The Laboratory Rat

Volume I
Biology and Diseases

AMERICAN COLLEGE OF LABORATORY ANIMAL MEDICINE SERIES

Steven H. Weisbroth, Ronald E. Flatt, and Alan L. Kraus, eds.:
The Biology of the Laboratory Rabbit, 1974

Joseph E. Wagner and Patrick J. Manning, eds.:
The Biology of the Guinea Pig, 1976

Edwin J. Andrews, Billy C. Ward, and Norman H. Altman, eds.:
Spontaneous Animal Models of Human Disease, Volume I, 1979;
Volume II, 1979

Henry J. Baker, J. Russell Lindsey, and Steven H. Weisbroth, eds.:
The Laboratory Rat, Volume I: Biology and Diseases, 1979

In preparation

Henry J. Baker, J. Russell Lindsey, and Steven H. Weisbroth, eds.:
The Laboratory Rat, Volume II: Research Applications

The Laboratory Rat

Volume I
Biology and Diseases

EDITED BY

Henry J. Baker
Department of Comparative Medicine
Schools of Medicine and Dentistry
University of Alabama in Birmingham
Birmingham, Alabama

J. Russell Lindsey
Department of Comparative Medicine
Schools of Medicine and Dentistry
University of Alabama in Birmingham
and the Veterans Administration Hospital
Birmingham, Alabama

Steven H. Weisbroth
AnMed Laboratories, Inc.
New Hyde Park, New York

ACADEMIC PRESS   1979
A SUBSIDIARY OF HARCOURT BRACE JOVANOVICH, PUBLISHERS
New York   London   Toronto   Sydney   San Francisco

ACADEMIC PRESS, INC.
111 Fifth Avenue, New York, New York 10003

United Kingdom Edition published by
ACADEMIC PRESS, INC. (LONDON) LTD.
24/28 Oval Road, London NW1 7DX

Library of Congress Cataloging in Publication Data
Main entry under title:

The Laboratory rat.

Includes bibliographical references and indexes.
CONTENTS: v. 1. Biology and diseases.
1. Rats as laboratory animals. 2. Rattus
norvegicus. I. Baker, Henry J. II. Lindsey,
James Russell, 1933–   III. Weisbroth, Steven H.
QL737.R666L3   619'.93   79–51688
ISBN 0–12–074901–7

PRINTED IN THE UNITED STATES OF AMERICA

79 80 81 82   9 8 7 6 5 4 3 2 1

Henry Herbert Donaldson (1857–1938)

From 1906 to 1938 this great scientist directed a multidisciplinary research program at the Wistar Institute in Philadelphia which laid the initial foundations for use of the rat in research. Thus, he is rightfully considered to be the originator of the laboratory rat. But Donaldson was more than a great scientist; he was a man of exemplary personal qualities, as revealed in the following description by Conklin:* "Anyone who had once seen him could never forget his magnificent head, his steady sympathetic eyes, his gentle smile. With these were associated great-hearted kindness, transparent sincerity, genial humor . . . orderliness, persistence, serenity. His laboratory and library were always in perfect order, his comings and goings were as timely as the clock, he never seemed hurried and yet he worked *Ohne Hast, Ohne Rast* [no haste, no rest]." For these reasons, the editors dedicate this text to the memory of H. H. Donaldson. (Photo by permission *J. Comp. Neurol.*)

*E. G. Conklin, "Biographical Memoir of Henry Herbert Donaldson, 1857–1938." Biogr. Mem., Vol. XX. 8th Mem., pp. 229–243. Natl. Acad. Sci., Washington, D.C., 1939.

Contents

List of Contributors

Numbers in parentheses indicate the pages on which the authors' contributions begin.

Norman H. Altman (333), Papanicolaou Cancer Research Institute, Miami, Florida 33136

Miriam R. Anver (377), Unit for Laboratory Animal Medicine, University of Michigan Medical School, Ann Arbor, Michigan 48109

Dennis E. J. Baker (153), Department of Animal Science, State University Agricultural and Technical College, Delhi, New York

Henry J. Baker (169, 243, 411), Department of Comparative Medicine, Schools of Medicine and Dentistry, University of Alabama in Birmingham, Birmingham, Alabama 35294

Pravin N. Bhatt (271), Section of Comparative Medicine, Yale University School of Medicine, New Haven, Connecticut 06510

W. Sheldon Bivin (73), Department of Anatomy, School of Veterinary Medicine, Louisiana State University, Baton Rouge, Louisiana 70803

N. R. Brewer (73), 5526 Blackstone Avenue, Chicago, Illinois 60637

Gail H. Cassell (243), Departments of Microbiology and Comparative Medicine, Schools of Medicine and Dentistry, University of Alabama in Birmingham, Birmingham, Alabama 35294

Bennett J. Cohen (377), Unit for Laboratory Animal Medicine, University of Michigan Medical School, Ann Arbor, Michigan 48109

M. Pat Crawford (73), Department of Physiology, Pharmacology, and Toxicology, School of Veterinary Medicine, Louisiana State University, Baton Rouge, Louisiana 70803

Lyubica Dabich (105), Department of Internal Medicine, University of Michigan Medical School, Ann Arbor, Michigan 48109

Jerry K. Davis (243), Departments of Microbiology and Comparative Medicine, Schools of Medicine and Dentistry, University of Alabama in Birmingham, Birmingham, Alabama 35294

Michael F. W. Festing (55), Medical Research Council, Laboratory Animals Centre, Carshalton, Surrey, England

Estelle Hecht Geller (401), Animal Institute and Department of Pathology, Albert Einstein College of Medicine, Yeshiva University, Bronx, New York 10461

Dawn G. Goodman* (333), Tumor Pathology Branch, Carcinogenesis Testing Program, National Cancer Institute, National Institutes of Health, Bethesda, Maryland 20014

Chao-Kuang Hsu (307), Division of Comparative Medicine, Georgetown University Medical Center, Washington, D.C. 20007

Robert O. Jacoby (271), Section of Comparative Medicine, Yale University School of Medicine, New Haven, Connecticut 06510

Albert M. Jonas (271), Tufts University School of Veterinary Medicine, Boston, Massachusetts 02111

Sam M. Kruckenberg (413), Department of Pathology, College of Veterinary Medicine, Kansas State University, Manhattan, Kansas 66502

J. Russell Lindsey (1, 169, 243, 411), Department of Comparative Medicine, Schools of Medicine and Dentistry, University of Alabama in Birmingham and the Veterans Administration Hospital, Birmingham, Alabama 35294

*Present Address: Division of Pathology, Clement Associates, Inc., Washington, D.C. 20007

Daniel H. Ringler (105), Unit for Laboratory Animal Medicine, University of Michigan Medical School, Ann Arbor, Michigan 48109

Roy Robinson (37), St. Stephens Road Nursery, London W13 8HB, England

Adrianne E. Rogers (123), Department of Nutrition and Food Science, Massachusetts Institute of Technology, Cambridge, Massachusetts

Steven H. Weisbroth (169, 193, 411), AnMed Laboratories, Inc., New Hyde Park, New York 11040

Preface

The American College of Laboratory Animal Medicine (ACLAM), founded in 1957, is the board of the American Veterinary Medical Association for specialists in laboratory animal medicine. Along with its mission of certifying competent professionals in the field, the college maintains an aggressive continuing education program. This book is a product of that program.

In 1969 ACLAM embarked on a bold and exciting project to develop a series of comprehensive scientific texts on laboratory animals. "The Biology of the Laboratory Rabbit" was published in 1974 and "The Biology of the Guinea Pig" in 1976.

An authoritative reference work on the laboratory rat to meet the needs of modern science has been long overdue. Relatively few books dedicated exclusively to this species have ever been printed. Thirty years have now elapsed since Edmond Farris and John Griffith, assisted by twenty-nine contributors, published the second (and last) edition of "The Rat in Laboratory Investigation." Thus, this book is intended to fill a void of thirty years during which no similar text appeared despite phenomenal progress in the biomedical sciences and a parallel expansion in knowledge of the laboratory rat.

The division of this work into two volumes was dictated by the large volume of material to be included. The subjects sorted readily into two major groupings, Biology and Diseases and Research Applications, the focal points for Volumes I and II, respectively.

Volume I details the more fundamental aspects of *Rattus norvegicus* as a species, its biology and diseases. Chapter 1, for the first time, brings together in a single narrative those events and personalities involved in the development of the species as a laboratory animal. The basic biology of the rat is described in chapters on taxonomy, genetics, anatomy,

physiology, and hematology and clinical biochemistry. It will be apparent that these chapters emphasize information of the greatest importance to research applications. Chapters on nutrition, reproduction, and husbandry include aspects of basic biology but emphasize matters of practical importance in the management of rats for laboratory investigation. Considerable emphasis is given to spontaneous diseases because of their importance as complications in the use of rats as research subjects and because they may present unique models for the study of disease mechanisms. Finally, the important zoonotic hazards associated with the use of laboratory rats are described.

Volume II chronicles topics of importance to research applications of the rat. Some chapters such as Research Methodology, Gnotobiology, and Wild Rats in Research are of general interest. Others focus on the use of rats in specific areas of research, ranging from dental research to toxicology. A few research specialties in which rats are important research subjects, such as endocrinology and behavior, will be obvious by their omission. These topics were not intentionally overlooked and will, we hope, be included in another volume.

Some chapter topics were presented at symposia sponsored by ACLAM. Chapters on taxonomy, morphophysiology, hematology and clinical chemistry, nutrition, reproduction, and housing were presented at a symposium entitled "The Laboratory Rat: Biology and Use in Research" held in conjunction with the 26th Annual Scientific Session of the American Association for Laboratory Animal Science on November 19, 1975, in Boston, Massachusetts. Chapters on bacterial disease, mycoplasmoses, parasitic diseases, lesions associated with aging, and spontaneous tumors were presented at the symposium "Spontaneous Diseases of Laboratory Rats as Complications of Research" held during the 113th Annual

Meeting of the American Veterinary Medical Association on July 20, 1976, in Cincinnati, Ohio.

This book is offered to a wide range of individuals concerned with the use of rats in research. It is hoped that students in graduate and professional curricula will find the broad coverage of material useful for rapid introduction to complex subjects. Commercial and institutional organizations involved in producing rats for research use will find these volumes a rich source of practical information. Specialists in laboratory animal science will welcome its addition to the volumes on rabbits and guinea pigs. Animal care and research technicians should find topics on husbandry, reproduction, and research methodology of particular interest. Above all, investigators will find this broad-based reference work of great value.

The editors wish to express special appreciation to the contributors. They were selected from the most knowledgeable and experienced scientists in the world. As with all other texts in this series, each author has contributed all publication royalties to the American College of Laboratory Animal Medicine for the purposes of continuing education. Officers and members of the College are acknowledged for their enthusiastic support of this project. Finally, we are grateful for the patience and assistance given by those on the staff of Academic Press who have contributed so much to development of this work and to the success of the series.

Henry J. Baker
J. Russell Lindsey
Steven H. Weisbroth

Contents of Volume II
Research Applications

Chapter 1

Historical Foundations

J. Russell Lindsey

I.   ORIGIN OF THE LABORATORY RAT

The purpose of this chapter is to retrace, for the first time, one of the most fascinating stories of biomedical history—that of the early events and personalities involved in establishment of *Rattus norvegicus* as a leading laboratory animal. From the onset, three features of the story deserve acknowledgment. First, it will be possible in the brief space of this chapter to touch only the highlights (but these cannot be treated equally because many important details already have been lost). Second, it is almost exclusively an American story as virtually all of the modern strains of rats in use today the world over trace their ancestry to stocks in the United States. Third, the story is literally one of ascendancy from the gutter to a place of nobility, for what creature is more lowly than the rat as a feral pest or more noble than the same species which has contributed so much to the advancement of knowledge as the laboratory rat!

A.   Earliest Records

The earliest records of *R. norvegicus* are a bit sketchy, but there appears to be good agreement on the major events concerning the species (21,40,41,127,132). Its original natural habitat is thought to have been the temperate regions of Asia, specifically the area between the Caspian Sea and Tobolsk (U.S.S.R.) and perhaps, extending as far east as Lake Baykal (U.S.S.R.). With the coming of modern civilization, a suitable ecological niche became available to the species as an economic pest, allowing its numbers to increase rapidly and spread over the world in close association with man. It is said to have reached Europe early in the eighteenth century, England between 1728 and 1730, and northeastern United States by 1775. That the species did at one time in history spread through Norway is readily accepted, but the name "Norway rat" or "Norwegian rat" actually has no meaning other than possibly to reflect the species name, i.e., *norvegicus*.

Richter (127) has summarized the likely sequence of events in the domestication of *R. norvegicus,* as follows.

> It is quite likely that Norway rats came into captivity as albinos. We know that rat-baiting was popular in France and England as early as 1800, and in America soon afterward. This sport flourished for seventy years or more, until it finally was stopped by decree. In this sport 100 to 200 recently trapped wild Norways were placed at one time in a fighting pit. A trained terrier was put into the pit. A keeper measured the time until the last rat was killed. Sportsmen bet on the killing times of their favorite terriers. For this sport many Norway rats had to be trapped and held in pounds in readiness for contests. Records indicate that albinos were removed from such pounds and kept for show purposes and/or breeding. It is thus very likely that these show rats, that probably had been tamed by frequent handling, found their way at one time or another into laboratories.

B.   Earliest Experiments

Regardless of the exact details, the Norway rat became the first mammalian species to be domesticated primarily for scientific purposes (128). There is some evidence to suggest that rats probably were used sporadically for nutrition experiments in Europe prior to 1850 (150). However, the work generally recognized as the first use of the rat for experimental purposes was a study on the effects of adrenalectomy in albino rats by Philipeaux (120) published in France in 1856. Soon afterward in 1863, Savory (135), an English surgeon, published what is thought to be the first attempt to evaluate nutritional quality of proteins in a mammal. His rats were of mixed coat colors, including black, brown, and white. The first known breeding experiments with rats employed both albino and wild animals, and were performed in Germany from 1877 to 1885 by Crampe (29–32).

The first experiments known to use rats in the United States were neuroanatomical studies performed during the early 1890's by S. Hatai and other faculty members in H. H. Donaldson's Department of Neurology at the University of Chicago (27). It seems likely, as suggested by Richter (129), that these albino rats had been brought from the Department of Zoology at the University of Geneva by Adolf Meyer, the Swiss neuropathologist, shortly after he immigrated to the United States and joined Donaldson's faculty around 1890. In 1894, Stewart at Clark University (in Worcester, Massachusetts) initiated studies on the effects of alcohol, diet, and barometric phenomena on activity (146). He began with wild rats, but in 1895 switched to albinos, possibly also provided by Adolf Meyer who took a job at the mental hospital in Worcester, Massachusetts in 1895 (152). However, the precise origin of the albino rats used in research in the United States must remain in some doubt as Donaldson stated explicitly in two separate articles that it was not clear whether the North American albinos were descendants of those stocks which had been used earlier in research laboratories in Europe, or mutants selected from wild rats captured in the United States (40,41).

II. THE WISTAR INSTITUTE

A. Background

The Wistar Institute of Philadelphia occupies a special place in the annals of biomedical history. First of all, it is the oldest independent research institute in the United States. Second, and particularly germane to the subject of this chapter, it provided a major share of the foundation on which the rat came to be established as an important laboratory animal.

The Institute was named in honor of a physician, Dr. Caspar Wistar (1761–1818). While Professor of Anatomy at the University of Pennsylvania School of Medicine, he authored the first standard text of human anatomy in America and developed a magnificent collection of anatomical specimens. These specimens were brought together as "The Wistar Museum" in 1808, and given to the University of Pennsylvania after his death. Additional material subsequently was contributed by his colleagues, particularly Dr. George Horner. Interest in the collection declined over the next 60 years until Dr. James Tyson, Dean of the Medical School, became keenly interested in assuring future preservation of the Museum. He approached General Isaac Wistar (1827–1905), Philadelphia lawyer, great nephew of Caspar Wistar and former Brigadier General in the Union Army, who initially contributed only $20.00. Later, under the General's direction, however, the Wistar Institute of Anatomy and Biology was incorporated in

1892, and funds were set aside for a "fire-proof museum building" and an endowment. The original building, which still stands today (Fig. 1), was built at a cost of $125,000 and formally opened in May, 1894. The University of Pennsylvania gave the land (123,158).

B. Milton J. Greenman

The first two directors of the Institute served only briefly. A physician, Milton Jay Greenman (1866–1937) (Fig. 2), became the third director in 1905, and held that position for 32 years until his death (2,45,158). Although his role was chiefly in administration, Greenman's era was clearly a "golden age" in the life of the Institute. Henry Donaldson (44) characterized him as "the Institute's real scientific founder..." and as a "genial, alert man, trained in biology, gifted to an unusual degree with mechanical and inventive abilities, with business capacity and good judgement, based on the imagination needed for an administrator."

Greenman wanted the Institute to be a center of scientific investigation, not merely a museum. Thus, immediately upon being made director, he established a Scientific Advisory Board composed of ten professors of anatomy and zoology from the nation's leading universities (Fig. 3). In April, 1905 this Board, at its first meeting, voted unanimously in favor of the Institute being devoted primarily to research, initially

Fig. 1. Original building of The Wistar Institute of Anatomy and Biology in Philadelphia built at a cost of $125,000 and formally opened in May, 1894. The street (Woodland Avenue) to the right of the building in this photo was subsequently closed and is today a small park. (Courtesy of The Wistar Institute.)

Fig. 2. Milton Jay Greenman, Director of The Wistar Institute from 1905 until 1937, and characterized by Henry Donaldson as "the Institute's real scientific founder." (Courtesy of The Wistar Institute.)

focusing its efforts in the areas of "neurology, comparative anatomy and embryology." Further, the Board unanimously recommended one of its own number, Henry Donaldson, to be the Institute's first Scientific Director (2,158).

In 1908, the Institute acquired a printing press and the rights to five leading biological journals. Others have been added since that time, and the group is still being published today by the Wistar Press (123).

C. Henry H. Donaldson

Henry Herbert Donaldson (1857–1938) was one of the truly remarkable men of early American science, undoubtedly one of its greatest heroes. In essence, Donaldson did for the laboratory rat what Clarence Cook Little (1888–1971) was to do later for the laboratory mouse at the Jackson Laboratory (established in 1929). Donaldson and his team of investigators at the Wistar began efforts in 1906 to standardize the albino rat. Initially, the main intent was to produce reliable strains for their studies of growth and development of the nervous system. In reality, the work directly gave the broad foundation for use of the rat in nutrition, biochemistry, endocrinology, genetic, and behavioral research, and indirectly, in many other fields of investigation.

Following completion of the B.A. degree at Yale and a year of graduate work in physiology with R. H. Chittenden, Donaldson enrolled in the College of Physicans and Surgeons

Fig. 3. General Isaac Wistar and Milton Greenman with the first Scientific Advisory Board of The Wistar Institute during their meeting of April 11 and 12, 1905. On the advice of this group, The Wistar Institute changed its primary mission from that of a museum to scientific research. From left to right: Simon H. Gage, Franklin P. Mall, Isaac J. Wistar, Milton J. Greenman, Edwin G. Conklin, Charles S. Minot, George A. Piersol, Lewellys F. Barker, J. Playfair McMurrich, Henry H. Donaldson, Carl Huber, and George S. Huntington (158). (Courtesy of The Wistar Institute.)

in New York, but only one year of study convinced him that he really wanted a career in basic research rather than clinical medicine. He entered graduate school at the Johns Hopkins University in 1881 and was awarded the Ph.D. in 1885. During the next few years the work of his dissertation research was continued at different institutions in Europe and the United States, work which culminated in a book, "The Growth of the Brain," published by Scribners in 1895. From 1889 to 1892 he was Assistant Professor of Neurology at Clark University. Donaldson was Dean of the Ogden School of Science at the University of Chicago for six years, and Professor of Neurology there from 1892 until 1906 (during this interval he developed tuberculosis of the hip which left him permanently crippled) when, at the invitation of Milton Greenman, he became Professor of Neurology and Director of Research at the Wistar Institute(27).

Donaldson brought with him from Chicago one colleague, Dr. Shinkishi Hatai, and four pairs of albino rats. Although Donaldson up to this point in his career had studied primarily the nervous systems of man and frogs, he was convinced (by Adolf Meyer) that the albino rat was the best available animal for laboratory work on problems of growth. Accordingly, he recruited to Wistar a team of scientists who, for their day, launched a colossal multidisciplinary research program (27). Donaldson and his group of collaborators carried out the following studies using albino rats [from a summary of Donaldson's life's work by McMurrich and Jackson (105)].

1. Determined growth curves for body length and body weight
2. Correlated growth curves of brain and cord with body length and weight
3. Studied various conditions which might affect the nervous system:
   a. Domestication, age, exercise, inbreeding (H. D. King)*
   b. Gonadectomy (J. M. Stotsenberg and S. Hatai)
   c. Growth of cerebral (N. Sugita) and cerebellar (W. H. F. Addison) cortex
   d. Blood supply to different portions of the brain (E. H. Craigie)
   e. Chemistry of the brain solids (W. and M. L. Koch)
   f. Relation of size of nerve cells to length of their neurites, the spacing of nodes in nerve fibers (S. Hatai)
   g. Regeneration of nerve fibers (M. J. Greenman)
4. Growth curves:
   a. Skull
   b. Skeleton
   c. Viscera (S. Hatai)
   d. Individual organs, the submaxillary glands, the hypophysis (W. H. F. Addison)

*Names of Donaldson's collaborators are given in parenthesis.

e. Thymus, adrenals (J. C. Donaldson)
f. Thyroid and effects of thyroidectomy (F. S. Hammett)

Donaldson himself published nearly 100 papers and books, the most famous being "The Rat: Data and Reference Tables for the Albino Rat (*Mus norvegicus albinus*) and the Norway Rat (*Mus norvegicus*)" (1st ed., 1915; 2nd ed., 1924). This book (42,43) was truly a remarkable compendium of data pertaining to the rat; the second edition, published in 1924, contained 469 pages which included 212 tables, 72 charts and over 2050 references! [These references combined with a later list for the period 1924–1929 by Drake and Heron (47) provide a complete bibliography on the rat to 1930.] An appraisal of Donaldson's contribution was made by McMurrich and Jackson (105); it reads: "Donaldson's work then, is fundamental: it gives a paradigm of growth with which other investigations may be compared and is an outstanding monument of perseverance, thoroughness, accuracy and cooperation." It remains today the all time classic work on the laboratory rat.

Donaldson's choice of the rat as a laboratory animal was no mere accident. The depth of his reasoning can be seen in the following quotes from his book (43):

In enumerating the qualifications of the rat as a laboratory animal, and in pointing out some of its similarities to man, it is not intended to convey the notion that the rat is a bewitched prince or that man is an overgrown rat, but merely to emphasize the accepted view that the similarities between mammals having the same food habits tend to be close, and that in some instances, at least, by the use of equivalent ages, the results obtained with one form can be very precisely transferred to the other.... I selected the albino rat as the animal with which to work. It was found that the nervous system of the rat grows in the same manner as that of man—only some 30 times as fast. Further, the rat of 3 years may be regarded as equivalent in age to a man of 90 years, and this equivalence holds through all portions of the span of life, from birth to maturity. By the use of the equivalent ages observations on the nervous system of the rat can be transferred to man and tested.

It appears that Donaldson's work impacted particularly on progress in nutrition. Addison (1) gives the following quote from an address presented by H. Gideon Wells, [Chairman of the Department of Pathology at the University of Chicago and Director of its affiliated Otha S. A. Sprague Memorial Institute (26,117)] in 1938 on the subject "War Time Experience in Nutrition:"

If it had not been for Henry Donaldson's rats the knowledge of vitamins would not have come in time to be applied when it was most acutely needed. It had taken years for Donaldson to learn the normal curve of growth of white rats and to establish strains of healthy rats that were reliable material for research. These rats and their growth curves were basic factors in nutrition research, and without them this research would have been delayed until nutritionists did what Donaldson had already done for them, and that delay would have been serious for the feeding of starving nations from 1915 to 1919.

D. Contributions of Donaldson's Colleagues

By the time of Donaldson's death in 1938, his students, assistants and associates at the Wistar had published more than 360 articles and books, almost exclusively on the rat (27). Some of them made major contributions to establishment of the rat as a laboratory animal. A few highlights of these contributions follow.

Helen Dean King (1869–1955) (Fig. 4) was an extremely productive scientist at the Wistar from 1907 until her retirement in 1949. In 1909 she began inbreeding the albino rats which Donaldson had brought from Chicago. [It is interesting that 1909 was also the year in which Clarence Cook Little began inbreeding the oldest known strain of mouse, the DBA (84).] This was an enormous undertaking as some of her many papers amounted to summaries of observations made on 25,000 rats (90–93)! By 1920 these rats were maintained as two separate lines which had reached generation 38 of brother × sister matings, prompting a note by Dr. King in the annual report that Wistar was "11 years ahead of all other attempts" to inbreed rats. One of these lines was carried to the 135th generation of inbreeding by Dr. King, and eventually became known as the King Albino (today designated the PA strain). Many other modern strains of rats (Table III) take their origin from this and other stocks at the Wistar (8). For example, the Lewis strain was produced from animals selected and inbred at Wistar by Margaret Reed Lewis; in 1956 this strain reached its eighth generation of inbreeding. King also captured some wild Norways in the vicinity of Philadelphia and proceeded to produce an inbred strain of them. By 1934 they had reached generation 35 of inbreeding. This line became the strain known today as the Brown Norway (BN) rat. Several mutations were identified and studied by Dr. King (2,21,132).

Fig. 4.   Helen Dean King during the early years of her career as a scientist at The Wistar Institute. In 1909 she initiated the first inbred strain of rats, now known as the PA strain. (Courtesy of The Wistar Institute.)

Another important milestone in the history of the laboratory rat came in 1935 when Eunice Chace Greene (1895–1975) published the first comprehensive treatise on gross anatomy of this species, entitled "Anatomy of the Rat" (75). This book (with several reprintings by Hafner Publishing Co.) remains today a standard reference on the rat. She also published a chapter in "The Rat in Laboratory Investigation" (69,79). These two publications comprise her total bibliography. At the time of her death in 1975, she and her husband, Walter F. Greene, formerly Professor of Biology at Springfield College in Massachusetts, lived in Jaffrey Center, New Hampshire (64,76).

E. The Wistar Rat Colony

It seems impossible to overstate the contribution made by The Wistar Institute to husbandry of the laboratory rat. The Wistar rat colony served as the initial testing ground for developing satisfactory cages and ancillary equipment, diets, breeding practices, and facilities for rats. It must also be emphasized that much of the credit for these innovations and advances goes to none other than the Director of the Institute, Milton Greenman. He gave the colony a great deal of his personal attention and, with the help of an able assistant, Miss F. Louise Duhring, rapidly evolved a highly successful set of husbandry practices. That the two of them took this responsibility seriously is clearly revealed in their book. "Breeding and Care of the Albino Rat for Research Purposes" (Greenman and Duhring, 1st ed., 1923; 2nd ed., 1931). The preface to the first edition states, "It became more and more evident that clean, healthy, albino rats were essential for accurate research and that their production was a serious, difficult, and worthwhile task" (77,78).

The first rats brought to the Wistar in 1906 by Donaldson were presumably housed in the museum building (Fig. 1). In 1913, however, the colony was moved into a 30 × 90 ft, three story building referred as The Annex, which had formerly been a police station (Fig. 5). In 1918 there arose a series of circumstances which seriously threatened the colony's productivity and continued existence. Cottonseed meal was introduced into the diet, and many rats died or ceased breeding as result of gossypol toxicity. In order to meet demands of the Institute's investigators, rats were purchased from outside sources and brought into the facility. Consequently, many further deaths were incurred because of an epizootic of infectious disease(s). Greenman found himself increasingly involved with the details of operating the colony. As a result of this experience, Miss Duhring was given charge of the colony and Greenman began planning for a new building. Plans were developed over a period of two years and, in 1921 a member of the Institute's Board of Managers, Mr. Samuel S. Fels, contributed $30,000 toward its construction. The "Colony Building" (Figs. 6–8),

Fig. 5. An early picture of the rat colony at Wistar Institute. The building is thought to be ''The Annex,'' a former police station which housed the Wistar rat colony from 1913 to 1922. (Courtesy of The Wistar Institute.)

Fig. 6

Fig. 7

with 12 modular rooms measuring 14 × 22 ft each, for a total of approximately 4635 net ft², was built at a cost of $33,458.66! The first rats were moved in on May 5, 1922 and the last ones entered on August 9, 1922, each rat being treated with tincture of larkspur seed by Miss Duhring to ''rid the colony of ectoparasites!'' At this time, the colony had a total population of approximately 6000 rats (2).

The Wistar Colony Building was remarkable because of its forward looking design characteristics. It was ''constructed of brick, concrete, steel, and glass'' because ''these materials offer less harbor for dirt and vermin if properly put together.'' The floor plan (Fig. 8) provided for separation of animal facility functions: rooms for housing animals, an office for records, a surgery, a cage washing room, a diet kitchen and lavatories for personnel, If one accepts the central hallway as the ''clean'' corridor and the outside concrete walk as the ''dirty'' corridor, it qualifies as the first animal facility of the ''clean–dirty corridor'' concept. It was recognized that ''the work of a colony is greatly facilitated if the entire colony is located on one floor level, preferably the ground floor.'' Also, and possibly the most important principle of all, was the idea that ''no space unfit for human habitation is desirable for an albino-rat colony'' (77,78).

Husbandry practices for the colony in the early years of the new colony building were documented in detail in the book of Greenman and Duhring (77,78). Most of the cages were constructed of white pine with removable galvanized wire floors (No. 22 wire, ⅛-inch mesh) mounted over steel trays (Figs. 9–12). (Both editions of Greenman and Duhring's book contained a drawing giving details for construction of these cages

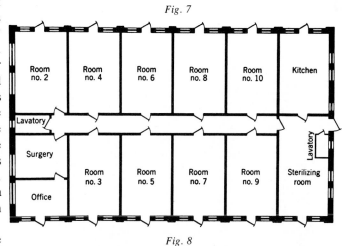

Fig. 8

Figs. 6–8. The Wistar Colony Building completed in 1922 and referred to by Donaldson in 1927 (44) as ''a home for the friendless rat.'' The building is shown here in three perspectives. Figure 6 is a view looking across the court-yard from the main Institute building. Figure 7 is a view from the Woodland Avenue side of the Institute's triangular lot looking back toward the main Institute building. Figure 8 is the floor plan of the Colony Building. (Courtesy of The Wistar Institute.)

Fig. 9.   A typical animal room in The Wistar Colony Building. The cages constructed of white pine were suspended from the ceiling on either side. (Courtesy of The Wistar Institute.)

Fig. 10.   Close up view of suspended cages (Fig. 9) in The Wistar Colony Building. This photo probably was made in the 1940s after pelleted diet came into use.

as a large fold-out frontpiece.) The floors and pans were washed by hand (Figs. 13 and 14). The diet, up until about 1920, had been scraps from local restaurants. But, with the diet kitchen in the new colony building, Greenman and Duhring developed a great diversity of recipes, including "wheat and peas with milk," "barley, salmon, and eggs," "hominy grits, vegetables, and eggs," and many others. Such homemade mixtures apparently were continued until the 1940's when dry commercial dog and fox diets came into use (2).

The Annual Reports of the Director (1911–1956) at Wistar give many valuable insights into realities of daily operations of the rat colony (2). First of all, the preeminence of the colony is revealed in the fact that every report from 1911 to 1956 contains a major section entitled "The Animal Colony." This section is a summary of each year's activities, including rat production, a listing of recipients of animals, progress of the inbreeding program under Dr. King, budget information, personnel matters, and comments on health problems encountered in the colony.

 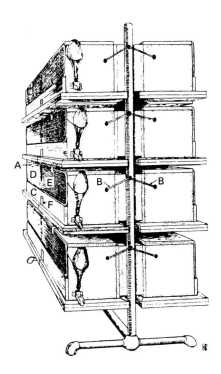

Figs. 11 and 12. "Cage and cage support, one of the older types used in The Wistar Institute Colony" (77). [Figure 11, courtesy of The Wistar Institute; Figure 12, from Greenman and Duhring (77).]

Fig. 13. The "sterilizing room" in The Wistar Colony Building. The buckets on the floor to the right of the scrub tank had hinged lids and holes in the sides, and were used in transporting small groups of rats around the Institute. (Courtesy of The Wistar Institute.)

Fig. 14. Hand scrubbing the galvanized wire floors used in cages in The Wistar Colony Building. (Courtesy of The Wistar Institute.)

Table I

Rat Production at The Wistar Institute, 1911–1928[a]

Year	Used in research at Wistar	Supplied to other institutions	Number of institutions supplied	Income from rat sales ($)	Expenditure ($) on rat colony[b]
1911	1,646	0	0	0	645.00
1912	1,737	2,135	—[c]	0	400.00
1913	8,000	3,000	—	0	293.00
1914	12,043	1,197	21	0	919.00
1915	1,091	5,037	21	831.00	1526.00
1916	1,462	6,025	—	1218.00	1021.00
1917	—	~7,000	—	1348.00	1213.00
1918	3,313	11,014	—	3292.00	5593.00[d]
1919	1,693	5,572	—	5577.00	5103.00
1920	1,272	2,517	—	1594.00	3815.00
1921	1,450	5,222	59	3294.00	5093.00
1922	1,262	2,217	56	2594.00	2626.00
1923	1,300	3,093	54	2819.00	4809.00
1924	784	1,686	60	2817.00	3037.00
1925	522	2,447	55	—	—
1926	727	2,809	~50	4,500.00	—
1927	—	3,443	120	7231.00	—
1928	787	2,524	121	8427.00	7669.00

[a] Compiled from Annual Reports of the Director, The Wistar Institute of Anatomy and Biology, Philadelphia, Pennsylvania.

[b] Amounts shown presumably represented costs of consumable supplies only; salaries (usually slightly more each year than the figures here) and a major cage purchase ($1596 in 1916) were reported separately.

[c] No data.

[d] Includes cost of mice produced for the U.S. Army to be used "in the diagnosis of pneumonia," no doubt pneumonia secondary to influenza as this was the year of the pandemic.

Table I gives the available data on rat production at the Wistar from 1911 through 1928. During this period, albino rats were supplied to many institutions throughout the United States and in many foreign countries. In fact, more animals were supplied to other institutions than to investigators in the Wistar. Also, it will be noted that income from rat sales usually covered most of the expenditures for the rat colony except for salaries of personnel. The data after 1928 are less complete (2).

Parenthetically, it must be emphasized that there has been enormous confusion about the genetic background of the rats which Wistar sold to users all over the world until 1960 (see below). Even today, one still sees references to "*the* Wistar rat" as if all rats ever sold by them were absolutely identical. From its inception, the Wistar apparently maintained a random bred (heterogeneous) colony, in addition to those which were maintained by strict brother × sister matings (e.g., the PA, BN and LEW strains). It was this random bred stock which served as the commercial colony. It presumably gave origin to the LEW strain (8). Whether an albino line(s) other than the one brought from Chicago in 1906 were introduced into the commercial colony is unknown, but this may have happened. It is known that outside breeders were brought in during 1918 to boost production of the breeding colony (2).

In the early years of the colony, disease problems were common (2). Some years (excluding 1918 when many died of gossypol toxicity) natural deaths exceeded 10%. Respiratory disease was mentioned frequently and, indeed, the first known report of the disease known today as murine respiratory mycoplasmosis (MRM) was published by the Chicago pathologist, Ludwig Hektoen (82), using 328 rats shipped to him from the Wistar in 1915 (2). In 1931 it was noted that 75% of the rats had middle ear disease, and attempts were being made to breed a resistant strain (2). In 1950 (68) the following was observed:

> The incidence of respiratory infections greatly increases when the following conditions exist in an animal colony:* (a) the air in the room is stagnant and heavily laden with fumes from the excreta; (b) the debris pans are not changed often enough; (c) the bedding needs changing; (d) the cages need cleaning; (e) the room temperature is unsatisfactory; (f) the atmosphere is dusty.

[These conditions strongly suggest, among other possibilities, what has been demonstrated in recent years, i.e., environmental ammonia plays a role in MRM (12)]. The annual report of 1945 indicates that the disease known as "ringtail" had been eliminated by maintaining the animal rooms at 50% or greater relative humidity (2).

In 1928, the Wistar Institute was given a 150-acre farm by a

*Numbering of this series has been changed by author.

relative of General Wistar. The Effingham B. Biological Farm located near Bristol, approximately 30 miles from Philadelphia, was the home of the Director from 1928 until his death. In addition, facilities at the farm were used for production of rats, amphibia, and opossums (the latter being an interest of Greenman's). After Greenman's death in 1937, the farm was sold and all rats were returned to the colony building at the Institute (158).

F. Wistar under Edmund Farris

Edmund J. Farris (1907–1961) became Acting Director in 1937, and was promoted in 1939 to Executive Director, a position he held until 1957. In contrast to the previous sharply focused research program of the Institute, Farris' tenure as Director was marked by diversification, including anatomy, aging, microbiology, cytology, and chemistry. Two editions of the book entitled "The Rat in Laboratory Investigation" were published in 1942 (79) and 1949 (69). A second text, "The Care and Breeding of Laboratory Animals," was published in 1950 (68). During Farris' tenure, the Institute continued to supply rats to other institutions, averaging 4359 rats to approximately 50 institutions each year. Annual use of rats within the Institute was 6643, and the average monthly census of rats at the Institute was 2977 (2).

G. The Modern Era at Wistar

Dr. Hilary Koprowski, a virologist, became Director of the Wistar in 1957. Under his leadership, the major thrust of the Institute was redirected toward the viral, degenerative, and neoplastic diseases, fields in which it is now recognized as a world leader. Major contributions during this era have included development of improved vaccines against rabies and rubella viruses (123).

In 1964, the colony building was renovated and an additional floor placed on top of it (the colony building was stressed for an additional floor when it was built in 1922!). In 1975, two additional floors were added (using columns of reinforced concrete positioned along the outer walls of the old building) to provide laboratories and animal facilities for a greatly expanded cancer research program. The Institute continued to supply the Wistar rats to other laboratories until 1960 when all of the breeding stocks and rights to their perpetuation were sold to a commercial firm (133).

III. NUTRITION AND BIOCHEMISTRY

The Norway rat probably was first used as an experimental animal in a few nutrition studies conducted in Europe prior to 1850. Its popularity for this purpose continued upward as it was used increasingly in several nutrition laboratories in Europe during the next 50 years (150).

One of the more notable early studies was an attempt in 1863 to measure the quality of dietary proteins in rats by the English surgeon, Savory (135), who explained, "Rats were chosen as subjects for these experiments because they are omnivorous, and will readily feed on almost any kind of diet. Moreover, from their size, they are very convenient to manage." Many of his rats died on a diet of arrowroot starch, sago, tapioca, and lard or suet. Those on diets of meat with fat and starch survived and experienced some growth. In attempts to perform nitrogen balance studies, the urine volume and nitrogen content were found to be low in animals fed wheat alone, but high in those fed meat.

It appears that the majority of early investigators to use rats in nutrition research were interested in dietary proteins and growth (150). Their choice of the rat as an experimental animal, therefore, takes on special significance as one considers the many possible reasons why rats have excelled in nutrition research over the intervening years. The major reason is not, as commonly thought, that the rat is a good model for man. As pointed out by Hegsted (81),

> The very characteristics which have made the rat so attractive for nutritional studies contrast with those found in man. The young rat weighing 50 to 60 gm will grow some 5 to 6 gm/day, a rate of about 10% of his body weight per day. He continues to grow rapidly until reaching a weight of 200 to 300 gm. This results in a great dilution of the body stores of nutrients and makes it very easy to produce a variety of nutritional deficiencies. In contrast, a child of 4 to 5 years of age also grows about 5 gm/day but weighs 20 kg. The growth rate is negligible compared to body size.

Although probably not fully appreciated by early scientists, these principles were exploited intensively by them.

The advances made through early animal feeding experiments naturally led increasingly to more and more biochemical emphasis. Thus, nutrition and biochemistry are inseparable because of shared information. It may be more revealing, however, to consider that the fields of nutrition and biochemistry and the laboratory rat share much in historical background, including many of the same pioneering personalities.

A. McCollum and the McCollum-Pratt Institute

Elmer Verner McCollum (1879–1967) (Fig. 15) has been called "The Abe Lincoln of science" (144). As a young man, he was 6-ft tall and weighed only 127 pounds. Furthermore, he was one of the most unusual and distinguished men of early American science—a true pioneer in every sense of the word, including the fact that it was he who pioneered the use of the rat in nutrition research.

McCollum was born and reared on a farm in Kansas where

Fig. 15. Elmer Verner McCollum at his desk in Baltimore, 1955. McCollum's pioneering experiments with rats opened a new era of scientific advancement in nutrition and biochemistry. [From McCollum (101), by permission.]

he conducted his "first nutrition experiment—collaboratively with Mother." When she discontinued breast feeding him at the age of 7 months, he had developed scurvy, and she had treated him successfully by feeding him apples. He received his Bachelor of Arts degree in 1903 and his Master of Science degree in 1904 from the University of Kansas at Lawrence. He entered graduate school at Yale and obtained the Ph.D. in chemistry under T. B. Johnson in 1906, then remained at Yale an additional year as a postdoctoral fellow in L. B. Mendel's laboratory. For several months in 1906, he worked as an assistant in the laboratory of T. B. Osborne (101,130).

In 1907, on the advice of Mendel, McCollum accepted a position as instructor in agricultural chemistry at the University of Wisconsin. The salary was $1200 per year. After a few months' work on a nutrition research project using cows, he became convinced that "the most promising approach to the study of nutritional requirements of animals was through experimenting with small animals fed simplified diets composed of purified nutrients." Despite considerable resistance to this idea from his superiors, McCollum finally obtained their reluctant approval and set about his work by trapping 17 wild rats in the College's horse barn. He next "learned that the rats were too wild, too much alarmed, and too savage to be satisfactory for breeding and experimental work." Undaunted by such problems, he then paid a Chicago pet dealer $6 of his own money for 12 young albino rats. These served as the foundation stock for a colony from which the first experiment was initiated in January 1908. The following year, in 1909, McCollum was joined by a voluntary assistant who took charge of the rat colony, Miss Marguerite Davis, a recent graduate from the University of California at Berkeley. McCollum's idea of experimenting with rats was more than successful—it proved to be a major breakthrough! By the end of his 10-year sojourn in

Madison, he and Davis had published numerous papers, including one in 1913 (102) announcing the discovery of "fat-soluble A," and another in 1915 (103) on the discovery of "water-soluble B." It was probably his early successes, communicated through frequent correspondence with his former mentor, Mendel, that led Osborne and Mendel to establish their rat colony and begin their famous studies in 1909 (101).

In 1917, McCollum accepted the position of Professor and Chairman of the Department of Chemical Hygiene in the new School of Hygiene and Public Health at The Johns Hopkins University. Fifty female and 10 male rats were forwarded to Baltimore for the establishment of a breeding colony in the new Department. Miss Nina Simmonds was the assistant placed in charge of the colony. Once again, the research program using purified diets was underway, only larger than ever before. During his active career at Hopkins from 1917 to retirement in 1946, a total of approximately 150 papers was published from his laboratory. The topics covered a wide spectrum, including the dietary roles of calcium, phosphorus, potassium, sodium, numerous trace minerals, thiamine, riboflavin, and vitamin E. An illustration of the feeding device he used in his nutrition experiments with rats at Hopkins is shown in Fig. 16. The most famous work of this period was the

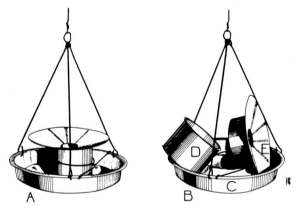

Fig. 16. Illustration of the feeding device used by E. V. McCollum, taken from the 1923 edition of Greenman and Duhring's book (77). (A) shows the device fully assembled, and (B) shows its components. As described by Greenman and Duhring, "It consists of an outer tin pan (C)—an ordinary cake pan—8 in. in diameter and 1¼ in. deep. In the middle of this pan a plain tin cup (D), 3½ in. in diameter and 2½ in. deep, is held in place by a wire loop attached at two points to the edge of the pan. This wire loop permits the cup to be removed for cleaning. The cup is covered by a circular disc of tin, 6 in. in diameter, with a central hole 1⅛ in. in diameter (E). This tinc disc is slightly convex, dipping toward the central hole, forming a funnellike top to the cup. A circular flange 1¼ in. wide is soldered to the underside of the circular disc. This flange fits accurately into the cup and holds the disc in place. The whole apparatus is suspended by three wires attached to the edge of the pan and brought together in a ring 10 in. above the pan. From this ring a single wire suspends the apparatus from the ceiling of the cage. Food is placed in the cup. The rats may stand on the disc and take food from the cup through the central opening in the disc. Food not eaten tends to fall back into the cup or if scattered over the edge of the disc is caught by the pan below."

discovery of vitamin D as the cause of rickets in 1922 (38,104,122). He authored several books, including "A History of Nutrition" (100) and his autobiography "From Kansas Farmboy to Scientist" (101). McCollum received many honors during his illustrious career. He served as president of the American Society of Biological Chemists in 1928 and 1929 (26).

For several months in 1943, McCollum served as a consultant to the U.S. Lend-Lease Administration in Washington. During some of the committee meetings there he met Mr. John Lee Pratt, a former executive of General Motors and the DuPont Co. who had retired to farming at Fredericksburg, Virginia. Their mutual interests in nutrition led to the development of a warm personal friendship. In 1947, Mr. Pratt contributed $500,000 to the Hopkins for research on trace elements in nutrition. This led to establishment of the McCollum-Pratt Institute on Hopkins' Homewood Campus (101).

It seems that most assessments of McCollum's life's work emphasize his scientific contributions without recognizing the impact his introduction of the rat had on nutrition research generally. This is a serious mistake, because the new trail he blazed with the rat directly opened a new era of discovery in nutrition and biochemistry—one of unprecedented advancement spanning the period from 1910 to 1940.

The stock of rats which McCollum began in 1907 at Madison, Wisconsin is still perpetuated today in the Department of Biochemistry at The Johns Hopkins University's School of Hygiene and Public Health in Baltimore. The initial rats obtained in 1907 were said to be albinos (101). Other bloodlines were subsequently brought in, including "a dozen young rats from Henry H. Donaldson" (Wistar) introduced in 1913, and some of Castle's "ruby-eyed, yellow-coated rats," and "pied (black and white) rats" introduced in 1915 (99,162). Today, three color patterns are represented: black, white, and albino. Over the years the colony has always been random-bred (131).

B. Osborne and Mendel

Thomas Burr Osborne (1859–1929) (Fig. 17) was born in New Haven, Connecticut. Thus, it was only natural that he attend Yale where he trained in chemistry, receiving a Bachelor of Arts degree in 1881 and the doctorate in 1885. From 1883 to 1886 he was an assistant in analytical chemistry and published several papers on analytical methods for metals and soils. In 1886 he became a member of the scientific staff (and son-in-law of the Director, Dr. S. W. Johnson) of the Connecticut Agricultural Experiment Station, a position he retained until his death. He was a charter member of the American Society of Biological Chemists and served as the fifth president of the Society (26). He authored and coauthored over

Fig. 17. Thomas Burr Osborne, biochemist at the Connecticut Agricultural Experiment Station from 1886 to 1929 and major contributor to understanding the chemistry of vegetable proteins. He and his chief collaborator, Mendel (Fig. 18), established the Osborne-Mendel strain of rat. (Courtesy of the National Library of Medicine.)

250 scientific publications during his career. In addition, he was a noted authority on birds and served for many years as director of a New Haven bank (28,151).

Osborne's impressive credentials and the innumerable awards he received, however, fail to reveal his real character. The following was written of Osborne by H. B. Vickery and L. B. Mendel, two of his closest associates:

> To those who were privileged to be associated with him in his work he was a rare stimulus, a formidable opponent in argument and an ever genial but just critic. He frequently closed a discussion with the remark that facts were to be found in the laboratory, not in the books. Naturally shy and retiring, the delivery of a public address or of a paper was a severe trial to which he looked forward with trepidation. But among a small group of friends he showed himself as a gifted conversationalist (28).

In short, he was a brillant and shy man, a perfectionist in the chemical laboratory where his true genius found its most natural expression.

Osborne's major scientific interest during the early years of his career was the chemistry of vegetable proteins. These studies consumed most of his effort from 1886 to 1909 and culminated in extensive amino acid analyses of many plant proteins, particularly those of major economic importance: gliadin (wheat) and zein (corn). These studies were summarized in his renowned monograph entitled "The Vegetable Proteins" (111,112), first published in 1909, and extensively revised in 1924. This work, which spanned the first half of Osborne's career, set the stage for the remainder of his career and his important collabortive studies with L. B. Mendel.

Fig. 18. Lafayette Benedict Mendel of Yale, a great teacher and early leader in biochemistry and nutrition research. (Courtesy of the National Library of Medicine.)

Lafayette Benedict Mendel (1872–1935) (Fig. 18) was a native of Delhi, New York. Like Osborne, he received his undergraduate and graduate training at Yale and spent his entire professional career as a member of the Yale faculty. He received his Ph.D. in physiological chemistry at the age of 21 (in 1893) under R. H. Chittenden, one of the most noted early American biochemists. He was first appointed to the Yale faculty in 1892 as Assistant in Phyioslogical Chemistry, and subsequently advanced through the ranks to the position of Professor in 1903. From 1911 to 1935 he held one of the first endowed chairs at Yale, as Sterling Professor of Physiological Chemistry. Mendel was the model university professor. He excelled as an administrator, teacher, and experimentalist, and had an enormous capacity for accomplishment. He authored or coauthored a total of 340 publications in his career. Furthermore, he was a most gifted teacher. In the words of one of his students, "He had a wonderful talent for packing thought close, and rendering it portable" (134). During his career, 92 students received the Ph.D. under his direction, and 237 graduate students and 96 advanced research fellows received part of their training under his supervision. It is said that Mendel trained more departmental chairmen in biochemistry than has any other professor in the United States. Thus, his influence in biochemistry has been profound. He was a charter member of the American Society of Biological Chemists, and at one time or another, held every office of that organization (25,26,134,137,142).

In 1909 Osborne ("the shy, retiring servant") and Mendel ("the sociable extrovert scholar") began 20 years of collaborative studies using rats. The work was supported by funds from the Carnegie Institution and yielded an average of eight papers per year from 1911 to 1927 (142). They were convinced that only through the use of purified diets would it be possible to obtain definitive data on the nutritive value of foods. The striking differences in amino acid composition of plant proteins which had been documented by Osborne, suggested that possible differences might exist in their biological value. The nutritive values of various purified proteins from cereal grains and other plant sources were compared for growth and maintenance in rats. This led to supplementation of "incomplete proteins" with those amino acid(s) limiting each foodstuff's "biologic quality" (e.g., tryptophan and lysine in corn). Casein was found to be a "complete protein," thus paving the way for the use of this protein in modern rat diets. Within a few years, it was possible to list the "essential" and "nonessential" amino acids. These studies necessitated development of a technique (Fig. 19) for feeding individual small animals to permit accurate measurements of food intake (70,113). During the course of these studies, it was noted that coprophagy could be an important variable in use of the rat in nutrition research, and this prompted subsequent investigators in the field to house their rats on wire floors (100). They also developed one of the first metabolism cages (Fig. 19) used in animal experiments (113).

Osborne and Mendel's work with rats also provided important contributions in other branches of nutrition. Their independent discovery (114) of vitamin A came only weeks after it was first reported in 1913 by McCollum and Davis (102). They were the first to recognize cod liver oil as an important source of vitamin A and to show that it was curative of xerophthalmia in deficient rats and children (115). Along with McCollum's group, they did considerable work with the "water-soluble B" vitamins. In 1918 they showed that phosphorus was an essential nutrient (116). Osborne performed several collaborative studies with H. Gideon Wells of the University of Chicago on the anaphylactogenic properties of plant proteins given to animals (28).

The origin of the rats used by Osborne and Mendel is uncertain. Presumably, this ablino stock was established about 1909 (101) and maintained as a random-bred colony for all investigators at the Connecticut Agricultural Experiment Station and Mendel's laboratory. It is known to have been maintained as a breeding colony of about 200 animals by Miss Edna L. Ferry (70), and operated as a closed colony until at least 1935 (106). At times, they were referred to as the "Yale strain" (162), and were noted for their large size (162). Sublines of this stock were established at other institutions around 1915 and used by other famous investigators in the early years of nutrition research, including C. M. McCay and L. A. Maynard at Cor-

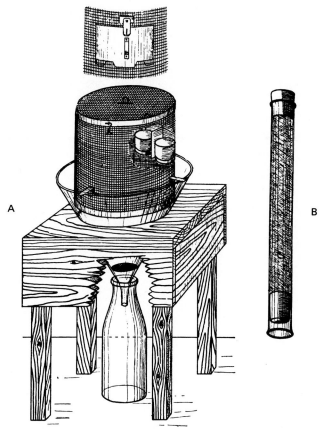

Fig. 19. Metabolism cage and food dispenser of Osborne and Mendel, from their publication of 1911, "Feeding Experiments with Isolated Food-Substances" (112). Their description reads as follows. "(A) Sketch of cage used for feeding and collection of urine and feces. Upper figure shows outer view of food-and-water-receptacle. (Reduced to one-twelfth natural size). (B) Illustration of tube from which daily ration is discharged during each diet period. (Reduced to one-fourth natural size)."

nell, and H. C. Sherman, T. F. Zucker and L. M. Zucker at Columbia University (85,162). A subline established at Vanderbilt University in 1927 by K. E. Mason became known as the "Vanderbilt strain" (160).

C. Henry C. Sherman

Henry Clapp Sherman (1875–1955) (Fig. 20) was born on a farm near Ash Grove, Virginia. He received a Bachelor of Science degree from Maryland Agricultural College (now University of Maryland) and remained there two additional years as a graduate student and assistant in chemistry. He then received a fellowship in chemistry from Columbia University where he was awarded a Master of Science degree in 1896 and the Ph.D. in 1897. From his first appointment in 1895 as Fellow in Analytical Chemistry, he rose within the faculty at Columbia to Professor of Organic Analysis in 1907. His title

Fig. 20. Henry Clapp Sherman of Columbia University, "an analytical chemist, nutritionist, experimental biologist and great humanitarian. "He established a subline of the Osborne-Mendel stock of rat which still bears his name today. (Courtesy of the National Library of Medicine.)

was changed to Professor of Food Chemistry in 1911, and again in 1924 to Mitchell Professor of Chemistry which remained his title until he retired in 1946. He was executive officer of the Department of Chemistry from 1919 to 1939. The University awarded him an honorary doctorate in 1929 (4,39,89).

According to Day (39), Sherman's career progressed through four fairly distinct phases: analytical chemist, nutritionist, experimental biologist, and great humanitarian. The last phase was the natural product of Sherman's real motivations in life: a deep religious faith and a burning desire to achieve long-term improvements in the health of people of all nations. "He was self-effacing because he believed his task transcended self." Thus, he was a quiet, shy person, an introvert whose life was characterized by discipline, courtesy, honesty and kindliness (89). He served in many capacities dedicated to improved human nutrition, ranging from the local level (e.g., the Association for Improving the Condition of the Poor in New York City) to international (e.g., member of a Red Cross team which evaluated the postrevolution food status of Russia in 1917). He received many honors. He served as president of the American Society of Biological Chemists during 1926 (26).

Sherman's career as a scientist was distinguished by its unusual breadth and depth of contribution rather than by an outstanding individual discovery. He and his students often provided the base of knowledge or the analytical techniques which were exploited by many others to advance the field of nutrition. He was by training a chemist, and this background greatly influenced his entire professional career. His early work brought much attention to accurate analysis of foods. During the period from 1895 to 1910, he and his students published about 34 papers on chemical analyses of a wide variety of natural foodstuffs, as well as a few studies dealing with other materials. About 1910, he became interested in digestion and, from 1911 to 1934, published an important series of approximately 50 papers on digestive enzymes, particularly amylases. He and his students clearly established the protein nature of enzymes (39,89).

As the evidence began to accumulate on the importance of dietary deficiencies or imbalances of such things as amino acids, vitamins, iron, and calcium, Sherman began to conduct short-term rat experiments, the first being published in 1919 (138). Within a few years, however, he began emphasizing the two types of life-span studies which he made most famous: (a) comparisons of experimental diets having definite quantities of ingredients and fed to rats over several generations and (b) studies of the improvements in longevity from adding known quantities of an ingredient to diets deficient or marginally adequate in that ingredient. Some of these studies using precisely formulated diets continued beyond 40 generations of rats! Sherman also developed many assay methods for food substances, including vitamins A, B_1, B_2, and C. He was a pioneer in the use of statistical methods in the evaluation of biological data. Altogether, Sherman authored or coauthored well over 300 articles, and published 10 books, many of which went through several editions. One of his collaborators and coauthors was his daughter, Caroline (Mrs. Oscar Lanford, Jr.), who received her Ph.D. in biochemistry at Yale (3,39,89).

The first of Sherman's studies to use rats was published in 1919 (138). From that time forward, rats were considered as essential to the accomplishment of his goals as the chemical laboratory itself. When asked or chided about devoting so much effort to animal research, he replied, ''These animals are my burettes and balances. They give quantitative answers in chemical terms to many of man's greatest problems!'' (89).

Complete details of the albino rats used by Sherman are not available. However, it appears clear that he obtained his initial breeding stock from Osborne and Mendel shortly before 1919 (85,138,139). Furthermore, the methodology for breeding and maintaining the colony was obtained from the Osborne-Mendel group at Yale and the McCollum group at Hopkins (139). Descendants of this presumably random-bred stock eventually were established in other laboratories where sublines are still maintained today.

IV. ENDOCRINOLOGY AND REPRODUCTIVE PHYSIOLOGY

The first known experimental use of rats in an endocrinologic study was an assessment of the effects of adrenalectomy in albino rats by the Frenchman, Philipeaux in 1856 (120). This study was prophetic of one of the major uses to be made of rats, experimental ablation of endocrine glands followed by assessment or restoration of altered function(s) as tools in endocrinologic investigation.

Other examples of surgical ablations included studies by J. M. Stotsenburg at the Wistar on the effects of castration (148) and spaying (149) on growth of the rat. Frederick S. Hammett (80) at the Wistar published in the early 1920s approximately 20 studies involving thyroidectomies and parathyroidectomies in rats. The most notable achievement in this succession, however, came in 1927 when Phillip E. Smith succeeded in hypophysectomizing rats by the parapharyngeal approach (143). These techniques, as well as others, served as a pivotal base for the establishment of the rat as the prime animal subject in experimental endocrinology.

Long and Evans

Sometime between 1915 and 1920, two faculty members at the University of California at Berkeley, Joseph Long of the Department of Zoology and Herbert Evans of the Department of Anatomy (Fig. 21), commenced a collaborative research program of major historical and scientific importance, and launched one of the leading strains of rats, the Long-Evans. Dr. Long had been quietly studying the estrous cycle of the rat for quite a while, but had not published any papers. The combined talents of Long and Evans, however, proved catalytic as they authored during a 3-year period (1920 through 1922) a total of 25 publications dealing with various aspects of reproduction in the rat. These efforts culminated in 1922 with publication of their classic monograph, ''The Oestrous Cycle in the Rat and Its Associated Phenomena'' (97). Their collaboration was over, but a monumental legacy had been given to endocrinology and reproductive physiology (6, 73, 74).

Joseph Abraham Long (1879–1953) received the Bachelor of Science (1904), Master of Arts (1905), and Ph.D. (1908) degrees from Harvard. His doctoral dissertation was on maturation of the egg in the mouse (under E. L. Marks who had previously been advisor on W. E. Castle's doctorate). Immediately upon completion of his doctorate in 1908, he joined the faculty at the University of California as Instructor in Zoology. Through a number of promotions, he became Professor of Embryology in 1939. He became Emeritus Professor of Embryology in 1950. Professor Long was noted as a teacher and as an inventor-master technician in the embryological research laboratory.

Fig. 21. Herbert McClean Evans and Joseph Abraham Long, 1945. These two investigators initiated the Long-Evans stock of rats at Berkeley, California, about 1915. (Courtesy of The Bancroft Library, University of California, Berkeley.)

For example, he invented the first glass wheel-type sharpener for microtome knives. He did some research in endocrinology, including an evaluation of the effects of hypophysectomy on gestation in the rat (with a student Richard I. Pencharz). He also contributed to the field of organ culture by inventing several devices to circulate a nutrient medium through excised organs and embryos. In the later years of his career he was an affiliate of the Institute of Experimental Biology of which Herbert Evans was Director (60,61).

Herbert McLean Evans (1882–1971) was born in Modesto, California where his father was a well-known physician. He obtained a Bachelor of Science degree at the University of California in 1904 and the M.D. degree at the Johns Hopkins University in 1908. As a medical student, he published a number of papers, including one on anatomy of the parathyroid glands with William S. Halsted. He remained at Hopkins as a member of the faculty in the Department of Anatomy from 1908 to 1915, publishing a number of anatomical and embryological papers, while reaching the rank of Associate Professor. During this interval, as a Research Associate of the Carnegie Institution of Washington, he became interested in vital staining of tissues with benzidine dyes, studies which ultimately led to development of the well known Evans blue (6,65).

In 1915, Evans became Professor and Chairman of Anatomy at the University of California in Berkeley, began his collaborative work with Dr. Long, and proceeded along the course of becoming one of the most distinguished scientists and scholars of all time. While most men are content to make a significant contribution in a single discipline, Evans made highly important contributions in no less than six fields: anatomy, embryology, reproduction, endocrinology, nutrition, and the history of science and medicine (6,73,74)!

Evans' work with Long on the estrous cycle of the rat yielded more than an understanding of a few basic phenomena of reproduction in the rat. In the words of a later student of Evans', Dr. Leslie Bennett, "It defined clearly for the first time the duration of the estrous cycle of the rat and showed how the changes could be followed clearly and simply by observing the succession of cell types which are thrown off within the vagina. The nature of the vaginal smear was correlated in great detail with histological studies of all parts of the reproductive system." Further, their work "was the launching pad for the chemistry and biology of the anterior lobe hormones" (6).

Evans and his associates (particularly, Miriam E. Simpson, Olive Swezy, C. H. Li, and others) went on to characterize the hormones of the anterior pituitary and define the interactions between · this gland, the ovary, and the uterus. Many hypophysectomized rats were required in these studies. Initially, they were prepared by Phillip Smith, but he left the department, and this activity as performed by Richard Pencharz, a former student of Dr. Long's (6). By the early 1930s a number of hormones of the anterior pituitary were well known, and a detailed monograph entitled "The Growth and Gonad—Stimulating Hormones of the Anterior Pituitary" was published in 1933. By the end of the 1940s they had separated the luteotropic and follicle stimulating fractions of the anterior pituitary. They next investigated adrenocorticotropic and thyrotropic hormones of the anterior pituitary (73,74).

During the period from 1922 into the early 1940s, Evans and his colleagues (particularly, Katherine Scott Bishop, George Oswald Burr, Gladys A. Emerson, and Oliver H. Emerson) also conducted a major nutrition research program which lead to the recognition of vitamin E (67), characterization of the muscle and infertility problems in the rat due to vitamin E deficiency, and the chemical isolation of α-tocopherol (66). In 1927, Evans and Burr published their monograph, "The Antisterility Vitamin: Fat Soluble E" (66,98).

In 1930, Evans was made Herzstein Professor of Biology and Director of the Institute of Experimental Biology, established on the Berkeley campus for his every-expanding research group (Fig. 22). Evans retired in 1953. During his career he authored or coauthored well over 350 articles and monographs, the majority of which concerned studies using rats (118).

Long and Evans (97) gave a brief account of the establishment and maintenance of their rat colony in their monograph, "The Oestrous Cycle in the Rat and Its Associated Phenomena." The rats were "descendants of a cross" made about 1915 "between several white females and a wild gray male caught in Berkeley." The following statement by Dr. Leslie Bennett more precisely identifies the origin of the Long-Evans rat. "On many occasions I heard Dr. Evans say that the strain was developed by Dr. Long through crossing the Wistar Institute white rat with the wild Norwegian gray rat which

Fig. 22. Professional staff of the Institute of Experimental Biology, posed in 1947 at one of the entrances of the Life Sciences Building (which housed the Institute along with several departments) on the Berkeley campus of the University of California. From left to right: J. A. Long, Hermann Becks, Marjorie Nelson, Alexi A. Knoneff, H. M. Evans, Miriam E. Simpson, Choh H. Li, and Leslie L. Bennett. (Courtesy of The Bancroft Library, University of California, Berkeley.)

Dr. Long had trapped in the banks of Strawberry Creek as it ran through the Berkeley campus of the University of California" (7).

The coat colors represented in the colony of Long and Evans (97) were black, gray, and hooded, but there was no "difference to be observed between these different colored rats with respect to the oestrous cycle." The colony was fed table scraps from a large hotel, except for animals on experiment whose diets were supplemented with whole milk and occasionally, other foods such as raw liver and greens (Fig. 23).

A number of observations on other husbandry practices were made through their early experiences (97). Crowding of animals in cages was associated with "respiratory inadequacy" (the author suspects this was murine respiratory mycoplasmosis), necessitating "separation of animals into pairs or at most into fours...." Maintenance of room temperature (65% to 68°F) by an electric heater was considered desirable. Experience clearly taught the "inestimable advantages of direct and frequent handling of his stock on the part of the investigator" as opposed to the use of "tongs or other rough devices" for handling rats.

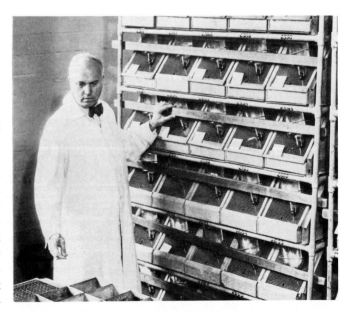

Fig. 23. Herbert Evans working in his rat colony, about 1920. (Courtesy of The Bancroft Library, University of California, Berkeley.)

V. CANCER RESEARCH INSTITUTES

A. The Crocker Institute of Cancer Research

A series of key events in the establishment of the rat for research purposes took place in New York City. In 1912, Mr. George Crocker, a businessman, donated $1,600,000 ("The George Crocker Special Research Fund") to Columbia University for the purpose of studies to find "the cause and cure of cancer" (63,161). [Elsewhere the figure has been given as $2,500,000 (49).] Unlike many philanthropists who delight in architectural edifices bearing their own names, Mr. Crocker specifically forbid that any of his contribution be used for a building. For this reason, the cancer program had to be housed in "borrowed space" in the Zoology Department, Schermerhorn Hall, until the University was able to raise the additional support. Eventually, the sum of $40,000 was provided and used to erect, at the corner of Amsterdam Avenue and 116 Street, "the largest possible building, with the sole requirements of the best illumination and the most floor space" for the money. It had a basement and three floors but, because of lack of funds, initially had no partitions. It was opened December 15, 1913 and named "Crocker Research Laboratory" (Fig. 24) but, in jest, called "The Workshop of Unrivaled Plainness" (161) or, by Columbia students, "The Canker Fund" (63). Nevertheless, the name "Institute of Cancer Research" became most widely accepted (159).

Francis Carter Wood (1869–1951), a pathologist who had authored the text "Clinical Diagnosis" in 1899, was chosen as the first Director of the Crocker Laboratory in 1912. Wood had major service commitments at surrounding hospitals, particularly St. Luke's located two blocks from the Crocker Laboratory. Nevertheless, he and other managers of the Crocker Fund maintained that "In view... of the failure of the medical profession to discover, after two thousand years of observation, the nature and cause of cancer in man, it is considered... inadvisable to expend much energy in investigations along that line" (i.e., clinical studies alone). Thus, Wood was cast throughout most of his 28 years as Director, in the difficult role of conserving the Crocker Fund for "purely experimental purposes" (basic research) against equally strong-willed administrators more interested in clinical medicine. The Fund suffered severe damage in the stock market crash of 1929, funding for cancer research generally was extremely meager over the next decade, and Wood went into semiretirement in 1936 (his official retirement was announced in 1940) at which time the cancer research program was moved into the College of Physicians and Surgeons, where it became the "Department of Cancer Research" (159).

Despite the turbulence of the times, Dr. Wood managed to mount a sizable research program. Early additions to his staff included three other pathologists, Frederick Dabney Bullock

Fig. 24. The Crocker Research Laboratory of Columbia University, built in 1913 at a cost of $40,000. It was here, on the third floor, that inbreeding of six major bloodlines of rats began: August, Copenhagen, Fischer, Marshall, Zimmerman, and Avon. [Courtesy of Ms. Betty Moore, Columbia University; from Wood (161).]

Fig. 25. Frederick Dabney Bullock, pathologist and scientist at the Crocker Institute of Cancer Research from 1913 to 1937. (Courtesy of the National Library of Medicine.)

(1878–1937) (Fig. 25), William H. Woglom, (1879–1953), and George Louis Rohdenburg (1883–1967), and a parasitologist, Gary N. Calkins. Bullock and Rohdenburg held appointments at Lenox Hill Hospital (63,108).

A major problem confronting cancer research at the time the Crocker Laboratory opened was the need for a reproducible animal model of cancer. It was known that Borrel (11), a French worker, had reported the production of sarcomas in rats fed tapeworm eggs. Bullock and Rohdenberg attempted to repeat Borrel's work, but failed. At about this time (1916), the group was joined by Dr. M. R. Curtis (50).

Maynie Rose Curtis (1880–1971) (Fig. 26) was a native of Mason, Michigan who had obtained undergraduate and advanced degrees (B.S. 1905, M.A. 1908, Ph.D. 1913) in Biology at the University of Michigan and had subsequently worked for eight years as a research associate studying poultry reproduction at the Maine Agricultural Experiment Station at Orono. It was her belief that Bullock and Rohdenburg had failed in producing cancers by Borrel's method because their experimentally infected rats simply did not live long enough. Most of their rats had died by 6 months of age. This Curtis attributed to poor housing conditions and diet. The standard practice had been to keep the rats in small boxes and to feed each rat a piece of dry bread and some vegetable (their sole source of water) every day. Curtis instituted an improved caging system and the practice of dipping the bread in whole milk before it was fed. (These pieces of milk-soaked bread were called "soggies".) With these improvements, the rats lived

much longer, and Bullock and Curtis reported their first successes in producing sarcomas in 1920 (14). The minimal time for sarcoma induction proved to be 8 months (50).

Encouraged by their early successes in producing cancers, Bullock and Curtis sought to expand their work. It was soon noted that rats from some vendors developed a higher incidence of sarcomas than those from other sources, when infected with *Taenia* eggs. Dr. Curtis then persuaded the group that the best way to find out whether these differences were truly genetic was to develop inbred strains of rats, and further, that these strains should be selected for variations in visible characteristics which might be related (linked?) in some way to the incidence of cancer (50). In 1919 (15), she purchased a few breeding pairs of rats from each of four local breeders whose names were: Fischer, Zimmerman, Marshall, and August. The rats from Marshall were always albinos. Fischer and Zimmerman each had rats which were black non-agouti piebald, but carriers of the albino gene. August had the most varied rats including some of the pink-eyed dilutes which Thomas Hunt Morgan (Nobel Laureate in Physiology and Medicine, 1931) had given to him after bringing them back from England some years before (50). A few rats also were obtained in 1920 from Dr. Jacob Rosenstirn (36). This stock he had originally obtained from a breeder in Copenhagen who supplied rats to Dr. Johannes Fibiger (Nobel Laureate for cancer research in 1927) (72, 157). Many years later, in 1941, an interesting group of

Fig. 26. Maynie Rose Curtis, biologist, cancer scientist, and chief developer of the inbred rat colony at the Crocker Institute of Cancer Research. (Courtesy of Dr. W. F. Dunning.)

Table II

Inbred Strains of Rats Which Originated at the Crocker Research Institute
of Columbia University

Strain[a]	Date of first mating	Generations of brother × sister matings in 1953[b]
Fischer 230	Sept. 1920	49
Fischer 344	Sept. 1920	51
Zimmerman 61	June 1920	56
Marshall 520	Nov. 1920	51
August 990	Feb. 1921	46
August 7322	Nov. 1925	53
August 28807	Feb. 1936	23
August 35322	Oct. 1942	12
Copenhagen 2331	Aug. 1922	43
A × C 9935	Dec. 1926	38
Avon 34986	Dec. 1941	14

[a] Most of these strains are still maintained by Dr. W. F. Dunning at the Papanicolaou Cancer Research Institute, Miami, Florida.

[b] From Heston et al. (85). Many more generations have been added over the years in the parent lines (Dunning) and in sublines maintained by others (see Chapter 3).

Fig. 27. Wilhelmina Francis Dunning, geneticist, cancer scientist, and chief preserver of the inbred rat strains began at the Crocker Institute of Cancer Research. (Courtesy of Dr. W. F. Dunning.)

rats with "ruby eye" were obtained from a breeder in Avon, Connecticut (50). These purchases provided the seed stocks for brother × sister matings and the development of several of today's more important inbred strains of rats (Table II). The first litter of pedigreed rats in the Institute was from mating number 344, and this was the beginning of the Fischer 344 strain (50).

Wilhelmina Francis Dunning (Fig. 27) (born in 1904 at Topsham, Maine) joined Drs. Curtis and Bullock as a part-time student assistant in 1926. She had completed her Bachelor of Arts degree at the University of Maine and had been admitted as a graduate student with Dr. Thomas Hunt Morgan in Zoology at Columbia. She completed the doctoral program at Columbia in Genetics and Cytology in 1932, and elected to join Drs. Curtis and Bullock full time. Her major role initially was to analyze the large amount of data which Dr. Curtis and Dr. Bullock had accumulated on sarcoma induction in the several strains of rats. By 1933, data was available on 3669 rats in which sarcomas had been induced experimentally. Two major factors emerged as being of paramount importance in sarcoma induction: (a) the relative susceptibility of the strains of rats to the parasite infection and (b) the longevity of the different strains of rats (37,52). Simultaneously, data were collected on a large number of spontaneous neoplasms from rats of the various lines (15,33), a transplantable lymphosarcoma was described (35), and a number of coat color mutations were documented (34).

By 1940, a great deal of animal research history had been made using the inbred rats at Columbia. In a retrospective summary of events for the 20 years from 1920 to 1940 written in 1940 by Dunning (51), the following recap appeared.

A rat village of some 10,000 individuals was built up, composed of as many divergent strains as was practicable and with the greatest attainable biological uniformity within each line. This colony has been under observation for more than twenty years. The rats were uniformly housed in clean wooden cages on hangers suspended from the ceiling [Fig. 28] in a room well suited to such a society. They were uniformly exposed to a constant supply of clean water and an adequate varied diet consisting of bread, cereal, whole milk, fresh vegetables, and raw beef. Pregnant females were removed from breeding cages and reared their young in isolation boxes making it possible to record their birth date and pedigree. Each rat was autopsied and notes were taken on the presence or absence of gross spontaneous or experimentally induced tumors. All gross lesions suggestive of cancer were examined microscopically. To date more than 100,000 rats have been autopsied and there are available a complete pedigree and case history for each of more than 14,000 bearers of experimentally induced and spontaneous tumors.

The animal colony was housed on the third floor of the Crocker Research Laboratory.

Following the reorganization of the cancer research program at Columbia as the Department of Cancer Research within the Medical School, Dr. W. H. Woglom served as Acting Director from 1940 to 1946 (63). Subsequently, the program was administered by the "Cancer Research Coordinating Committee" with Dr. Alfred Gellhorn serving as chairman from 1957 to 1968. Dr. Sol Spiegelman became Director in 1968. In

Fig. 28. A photograph of the "rat village" at the Crocker Institute of Cancer Research in the late 1930s. Wood cages were kept on wood shelves suspended from the ceiling. (Courtesy of Dr. W. F. Dunning.)

1972, the program was once again reorganized as the "Cancer Research Center" under Dr. Paul A. Marks, Vice President for Health Sciences. It includes both the Institute of Cancer Research and the Clinical Cancer Facility in Presbyterian Hospital (108).

The Crocker Research Laboratory building (Fig. 24) was razed about 1960 to make way for new schools of Law and International Affairs which share the site today (108).

B. The Detroit Institute for Cancer Research

The reorganization of the Columbia cancer research program, which began in 1936 when the program was made a department in the School of Medicine, continued well into the 1940s. The combined pressures of changing faculty research interests, shortages of laboratory space, the increasing expense of the rat colony, and the general emergency associated with World War II seriously threatened the continued existence of the inbred rat colony. Dr. W. C. Rappleye, the Dean of Medicine, made serious efforts to relocate the colony to a farm site, but these attempts proved unsuccessful. Thus, the environment for continued research at Columbia became unacceptable to Dr. Dunning, particularly after Dr. Curtis' retirement from that institution in 1941 (50,108).

Dr. Dunning and Dr. Curtis soon received an invitation from Dr. Rollin Stevens, Radiologist and Chairman of the Cancer Committee of the Wayne County Medical Association, and Dr. J. Edgar Norris, Chairman of Pathology and Acting Dean of the Wayne University College of Medicine, to move to Detroit and serve as the nucleus of scientists for a new cancer institute. Through the help of Dr. Wood, Dr. Dunning obtained permission to move the pedigreed rat stocks to Detroit. On Memorial Day of 1941, the entire colony of rats, which had been reduced to approximately 1000, was transported on the train known as the "Commodore Vanderbilt" from New York City through Canada to Jacksonville, Michigan. From Jacksonville, they were carried by truck approximately 25 miles to the Curtis family farm at Mason, Michigan where Dr. Curtis had grown up and which she had now inherited. Work had begun on renovating a swine barn for the rats, but this was incomplete so the rats initially had to be housed in the basement of the farm house. Because of the war, commercial dog food (Carnation's "Friskies" and Wayne's "Lucky" brands were used) was not always available, so yellow corn and vegetables grown on the Curtis farm were at times fed to the rats. Initially, the laboratory space provided Dr. Dunning at Wayne University was very limited, and it was necessary to maintain the breeding colony at the farm. Dr. Dunning spent the weekends at the farm and, as they were needed for experiments, carried a hundred or so rats by car to the laboratory some 84 miles away (50).

Within a few years, the Southeastern Michigan chapter of the Women's Field Army (now American Cancer Society) had raised approximately $250,000 to be used for a building to house the Detroit Institute for Cancer Research. A two-story building which had been used by the Ford Motor Company's Sales and Service Division and located at 4811 John R (Street) in Detroit was purchased and renovated. The top floor was used for housing the animal colony and for the research laboratories. It was here that Dunning and Curtis (53) demonstrated that the intraperitoneal injection of washed and ground larvae of the tapeworm would produce multiple intraperitoneal sarcomas in rats. Further strain differences in neoplastic diseases of the rats also were elaborated (54,59).

The Women's Field Army occupied the lower floor of the Institute's building for office space. The frequent (often weekly) visits of up to 200 women into the animal rooms and laboratories proved to be detrimental to the research program, and Dr. Dunning once again began searching for a more favorable environment. The Detroit Institute for Cancer Research later became the Michigan Cancer Foundation which continues a cancer research program at this same site today (50,62).

C. The University of Miami and the Papanicolaou Cancer Research Institute

For some time the University of Miami had been interested in building a medical school and attracting an affiliated Veterans Administration Hospital to its campus. In order to be successful, the administration felt that an essential first step was to begin a research program. As part of this effort, the offer of a faculty position was extended to Dr. Dunning along with a promise of support for developing laboratory space and renovating building No. 29 (Fig. 29) to house the rats on the South Campus (a former naval base for blimps at Perrine, Florida). The sum of $50,000 had been provided for renovations by the Damon Runyon Fund (50).

After completion of the necessary renovations, in June of 1950, Dr. Dunning, accompanied by her close friend and collaborator, Dr. Curtis (who sold her farm in Michigan), moved the colony of inbred rats to the University of Miami. The rats

Fig. 29. "Building No. 29," on the South Campus of the University of Miami (Florida). After the inbred rats from Columbia had been moved to Michigan in 1941, they were moved into this building in 1950 and later moved to the Papanicolaou Cancer Research Institute in Miami. (Courtesy of Dr. W. F. Dunning.)

were placed in wooden cheese boxes with holes on the sides covered by screen wire, and carried by truck to the Detroit Airport where they were accepted by a well-known commercial airline and loaded on an aircraft bound for Miami. Having now moved into the modern era of transportation, about half of the rats succumbed on the way. The water bottles arrived separately so that an improvised watering system had to be provided. The Marshall 520 strain was reduced to one pregnant female which later whelped to provide the only breeders for perpetuating that line. The Fischer 230 strain was eventually lost, but this may not have been due to problems in transporting the colony (50).

At the University of Miami the research program on the larval tapeworm-induced sarcoma in the inbred rats was again very active. Efforts by Dr. Dunning and Dr. Curtis (57) to isolate the carcinogen of *Taenia* larvae demonstrated that the active agent is associated with the calcareous corpuscles of the parasite. In addition, the inbred rat strains were of crucial importance in studies of transplantable neoplasms (58) and in experimental carcinogenesis (55,56). Also, many inbred rats (and mice) were produced under contract with the National Cancer Institute. Dr. Curtis continued to be active in the research, "working 7 days a week in the laboratory until within 2 days of her death on April 13, 1971 at 91 years of age" (50).

In 1971, the inbred rat colony was moved to the Papanicolaou Cancer Research Institute where it is maintained today. Dr. Dunning continued an active research program there until her retirement July 1, 1977. She still resides in Miami and does part time research at the Papanicolaou.

VI. BEHAVIORAL RESEARCH

The use of the rat in behavioral research began in the Department of Neurology at Clark University in Worcester, Massachusetts shortly after Donaldson left that department in 1892 for the University of Chicago (27,109,152).

A. Doctoral Students at Clark University

The first behavioral studies using rats were done by three doctoral students at Clark University: Stewart, Kline, and Small. According to Munn (109), Colin C. Stewart at Clark University in 1894 began working "on an investigation of the effect of alcohol, diet, and barometric changes on animal activity (146). He placed wild gray rats in revolving drums and measured their activity in terms of the number of revolutions of the drums (Fig. 30A). Wild gray rats were difficult to handle and so Stewart changed in 1895 to white rats." The source of the white rats is uncertain, but it seems probable they were given to Stewart by Adolf Meyer who joined the staff of the Worcester State Hospital in 1895 (107). Further contributions

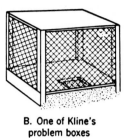

A. Stewart's activity wheel

B. One of Kline's
problem boxes

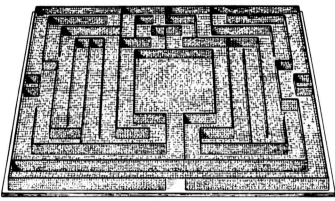

C. Small's Hampton Court maze

Fig. 30. "Some historically significant devices used in early research on rat behavior. (A) Stewart's activity wheel. The rat lives in the cage at the center of the wheel. Part of the recording mechanism is at the right. (B) One of Kline's problem boxes. Escape is effected by burrowing through the sawdust. (C) Small's Hampton Court Maze. Food was in center." [From Munn (109), used by permission.]

from Clark University included Linus W. Kline's work (94) with problem boxes (Fig. 30B) and Willard S. Small's studies (140,141) with the "Hampton Court Maze" (patterned after the maze at Hampton Court Palace near London, England) (Fig. 30C) in which rats were trained to run through the maze to obtain food (136).

B.   Watson and Meyer

Following the initial studies using rats at Clark University, the focus of attention in animal behavior quickly shifted to one of Donaldson's doctoral students at the University of Chicago, John Broadus Watson (1878–1958) (Fig. 31), who was destined to become one of the great leaders in the field. Munn (109) summarized his contributions made with the rat as follows.

Although the use of white rats for psychological investigation began at Clark University, research with rats was given its greatest impetus by the Chicago University investigations of Watson and Carr. In 1901 Watson was investigating the relation between myelinization and learning ability in white rats. This work was published in his "Animal Education" in 1903 (153). In 1907 Watson (154) published his very significant work on the role of kinesthetic and organic processes in maze learning by white rats. This work was followed in 1908 by Carr and Watson's study of

Fig. 31.   John Broadus Watson, psychologist of the University of Chicago and The Johns Hopkins University, gave great impetus to early use of the rat in behavioral research. [From Watson (155a), used by permission.]

orientation in the maze (16). These experiments and others which soon followed from the Chicago Laboratory firmly established the white rat as a subject for psychological investigation.

Watson moved in 1908 to the Johns Hopkins University School of Medicine to become Professor of Experimental and Comparative Psychology, and Director of the new Psychological Laboratory there (107). The next year, in 1909, Adolf Meyer also moved to the Hopkins as Professor of Psychiatry and Director of the nearby Henry Phipps Psychiatric Clinic. The combined presence of these two great men provided an unusually fertile environment for behavioral research. However, Watson's work, in particular, was to give further impetus to use of the rat in research. According to Munn (109), "the most influential book in bringing about the rapid growth of animal psychology was Watson's "Behavior: An Introduction to Comparative Psychology," published in 1914. Although Watson believed that animal research would contribute much to the understanding of fundamental processes underlying human behavior, he also believed that animal research had a right to exist for its own sake." He stated, "the range of responses, and the determination of effective stimuli, of habit formation, persistency of habits, interference and reinforcement of habits, must be determined and evaluated in and for themselves, regardless of their generality, or of their bearing upon such laws in other forms, if the phenomena of behavior are ever to be brought within the sphere of scientific control" (155,155a).

Watson and Meyer had two graduate students at Hopkins who subsequently proved to be major contributors to the field. Karl Lashley began his work in 1912 and subsequently published on a wide diversity of topics concerning rats, including: form and size discrimination and the effects of drugs and experimentally induced brain lesions on learning (107). The second student was Curt Paul Richter (Fig. 32) who joined Watson in 1919 shortly after the Psychological Laboratory was transferred to the Phipps Clinic. Watson resigned from the Hopkins in 1920 as result of a controversy surrounding his divorce. A year later, Curt Richter was appointed director of the laboratory, and its name was changed to the "Psychobiological Laboratory" (107).

From 1919 to 1977, Richter conducted a steady stream of research projects on psychobiological phenomena of the rat, including spontaneous activity, biological clocks, physiologic effects of adrenalectomy, self-selection of nutrients, poisoning, stress, and domestication. His laboratory, the direct successor of the laboratory began by Watson, produced approximately 250 research papers and one book of broad fundamental importance to use of the rat in behavioral research (9).

Richter's research, throughout his career, was heavily dependent on a rat colony of 400–600 animals he established at Hopkins in 1922 on the third floor of the Phipps Clinic. The foundation stock for this colony was albinos from the Wistar Institute. "A few hooded, tan and black rats from the colony of

Fig. 32. Curt Paul Richter, immediate successor of Watson as Director of the Psychological Laboratory (changed to "Psychobiological Laboratory" in 1919) at The Johns Hopkins University School of Medicine. (Courtesy of the National Library of Medicine, Bethesda, Maryland; used with permission of Bachrach Studios, Watertown, Massachusetts.)

Dr. E. V. McCollum were introduced in 1927." The diet always consisted of graham flour 72.5%, casein 10.0%, butter 5.0%, skim milk powder 10.0%, calcium carbonate 1.5%, and sodium chloride 1.0% (126).

The pioneering work of the group at Clark University, Watson, Richter, and others touched off a major wave of behavioral research using the rat in the 1920s and 1930s. By the 1940s a major share of the methodologic approaches used in this field had been developed (95,107,109,119,125,152).

VII. GENETICS

The earliest studies of genetics in the Norway rat were published in Germany by Crampe from 1877 to 1885 (29–32). Albino mutants crossed with wild gray (agouti) rats resulted in offspring of 3 colors (agouti, black, and albino) and two patterns (uniform pigmentation and white spotting). Following the rediscovery of Gregor Mendel's laws in 1900, Crampe's data were reexamined by Bateson (5) and his student Doncaster (46) who found the coat color ratios to conform with Mendelian expectation (22,156).

Other significant events concerning genetics of the rat came with the establishment of the Wistar Institute and the beginning of an inbred colony by H. D. King in 1909. Over the next few decades, genetics of the rat increasingly gained the attention of one of this nation's first mammalian geneticists, W. E. Castle.

A. Castle and the Bussey Institution of Harvard

William Ernest Castle (1867–1962) (Fig. 33) was born in 1867 on his father's farm near Alexandria, Ohio. His first college experience was at Denison University in Ohio, a sectarian college emphasizing the classics and ancient languages, from which he obtained a Bachelor of Arts degree. After 3 years of teaching Latin at a college in Kansas, he entered Harvard University and took a second Bachelor of Arts degree, graduating Phi Beta Kappa in 1893. He then became a laboratory assistant in Zoology at Harvard where he received the Master of Arts in 1894 and the Ph.D. in 1895 (under E. L. Marks). Subsequently, he held brief appointments as Instructor of Zoology at the University of Wisconsin, Knox College, and Harvard. He remained at Harvard where he was promoted to Assistant Professor in 1903, to Professor in 1908, and became Emeritus Professor of Genetics in 1936. Subsequently, he was Research Associate in Mammalian Genetics at the University of California in Berkeley where he continued an active professional career until his death in 1963 (48).

The rediscovery of Gregor Mendel's principles of inheritance around 1900 gave birth to the new science of genetics. Out of the group of bright young biologists who responded to

Fig. 33. William Ernest Castle of Harvard, first full time mammalian geneticist in the United States and heavy contributor to early genetic foundations of the rat. [Courtesy of the National Library of Medicine; see Dunn (48).]

Fig. 34. The Bussey Institution for Applied Biology (The Harvard Graduate School of Biology from 1908 to 1936). It was here that Castle "for the first time had space and freedom to develop an extensive program in mammalian genetics" (48). (Courtesy of the Harvard Archives.)

this new challenge in the United States, there emerged a small nucleus of scientists who were to serve as this nation's first generation of leaders in genetics: T. H. Morgan, C. B. Davenport, W. E. Castle, H. S. Jennings, B. M. Davis, R. A. Emerson, G. H. Schull, and E. M. East. Castle was the first of this illustrious group to devote himself exclusively to genetics, and he had the longest active career—almost 60 years! His first paper on genetics appeared in 1903, and his last (bringing his total bibliography to 246 titles) in 1961 (48). He published a number of important books, including "Genetics and Eugenics" (18), "The Genetics of Domestic Rabbits" (19), and "Mammalian Genetics" (20).

Following Castle's appointment as Assistant Professor at Harvard in 1903, his career in genetics soared. He was promoted to the rank of Professor in 1908 and elected to membership in the National Academy of Sciences in 1915. One of the key developments in his career came, however, in 1908 with the establishment of the Bussey Institution for Applied Biology. "The Bussey," as it was known, was the graduate school of biology at Harvard and the major base of operations for Castle from 1908 to 1936.

The main building of grey stone [Fig. 34] had been built some forty years before in the midst of the fields adjacent to the Arnold Arboretum in that part of Jamaica Plain known as Forest Hills, some ten miles distant from Cambridge. There were in addition greenhouses, barns and outhouses, and a frame dwelling used by the graduate students as a dormitory. Castle moved his rabbits into the basement of the main building, his rats into the large west room on the first floor, and his guinea pigs into an out building known as the pigeon house. Soon a mouse-room was established in what had been a greenhouse attached to the front of the main building. Here Castle for the first time had the space and freedom to develop an extensive program in mammalian genetics. He was fortunate in having the help of a foreman who acted also as janitor—Mr. Patch—a former Maine farmer who tilled the fields, grew food for the animals, and invented devices for feeding, watering, and keeping the animals securely in their cages [Fig. 35]. Here he received as fellow-workers the first of the graduate students who took their degrees under his direction (48).

Genetics as taught at the Bussey encompassed both mammalian genetics (under Castle) and plant genetics (under E. M. East). "Both [men] were strong and positive personalities and in temperament and general views they were destined to be often in disagreement." Nevertheless, they spearheaded a cohesive program which by 1936 had prepared 40 doctoral candidates, among them: C. C. Little, Sewall Wright, L. C. Dunn, C. E. Keeler, G. D. Snell, and P. B. Sawin (48). The importance of these teaching activities to modern mammalian genetics can be more fully appreciated by studying the Castle "intellectual family tree" recently published by Morse (108a).

Castle published papers on genetics, particularly inheritance of coat colors, in many animal species. However, the rat proved crucial in answering the question which dominated his early career up to about 1919, i.e., "whether Mendelian characters are, as generally assumed, incapable of modification by selection...." ("Mendelian characters" were referred to by Castle as "unit characters";... they later became

Fig. 35. "Rat cage of Professor William E. Castle, of the Bussey Institution, Harvard Institution," from the 1923 edition of the book by Greenman and Duhring (77). The cage almost certainly was designed and built by Mr. Patch. Castle's remarkable janitor, animal technician, and generally versatile assistant. As described by Greenman and Duhring, "The cage consists of a galvanized sheet-metal pan 17 inches long, 15 inches wide and 3 inches deep. In this pan rests the cage proper which is a completely closed box constructed of ½ inch mesh no. 18 galvanized wire cloth. This box is 16 inches long, 14 inches wide, 8½ inches high at one end and 4 inches high at the opposite end. For a space of 4 inches wide across the high end, the top is flat, the remaining area of the top slopes from a height of 8½ inches to a height of 4 inches. In the sloping portion of the top is the only entrance to the cage, 5 inches × 5 inches, closed by a wire-cloth door. Upon the door rests the water bottle.... The weight of the water bottle is not always sufficient to hold the door closed, an ingenious clip is therefore provided to hold it securely in place. These cages will accommodate five or six adult rats. The cleaning process is reduced to a minimum, as the cage containing the rats may be lifted from the pan which catches the litter falling from the cage."

known as "genes.") In order to answer this question, extensive studies were made of coat color inheritance in rats.

Castle with a succession of collaborators, chiefly J. C. Phillips, proved that a black-and-white spotting pattern, called "hooded" known to segregate as a "unit-character" in rats, could be modified by selection in both directions toward more black and toward more white and that crossing with wild rats increased the amount of white in the hooded pattern of the spotted animals extracted from the cross. This happened to a conspicuous extent when the line selected toward more white was outcrossed, but there was slight effect of this kind when the dark-selected line was used. This seemed to contradict the dogma of "purity of the gametes," so an extensive study involving some 50,000 rats was carried out between 1907 and 1919. This showed that the character was modifiable, but the crucial test (suggested by Castle's student Sewall Wright) showed that the modifications were due to genes separable in crosses from the hooded gene, which had not itself been changed. The results were shown to be due to the operation of multiple modifying factors... (23,48).

Thus, a basic understanding of the hooded pattern in rats had been developed by 1919 (17), laying a firm foundation for further understanding of coat colors in rats (Figs. 36–39) in subsequent years (22). After about 1920, Castle's main interests shifted to analysis of inheritance of body size in mammals and the construction of chromosome maps for the rabbit, rat, and mouse (48).

Hooded rats

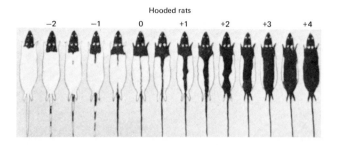

Fig. 36. Castle's grading scale for hooded rats, largely worked out by Castle and Phillips in 1914 (23). [From Castle (22).]

B. Castle's Second Career

The year 1936 was another major turning point in Castle's life. With the completion of the new Biological Laboratories on the main campus, the Bussey closed its doors and Castle went into retirement, both on July 1, 1936. In the words of Dunn (48), "... the rich odors of an old building which housed hundreds or thousands of rats and mice and rabbits and guinea pigs, and the spaciousness (and often the low temperature) of its high-ceilinged rooms failed in the end to compensate for its physical separation from the main center of the University." Despite his disappointment, Castle faced these new realities squarely as expressed in a letter to Dunn in February of 1936. "I am grateful for the long continued opportunities which I

Fig. 38

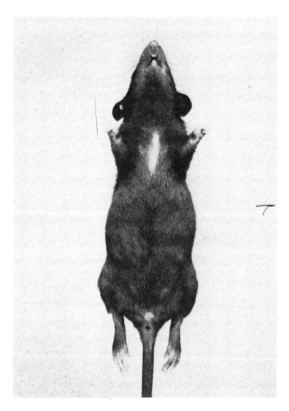

Fig. 37

Fig. 39

Figs. 37–39. Rats representing 3 alleles at the H locus, from the work of Castle. Fig. 37. Irish ($h^i h^i$) rat of grade + 5¾. Fig. 38. Hooded rat (hh) of grade + 2. Fig. 39. Notch rat ($h^n h^n$) of grade −3. [From Castle (22), by permission.]

have enjoyed for scientific research, which is indeed a privileged status. Now I shall take a vacation and look around for something worth doing while I continue so damned healthy and so am unable to die, as I should.'' He subsequently became Research Associate in Mammalian Genetics at the University of California in Berkeley and, in essence, launched a second career in genetics which spanned 26 years and produced 55 publications. It was during this interval that he and H. D. King did most of their ten collaborative linkage studies in the rat (48). By 1947, he had identified a total of 23 mutations in the rat (21). These and his many other fundamental contributions have served as the foundation for modern genetics of the rat as well as other laboratory animals—indeed for mammalian genetics.

VIII.   OTHER STOCKS OF RATS

A.   Sprague-Dawley

The precise origin of the so-called ''Sprague-Dawley rat'' appears to be lost in uncertainty. The original stock was reportedly established about 1925 by Mr. Robert Worthington Dawley (1897–1949), a physical chemist at the University of Wisconsin (83). In naming the strain, he simply combined the maiden name (Sprague) of his first wife and his own name to form Sprague-Dawley. He subsequently established in Madison, Wisconsin the commercial firm known as Sprague-Dawley, Inc. dedicated exclusively to the advancement and sale of his rats.

The following excerpt from a letter, dated July 22, 1946, to Mr. S. M. Poiley (121) of the National Institutes of Health from Sprague-Dawley, Inc. summarizes the meager information available on origin of their rat stock.

> Regarding the origin of the strain, this strain as developed by Mr. Dawley, started originally with a hybrid hooded male rat of exceptional size, and vigor, which genetically was half-white. He was mated to a white female and subsequently to his white female offspring for seven successive generations. The origin of this male is unknown. The original white female was of the Douredoure strain which probably was from Wistar. After his death, his white offspring were inbred in a number of different lines from which the best ten were combined. Selection was made to retain or acquire characteristics of high lactation, rapid growth, vigor, good temperament, and high resistance to arsenic trioxide.

Mr. Dawley died in 1949. His original company continues today under the name of ARS/Sprague-Dawley.

Mr. Evan Carl Holtzman, who for many years was an employee of Sprague-Dawley, Inc., left that company in the 1940s and established his own business (also in Madison, Wisconsin) using Sprague-Dawley animals as seed stock. Thus, came into being the so-called ''Holtzman rat'' (83,121).

Many sublines of these stocks exist today and, indeed, Sprague-Dawley descended animals are among the more popular rats used experimentally. Virtually all of them are randombred, perhaps the most distinctive feature of the geneology of these stocks.

B.   Albany Strain

In October of 1930, Professor Arthur Knudson of the Department of Biochemistry at the Albany Medical College (Albany, New York) obtained 11 female and 5 male rats from Dr. C. E. Bills of the Mead Johnson Company of Evansville, Indiana. (The origin of Bills' stock is unknown.) Professor Knudson maintained a breeding colony of these animals until 1936 when they were moved to other quarters at the Albany Medical College. Shortly thereafter, they were discovered to have a high incidence of spontaneous mammary tumors and designated the ''Albany strain'' (13). Subsequently, these animals were studied extensively because of their high incidence of tumors, comparisons usually being made with the ''Vanderbilt strain,'' a subline of Osborne and Mendel's stock which had been established at Vanderbilt by K. E. Mason in 1927 (160). In 1950, representatives of the Albany stock were transferred from J. M. Wolfe and A. M. Wright at Albany Medical College to the National Institutes of Health where inbreeding was begun (71).

C.   Hunt's Caries-Susceptible and Caries-Resistant Strains

In 1937, at the suggestion of Dr. Morris Steggarda of the Carnegie Institution of Washington, Dr. H. R. Hunt, Dr. C. A. Hoppert and their associates at Michigan State College (now Michigan State University) initiated a long series of studies of hereditary influences in dental caries of rats. Their strategy and execution of this project are revealed in the following excerpt from a paper by Hunt, Hoppert, and Erwin (87).

> The first objective of the investigation was to determine whether there is an inheritance factor in susceptibility and resistance to dental caries in the albino rat. We therefore undertook to build a caries susceptible line, and a caries resistant line, using phenotypic selection, brother × sister inbreeding, and progeny testing of breeders as genetic techniques to effect this purpose. It was desirable to start with a considerable number of animals from a variety of sources to increase the chance of securing at the outset as large an assortment as possible of genes for susceptibility and resistance. The following rats comprised our first generation: 18 from the colony of the Psychology Department at Michigan State College, 17 from the Nutrition Laboratory of the Home Economics Division, and 84 from the rat laboratory of the Chemistry Department. Selection was fairly stringent at the outset, for only 16 animals (13.4%) chosen as breeders produced young that were themselves mated. These were: susceptibles, 2 males (1 from the Psychology laboratory, 1 from the Chemistry laboratory), 5 females (2 from Psychology, 2 from Chemistry, and 1 from Home Economics); resistants, 3 males (all from the Chemistry colony), and 6 females (1 from Psychology and 5 from Chemistry). The object in selecting and intensively inbreeding by brother × sister matings was not only to

build a susceptible and a resistant line, and thereby to demonstrate the inheritance factor, but also to create homozygous susceptible and resistant types which would be crossed, so that genetic segregation could be observed among the grandchildren from such matings, supplying information by which we might determine the number of pairs of genes that are involved.

The studies in Hunt's laboratory were initially based on use of the cariogenic diet of Hoppert, Webber, and Canniff (86), which consisted of coarsely ground hulled rice 66%, whole milk powder 30%, alfalfa leaf meal 3%, and NaCl 1%. Although caries-susceptible and -resistant strains were developed using this diet (88), the same differences were later demonstrated using a diet containing 57% sucrose (147). Intensive research in this field in recent years by many investigators has led to the belief that the strain differences observed by Hunt and associates are attributable not to Mendelian factors, as originally thought, but to multiple factors, particularly microbial flora of the oral cavity and dietary factors (96,110). Nevertheless, these pioneering studies led to the establishment of two inbred strains of rats, Hunt's caries-susceptible (CAS) and caries-resistant (CAR).

IX. GENEALOGY OF MAJOR RAT STOCKS AND STRAINS

Figure 40 shows the pedigree relationships of the major bloodlines of laboratory rats in use today. They comprise only about a dozen more or less distinct families.

Donaldson brought an albino stock from the University of Chicago and established it at the Wistar Institute in 1906. These animals gave rise to two lines: the inbred "King Albino" (now PA) bred by H. D. King and an outbred commercial colony. Other breeding stock from an outside source may have been introduced into the commercial colony in 1918. Rats, referred to by investigators for 50 years as "*the* Wistar rat," were disseminated from this commercial colony all over the world beginning at least as early as 1911 (Table I). Thus, this commercial colony contributed genetically to a major proportion of the established strains of rats. Of the 111 strains listed in the January, 1978 issue of *Rat News Letter* (71), at least 45 are known either to have descended directly from this commercial stock or to have otherwise received genetic input from it (Table III). No doubt many others also originated in part or exclusively from commercial Wistars. Regardless, the Wistar bloodline has contributed far more strains of rats than any other single line. In fact, the contribution of the commercial Wistar colony to the total gene pool of all modern strains and stocks of rats may exceed the combined influence of all other bloodlines!

McCollum (101) purchased a stock of albino rats from a Chicago dealer in 1907, and subsequently introduced into his

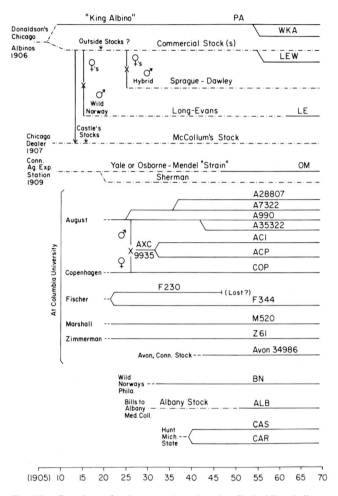

Fig. 40. Genealogy of major rat stocks and strains. Dashed lines indicate initiation of b × s matings; solid lines identify inbred strains; lines of dashes and dots designate outbred stocks.

Table III

Rat Strains and Stocks Originated at Least in Part from Wistar Bloodlines

A. Inbred line began by Helen Dean King in 1909 and now known as the PA strain, one subline: WKA

B. Direct descendants of Wistar Commerical Stock(s)

AS	K	MNR	SHR	WE/Cpb
B	KYN	MNRA	W	WF
BROFO	LEW	MR	WA	WKY/N
BUF	LOU/C	OKA/Wsl	WAB	WM
GH	LOU/M	R	WAG	WN

C. From crosses between Wistar commercial stock and other rats

BD II	BD VI	BD X	IS	McCollum
BD III	BD VII	BDE	LE	Sprague-Dawley
BD IV	BD VIII	BS	LGE	
BD V	BD IX	GHA	MAXX	

breeding colony Wistar's (in 1913) and two strains of animals from Castle (in 1915). This remains an outbred stock to the present day (131).

Joseph Long probably began his work on the estrous cycle of the rat with Wistar animals (although the Annual Reports at Wistar do not record a purchase by him) (124). On the belief that these rats were effete, a number of Wistar females were mated about 1915 to a wild Norway caught in Berkeley (6). The stock came to be known as "Long-Evans." Muhlbock at the Netherlands Cancer Institute began inbreeding a line of these rats in 1960; representatives were transferred to the National Institutes of Health in 1973 (71).

Robert W. Dawley began the Sprague-Dawley stock about 1925 using females of probable Wistar descent and a "hybrid" male of unknown origin. There are many Sprague-Dawley descended stocks in existence today, and practically all are continued as outbred populations (121).

Possibly as a result of McCollum's early research successes using rats (101), Osborne and Mendel decided to establish their rat colony at the Connecticut Agricultural Experiment Station about 1909, using stock from an unknown source. Inbreeding of this line was initiated in 1946 by Heston (71). Sherman established his subline of this stock about 1918 (138).

Curtis and Bullock at Columbia University initiated five inbred lines in 1921, using rats from four local vendors (August, Fischer, Marshall, and Zimmerman) and one vendor in Copenhagen, Denmark (from whom rats had been brought to the United States by Rosenstirn) (15). An accidental mating between an August male with an Irish coat and a COP (originally called Copenhagen 2331) female in 1926 resulted in establishment of the ACI (August × Copenhagen Irish) and ACP (August × Copenhagen Piebald) strains. Inbreeding of a sixth stock, the Avon, began in 1941 with rats obtained from a breeder in Avon, Connecticut. One strain, the F230, was lost in 1945, leaving a total of 11 strains. Dr. Wilhelmina F. Dunning joined Curtis and Bullock in 1926 and helped to develop these strains. She continues to maintain most of them at the Papanicolaou Cancer Research Institute in Miami (50).

The origin of the ALB strain can be traced back only to 1930 when a biochemist, Dr. Arthur Knudson, at the Albany Medical College obtained some breeders from Dr. C. E. Bills of the Mead Johnson Company of Evansville, Indiana. Inbreeding of this line began in 1950 at the National Institutes of Health (13,71).

The BN strain was developed by Silvers and Billingham from a pen-bred stock maintained by King and Aptekman at Wistar. This stock presumably had been selected out of the wild Norways trapped around Philadelphia about 1930 and inbred for a while by King (2,8).

Hunt at Michigan State University in 1937 initiated development of a "caries-susceptible" (CAS) and a "caries-resistant" (CAR) strain of rats, using as foundation stock albino rats from three local departmental (psychology, home economics and chemistry) colonies. Although these stocks are no longer accepted as being genetically susceptible or resistant (this can now be explained on the basis of oral flora and factors other than genetics), the two strains do represent two branches of a separate and distinct bloodline in the laboratory rat's genealogy (87).

X. SOME KEY EVENTS AFTER WORLD WAR II

Following World War II, the tremendous upsurge of research employing rats as well as other animals in the United States demanded that measures be taken to improve: (a) the dissemination and exchange of information on laboratory animals, (b) mechanisms of support and preservation of valuable animal stocks, (c) the nomenclature and genetic standards for various strains and stocks of animals, and (d) many other aspects of assuring adequate supplies of high quality animals for modern research. Numerous organizations, including scientific, commercial, and governmental, arose to meet these needs. A few are of particular historical interest and importance to the laboratory rat and will be summarized below. Similar events also occurred in England.

A. Institute of Laboratory Animal Resources

Realizing the enormous need for improving the standards and logistics of meeting the national requirements of laboratory animals, the Division of Biology and Agriculture of the National Research Council–National Academy of Sciences established, in 1952, the Institute of Laboratory Animal Resources (ILAR). The "first step" of that organization toward meeting its charge was the production of the "Handbook of Laboratory Animals" (83). This publication presented the "state of the art" in terms of genetic standards, nutrition, diseases, and uses of laboratory animal species. It also gave the first national listing of animal vendors and users. At that time there were 25 commercial producers of rats. Of that group, only about 5 are still in business today.

The "Handbook of Laboratory Animals" was the direct forerunner of a wide diversity of publications and services available today from ILAR (Dr. Wayne Grogan, Executive Secretary, Institute of Laboratory Animal Resources, National Academy of Sciences, 2101 Constitution Ave., Washington, D.C. 20418).

B. NIH Rodent Repository

Beginning in the 1940s, the Laboratory Aids Branch of the National Institutes of Health (NIH), under the direction of Dr.

Walter E. Heston, began establishing breeding colonies of selected strains of rodents at the NIH in Bethesda, Maryland (85). At the time, Dr. George E. Jay, Jr. was geneticist and Mr. Sam E. Poiley was animal husbandman. By 1953, seven inbred strains of rats had been established: AxC 9935, Buffalo, Fischer 344, Marshall 520, O'Grady, Osborne-Mendel, and Wistar (83,88a). Random-bred colonies at the NIH at this time included rats of Sprague-Dawley, Holtzman-Rolfsmeyer, and NIH black stocks (121).

Over the years, animal breeding activities at the NIH continued to expand. The name of the unit in charge of the animal colonies was changed to the Veterinary Resources Branch (Division of Research Services). Dr. Robert Whitney is currently chief of this branch. Dr. Carl T. Hanson joined the staff of the Veterinary Resources Branch in 1964, and subsequently established within the branch "The NIH Rodent Repository." This repository is "a collection of strains and stocks of rodents and lagomorphs of known genetic characteristics—maintained to provide a defined source of breeders." The Repository includes "foundation colonies of inbred and congenic strains and nucleus colonies of outbred and mutant stocks of rodents." At present, it maintains 45 strains or stocks of rats [24 inbred strains, 15 mutant stocks, 4 outbred stocks and 2 other species: the cotton rat (*Sigmodon hispidus hispidus*) and the rice rat (*Oryzomous palustris*)] (24).

The NIH Rodent Repository serves both as a national and international resource, by virtue of its designation as a World Health Organization (WHO) Collaborating Center for Defined Laboratory Animals. The Repository supplies breeding stocks for scientific purposes to universities, NIH contractors, and commercial breeders. (Inquiries should be addressed to Dr. Robert A. Whitney, Jr., Chief, Veterinary Resources Branch, DRS Building 14G, Room 102, National Institutes of Health, Bethesda, Maryland 20014.)

C.   Laboratory Animals Centre

The Laboratory Animals Centre (LAC) in the United Kingdom was founded (as the Laboratory Animals Bureau) in 1947 with the general aim of improving the quality and availability of laboratory animals used in the United Kingdom. Although financed by the Medical Research Council, the Centre has been run as a national service to all biomedical research workers in the United Kingdom, and it has recently been designated a WHO reference centre for the supply of defined laboratory animals. The first director was Dr. R. E. Glover (1947–1949) followed by Dr. W. Lane-Petter (1949–1965) and Dr. John Bleby (from 1965). Breeding colonies have been under the control of Dr. M. F. W. Festing, who joined the staff as Geneticist in 1966. The work of the LAC has evolved so that it now covers four main areas: information, supply of animals,

training, and research (10). Only the first two of these will be summarized here.

The Centre acts as a clearing house for information on all aspects of laboratory animal science. Apart from dealing with a wide range of individual queries, it publishes *LAC News,* a range of LAC manuals, and distributes *Mouse News Letter* (edited by Dr. A. G. Searle, Harwell), *Guinea-Pig News Letter* (edited by Dr. M. F. W. Festing, LAC), and more recently, *Rat News Letter.* The latter was established in 1977 following a request from Dr. T. J. Gill, III of the University of Pittsburgh (see Volume II, Chapter 210), who acts as an advisor. It is edited by Dr. M. F. W. Festing with the assistance of Dr. J. C. Howard (alloantigen news) and Roy Robinson (mutant lists), and is distributed internationally. (Subscription requests for *Rat News Letter* should be directed to Dr. Festing at the LAC). The LAC also produces the *International Index of Laboratory Animals* (editor Dr. M. F. W. Festing with the assistance of Mrs. W. Butler) which is a world-wide listing of more than 4000 animal colonies, aimed at assisting people to locate particular stocks of any laboratory species.

The LAC maintains a comprehensive collection of genetically defined strains of rodents, and a few other species, maintained in germ-free, specific pathogen-free, and conventional conditions for supply as high quality breeding nuclei. Currently, the LAC maintains seven inbred, two mutant, and one outbred stock of rats. Commercial breeders are responsible for the large-scale production of laboratory animals in the United Kingdom. Most breeders belong to the Accreditation Scheme, run by the LAC. Such breeders must exceed certain minimum standards of husbandry, and the animals are checked routinely for the presence of designated parasites, bacteria, and viruses, with the colonies being graded according to the types of microorganisms that they harbor (145).

Enquiries on any aspect of the Centre's work should be addressed to: Dr. John Bleby, Director, Medical Research Council Laboratory Animals Centre, Woodmansterne Rd., Carshalton, Surrey SM5 4EF, England.

XI.   INFORMATION REQUESTED

While doing the research for this chapter, the author was repeatedly impressed that most of the generation of scientists responsible for early development of the laboratory rat is now deceased by only a few years or decades. Unfortunately, this means that many important facts and memorabilia from the early history of the rat have been lost forever. In order to try and preserve as much as possible of this glorious history, the author hereby invites correspondence on this subject from past associates and relatives (locating close relatives of these early scientists is a serious problem) and from other interested parties. All correspondance should be addressed to Dr. J. Russell Lindsey, Department of Comparative Medicine, University of Alabama Medical Center, Birmingham, Alabama 35294 (U.S.A.).

ACKNOWLEDGMENTS

Supported in part by research funds of the Veterans' Administration, United States Public Health Service Grant RR00463 and NCI Contract N01-CM-6708. The author gratefully acknowledges the invaluable contributions of all who assisted in preparation and verification of information in this chapter. In particular, appreciation is due the following (by institution, historical personality, or rat stock): The Wistar Institute—Dr. Robert Roosa, Mr. William J. Purcell, Mr. Martin Cohn, Mr. Dick Walsh, and Dr. Walter F. Greene; McCollum—Ms. Agatha A. Rider; Sherman—Dr. Paul L. Day and Dr. Charles G. King; Long and Evans—Dr. Leslie L. Bennett, Mr. J. R. K. Kantor, Dr. H. W. Magoun, and Dr. George O. Burr; Cancer Research Institutes—Dr. Wilhelmina F. Dunning and Ms. Betty R. Moore; Castle and The Bussey Institution—Mr. Bill Whalen; Albany Strain—Dr. J. M. Wolfe; and Laboratory Animals Centre—Dr. Michael F. W. Festing. Photographic credits are due Mr. Phil Foster of the University of Alabama in Birmingham and Ms. Lucinda Keister of The National Library of Medicine. Mrs. Hilda Harris of the Lister Hill Library at the University of Alabama in Birmingham provided invaluable assistance in library research.

REFERENCES

1. Addison, W. H. F. (1939). Henry Herbert Donaldson (1857–1938). *Bios.* **10,** (Madison, N. J.) 4–26.
2. Annual Reports of the Director (1911–1956). Wistar Institute of Anatomy and Biology, Philadelphia, Pennsylvania.
3. Anonymous (1948). "Selected Works of Henry Clapp Sherman." Macmillan, New York.
4. Anonymous (1962). "The National Cyclopaedia of American Biography," Vol. 45, 170–171.
5. Bateson, W. (1903). The present state of knowledge of colour heredity in mice and rats. *Proc. Zool. Soc. London* **2,** 71–93.
6. Bennett, L. L. (1975). Endocrinology and Herbert M. Evans. *In* "Hormonal Proteins and Peptides" (C. H. Li, ed.), Vol. 3, Chapter 5, pp. 247–272. Academic Press, New York.
7. Bennett, L. L. (1978). San Francisco, California (personal communication).
8. Billingham, R. E., and Silvers, W. K. (1959). Inbred animals and tissue transplantation immunity. *Plast. Reconstr. Surg.* **23,** 399–406.
9. Blass, E. M., ed. (1976). "The Psychobiology of Curt Richter." York Press, Baltimore, Maryland.
10. Bleby, J. (1967). The function and work of the United Kingdom Laboratory Animals Centre. *Lab. Anim. Care* **17,** 147–154.
11. Borrel, A. (1906). Tumeurs cancéreuses et helminthes. *Bull. Acad. Med. Paris* **56,** 141–145.
12. Broderson, J. R., Lindsey, J. R., and Crawford, J. E. (1976). The role of ammonia in respiratory mycoplasmosis of rats. *Am. J. Pathol.* **85,** 115–130.
13. Bryan, W. R., Klinck, G. H., Jr., and Wolfe, J. M. (1938). The unusual occurrence of a high incidence of spontaneous mammary tumors in the Albany strain of rats. *Am. J. Cancer* **33,** 370–388.
14. Bullock, F. D., and Curtis, M. R. (1920). The experimental production of sarcoma of the liver of rats. *Proc. N.Y. Path. Soc.* **20,** 149–175.
15. Bullock, F. D., and Curtis, M. R. (1930). Spontaneous tumors of the rat. *J. Cancer Res.* **14,** 1–115.
16. Carr, H. A., and Watson, J. B. (1908). Orientation in the white rat. *J. Comp. Neurol. Psychol.* **18,** 27–44.
17. Castle, W. E. (1919). Piebald rats and selection, a correction. *Am. Nat.* **53,** 370–375.
18. Castle, W. E. (1924). "Genetics and Eugenics." Harvard Univ. Press, Cambridge, Massachusetts. (1st ed. 1916, revised 1920, 1924, and 1930).
19. Castle, W. E. (1930). "The Genetics of Domestic Rabbits." Harvard Univ. Press, Cambridge, Massachusetts.
20. Castle, W. E. (1940). "Mammalian Genetics." Harvard Univ. Press, Cambridge, Massachusetts.
21. Castle, W. E. (1947). The domestication of the rat. *Proc. Natl. Acad. Sci. U.S.A.* **33,** 109–117.
22. Castle, W. E. (1951). Variation in the hooded pattern of rats and a new allele of hooded. *Genetics* **36,** 254–266.
23. Castle, W. E., and Phillips, J. C. (1914). Piebald rats and selection. *Carnegie Inst. Washington Publ.* **195.**
24. Chase, H. B., chairman (1975). "NIH Rodent Repository," Report of the Committee on Maintenance of Genetic Stocks. Inst. Lab. Anim. Resour., NAS, Washington, D.C.
25. Chittenden, R. H. (1938). "Biographical Memoir of Lafayette Benedict Mendel, 1872–1935," Biogr. Mem., Vol. XVIII, 6th mem., pp. 123–155. Natl. Acad. Sci., Washington, D.C.
26. Chittenden, R. H. (1945). "The First Twenty-five Years of the American Society of Biological Chemists." Waverly Press, Baltimore, Maryland.
27. Conklin, E. G. (1939). "Biographical Memoir of Henry Herbert Donaldson, 1857–1938," Biogr. Mem., Vol. XX, 8th Mem., pp. 229–243. Natl. Acad. Sci., Washington, D.C.
28. Connecticut Agricultural Experiment Station (1930). Thomas B. Osborne, a Memorial. *Conn., Agric. Exp. Stn., New Haven, Bull.* **312,** 281–394.
29. Crampe, H. (1877). Kreuzungen zwischen wanderratten verschiedener Farbe. *Landwirtsch. Jahrb.* **6,** 385–395.
30. Crampe, H. (1883). Zuchtversuche mit zahmen Wanderratten. I. Resulte der Zucht in Verwandtschaft. *Landwirtsch. Jahrb.* **12,** 389–449.
31. Crampe, H. (1884). Zuchtversuche mit zahmen Wanderratten. II. Resultate der Kreuzung der zahmen Rtten mit Wilden. *Landwirtsch. Jahrb.* **13,** 699–754.
32. Crampe, H. (1885). Die Gesetze der Bererbung der Farbe. *Landwirtsch. Jahrb.* **14,** 539–619.
33. Curtis, M. R., Bullock, F. D., and Dunning, W. F. (1931). A statistical study of the occurrence of spontaneous tumours in a large colony of rats. *Am. J. Cancer* **15,** 67–121.
34. Curtis, M. R., and Dunning, W. F. (1937). Two independent mutations of the hooded or piebald gene of the rat. *J. Hered.* **18,** 382–390.
35. Curtis, M. R., and Dunning, W. F. (1940). Transplantable lymphosarcomata of the mesenteric lymph nodes of rats. *Am. J. Cancer* **40,** 299–309.
36. Curtis, M. R., and Dunning, W. F. (1940). An independent recurrence of the blue mutation in the Norway rat. *J. Hered.* **31,** 219–222.
37. Curtis, M. R., Dunning, W. F., and Bullock, F. D. (1933). Genetic factors in relation to the etiology of malignant tumors. *Am. J. Cancer* **17,** 894–923.
38. Day, H. G. (1975). Contributions of Elmer Verner McCollum. *In* "Nutrition and Public Health—A Symposium Celebrating the Johns Hopkins University Centenial and Honoring Elmer V. McCollum" (R. M. Herriott, ed.), pp. 3–21. Johns Hopkins School of Hygiene and Public Health, Baltimore, Maryland.

39. Day, P. L. (1957). Henry Clapp Sherman. *J. Nutr.* **61,** 3–11.

40. Donaldson, H. H. (1912). A comaprison of the European Norway and albino rats *Mus norvegicus albinus*) with those of North America in respect to the weight of the central nervous system and to cranial capacity. *J. Comp. Neurol.* **22,** 71–77.

41. Donaldson, H. H. (1912). The history and zoological position of the albino rat. *J. Acad. Nat. Sci. Philadelphia* **15,** 365–369.

42. Donaldson, H. H. (1915). "The Rat. Reference Tables and Data for the Albino Rat (*Mus norvegicus albinus*) and the Norway Rat (*Mus norvegicus*)," 1st ed., Mem. No. 6. Wistar Inst. Anat. Biol., Philadelphia, Pennsylvania.

43. Donaldson, H. H. (1924). "The Rat. Reference Tables and Data for the Albino Rat (*Mus norvegicus albinus*) and the Norway Rat (*Mus norvegicus*)," 2nd ed., Memoirs No. 6. Wistar Inst. Anat. Biol., Philadelphia, Pennsylvania.

44. Donaldson, H. H. (1927). The museum of Caspar Wistar. Address given before Sigma Xi, Philadelphia, Pa., January 19, 1927. *In* "The Donaldson Memoirs," Library of the Wistar Institute, Philadelphia, Pennsylvania.

45. Donaldson, H. H. (1937). Milton Jay Greenman (1866–1937). *Anat. Rec.* **68,** 262–265.

46. Doncaster, L. (1906). On the inheritance of coat color in rats. *Proc. Cambridge Philos. Soc.* **13,** 215–227.

47. Drake, L. E., and Heron, W. T. (1930). The rat. A bibliography 1924–1929. *Psychol. Bull.* **27,** 141–239.

48. Dunn, L. C. (1967). "Biographical Memoir of William Ernest Castle," Biogr. Mem., Vol. XXXVIII, pp. 33–80. Natl. Acad. Sci., Washington, D.C.

49. Dunning, W. F. (1951). In memorium—Francis Carter Wood (1869–1951). *Cancer Res.* **11,** 296.

50. Dunning, W. F. (1977). The Papanicolaou Cancer Research Institute, Miami, Florida (personal communication).

51. Dunning, W. F. (1978). Unpublished manuscript prepared in 1940, made available to J. R. Lindsey in 1978.

52. Dunning, W. F., and Curtis, M. R. (1941). Longevity and genetic specificity as factors in the occurrence of spontaneous tumors in the hybrids between two inbred lines of rats. *Genetics* **26,** 148–149.

53. Dunning, W. F., and Curtis, M. R. (1946). Multiple peritoneal sarcoma in rats from intraperitoneal injection of washed, ground *Taenia* larvae. *Cancer Res.* **6,** 668–670.

54. Dunning, W. F., and Curtis, M. R. (1946). The respective roles of longevity and genetic specificity in the occurrence of spontaneous tumors in the hybrids between two inbred lines of rats. *Cancer Res.* **6,** 61–81.

55. Dunning, W. F., and Curtis, M. R. (1952). Diethylstilbestrol-induced mammary cancer in reciprocal F₁ hybrids between negative and positive inbred lines of rats. *Cancer Res.* **12,** 257–258.

56. Dunning, W. F., and Curtis, M. R. (1952). The influence of diethylstilbestrol-induced cancer in reciprocal F₁ hybrids obtained from crosses between rats of inbred lines that are susceptible and resistant to the induction of mammary cancer by this agent. *Cancer Res.* **12,** 702–706.

57. Dunning, W. F., and Curtis, M. R. (1953). Attempts to isolate the active agent in *Cysticercus fasciolaris*. *Cancer Res.* **13,** 838–843.

58. Dunning, W. F., and Curtis, M. R. (1957). A transplantable acute leukemia in an inbred line of rats. *J. Natl. Cancer Inst.* **19,** 845–852.

59. Dunning, W. F., Curtis, M. R., and Madsen, M. E. (1951). Diethylstilbestrol induced mammary gland and bladder cancer in reciprocal F₁ hybrids between two inbred lines of rats. *Acta Unio Int. Cancrum* **7,** 238–244.

60. Eakin, R. M. (1956). History of zoology at the University of California, Berkeley. *Bios. (Madison, N.J.)***27,** 67–90.

61. Eakin, R. M., Evans, H. M., Goldschmidt, R. B., and Lyons, W. R. (1959). Joseph Abraham Long, 1879–1953. "In Memorium," April issue, pp. 40–42. University of California.

62. Edward, A. G. (1978). Department of Comparative Medicine, Wayne State University, Detroit, Michigan (personal communication).

63. Eisen, M. J. (1954). Obituary—William H. Woglom (1879–1953). *Cancer Res.* **14,** 155–156.

64. Erikson, G. E. (1977). Eunice Chace Greene, 1895–1975. *Anat. Rec.* **189,** 306–307.

65. Evans, H. M. (1914). Vital straining of protoplasm. *Science* **39,** 843–844.

66. Evans, H. M. (1962). The pioneer history of vitamin E. *Vitam. Horm. (N.Y.)* **20,** 379–387.

67. Evans, H. M., and Bishop, K. S. (1922). On the existence of a hitherto unrecognized dietary factor essential for reproduction. *Science* **56,** 650–651.

68. Farris, E. J. (1950). The rat as an experimental animal. *In* "The Care and Breeding of Laboratory Animals" (E. J. Farris, ed.), Chapter 2, pp. 43–78. Wiley, New York.

69. Farris, E. J., and Griffith, J. Q., Jr., eds.) (1949). "The Rat in Laboratory Investigation," 2nd ed. Lippincott, Philadelphia, Pennsylvania.

70. Ferry, E. L. (1919). Nutrition experiments with rats. A description of methods and technic. *J. Lab. Clin. Med.* **5,** 735–745.

71. Festing, M. F. W. (1978). "Rat News Letter," No. 3, pp. 18–35. Med. Res. Counc. Lab. Anim. Cent., Woodmansterne Rd., Carshalton, Surrey SM5 4EF, England.

72. Fibiger, J. A. G. (1965). Investigations on *Spiroptera carcinoma* and the experimental induction of cancer: Nobel Lecture, December 12, 1927. *In* "Nobel Lectures Including Presentation Speeches and Laureates' Biographies, Physiology and Medicine, 1922–1941," pp. 122–150. Am. Elsevier, New York, 1965.

73. Friends of H. M. Evans (1943). Herbert M. Evans—Biographical sketch and Bibliography of Herbert McLean Evans (1804–1902). *In* "Essays in Biology in Honor of Herbert M. Evans," pp. ix–x and xii–xxvii. Univ. of California Press, Berkeley.

74. Gorski, R. A., and Whalen, R. E., eds. (1963). "The Brain and Gonadal Function," UCLA Forum for Medical Sciences, No. 3. Univ. of California Press, Berkeley.

75. Greene, E. C. (1935). "Anatomy of the Rat," Trans. Philos. Soc. Philos. Soc., Philadelphia, Pennsylvania.

76. Greene, W. F. (1978). Jaffrey Center, New Hampshire (personal communication).

77. Greenman, M. J., and Duhring, F. L. (1923). "Breeding and Care of the Albino Rat for Research Purposes." Wistar Inst. Anat. Biol., Philadelphia, Pennsylvania.

78. Greenman, M. J., and Duhring, F. L. (1931) "Breeding and Care of the Albino Rat for Research Purposes," 2nd ed. Wistar Inst. Anat. Biol., Philadelphia, Pennsylvania.

79. Griffith, J. Q., Jr., and Farris, E. J., eds. (1942). "The Rat in Laboratory Investigation," 1st ed. Lippincott, Philadelphia, Pennsylvania.

80. Hammett, F. S. (1924). Studies of the thyroid apparatus. XX. The effect of thyro-parathyroidectomy and parathyroidectomy at 75 days of age on the growth of the brain and spinal cord of male and female albino rats. *J. Comp. Neurol.* **37,** 15–30.

81. Hegsted, M. (1975). Relevance of animal studies to human disease. *Cancer Res.* **35,** 3537–3539.

82. Hektoen, L. (1915–1916). Observations on pulmonary infections in rats. *Trans. Chicago Pathol. Soc.* **10,** 105–109.

83. Herrlein, H. G., Coursen, G. B., Randall, R., and Slanetz, C. A. (1954). "Handbook of Laboratory Animals," Inst. Anim. Resour., Natl. Acad. Sci.—Natl. Res. Counc., Washington, D.C.

84. Heston, W. E. (1972) Obituary—Clarence Cook Little. *Cancer Res.* **32,** 1354–1356.

85. Heston, W. E., Jay, G. E., Jr., Kaunitz, H., Morris, H. P., Nelson, J. B., Poiley, S. M., Slanetz, C. A., Zucker, L. M., and Zucker, T. F. (1953). "Rat Quality—A Consideration of Heredity, Diet and Disease." Natl. Vitam. Found., Inc., New York.

86. Hoppert, C. A., Webber, P. A., and Conniff, T. L. (1932). The production of dental caries in rats fed an adequate diet. *J. Dent. Res.* **12**, 161–173.

87. Hunt, H. R., Hoppert, C. A., and Erwin, W. G. (1944). Inheritance of susceptibility to caries in albino rats (*Mus norvegicus*). *J. Dent. Res.* **23**, 385–401.

88. Hunt, H. R., Hoppert, C. A., and Rosen, S. (1955). Genetic factors in experimental rat caries. *Adv. Exp. Caries Res., Symp. 1953* pp. 66–81.

88a. Jay, G. E., Jr. (1953). The use of inbred strains in biological research. *In* "Rat Quality—A Consideration of Heredity, Diet and Disease" (W. E. Heston *et al.*, eds.), pp. 98–103. Natl. Vitam. Found., New York.

89. King, C. G. (1975). "Biographical Memoir of Henry Clapp Sherman, 1875-1955," Biogr. Mem., Vol. XLVI, pp. 397–429. Natl. Acad. Sci., Washington, D.C.

90. King, H. D. (1918). Studies on inbreeding. I. The effects of inbreeding on the growth and variability in the body weight of the albino rat. *J. Exp. Zool.* **26**, 1–54.

91. King, H. D. (1918). Studies on inbreeding. II. The effects of inbreeding on the fertility and on the constitutional vigor of the albino rat. *J. Exp. Zool.* **26**, 335–378.

92. King, H. D. (1918). Studies on inbreeding. III. The effects of inbreeding, with selection, on the sex ratio of the albino rat. *J. Exp. Zool.* **27**, 1–35.

93. King, H. D. (1919). Studies on inbreeding. IV. A further study of the effects of inbreeding on the growth and variability on the growth and variability in the body weight of the albino rat. *J. Exp. Zool.* **29**, 135–175.

94. Kline, L. W. (1899). Methods in animal psychology. *Am. J. Psychol.* **10**, 256–279.

95. Kreezer, G. L. (1949). Technics for the investigation of behavioral phenomena in the rat. *In* "The Rat in Laboratory Investigation" (E. J. Farris and J. Q. Griffith, Jr., eds.), Chapter 10, pp. 203–277. Lippincott, Philadelphia, Pennsylvania.

96. Larson, R. H., Amsbaugh, S. M., Navia, J. M., Rosen, S., Schuster, G. S., and Shaw, J. H. (1977). Collaborative evaluation of a rat caries model in six laboratories. *J. Dent. Res.* **56**, 1007–1012.

97. Long, J. A., and Evans, H. M. (1922). "The Oestrous Cycle in the Rat and Its Associated Phenomena," Univ. Calif., No. 6. Univ. of California Press, Berkeley.

98. Mason, K. E. (1977). The first two decades of Vitamin E. *Fed. Proc., Fed. Am. Soc. Exp. Biol.* **36**, 1906–1910.

99. McCollum, E. V. (1948). Letter dated May 4, 1948 from E. V. McCollum to W. D. Salmon of Auburn University, Auburn, Alabama (copy made available by Ms. Agatha Rider, Department of Biochemistry, School of Hygiene and Public Health, The Johns Hopkins University, Baltimore, Maryland).

100. McCollum, E. V. (1957). "A History of Nutrition—The Sequence of Ideas in Nutrition Investigation." Houghton, Boston, Massachusetts.

101. McCollum, E. V. (1964). "From Kansas Farm Boy to Scientist—The Autobiography of Elmer Verner McCollum." Univ. of Kansas Press, Lawrence.

102. McCollum, E. V., and Davis, M. (1913). The necessity of certain lipids in the diet during growth. *J. Biol. Chem.* **15**, 167–175.

103. McCollum, E. V., and Davis, M. (1915). The essential factors in the diet during growth. *J. Biol. Chem.* **23**, 231–246.

104. McCollum, E. V., Simmonds, N., Becker, J. E., and Shipley, P. G. (1922). Studies on experimental rickets. XXI. An experimental demonstration of the existence of a vitamin which promotes calcium deposition. *J. Biol. Chem.* **53**, 293–312.

105. McMurrich, J. P., and Jackson, C. M. (1938). Henry Herbert Donaldson (1857-1938). *J. Comp. Neurol.* **69**, 173–179.

106. Mendel, L. B., and Hubbell, R. B. (1935). The relation of the rate of growth to diet. *J. Nutr.* **10**, 557–563.

107. Miles, W. R. (1930). On the history of research with rats and mazes. *J. Gen. Psychol.* **3**, 324–337.

108. Moore, B. R. (1978). Library Service Coordinator for Cancer Research Center, Columbia University, New York (personal communication).

108a. Morse, H. C., III (1978). Introduction. "Origins of Inbred Mice," H. C. Morse, III, ed. Academic Press, New York, p. 9.

109. Munn, N. L. (1950). "Handbook of Psychological Research on the Rat," pp. 2–5. Houghton, Boston, Massachusetts.

110. Navia, J. M. (1977). Experimental dental caries. *In* "Animal Models in Dental Research," Chapter 13, pp. 257–297. Univ. of Alabama Press, Birmingham.

111. Osborne, T. B. (1909). "The Vegetable Proteins." Longmans, Green, New York.

112. Osborne, T. B. (1924). "The Vegetable Proteins," 2nd ed. Longmans, Green, New York.

113. Osborne, T. B., and Mendel, L. B. (1911). "Feeding Experiments with Isolated Food-Substances," Parts I and II. Carnegie Inst. Washington, Washington, D.C.

114. Osborne, T. B., and Mendel, L. B. (1913). The influence of butter-fat on growth. *J. Biol. Chem.* **16**, 423–437.

115. Osborne, T. B., and Mendel, L. B. (1914). The influence of cod liver oil and some other fats on growth. *J. Biol. Chem.* **17**, 401–408.

116. Osborne, T. B., and Mendel, L. B. (1918). The inorganic elements in nutrition. *J. Biol. Chem.* **34**, 131–139.

117. Otha S. A. Sprague Memorial Institute of Chicago (1911-1941). "Collected Papers," Vols. 1–25.

118. Parkes, A. S. (1969). Herbert McLean Evans. *J. Reprod. Fertil.* **19**, 1–29.

119. Peterson, G. M. (1946). The rat in animal psychology. *In* "The Encyclopedia of Psychology" (P. L. Harriman, ed.), pp. 765–798. Philosophical Library, New York.

120. Philipeaux, J. M. (1856). Note sur l'extirpation des capsules survenales chez les rats albios (*Mus rattus*). *C. R. Habd. Seances Acad. Sci.* **43**, 904–906.

121. Poiley, S. M. (1953). History and information concerning the rat colonies in the animal section of the National Institutes of Health. *In* "Rat Quality—A Consideration of Heredity, Diet and Disease" (W. E. Heston *et al.*, eds.), pp. 86–97. Natl. Vitam. Found., New York.

122. Proceedings of the Borden Centennial Symposium on Nutrition (1958). "The Nutritional Ages of Man—Nutrition: Past, Present, and Future," pp. 120–126. Borden Company Found., Inc., New York.

123. Purcell, W. L. (1968-1969). "An Outline of the History of the Wistar Institute," Biennial Res. Rep., pp. xi-xiii. Wistar Inst. Anat. Biol., Philadelphia, Pennsylvania.

124. Purcell, W. L. (1978). Librarian, Wistar Institute, Philadelphia, Pennsylvania (personal communication).

125. Richter, C. P. (1949). The use of the wild Norway rat for psychiatric research. *J. Nerv. Ment. Dis.* **110**, 379–386.

126. Richter, C. P. (1950). Domestication of the Norway rat and its implications for the problem of stress. *Res. Publ., Assoc. Res. Nerv. Ment. Dis.* **29**, 19–47.

127. Richter, C. P. (1954). The effects of domestication and selection on the behavior of the Norway rat. *J. Natl. Cancer Inst.* **15**, 727–738.

128. Richter, C. P. (1959). Rats, man, and the welfare state. *Am. Psychol.* **14**, 18–28.

129. Richter, C. P. (1968). Experiences of a reluctant rat-catcher. The common Norway rat—Friend or enemy? *Proc. Am. Philos. Soc.* **112**, 403–415.

130. Rider, A. A. (1970). Elmer Verner McCollum—A biographical sketch. *J. Nutr.* **100,** 1–10.
131. Rider, A. A. (1978). Department of Biochemistry, Johns Hopkins University School of Hygiene and Public Health, Baltimore, Maryland (personal communication).
132. Robinson, R. (1965). "Genetics of the Norway Rat." Pergamon, Oxford.
133. Roosa, R. (1977). Wistar Institute, Philadelphia, Pennsylvania (personal communication).
134. Rose, W. C. (1936). Lafayette Benedict Mendel—An appreciation. *J. Nutr.* **11,** 607–613.
135. Savory, W. S. (1863). Experiments on food; its destination and uses. *Lancet* **2,** 381–383.
136. Scott, T. R. C. (1931). The Hampton Court maze. *J. Genet. Psychol.* **39,** 287–289.
137. Sherman, H. C. (1936). Lafayette Benedict Mendel. *Science* **83,** 45–47.
138. Sherman, H. C., Rouse, M. E., Allen, B., and Woods, E. (1919). Growth and reproduction upon simplified food supply. *Proc. Soc. Exp. Biol. Med.* **17,** 9–10.
139. Sherman, H. C., Rouse, M. E., Allen, B., and Woods, E. (1921). Growth and reproduction upon simplified food supply. I. *J. Biol. Chem.* **46,** 503–419.
140. Small, W. S. (1900). An experimental study of the mental processes of the rat. *Am. J. Psychol.* **11,** 133–165.
141. Small, W. S. (1901). Experimental study of the mental processes of the rat. *Am. J. Psychol.* **12,** 206–239.
142. Smith, A. H. (1956). Lafayette Benedict Mendel. *J. Nutr.* **60,** 3–12.
143. Smith, P. E. (1927). The disabilities caused by hypophysectomy and their repair. *J. Am. Med. Assoc.* **88,** 158–161.
144. Snyder, E. K., and Jones, E. A. (1968). Elmer Verner McCollum. *J. Am. Diet. Assoc.* **52,** 49.
145. Sparrow, S. (1976). The microbiological and parasitological status of laboratory animals from accredited breeders in the United Kingdom. *Lab. Anim.* **10,** 365–373.
146. Stewart, C. C. (1898). Variations in daily activity produced by alcohol and by changes in barometric pressure and diet, with a description of recording methods. *Am. J. Physiol.* **1,** 40–56.
147. Stewart, W. H., Hoppert, C. A., and Hunt, H. R. (1952). The incidence of dental caries in caries-susceptible and caries-resistant albino rats (*Rattus norvegicus*) when fed diets containing granulated and powdered sucrose. *J. Dent. Res.* **32,** 210–221.
148. Stotsenberg, J. M. (1909). On the growth of the albino rat (*Mus norvegicus* var. *albus*) after castration. *Anat. Rec.* **3,** 233–244.
149. Stotsenberg, J. M. (1913). The effect of spaying and semi-spaying young albino rats (*Mus norvegicus albinus*) on the growth in body weight and body length. *Anat. Rec.* **7,** 183–194.
150. Verzar, F., ed. (1973). "Clive M. McCay's Notes on the History of Nutrition Research." Huber, Bern.
151. Vickery, H. B. (1932). "Biographical Memoir of Thomas Burr Osborne, 1859–1929," Biogr. Mem., Vol. XIV, 8th mem., pp. 261–304. Natl. Acad. Sci., Washington, D.C.
152. Warden, C. J. (1930). A note on the early history of experimental methods in comparative psychology. *J. Genet. Psychol.* **38,** 466–471.
153. Watson, J. B. (1903). "Animal Education. An Experimental Study on the Psychical Development of the White Rat, Correlated with the Growth of its Nervous System." Univ. of Chicago Press, Chicago, Illinois.
154. Watson, J. B. (1907). Kinaesthetic and organic sensations: Their role in the reactions of the white rat. *Psychol. Rev. Monog.* **8,** No. 33, 1–100.
155. Watson, J. B. (1914). "Behavior: An Introduction to Comparative Psychology." Holt, New York.
155a. Watson, J. B. (1936). John Broadus Watson. *In* "A History of Psychology in Autobiography" (C. Murchison, ed.), Vol. III, pp. 271–281. Clark Univ. Press, Worcester, Massachusetts.
156. Weisbroth, S. H. (1969). The origin of the Long-Evans rat and a review of the inheritance of coat colors in rats (*Rattus norvegicus*). *Lab. Anim. Care* **19,** 733–737.
157. Wernstedt, W. (1965). Presentation speech, Nobel prize in physiology or medicine, 1926. *In* "Nobel Lectures Including Presentation Speeches and Laureates' Bibliographies Physiology and Medicine, 1922–1941," pp. 119–121. Am. Elsevier, New York.
158. Wistar, I. J. (1937). "Autobiography of Isaac Jones Wistar (1727–1905), Half a Century in War and Peace," with an Appendix entitled "The Wistar Institute of Anatomy and Biology" by Milton J. Greenman (pp. 503–516) and an Addendum by Edmond J. Farris (pp. 517–518). Wistar Inst. Anat. Biol., Philadelphia, Pennsylvania.
159. Woglom, W. H. (1951). Francis Carter Wood (1869–1951). *Am. J. Roentgenol. Radium Ther.* [n.s.] **65,** 955–959.
160. Wolfe, J. M., Bryan, W. R., and Wright, A. W. (1938). Histologic observations on the anterior pituitaries of old rats with particular reference to the spontaneous appearance of pituitary adenomata. *Am. J. Cancer* **34,** 352–372.
161. Wood, F. C. (1914–1915). The Crocker Research Laboratory. *Columbia Univ. Q.,* **17,** 82–86.
162. Zucker, T. F. (1953). Problems in breeding for quality. *In* "Rat Quality—A Consideration of Heredity, Diet and Disease" (W. E. Heston *et al.,* eds.), pp. 48–76. Natl. Vitam. Found., New York.

Chapter 2

Taxonomy and Genetics

Roy Robinson

I. INTRODUCTION

The Norway rat should be regarded as a useful foil to the house mouse in the field of genetics. The rat is more expensive to house and feed, admittedly, but its reproductive prolificacy is on a par with that of the mouse. The larger size of the animal, for instance, is a decided advantage for certain experiments. However, the real value of the rat resides in extending the generality of the findings of mouse genetics. Other mammalian species must also feature in this type of investigation, but few can compete with the rat or mouse in ease of laboratory management.

A broadly based review of the literature on the laboratory rat was completed in 1962 (137); the present summary, however, is more restrictive in scope but includes more recent reviews. Greater emphasis will be placed upon monogenic variation because of space limitations, but there is ample evidence that every trait or process of rat physiology is subject to omnipresent polygenic variation (137).

II. TAXONOMY AND GEOGRAPHICAL DISTRIBUTION

The Rodentia are by far the most numerous of mammalian orders, as judged by sheer numbers or by recognized species. A conservative estimate places the number of species as 1729 (112), but this could be an underestimate. The genus *Rattus,* itself, is comprised of some 137 species, according to the same conservative source, although Ellerman (41) speaks of over 500 named forms, a figure which includes a great many subspecies of doubtful validity. Missonne (108) has considered the classification of the Muridae from an evolutionary point of view, and his work apparently is the most recent study of this complex family. He has divided Ellerman's vast *Rattus* genus into several genera of seemingly equal standing. Only future work will establish the propriety of this division.

The two most interesting *Rattus* species for the purpose of this review are *Rattus norvegicus* and *Rattus rattus* (black rat) of which a large number of subspecies and forms have been described by Ellerman (41), to which reference should be made. Ellerman also recognizes five forms of *Rattus norvegicus: R. n. norvegicus, R. n. caraco, R. n. praestans, R. n. primarius,* and *R. n. socer.* The existence of these forms may be taxonomically valid but, beyond this, few details seem to be available. The common brown or Norwegian rat is *Rattus norvegicus,* and it is a cosmopolitan species, relatively recently distributed by man. For this reason, Norway rats from any part of the world will interbreed freely and are regarded as a single species. If aboriginal *norvegicus* do exist, they must be as isolated populations in central Asia or Northern China, the presumed center of dispersal of the species.

The early history of the rat is vague (34,73). There is no real evidence that the Greeks or Romans were acquainted with the animal. The spread of both *R. rattus* and *R. norvegicus* began with the opening up of trade routes with the East. *Rattus rattus* was not known to occur in Europe in appreciable numbers before the twelfth century, subsequently reaching the Americas sometime in the sixteenth century. *Rattus norvegicus* appeared later, arriving in Europe in the early eighteenth century, but spreading very rapidly. It apparently reached America by the latter part of the century. An interesting aspect is the efficiency by which *R. norvegicus* can supplant *R. rattus* in almost every environment except the warmer zones. In Africa, southeast Asia, and Oceania, *R. norvegicus* is present but is largely coastal as an introduced species and unable to displace the indigenous *R. rattus* species or subspecies.

Southeast Asia is depicted as the main center of origin of murid rodents (108). India, the Malayan peninsula, and the larger islands are particularly rich in species. *Rattus rattus* has been especially successful in the region, colonizing most of the islands, largely inadvertently by man. *Rattus norvegicus* is

Table I
Taxonomic Classification of the Laboratory Rat[a]

Class: Mammalia
 Subclass: Theria
  Infraclass: Eutheria
   Order: Rodentia
    Suborder: Myomorpha
     Superfamily: Muroidea
      Family: Muridae
       Subfamily: Murinae
        Genus: *Rattus*
         Species: *norvegicus*

[a]From Ellerman (41), Missonne (108), Morris (112), and Simpson (145).

clearly a colder climate species, and it is generally assumed to have evolved in either central Asia or China. The spread of both species is undoubtedly due to their commensal habits, a feature shared with a few other murids, notably *Rattus exulans, R. (Praomys) natalensis,* and *Mus musculus.*

III. ADAPTATION TO THE LABORATORY

The rat appears to have entered captivity as a consequence of rat-baiting, a popular pastime of the early 1800s. Large numbers of rats had to be procured for these spectacles, and the odd mutant form was probably preserved as a curiosity. Rat breeding as a hobby was established in the latter part of the century. The "official" introduction of the rat to the laboratory may be taken to date from the establishment of breeding colonies by Crampe (ca. 1877) and Donaldson (ca. 1893). Earlier experimental work has been reported, but it is unknown if these represented more than casual employment of animals purchased from dealers (137) (see Chapter 1).

Profound changes have occurred as a consequence to domestication, affecting size and function of many organs, reproductive performance, and behavior (135,137). These seem to be mediated by changes in size and output of the major endocrine glands. The adrenals have been sharply reduced in size, especially the cortex, together with the preputials. The pituitary, thymus, and thyroid are relatively unchanged. While the ovaries, testes, and secondary sex glands are unchanged in size, they probably mature earlier and function more continuously. The laboratory rat attains puberty earlier than the wild animal, displays little sign of an annual sexual cycle, and is more prolific. If the animal matures earlier, it probably dies sooner. Wild rats probably live as long as 4 or 5 years, whereas it is unusual for the laboratory rat to exceed a lifespan of 3 years. There has been an overall reduction in size, but no other obvious morphological changes have occurred, with the exception of glandular modifications.

IV. GENETICS

In this section some of the more important inherited diseases and variations of the laboratory rat are described (also see Table II).

A. Color and Coat Variation

The wild-type rat is a typical light-bellied agouti, i.e., agouti on the dorsum with light cream or whitish stomach. The color mutants which modify the type may be divided into those which change the color and those which induce white spotting. A full account of the older mutants may be found in Robinson (137).

The agouti locus has two known alleles: **non-agouti** (*a*), which produces a uniform black coat, and **agouti-melanic** (*a^m*). Phenotypically, agouti-melanic is almost black, but the hairs are tipped with yellow and the stomach fur is light. Preliminary data infer that a^m is recessive to type but dominant to *a* (100).

The normal black pigmentation of agouti and self black may be changed to dark chocolate by the **brown** (*b*) mutant, resulting in the cinnamon ($++bb$) and self chocolate (*aabb*), respectively.

Three mutant alleles are known for the albino series. Full **color** (*C*) is the wild type, with mutants designated as **ruby-eyed dilute** (c^d), **Himalayan** (c^h), and **albino** (*c*). Ruby-eyed dilute rats have no phaeomelanin, and the eumelanin is degraded to pale sepia. The eyes are ruby colored. The non-

agouti form ($aac^d c^d$) is pale sepia. Himalayanism is a recent discovery (114), the coat is cream initially but becomes whitish while, conversely, the extremities darken with age, the phenotype eventually becoming an acromelanic albino. The complete albino (*c*) has a white coat and pink eyes.

The **dilution** gene (*d*) produces a bluish-cream in combination with agouti ($++dd$) and a uniform slate-blue with non-agouti (*aadd*). In conjunction with *b*, a somewhat paler bluish-cream ($++bbdd$) is produced and a softer toned blue-grey or lilac (*aabbdd*).

The effects of **fawm** (*f*) on the coat color has not been worked out fully. In company with non-agouti and blue dilution, the following phenotypes have been determined: black ($aa++++$), blue ($aadd++$), coffee brown ($aa++ff$), and lavender (*aaddff*). Gene *f* could be a second brown-type mutant but of somewhat stronger action to produce a lighter phenotype (coffee versus chocolate brown).

Pink-eyed dilution (*p*) sharply degrades eumelanism to a drab beige but leaves phaeomelanin unaffected. With agouti ($++pp$), the effect is an orange-fawn, but with non-agouti (*aapp*), a uniform beige. The eyes are pink. Pink-eyed dilute is effectively epistatic to *b* because the phenotypes $++pp$ and *bbpp* are indistinguishable. An allele (p^m) of *p* has been reported (175) in which the eye color is ruby and the effect on eumelanin is not so drastic.

A similar gene to *p*, but not allelic to it, is **red-eyed dilution** (*r*). This mutant produces an orange-fawn phenotype in combination with agouti but the eyes are a dark red.

The mutant designated as **silver** (*s*) by Castle (24) would appear to be more aptly described as a form of achromia. The juvenile coat is normally pigmented but in succeeding molts the hairs become progressively lighter at the base. In the adult coat, only the distal part of the guard hairs are pigmented, the coat appearing predominately white.

Macy and Stanley (100) have described a mutant **sand** (*sd*), which is brownish-yellow in combination with agouti. With non-agouti (*aasdsd*), a slate-colored rat is engendered, said to be darker than blue (*aadd*). Pink-eyed yellow segregants (*ppsdsd*) from matings of sand animals are stated to be identical to $pp++$. Hence, *sd* may only dilute eumelanin but not phaeomelanin. (The authors employ the symbol *s* in their report but this has been preempted; here *sd* is used).

A **yellow,** or nonextension of eumelanin, mutant has not been genetically analyzed, but a number of specimens of this phenotype have been trapped by Figala (44) in a wild population (Mydlarka, Bohemia). The eyes are dark, and the coat is described as flavus, with no trace of eumelanic pigmentation. The homology with yellow mutants of other rodent species is apparent. Provisionally, the mutant is symbolized as *e*.

A gene producing **white belly** (*wb*) fur in agouti individuals has been reported (160). The gene (*wb*, revised symbol proposed by L. Vasenius) is inherited as a recessive and shows

Table II

Summary of Linkage Groups

	Locus relationships	References
I	$\begin{array}{cccccccc} p & fz & c & r & Hbb & lg & Rw & he & w \\ \overline{0} & 2 & 19 & 20 & 28 & 30 & 41 & 42 & 43 \end{array}$	17,46,51,114,138
II	$\begin{array}{cccccc} Sh & Cu\text{-}1 & an & in & s & b \\ \overline{0} & 4 & 6 & 20 & 44 & 52 \end{array}$	138
III	$\begin{array}{cc} k & st \\ \overline{0} & 26 \end{array}$	138
IV	$\begin{array}{ccc} a & svp\text{-}1 & f \\ \overline{0} & 6 & 45 \end{array}$	49,114,138
V	$\begin{array}{ccc} Ag\text{-}C & Es\text{-}2 & Es\text{-}1 \\ \overline{0} & 6 & 11 \end{array}$	52,166
VI	$\begin{array}{cc} Gl\text{-}1 & h \\ \overline{0} & 12 \end{array}$	115
VII	$\begin{array}{cc} H\text{-}5 & lx \\ \overline{0} & 5 \end{array}$	88

phenotypic interaction with non-agouti. Homozygous agouti white belly (*AAwbwb*) have white venters, but heterozygous agouti (*Aawbwb*) tends to be intermediate between white belly and the grey-bellied type. The difference is most apparent at about 30 days of age. Non-agouti is epistatic to white belly.

Vasenius (159) has described a dominant gene, **microphthalamia** (*Mi*, revised symbol proposed by L. Vasenius), which eliminates yellow pigment from the coat and produces minor white spotting in the forehead of heterozygotes. In conjunction with +*h*, the amount of white is sharply increased, the ventrum is almost completely white, and the spot on the forehead is enlarged. With *hh*, the animal is entirely white except for light patches of colored fur around each eye and extending to the nose. The homozygote is white, with pink microphthalmic eyes. The teeth fail to erupt, and the animal eventually dies from emaciation after weaning (L. Vasenius, personal communication, 1976).

Only one locus for white spotting is known, although four alleles have been reported (including one for the singularly characteristically hooded pattern of the rat). Three of the alleles are conventionally regarded as recessive to type, designated as **Irish** (*h^i*), **hooded** (*h*), and **notch** (*h^n*). "Irish" homozygotes have chest and mid-venter spotting, while hooded homozygotes are white except for the head and shoulders (the "hood") and a spinal stripe. "Notch" homozygotes are similar to hooded except that the area of pigmentation is further restricted. The hood is small and there is no spinal stripe. All of the heterozygotes display white spotting roughly intermediate to that shown by the corresponding homozygotes. Type animals heterozygous for any of the above alleles usually, but not invariably, show some ventral white. Wendt-Wagener (162) states that the white pattern is due to failure of melanoblast migration.

The most interesting allele is **restricted** (*H^{re}*) which produces white spotting, male sterility, and lethality when homozygous (60,61,67,128,141). The allele produces minor white spotting on the forehead and venter by itself (*H^{re}*+), but extensive spotting in conjunction with *h^i* or *h*. *$H^{re}h^i$* is high grade hooded white, and *$H^{re}h$* is almost all white except for small areas of pigmentation around the eyes and ears. These two genotypes show histological irregularities of eye pigmentation which reaches the macroscopic level in combination with *r*. In *$H^{re}h^irr$* and *$H^{re}hrr$*, the usual dark red eye (due to *r*) may be modified to pink, either unilaterally or bilaterally depending upon the magnitude of the effect of *H^{re}*. *H^{re}* males display normal copulatory behavior but usually are sterile or are fertile until 3 months of age soon thereafter becoming sterile. The testes are flaccid and undersized. Development of the tubular germinal epithelium is grossly impaired and ultimately ceases. Death of the homozygote *$H^{re}H^{re}$* occurs near or immediately after parturition as a result of cessation of hematopoiesis and severe macrocytic anemia.

Four independent rexoid mutants are known: three dominant [**curly-1** (*Cu-1*), **curly-2** (*Cu-2*), and **shaggy** (*Sh*)] and one recessive [**kinky** (*k*)]. All four are of the same general phenotype, with rough, unkempt coats and curved vibrissae (137). A fifth dominant rexoid mutant [**rex** (*Re*)] of typical appearance has appeared recently. *Re* is inherited independently of *b* (R. Robinson, unpublished). This implies that *Re* is not a repeat of *Cu-1* (which is linked to *b*) but it may be a repeat of either *Cu-2* or *Sh*.

Cowlick (*cw*) causes a whorl of hair situated along the dorsal midline, frequently in the middle of the back, but it may occur as far forward as between the ears. There is some variation of expression, but the full cowlick is a radiation of hairs, pointing in all directions from a center point.

Several instances of hypotrichosis have been reported. Two early discoveries were **hairless** (*hr*) and **naked** (*n*). In each, there is an initial growth of hair at successive molts, which is subsequently lost. The amount of new growth diminishes steadily. The skin of *hrhr* eventually becomes thickened and wrinkled, whereas that of *nn* remains pliable. Hairless and naked are genetically distinct (25,136).

Additional hairless mutants are **hypotrichotic** (*hy*) and **atrichis** (*at*), described by McGregor (104) and Lefebvres-Boisselot (95), respectively. There is a similar general picture of initial erupture of hair for each cycle of growth but subsequent loss. **Vibrissaeless** (*vb*) is reported by Lutzner and Hansen (99). There is almost complete loss of hair, including the vibrissae. This form can be identified at birth because of shortened and reduced number of vibrissae. The skin develops blisters due to intraepidermal separation just above the basal cell layer. These form crusts but heal without scarring. Body size is reduced. The females cannot nurse because of blistering of the nipples. The claws are easily detached and are often shed spontaneously. A further hair deficiency mutant was briefly noted by Palm (126) as **fuzzy** (*fz*). Fuzzy individuals are covered by grossly abnormal hair to about 6 to 8 weeks when it is lost. Adult fuzzy rats are hairless (J. Palm, personal communication, 1971). The relationships of the more recent genes to *hr*, or *n*, or to each other is unknown.

An alopecia, associated with a brownish exudate, chronic inflammation of the eyelids, and marked susceptibility to infection by *Pasteurella pneumotropica* has been described (82,156). The syndrome is due to a recessive gene, **masked** (*mk*), so called from a characteristic brownish crust which forms around the eyes and nose. The alopecia commences at about the fourth week and persists throughout life. There is regrowth of hair, but this is subsequently lost. The blepharitis is apparent by 21 days and becomes chronic. There is a thickening of the eyelids and some hyperemia. The eyes are usually closed. The brownish crust is found adhering to the skin and hair but can be removed by wet cotton swabs. Treatment with antibiotics failed to abolish the persistent pasteurellosis.

B. Growth and Reproduction

Differences in rats of growth and weight/size at various ages, attainment of puberty, litter size, and many other parameters of growth and reproductive physiology are largely polygenic, as far as the genetic component of the variation is concerned. The earlier literature has been previously summarized (137). The more recent work has dealt with response to selective breeding (6,7,140) and with heterotic effects (121) from crosses between inbred and partially inbred strains.

Growth, as measured by efficiency of food conversion and basal metabolism, has been shown to exhibit interstrain variation and, in some instances, evidence of polygenic heredity. Of equal importance has been the many studies on specific nutritional requirements. These aspects have also been reviewed previously (137). However, further studies have been made (144), and two recent reviews may be usefully cited (101,106).

The dwarfism (**dwarf-1,** *dw-1*) of Lambert and Schinchetti (92) appears to be basically endocrine in origin, although pituitary therapy failed to correct the disease. At birth, the dwarfs are of normal size, but growth impairment is obvious by the fifth day. The coat is thin, activity is curtailed, and the sexes do not become differentiated by size. No sexual activity was observed, sterility was the rule, and the dwarfs are shorter-lived than normal.

A second dwarfism (**dwarf-2,** *dw-2*) was found to differ in several respects from the above (170). The growth retardation is not observed until about 2 months of age; the males are sterile but the females may have one or two litters. The genetic relationship between the two dwarfs has not been investigated.

Fatty (*fa*) confers a noticeably obese condition from about the fifth week of life (180). The rat becomes grossly rotund. The blood serum assumes a milky appearance, with the fatty acids about ten times the normal level and cholesterol and phosphatides about four times normal. The blood lipids remain high even on a restricted diet. Females are sterile, but some males are fertile.

A second obese condition due to a recessive gene (*cp*, **corpulent**) has been described by Koletsky (86). The obesity can be detected by about the fifth week, and by 5 months the animal is excessively rotund, with copious fat deposits. The blood serum has a milky appearance, and there is marked hyperlipemia. The adrenals and islets of Langerhans are enlarged. Corpulent animals of both sexes failed to reproduce, but this is credited to mechanical problems, rather than to infertility since spermatogenesis was observed in the male and graafian follicles are well developed in the female. Koletsky designated the gene as *f*, but this has been changed to *cp* because *f* has been preempted (C. T. Hansen, private communication, 1976).

With **Kon's lethal** (lk), growth is normal until about the tenth day, when a steady decline is evident. Afflicted animals appear more restless than usual, and the suckling reflex is maintained but death ensues from apparent inanition. There is an absence of fat in the body, and some dehydration may be present (31).

Two recessive mutants affecting growth processes were observed by Taylor (156) in his experiments on X-ray-induced mutation. These are **skinny** (*sk*) and **runt** (*rt*). Only brief descriptions are given. Skinny individuals are small and emaciated at birth, usually dying within a few days, although a number manage to survive to weaning. Runt animals are noticeably smaller than their littermates at birth and die almost immediately.

The pseudohermaphrodite rat has a XY karyotype but a phenotype of feminine number and positioning of nipples, and a short vagina which ends blindly. There is an absence of reproductive organs other than inguinal testes with arrested spermatogenesis. The condition is transmitted by unaffected females to half of the male offspring. This is consistant with the inheritance of a dominant gene (**pseudohermaphrodite,** *Ps*), either autosomally with male-limited expression or X borne. The latter explanation is favored at this time (1,28,59).

C. Skeleton and Viscera

The gene **stub** (*st*) is named after the more obvious effect on the tail which is shortened, bent, or twisted, due to ankylosed vertebrae (133). Much of the lower vertebral column may be grossly abnormal, as well as the hindlimbs, lower viscera, and urogenital tract. There is high mortality at birth or shortly afterward, while the survivors are undersized. Most of the females are sterile. There is evidence of thyroid hyperplasia.

Gruneberg (58) has studied in detail a gene (**Gruneberg's lethal,** *lg*) causing a cartilage hypertrophy. At birth, homozygotes are of normal weight but fail to grow, and death usually occurs by day 14. The excessive growth of cartilageous tissue results in a wide variety of skeletal defects, all of which lead to eventual death. Gruneberg discusses the progression of the defects in some detail.

The mutant **incisorless** (*in*) may be readily classified by the absence of both sets of incisors (57). The anomaly has been investigated in detail (137). The lack of incisors is merely an indication of a more fundamental skeletal disturbance. Most of the skeleton appears to undergo a transient osteosclerosis. Early administration of parathyroid hormone brings about some alleviation of the condition, producing near normal bone growth and, in some instances, eruption of incisors.

A similar condition is **osteopetrosis** (*op*), even to the extent of noneruption of the incisors. This anomalous condition is more severe in that the normal medullary cavities in the diaphysis of long bones of *inin* animals commence to appear by 10 days, whereas no cavities occur in *opop* animals. By 20 to 30 days, the contrast between the two genotypes is sufficiently great for this feature to be used as a means of identification. Parabiotic union between *opop* and normal rats can enable the former to survive and breed. The *ininopop* individual is identical to *opop* in lacking diaphyseal cavities (118).

A third congenital osteopetrosis has been described by Cotton and Gaines (30). The causative gene (*tl*) is named **toothless** because of the unerupted incisor and molar dentition. Toothless can usually be distinguished from normal by the lack of incisors, shorter snout, and encrustation about the eyes. Many of the bones are abnormally thick and lack characteristic cortical and medullary areas. No obvious signs of anemia were observed. Evidently there is increased extramedullary hemopoiesis (splenomegaly was in fact observed in a few cases). Longevity of *tltl* rats did not seem to be impaired.

The **hypodactyly** (*hd*) studied by Moutier *et al.* (116) causes a reduction in number of toes of both the front and rear feet. Body growth is not affected, and the females are fertile. The males, on the other hand, are regularly sterile. Testes size is sharply reduced to about one-third of normal at 12 weeks of age. There is an almost complete absence of spermatozoa in the testes.

The **polydactly-luxate** syndrome observed by Kren (88) is due to a recessive gene *lx*. There is considerable variability of expression, according to the genetic background. In some strains, the gene behaved as a recessive but in others as a semidominant with poor penetrance. The severity of the defect in homozygotes was correlated with the penetrance of the heterozygotes. In severe cases, the front as well as the rear legs were affected. The polydactyly was of the preaxial form (resulting in six, sometimes seven digits), with reduction in size of the tibia and distortion of both tibia and fibula leading to luxation.

Two recessive mutants affecting skeleton growth have been reported by Taylor (156), namely, **chubby** (*cb*) and **stubby** (*sb*). Chubby animals are normal until 2 weeks of age when growth is retarded. Opening of the eyes is delayed by several days. At weaning (about 3 weeks), they are undersized, have small ears and limbs, and are lethargic. They are called chubby because of their appearance due to reduced skeletal size. They rarely live beyond 35 days. Stubby can be recognized by their very short legs of affected rats at birth. They seem to resemble the "bulldog" mutants of larger species. Death usually occurs within a few days.

An interesting array of gross anomalies of various parts of the skeleton have been described. In most cases, a clear-cut indication of monogenic causation is lacking, but many of the conditions appear to be due to polygenically mediated threshold characters (137). In some, the genetic influence can be modified by an overt environmental factor. The more significant conditions are congenital taillessness (37), spontaneous tail amputation (169), bent nose (72), malocclusion of incisors (81), umbilical hernia (109), and diaphragmatic hernia (2).

There has been an impressive amount of work on the etiol-

ogy of dental caries in the rat (77,78,137) but not necessarily with clear-cut results. Apparent genetic differences were demonstrated, but it was difficult to decide if these were due to differences between rats or to differences between the cariogenic microflora interacting with tooth morphology. The most long term of the many experiments which have been conducted is discussed by Chai *et al.* (27). They concluded that, despite the obvious influence of certain environmental factors, an appreciable portion of the resistance to caries is genetic.

The congenital **hydronephrosis** discovered by Lozzio *et al.* (98) is due to a dominant gene (*Ne*), which is lethal when homozygous. The kidneys of affected rats displayed various lesions, but hydronephrosis was the most common anomaly. There is variation of onset and expression of the condition.

Renal tumors, probably due to a dominant gene, were described by Eker and Mossige (40). They were rarely detected before 10 to 12 months of age and varied in size. They probably originated from renal tubules, since none have been found to contain glomeruli. The gene **renal tumor** is symbolized as *Tu* (J. C. Morsige, personal communication, 1976). There is evidence that the *TuTu* homozygote is lethal.

D. Sense Organs

Retinal dystrophy (15) is a well known eye anomaly of rats and is due to a recessive gene (*rdy*). The anomaly has been well researched (94), and this work is continuing (93). The primary effect of the gene is a progressive degeneration of the retina, commencing at about 3 weeks of age. However, it appears that anomalous differentiation of the retina is occurring as early as day 12 (36). Notable concurrent defects are a persistent hyaloid artery, unusual lens reflection, and cataract in the majority of animals. The cataract is the most conspicuous effect of the anomaly. Rearing the animals in darkness or combining the *rdy* gene with a pigmented, as opposed to the transparent albino, eye results in a delay in the onset of the degenerative process. The site of action of the *rdy* gene appears to be in the pigment epithelial cell layer of the retina (119). A parallel is drawn between the present anomaly and certain heritable retinal dystrophies in man.

A cataract due to a dominant gene **cataract** (*Ca*) has been described (149). The anomaly was observable from the moment the eyes are opened in most individuals and was regularly and bilaterally manifested. The lens was abnormal, smaller in size, and lacked the normal plumb shape. The cause of the opacity is unknown.

A second cataractous condition has been reported by Leonard and Maisin (96), behaving as a recessive gene **cataract lens** (*cat*). Approximately 25% of *catcat* animals do not have cataract at weaning, but the condition is fully expressed bilaterally by 70 days. The cataractous eye is smaller than normal, and the weight of the lens is merely one-fifth to one-tenth of normal at 70 days.

Yet a third monogenic cataract has been observed by Smith *et al.* (148). The anomaly was stated to be inherited as a recessive. No symbol has been ascribed to the gene, but *rc* is employed here. The cataract was associated with abnormal skull shape which seemed unusually flattened. Wartlike excrescences could be seen on the inner surface of the interior lens capsule, and there appeared to be excessive proliferation of the lens epithelium.

Several other eye defects have been described (137), although the genetic basis for many of them is obscure. Two exceptions are the microphthalmia of Browman (22) and the buphthalmos of Young *et al.* (177). Both of these anomalies appear to be inherited as threshold characters with a marked maternal influence complicating matters for microphthalmia. The microphthalmia appears to arise from arrested development of the eye. The buphthalmos is due to a persistent pupillary membrane interfering with proper drainage of the aqueous humor.

E. Nervous System

Several mutants which have profound effects on behavior are known. Three genes which have been reported upon in detail are waltzer (*w*), wobbly (*wo*), and shaker (*sr*). The abnormal behavior associated with each gene is characteristically different. Waltzers tend to circle vigorously, either to the left or right (83). Wobblies walk in a jerky manner, with the head and tail held high instead of in line with the body (26). Shakers are nervous and walk with a curious shaking movement. If disturbed, they rush around in an excitable manner (25).

Two major behavior mutants were briefly described by Taylor (156). These are **crawler** (*cr*) and **spastic** (*sp*). Crawler homozygotes show trembling at about 2 weeks of age, followed by a paralysis affecting the hind limbs. The animal can only move by pulling itself forward with the front legs. Death ensues by approximately the fourth week. Assortment data indicate a deficiency of *crcr* individuals. Spastic rats are incapable of normal motility. They fall on their sides and hold their legs out rigidly, with the back arched. This behavior is observable immediately when they leave the nest, and death occurs by about 3 to 4 weeks of age.

C. T. Hansen (personal communication, 1976) has contributed details of two hitherto unreported mutants affecting behavior. The first is the recessive gene **spinner** (*sr*). Heterozygotes are of normal behavior, but the homozygote displays a mild hydrocephalus and a propensity to circle, mostly in a clockwise direction. The second is **trembler** (*tr*), also recessive. The affected animal commences to show a tremor at about 3 weeks of age, which becomes progressively worse until death intervenes at about 6 weeks.

Some strains of rats seem to be particularly liable to audiogenic seizures or electroshock convulsions. A considerable amount of effort has been made to unravel the genetics of this condition. These have involved both strain comparisons and selective breeding of susceptible and nonsusceptible lines (137). The experiments have shown that genetic variation certainly exists and that this may be regarded as polygenic.

F. Physiology and Biochemistry

An acholuric **jaundice** (*j*) was described by Gunn (62). Affected young are icteric at birth when born to a jaundiced mother but are normal at birth when born to a heterozygous mother, becoming icteric some 12 hr later. Growth rate is subnormal. Although the condition is formally regarded as a recessive, some 50% of heterozygotes (+*j*) display increased fragility of the erythrocytes and a reticulocytosis. The most obvious feature of the jaundice is a persistent bilirubinaemia throughout life. A "wobbly" gait may develop due to kernicterus. The condition appears to be traceable to a deficiency of the enzyme glucuronyltransferase. For a review of the pathogenesis see Blanc and Johnson (13).

A lethal anemia (**anemia-1,** *an*) has been analyzed by Smith and Bogart (150). The young rat shows a pallor at about 2 to 3 days and death usually ensues by day 14. There is loss of weight and an accompanying jaundice. The anemia appeared to be microcytic in form and to be derived from a dyshemopoiesis of the bone marrow. There was a deficiency of erythrocytes, although the number of leukocytes were about normal.

A second anemia (**anemia-2,** *am*) was described by Sladic-Simic *et al.* (147) as characterized by hypochromia, persistent microcytosis and poikilocytosis, and heightened osmotic resistance of erythrocytes. Newborn anemic rats were pale yellow and showed retarded growth. The majority died soon after birth, but a few survived until about 6 months. Administration of iron dextran prolonged the life of anemic animals but did not change the blood picture.

The vascular thrombosis described by Dunning and Curtis (38) appears to be due to a recessive gene **hematoma** (*he*). Spontaneously thrombosed arteries arise in the liver (usually) and lungs (rarely) of afflicted individuals, resulting in death.

Resistance to warfarin poisoning has been shown (55) to be due to a dominant gene (*Rw*). Both *he* and *Rw* are linked to *p*, with similar cross-over values. This had led to speculation that *he* and *Rw* might be allelic (or pseudoallelic) and concerned in the same hemorrhagic or blood clotting processes (56). A search for the mechanism which confers resistance has been unproductive (129).

A form of hypothalmic **diabetes insipidus** (142) is due to a recessive gene (*di*). Affected rats have an excessive intake of water and subsequent urination on the order of 20 to 125% of body weight per day, compared with 10% or less for the normal individual. There also is a difference of urine osmolarity. The severity of the disease (as measured by amount of urine voided) was variable and appeared to be familial, suggesting some degree of genetic modification.

Tschopp and Zucker (157) have described a strain of rats with an inherited mild hemorrhagic diathesis. The mode of heredity has not been determined.

The inability to react to intraperitoneal injection of dextran has been shown (70) to be inherited as a monogenic recessive trait (*dx*). The typical reaction is hyperemia, edema, and pruritis of the head and paws, while the nonreactor *dxdx* either failed to show any or showed only mild symptoms.

Rapp *et al.* (130,131) have demonstrated a difference of adrenal metabolism for two rat strains selected for susceptibility and resistance to salt-induced **hypertension.** The criterion is rate of conversion of ^{14}C-labeled deoxycorticosterone to 18-hydroxydeoxycorticosterone for homogenized adrenal tissue. Despite the individual variation, the results indicated the segregation of two codominant alleles of a locus *Hyp-1* (symbol proposed by J. P. Rapp).

Ordinarily, the inheritance of essential hypertension would be regarded as polygenic, but it seems possible that the number of genes involved are few rather than many. It is possible to produce both hypertensive and hypotensive strains fairly easily by selective breeding (39,132). Yen *et al.* (173) has demonstrated the apparent segregation of a major gene for hypertension by the use of a discriminant function. Rapp *et al.* (132) found that allelic differences at the *Hyp-1* locus (governing the rate of conversion of ^{14}C-labeled deoxycorticosterone) is influential in determining hypertension. These genes accounted for about 16% of the difference between two strains differentially selected for salt-induced hypertension.

The rat has featured in the current interest in biochemical variation as revealed by elegant electrophoretic analyses. Various studies of esterases have been particularly rewarding. Variation for five different esterases has been isolated, as judged by their electrophoretic characteristics and tissue distribution (114,117,168). These have been designated as *Es-1, Es-2, Es-3, Es-4,* and *Es-5,* and each has two codominant alleles except for *Es-3* which probably has three and *Es-2* which may have as many as six (18). A sex-influenced esterase has also been detected, but the independent status of the variation is unsettled (52). All of the above appear to be on the same chromosome, several seeming to be very tightly linked. An esterase (*Es-12*) derived from heart tissue displays variation which is due to two codominant alleles (79), but the relationships with the above group as yet has to be determined.

All of the esterase variability discussed above is concerned with electrophoretic mobility (distance traveled) of the various

alternative forms. However, a difference in activity (degree of staining for a band of the electrophoretic spectrum) has been reported (178) for a hepatic esterase. High activity behaves as a dominant to low, and the locus is denoted as *EsC* (**esterase concentration**) with two alleles.

Variation has been observed (23,85) for **phosphogluconate dehydrogenase** (PGD) and population studies have revealed the presence of two codominant alleles of a locus *Pgd*.

The seminal vesicle protein (*Svp-1*) has been shown to exist as two allelic forms (40).

Three electrophoretic patterns of plasma proteins have been described by Moutier *et al.* (114). These are produced by two codominant alleles of a locus *Gl-1*.

Jimenez-Marin (179) has shown that variation of plasma **alkaline phosphatase** (ALP) could be ascribed to two codominant alleles of a locus *Alp*.

Rat hemoglobulin shows a consistent variation of electrophoretic pattern which is determined by a *Hbb* locus with two codominant alleles (16,17,46). It is proposed that the locus is concerned with differences of composition of a minor B chain. Comparable variations probably have been discovered by Martinovic *et al.* (102).

Despite the electrophoretic diversity which has been genetically analyzed to date, there would seem to be further variation waiting to be studied in detail (33,80). General surveys have uncovered individual and intrastrain differences for ceruloplasmins, haptoglobins, transferrins, glucose-6-phosphate dehydrogenase, and 6-phosphogluconate dehydrogenase, together with a few esterases which may be different from those previously investigated.

G. Immunogenetics

The early studies on blood groups and immune reaction were unfortunately rather disconnected, and the results were sometimes unclear from a genetic standpoint. The situation was not helped by a lack of appreciation of the necessity to preserve stocks of rats carrying definable antigens. In view of this, it is convenient to consider the papers of Palm (124) and Palm and Black (127) as initiating modern rat immunogenetics.

It seems worthwhile to distinguish between those antigens whose function appears to be mainly serological and those which may determine serotype differences but are also concerned in histocompatibility activity. Three blood group loci have been recognized (127) and designated *Ag-A, Ag-C,* and *Ag-D,* each with two alleles. One of the alleles of *Ag-A* appears to be uncommon, but the alternative alleles of the other two loci appear to have proceeded to fixation in different inbred strains more or less at random (127).

Two loci, *H-2* and *H-3,* each with serological and tissue antigenic properties, have been described by Kren *et al.* (89).

Each locus has two alleles. *H-3* could be identical with *Ag-C,* since both display linkage with *Es-2,* with a cross-over percentage of 5.1 ± 2.5% (V. Kren, personal communication, 1976) and 5.6 ± 3.9% (52), respectively.

A locus (*Ag-F*) was defined by DeWitt and McCullough (32) as controlling the expression of a lymphocyte surface antigen. This locus has at least two alleles and possibly as many as four. Close linkage has been found with albinism.

A major histocompatibility locus has emerged for the rat, comparable to the *H-2* locus of the house mouse. The locus has definite serological aspects but attention has been focused upon the histocompatibility properties as being of greater fundamental importance. The locus has been designated *Ag-B* in the terminology of Palm (127) but also as *H-1* to emphasize the histocompatibility aspect and the apparent homology to *H-2* of the mouse (89,151). The amount of work devoted to analysis of the antigenic properties of the locus is impressive. Thus far, ten alleles have been identified, characterized by 22 different antigenic specificities (151). It is evident that other alleles remain to be found. At this time, few experiments have been performed with the objective of detecting cross-overs within the *H-1* complex. The relationship of the six *H-1* alleles isolated by Palm to the ten *H-1* alleles of the Prague school of workers have been defined by Kren *et al.* (89). More recently, two new *Ag-B* alleles have been found by Kunz and Gill (90) which are distinct from the six Palm alleles.

The distribution of *H-1* alleles has been determined for some 20 or more American (90,125,127), 36 European (151,152) and 28 British (87) inbred strains and stocks. The distribution of the various alleles is not random. The allele $H-1^w$ occurs more frequently (about 22 cases) than the next two ($H-1^d$, $H-1^1$; about 12–15 cases), with the other alleles occurring rarely. There has been speculation as to whether this may indicate selective advantage of certain alleles or if the distribution merely reflects common origin of the strains. Since the origin of many extant strains is obscure, it is difficult to decide between the two alternatives. *H-1* polymorphisms exist in random-bred stocks, hence the possibility of selective advantage cannot be ruled out.

The histocompatible locus detected by Michie and Anderson (105) has been symbolized as *Ag-E* by Palm (125). She (123) considers that the skin graft rejection observed by Michie and Anderson is not due to differences at the *H-1* locus because the survival times are different. However, there is the possibility that this aspect might be mediated by the strain background.

A differential survival of individuals heterozygous for *H-1* and *Ag-E* alleles over the corresponding homozygote has been described (105,123). The reason for the disparity is not obvious. Palm suggests that an allelic *H-1* difference between mother and fetus might protect the latter from attack by a non-*H-1* incompatibility. Michie and Anderson suggest that the heterozygote may enjoy an initial advantage of rate of early

development or that a form of selective fertilization may be operating by means of an immunological reaction between surface membranes of sperm and egg. Beer *et al.* (9) have concluded that implantation is probably the critical time for manifestation of the phenomenon.

The existence of weak H-Y (71,120) and H-X (143) antigens have been demonstrated. No enhancement of the antigenicity was observed by the few *H-1* alleles which also were present in the experimental material.

Kren *et al.* (89) have described two loci, each with two alleles, which function as weak histological antigens (in comparison with *H-1*). The loci are designated as *H-4* and *H-5.* The relationship of these loci to those of the *Ag* series is unknown at present. However, *H-4* is linked to albino (*c*) and *H-5* to polydactyly-luxate (*lx*).

Studies on the inheritance of immunoglobulin allotypes have uncovered a locus *RI-1* involved in the determination of properties of the light chain (3,66). Two codominant alleles could be demonstrated. A second locus *Iα* (*la*) is concerned with variants of the heavy α-chain (8). A pair of codominant alleles could be demonstrated.

A picture of a complex immune response locus apparently closely linked to *H-1* is slowly being constructed. Wurzburg *et al.* (171, 172) have shown that high responsiveness to **lactic dehydrogenase** (LDH-A) is due to a dominant gene. High responsiveness was observed to be associated with the $H-1^l$ allele and low responsiveness with $H-1^a$ in crosses. Examination of the response of inbred strains with known *H-1* alleles implied that high response is also associated with the $H-1^d$, $H-1^e$, $H-1^f$, $H-1^n$, and $H-1^w$ alleles.

Resistance or susceptibility of the rat to experimental allergic encephalomyelitis (EAE) was thought to be mediated by a pair of genes linked to *H-1* (50,166). Resistance appeared to be associated with the $H-1^n$ allele and susceptibility to $H-1^l$ and $H-1^a$. Susceptibility was depicted as due to a dominant gene *Ir-EAE,* linked to *H-1* with a cross-over frequency of about 10%. However, subsequent work indicates that the situation is more complex (51). Allele $H-1^n$ still confers complete resistance, but not all susceptible animals developed obvious lesions. It is undecided whether the variable results were due to unknown technical or genetic factors. In the light of this, the previous evidence for cross-over between *H-1* and the postulated *Ir-EAE* gene was considered to be unreliable.

Some interesting work is shaping up with immune response to synthetic polypeptides. High responsiveness to (T,G)-A-L is governed by a dominant gene linked to the $H-1^a$ allele while low response was associated with $H-1^w$ (65). Examination of the response by a battery of inbred strains carrying known *H-1* alleles revealed that low response also was associated with $H-1^b$, $H-1^d$, $H-1^e$, $H-1^l$, and $H-1^n$ (64). The $H-1^f$ allele gave a medium response. The same battery of inbred strains was challenged with (H,G)-A-L and (Phe,G)-A-L. Each antigen gave a different pattern of high and low response for the various *H-1* alleles. High responsiveness was stated to be dominant to low responsiveness in each case. Allelic complementation and "overdominance" of response to (H,G)-A-L has been demonstrated (63). Alleles with nil or poor response when homozygous can give high response in certain heterozygous combinations.

Armerding *et al.* (5) have shown that high response to the synthetic copolymers GA and GT is due to two distinct dominant genes which they called *Ir-GA* and *Ir-GT*, respectively. Both genes are linked to different *H-1* alleles, *Ir-GA* to *H-1^w* and *Ir-GT* to *H-1^a*; *H-1^a* and *H-1^w* display linkage to the obverse low response allele of the respective immune gene. In an earlier paper (4), working with inbred strains, high response to GA was associated with *H-1^b*, *H-1^c*, and *H-1^l* and low response with *H-1^n*, while high response to GT was associated with *H-1^c* and low response with *H-1^b*, *H-1^l*, and *H-1^n*.

The extensive studies by Gill *et al.* (53,54) of response to poly(Glu^{52}Lys^{52}Tyr^{15}) between two strains could not be explained in simple monogenic terms. At least two genes appeared to be involved and probably more. A later study (91) revealed that the magnitude of the response was associated with *H-1* alleles. Alleles *H-1^a* and *H-1^c* with high response and *H-1^b*, *H-1^l*, and *H-1^n* with low. *H-1^w* gave an intermediate response.

Differences of susceptibility to experimental autologous immune complex glomerulonephritis appear to be related to the *H-1* status (153). These observations are preliminary but the allele *H-1^n* seemed to confer resistance under certain conditions, whereas alleles *H-1^d* and *H-1^l* did not.

The foregoing would portent that the *H-1* locus could be a complex of histogenic and immune response sites comparable to the *H-2* locus of the mouse. Just how comparable will depend upon the fine structure. This indicates that the next step in the analysis should be primarily genetic. That is, while the immunogenetic description should continue, emphasis should be placed on mapping the units of the region. Thus far, the number of animals examined have been just sufficient to establish linkage of the *H-1* and the *Ir* loci, but certainly insufficient for assessment of closeness of the linkage and to detect rare cross-overs. The latter are necessary if any idea is to be obtained of the number of subloci, their linear arrangement, and potential for recombination to engender detectable *H-1* and *Ir-1* alleles.

Koch (84) has observed that high immune response to high or low (LW) molecular weight (G,T)-A-L was inherited as a dominant to low response. However, while high response to LW (T,G)-A-L was linked to *H-1^n* (low response to *H-1^l*), high response to LW (G,T)-A-L was not.

The interstrain difference of level of response to sheep red blood cells (SRBC) and to bovine γ-globulin (BGG) has been shown to be monogenic (155). High response in each case was due to independent dominant genes designated as *Ir-SRBC* and *Ir-BGG*, respectively. No linkage was found between these loci and *H-1*. It remains to be seen if these sites determine specific or general immune responsiveness.

A theta-like antigen was described by Douglas (35) for the rat, but no variants were detected for six inbred strains.

A technique for the discovery of minor antigenic differences, described by Kren *et al.* (89), deserves brief comment. Individuals with a dominant gene are repeatedly backcrossed to an inbred isogenic stock and challenged by a battery of tests to reveal antigenic differences which could be attributed to the segment of chromosome tagged by the gene. Homozygosity at the major histogenic *H-1* and other known antigenic loci must be achieved in the early generations so as to reveal any remaining differences in later generations. As the backcrossing continues, the genome becomes progressively homozygous except for the tagged segment of chromosome. Two minor histogenic loci have been isolated by this method.

H. Psychogenetics

There is a considerable volume of material on rat psychogenetics or behavioral differences. No major genes have been discovered (genes *cr, sn, sp, sr, tr, w,* and *wo* may be excluded as representing grossly abnormal behavior), but ample evidence exists for polygenic variation in a wide variety of behavioral traits. Two methods of analysis have disclosed the genetic component of the observed variation. The first is bidirectional selection or selective breeding for high and low expression which, when successful, results in two steadily diverging phenotypic strains. The second is a biometrical analysis by means of diallel crosses employing inbred strains. The two approaches have been combined on occasion. The methodology may be different, but both approaches are powerful tools for the genetic analysis of behavior.

Genetic differences have been demonstrated for the following patterns of behavior or responses to specific experimental situations. Cognitive ability (maze learning of several different types, problem solving, avoidance learning and various discrimination tests), activity (spontaneous, exploratory, ambulatory and geotropic), temperament (wildness, aggressiveness, emotionality, and conditioned emotionality), preferences (alcohol, luminance, morphine and saccharine) and miscellaneous traits (dominance behavior, grooming activity, hoarding instinct, neophobia, social behavior, startle response, toxiphobia and wood gnawing). This work has been reviewed fully and documented elsewhere (137).

With a few exceptions, the more recent research has been much on the same lines as before, developing and consolidating previous observations rather than opening up completely fresh avenues. Differences of conditioned avoidance learning have been studied by Bignami (11,12,20) and others (45,74,163), and emotional behavior by Broadhurst and associates (21,47,74,164,165) and Harrington (69). Other investigations of genetic interest have involved maze learning (68), stress-induced stomach ulceration (146), water intake (139), and preferences for alcohol (42,163) or ethanol (19).

V. CHROMOSOME MAPPING

The allocation of genes to the various known linkage groups has been critically evaluated elsewhere (138). A summary of the conclusions, together with the results of more recent findings is shown by Table III (see also Fig. 1). Seven groups may be regarded as definitely established. The map lengths are approximate and the best that can be estimated from the available data.

A few genes are known to belong to certain groups, but their exact positions are uncertain at present. *Ag-F* is linked to *c* (32) with about 2% of crossing over but has not been tested against any other locus in group I. Also, *H-4* is stated (89) to be linked to *c*. Close linkage is reported between the esterase loci *Es-2, Es-3, Es-4,* and *Es-5,* with no crossing over in moderate sized samples (117,166). The existence of a histogenic antigen carried by the *Y* chromosome (if antigenicity is not a property of the *Y* chomosome per se) may be regarded as substantiated (71).

VI. CYTOGENETICS

The rat is a favorite species for chomosome studies, and there is an extensive literature. This has been reviewed to 1968 (137,138) and, in general, only more recent contributions need be noted. The haploid number of chromosomes is 21 and only with the event of differential banding techniques has it been possible to identify the chromosomes individually. Grossly, there are two large chromosomes, one telocentric and the other subtelocentric. From then onward, the elements decrease steadily in size, few of which can be consistently identified on an individual basis. Almost all of the medium-sized chromosomes are either telocentric or subtelocentric, while most of the smaller ones are metacentric. The X is a medium-sized subtelocentric (sometimes acrocentric) and the Y is a small telocentric. As yet, no detectable differences have emerged between the karyotypes of laboratory and wild *R. norvegicus,* nor is there reason to anticipate any.

Detailed descriptions of the banded chromosomes, as revealed by fluorescent or Giemsa techniques, have been presented by a number of investigators (48,107,110,143,158,167,

176) (See Fig. 2). There is some variation of details, and a different system of numbering the chromosomes was employed in each case. This unsatisfactory state of affairs was rectified by a committee which has defined a mutually agreed nomenclature based upon decreasing size of the chromosomes. The committee noted that, although each chromosome can now be identified, certain elements (particularly numbers 16 to 18) may only be distinguishable in good preparations (29).

The question of nomenclature has been further considered by Levan (97), who proposed a system based on human chromosome methodology. This would enable individual band and the lighter staining interband regions to be defined precisely. Levan's portrayal of the banding pattern for individual chromosome differs from that of the standardized karyotype in a number of aspects. He gives the most detailed description as yet of the banding pattern for individual chromosomes. The published idiograms are also critically compared.

Comparisons have been undertaken between the banding patterns of individual chromosomes belonging to seven species of rats (111,176). There appears to be a fundamental similarity of pattern for the various species, even when they differed in haploid number. It is suggested that this similarity supports the concept that the major interspecies differences are the result of fusion or fission of the larger chromosomes. Pericentric inversions seem to be a feature of the smaller elements.

Nesbitt (122) has studied the banding patterns of the rat and mouse and found that about 40% of the genome could be matched for the two species. This estimate may turn out to be optimistic, but it is patent that this sort of cytogenetic analysis can be extended with the promise of interesting results, especially if the banding techniques can be improved to give higher resolution of bands.

A number of polymorphisms of various chromosomes have been observed (138). The third largest chromosome (No. 3) appears to be either telocentric or subtelocentric (or satellited) in different stocks (75,134,174). Chromosome No. 3 is distinctive enough for the difference to be useful as a marker for certain experiments. Similar variation has been reported for several of the smaller chromosomes (134), in particular for chromosome No. 13 (10,103,130). Lately, variation of banding pattern has been exhibited for one of the small subtelocentrics (48).

The large X chromosome has displayed heteromorphism, being either telocentric or subtelocentric (138). Zidek (179) has presented a list of rat strains in which one or the other chromosome has been observed. The difference has been utilized by Wake *et al.* (161) to demonstrate that while the maternal or paternal derived X chromosome apparently is inactivated at random in the rat embryo, the paternal X is preferentially inactivated in the yolk sac. The Y chromosome has given some evidence of variation in size, being extremely small in some preparations but distinctly larger in others (76).

Table III

List of Defined Genes in the Rat

Linkage group	Symbol	Gene designation	Primary characteristic	Reference
IV	a^m	Agouti-melanic	Coat color	100
	a	Non-agouti	Coat color	137
	$Ag\text{-}A$	Agglutinogen-A	Antigenicity	127
V	$Ag\text{-}C$	Agglutinogen-C	Antigenicity	127
	$Ag\text{-}D$	Agglutinogen-D	Antigenicity	127
	$Ag\text{-}E$	Histocompatibility	Antigenicity	125
I	$Ag\text{-}F$	Lymphocytotoxin	Antigenicity	32
	Alp	Alkaline phosphatase	Electrophoresis	79
	am	Anemia-2	Blood	147
	an	Anemia-1	Blood	150
	at	Atrichis	Atrichosis	95
II	b	Brown	Coat color	137
I	c^d	Ruby-eyed dilute	Coat color	137
I	c^h	Himalayan	Coat color	113
I	c	Albino	Coat color	137
	Ca	Cataract	Lens anomaly	22
	cat	Cataract lens	Lens anomaly	46
	cb	Chubby	Skeleton	156
	cp	Corpulent	Physiology	86
	cr	Crawler	Behavior	156
	ct	Cataractous	Lens anomaly	148
II	$Cu\text{-}1$	Curly-1	Hair anomaly	137
	$Cu\text{-}2$	Curly-2	Hair anomaly	137
	cw	Cowlick	Hair anomaly	137
	d	Dilute	Coat color	137
	di	Diabetes insipidus	Physiology	142
	dx	Anaphylactoid reaction	Physiology	70
	$dw\text{-}1$	Dwarf-1	Body size	92
	$dw\text{-}2$	Dwarf-2	Body size	168
V	$Es\text{-}1$	Esterase-1	Electrophoresis	114
V	$Es\text{-}2$	Esterase-2	Electrophoresis	52
V	$Es\text{-}3$	Esterase-3	Electrophoresis	114
V	$Es\text{-}4$	Esterase-4	Electrophoresis	114
V	$Es\text{-}5$	Esterase-5	Electrophoresis	114
	$Es\text{-}12$	Esterase-12	Electrophoresis	79
	EsC	Esterase concentration	Electrophoresis	178
	$e(?)$	Yellow	Coat color	44
IV	f	Fawn	Coat color	137
	fa	Fatty	Physiology	180
I	fz	Fuzzy	Atrichosis	126
VI	$Gl\text{-}1$	Plasma protein	Electrophoresis	66
VI	H^{re}	Restricted	White spotting	60
VI	h^i	Irish	White spotting	137
VI	h	Hooded	White spotting	137
VI	h^n	Notch	White spotting	137
	$H\text{-}1\ (Ag\text{-}B)$	Histocompatibility-1	Antigenicity	89, 127
	$H\text{-}2$	Histocompatibility-2	Antigenicity	89
V	$H\text{-}3$	Histocompatibility-3	Antigenicity	89
	$H\text{-}4$	Histocompatibility-4	Antigenicity	89
VII	$H\text{-}5$	Histocompatibility-5	Antigenicity	89
	$H\text{-}Y$	Histocompatibility-Y	Antigenicity	71
I	Hbb	Hemoglobin-B	Blood	46
	hd	Hypodactyly	Skeleton	116
	he	Hematoma	Blood	38
	hr	Hairless	Atrichosis	25
	hy	Hypotrochosis	Atrichosis	104

(continued)

Table III (*Continued*)

Linkage group	Symbol	Gene designation	Primary characteristic	Reference
	Hyp-1	Hypertension-1	Physiology	131
	Iα(1a)	Immunoglobulin-α	Immune reactivity	8
II	*in*	Incisorless	Skeleton	35
	(Ir-BGG)ᵃ	Immune response to BGG	Immune reactivity	155
	(Ir-EAE)	Immune response to EAE	Immune reactivity	51
	(Ir-GA)	Immune response to GA	Immune reactivity	5
	(Ir-GT)	Immune response to GT	Immune reactivity	5
	(Ir-LDHA)	Immune response to LDHA	Immune reactivity	170
	(Ir-SRBC)	Immune response to SRBC	Immune reactivity	155
	j	Jaundice	Physiology	62
	k	Kinky	Hair anomaly	137
	lg	Gruneberg's lethal	Skeleton	58
	lk	Kon's lethal	Growth	31
VII	*lx*	Polydactly-luxate	Skeleton	88
	Mi	Microphthalmia	Coat color	171
	mk	Masked	Atrichosis	82
	n	Naked	Atrichosis	136
	Ne	Hydronephrosis	Physiology	98
	op	Osteropetrosis	Skeleton	118
I	*p*	Pink-eyed dilution	Coat color	137
I	*pᵐ*	Ruby-eyed dilution	Coat color	175
	Pgd	Phosphogluconate dehydrogenase	Electrophoresis	85
	Ps	Pseudohermaphroditism	Sex organs	1
I	*r*	Red-eyed dilute	Coat color	137
	rdy	Retinal dystrophy	Retinal anomaly	94
	Re	Rex	Hair anomaly	Section IV,A
	RI-1	Immunoglobulin-1	Immune reactivity	66
	rt	Runt	Growth	156
I	*Rw*	Warfarin resistance	Physiology	55
II	*s*	Silvering	Coat color	24
	sd	Sand	Coat color	100
	sb	Stubby	Skeleton	156
II	*Sh*	Shaggy	Hair anomaly	137
	sk	Skinny	Growth	156
	sn	Spinner	Behavior	Section IV, E
	sp	Spastic	Behavior	156
	sr	Shaker	Behavior	25
III	*st*	Stub	Skeleton	133
IV	*Svp-1*	Seminal vesicle protein	Electrophoresis	49
	tl	Toothless	Skeleton	30
	tr	Trembler	Behavior	Section IV, E
	Tu	Renal tumor	Physiology	40
	vb	Vibrissaeless	Atrichosis	99
I	*w*	Waltzing	Behavior	83
	wo	Wobbly	Behavior	26
	wb	White belly	Coat color	172

ᵃDescriptive notations at this time. See also Section IV, G (pp. 44–46).

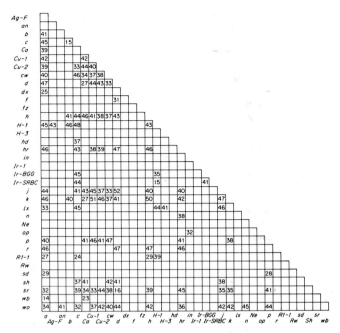

Fig. 1. Summary of tests for independent segregation of genes and the closest linkage compatible with the assortment data.

A spontaneous reciprocal translocation has been described by Bouricius (14), inducing partial sterility of carriers. The translocation appeared to involve a large and small chromosome.

Differential staining of sister chromatids have been described

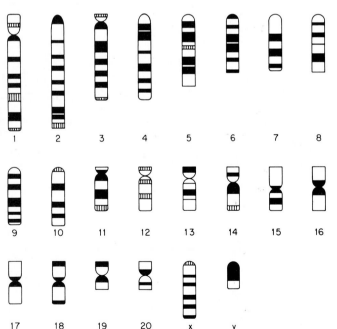

Fig. 2. Idiogram of the banded karyotype of the rat [based upon the observations of Levan (163)].

for rat chromosomes (154), and the technique will doubtless prove to be a useful experimental tool.

Chiasma frequency per bivalent has been intensively studied (138). A point of interest is a comparison of chiasma frequencies between the sexes. The number at diplotene for the male ranged from 1 to 4 per bivalent, with a mean of 2.09 ± 0.06, compared with a range of 1 to 5 and a mean of 2.70 ± 0.11 for the female. These values imply a higher rate of crossing over for female gametogenesis. This prediction is supported by the majority of intersex comparisons of crossing over (138). A slow decline in chiasmata with age was observed for the male, but the decline was not marked except in very old animals when it was accompanied by a fall in spermatogenesis.

VII. INBRED STRAINS

Numerous inbred strains have been developed from time to time, but there have been few attempts to ensure that these will survive on a permanent basis. The most famous stock of rats is the Wistar albino, established by Donaldson about 1900. The original animals were not inbred but selected for vigor and prolifacy. Numerous substocks have been produced as a consequence of the original animals beings distributed to many laboratories (43,137) (also, see Chapter 1).

Several surveys of existing inbred strains have been made. These are mainly of historical interest, but they indicate how the strains have developed. A recent survey (43) not only lists current strains but attempts to place the designation of strains on a rational basis (see Chapter 3). Rules are presented for symbolization of strains and for indicating their inbred status. These are based upon the proven rules for mice. The survey is comprehensive and lists 100 strains and substrains, giving brief summaries of their origin, known characteristics, and location. The problem with these compilations is that the information slowly becomes dated; however, it is hoped that periodic surveys will be conducted.

It is generally appreciated that highly inbred strains of mammals are valuable assets because of the effort expended on their development and the useful properties which many of them come to possess. Their loss in many cases would be irreplaceable. There is urgent need for a scheme to safeguard both mutant genes and inbred strains of rats against loss as exists for the mouse.

REFERENCES

1. Allison, J. E., Stanley, A. J., and Gumbreck, L. G. (1965). Sex chromatin and idiograms from rats exhibiting anomalies of the reproductive organs. *Anat. Rec.* **153,** 85–92.

2. Andersen, D. H. (1949). Effect of diet during pregnancy upon the incidence of congenital hereditary diaphragmatic hernia in the rat. *Am. J. Pathol.* **25,** 163–186.

3. Armerding, D. (1971). Two allotypic specificities of rat immunoglobulin. *Eur. J. Immunol.* **1,** 39–45.

4. Armerding. D., Katz, D. H., and Benacerraf, B. (1974). Immune response genes in inbred rats. I. Analysis of responder status to synthetic polypeptides and low doses of bovine serum album. *Immunogenetics* **1,** 329–339.

5. Armerding, D., Katz, D. H., and Benacerraf, B. (1974). Immune response genes in inbred rats. II. Segregation studies of the GT and GA genes and their linkage to the major histocompatibility locus. *Immunogenetics* **1,** 340–351.

6. Baker, R. L., and Chapman, A. B. (1975). Correlated responses to selection for postweaning gain in the rat. *Genetics* **80,** 191–203.

7. Baker, R. L., Chapman, A. B., and Wardell, R. T. (1975). Direct response to selection for postweaning gain in the rat. *Genetics* **80,** 171–189.

8. Bazin, H., Beckers, A. Vaerman, J. P., and Hereman, J. F. (1974). Allotypes of rat immunoglobulins. I. An allotype at the a-chain locus. *J. Immunol.* **112,** 1035–1041.

9. Beer, A. E., Billingham, R. E., and Scott, J. R. (1975). Immunogenetic aspects of implantation, placentation and feto-placental growth rates. *Biol. Reprod.* **12,** 176–189.

10. Bianchi, N. O., and Molina, O. (1966). Autosomal polymorphism in a laboratory strain of rat. *J. Hered.* **57,** 231–232.

11. Bignami, G. (1965). Selection for high rates and low rates of conditioning in the rat. *Anim. Behav.* **13,** 221–227.

12. Bignami, G., and Bovet, D. (1965). [A selection experiment on conditioned avoidance reaction in the rat.] *C. R. Hebd. Seances Acad. Sci.* **260,** 1239–1244.

13. Blanc, W. A., and Johnson, L. (1959). Studies in kernicterus. Relationship with sulfonamide intoxication; report on kernicterus in rats with glucuronyl transferase deficiency and review of pathogenesis. *J. Neuropathol. Exp. Neurol.* **18,** 165–189.

14. Bouricius, J. K. (1948). Embryological and cytological studies in rats heterozygous for a probable reciprocal translocation. *Genetics* **33,** 577–587.

15. Bourne, M. C., and Gruneberg, H. (1939). Degeneration of the retina and cataract. A new recessive gene in the rat. *J. Hered.* **30,** 131–136.

16. Brdicka, R. (1967). Genetics of the rat haemoglobin. *Polymorphismes Biochim. Anim., Congr. Eur. Groupes Sang. Polymorphisme Biochim. Anim., 10th, 1966* pp. 407–411.

17. Brdicka, R. (1968). The chromosome I of the laboratory rat. *Acta Univ. Carol., Med.* **14,** 93–98.

18. Brdicka, R. (1974). Alleles of the second plasma esterase locus in the laboratory rat and their distribution among some inbred strains. *Folia Biol. (Prague),* **20,** 350–353.

19. Brewster, D. J. (1968). Genetic analysis of ethanol preference in rats selected for emotional activity. *J. Hered.* **59,** 283–286.

20. Broadhurst, P. L., and Bignami, G. (1965). Correlative effects of psychogenetic selection. A study of the Roman high and low avoidance strains of rats. *Behav. Res. Ther.* **2,** 273–280.

21. Broadhurst, P. L., and Jinks, J. L. (1966). Stability and change in the inheritance of behaviour in rats: A further analysis of statistics from a diallel cross. *Proc. R. Soc. London, Ser. B* **165,** 450–472.

22. Browman, L. G. (1954). Microphthalmia and maternal effect in the white rat. *Genetics* **39,** 261–265.

23. Carter, N. D., and Parr, C. W. (1969). Phosphogluconate dehydrogenase polymorphism in British wild rats. *Nature (London)* **224,** 1214.

24. Castle, W. S. (1953). Silver, a new mutation of the rat. *J. Hered.* **44,** 205–206.

25. Castle, W. S., Dempster, S. R., and Shurragor, H. C. (1955). Three new mutations of the rat. *J. Hered.* **46,** 9–14.

26. Castle, W. S., King, H. D., and Daniels, A. L. (1941). Linkage studies of the rat. IV. *Proc. Natl. Acad. Sci. U.S.A.* **27,** 250–253.

27. Chai, C. K., Hunt, H. R., Hoppert, C. A., and Rosen, S. (1968). Hereditary basis of caries resistance in rats. *J. Dent. Res.* **47,** 127–138.

28. Chan, F., and Allison, J. E. (1969). The mutant gene for male pseudohermaphroditism in rats. *Anat. Rec.* **163,** 167.

29. Committee for a Standardized Karyotype of *Rattus norvegicus* (1973). Standard karyotype of the Norway rat. *Cytogenet. Cell Genet.* **12,** 199–205.

30. Cotton, W. R., and Gaines, J. F. (1974). Unerupted dentition secondary to congenital osteopetrosis in Osborne-Mendel rats. *Proc. Soc. Exp. Biol. Med.* **146,** 554–561.

31. Crew, F. A. E., and Kon, S. K. (1934). A lethal in the rat. *J. Genet.* **28,** 25–31.

32. DeWitt, C. E., and McCullough, M. (1975). *Ag-F:* Serological and genetic identification of a new locus in the rat governing lymphocyte membrane antigens. *Transplantation* **19,** 310–317.

33. Dolezalova, V., and Brada, Z. (1968). Serum protein heterogeneity in some inbred rat strains studied by electrophoretic methods. *Comp. Biochem. Physiol.* **26,** 301–309.

34. Donaldson, H. H. (1924). "The Rat." Wistar Inst., Philadelphia, Pennsylvania.

35. Douglas, T. C. (1972). Occurrence of a theta-like antigen in rats. *J. Exp. Med.* **136,** 1054–1062.

36. Dowling, J. E., and Sidman, R. L. (1962). Inherited retinal dystrophy in the rat. *J. Cell Biol.* **14,** 73–109.

37. Dunn, L. C., Gluecksohn-Schoenheimer, S., Curtis, M. R., and Dunning, W. F. (1942). Heredity and accident as factors in the production of taillessness in the rat. *J. Hered.* **33,** 65–67.

38. Dunning, W. F., and Curtis, M. R. (1939). Linkage in rats between factors determining a pathological condition and a coat colour. *Genetics* **24,** 70.

39. Dupont, J., Dupont, J. C., Froment, A., Milon, H., and Vincent, M. (1973). Selection of three strains of rats with spontaneous different levels of blood pressure. *Biomedicine* **19,** 36–41.

40. Eker, R., and Mossige, J. (1961). A dominant gene for renal adenomas in the rat. *Nature (London)* **185,** 858–859.

41. Ellerman, J. R. (1941). "The Families and Genera of Living Rodents," Vol. II. British Museum (Natural History), London.

42. Eriksson, K. (1968). Genetic selection for voluntary alcohol consumption in the albino rat. *Science* **149,** 739–741.

43. Festing, M., and Staats, J. (1973). Standardized nomenclature for inbred strains of rats. Fourth listing. *Transplantation* **16,** 221–245.

44. Figala, J. (1963). Yellow mutant in the brown rat. *Vestn. Cesk. Spol. Zool.* **27,** 209–210.

45. Fleming, J. C., and Broadhurst, P. L. (1975). The effects of nicotine on two-way avoidance conditioning in bidirectional selected strains of rats. *PsychoPsychopharmacologia* **42,** 147–152.

46. French, E. A., Roberts, K. B., and Searle, A. G. (1971). Linkage between a hemoglobin locus and albinism in the Norway rat. *Biochem. Genet.* **5,** 397–404.

47. Fulker, D. W., Wilcock, J., and Broadhurst, P. L. (1972). Studies on genotype-environment interaction. I. Methodology and preliminary multi-variate analysis of a diallel cross of eight strains of rats. *Behav. Genet.* **2,** 261–287.

48. Gallimore, P. H., and Richardson, C. R. (1973). An improved banding technique exemplified in the karyotype analysis of two strains of rat. *Chromosoma* **41,** 259–263.

49. Gasser, D. L. (1972). Seminal vesicle protein in rats: A gene in the fourth linkage group determining electrophoretic variants. *Biochem. Genet.* **6,** 61–63.

50. Gasser, D. L., Newlin, C. M., Palm, J., and Gonatas, N. K. (1973). Genetic control of susceptibility to experimental allergic encephalomyelitis in rats. *Science* **181**, 872–873.

51. Gasser, D. L., Palm, J., and Gonatas, N. K. (1975). Genetic control of susceptibility to experimental allergic encephalomyelitis and the *Ag-B* locus of rats. *J. Immunol.* **115**, 431–433.

52. Gasser, D. L., Silvers, W. K., Reynolds, H. M., Black, G., and Palm, J. (1973). Serum esterase genetics in rats: Two new alleles at *Es-2,* a new esterase regulated by hormonal factors and linkage of these loci to the *Ag-C* blood group locus. *Biochem. Genet.* **10**, 203–217.

53. Gill, T. J., Kunz, H. W., Stechschulte, D. J., and Austen, K. F. (1970). Genetic and cellular factors in the immune response. I. Genetic control of the antibody response to poly $Glu^{52}Lys^{33}Tyr^{15}$ in the inbred rat strains ACI and F344. *J. Immunol.* **105**, 14–28.

54. Gill, T. J., and Kunz, H. W. (1971). Genetic and cellular factors in the immune response. II. Evidence for the polygenic control of the antibody response from further breeding studies and from pedigree analyses. *J. Immunol.* **106**, 980–992.

55. Greaves, J. H., and Ayres, P. (1967). Heritable resistance to warfarin in rats. *Nature (London)* **215**, 877–878.

56. Greaves, J. H., and Ayres, P. (1969). Linkages between genes for coat colour and resistance to warfarin in *Rattus norvegicus. Nature (London)* **224**, 284–285.

57. Greep, R. O. (1941). An hereditary absence of the incisor teeth. *J. Hered.* **32**, 297–298.

58. Gruneberg, H. (1939). An analysis of the "pleiotropic" effects of a new lethal mutation in the rat (*Mus norvegicus*). *Proc. R. Soc. London, Ser. B* **125**, 123–144.

59. Gumbreck, L. G. (1964). New genetic factors that effect fertility in the male rat. *Proc. Int. Congr. Anim. Reprod. Artif. Insem., 5th, 1963* Vol. 2, 319–325.

60. Gumbreck, L. G., Stanley, A. J., Allison, J. S., and Easley, R. B. (1972). Restriction of color in the rat with associated sterility in the male and heterochromia in both sexes. *J. Exp. Zool.* **180**, 333–349.

61. Gumbreck, L. G., Stanley, A. J., Macy, R. M., and Peeples, E. E. (1971). Pleiotropic expression of the restricted coat color gene in the Norway rat. *J. Hered.* **62**, 356–358.

62. Gunn, C. K. (1938). Hereditary acholuric jaundice in a new mutant strain of rats. *J. Hered.* **29**, 137–139.

63. Gunther, E., and Rude, E. (1975). Genetic complementation of histocompatibility linked *Ir* genes in the rat. *J. Immunol.* **115**, 1387–1393.

64. Gunther, E., Rude, E., Meyer-Delius, M., and Stark, O. (1973). Immune response genes linked to the major histocompatibility system in the rat. *Transplant. Proc.* **5**, 1467–1469.

65. Gunther, E., Rude, E., and Stark, O. (1972). Antibody response in rats to the synthetic polypeptide (T,G-A-L) genetically linked to the major histocompatibility system. *Eur. J. Immunol.* **2**, 151–155.

66. Gutman, G. A., and Weissman, I. L. (1971). Inheritance and strain distribution of a rat immunoglobulin allotype. *J. Immunol.* **107**, 1390–1393.

67. Harkins, R. K., Allison, J. E., and Macy, R. M. (1974). Ocular sites of gene expression in the Norway rat. *J. Hered.* **65**, 273–276.

68. Harrington, G. M. (1968). Genetic-environmental interaction in intelligence. I. Biometric genetic analysis of maze performance of *Rattus norvegicus. Dev. Psychobiol.* **1**, 211–218.

69. Harrington, G. M. (1972). Strain differences in open-field behavior in the rat. *Psychonom. Sci.* **27**, 51–53.

70. Harris, J. M., Kalmus, H., and West, G. B. (1963). Genetic control of the anaphylactoid reaction in rats. *Genet. Res.* **4**, 346–355.

71. Heslop, B. F. (1973). The male antigen in the rat. *Transplantation* **15**, 31–35.

72. Heston, W. E. (1938). Bent-nose in the Norway rat. *J. Hered.* **29**, 437–448.

73. Hinton, M. A. C. (1920). Rats and mice as enemies of mankind. *Br. Mus. (Nat. Hist.), Econ. Ser.* **8**.

74. Holland, H. C., and Gupta, B. O. (1966). Some correlated measures of activity and reactivity in two strains of rats selectively bred for differences in the acquisition of a conditioned avoidance response. *Anim Behav.* **14**, 574–580.

75. Howard, J. C. (1971). A histocompatible chromosome marker system in the laboratory rat. *Transplantation* **12**, 95–97.

76. Hungerford, D. A., and Nowell, P. C. (1963). Sex chromosome polymorphism and the normal karyotype in three strains of the laboratory rat. *J. Morphol.* **113**, 275–286.

77. Hunt, H. R., Hoppert, C. A., and Rosen, S. (1955). Genetic factors in experimental rat caries *Adv. Exp. Caries-Res., Symp., 1953.*

78. Hunt, H. R., Hoppert, C. A., and Rosen, S. (1956). The role of heredity in the causation of dental caries in rats. *In* "A Symposium on Preventive Dentistry" (J. C. Muhler and M. K. Hine, eds.), Mosby, St. Louis, Missouri.

79. Jimenez-Marin, D. (1974). Enzyme inheritance in the laboratory rat. *J. Hered.* **65**, 235–237.

80. Jimenez-Marin, D., and Dessauer, H. C. (1973). Protein phenotype variation in laboratory populations of *Rattus norvegicus. Comp. Biochem. Physiol. B* **46**, 487–492.

81. Jones, E. E. (1925). The occurrence of an eye and of a tooth abnormality in a line of albino rats. *Am. Nat.* **59**, 427–440.

82. Kent, R. L., Lutzner, M. A., and Hansen, C. T. (1976). The masked rat. *J. Hered.* **67**, 3–5.

83. King, H. D. (1936). A waltzing mutation in the white rat. *J. Mammal.* **17**, 157–163.

84. Koch, C. (1974). Inheritance in the rat of the antibody response to different determinants of (TG)-A-L. *Immunogenetics* **1**, 118–125.

85. Koga, A., Harada, S., and Omoto, K. (1972). Polymorphisms of erythrocyte 6-phosphogluconate dehydrogenase and phosphoglucomutase in *Rattus norvegicus* in Japan. *Jpn. J. Genet.* **47**, 335–338.

86. Koletsky, S. (1973). Obese spontaneously hypertensive rats. *Exp. Mol. Pathol.* **19**, 53–60.

87. Kren, V. (1974). The major histocompatibility system (*H-1*) alleles of some British rat strains. *Transplantation* **17**, 148–152.

88. Kren, V. (1975). Genetics of the polydactly-luxate syndrome in the Norway rat. *Acta Univ. Carol. Med., Monogr.* **68**, 1–103.

89. Kren, V., Stark, O., Bila, V., Frenzl, B., Krenova, D., and Krisakova, M. (1973). Rat alloantigen systems defined through congenic strain production. *Transplant. Proc.* **5**, 1463–1466.

90. Kunz, H. W., and Gill, T. J. (1974). Genetic studies in inbred rats. I. Two new histocompatibility alleles. *J. Immunogenet.* **1**, 413–420.

91. Kunz, H. W., Gill, T. J., and Borland, B. (1974). The genetic linkage of the immune response to poly($Glu^{52}Lys^{33}Tyr^{15}$) to the major histocompatibility locus in inbred rats. *J. Immunogenet.* **1**, 277–287.

92. Lambert, W. V., and Schinchetti, A. M. (1935b). A dwarf mutation in the rat. *J. Hered.* **26**, 91–94.

93. Lavail, M. M., and Battells, B. A. (1975). Influence of eye pigmentation and light deprivation on inherited retinal dystrophy in the rat. *Exp. Eye Res.* **21**, 167–192.

94. Lavail, M. M., Sidman, R. L., and Gerhardt, C. O. (1975). Congenic strains of RCS rats with inherited retinal dystrophy. *J. Hered.* **66**, 242–244.

95. Lefebvres-Boisselot, J. (1968). Effets complexes de la mutation Atrichis chez le Rat. *C. R. Hebd. Seances Acad. Sci., Ser. D* **267**, 1636–1638.

96. Leonard, A., and Maisin, J. R. (1965). Hereditary cataract induced by X irradiation of young rats. *Nature (London)* **205**, 615-616.

97. Levan, G. (1974). Nomenclature for G bands in rat chromosomes. *Hereditas* **77**, 37-52.

98. Lozzio, B. B., Chernoff, A. I., Machado, E. R., and Lozzio, C. B. (1967). Hereditary renal disease in a mutant strain of rats. *Science* **156**, 1742-1744.

99. Lutzner, M. A., and Hansen, C. T. (1975). Skin blisters and hair loss in a rat mutant called vibrissaeless. *J. Invest. Dermatol.* **65**, 212-216.

100. Macy, R. M., and Stanley, A. J. (1973). Light and dark color genes in a wild population of Norway rats. *J. Hered.* **64**, 98.

101. Marshall, M. W., Durand, A. M. A., and Adams, M. (1971). Different characteristics of rat strains: Lipid metabolism and response to diet. *Proc. Symp. Int. Comm. Lab. Anim., 4th, 1971* pp. 381-413.

102. Martinovic, J. V., Martinkovic, D. V., Kanaizir, D. T., and Martinovitch, P. N. (1970). Inherited differences in the electrophoretic pattern of hemoglobin revealed in a local random bred colony of albino rats. *Blood* **35**, 447-450.

103. Masuji, H. (1970). Autosomal polymorphism in Donryu rat strain. *Acta Med. Okayama* **24**, 81-91.

104. McGregor, J. F. (1974). Possible usefulness of hypotrichotic rats for studies on skin tumorigenesis. *Can. J. Genet. Cytol.* **16**, 341-348.

105. Michie, D., and Anderson, N. F. (1966). A strong selective effect associated with a histocompatibility gene in the rat. *Ann. N.Y. Acad. Sci.* **129**, 88-93.

106. Mickelsen, O., Schemmel, R., and Gill, J. L. (1971). Influence of diet, sex and age in skeleton size in seven strains of rats. *Growth* **35**, 11-22.

107. Miller, D. A., Dev, V. G., Borek, C., and Miller, O. J. (1972). The quinacrine fluorescent and Giemsa banding karyotype of the rat and banded chromosome analysis of transformed and malignant rat liver cell lines. *Cancer Res.* **32**, 2375-2382.

108. Missonne, X. (1969). African and Indo-Australian Muridae; evolutionary trends. *Ann. Mus. R. Afr. Cent. Zool.* **172**, 1-219.

109. Moore, L. A., and Schaible, P. J. (1936). Inheritance of hernia in rats. *J. Hered.* **27**, 272-280.

110. Mori, M., and Sasaki, M. (1973). Fluorescence banding patterns of the rat chromosomes. *Chromosoma* **40**, 173-182.

111. Mori, M., Sasaki, M., and Takagi, N. (1973). Chromosomal banding patterns in three species of rats. *Jpn. J. Genet.* **48**, 381-383.

112. Morris, D. (1965). "The Mammals." Hodder & Stoughton, London.

113. Moutier, R., Toyama, K., and Charrier, M. F. (1973). Himalayan allele of the albino locus in the Norway rat. *J. Hered.* **64**, 303-304.

114. Moutier, R., Toyama, K., and Charrier, M. F. (1973). Biochemical polymorphism in the rat, *Rattus norvegicus*: Genetic study of four markers. *Biochem. Genet.* **8**, 321-328.

115. Moutier, R., Toyama, K., and Charrier, M. F. (1973). Linkage of a plasma protein marker (*Gl-1*) and the hooded locus in the rat. *Biochem. Genet.* **10**, 395-398.

116. Moutier, R., Toyama, K., and Charrier, M. F. (1973). Hypodactyly, a new recessive mutation in the Norway rat. *J. Hered.* **64**, 99-100.

117. Moutier, R., Toyama, K., and Charrier, M. F. (1973). Evidence for linkage between four esterase loci in the rat. *Biochem. Genet.* **9**, 109-115.

118. Moutier, R., Toyama, K., and Charrier, M. F. (1974). Genetic study of osteopetrosis in the Norway rat. *J. Hered.* **65**, 373-375.

119. Mullen, R. J., and LaVail, M. M. (1976). Inherited retinal dystrophy: Primary defect in pigment eipthelium determined with experimental rat chimeras. *Science* **192**, 799-801.

120. Mullen, Y., and Hildemann, W. H. (1972). X and Y linked transplantation antigens in rats. *Transplantation* **13**, 521-529.

121. Naso, R. B., Baker, F. T., Butcher, D., Kaczmarcyk, W., and Ulrich, V. (1975). Weight gain and heterosis at different stages of development of the rat. *Growth* **39**, 345-361.

122. Nesbitt, M. N. (1974). Evolutionary relationships between rat and mouse chromosomes. *Chromosoma* **46**, 217-224.

123. Palm, J. (1970). Maternal-foetal interactions and histocompatibility antigen polymorphisms. *Transplant. Proc.* **2**, 162-173.

124. Palm, J. (1971). Immunogenetic analysis of Ag-B histocompatibility antigens in rats. *Transplantation* **11**, 175-183.

125. Palm, J. (1971). Classification of inbred rat strains for Ag-B histocompatibility antigens. *Transplant. Proc.* **3**, 169-181.

126. Palm, J. (1975). The laboratory rat. *Handb. Genet.* **4**, 243-254.

127. Palm, J., and Black, G. (1971). Interrelationships of inbred rat strains with respect to Ag-B and non-Ag-B antigens. *Transplantation* **11**, 184-189.

128. Palmer, M. L., Allison, J. E., Peeples, E. E., and Whaley, G. D. (1974). Coat colour restriction gene in rats: Its effect in the homozygous condition. *J. Hered.* **65**, 291-296.

129. Pool, J. G., O'Reilly, R. A., Schneiderman, L. J., and Alexander, A. (1968). Warfarin resistance in the rat. *Am. J. Physiol.* **215**, 627-631.

130. Rapp, J. P., and Dahl, L. K. (1972). Mendelian inheritance of 18- and 11b-steroid hydroxylase activities in the adrenals of rats genetically susceptible or resistant to hypertension. *Endocrinology* **90**, 1435-1446.

131. Rapp, J. P., and Dahl, L. K. (1976). Mutant forms of cytochrome *P-450* controlling both 18- and 11b-steroid hydroxylation in the rat. *Biochemistry* **15**, 1235-1242.

132. Rapp, J. P., Knudsen, K. D., Iwai, J., and Dahl, L. K. (1973). Genetic control of blood pressure and corticosteroid production in rats. *Circ. Res.* **33**, Suppl. 1, 139-149.

133. Ratcliffe, H. L., and King, H. D. (1941). Developmental abnormalities and spontaneous diseases found in rats of the mutant strain, stub. *Anat. Rec.* **81**, 283-305.

134. Rees, E. D., Shuck, A. E., Christian, J. C., and Pugh, J. R. (1968). Karyotypes of rats from strains of different susceptibility to mammary cancer induction. *Cancer Res.* **28**, 823-830.

135. Richter, C. P. (1954). The effects of domestication and selection on the behavior of the Norway rat. *J. Natl. Cancer Inst.* **15**, 727-738.

136. Roberts, E., Quisenberry, J. H., and Thomas, L. C. (1940). Hereditary hypotrichosis in the rat. *J. Invest. Dermatol.* **3**, 1-29.

137. Robinson, R. (1965). "Genetics of the Norway Rat." Pergamon, Oxford.

138. Robinson, R. (1972). "Gene Mapping in Laboratory Mammals," Part B. Plenum, New York.

139. Roubicek, C. B. (1975). Selection for water intake in the rat. *Genet. Lect.* **4**, 25-38.

140. Roubicek, C. B., and Ray, D. S. (1971). Direct and transfer response of rats selected for growth at two environmental temperatures. *Growth* **35**, 1-10.

141. Russell, L. D., and Gardner, P. J. (1974). Testicular ultrastructure and fertility in the restricted color rat. *Biol. Reprod.* **11**, 631-643.

142. Saul, G. B., Garrity, E. B., Benirschke, K., and Valtin, H. (1968). Inherited hypothalamic diabetes insipidus in the Brattleboro strain of rats. *J. Hered.* **59**, 113-117.

143. Schnedl, W., and Schnedl, M. (1972). M banding patterns in rat chromosomes. *Cytogenetics* **11**, 188-196.

144. Shiehzadeh, S. A., Harbers, L. H., and Schalles, R. R. (1972). Inheritance of response to lysine deficient diet by rats. *J. Hered.* **63**, 119-121.

145. Simpson, G. G. (1945). The principles of classification and a classification of mammals. *Bull. Am. Mus. Nat. Hist.* **85**, 1-350.

146. Sines, J. O., and McDonald, D. G. (1968). Heritability of stress-ulcer susceptibility in rats. *Psychosom. Med.* **30**, 390-394.

147. Sladic-Simic, D., Zivkovic, N., Pavic, D., Marin-Kovic, D., Mar-

tinovic, J., and Martinovitch, P. N. (1966). Hereditary hypochromic microcytic anaemia in the laboratory rat. *Genetics* **53**, 1079–1089.

148. Smith, R. S., Hoffman, H., and Cisar, C. (1969). Congenital cataract in the rat. *Arch. Ophthalmol.* **81**, 259–264.

149. Smith, S. E., and Barrentine, B. F. (1943). Hereditary cataract, a new dominant gene in the rat. *J. Hered.* **34**, 8–10.

150. Smith, S. E., and Bogart, R. (1939). The genetics and physiology of lethal anaemia in the rat. *Genetics* **24**, 474–493.

151. Stark, O., Kren, V., and Gunther, E. (1971). RtH-1 antigens in 39 rat strains and six congenic lines. *Transplant. Proc.* **3**, 165–168.

152. Stark, O., Kren, V., and Gunther, E. (1972). Ten alleles of the *RtH-1* system in 34 inbred strains and two random bred populations of laboratory rats. *Eur. Conf. Anim. Blood Groups Biochem. Polymorphismo.* [*Proc.*], *12th, 1970* pp. 627–630.

153. Stenglen, B., Thoenes, C. H., and Gunther, E. (1975). Genetically controlled autologous immune complex glomerulonephritis in rats. *J. Immunol.* **115**, 895–897.

154. Sugiyama, T., Goto, K., and Kano, Y. (1976). Mechanism of differential Giemsa method for sister chromatids. *Nature (London)* **259**, 59–60.

155. Tada, N., Ikakura, K., and Aizawa, M. (1974). Genetic control of the antibody response in inbred rats. *J. Immunogenet.* **1**, 265–275.

156. Taylor, B. A. (1968). The frequency of X-ray induced recessive visible mutations in the rat. *Genetics* **60**, 559–565.

157. Tschopp, T. B., and Zucker, M. B. (1972). Hereditary defect in platelet function in rats. *Blood* **40**, 217–226.

158. Unakul, W., and Hsu, T. C. (1972). The C and G banding patterns of *Rattus norvegicus* chromosomes. *J. Natl. Cancer Inst.* **49**, 1425–1431.

159. Vasenius, L. (1971). A new dominant coat colour gene in the Norway rat. *Acta Vet. Scand.* **12**, 109–110.

160. Vasenius, L. (1972). Genetics of the white bellied agouti colour in the Norway rat. *Z. Versuchstierk.* **14**, 142–147.

161. Wake, N., Takagi, N., and Sasaki, M. (1976). Non-random inactivation of the X chromosome in the rat yolk sac. *Nature (London)* **262**, 581–582.

162. Wendt-Wagener, G. (1961). Unterunchungen uber die Ausbreitung der Melanoblasten bei einfarbig schwarzen Ratten und bei Haubenratten. *Z. Vererbungsl.* **92**, 63–68.

163. Whitney, G., McClearn, G. E., and Defries, J. C. (1970). Heritability of alcohol preference in laboratory mice and rats. *J. Hered.* **61**, 165–169.

164. Wilcock, J. (1968). Strain differences in response to shock in rats selectively bred for emotional elimination. *Anim. Behav.* **16**, 294–297.

165. Wilcock, J., and Broadhurst, P. L. (1967). Strain differences in emotionality: Open field and conditioned behaviour in the rat. *J. Comp. Physiol. Psychol.* **63**, 335–338.

166. Williams, R. M., and Moore, M. J. (1973). Linkage of susceptibility to experimental allergic encephalomyelitis to the major histocompatibility locus in the rat. *J. Exp. Med.* **138**, 775–783.

167. Wolman, S. R., Phillips, T. F., and Becker, F. F. (1972). Fluorescent banding patterns of rat chromosome in normal cells and primary hepatocellular carcinomas. *Science* **175**, 1267–1269.

168. Womack, J. E. (1973). Biochemical genetics of rat esterases: Polymorphism, tissue expression and linkage of four loci. *Biochem. Genet.* **9**, 13–24.

169. Woolley, G. W., and Cole, L. J. (1938). Spontaneous tail amputation in the Norway rat. *J. Hered.* **29**, 123–127.

170. Woolley, G. S., and Cole, L. J. (1939). A new dwarf mutation (dw_2) in *Rattus norvegicus*. *Genetics* **24**, 111.

171. Wurzburg, U. (1971). Correlation between the immune response to an enzyme and histocompatibility type in rats. *Eur. J. Immunol.* **1**, 496–497.

172. Wurzburg, U., Schutt-Gerowitt, H., and Rajewsky, K. (1973). Characterization of an immune response gene in rats. *Eur. J. Immunol.* **3**, 762–766.

173. Yen, T. T., Hu, P. L., Roeder, H., and Willard, P. W. (1974). Genetic study of hypertension in Okamoto-Aoki spontaneously hypertensive rats. *Heredity* **33**, 309–316.

174. Yosida, T. H., and Amano, K. (1965). Autosomal polymorphism in laboratory bred and wild Norway rats. *Chromosoma* **16**, 658–667.

175. Yosida, T. H., Kurita, Y., and Taneda, S. (1961). Genetic study in two wild mutants of rats. *Bull. Exp. Anim.* **10**, 20–22.

176. Yosida, T. H., and Sagai, T. (1973). Similarity of Giemsa banding patterns of chromosomes in several species of the genus *Rattus*. *Chromosoma* **41**, 93–101.

177. Young, C., Festing, M. F. W., and Barnett, K. C. (1974). Buphthalmos (congenital glaucoma) in the rat. *Lab. Anim.* **8**, 21–31.

178. Zemaitis, M. A., Hill, R. N., and Greene, F. S. (1974). Autosomal genetic control of the activity of an esterase in rat hepatic microsomes. *Biochem. Genet.* **12**, 295–308.

179. Zidek, Z. (1968). Karyotypes of four inbred strains of rats. *Folia Biol. (Prague)* **14**, 74–79.

180. Zucker, L. M., and Zucker, T. F. (1961). Fatty, a new mutation in the rat. *J. Hered.* **52**, 275–278.

Chapter 3

Inbred Strains

Michael F. W. Festing

I. INTRODUCTION

Inbred strains of laboratory animals have already made a substantial contribution to biomedical research. In cancer research, transplantable tumors have been widely used in studies ranging from the biochemistry, immunology, and growth of tumors, to the screening of potential anticancer drugs. Many tumors can only be transplanted to a host genetically identical with the animal in which it first arose, i.e., within an inbred strain or F1 hybrid.

Progress in immunological research has been rapid in the past two decades. Much of this research has been dependent first on the availability of inbred strains, and more recently on the widespread use of congenic lines. The latter have been developed by backcrossing a mutant or variant, such as a major histocompatibility haplotype, on to an inbred strain genetic background. Recent progress in understanding the major histocompatibility complex of rats and mice would have been impossible without the use of inbred and congenic strains.

Inbred strains have also been used in other disciplines, although in many cases their use is not yet widespread. This is largely because workers are not aware of the benefits of using such animals. The properties of inbred strains which make them such valuable research tools are discussed in detail by Festing (33). Very briefly, these include the following.

1. Long-term genetic stability: Inbred strains stay geneti-

cally constant for long periods of time so that background data on strain characteristics may be useful for many years. Some genetic change occurs as a result of mutation, but this is probably much less than occurs in an outbred stock which can also be changed as a result of selection and random drift of gene frequency.

2. Isogenicity: All individuals of a strain are genetically identical. Thus skin grafts and tumors may be transplanted within an inbred strain without immunological rejection. Similarly, genetically determined biochemical and immunological polymorphisms may be studied in a single animal in order to type the whole strain. In this way, data may be accumulated on, for example, the haplotype at the major histocompatibility complex in a way that is impossible in an outbred stock.

3. Homozygosity: Inbred animals are homozygous at virtually all loci, and thus will breed true within the strain.

4. International distribution: Many of the widely used inbred strains of mice and rats have an international distribution, so international comparisons of research results are possible using genetically identical animals.

5. Identifiability: Many inbred strains can be identified by their immunological and biochemical characteristics, thus reducing the chance of unnoticed genetic contamination. In contrast, there is no way, except possibly from morphological characteristics, that outbred stocks may be identified (32).

6. Uniformity: The elimination of genetic variation which occurs during inbreeding means that for many highly inherited polygenic characters, inbred strains are more uniform than outbred stocks. This means that fewer inbred than outbred animals are needed in an experiment to achieve a given statistical precision.

7. Individuality: Each inbred strain represents a unique genotype which often leads to a phenotype of biomedical interest. Inbred strains may be found with a very high or a very low tumor incidence. Many strains develop diseases which may be used as models of disease in humans or domesticated animals. Each strain has its own patterns of growth; development; behavior; response to infections, antigens, and toxins; and occurrence of spontaneous disease. Contrasts between strains which differ in these characteristics can lead to a better understanding of the biology of mammals.

This chapter lists the inbred strains of rats which are currently available, with an outline of their history and characteristics.

The first listing of inbred strains of rats was by Billingham and Silvers (12). This is the fifth listing (Table I) and has been substantially revised, although it is heavily dependent on previous listings (35). In contrast with previous listings an attempt has been made to cite the original reference for any given strain characteristics. Particular attention has also been given to studies involving several inbred strains, and where possible, the strains have been ranked on their characteristics with ap-

proximately the top quartile being classified as "high" and the bottom quartile as "low." Thus "low" litter size (10/12) indicates that the strain in question ranked tenth out of 12 strains examined for litter size. It should be strongly emphasized that the ranking may be dependent on environmental as well as genetic factors. In a different environment strains could rank differently. Known polymorphisms are given in Table II. Chromosomal polymorphisms are given in Table III, and coat color genes carried by albino strains in Table IV.

Strain synonyms also represent a problem. Users of inbred rat strains are not yet in the habit of using standardized nomenclature and continue to use obsolete synonyms. Accordingly, a table of strain synonyms whose sole function is to assist in the "translation" of older papers is included (Table V) and the list of modern strain designations is given in the main text without these old synonyms. Strains may be located using Table VI, and a list of congenic lines is given in Table VII.

Finally, it is clear from this listing that there are already a very large number of inbred strains of rats, but virtually nothing is known about the characteristics of many of them. Future priority should go to the study and characterization of those strains that are already available, rather than the development of new general purpose inbred strains. Comparative studies involving six or more inbred strains are particularly valuable, and if possible such studies should include F344 and/or LEW, which seem likely to emerge as the most widely used of all inbred rat strains. New inbred strains with diseases or conditions of biomedical interest may well be worth developing, and there may be a need for a much wider range of congenic lines.

II. NOMENCLATURE RULES

The following rules are published without alteration and with permission from Festing and Staats (35). Rules for nomenclature of inbred strains of mice are currently being revised, and it is probable that rules for the nomenclature of rat strains will have to be revised in the near future. In the meantime, the rules published here should be adhered to as far as possible.

Abbreviated Rules for Symbols to Designate Inbred Strains of Rats

1. Definition of Inbred Strain

A strain shall be regarded as inbred when it has been mated brother × sister (hereafter called b × s) for 20 or more consecutive generations. Parent × offspring matings may be substituted for b × s matings provided that in the case of consecutive parent × offspring matings, the mating in each case is to the younger of the two parents.

2. Symbols for Inbred Strains

Inbred strains shall be designated by a capital letter or letters in Roman type. It is urged that anyone naming a new stock consult this article to avoid duplication. Brief symbols are preferred. An exception is allowed in the case of stocks already widely used and known by a designation which does not conform.

3. Definition of Substrain

The definition of substrain presents some of the same problems as the definition of species. In practice, the determination of whether two related strains should be treated as substrains and whether, in published articles, substrain symbols should be added to the strain symbol must rest with the investigators using them. The following rules, however, may be of assistance.

Any strains separated after 8–19 generations of b × s inbreeding and maintained thereafter in the same laboratory without intercrossing for a further 12 or more generations shall be regarded as substrains. It shall also be considered that substrains have been constituted (a) if pairs from the parent strain (or substrain) are transferred to another investigator or (b) if detectable genetic differences become established.

4. Designation of Substrains

A substrain shall be known by the name of the parent strain followed by a slant line and an appropriate substrain symbol. Substrain symbols may be of two types.

a. Abbreviated Name as Substrain Symbol. The symbol for a substrain should usually consist of an abbreviation of the name of the person or laboratory maintaining it. The initial letter of this symbol should be set in roman capitals; all other letters should be in lower case. Abbreviations should be brief, should as far as possible be standardized, and should be checked with published lists to avoid duplication, for example, F344/N (substrain of F344 maintained at the Laboratory Aids branch of the National Institutes of Health).

When a new substrain is created by transfer, the old symbol may be retained and a new one added. The accumulation of substrain symbols in this fashion provides a history of the strain. If the substrain symbols are not accumulated, a history of transfers should be kept, for example, PVG/ComLac (a subline of strain PVG maintained at the MRC Laboratory Animals Centre, and previously at the ARC Unit Compton).

b. Numbers or Lower Case Letters. Numbers or lower case letters may be used as substrain symbols in certain circumstances. The position of these relative to other parts, if any, of the substrain symbol should be suggestive of an historic or time sequence (104).

5. Coisogenic and Congenic Stocks

Coisogenic stocks produced by the occurrence of a single major mutation within an inbred strain, or congenic stocks produced by the introduction of a gene into an inbred background by a series of crosses, shall be designated by the strain symbol and, where appropriate (see rule 7), substrain symbol, followed by a hyphen and the gene symbol (in italics in printed articles). When the mutant or introduced gene is maintained in the heterozygous condition, this may be indicated by including a plus in the symbol. *Example*: A35322-D*h*/+. (The term congenic is now used to designate stocks which approximate the coisogenic state but which, because they are derived through a limited series of crosses rather than by mutation, may be presumed to carry, in addition to the distinguishing foreign allele, some other contaminant genes.)

When a congenic strain is produced by inbreeding with forced heterozygosis, indication of the segregating locus is strictly optional.

Nomenclature for strains congenic at alloantigen loci has not yet been officially agreed, but it should indicate first the inbred partner strain. This is followed either by a hyphen and the differential locus with the donor strain in parentheses, e.g., PVG-RT1^a(DA), or by a dot followed by the designation of the donor strain with the differential locus in parentheses, e.g., PVG.DA(RT1^a). Optionally, the donor strain designation (particularly if the donor was an undefined stock) or the differential locus may be omitted, provided this information is readily available from published lists. Similarly, the differential locus may be abbreviated, e.g., PVG-1^a or even PVG.1A. The final rules should be agreed soon.

In the case of congenic stocks produced by repeated crosses of a gene into a standard inbred strain, it may be desirable to indicate the number of backcross generations. Strains shall be regarded as congenic when at least seven such crosses have been made, though in draft revised rules 12 generations will be required. The first hybrid or F_1 generation should be counted as generation 1 (N_1), the first backcross generation as generation 2 (N_2), etc.

6. Substrains Developed through Foster Nursing, Ova Transfer, or Ovary Transplant

See Staats (104).

7. Compound Substrain Symbols for Stocks of Complex Origin

See Staats (104).

8. Indication of Inbreeding

When it is desired to indicate the number of generations of b × s inbreeding, this shall be done by appending, in parentheses, an F followed by the number of inbred generations. If, because of incomplete information, the number given represents only part of the total inbreeding, this should be indicated by proceeding it with a question mark and plus sign.

9. Priority in Strain Symbols

If two inbred strains are assigned the same symbol, the symbol to be retained shall be determined by priority in publication.

Table I
Inbred Strains[a]

ACI	*Color:* black with white belly and feet (a,h^i). Inbr. > F100. *Origin:* Curtis and Dunning at the Columbia University Institute for Cancer Research, 1926. To Heston, 1945 at F30, to NIH in 1950 at F41. Subsequent sublines from either Dunning or the NIH. *Characteristics:* long latency to emerge into familiar (11/12 male, 9/12 female) and novel environment (12/12 male, 11/12 female) (48). Median survival 113 weeks in males, 108 weeks in females. Spontaneous tumors in males: 46% testes, 16% adrenal, 5% pituitary, 6% skin and ear duct, and fewer of other types. In females: 21% pituitary, 13% uterus, 11% mammary gland, 6% adrenal and fewer other types. Other conditions include 28% of males and 20% of females with absent, hypoplastic or cystic kidneys on one side sometimes associated with an absent or defective uterine horn or atrophic testes on the same side (65). These abnormalities have a polygenic mode of inheritance (26). Spontaneous adenocarcinomas of ventral prostate seen in 7/41 untreated males at 34–37 months of age (97). High early prenatal mortality (2/8 = 11%) and high incidence of congenital malformations (1/8 = 10%) (99). Low serum thyroxine (5/5) (31). Low systolic blood pressure (17/17) (47). High hepatic metabolism of aniline in females (2/10) (82). Absorbs diethylstilbestrol at intermediate rate, leading to a high incidence of mammary tumors (30). High dose of pentobarbital sodium (120 mg/kg) required for LD_{50}(1/7) (98). Poor reproductive performance (9/12) and low litter size (11/12) (47). High *in utero* embryo mortality which depends on maternal genotype (26). Will grow transplantable tumors: M-C961, 970, R3234, R3559.
ACP	*Color:* black with white belly and feet (h^i). Inbr. F64. *Origin:* Dunning to National Cancer Institute 1967 at F54.
AGA	*Color:* black (a). Inbr. F20. *Origin:* Nakic, Zagreb (107). Used for immunological studies. No further information.
AGUS	*Color:* albino (a,c,H). Inbr. F35. *Origin:* germ-free strain developed by Gustafsson from stock (Sprague-Dawley ?) hand-reared in 1948 at F10. To Laboratory Animals Centre, Carshalton in 1968 at F26. *Characteristics:* Susceptible to experimental allergic encephalomyelitis (55). Relatively resistant to infection by *Entamoeba histolytica* (78). Good breeding performance, though sensitive to environmental influences. About 1–4% tailless young, which are infertile (M. F. W. Festing, original observation).
ACH	*Color:* black hooded (a,h). Inbr. F73. *Origin:* Curtis and Dunning, 1926 at the Columbia University Institute for

Table I (*Continued*)

	Cancer Research. *Characteristics:* high incidence of spontaneous lymphosarcoma of ileocecal mesentary. Will grow transplantable tumors R2788, IRS6820, B-P839.
ALB	*Color:* dilute brown (a,b,d ?). Inb. F41. *Origin:* Wolf and Wright, Albany Medical College, to NIH in 1950. No inbreeding records prior to transfer. *Characerics:* docile behavior. Some mammary fibroadenoma. Variable frequency of fetal resorption which affects reproductive performance (47). Low hepatic metabolism of aniline in females (10/10) (82). Poor reproductive performance (11/12) and small litter size (10/12) (47).
AM	*Color:* yellow coat (genetics unknown). Inbr. F26. *Origin:* Torres, Rio de Janeiro, from outbred stock.
AMDIL	*Color:* dilute yellow coat (genetics unknown). Inbr. F25. *Origin:* Torres (see AM).
AO	*Color:* albino (A,c,h). Inbr. F? + 48. *Origin:* From ARC Compton, probably as "WAG" to Gowans, Oxford in 1957. Appears to differ from some other WAG sublines in having *A* at the agouti locus.
AS	*Color:* albino (c). Inbr. F60+. *Origin:* University of Otago from Wistar rats imported from England in 1930. May be a subline of GH, with which it is histocompatible (52). *Characteristics:* hypertensive, though not to such an extent as GH (52). Susceptible to development of experimental allergic encephalomyelitis (55). Susceptible to induction of autologous immune complex glomerulonephritis, associated with the major histocompatibility complex (111). Good reproductive performance.
AS2	*Color:* albino (c). Inbr. F60+. *Origin:* Outbred rats at the University of Otago Medical School to Department of Surgery 1963 at F22-24. High secondary antibody response to the polypeptide (T,G)-Pro-L (3/20) (49).
AUG	*Color:* dilute hooded (h,p). (i.e., yellow). Inbr. F? + 34. *Origin:* Derived from one of the U.S. "August" sublines in 1951 and distributed by the Chester Beatty Institute, Pollards Wood, England. *Characteristics:* susceptible to experimental allergic encephalomyelitis (55). Resistant to induction of autoimmune thyroiditis (91).
AVN	*Color:* albino (c). Inbr. F44. *Origin:* Unknown. Will grow ferridextran-induced sarcoma FEDEX-AVN (59). Low secondary antibody response to the polypeptide (T,G)-Pro-L (19/20) (44).
A990	*Color:* agouti or non-agouti hooded (A,C,h or a,C,h). Inbr. F87. *Origin:* Curtis 1921 at the Columbia University Institute for Cancer Research. *Characteristics:* life span 14 ± 1 months. Resistant to cysticercus. Susceptible to estrogen-induced mammary and adrenal tumors. Will grow transplantable tumors IRC 855 and R3409. Good reproduction. High open-field defecation (2/12 males, 3/12 females) and low ambulation (11/12) (50). Low wheel activity (12/12) (49).
A7322	*Color:* pink-eyed dilute hooded (h,p). Inbr. F64. *Origin:* Curtis 1925 at the Columbia University Institute for Cancer Research. *Characteristics:* life span 14 ± 1 months. Spontaneous mammary tumors frequent. Resistant to cysticercus. Will grow transplantable tumors R2857, R2737, R2426, R3442.
A28807	*Color:* pink-eyed hooded, dilute (h, ?). Inbr. F53. *Origin:* Subline of A7322 derived from a half-brother × sister mating at F15. *Characteristics:* similar to A7322.

(continued)

Table I (*Continued*)

A35322	*Color:* black hooded (a,h). Inbr. F50. *Origin:* Curtis and Dunning 1942, from a mutation originating in an aunt × nephew cross at F27 of animals of strain A990. *Characteristics:* vaginal prolapse frequent. Will grow transplantable tumor R3280 (bronchiogenic carcinoma) and R3371. Short latency to emerge from cage into familiar (2/12) and novel environment (4/12) (48). High open-field defecation (3/12 males, 1/12 females) (50). High wheel activity (3/12 females, 4/12 males) (99).
B	*Color:* albino (c). Inbr. F69. *Origin:* Dr. P. Swanson from Wistar stock to E. Dempster at F43, *Characteristics:* large body size (68 gm at 28 days). Poor maternal instincts; fertile.
BDE	*Color:* black hooded (a,h). Inbr. F35. *Origin:* Zentralinstitut für Versuchstierzucht Hannover, from a cross between BD VII and E3. *Characteristics:* low incidence of megaesophagus/aperistalsis and microphthalmia.
BDI	*Color:* yellow, pink-eyed (A,C,H,p). Inbr. F? + 50. *Origin:* Druckrey 1937 from a yellow, pink-eyed strain. Inbred and reduced to one pair after World War II. Crosses with Wistar stock and subsequent inbreeding led to the development of BD II. According to Druckrey (29) strains BD III – BD X were then developed from a cross of a single BD I × BD II mating pair, with subsequent selection for coat color alleles. However, the strains have 4 different *H-1 (Ag-B)* haplotypes (H-1^d, H-1^w, H-1^l, and H-1^e) rather than the two that would be expected from such a cross (109). The strains cannot be regarded as a set of recombinant inbred strains as defined by Bailey (3) though their definition by coat color alleles makes the set easily identifiable, and should help to ensure authenticity. According to Druckrey (29) all strains have a low tumor incidence, with a median life span of 700–950 days depending on strain.
BD II	*Color:* albino (a,c,h,P). Inbr. F? + 50. *Origin:* see BD I. Frequent microphthalmia.
BD III	*Color:* pink-eyed, yellow, hooded (A,C,h,p). Inbr. F? + 50. *Origin:* see BD I.
BD IV	*Color:* black hooded (a,C,h,P). Inbr. F? +50. *Origin:* see BD I.
BD V	*Color:* pink-eyed, non-agouti, hooded (a,C,h,p). Inbr. F? + 50. *Origin:* see BD I.
BD VI	*Color:* black (a,C,H,P). Inbr. F? + 50. *Origin:* see BD I.
BD VII	*Color:* pink-eyed sandy (a,C,H,p). Inbr. F? + 50. *Origin:* see BD I. Low secondary antibody response to polypeptide (T,G)-Pro-L (20/20) (44).
BD VIII	*Color:* agouti hooded (A,C,h,P). Inbr. F? + 50. *Origin:* see BD I. Occasional vaginal atresia.
BD IX	*Color:* agouti. (A,C,H,P). Inbr. F? + 50. *Origin:* see BD I.
BD X	*Color:* albino (a,c,h,p). Inbr. F? + 50. *Origin:* see BD I.
BIRMA	*Color:* albino (c). Inbr. F27. *Origin:* A. M. Mandl, 1952 from albino rats purchased from Birmingham market.
BIRMB	*Color:* albino (c). Inbr. F25. *Origin:* as for BIRMA.
BN	*Color:* brown (a,b,h^l). Inbr. F35. *Origin:* Silvers and Billingham, 1958 from a brown mutation maintained by D. ₁H. King and P. Aptekman in a pen-bred colony. *Characteristics:* endocardial disease 7% at an average of 31 months of age (13). Tumors of epithelium 28% in males, 2% in females. Ureter tumors 20% in females, 6% in

males. Estimated median life span more than 24 months in males and more than 25 months in females (14). Congenital hydronephrosis 30% (24). Survival curves based on 236 females and 74 males show a median life span of 29 months in males, and 31 months in females. Most common neoplastic lesions in males were urinary bladder carcinoma 35%, pancreas islet adenoma 15%, pituitary adenoma 14%, lymphoreticular sarcomas 14%, adrenal cortex adenoma 12%, medullary thyroid carcinoma 9%, adrenal pheochromocytoma 8%. Four other types of tumors were observed. In females: pituitary adenoma 26%, ureter carcinoma 22%, adrenal cortical adenoma 19%, cervix sarcoma 15%, mammary gland fibroadenoma 11%, islet adenoma 11%. Twelve other tumor types were observed (20). Chance of death from metastases increases with age in females, but reaches a peak at 25–30 months in males (21). The cervical and vaginal tumors have been studied in more detail (20) and further details of the aging colony are given by Hollander (53) and Burek and Hollander (22). Vaginal and cervical tumors, mostly sarcomas but also 7 squamous cell carcinomas and 4 leiomyomas were seen in 20% of animals that died naturally (23). Intermediate susceptibility to pentobarbital sodium (3/7) with LD$_{50}$ of 90 mg/kg (98). Resistant to experimental allergic encephalomyelitis (1/7) (39,68). Resistant to induction of autologous immune complex glomerulonephritis (111). Low fertility.

BP	*Color:* black Irish hooded (a,h^l). Inbr. F25. *Origin:* from Professor Sekla. Strain selected for resistance to Walker 256 tumor. *Characteristics:* low fertility. Adult spleen cells highly effective in the induction of GVH reaction in allogeneic newborn, but BP newborn resistant to GVH reaction.
BROFO	*Color:* albino (c). Inbr. F28. *Origin:* Medical Biological Lab., Defence Res. Org., Netherlands. *Characteristics:* large Wistar type of rat maintained SPF and germ-free.
BS	*Color:* black (a). Inbr. F41. *Origin:* University of Otago Medical School from a cross of wild rats × Wistar stock, with the black phenotype backcrossed to the Wistar (118). *Characteristics:* docile. Fair reproduction. Low incidence of hydrocephalus.
BUF	*Color:* albino (c). Inbr. F58. *Origin:* Heston, 1946 from Buffalo stock of H. Morris. To NIH in 1950 at F10. *Characteristics:* low incidence of dental caries. Will grow hepatoma 5123 (which has some enzymes similar to normal liver), Yoshida ascites sarcoma (16%), and also Morris hepatoma and pituitary tumor. Spontaneous tumors of anterior pituitary 30% and adrenal cortex 25% in older animals. Survival 58% at 2 years of age with thyroid carcinoma 25% (64). High incidence of prenatal mortality (1/8 = 18%) early and (2/8 = 11%) late but no congenital abnormalities detected (99). Autoimmune thyroiditis seen in 6/11 males at 36 weeks of age, but not seen in 5 other strains. Spontaneous autoimmune thyroiditis with mononuclear cell infiltration of the thyroid in 26% of animals over one year (79). Autoimmune thyroiditis develops spontaneously and after ingestion of 3-methylcholanthrene, but reaches nearly 100% after neonatal thymectomy (102). Poor immune response to sheep red blood cells (6/7) (112). Low hepatic

(*continued*)

Table I (*Continued*)

metabolism of ethylmorphine (7/10 in males 10/10 females), but high metabolism of aniline (3/10 males, 1/10 females) (82). Intermediate breeding performance (5/12) and litter size (8/12) (47).

B3   *Color:* Inbr. F10 (i.e., only part-inbred). *Origin:* From stock bred for homozygosity of coat color genes, as a natural recombinant of the major histocompatibility complex (40). Strain has the serologically defined *RT1ⁿ* haplotype together with the mixed lymphocyte reactivity-4 (*MLR4*) locus.

CAP   *Color:* albino (a,h). Inbr. F30+. *Origin:* Polish Academy of Sciences, Krakow (106).

CAR   *Color:* albino (c). Inbr. F77. *Origin:* Hunt 1937, developed for resistance to dental caries. Poor reproductive performance (10/12) and low litter size (9/12) (47).

CAS   *Color:* albino (c). Inbr. F38. *Origin:* Hunt 1937. Developed for a high incidence of dental caries. Poor reproductive performance (12/12) and low litter size (12/12) (47).

COP   *Color:* black hooded (a,h). Inbr. F73. *Origin:* Curtis in 1921 at the Columbia University Institute for Cancer Research. *Characteristics:* Mean life span 20 ± 0.2 months. Spontaneous tumors of thymus. Slow absorption of diethylstilbestrol leading to death from bladder calculi and papillomas. Resistant to mammary tumor induction. Small pituitaries (30). Resistant to cysticercus. Will grow transplantable tumors IRS 4337 and R3327 prostate adenocarcinoma, which is a model of human prostate cancer.

CPBB   *Color:* Agouti hooded. Inbr. F65. *Origin:* Hagedoorn, Holland to CPB in 1949 at F15. *Characteristics:* susceptible to audiogenic seizures (113).

DA   *Color:* Agouti (A,B,C). Inbr. ? + 50. *Origin:* not known, but Palm and Black (83) suggest that it may be related to COP. *Characteristics:* susceptible to induction of autoimmune thyroiditis (91).

DONRYU   *Color:* albino (c). Inbr. F59. *Origin:* R Sato in 1950 by inbreeding Japanese albino rats. *Characteristics:* mild nature, relatively small size. Will accept 100% of transplantable Yoshida sarcoma and ascites hepatoma (Dr. S. Sato, personal communication, 1977).

E3   *Color:* non-agouti, yellow-brown hooded (a,C,h). Inbr. F55. *Origin:* Kröning, Göttingen in 1949 from rats of unknown origin, to Hannover in 1957.

F344   *Color:* albino (a,c,h). Inbr. F86. *Origin:* Curtis in 1920 at the Columbia University Institute for Cancer Research. To Heston 1949 then to NIH, Bethesda in 1950 at F51. Subsequent sublines either from Dunning or the NIH colonies. *Characteristics:* low open-field defecation (10/12) in males (50). Low wheel activity (11/12 females, 8/12 males) (49). Median life span about 31 months in males and 29 months in females with about 87% survival to 24 months in both sexes. Mammary tumors 41% in females and 23% in males, pituitary adenomas 36% in females and 24% in males, testicular interstitial cell tumors 85% in males. Other tumor types less common (95). Thyroid carcinoma 22% (64). Mean life span 24 months in both sexes, in presence of severe pulmonary infection. Interstitial cell testicular tumors 65%. Mononuclear cell

leukemia 24%, mammary fibroadenoma 9% in females. Both sexes have a 5% incidence of nodular hyperplasia of the liver (28). Mean life span 675 days in females and 725 days in males. Testicular interstitial cell tumors in males 68%, uterine polyploid tumors of endometrial origin 21% in females (56). Median life span in SPF females was 25 months. A unique mononuclear cell leukemia was developed in 21/86 animals with uniform involvement of the liver and spleen (71). Incidence of tumors under germ-free conditions: Leukemia 26% in males and 36% in females, mammary tumors 12% in males and 20% in females, all other tumors 9% in males and 5% in females (93). Low serum insulin (5/5) (31). High hepatic metabolism of ethylmorphine and aniline in males (2/10 in each case) (82). Resistant to the development of salt hypertension (46). High specific activity but low inducibility of NADPH-cytochrome c reductase compared with oubred Sprague-Dawley rats (41). Large pituitaries (30). Susceptible to cysticercus infection. Rapidly absorbs implanted diethylstilbestrol pellets leading to death. Fatty liver most common autopsy findings (30). Low LD_{50} of pentabarbital sodium (5/7 = 70 mg/kg) (98). Low primary and secondary immune response to sheep red blood cells (7/7) (112). Good breeding performance (1/12) and large litter size (1/12) (47). Low incidence of nephropathy. Will grow the following transplantable tumors: Dunning hepatoma, hepatoma LC-18, Novikoff hepatoma, mammary carcinomas HMC and R-3230, pituitary tumors MtT and MtTf4, Walker 256 carcinosarcoma, Dunning leukemia, leukemias HLF1, IRC-741 and R3149, lymphosarcoma R-3251, mammary fibroma F-609, fibrosarcoma R-3244, sarcomas IRS 9802 and R13259, uterine sarcoma F-529 and also leukemias R-3323, 3330, 3399 and 3432 (47).

G/Cpb   *Color:* albino. Inbr. F86. *Origin:* Gorter, Holland to Hagedoorn to CPB at F35 (113).

GH   *Color:* albino (c). Inbr. F31. *Origin:* University of Otago Medical School from rats of Wistar origin imported from England in 1930. Selection for high blood pressure started by Smirk, 1955. A number of sublines have been developed. Closely related to strain AS (52). *Characteristics:* hypertension, cardiac hypertrophy and vascular disease (85). Heart rate about 20% greater, lower body fat and heart weight about 50% greater than in normotensive strains. Genetic hypertension in GH (but not SHR) may be associated with a defect in renal prostagladin catabolism (2). Strain characteristics in relation to SHR reviewed by Simpson *et al.* (103).

GHA   *Color:* ginger hooded (?, h). Inbr. F 20. *Origin:* The Queen Elizabeth Hospital, Woodville, S. Australia from mixed Wistar, LEW, and coloured stock.

HCS   *Origin:* Harvard to Liverpool, U.K. 1960. May be a subline of CAS (?).

HO   See PVG.

HS   *Color:* black hooded (a,h). Inbr. F20 +. *Origin:* Probably from same cross as BS (118). *Characteristics:* docile, fair reproduction. Approximately 12% hydrocephalus.

INR   *Color:* black hooded (a,h). Inbr. F21. *Origin:* Harrington, 1962 from a stock selected by Hall for low open-field defecation. *Characteristics:* Short emergence la-

Table I (*Continued*)

tency from cage into both familiar and novel environment (1/12) (48). Low open-field defecation in females (12/12) and high open-field ambulation in both sexes (1/12) (50). Low wheel activity, and poor complex maze learning.

IR   *Color:* pale cinnamon beige hooded (a,h,c^d?). Inbr. F20. *Origin:* Harrington, 1962 from a mutation of a Michigan stock. *Characteristics:* vigorous and healthy, but high incidence of physical anomalies. Long latency to emerge into familiar environment (12/12 males, 10/12 females) (48). High open-field exploration in females (3/12) (50). Low wheel activity (11/12 males, 10/12 females) (49).

IS   *Color:* agouti (+). Inbr. F23. *Origin:* from a cross between a wild male and a Wistar female, with sib mating since 1968 (55a). *Characteristics:* malformations of the thoracicolumbar vertebrae leading to kyphoscoliosis with restricted spinal canals and compressed spinal cords. The malformations occur from the twelfth thoracic to the sixth lumbar vertebrae. Usually two to four vertebrae, but sometimes as many as seven are affected, and some degree of abnormality occurs in about 90% of individuals.

K   *Color:* albino. Inbr. F40. *Origin:* Dr. E. Matthies, Halle-Wittenberg, 1958, from outbred Wistar stock. *Characteristics:* low spontaneous tumor incidence (less than 0.5%). Good breeding performance. Weight at 100 days is 290 gm in males and 200 gm in females. Developed by selection for resistance to a range of transplantable tumors (67).

KGH   Inbr. F43. *Origin:* Gill from animals supplied by the Animal Research Centre, Harvard University (61,62).

KX   *Color:* albino (formerly called NEDH). Inbr. F45. *Origin:* Developed from Slonaker colony (University of Chicago ca. 1928). Sublines carrying *ic* (infantile ichthyosis) and color genes *C* and *H* are also kept (58).

KYN   *Color:* black with some white on ventral surface (a,h^n). Kyoto-Notched. Inbr. F ?. *Origin:* Makino, Hokkaido University 1960 from stock carrying the "notched" character isolated by Nakata from wild rats in Kyoto, and from a cross involving WKA (99). Abnormalities of the urogenital organs 11% (100).

LA/N   *Color:* black (a,b?). Inbr. F23. *Origin:* From a cross between ALB/N and a hooded stock of unknown origin (47).

LE   *Color:* black hooded. Inbr. F43. *Genetics:* CCaahh, H-1^w?, H-6^d. *Origin:* Mühlbock, The Netherlands Cancer Institute in 1960 from pen-bred Orl:LE.

LEJ   *Color:* black hooded (a,h). Inbr. F39. *Origin:* From Pacific Farms to Nat. Institute of Genetics, Misima, Japan in 1956.

LEP   *Color:* black hooded (a,h,?). Inbr. F23. *Origin:* Czechoslovakia, Prague from stock of unknown ancestry. Low secondary antibody response to the polypeptide (T,G)-Pro-L(18/20) (44).

LEW   *Color:* albino (a,h,c). Inbr. F67. *Origin:* Lewis from Wistar stock to Aptekman and Bogden 1954 at F20, to Silvers, 1958 to F31. Subsequently distributed by Silvers. Used as the inbred partner for a number of congenic strains at the major histocompatibility complex (108). *Characteristics:* docile. High fertility. Survival: 26% at 2 years of age (69). High serum thyroxine (1/5), high serum insulin (1/5) and high serum growth hormone (1/5) (31). Becomes obese on a high fat diet (2/7) (96). High hepatic

Table I (*Continued*)

metabolism of ethylmorphine in females (3/10) (82). Susceptible to the development of experimental allergic encephalomyelitis after challenge with guinea pig myelin basic protein (contrast BN) (55,168). Highly susceptible to development of induced autoimmune myocarditis (37). Susceptible to induction to autologous immune complex glomerulonephritis (this is linked to the major histocompatibility complex) (111). Susceptible to induction of experimental allergic encephalitis and adjuvant-induced arthritis (84). Host for lymphoma 8, kidney sarcoma, fibrosarcomas MC-39, ML-1, ML-7, carcinoma No. 10 Lewis, and sarcoma No. 3 Lewis (47).

LGE   *Color:* black hooded. Inbr. F18 +. *Origin:* Long-Evans stock from Poiley to Gill in 1971.

LOU/C   *Color:* albino (?). Inbr. F20 + ?. *Origin:* Bazin and Beckers from rats of presumed Wistar origin kept at the University of Louvain. From 28 parallel sublines LOU/C was selected for its high incidence of plasmacytomas, and LOU/M for its low incidence. The two are histocompatible, and this is maintained by selection of LOU/M males on the basis of acceptance of skin grafts from LOU/C animals (5). *Characteristics:* develop spontaneous plasmacytomas after about 8 months of age, with an incidence of about 30% in males and 16% in females. Tumors develop rapidly, and may be detected by palpation. They usually develop in the ileocecal lymph nodes, and about 60% of them synthesize monoclonal immunoglobulins (Bence-Jones proteins) of IgC_1 (35%), IgE (36%) or IgA classes. Tumors are transplantable in solid or ascites form, and retain their secretory properties through successive passages (4,6,7,9). Amino acid sequence of a **k** Bence-Jones protein has been studied by Starace and Quérinjean (105). More than 800 transplantable immunocytomas are now available.

LOU/M   *Color:* albino (?). Inbr. ?. *Origin and characteristics:* see LOU/C. Good IgE antibody response to ovalbumin and DNP hapten (8).

MAXX   *Color:* black hooded (a,h). Inbr. F20 +. *Origin:* From a cross of BN × LEW, with subsequent inbreeding.

MNR   *Color:* albino (A or a,c,h). Inbr. F44 +. *Origin:* P. L Broadhurst, 1954, from a commercial Wistar stock, with selection for low defecation response in an open field. A number of parallel sublines are in existance. These differ at least at the agouti (A) and the major histocompatibility (*Ag-B* or *H-1*) loci. *Characteristics:* long latency to emerge into a familiar (11/12) and novel (12/12) environment in females (48). Low open-field defecation (12/12 in males, 11/12 to 9/12 depending on subline) (50). High wheel activity (1/12) (49). Low open field defecation (8/8) in hybrid offspring (87). Good breeding performance (3/12) and large litter size (3/12) (47). See also under strain MR. An extensive review of the difference between MR and MNR is given by Broadhurst (19).

MNRA   *Color: albino (a,c,h). Inbr. ?. Origin:* subline of MNR which differs in a number of characteristics, and carries the *a* allele at the agouti locus (G. M. Harrington, personal communication, 1977).

MR   *Color: albino (a,c,h). Inbr. F40. Origin:* P. L. Broadhurst, 1954 from commercial Wistar stock, with

(*continued*)

Table I (*Continued*)

selection for high defecation response in open field (see MNR). *Characteristics:* compared with MNR the strain has a high open field defecation, low ambulation, low rearing, low shock avoidance conditionability and tends to be more "emotional" in a wide range of behaviour (19). Long emergence latency into novel environment in females (12/12) and also into familiar environment in males (11/12) (48). High open field defecation (1/12 males, 2/12 females) (50). Good breeding performance (4/12) and large litter size (2/12) (47). See also under MNR.

MSUBL/Icgn.   *Color:* black (a). Inbr. F23. *Origin:* Stroeva Institute of Developmental Biology, Moscow, from a cross of wild rats × MSU microphthalmic rats obtained from Dr. Brouman, Montana State University. *Characteristics:* high incidence of microphthalmia and anophthalmia (16). High reactivity of adrenals to ACTH (1/4). Immunoglobulin (**k** chain) allotype reported by Rokhlin and Nezlin (90).

M520   *Color:* albino (a,c,h). Inbr. F86. *Origin:* Curtis, 1920, at the Columbia University Institute for Cancer Research, to Heston in 1949 at F49, to NIH in 1950 at F51. *Characteristics:* low systolic blood pressure (16/17) (47). Low hepatic metabolism of aniline (9/10 in males, 7/10 in females), but high metabolism of ethylmorphine (1/10) (82). Good breeding performance (2/12) and intermediate litter size (6/12) (47). Susceptible to cysticercus. NIH subline has 21–25% incidence of adrenal medulla tumors. Susceptible to induction of tumors by 2-acetylaminofluorine. Low frequency of pancreatic exocrine tumors. Highly susceptible to nephritis. Tumors of the uterus, anterior pituitary, adrenal medula and cortex, and interstitial cell tumors range from 0–10% in animals under 18 months, but after 18 months 12–50% of females show uterine tumors, 35% of virgin males show interstitial cell tumors, 60–85% and 20–45% show tumors of adrenal medulla and cortex, respectively, and 20–40% have tumors of the anterior pituitary. Will grow Jensen sarcoma and Yoshida ascites tumors, and 75% will grow hepatomas 7974 and 130. Host for mammary tumor BICR/MI, osteogenic sarcoma 344, carcinoma 343, Harderian gland carcinoma 2226, carcinoma 338 and sarcoma E-2730 (47).

M14   *Color:* albino (c). Inbr. F40 +. *Origin:* A. B. Chapman, 1940 from Sprague-Dawley stock, with selection for low ovarian response to pregnant mare's serum.

M17   *Color:* albino. Inbr. F40 +. *Origin:* A. B. Chapman, 1940, from Sprague-Dawley stock, with selection for high ovarian response to pregnant mare's serum.

NBR   *Color:* black (a). Inbr. F20 +. *Origin:* Poiley 1966 from heterogeneous stock. *Characteristics:* good reproduction, active, small, S5B may be closely related (see S5B).

NIG-III   *Color:* pink-eyed yellow (?). (aaBBCCHHp^{m}p^m). *Origin:* From a mating between a wild rat trapped in Misima City and Castle's black rat.

NSD   *Color:* albino (c). Inbr. F50. *Origin:* NIH, Bethesda, 1964, from non-inbred (Sprague-Dawley) stock. *Characteristics:* used for studies of mammary carcinomas. High systolic blood pressure (4/17). Intermediate breeding performance (6/12) and litter size (5/12) (47).

NZR/Gd   *Color:* albino (c). Inbr. F32 +. *Origin:* developed as a

Table I (*Continued*)

subline of AS2 at F32. *Characteristics:* tumors (atriocaval epithelial mesotheliomas) of right atrium or inferior vena cava occur in approximately 20% of animals of both sexes over 1 year of age. These tumors are slow growing, but apparently malignant, and closely resemble tumors found in humans (42).

OKA/Wsl   *Color:* albino. Inbr. F20 + ?. *Origin:* from Faculty of Medicine, Kyoto, Japan to Dr. J. Roba, Machelen, Belgium in 1970. To Dr. H. Bazin 1971. Should probably be regarded as a subline of SHR, though skin grafts between OKA and SHR are rejected after about 30–45 days.

OM   *Color:* albino (c). Inbr. F30 +. *Origin:* Heston 1946 from non-inbred Osborne-Mendel stock obtained from J. White, to NIH at F10. *Characteristics:* spontaneous thyroid carcinoma 33% (64). Adrenal cortical tumors 94% over 18 months of age, pituitary tumors 8–18% at 18 months (47). Mammary tumors 26–30%. High incidence of retinal degeneration in animals over 10 months of age (94). High systolic blood pressure (2/17) (47). Becomes obese on a high fat diet (1/7) (96). Intermediate breeding performance (8/12) and good litter size (4/12) (47).

PA   *Color:* albino (c). Inbr. F180. *Origin:* King, 1909 from Wistar Institute stock to Aptekman 1946, at F135, to Bogden, 1958 at F155. The oldest inbred strain of rats. WKA is probably a subline of this strain. *Characteristics:* vigorous (and vicious), healthy, good reproduction. Will grow ascites tumor 9A, leukemia LK2, carcinoma 5, lymphoma 6.

PETH   *Color:* pink-eyed hooded (a,h,p). Inbr. F39. *Origin:* Bourne 1938 to Sidman to NIH in 1966 at F9N1F18. Should probably be regarded as a subline of RCS. *Characteristics:* carries gene for retinal dystophy (*re*), causing cataracts at 3 months of age. Secondary effects include unusual lens reflection and persistent hyaloid artery. Low systolic blood pressure (13/17) (47).

PVG   *Color:* black hooded (a,h). Inbr. F70 +. *Origin:* Kings College of Houseold Science to Lister Institute, to Virol, to Glaxo in 1946. Inbred by Glaxo. *Characteristics:* docile and good breeding performance. Low defecation and activity in open field in Broadhurst's subline. Resistant to induction of autoimmune thyroiditis (91). Susceptible to infection by *Entamoeba histolytica* (78). Lower incidence of polyspermia than WAG (17). A subline with a chromosome marker analogous to the CBA-T6 strain has been developed by Howard (54), and it is also being used as the background strain in the development of a number of congenic lines (W. L. Ford, personal communication).

R   *Color:* albino (c). Inbr. F78. *Origin:* Mühlbock from a Wistar stock in 1947.

RCS   *Color:* pink-eyed hooded (a,h,p). Inbr. F20 + ?. *Origin:* developed prior to 1965 by Sidman from a stock obtained from Sorsby of the Royal College of Surgeons, London (101). Presumed to be very similar to PETH. *Characteristics:* carries gene for retinal dystrophy. Cataracts develop in 24% and there is a 4% incidence of microphthalmia. A congenic strain RCS-p/+ has been developed in which the retinal dystrophy develops more slowly. Mean litter size 6.8 ± 0.1 (SEM). Body weight plateaus at 5–6 months at about 185 gm in females and 275 gm in males (53).

(*continued*)

Table I (Continued)

Table I (Continued)

RHA/N	Color: albino (c). Inbr. F23. Origin: Bignami selected for high avoidance conditioning with light as a conditioned stimulus and electric shock as an unconditioned stimulus. This outbred stock to NIH in 1968, where brother × sister mating was initiated. (Note: the original outbred stock, and possibly some independently derived inbred strains are still in existence so it is important to retain the ''N'' designation in describing this strain).
RLA/N	Color: albino (c). Inbred. F33. Origin: as for RHA, but original stock selected for low avoidance conditioning.
SD	Color: albino (?). Inbr. F25. Origin: Halber, University of Minnesota to NIH at F25. Characteristics: grows Walker 256 tumor.
SEL	Color: black (a). Inbr. F28. Origin: Dunning 1948. Characteristics: will grow transplantable tumors R3401 and R3478.
SHR	Color: albino (c). Inbr. F23 +. Origin: Okamoto in 1963 from outbred Wistar Kyoto rats. From a colony of many animals one male with spontaneous hypertension, and a female with elevated blood pressure were mated. Brother × sister mating with continued selection for spontaneous hypertension was then started (80). A number of sublines have also been developed with a tendency to develop cerebrovascular lesions and stroke (77). Characteristics: incidence of hypertension is high and there are no obvious primary organic lesions in kidneys or adrenal glands. Hypertension is very severe, with blood pressure frequently over 200 mm. There is a high incidence of cardiovascular disease (81). Genetic analysis indicates that the condition is controlled by 3–4 genetic loci, one of which may be a ''major'' locus (117). It is suggested that the blood pressure maintenance in the hypothalamus is deranged. Alloxan diabetes further increases blood pressure, but the animals respond to antihypertensive drugs (80). Roba (89) has concluded that the strain is a suitable model for screening drugs for antihypertensive action. In young SHR rats the plasma levels of both noradrenaline and dopamine b-hydroxylase were increased over control WKA rats, but total catecholamines were not significantly different. Catecholamine content of the adrenals was reduced (43). Circulating thyrotropin levels were markedly elevated over two control strains (114). There was a reduced ^{131}I metabolism and increased thyroid weight relative to Wistar controls (36). The ''Committee on the Care and Use of Spontaneously Hypertensive (SHR) Rats'' (25) has issued guidelines for the breeding, care and use of this strain.
S5B	Color: albino (a,c). Inbr. F38. Origin: Poiley, 1955 from a cross of outbred NBR rats × Sprague-Dawley, with 5 generations of backcrossing of albino gene followed by sib mating. Characteristics: high tolerance of toxic compounds (?). Good learning. Resistant to dietary induction of obesity (1/7) (96).
TMB	Color: agouti (+). Inbr. F23 +. Origin: inbred by P. L. Broadhurst from stock selected by Tryon for good maze learning performance. Although TMB and TS1 are derived from the same outbred selected stock they were inbred independently and should therefore be regarded a different inbred strains. Characteristics: developed from stock selected for behavioral characteristics but shuttle-

	box avoidances, ambulation, defecation, and brain GABA production lie in the normal range (88).
TMD	Color: black (a). Inbr. F23 +. Origin: see TMB, but inbred from stock selected for poor maze learning performance. Should also be regarded as a different strain from TS3. Characteristics: various measures of behavior lie in the normal range (88).
TO	Color: albino (c). Inbr. F38. Origin: a breeder in Tokyo to Hokkaido University 1952. Characteristics: will grow Yoshida sarcoma.
TS1	Color: agouti with some white on ventral surface (A,h^i). Inbr. F20 +. Origin: inbred by G. M. Harrington about 1965 from stock developed by Tryon about 1929 by selective breeding for good maze learning performance. Characteristics: high voluntary alcohol acceptance concentration (1/4) (92). Low open field activity (8/8) (87). Low open field activity (12/12) (50). Low wheel activity in males (10/12) (49).
TS3	Color: black (a). Inbr. F20 +. Origin: see TS1, but from stock selected by Tryon for poor maze performance. Characteristics: low voluntary alcohol final acceptance level (4/4) (92), low shock avoidance (7/8) (88), high brain GABA (3/8) (88), high open field ambulation (4/12) and low defecation (9/12 males, 10/12 females) (50), short latency to emerge into novel (3/12 male, 2/12 female) and familiar (3/12 male, 4/12 female) environments (48).
U	Color: Inbr. F48. Origin: Zootechnical Institute, Utrecht, 1958.
W	Color: albino. Inbr. F65. Origin: Wistar Institute to Tokyo University to Hokkaido University 1944. Inbred by Makino. Characteristics: congenital cleft palate 0.5% (99).
WA	Color: albino (a,h,c). Inbr. F66. Origin: Wistar stock to St. Thomas's Hospital to Laboratory Animals Centre in 1964 at F43. Characteristics: Mean life span 749 ± 40 days in females and 645 ± 30 days in males (based on 22 males and 25 females maintained in SPF conditions). High incidence of polycystic nephrosis (34). Good breeding performance.
WAB	Color: albino. Inbr. F81. Origin: from the same stock as WAG, but separated in 1926, prior to inbreeding. Characteristics: benign thymoma in 23% of individuals over 2 years of age, with 50% incidence in castrated males, and 57% in spayed females.
WAG	Color: albino (a,c,h; A,c,h, or A,c,H depending on subline). Inbr. F101. Origin: A. L. Bacharach 1924 from Wistar stock. The presence of different coat color alleles implies that this strain, or some sublines, may have become genetically contaminated at some point in the past. It is, therefore, important that the subline should be stated carefully in published work. Most common sublines are a,c,h. Characteristics: short latency to emerge from cage into familiar environment (4/12 in males, and 3/12 in females) and novel environment (2/12 males and 3/12 females) (48). High open field ambulation (2/12) (50). High wheel activity (2/12) (49). Low open field defecation (7/8) but high ambulation and shock avoidance learning (1/8) (87). Low brain GABA (8/8) (88). Endocardial disease 4% at an average age of 31 months (15). Tumors:

(continued)

Table I (Continued)

	Medullary thyroid carcinoma 27%, pheochromocytoma 2%, islet cell adenoma 1%. Mean life span more than 31 months in females and 22 months in males (15). Mean life span 31 months in females, with adenoma found in 69% of females. Other common tumors include medullary thyroid carcinoma 40%, adrenal cortical adenoma 29%, fibroadenoma of breast 21% with 19 other types of tumor found in 290 animals (13). Further details of the aging colony given by Hollander (53). Resistant to experimental allergic encephalomyelitis (55). Some sublines may carry recessive gene *dx* preventing anaphylactoid reaction to dextran (51). Susceptible to induction of autoimmune thyroiditis (91). Higher incidence of polyspermia than in PVG (17). Susceptible to iron deficiency. Low secondary antibody response to the polypeptide (T,G)-Pro-L (17/20) (44).
WE/Cpb	*Color:* light beige. Inbr. F38. *Origin:* outcross involving strains B, WAG and others in 1956, followed by inbreeding. *Characteristics:* relatively aggressive (113).
WF	*Color:* albino (c). Inbr. F27 +. *Origin:* J. Furth, 1945, from a commercial Wistar stock in an attempt to develop a high leukemia rat strain. *Characteristics:* carcinoma of colon found in 8/21 males and 8/29 females (69). In females tumors include 27% pituitary, 21% mammary, 3% adrenal, 9% leukemia (generalized), 7% lymphoma (malignant), 3% lipomas, and 4% unclassified (57). Mean life span 21 months in females and 23 months in males. Leukemia 15–22% characterized by an unusual type of mononuclear cell containing reddish granules. Administration of 3-methylcholanthrene to young increased, but X rays decreased leukemia (70). Low serum growth hormone (5/5) (31). Resistant to adrenal regeneration hypertension (72). Strain carries a distinctive heteropycnotic Y chromosome which may be used as a cellular marker (119).

Table I (Continued)

WKA	*Color:* albino. Inbr. F201. *Origin:* King 1909 from Wistar Institute stock to Aptekman 1946 at F135 to Hokkaido University 1953 at F148. This should probably be considered a subline of PA. *Characteristics:* congenital clubfoot and polydactyly 2% (99).
WKY/N	*Color:* albino (c). Inbr. F15 (i.e., still in development). *Origin:* outbred Wistar stock from Kyoto School of Medicine to the NIH, Bethesda in 1971. Inbred as a normotensive control strain for SHR (47).
WM	*Color:* albino. Inbr. F60. *Origin:* From Wistar Institute to Tokyo University, 1938, to Hokkaido 1944, to National Institute of Genetics, Misima 1951.
WN	*Color:* albino (c). Inbr. F34. *Origin:* Heston 1942 from Wistar stock of Nettleship, to NIH in 1950 at F15. *Characteristics:* incidence of spontaneous malignant and benign mammary tumors 30–50%. No tumors of uterus or interstitial cells found at any age. Anterior pituitary tumors 0–35% at less than 18 months of age, and 40–93% over 18 months. Tumors of adrenal cortex and medulla 0–10% at less than 18 months, but adrenal cortical tumors 25–50% after 18 months of age. Squamous cell metaplasia and hyperplasia of the thyroid has been found. Low systolic blood pressure (13/17), intermediate breeding performance (7/12) and litter size (7/12) (47).
WR	*Color:* Inbr. F30 +. *Origin:* Sykora, Rosice (107). No further information.
YO	*Origin:* Poiley, National Cancer Institute. No further information.
Y59	Strain developed in Zagreb, Yugoslavia.
Z61	*Color:* albino (a,c,h). Inbr. F70. *Origin:* Curtis 1920 at the Columbia University Institute for Cancer Research. *Characteristics:* susceptible to cysticercus and to estrogen-induced tumors and 2-acetylaminofluorine tumors: will grow Jensen sarcoma R3449 and R92.

[a] See also Fig. 40 and Table III.

Table II

Polymorphisms in Inbred Rats

Locus	Allele	Strains[a]	Reference
RT1 or (Ag-B, H-1)	$RT1^1 = Ag\text{-}B^1 = H\text{-}1^1$	AGA,AGUS,BD III,BS,CAR,CAS,CDF,F344,HS,LEW,NBR/1, S5B	35,61,86,106,107,110
	$RT1^u = Ag\text{-}B^2 = H\text{-}1^w$	AO,BDE,BD II,E3,LEP,LGE,LOU/C/Wsl,LOU/M,NBR,OM, PETH,R,RHA,WAG,WF,WP,YO	
	$RT1^n = Ag\text{-}B^3 = H\text{-}1^n$	BN,MAXX	
	$RT1^a = Ag\text{-}B^4 = H\text{-}1^a$	ACI,ACP,AVN,DA,COP,MNR/N	
	$RT1^c = Ag\text{-}B^5 = H\text{-}1^c$	A28807,A7322,AUG,CAP,HO,MNR/Brh,PD,PVG/c,Y59	
	$RT1^b = Ag\text{-}B^6 = H\text{-}1^b$	ALB,BP,BUF LA,M520,NSD	
	$RT1^g = Ag\text{-}B^7 = H\text{-}1^g$	KGH	
	$RT1^k = Ag\text{-}B^8 = H\text{-}1^k$	WKA,OKA,SHR	
	$RT1^d = Ag\text{-}B^9 = H\text{-}1^d$	BD I, BD IV, BD V, BD VI, BD IX, BD X, MR/Psy	
	$RT1^f = Ag\text{-}B^{10} = H\text{-}1^f$	AS2	
	$RT1^e = \quad H\text{-}1^e$	BD VII	
	$RT1^h = \quad H\text{-}1^h$	HW	

(continued)

Table II (*Continued*)

Locus	Allele	Strains[a]	Reference
RT2 (Ag-C)	*a*	ACP,AUG,A28807,BN,B3,CAR,F344,HW,LEW,MAXX,MNR, M520,OM,RHA,WKA,Z61	86; T.J. Gill, III, personal communication
	b	ACI,ALB,AS2,BDV,BDVII,CAS,COP,DA,HO,KGH,LA,LGE, LOU/C,NBR/1,NBR/2,NSD,OKA,PETH,PVG,RLA,SD,SHR, S5B,WAG,WF,YO	
Ag-F	*a*	ACI,A28807,DA,F344,M520,WF	35
	b	LEW	
	c	BN,MAXX	
	d	BUF	
Es-1	*a*	AUG,AVN BDVII,BDX,BN,BP,CAP,COP,DA,LEW	18,116
	o	ACI,AS,AS2,A28807,BDV,BS,BUF,DA,LEP,WAG/Rij	
Es-2	*a*	BDX	18
(see footnote *c*)	*b*	AVN,AUG,BD VII,BN,DA,LEP	
	c	LEW,WAG/Rij	
	d	CAP	
	l	BDV	
	t	BP	
Es-3	*a*	AUG,WAG	76
(see footnote *c*)	*b*	LE	
	c	BN,LEW	
Es-4	*a*	ACI,A28807,AUG,AVN,BDV,BD VII,BS,BUF	11,76
	b	AS,AS2,BN,LEW,LE	
Es-5	*a*	AS,BN,LEP,LEW	11
	b	ACI,AS2,A28807,AVN,BDV,BD VII,BS,BUF,COP,DA,WAG	
Gl-1	*a*	AUG,BN,LEW,WAG	76
	b	LE	
Hb	*A*	AVN,BD II,BDV,BD VII,CDF,WAG,Y59	
	B	AGA,AUG,BP,CAP,LEP,LEW	
Ig-1	*Ig-1*[a]	ALB,AS AS2,BN,BS,BUF,CAP,F344,LEW,LOU/C,LOU/M, M520,MSU,PVG/c,WAG,WF,YOS	10
	Ig-1	ACI,ACJ,ACP,AGUS,AVN,ACH,BD V,BD X,COP,DA,LEP, LOU/C/IH[b],NBR,SHR,OM,S5B,Z61,OKA	
Ig (kappa chain) (see footnote *d*)	*Ig-1a = RL-1 = SD-1 = RI-1*	ACI,ACJ,ALB,BD 1,BD V,CDF,COP,DA,F344,MSUBL,OFA, OKA,SHR,YOS	
	Ig-1b = RL-2 = W-1 = lk(ln)	ACP,AGUS,AO,AS,AS2,AUG,AVN,A28807,BDE,BD II, BD VII,BD X,BiCR/MIR (Marshall),BN,BS,BUF,CAP,E3, HW,HS,HO,LEW,LEP,LOU/c,LOU/M,M520,PB,PVG,WAG, WF,WP,Z61	1,10,45,90,115
MLR	*1*	KGH,LEW	27
	2	YOS,WF	
	3	BN,MAXX	
	4	ACI,DA	
	5	PVG,AUG,HO	
	6	ALB,BUF	
	7	WKA	
Svp-1	*a*	ACI,AS,AS2,A28807,AVN,AUG,BDV,BS,COP,DA,F344, WAG	10,38,73,75
	b	LEW,BN,LEP	
Tf (transferrin)	*F*	BS	
	S	ACI,AS,AS2,A28807,AVN,BDV,BD VII,BN,BUF,COP,DA, LEP,LEW,WAG	

(*continued*)

Table II (*Continued*)

Locus	Allele	Strains[a]	Reference
Pgd-1	F	LEP	11,60
	S	ACI,AS,AS2,A28807,AVN,BDV,BD VII,BN,BS,BUF,COP, DA,LEW,WAG	

[a]Congenic lines not included. See Table VII.
[b]This is a congenic line, deliberately bred to carry the blank allele of OKA rats.
[c]It is not clear at this stage how the various esterase loci correspond with those given in 60.
[d]Nomenclature not yet standardized.

Table III

Chromosome Polymorphisms: Combinations of Polymorphic Chromosome Markers in Twelve Strains of Inbred Rats[a,b]

						Strain						
Chromosome	ACI	ALB	BUF	F344	KYN	LEJ	NIG	SD[c]	T[d]	TO	W	WKA
No. 3	st	st	st	st	st	st	t	t	t	st	st	st
X	t	t	t	t	st	t	t	t	st	t	st	st
No. 4	−	−	+	+	−	−	+	−	−	−	+	+
No. 5	−	−	−	−	+	+	+	−	−	−	−	−
No. 7	−	−	+	−	+	+	+	−	−	+	+	+
No. 9	+	+	+	+	−	−	−	+	+	−	−	−

[a]From Sasaki (personal communication, 1977).
[b]Key to abbreviations: st, subtelocentric; t, telocentric; +, large C band; −, no or very small C band.
[c]Sprague-Dawley.
[d]Maintained by Dr. K. Takewaki, University of Tokyo.

Table IV

Coat Color Genes of Inbred Albino Rat Strains

	Coat color locus						Coat color locus				
Strain	A	B	H	P	Other	Strain	A	B	H	P	Other
AGUS	a	+	+	+		M520	a	+	h	+	+
AO	A	+	h	+		M14					
AS						M17					
AS2						NEDH					
AVN						NSD					
B						OM					
BDII	a	+	h			PA					
BDX	a	+	h	p	+	R					
BIRMA						SHR					
BROFO						S5B	a				
BUF		+				TO					
CAR						W					
CAS						WA	a	+	h	+	+
F344	a	+	h	+	+	WAG	a/+[b]	+	h/+[b]	+	+
GH						WF					
HCS						WKA					
LEW	a	+	h	+	+	WM					
MNR	+/a[a]	+	h	+	+	WN					
MR	a	+	h	+	+	YOS					
						Z61	a	+	h	+	+

[a]Sublines differ and one colony is segregating at this locus. Blanks indicate that information is not available.
[b]Depending on subline.

Table V

Synonyms for "Translation" of Older Strain Names[a]

Synonym	Current designation	Synonym	Current designation
ACP 9935 Irish piebald	ACP	Kyoto notched	KYN
Albany	ALB	Long Evans	LE
August	AUG	Long-Evans Praha	LEP
August 990	A990	Louvain	LOU
August 7322	A7322	Lewis	LEW
August 28807	A28807	Lister hooded	LIS
August 35322	A35322	Maudsley nonreactive	MNR
AxC 9935 Irish	ACI	Maudsley reactive	MR
AxC 9935 Piebald	ACH	Marshall 520	M520
B (black hooded)	PVG	NIH black	NBR
Berkeley S1	TMB or TS1	Okamoto (hypertensive)	SHR,
Berkeley S3	TMD or TS3		OKA/Wsl
Birmingham A	BIRMA	Osborne-Mendel	OM
Birmingham B	BIRMB	P.A. or King albino	PA
Black Praha	BP	SD/N	NSD
Brown Norway	BN	Selfed 36670	SEL
Buffalo	BUF	Slonaker	KX
CDF	F344	Spontaneously hypertensive rat	SHR
CPB-G	G	S1	TMB or TS1
CPB-WE	WE	S3	TMD or TS3
CPB-B	CPBB	Tokyo	TO
Copenhagen 2331	COP	Tryon Maze Bright	TMB or TS1
Fischer, Fischer 344	F344	Tryon Maze Dull	TMD or TS3
Genetic hypertension	GH	Wistar albino Boots	WAB
Harvard caries susceptible	HCS	Wistar albino Glaxo	WAG
Hunt's caries resistant	CAR	Wistar Furth	WF
Hunt's caries susceptible	CAS	Wistar King A	WKA
Iowa nonreactive	INR	YOS	DONRYU
Iowa reactive	IR	Zimmerman	Z61

[a] See also Fig. 40 and Table III.

Table VI

Some Holders of Inbred Strains of Rats[a]

Strain	Maintained by[b]	Strain	Maintained by[b]	Strain	Maintained by[b]	Strain	Maintained by[b]
ACI	N, PIT, MAX	A990	DU	BDVII	MAX, BU	CAR	N
ACP	PIT	A7322	DU	BDVIII	BU	CAS	N
AGA	?	A28807	PIT, DU	BDIX	BU	COP	MAX
AGUS	LAC	A35322	DU, CR	BDX	BU	CPBB	CPB
ACH	DU			BIRMA	BUA		
ALB	N, PIT, HOK	B	HAR	BIRMB	BUA	DA	HAN, PIT, IAP
AM	TOR	BDE	HAN	BN	N, HAN, ORL	DONRYU	SATO, DU
AMDIL	TOR	BDI	BU	BP	CUB		
AO	IAP	BDII	HAN, BU	BROFO	RIJ	E3	HAN
AS	OSU, MAX	BDIII	BU	BS	OSU, MAX	F344	N, LAC, HAN
AS2	OSU, MAX	BDIV	BU	BUF	N, PIT, MAX	G/Cpb	CPB
AUG	ORL, IAP	BDV	MAX, KIEL, BU	B3	PIT	GH	OMR
AVN	MAX	BDVI	BU	CAP	KIEL	GHA	QEH

(continued)

Table VI (*Continued*)

Strain	Maintained by[b]	Strain	Maintained by[b]	Strain	Maintained by[b]	Strain	Maintained by[b]
HCS	LHR	LGE	PIT	OKA/WSl	PIT	U	TOR
HO	IAP	LOU/C	WSL	OM	N, HAN		
HS	OSU	LOU/M	WSL	PA	CR	W	HOK, MS
		MAXX	PIT	PETH	N	WA	LAC
INR	HAR	MNR	N, BRH	PVG	LAC, PIT	WAB	BELL
IR	HAR	MNRA	HAR			WAG	LAC, ORL
IS	IS	MR	N, PIT, BRH	R	LN	WE/Cpb	CPB
K	HALLE	MSUBL	NOVO	RCS	HV	WF	PIT
KGH	PIT	M520	N, PIT	RHA/N	N	WKA	PIT, HOK, MS
KX	KNOX	M14	CP	RLA/N	N	WKY/N	N
KYN	HOK	M17	CP	SD	N ?	WM	MS
				SEL	DU	WN	N
LA/N	N	NBR	PIT	SHR	N, HAN, ORL	WR	?
LE	HAN, ORL	NIG-III	MS	S5B	CR		
LEJ	HOK, MS	NSD	N			YO	PIT
LEP	MAX	NZR/Gd	GD	TO	HOK	Y59	?
LEW	N, LAC, HAN			TMB	BRH	Z61	DU
				TMD	BRH		

[a]This list is not exhaustive. A maximum of three locations of any one stock have been given. Additional information may be given in current issues of *Rat News Letter* and the International Index of Laboratory Animals, both of which are obtainable from Medical Research Council, Laboratory Animals Centre, Woodmansterne Road, Carshalton, Surrey, SM5 4EF, England.

[b]Key to abbreviations: BELL, Professor D. Bellamy, Dept. of Zoology, University College, Cardiff, S. Wales; BRH, Professor P. L. Broadhurst, Dept. of Psychology, University of Birmingham, Birmingham B15 2TT, England; BU, Dr. Walter Burdette, University of Texas, M. D. Anderson Hospital and Tumor Institute, Houston, Texas 77025; BUA, Birmingham University, Dept. of Anatomy, The Medical School, Birmingham 15, England; CP, Dr. A. B. Chapman, Dept. of Genetics, University of Wisconsin, Madison, Wisconsin 53706; CPB, Dr. J. C. J. van Vliet, Centraal Proefdierenbedrig of TNO, P. O. B. 167, Zeist, The Netherlands; CR, Division of Cancer Treatment, National Cancer Institute, Bethesda, Maryland 20014; CUB, Charles University, Faculty of General Medicine, Dept. of Biology, Albertov 4, Prague 2, Czechoslovakia; DU, Dr. Wilhelmina F. Dunning, Papanicolaou Cancer Research Institute, 1445, N.W. 14th St., Miami, Florida 33136; GD, Dr. C. M. Goodall, National Cancer Research Laboratories, University of Otago, Dunedin, New Zealand; HALLE, Dr. E. Matthies, Martin-Luther-Universitat Halle-Wittenberg, Lehrstuhl für Industrietoxikologie, 402 Halle (Saale) Leninallee 4, DDR: HAN, Zentral institut für Versuchstiere Lettow-Vorbeck Allee 57, D-3000 Hanover 91, Germany; HAR, Dr. G. M. Harrington, University of Northern Iowa, Dept. of Psychology, Cedar Falls, Iowa 50613; HOK, Dr. M. Sasaki, Chromosome Research Unit, Faculty of Science, Hokkaido University, North 10 West 8, Sapporo, Japan; HV, Dr. R. Sidman, Children's Hospital Medical Centre, Dept. of Neuroscience, Boston, Massachusetts 02115; IAP, Dr. J. C. Howard, Dept. of Immunology, Agricultural Reseach Council, Institute of Animal Physiology, Babraham, Cambridge CB2 4AT, England; IS, Dr. M. Ishibashi, Laboratory of Animal Sciences, Azabu Veterinary College, Sagamichara, Kanagawa, Japan; KIEL, Prof. W. Müller-Ruchholtz, Division of Immunology, Hygiene Institute der University, Brunswikerstr. 2-6, D-2300 KIEL, D.D.R.; KNOX, Dr. W. E. Knox, New England Deacones Hospital, 194, Pilgrim Road, Boston, Massachusetts 02215; LAC, Medical Research Council, Laboratory Animals Centre, Woodmansterne Road, Carshalton, Surrey SM5-4EF, England; LHR, London Hospital Research Laboratories, Animal House, Ashfield Street, London E.1. England; LN, Dr. R. Leyten, Profdierencentrum KUL, de Croylaan 34, B-3030 Heverlee-Leuven, Belgium; MAX, Dr. E. Gunther, Max-Planck Institute für Immunologie, 78 Freiburg-Zahringen, Stubweg 51, West Germany; MS, Dr. T. H. Yosida, National Institute of Genetics, Misima, Sizouka-ken, Japan (411); N, N.I.H. Laboratory Animal Genetic Centre, National Institutes of Health, Bethesda, Maryland 20014; NOVO, Dr. P. M. Borodin, Institute of Cytology and Genetics, Siberian Branch of the Academy of Science, Novosibirsk, 630090, U.S.S.R.; OMR, University of Otago Medical School, Dept. of Medicine, Wellcome Medical Research Institute, P. O. Box 913, Dunedin, New Zealand; ORL, Centre de Selection et d'Elevage d'Animaux de Laboratoire C.N.R.S., 45045 Orleans Cedex, France; OSU, Dr. Barbara F. Heslop, Dept. of Surgery, University of Otago Medical School, P. O. Box 913, Dunedin, New Zealand; PIT, Professor T. J. Gill III, University of Pittsburgh, School of Medicine, Dept. of Pathology, 716A Scaife Hall, Pittsburgh, Pennsylvania 15261; QEH, The Queen Elizabeth Hospital, Woodville, S. Australia 5011; RIJ, Repgo Instituten TNO, Lange Kleiweg 151, Rijswijk (Z-H) The Netherlands; SATO, Dr. Sadako Sato, Nippon Rat Co. Ltd., 608-3, Negishi Urawa, Saitama 336, Japan; TOR, Dr. S. Torres, Instituto Nacional do Cancer Secao de Pesquisas, Pca, da Druz Vermelha 23, Rio de Janeiro, Guanabara, Brazil; WSL, Dr. Herve Bazin, Experimental Immunology Unit—UCL 3056, University of Louvain, Faculty of Medicine, 30 Clos Chapelle-aux-Champs, 1200 Bruxelles, Belgium.

Table VII

Some Congenic Strains[a]

Differential locus	Congenic line	Inbred background	Donor strain	Current status of breedng
$RT1^l(Ag$-$B1,H$-$1^l)$	BN.LEW	BN	LEW	N11
$RT1^u(Ag$-$B2,H$-$1^w)$	BN.WF	BN	WF	N11
$RT1^u(Ag$-$B2,H$-$1^w)$	LEW.1W	LEW	WP	N6F13
$RT1^u(Ag$-$B2,H$-$1^w)$	BN.YO	BN	YO38366	N7
$RT1^u(Ag$-$B2,H$-$1^w)$	PVG.B2=HO.B2	PVG=HO	AO	N6
$RT1^n(Ag$-$B3,H$-$1^n)$	LEW.BN	LEW	BN	N11
$RT1^n(Ag$-$B3,H$-$1^n)$	DA.B3	DA	B3	N5
$RT1^a(Ag$-$B4,H$-$1^a)$	BN.DA	BN	DA	N11
$RT1^a(Ag$-$B4,H$-$1^a)$	LEW.1A	LEW	AVN	N4F13
$RT1^a(Ag$-$B4,H$-$1^a)$	PVG.MNR	PVG	MNR	N4
$RT1^c(Ag$-$B5,H$-$1^c)$	BN.AUG	BN	AUG	N12F1
$RT1^b(Ag$-$B6,H$-$1^b)$	BN.BUF	BN	BUF	N11
$RT1^g(Ag$-$B7,H$-$1^g)$	BN.KGH	BN	KGH	N11
$RT1^g(Ag$-$B7,H$-$1^g)$	PVG.KGH	PVG	KGH	N11
$RT1^k(Ag$-$B8,H$-$1^k)$	BN.WKA	BN	WKA	N11
$RT1^k(Ag$-$B8,H$-$1^k)$	PVG.WKA	PVG	WKA	N11
$RT1^d(Ag$-$B9,H$-$1^d)$	LEW.1D	LEW	BD.V	N7F31N1F3
$RT1^f(Ag$-$B10,H$-$1^f)$	LEW.1F	LEW	AS.2	N9F11N1F3
$MLR\ 5^b$	DA.MNR	DA	MNR	N3
$MLR\ 4$	BN.B3	BN	B3	N3
Ag-$C2$	AUG.PVG	AUG	PVG	N4
Ag-$C1$	PVG.AUG	PVG	AUG	N4
RI-$1a$	PVG.1a	PVG	DA	N10
K^c	LOU/C.IK(OKA)	LOU/C	OKA	N7
H^d	LOU/C.IH(OKA)	LOU/C	OKA	N10
$K+H$	LOU/c.IH,K(OKA)	LOU/C	OKA	N7
$RT1^k$	LOU/C.RT1^k(OKA)	LOU/C	OKA	N5
H	LOU/C.IH(AUG)	LOU/C	AUG	N7
H	LOU/C.IH(AxC)	LOU/C	AxC[e]	N7
P	RCS-p/+	RCS	?	?

[a]T. J. Gill, personal communication (1977) and (108).

[b]Now discontinued.

[c]K, Kappa light chain of *Ig* locus.

[d]H, Heavy chains *Ig* loci.

[e]AxC, Probably ACH.

ACKNOWLEDGMENTS

I am indebted to numerous people for assistance in compiling this list of inbred rat strains. In some cases individuals have supplied me with data which has been incorporated without a specific acknowledgement. Particular thanks are due to the following: Professor M. Sasaki, Dr. P. M. Borodin, Dr. J. C. J. van Vliet, Dr. S. V. Hunt, Dr. H. Bazin, Mr. Roy Robinson, Dr. W. J. I. van der Gulden, Dr. R. Moutier, Professor P. L. Broadhurst, Dr. T. J. Gill III and his associates, Professor W. L. Ford, Dr. E. Günther, Dr. J. C. Howard, Dr. H. J. Hedrich, and Dr. G. M. Harrington. This listing has depended heavily on previous listings, particularly fourth listing (35) prepared in cooperation with Ms. Joan Staats and a substantial proportion of this chapter is reprinted, with permission, from Festing (33). Comments and/or corrections and additional information for a subsequent updating of this list will be welcomed.

REFERENCES

1. Armerding, D. (1971). Two allotypic specificities of rat immunoglobulin. *Eur. J. Immunol.* **1,** 39–45.

2. Armstrong, J. M., Blackwell, G. J., Flower, R. J., McGiff, J. C., Mullane, K. M., and Vane, J. R. (1976). Genetic hypertension in rats accompanied by a defect in renal prostaglandin catabolism. *Nature (London)* **260,** 582–586.

3. Bailey, D. W. (1971). Recombinant inbred strains: an aid to finding identity linkage and function of histocompatibility and other genes. *Transplantation* **11,** 325–327.

4. Bazin, H. (1974). Tumeurs sécrétant des immunoglobulines chez le rat LOU/Wsl. Etude portant sur 200 immunocytomes. *Ann. Immunol. Paris* **125 c,** 277–279.

5. Bazin, H. (1977), Inbred strains. *Rat News Lett.* **1,** 27.

6. Bazin, H., Beckers, A., Deckers, C., and Moriame, M. (1973). Transplantable immunoglobulin-secreting tumors in rats. V. Monoclonal immunoglobulins secreted by 250 ileocecal immunocytomas in LOU/Ws1 rats. *J. Natl. Cancer Inst.* **51,** 1359–1361.

7. Bazin, H., Deckers, C., Beckers, A., and Heremans, J. F. (1972). Transplantable immunoglobulin-secreting tumours in rats. I. General features of LOU/Ws1 strain rat immunocytomas and their monoclonal proteins. *Int. J. Cancer* **10,** 568–580.

8. Bazin, H., and Platteau, B. (1976). Production of circulating reaginic antibodies (IgE class) by oral administration of ovalbumin to rats. *Immunology* **30,** 679–683.

9. Bazin, H., Quérinjean, P., Beckers, A., Heremans, J. F., and Dessy, F. (1974). Transplantable immunoglobulin-secreting tumours in rats. IV. Sixty-three IgE-secreting immunocytoma tumours. *Immunology* **26,** 713–723.

10. Beckers, A., Quérinjean, P., and Bazin, H. (1974). Allotypes of rat immunoglobulins. II. Distribution of the allotypes of kappa and alpha chain loci in different inbred strains of rats. *Immunochemistry* **11,** 605–609.

11. Bender, K., and Günther, E. (1978). Screening of inbred rat strains for electrophoretic protein polymorphisms. *Biochem. Genet.* **16,** 387–398.

12. Billingham, R. E., and Silvers, W. K. (1959). Inbred animals and tissue transplantation immunity. *Transplant. Bull.* **6,** 399–406.

13. Boorman, G. A., and Hollander, C. F. (1973). Spontaneous lesions in the female WAG/Rij (Wistar) rat. *J. Gerontol.* **28,** 152–159.

14. Boorman, G. A., and Hollander, C. F. (1974). High incidence of spontaneous urinary bladder and ureter tumors in the brown Norway rat. *J. Natl. Cancer Inst.* **52,** 1005–1008.

15. Boorman, G. A., Zurcher, C., Hollander, C. F., and Feron, V. J. (1973). Naturally occuring endocardial disease in the rat. *Arch. Pathol.* **96,** 39–45.

16. Borodin, P. M. (1977). Strains. *Rat News Lett.* **1,** 49.

17. Braden, A. W. H. (1958). Strain differences in the incidence of polyspermia in rats after delayed mating. *Fertil. Steril.* **9,** 243–246.

18. Brdicka, R. (1974). Alleles of the second plasma esterase locus in the laboratory rat and their distribution among some inbred strains. *Folia Biol. (Prague)* **20,** 350–353.

19. Broadhurst, P. L. (1975). The Maudsley reactive and non-reactive strains of rats: A survey. *Behav. Genet.* **5,** 299–319.

20. Burek, J. D., and Hollander, C. F. (1975). "Studies of Spontaneous Lesions in Ageing BN/Bi Rats. I. Neoplastic and Non-neoplastic Lesions," pp. 235–237. Annu. Rep. Org. Health Res., TNO, Rijswijk, The Netherlands.

21. Burek, J. D., and Hollander, C. F. (1975). "Studies of Spontaneous Lesions in Ageing BN/Bi Rats. II Age-associated Incidence of Tumours and Tumour Metastases," pp. 238–241. Annu. Rep. Org. Health Res., TNO, Rijswijk, The Netherlands.

22. Burek, J. D., and Hollander, C. F. (1977). Incidence patterns of spontaneous tumors in BN/Bi rats. *J. Natl. Cancer Inst.* **58,** 99–105.

23. Burek, J. D., Zurcher, C., and Hollander, C. F. (1976). High incidence of spontaneous cervical and vaginal tumors in an inbred strain of Brown Norway rats (BN/Bi). *J. Natl. Cancer Inst.* **57,** 549–554.

24. Cohen, B. J., DeBruin, R. W., and Kort, W. J. (1970). Heritable hydronephrosis in a mutant strain of brown Norway rats. *Lab. Anim. Care* **20,** 489–493.

25. Committee on the Care and Use of Spontaneously Hypertensive (SHR) Rats (1976). Spontaneously hypertensive (SHR) rats: Guidelines for breeding, care and use. *ILAR News* **19,** G1–G20.

26. Cramer, D. V., and Gill, T. J., III (1975) Genetics of urogenital abnormalities in ACI inbred rats. *Teratology* **12,** 27–32.

27. Cramer, D. V., Shonnard, J. W., Davis, B. K., and Gill, T. J., III (1977). Polymorphism of the mixed lymphocyte response of wild Norway rats. *Transplant. Proc.* **9,** 559–562.

28. Davey, F. R., and Moloney, W. C. (1970). Postmortem observations on Fischer rats with leukemia and other disorders. *Lab. Invest.* **23,** 327–334.

29. Druckrey, H. (1971). Genotypes and phenotypes of ten inbred strains of BD rats. *Arzneim-Forsch.* 21, 1274–1278.

30. Dunning, W. F., Curtis, M. R., and Segaloff, A. (1947). Strain differences in response to diethylstilbestrol and the induction of mammary gland and bladder cancer in the rat. *Cancer Res.* **7,** 511–521.

31. Esber, H. J., Menninger, F. F., Jr., and Bogden, A. E. (1974). Variation in serum hormone concentrations in different rat strains. *Proc. Soc. Exp. Biol. Med.* **146,** 1050–1053.

32. Festing, M. F. W. (1972). Mouse strain identification. *Nature (London)* **238,** 351–352.

33. Festing, M. F. W. (1979). "Inbred Strains in Biomedical Research." Macmillan, New York.

34. Festing, M. F. W., and Blackmore, D. K. (1971). Lifespan of specified-pathogen-free (MRC category 4) mice and rats. *Lab. Anim.* **5,** 179–192.

35. Festing, M. F. W., and Staats, J. (1973). Standardized nomenclature for inbred strains of rats fourth listing. *Transplantation* **16,** 221–245.

36. Fregly, M. J. (1975). Thyroid activity of spontaneous hypertensive rats. *Proc. Soc. Exp. Biol. Med.* **149,** 124–132.

37. Friedman, I., Ron, N., Laufer, A., and Davies, A. M. (1970). Experimental myocarditis: Enhancement by the use of pertussis vaccine in Lewis rats. *Experientia* **26,** 1143–1145.

38. Gasser, D. L. (1972). Seminal vesicle protein in rats: A gene in the fourth linkage group determining electrophoretic variants. *Biochem. Genet.* **6,** 61–63.

39. Gasser, D. L., Palm, J., and Gonatas, N. K. (1975). Genetic control of susceptibility to experimental allergic encephalomyelitis and the *Ag-B* locus of rats. *J. Immunol.* **115,** 431–433.

40. Gill, T. J., III, Kunz, H. W., Shonnard, J. W., Davis, B. K., and Hansen, C. T. (1977). Immunogenetic studies of the recombinant B3 strain of rats. *Transplant. Proc.* **9,** 567–569.

41. Gold, G., and Widnell, C. C. (1975). Response of NADPH-cytochrome *c* reductase and cytochrome *P450,* in hepatic microsomes to treatment with phenobarbital differences in rat strains. *Biochem. Pharmacol.* **24,** 2105–2106.

42. Goodall, C. M., Christie, G. S., and Hurley, J. V. (1975). Primary epitheliel tumour in the right atrium of the heart and inferior vena cava in NZR/Gd inbred rats: Pathology of 18 cases. *J. Pathol.* **116,** 239–252.

43. Grobecker, H., Roizen, M. F., Weise, V., Saavedra, J. M., and Kopin, I. J. (1975). Sympathoadrenal medullary activity in young, spontaneously hypertensive rats. *Nature (London)* **258,** 267.

44. Günther, E., Mozes, E., Rude, E., and Sela, M. (1976). Genetic control of immune responsiveness to poly(L-Pro)-poly(L-Lys)-derived polypeptides by histocompatibility-linked immune response genes in the rat. *J. Immunol.* **117,** 2047–2052.

45. Gutman, G. A., and Weissman, I. (1971). Inheritance and strain distribution of a rat immunoglobulin allotype. *J. Immunol.* **107,** 1390–1393.

46. Hall, C. E., Ayachi, S., and Hall, O. (1976). Immunity of Fischer 344 rats to salt hypertension. *Life Sci.* **18,** 1001–1008.

47. Hansen, C. T., Judge, F. J., and Whitney, R. A. (1973). "Catalogue of NIH Rodents," DHEW Publ. No. (NIH) 74-606. Natl. Inst. Health, Bethesda, Maryland.

48. Harrington, G. M. (1971). Strain differences among rats initiating exploration of differing environments. *Psychon. Sci.* **23,** 348–349.

49. Harrington, G. M. (1971). Strain differences in rotating wheel activity of the rat. *Psychon. Sci.* **23,** 363–364.

50. Harrington, G. M. (1972). Strain differences in open-field behaviour of the rat. *Psychon. Sci.* **27**, 51–53.

51. Harris, J. M., Kalmus, H., and West, G. B. (1963). Genetic control of the anaphylactoid reaction in rats. *Genet. Res.* **4**, 346–355.

52. Heslop, B. F., and Phelan, E. L. (1973). The GH and AS hypertensive rat strains. *Lab. Anim.* **7**, 41–46.

53. Hollander, C. F. (1976). Current experience using the laboratory rat in aging studies. *Lab. Anim. Sci.* **26**, 320–328.

54. Howard, J. C. (1971). A histocompatible chromosome marker system in the laboratory rat. *Rattus norvegicus. Transplantation* **12**, 95–97.

55. Hughes, R. A. C., and Stedronska, J. (1973). The susceptibility of rat strains to experimental allergic encephelomyelitis. *Immunology* **24**, 879–884.

55a. Ishibashi, M. (1976). ''Ishibashi rat''—A new strain of rats with congenital spinal malformations. *Teratology* 14, 242 (abstract).

56. Jacobs, B. B., and Huseby, R. A. (1967). Neoplasms occurring in aged Fischer rats, with special reference to testicular, uterine and thyroid tumours. *J. Natl. Cancer Inst.* **39**, 303–307.

57. Kim, U., Clifton, K. H., and Furth, J. (1960). A highly inbred strain of Wistar rats yielding spontaneous mammosomatotropic pituitary and other tumours. *J. Natl. Cancer Inst.* **24**, 1031–1055.

58. Knox, W. E. (1977). *Rat News Lett.* **1**, 26.

59. Kren, V., Krenova, D., and Stark, O. (1970). Properties of sarcomas induced by ferridextran spofa in the inbred rat strain AVN. I. Transplantability of fedex—AVN tumours *in vivo* and karyological and morphological properties of the cell line after explantation *in vitro*. *Neoplasma* **17**, 329–337.

60. Krog, H. H. (1977). 6-phosphogluconate dehydrogenase (Pgd-1) in the rat and its possible linkage. *Hereditas* **85**, 207–210.

61. Kunz, H. W., and Gill, T. J., III (1974). Genetic studies in inbred rats. I. Two new histocompatibility alleles. *J. Immunogenet.* **1**, 413–420.

62. Kunz, H. W., Gill, T. J., III, and Borland, B. (1974). The genetic linkage of the immune response to poly C(Glu^{52}, Lys^{33}, Tyr^{15}) to the major histocompatibility locus in inbred rats. *J. Immunogenet.* **1**, 277–287.

63. LaVail, M. M., Sidman, R. L., and Gerhardt, C. O. (1975). Congenic strains of RCS rats with inherited retinal dystrophy. *J. Hered.* **66**, 242–244.

64. Lindsay, S., Nichols, C. W., Jr., and Chaikoff, I. L. (1968). Naturally occuring thyroid carcinoma in the rat. Similarities to human medullary carcinoma. *Arch. Pathol.* **86**, 353–364.

65. Maekawa, A., and Odashima, S. (1975). Spontaneous tumours in ACI/N rats. *J. Natl. Cancer Inst.* **55**, 1437–1445.

66. Martin, J. B., Harris, C. A., and Dirks, J. H. (1974). Evidence of G. H. deficiency in the Munich-Wistar rat. *Endocrinology* **94**, 1359–1363.

67. Matthies, E., and Ponsold, W. (1973). Zur Ubertragung der Resistenz gegen inpftamoren bei ratten. *Z. Krebsforsch.* **80**, 27–30.

68. McFarlin, D. E., Chung-Ling Hsu, S., Slemenda, S. B., Chou, S. C.-H., and Kibter, R. F. (1975). The immune response against an encephalitogenic fragment of guinea-pig basic protein in the Lewis and Brown Norway strains of rat. *J. Immunol.* **115**, 1456–1458.

69. Miyamoto, M., and Takizawa, S. (1975). Colon carcinoma of highly inbred rats. *J. Natl. Cancer Inst.* **55**, 1471–1472.

70. Moloney, W. C., Boschetti, A. E., and King, V. P. (1969). Observations on leukemia in Wistar Furth rats. *Cancer Res.* **29**, 938.

71. Moloney, W. C., Boschetti, A. E., and King, V. P. (1970). Spontaneous leukemia in Fischer rats. *Cancer Res.* **30**, 41–43.

72. Molteni, A., Nickerson, P. A., Gallant, S., and Brownie, A. C. (1975). Resistance of W/Fu rats to adrenal regeneration hypertension. *Proc. Soc. Exp. Biol. Med.* **150**, 80–84.

73. Moutier, R., Toyama, K., and Charrier, M. F. (1971). Controle genetique des proteines de la secretion des glandes séminales chez la souris et le rat. *Exp. Anim.* **4**, 7–18.

75. Moutier, R., Toyama, K., and Charrier, M. F. (1973). Biochemical polymorphism in the rat *Rattus norvegicus*. Genetic study of four markers. *Biochem. Genet.* **8**, 321–328.

76. Moutier, R., Toyama, K., and Charrier, M. F. (1973). Evidence for linkage between four esterase loci in the rat, *Rattus norvegicus*. *Biochem. Genet.* **9**, 109–115.

77. Nagaoka, A., Iwatsuka, H., Suzuoki, A., and Okamoto, K. (1976). Genetic predisposition to stroke in spontaneously hypertensive rats. *Am. J. Physiol.* **230**, 1354–1359.

78. Neal, R. A., and Harris, W. G. (1975). Proceedings: Attempts to infect inbred strains of rats and mice with *Entamoeba histolytica. Trans. R. Soc. Trop. Med. Hyg.* **69**, 429–430.

79. Nobile, B., Yoshida, T., Rose, N. R., and Bigazzi, P. E. (1976). Thyroid antibodies in spontaneous autoimmune thyroiditis in the Buffalo rat. *J. Immunol.* **117**, 1447–1455.

80. Okamoto, K. (1969). Spontaneous hypertension in rats. *Int. Rev. Exp. Pathol.* **7**, 227–270.

81. Okamoto, K., Yamovi, Y., Nosaka, S., Ooshima, A., and Hazama, F. (1973). Studies on hypertension in spontaneously hypertensive rats. *Clin. Sci. Mol. Med.* **45**, 115–145.

82. Page, J. G., and Vesell, E. S. (1969). Hepatic drug metabolism in ten strains of Norway rats before and after pretreatment with phenobarbital. *Proc. Soc. Exp. Biol. Med.* **131**, 256–261.

83. Palm, J., and Black, G. (1971). Interrelationships of inbred rat strains with respect to Ag-B and non Ag-B antigens. *Transplantation* **11**, 184–189.

84. Perlik, F., and Zideck, Z. (1974). The susceptibility of several inbred strains of rats to adjuvant-induced arthritis and experimental allergic encephalomyelitis. *Z. Immunitactsforsch., Exp. Klin. Immunol.* **147**, 191–193.

85. Phelan, E. L. (1968). The New Zealand strain of rats with genetic hypertension. *N. Z. Med. J.* **67**, 334.

86. Poloskey, P. E., Kung, H. W., Gill, T. J., III, Shonnard, J. W., Hansen, C. T., and Dixon, B. D. (1975). Genetic studies in inbred rats. IV. Xenoantisera against erythrocyte (Ag-C) antigens. *J. Immunogenet.* **2**, 179–187.

87. Rick, J. T., and Fulker, D. W. (1972). Some biochemical correlates of inherited behavourial differences. *Prog. Brain Res.* **36**, 105–112.

88. Rick, J. T., Tunicliff, G. Kerkut, G. A., Fulker, D. W., Wilcock, J., and Broadhurst, P. L. (1971). GABA production in brain cortex related to activity and avoidance behaviour in eight strains of rat. *Brain Res.* **32**, 234–238.

89. Roba, J. L. (1976). The use of spontaneously hypertensive rats for the study of anti-hypertensive agents. *Lab. Anim. Sci.* **26**, 305–319.

90. Rokhlin, O. V., Vengerova, T. I., and Nezlin, R. S. (1971). RL allotypes of light chains of rat immunoglobulins. *Immunochemistry* **8**, 525–538.

91. Rose, N. R. (1975). Differing responses of inbred rat strains in experimental autoimmune thyroiditis. *Cell. Immunol.* **18**, 360–364.

92. Russell, K. E., and Stern, M. H. (1973). Sex and strain as factors in voluntary alcohol intake. *Physiol. Behav.* **10**, 641–642.

93. Sacksteder, M. R. (1976). Occurrence of spontaneous tumours in the germ-free F344 rat. *J. Natl. Cancer Inst.* **57**, 1371–1373.

94. Sallman, L. von, and Grimes, P. (1972). Spontaneous retinal degeneration in mature Osborne-Mendel rats. *Arch. Ophthalmol.* **88**, 404–411.

95. Sass, B., Rabstein, L. S., Madison, R., Nims, R. M., Peters, R. L., and Kelloff, G. J. (1975). Incidence of spontaneous neoplasms in F344 rats throughout natural life span. *J. Natl. Cancer Inst.* **54**, 1449–1456.

96. Schemmel, R., Mickelsen, O., and Gill, J. L. (1970). Dietary obesity

in rats: body weight and body fat accretion in seven strains of rats. *J. Nutr.* **100**, 1041–1048.

97. Shain, S. A., McCullough, B., and Segaloff, A. (1975). Spontaneous adenocarcinomas of the ventral prostate of aged A × C rats. *J. Natl. Cancer Inst.* **55**, 177–180.

98. Shearer, D., Creel, D., and Wilson, C. E. (1973). Strain differences in the response of rats to repeated injections of pentobarbital sodium. *Lab. Anim. Sci.* **23**, 661–664.

99. Shoji, R. (1977). Spontaneous occurrence of congenital malformations and mortality in prenatal inbred rats. *Proc. Jpn. Acad. Sci.* **53**, 54–67.

100. Shoji, R., and Harata, M. (1977). Abnormal urogenital organs occuring spontaneously in inbred ACI and Kyoto-notched rats. *Lab. Anim.* **11**, 247–249.

101. Sidman, R. L., and Pearlstein, R. (1965). Pink-eyed dilution (p) gene in rodents: Increased pigmentation in tissue culture. *Dev. Biol.* **12**, 93–116.

102. Silverman, D. A., and Rose, N. R. (1975). Spontaneous and methylcholanthrene-enhanced thyroiditis in BUF rats. I. The incidence and severity of the disease, and the genetics of susceptibility. *J. Immunol.* **114**, 145–147.

103. Simpson, F. O., Phelan, E. L., Clark, D. W. J., Jones, D. R., Gresson, C. R., Lee, D. R., and Bird, D. L. (1973). Studies on the New Zealand strain of genetically hypertensive rats. *Clin. Sci. Mol. Med.* **45**, 155–215.

104. Staats, J. (1976). Standardized nomenclature for inbred strains of mice: Sixth listing. *Cancer Res.* **36**, 4333–4377.

105. Starace, V., and Quérinjean, P. (1975). The primary structure of a rat *K* Bence-Jones protein: Phylogenetic relationships of *V*- and *C*-region genes. *J. Immunol.* **115**, 59–62.

106. Stark, O., Frenzl, B., and Kren, V. (1968). Erythrocyte and transplantation antigens in inbred strains of rats. VII. *H-1* alleles of the LEP, CAP, BN, BDV, BD VII and BD X strains. *Folia Biol. (Prague)* **14**, 169–175.

107. Stark, O., Kren, V., Frenzl, B., and Krenova, D. (1968). Histocompatibility. I. Alleles in inbred rats of the BDII, WR, AGA, CDF, AS, BS, HS and DA strains. *Folia Biol. (Prague)* **14**, 425–432.

108. Stark, O., and Kren, V. (1969). Five congenic resistant lines of rats differing at the *Rt H-1* locus. *Transplantation* **8**, 200–203.

109. Stark, O., and Zeiss, I. (1970). Antigens des ersten histokompatibilitats systems (H-1) bei den BD (Berlin Druckrey) und BDE—ratteninzuchstammen. *Z. Versuchstierkd.* **125**, 27–40.

110. Stark, O., Kren, V., and Günther, E. (1971). Rt.H-1 Antigens in 39 rat strains and six congenic lines. *Transplant. Proc.* **3**, 165–168.

111. Stenglein, B., Thoenes, G. H., and Günther, E. (1975). Genetically controlled autologous immune complex glomerulonephritis in rats. *J. Immunol.* **115**, 895–897.

112. Tada, N., Itakura, K., and Aizawa, M. (1974). Genetic control of the antibody response in inbred rats. *J. Immunogenet.* **1**, 265–375.

113. Van Vliet, J. C. J. (1977). Strains/stocks/mutants. *Rat News Lett.* **1**, 63.

114. Werner, S. E., Manger, W. M., Radichevich, I., Wolff, M., and von Estorff, I. (1975). Excessive thyrotropin concentrations in the circulation of the spontaneously hypertensive rat. *Proc. Soc. Exp. Biol. Med.* **148**, 1013–1017.

115. Wistar, R., Jr. (1969). Immunoglobulin allotype in the rat: Localization of the specificity to the light chain. *Immunology* **17**, 23–32.

116. Womack, J. E. (1973). Biochemical genetics of rat esterases: polymorphism, tissue expression, and linkage at four loci. *Biochem. Genet.* **9**, 13–24.

117. Yen, T. T., Roeder, H., and Willard, P. W. (1974). A genetic study of hypertension in Okamoto-Aoki spontaneously hypertensive rats. *Heredity* **33**, 309–316.

118. Zeiss, I. M. (1966). The fate of simultaneous and successive male to female skin grafts in an inbred strain of rats. *Transplantation* **4**, 48–55.

119. Zieverink, W. D., and Moloney, W. C. (1965). Use of the Y chromosome in the Wistar Furth rat as a cellular marker. *Proc. Soc. Exp. Biol. Med.* **119**, 370–373.

Chapter 4

Morphophysiology

W. Sheldon Bivin, M. Pat Crawford, and N. R. Brewer

I. INTRODUCTION

Since the early 1920s, several textbooks and laboratory manuals have been published on the anatomy and physiology of the rat (27,38,43,60,72,132,134,140,147,165). This chapter will describe and illustrate the most important anatomical and physiological features of the laboratory rat, providing a summary of previous findings and updating the information where necessary.

New experimental methods now make it possible to observe blood pressure, spinal fluid pressure, blood volume, hormone levels, brain learning centers, and other parameters in the rat which were previously limited to larger animals. Therefore, an additional goal of this chapter is to provide current information available on those anatomical and physiological characteristics of the laboratory rat which make it of particular value as a research animal model.

II. GENERAL APPEARANCE

A. Head and Body

The laboratory rat, *Rattus norvegicus,* is characterized by having a fusiform body consisting of a tapered head which blends into a slender trunk. The long rasplike tail may constitute 85% of the length of the body, being somewhat longer in the female than in the male.

The integument consists of typical mammalian hair dispersed over the body surface except for the nose, lips, palms, and soles. The hair is arranged by two classes, that with and that without long hair shafts. Hair growth has been shown to be cyclic with the resting period and the growing stage each being about 17 days in length. For a more complete understanding of hair growth and patterns in the rat, the reader is referred to the works of Fraser (49), Butcher (22), and Greene (60).

In certain regions, the bristles are developed as specialized tactile hairs (pili tactiles or vibrissae). According to Pococks' terminology, these vibrissae are arranged in four groups: the buccal vibrissae which are subdivided into mystacial and submental, the superciliary, and genal, and the interramal.

It is normal for the eyes to bulge in the rat. One reason for this appearance is a greatly reduced nictitating membrane called the plica semilunaris. The eyelids are well developed and have very fine short eyelashes. Located within the eyelids are the large tarsal or meibomian glands. By frequent winking, the eye is kept moist with secretions from the laterally placed lacrimal and medially placed Harderian glands.

The external nares are shaped like inverted commas and open on the lateral aspect of a rather mobile nose. The upper lip is usually cleft in the center by a vertical groove called the philtrum which ends just below the nares.

The surface of the rat snout pad is ridged in a fashion similar to the dermatoglyphic pattern found in human and primate digital skin. These epidermal and dermal ridges almost certainly play a role in sensory discrimination. This relationship provides further evidence of the parallel between fingertip function in bipeds and snout tip function in quadrupeds (96).

Both fore- and hindlimbs have five digits. The first digit "thumb or polex" is much reduced on the forelimb and has a flattened nail unlike the more rounded nails of the other digits. Typical walking pads (tori) are present. The fore limb has five apical pads, three interdigital pads, and two basal pads.

The most dorsal of the caudal body openings is the anus. This structure is easily visible in the female, but may be partly obscured in the mature male by the large scrotum.

B. External Reproductive Genitalia

The vaginal orifice is located about 7 mm ventral to the anus. Approximately 4 mm cranioventral to the vagina is the small prepuce containing the clitoris. The urethral orifice is located at the base of the clitoris and is directed ventrocaudally. The teats of the rat are usually twelve in number, three pair in the pectoral region and three pair in the abdominal region.

The size of the scrotum is directly related to age and whether the testis have descended or not. Immediately cranial to the scrotum and on a midline is the prepuce containing the penis. The urethral orifice opens through the end of the penis which is directed ventrocaudally. On the ventral wall of the penis is a single cartilaginous or bony process, the os penis.

C. Skeletal Structure

The skeleton of the rat differs very little from that of other four-footed animals. The vertebral formula is C_7 T_{13} L_6 S_4 Cy^{27-30}. Most of those differences observed are the result of the stance or the rodent profile. One minor difference is that the rat is slower than most other mammals in bone maturation. Ossification is not complete until after the first year of life (156). The dorsal segment of the ribs is completely ossified, and the ventral segment is usually calcified. True costal cartilages are absent in the rat. The clavicle and its associated structures extend as a chain between the sternum and the scapula in the following order: osmosternum, proximal procoracoid piece, clavicle, distal procoracoid piece. Of these, the clavicle and the osmosternum are ossified and the others are cartilaginous (43).

In the forelimb, the scapula is normally in a rather horizontal position. The acromion and coracoid processes are quite large and form a deep socket for the head of the humerus. A large deltoid tuberosity is easily identified on the humerus. The hindlimb presents two differences. One is that the femur has extremely large greater, lesser, and third trochanters coupled with a very small head and neck. Second, the tibia and fibula are fused in their distal quarter. For more detail concerning the skeleton of the rat, the reader is referred to the works of Greene (60) and Chiasson (27).

D. Musculature

A comprehensive description of the muscles has been published in Greene (60) and updated by Chiasson (27). Although the descriptions of the muscles of the rat are well described and illustrated in the above texts, the terminologies used are often antiquated and should be clarified with nomina anatomica before being used in current presentations.

The growth pattern of the rat has been studied by several authors with differences being recorded in the rates of development depending on the strain. The maximum body weight of the Lister strain, for example, is attained later in life than in the black-hooded rat (73,74); while the weight appears to increase throughout the life in WAG-C rats (98).

One author reports body length to follow a direct relationship with the square root of body weight in the CFHB Wistar rats (125). Another reports body length to follow a direct relationship with the cube root of weight in Lister rats (129).

Researchers have also observed that prolonged exposure to ambient temperatures caused an overall reduction in the rate of growth in albino rats. The tail appears to suffer most, suggesting that reduction in its length and surface area may prevent excessive heat loss (88).

Such differences in growth rate between the various species may explain the variation in times at which, for example, muscles exhibit maturity in terms of their histochemical appearance (124,125).

E. Superficial Glands of the Head and Neck

This group comprises the orbital glands, lateral nasal glands, salivary and lymphatic glands of the neck and the multilocular adipose tissue or "hibernating gland."

Just ventral and rostral to the ear lies the exorbital lacrimal gland (Fig. 1). Its duct joins with that of the intraorbital lacrimal gland to open onto the conjunctiva in the dorsolateral region of the eye. The intraorbital gland occupies the caudal

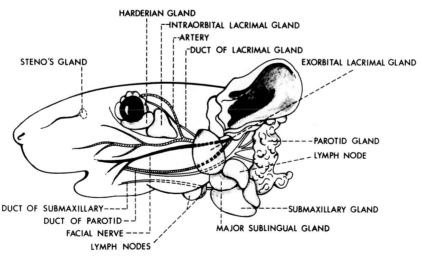

Fig. 1. Superficial glands of the head and neck (lateral view). [Redrawn from Greene (60).]

angle of the orbit and it is covered by a connective tissue sheath. The lacrimal duct epithelium in the rat is thought to transport sodium, potassium, and chloride ions. A study of the fine structure of this duct system has shown the basal layer of the intercalated ducts to consist of myoepithelial cells, the clavicle. The parotid duct is formed by the union of three and an electrolyte transporting function. The presence of nerve terminals in the epithelium observed within these ducts suggests that the duct system of the rat exorbital gland is active in the production of lacrimal fluid.

The Harderian gland is horseshoe-shaped and occupies a large portion of the orbit as it extends medially and very deep to encircle the optic nerve. It is thought that this gland is present in most mammals except Chiroptera and Simidae. The secretory cells of the Harderian gland have a small fat globule in their center which may account for the gland being well developed in swimming animals. Three cell types have been described for this gland, designated A, B and C. It is thought that the Harderian gland may not only have an apocrine (or merocrine) secretion, but also a holocrine mode of secretion (108).

Although the rat has several rather well-developed rostral nasal glands, the largest gland is the glandula nasalis lateralis or Steno's gland (Fig. 1). This gland is situated in the wall of the rostral portion of the maxillary sinus and has an excretory duct which courses to the vestibule along the root of the nasoturbinate. Steno's gland is characterized by cytological features quite similar to those described for the major serous salivary glands and is homologous with the salt gland found in marine birds. It produces a watery, nonviscous secretion which is discharged at the entrance to the nasal airway. At this point, the secretion must contribute to the humidification of the inspired air. Further, it may contribute to the maintenance of the proper viscosity of the mucous blanket covering the ciliated surface of the respiratory region. A great number of autonomic nerves have been described in close contact with the acinar cells and indicate that the gland of Steno is regulated by the nervous system in such a way that rapid adjustment of the secretory activity to changes in the humidity of the inhaled air or to airborne irritants is possible (104).

There are three pairs of salivary glands in the rat. The parotid gland is a diffuse structure which extends ventrorostrally from behind the ear to course along the caudal facial vein and end on the ventrolateral surface of the neck. At times the caudal border of the gland may reach the shoulder and cover a portion of the clavicle. The parotid duct is form by the union of three principle branches and crosses the lateral surface of the masseter muscle, in close association with the buccal and mandibular branches of the facial nerve, to open into the vestibule just opposite the molar teeth (Figs. 1 and 2).

In the rat, the parotid and the submaxillary salivary glands are of about equal size. Rat parotid saliva has a protein concentration of about 2%, and this is unique among animals so far tested (61).

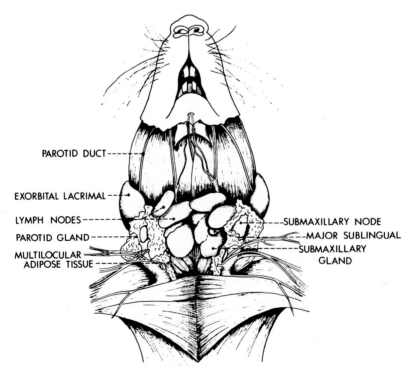

Fig. 2. Superficial lymph nodes and glands of the neck (ventral view). [Redrawn from Green (60).]

The second group is called the submaxillary or submandibular glands (Fig. 2). These are large elongated glands which lie just caudal to the angle of the mandible and extend caudally to the cranial extent of the manubrium. Two types of secretory granules occur in the submandibular gland of the adult rat; one type is found in acinar cells and the other in the granulated portion of secretory ducts. The granules of the acinar cells are of a mucous nature and those of the ducts are serous. Duct granules contain unusual substructures which are not found in the secretion granules of adult rats and suggest that these granules are transitory (82).

The sublingual glands are small, rounded glands located at the rostal edge of the submandibular glands and may be partially embedded in them. Their ducts parallel those of the sublingual glands to open on the plica sublingualis (75).

Other structures in the head region which may be confused with the salivary glands are the irregularly shaped, compact lymph nodes (Fig. 2). One lymph node is found partially embedded in the parotid salivary gland. Two other nodes may be found just rostral to the submaxillary gland. Additional smaller nodes may be seen on occasion accompanying the blood vessels of the head and neck.

F. Multilocular Adipose Tissue

Multilocular adipose tissue, sometimes referred to as the "hibernating gland" or brown fat is diffusely distributed throughout the ventral, lateral, and dorsal aspects of the neck and is a chief component found between the scapulae. This tissue contains a pigment which gives it a brown color. Microscopically, it is somewhat suggestive of an endocrine gland as the fat cells are filled with small lipid droplets giving it a multilocular appearance. Its significance as a hibernating gland is questionable in the rat, since the rat is not a hibernating animal. However, it has been shown that the multilocular tissue is critical to the life of the rat. One thought is that it may serve as an economizer of proteins by ensuring utilization of reserve carbohydrates and fats. In this regard, the brown fat plays a major role in thermogenesis during cold exposure. This thermogenic response is thought to be triggered by the innervation of brown adipose cells by sympathetic nerves (145). It also has the ability to modify the effects of various chemical agents on the body, accelerating some and retarding others (127,128). While the common fat tissue loses or accumulates neutral fat with changes in the nutritional condition of the animal, these factors do not seem to affect the multilocular adipose tissue. It has been claimed that this multilocular adipose tissue loses its lipids after hypophysectomy and that these changes can be prevented by injection of adrenocorticotropic hormone (44,45).

III. THE DIGESTIVE SYSTEM

No attempt has been made to present a detailed description of the viscera; however, special attention has been drawn to those areas in which the laboratory rat differs from the other familiar laboratory animals. Some of the more important differences include a lack of tonsils, no gallbladder, an extremely diffuse pancreas, and the presence of a number of accessory glands.

Rodents are semicontinuous feeders. Normally, a 400-gm rat eats about 20 gm of dry food daily (about 40 gm of diet including liquids) containing approximately 50 kcal. A 300-gm rat eats up to 4 gm of food at each feeding. They can be tube-fed 13 ml of water or be trained to take up to 26 ml at a time (72). Water turnover in a 300 gm rat is about 24 ml/24 hr (114 ml/kg/24 hr) in an ambient temperature of 25°C. Emotional stress increases drinking in laboratory rats. According to Chew (26), rats die sooner when they are deprived of food than when they are deprived of water.

The digestive system is described in sequential order from the oral cavity to the anus and rectum, and is followed by a discussion of the accessory organs associated with the alimentary tract.

A. Oral Cavity

Rostrally the oral cavity is bounded by a cleft upper lip, the philtrum, and an intact lower lip. The space between the lips, cheeks, and teeth is the vestibule. A single pair of well-developed incisor teeth is present on each jaw. Due to the absence of enamel on one surface, a sharp cutting edge is maintained as the teeth are used. The pulp cavity for these teeth remains open and growth continues, making it essential that the teeth are aligned to allow for proper wear. Canine teeth are not present, but caudal to the incisors is a space, the diastema, into which folds of mucosa extend. These cheek folds tend to separate the gnawing apparatus from the caudal part of the buccal cavity. This arrangement of the teeth is characteristic for rodents. The dental formula throughout the life of the animal is

$$2(I\ 1/1,\ C\ 0/0,\ Pm\ 0/0,\ M\ 3/3) = 16$$

The tongue is characterized by a single circumvallate papilla near the base of the tongue, fungiform papillae on the rostral surface, numerous filiform papillae over the dorsum of the tongue and scattered foliate papillae (152). Since each fungiform papilla in the rat has a single taste bud, the spatial distribution of fungiform papillae is equivalent to the location of taste buds on the rostral portion of the tongue. A mean total number of 187 fungiform papillae per tongue has been described for an average density of 3.4 papillae/mm². In the last two decades, the rat fungiform papillae have been used extensively as a model system of the mammalian taste receptor (103). Water taste receptors which have been identified in other animals (dog, cat, pig, etc.) have not been identified in the rat, and it is believed that rats cannot taste water (90). There is no medial fraenum, but two lateral ones extend from near the tip of the band which ties the lower lip to the gum. There are no sublingua; however, a small pair of salivary papillae lie close to the median line behind the incisors.

The roof of the oral cavity is composed of the hard palate and the soft palate. The hard palate is characterized by rows of palatine ridges located between small, horny, incisive papillae. These ridges are composed of stratified squamous epithelium which are replaced every 6–7 days and have, in addition, a most unusual feature. Small outgrowths of hair have been described sprouting from all sides of the papillae. Sebaceous glands have not been described as occurring with these hair follicles. It is assumed that these hairs play some role in the tactile sensitivity of the oral cavity (77).

The soft palate extends from the caudal border of the hard palate to the nasopharyngeal hiatus where it merges with the dorsal wall of the pharynx. Like the hard palate, it is covered by a stratified squamous epithelium which is replaced every 3½–4½ days (63). Laterally the soft palate is continuous with the cheeks and lateral wall of the pharynx. At the caudal extremity, there is no obvious tonsil, no uvula, and no rostral or caudal pillars of the fauces.

An unusual feature is the relationship of the larynx to the soft palate and nasopharyngeal hiatus. The epiglottis lies some 2–3 mm rostral to the nasopharyngeal hiatus while the caudal border of the larynx, namely, the arytenoid cartilages, actually lies within the nasopharyngeal hiatus rostral to its caudal border. A muscle, the nasopharyngeal sphincter, surrounds the nasopharyngeal hiatus and is closely related to the cartilage plates which lie on either side of it. There are no levator palati, musculus unulae or palatoglossus muscles (31).

The pharyngeal region presents no other unusual features and is rather typical of most mammalian species. The esophagus begins just dorsal to the glottis and extends caudally to connect the oral cavity with the stomach.

B.　Esophagus

The esophagus is accessible caudal to the diagram, permitting esophageal–intestinal anastomosis and gastrectomy with relative ease. As in all rodents, the epithelium of the esophagus is covered with a layer of keratin. The esophagus enters the stomach at the inner curvature through a fold of the limiting ridge that separates the forestomach from the glandular stomach. The fold of the limiting ridge makes it impossible for a rat to vomit.

C.　Stomach

On the left side of the abdominal cavity is the stomach which lies in contact with the liver. A prominent, highly vascularized mesentery attaches the stomach and other organs of the digestive tract to the dorsal body wall. The saclike mesentery which attaches to the greater curvature of the stomach and drapes like an apron over the stomach and intestines, is called the omentum.

The anatomy of the rat stomach is unique, and the terminology currently used is confusing. Since the rat is used extensively in gastroenterologic research, it is important that the nomenclature be standardized. In Fig. 3, the recommended name for each portion of the stomach is shown with the various synonyms used listed underneath. For a more complete discussion on proper terminology of the rat stomach, the reader is referred to the works of Robert (131).

A histological description of the stomach of the rat would not differ significantly from that of most rodents. The nonglandular forestomach has rumenlike mucosal folds covered with stratified squamous epithelium and serves as a reservoir. A glandular region, the corpus, is characterized by gastric pits

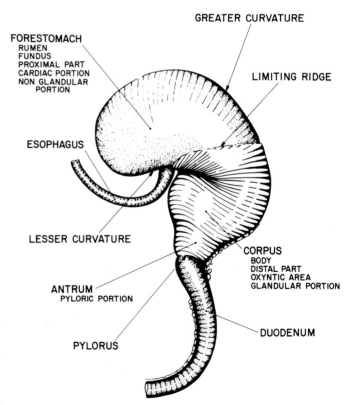

Fig. 3. This is a schematic drawing of the rat stomach. The correct name for each portion of the stomach is shown in large type with the various synonyms listed beneath. [Redrawn from Robert (131).]

lined with simple columnar epithelium. The gastric glands are composed primarily of parietal cells and chief cells, although argentaffinlike cells have been described on occasion. The pyloric portion of the stomach is again characterized by a mucosa of simple columnar epithelium lining the elongated gastric pits and beneath this lies the pyloric glands (147) (Fig. 4).

Gastric mucosal mast cells differ morphologically, histochemically, and pharmacologically from those elsewhere in the body. Histamine is believed to be involved in the secretion of acid by the stomach, and some of this histamine is derived from mast cells. Their true function still remains a mystery. However, recent studies indicate that they may be involved in producing local vasodilatation and increased capillary permeability during periods of increased secretory activity (65).

D. Small Intestine

The entire small intestine (duodenum, jejunum, and ileum) is about six times as long as the large intestine (colon and rectum, 15 cm). (Fig. 5). Approximately eighteen Peyers patches can be identified externally and internally along the distal jejunal and ileal portions of the small intestine. The ileum empties into the cecum at the ileocecocolic orifice. This structure is muscular and is easily identified when viewed from the opened cecum.

Marked changes in mucosal morphology and epithelial re-

placement occur in the rat intestine between birth and 60 gm body weight. The relation of these changes to the function of the intestine and the factors, dietary, nervous, or hormonal, which bring about these changes require further investigation (30).

Glands of Brunner are of interest because they are confined to mammals and their function is uncertain. Most investigators are of the opinion that the glands secrete an alkaline fluid containing mucin to protect the proximal duodenal mucosa against acid insult from gastric content. The cells in the guinea pig are reported to be typical mucous cells, whereas, in the cat and mouse, they include both serous and mucous types. In the rat and the rabbit, the general features of these cells characterize them as serous cells. Therefore, an antacid role for Brunner's gland secretion seems doubtful for the rat and the rabbit (5).

The rate of cell production in the crypts of Lieberkühn along the length of the small intestine is relatively constant and equals about 36 cells/crypt. The ratio of crypts to villi decreases from 27 in the duodenum to 10 in the terminal ileum (29).

The apices of the epithelial cells of the rat small intestine are the structures through which most nutrients are absorbed (Fig. 6). They possess as conspicuous specializations of their free surfaces, large numbers of vertically oriented, slender microvilli, known as "brush border." These and other components of the cell apex as a discrete unit without serious disrup-

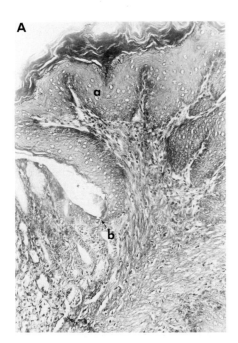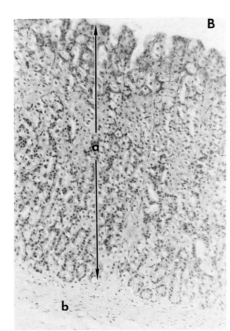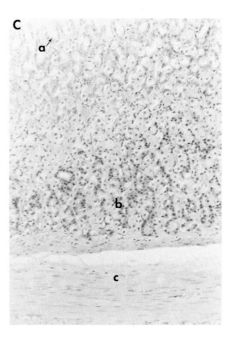

Fig. 4.   Photomicrographs of the mucosa of the stomach. (A) Forestomach: (a) Stratified squamous epithelium; (b) junction of forestomach with corpus. × 16. (B) Corpus: (a) Gastric glands composed of parietal cells, chief cells, and argentaffin cells; (b) outer smooth muscle layer. × 16. (C) Antrum: (a) Mucosal surface with gastric pits; (b) pyloric glands; (c) smooth muscle in wall. × 16.

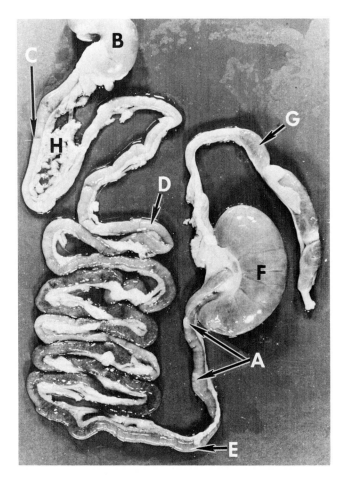

Fig. 5.   Intestinal tract of the rat. (A) Arrows point to peyers patches, (B) stomach, (C) duodenum, (D) jejunum, (E) ileum, (F) cecum, (G) colon, (H) pancreas.

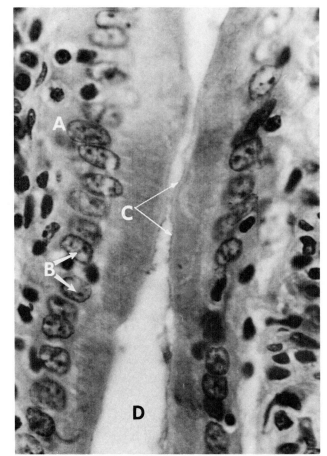

Fig. 6.   Photomicrograph of small intestine villi. (A) Villus, (B) epithelial cell nuclei, (C) brush border, (D) crypt of Lieberkühn. × 100.

tion of its overall structure. This would indicate that the mechanical strength of this area is different from that of the rest of the cytoplasm. Although early works attributed this to a well-defined filamentous or, terminal web, more recent electron microscopic studies have shown that the terminal web is a discontinuous, perforated, granular sheet in which the rootlets of the microvilli are embedded. In this way, the terminal web may provide rigidity to the cell apex and at the same time permit exchange of nutrients (20).

Values for length and surface area of the whole or parts of the absorptive gut in adult rats have been reported (18). The young rat, which is most often used in absorption and nutrition experiments, shows a mucosal area per unit of gut length ratio of about 2 : 1 (118).

Maximum absorption rates within various segments of the intestinal tract are depicted in Fig. 7. Glucose is one of the more readily absorbed substances within the gastrointestinal tract, and it is found to have its maximum absorption rate occurring within the jejunum and upper portion of the ileum

(11). Galactose, on the other hand, has its maximum absorption rate occurring in the middle third of the small intestine (47). The relative absorption rates of other carbohydrates in the intestine are shown in Table I. Whole protein and albumen are absorbed mainly in the lower segments of the small intestine by pinocytosis (28,116). The amino acids (especially, L-cystine, L-cysteine, and scL-tryptophan) are absorbed maximally in the duodenum and jejunum (109). Fat absorption has been shown to have its major absorption rate within the jejunum via the lacteals located within the microvilli of the "brush border" (151).

The fat-soluble vitamins, especially A and D, are all maximally, absorbed in the upper and middle third of the small intestine (111,122). Vitamins E and K are presumed to be absorbed primarily from the upper portion of the small intestine. Vitamin C is also absorbed in the upper small intestine by passive diffusion (150). The B vitamins are absorbed by passive diffusion almost entirely in the jejunum (17). Vitamin B_{12} requires intrinsic factor from the stomach to effect its absorption which is thought to occur maximally in the ileum (155). Iron and calcium are absorbed primarily in the duodenum, and magnesium is primarily transported in the ileum (39,133,136).

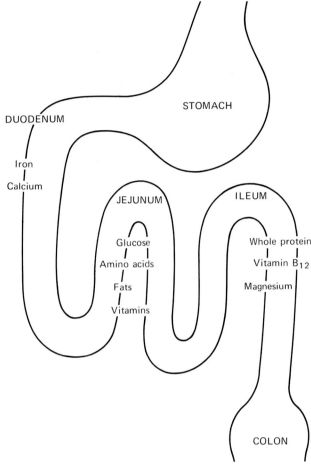

Fig. 7.   Sites of maximum absorption in the gastrointestinal tract of the rat.

Table I

Relative Absorption Rates of Carbohydrates (Glucose = 100)[a]

D-Glucose	100
D-Galactose	99
D-Allose	67
L-Xylose	46
D-Ribose	45
D-Lyxose	38
D-Xylose	37
N-Acetyl-D-glucosamine	34
D-Altrose	24
D-Talose	22
D-Glucosamine	22
D-Idose	20
L-Arabinose	17
D-Arabinose	16
D-Mannose	8
D-Gulose	6

[a] From Kohn (83).

E.   Cecum

The cecum is a large, thin-walled, blind pouch which is shaped somewhat like a comma and is lightly constricted about its middle. In the rat, it differs from that found in many other rodents as it is devoid of internal septa. Even though the cecum is not divided into cells, so commonly found in rodents, it does have a constriction which subdivides the cecum into two parts, an apical and a basal portion. The basal portion contains no lymphoid tissue (Fig. 8A). The apical portion, on the other hand, contains a distinct mass of lymphoid tissue in its lateral wall. This is thought to be analogous with the vermiform appendix of man (Fig. 8B) (14).

F.   Large Intestine

The colon runs cranially from the cecum as the ascending colon, crosses the duodenum, and then proceeds laterally toward the left side of the body as the transverse colon. The transverse colon extends for only a short distance before turning caudally as the descending colon to form the rectum in the pelvic region. The rectum is then continuous caudally with the anus which opens externally.

G.   Rectum and Anus

Some epithelial cells of the mucous membrane in the rectum of rats have the following peculiarities: (1) glycogen particles, which form extended accumulations in young animals, but are singly distributed in adults; (2) a brush border, the microvilli of which are larger and thicker than in the adjoining border cells; (3) bundles of filaments and rows of vesicles in the apical part of the cytoplasm. These glycogen-containing brush cells are very similar to the sensory cells described in the epithelium of the trachea of the rat; however, no connections have been demonstrated with dendrites (94).

The normal rectal temperature of the rat is about 37.5°C (range between 35.8° and 37.6°C). The critical temperatures (ambient temperatures within a range permitting the animal to maintain normal body temperature) are −10° and 32°C (149). The thermoneutrality zone (the range of temperatures over which the metabolic rate is lowest and constant in the unanesthetized animal) is between 28° and 30°C. With the ambient temperature at thermoneutrality, the use of O_2 by an adult rat is at the rate of 1169 ml/kg/hr; and heat is produced at a rate of 800 kcal/m²/24 hr (4). Thus, a 400-gm rat has a basal metabolic rate (BMR) of about 35 kcal/24 hr. With normal activity, the metabolic rate of a cage rat averages about 2.5 times the BMR over a 24-hr period, or about 85 kcal for a 400-gm rat/24 hr. Rodents have no sweat glands, cannot pant, and are poor regulators of core body temperature. Unlike ani-

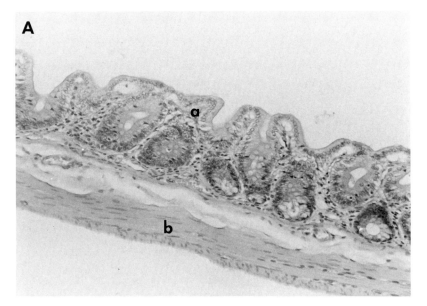
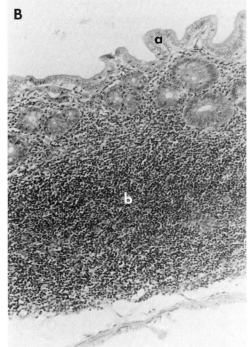

Fig. 8. Photomicrograph of rat cecum. (A) Basal portion of cecum without lymphoid tissue: (a) Mucosal surface; (b) smooth muscle in outer wall. (B) Apical portion of cecum with lymphoid tissue: (a) Mucosal surface; (b) lymphoid tissue in wall. × 16.

mals that salivate, pant, or sweat in response to a high ambient temperature (T_A), rats (and guinea pigs) do not increase water intake when the T_A becomes high. Heat actually seems to inhibit drinking in these animals (3). Rodents seek relief from heat by behavior patterns that cause them to seek shade and to burrow.

Rats adapt to cold much better than they do to heat. When exposed to an ambient temperature of 6°C for 3 or 4 weeks, nonshivering thermogenesis (NST) takes place. Electomyographic investigations in the rat reveal almost a complete cessation of shivering (14,35). At the same time, shivering is replaced by a mechanism that may lead to greater heat production than does shivering (38). After 3 to 6 weeks, the curarized rat is able to maintain its body temperature in the cold. The NST of an "adult" rat is higher than that found in any other animal (67). In the rat (and guinea pig), a sudden exposure to cold causes a lowering of ascorbic acid in the adrenals. The administration of vitamin C increases the survival time of rats exposed to low temperatures. Prolonged cold exposure causes adrenal hypertrophy, and vitamin C administration tends to prevent this (41). The neonatal rat does not develop thermoregulating mechanisms until the end of the first week (55). Unlike most other animals, the NST is not at its maximum in neonates, but it increases until, at about 7 days of age, it reaches 150% of the BMR (106). In the rat, muscle and liver glycogen stores are low in the neonate, and the body temperature (T_B) cannot be maintained. A reduced T_B leads to a reduced BMR, and the hypothermia tends to prolong life. In this it differs from the newborn lamb, foal, pig, dog, and human which are animals with larger glycogen reserves. In such animals, hypothermia will lead to a rapid depletion of glycogen reserves, and the ensuing hypoglycemia may be fatal (143).

H. Accessory Organs

Accessory organs associated with the viscera include the spleen, liver, and pancreas. The spleen is an elongated, flattened organ attached to the greater curvature of the stomach by mesentery. In the rat, it serves as one of the major sites for destruction of red blood cells. This is accomplished by the action of macrophages acting within the venous sinuses of the spleen. A specialized cell containing aldehyde-fuchsinophilic granules has also been identified. These cells are situated at the inner side of the marginal sinus, but they have been shown to move toward the center of the splenic follicles. As yet, no role has been identified for this cell in the antigen-trapping mechanism of the spleen (42). Other peculiarities of relationship or structure are lacking; however, it does vary greatly in size.

The pancreas is very diffuse and extends from the end of the duodenal loop to the left into the gastrosplenic omentum (Fig. 5). The pancreatic ducts vary in number from 15 to 40. They open into the lower part of the common bile duct. To collect pancreatic juice, the common bile duct is ligated near its origin and cannulated near its orifice.

The liver is divided into four major lobes: the median or cystic lobe which bears a deep fissure, the right lateral lobe which is partially divided into cranial and caudal lobes, a large left lobe, and a small caudate lobe fitting around the esophagus (Fig. 9). The rat has no gallbladder. Bile ducts from each lobe come together to form the ductus choledochus (common bile duct) which may be traced to its entrance into the descending duodenum about 25 mm from the pyloric sphincter of the stomach. The gallbladder is not only absent, but apparently the ducts have no ability to concentrate bile such as exists in the gallbladder of other rodents (102). Also, the small size of the ducts and a lack of tonus to the sphincter of Oddi would almost preclude any possibility for bile storage (97).

The mesentery in the rat is a double layer of peritoneal membrane extending from the dorsal body wall to the viscera. It is given a specific name according to the particular visceral organ to which it is attached and thus supports: mesogastrium, to the stomach; mesodudenum, to the duodenum; true mesentery or mesentery proper, to all of the small intestine exclusive of the duodenum; mesocolon, to the large intestine, mesorchium covers the testicle of the male; mesovarium, to the ovaries of the female; and broad ligament, to the fallopian tubes and uterus of the female.

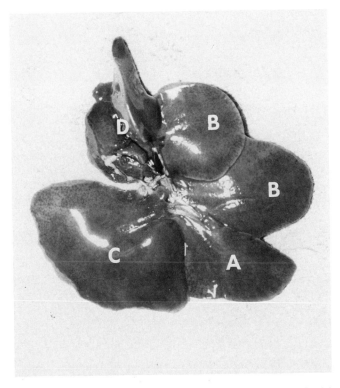

Fig. 9. Four major lobes of the liver. (A) Median or cystic lobe, (B) right lateral lobe, (C) left lobe, (D) caudate lobe.

IV. RESPIRATORY SYSTEM

A. Upper Respiratory System

In the adult rat, the upper portion of the respiratory system presents no unusual features. Rostrolateral external nares open caudally through the caudal nares into the nasopharynx. On the dorsolateral wall of the nasopharynx are the openings of the eustachian tubes.

B. Oropharynx and Trachea

In the oropharynx, the epiglottis guards the slitlike opening to the larynx. The trachea, as in all mammals, has its beginning ventral to the esophagus and extends caudally to form the two main branches or primary bronchi, each of which enters a lobe of the lung. Eighteen to twenty C-shaped cartilages form the framework of the trachea, which has a diameter ranging from 1.6 to 1.7 mm. Some areas of the epithelium have cells which appear to be cuboidal in appearance, with nuclei filling one-half to three-fourths of the cell. Parabasal cells are numerous and their nuclei are chiefly spherical. Numerous goblet cells with prominent cilia about 3–4 mm long are scattered uniformly between the columnar cells. Parabasal bodies of cilia are distinct, as is the basement membrane. Mucosal and submucosal layers are each relatively distinct, and together have a combined width of approximately 27.5 mm. Seromucous glands are abundant, but are limited to the proximal portion of the trachea (9).

C. Lower Respiratory System

At birth, the lung is immature and contains no alveoli and no alveolar ducts. In the newborn rat, lung gas exchange occurs in the smooth-walled channels and saccules, the prospective alveolar ducts, and alveolar sacs. The lung structure remains grossly unchanged between birth and day 4. Finally, an explosive restructuring of lung parenchyma occurs between days 4 and 7.

No respiratory bronchioles are present at birth, whereas from day 10 onward, they can easily be identified. The main changes occurring after day 13 are the expansion and thinning of the primary and secondary septa to form true alveolar septa (21).

The distribution and frequency of ten morphologically distinct cell types has been described in the surface epithelium of the rat intrapulmonary airways. Eight of these cell types are epithelial and two are migratory. One of these cell types, the epithelial serous cell, is thought to be unique for the rat as it has not been described in any other species. Recent studies indicate that this cell produces a secretion of lower viscosity than that of the mucous cell. This secretion may be partially responsible for the low viscosity periciliary liquid layer present at all levels of the respiratory tract (78).

Ciliated cells are described in this periciliary fluid of low viscosity. Lying immediately on top of these cells are discrete flakes or droplets of mucous. It is theorized that only the tips of the cilia, which are described as having a clawlike structure, penetrate the periciliary fluid and "claw" the mucus forward (78).

Goblet cells are present at all airway levels, but comprise less than 1% of the total cells. Ciliated cells increase in number progressively toward the periphery. Basal cells decrease and virtually disappear from small airways, and, at all airway levels, the remaining nonciliated cells form a high proportion (about 40%) of the total cells present (78).

The lungs lie within the thoracic cavity and are covered with a membrane called the visceral pleura. There are three lobes of the right lung (cranial, middle and caudal). A fourth deeper median lobe lies in contact with the diaphragm and apex of the heart and is notched to accommodate the caudal vena cava. This is sometimes called the postcaval lobe. The left lung has but one lobe (Fig. 10).

Respiration is the result of changes in the normally negative intrathoracic pressure at the rate of 85 breaths per minute producing an average tidal volume of 1.5 ml (21,54). The minute ventilation is about 100 ml/min (range 75–130 ml/min). The regulation of respiration is primarily in response to tissue CO_2 changes in the medullary respiratory center, but the carotid body or "glands" as they are sometimes called, also play a role. They are located, one on each side of the neck, in the bifurcation of the common carotid into external and internal carotid arteries. This area is very vascular and is abundantly supplied with nerves. These nerves, which comprise the carotid body, arise as branches of the glossopharyngeal and vagus trunks and the cranial cervical ganglion, and are accompanied by sympathetic ganglion cells. They are primarily sensitive to a low arterial oxygen saturation which increases the receptors firing rate markedly as the partial pressure of oxygen decreases below 100 mm Hg. Therefore, the carotid bodies function as chemoreceptors in the regulation of respiration. Similar chemoreceptors located in the aorta are called aortic bodies. The aortic body afferents travel via the vagi to the brain.

The respiratory bronchioles are short and lead almost immediately into alveolar ducts, each of which subdivides four or five times. The epithelial lining of respiratory bronchioles varies with the species. In the rat, the cells remain high cuboidal on the side adjacent to the accompanying arteriole, and the opposite side progressively decreases in height (84).

The lung weight is related to the body size, but the surface area of the lung is related to the O_2 consumption. The alveolar

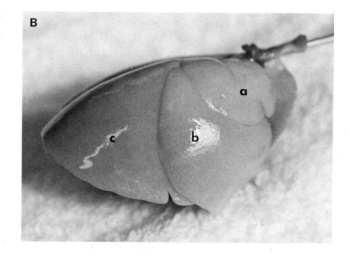

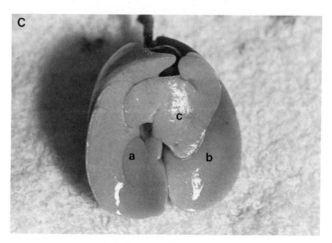

Fig. 10. Three views of the lobes of the lung. (A) Left lateral view (one lobe). (B) Right lateral view: (a) Cranial lobe, (b) middle lobe, (c) caudal lobe. (C) Diaphragmatic view: (a) Right caudal lobe, (b) left lobe, (c) postcaval lobe.

diameter is thus related to metabolic rate as measured by the O_2 utilization per unit of body weight (158). The mean alveolar diameter in the rat is about $70\,\mu$m. This can be compared with about $20\,\mu$m for a bat or a shrew, about $200-250\,\mu$m for a man or a whale, about $400\,\mu$m for a sloth, and about $1100\,\mu$m for a manatee. The total surface area of the lung of a 400-gm rat is about $7.5\,m^2$, compared with about $75\,m^2$ in a 70 kg man. The rat has a thin pleura. In animals with a thin pleura, there are no septa in the lungs, and branches of the brachial artery are confined.

The rat and the rabbit have been most extensively used to gain information about the connective tissue of the lung, of which collagen is the most important constituent. Information in this area is too new to have species differences defined well. In the rat and rabbit, the most important stimulant for lung growth is available space. Blood flow rate is not as important as it is in liver (33). A summary of the microscopic characteristics of the alveolar system is given in Table II.

It has been well established that the alveolar wall consists of three layers: the epithelial cell lining the alveolar space; a basement membrane, and the endothelial cell lining the capillary lumen (64,81). A specialized cell known as the great alveolar cell or type II pneumocyte is distinguished from neighboring components of the lung epithelium by its size and pleomorphic cuboidal shape. In addition, it has characteristic membrane-limited lamellar inclusion bodies which are thought to contain the physiologically important surface active system (surfactant) that facilitates inflation of the alveoli by lowering the surface tension at the alveolar–air interface (15,92,146). (Fig. 11).

Normally, all layers are relatively thin and continuous. The minimum thickness of the air–blood barrier is nearly identical in all mammalian species examined ($0.2\,\mu$m). The thicker portions are species specific, and these are mainly due to the connective tissue fibers. They are generally in direct propor-

Table II

Summary of Microscopic Observations of the Alveolar System[a]

Characteristic	Observation
Alveolar duct (μm)	
Length	288–624
Diameter	68–154
Number of branches alveolar duct	2–5
Atria, diameter (μm)	15–262
Alveoli, diameter (μm)	57–112

[a]From Babero et al. (9).

tion to the size of the lung, that is, to the animal (164). There is, however, little difference in the barrier thickness in animals with a thin pleura. Thus the mouse, rat, and rabbit have an average barrier thickness of about $1.5\,\mu$m. The alveolar spaces are clear of fluid or debris. Red cells and occasionally white cells are seen in the capillaries, but platelets are seldom observed in normal capillaries of rat lung (64,91).

The pulmonary artery of the rat is thinner than any other animal examined; whereas the pulmonary vein is thicker than in other animals. This peculiar feature of the pulmonary vein can be attributed to the presence of striated muscle fibers which are contiguous with those of the heart. These cardiac muscle fibers are found primarily in the longer intrapulmonary branches, but may be seen in smaller veins. Such a relationship allows for the possibility of infections to spread from the heart to the pulmonary veins and the lungs (32). It is suggested that conduction of impulses between cells is transmitted along the cardiac muscle of pulmonary veins in a manner analogous to that in muscle of the heart (95). It is difficult to differentiate a vein from an artery of a rat under the microscope. Precapillary anastomoses between the bronchial and pulmonary arteries have been demonstrated in the rat (126), as they have been in

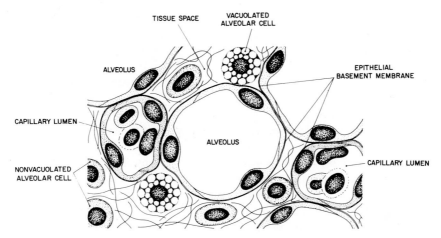

Fig. 11. Alveoli. Schematic thin section. [Redrawn from Bertalanffy, (15).]

man and the guinea pig (162). These anastomoses have been limited to the hilar region in the rat. Such anastomoses have not been demonstrated in the rabbit, sheep, horse, dog, mouse, cat, or monkey (100).

There is a species variation in the response of the pulmonary vessels to acetylcholine (ACh). In the rat, 0.2 and 5 μg ACh causes vasoconstriction.

Innervation of the lung is complex, with wide variation between species. The density of nerves is high in the rat, as it is in the calf, mouse, and guinea pig (157). It is low in the rabbit. Information about pulmonary lung activity from anatomic sources has been scarce. Most of the information comes from physiologic and pharmacologic sources. The rat (and the rabbit) do not have an adrenergic nerve supply to the bronchial muscle (66). In these two species, the bronchoconstrictor bronchioles are controlled by vagal tone. In the human, dog, cat, sheep, pig, and the calf, the bronchioles are controlled by sympathetic tone.

The rat lung has a high activity of serotonin or 5-hydroxtryptamine (5HT). Other animals with high 5HT activity are the mouse, the hamster, and the rabbit. The guinea pig lung has about zero 5HT activity, and the cat and dog lungs are low in the activity of this strong vasoconstrictor. The rat lung has a low concentration of histamine, as do the lungs of the rabbit, the hamster, and the mouse (117). Animals with a high concentration of histamine in the lungs are the guinea pig, the cat and the dog.

V. CARDIOVASCULAR SYSTEM

In many respects the anatomy of the cardiovascular systems of the human and rat are similar. No attempt can or need be made in this limited space to describe the complete cardiovascular system. A comprehensive illustration of the major blood vessels in the female is shown in Fig. 12 and emphasizes those vascular areas frequently involved in experimental work.

A. Heart

The heart is located on a midline in the thoracic cavity. Its apex rests near the diaphragm and its lateral aspects are bounded for the most part by the lungs. Due to the small size of the left lung, the heart is exposed to the thoracic wall for a considerable distance. This factor makes the cardiac puncture bleeding technique a relatively simple task if performed between ribs 3 and 5. As in other mammals, it is four-chambered heart with two artia and two ventricles. Aortic and pulmonary values each have three leaflets, whereas, the mitral and tricuspid valves have two major and minor accessory leaflets.

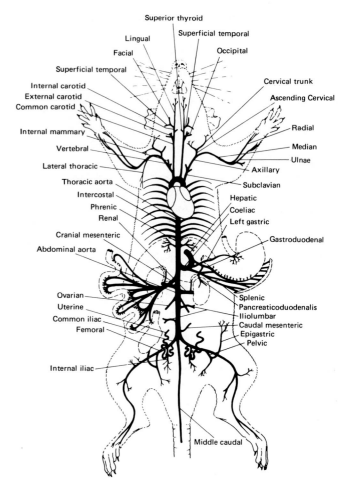

Fig. 12. Schematic of the major arteries of the female rat [Redrawn from Chiasson (27).]

The sinoatrial node in the adult rat has been studied extensively with the light microscope (62,107). It has also been studied in rat embryos where the primordium of the node appears at 12 days, located in the ventromedial wall of the right common cardinal vein, just above its entrance into the sinus venosus. Knowledge of the precise location of the node in newborn rats, a cellular package located in the ventromedial wall of the right upper vena cava at the atriocaval junction, has enabled investigators to isolate the structure for electron microscopy. These studies have shown that embryonic nodal cells are very similar to adult "P" cells which indicates that the node begins to function very early in life (37).

The aortic arch bends to the left as in other mammals and gives rise to the brachiocephalic (innominate), the left common carotid, and the left subclavian arteries. The brachiocephalic trunk divides at the sternoclavicular joint into right common carotid and right subclavian arteries. The aorta and its systemic branches are not described here as they are well illustrated by Greene (60). One exception which should be mentioned is that there is an extracoronary myocardial blood supply in the rat. This is true mainly for the atria as they receive a major input from branches of the internal mammary and subclavian arteries. The right cardiac arteries supply right and left atria, whereas the left cardiac arteries only supply a small portion of the left atrium. These characteristics make the blood supply to the heart similar to that found in some fish and is not like the circulatory system described for higher mammals.

The vena cava presents a slightly different appearance than in other mammals. Two precavae are present in the rat. The right precava empties into the right atrium directly. The left precava runs further caudally where it is joined by the azygous vein before uniting with the postcava to enter the right atrium.

B. Peripheral Circulation

The peripheral circulatory system does not differ significantly from that described for other small mammals.

Table III shows some of the hemodynamic parameters of the rat. Blood pressure is measured indirectly using the tail plethysmograph as seen in Fig. 13. Systolic pressure is most easily measured by this method, since the end point for determining diastolic pressure is not very discernable (50).

Regional blood flow can be measured by the technique of radioactive microsphere injection (135) or by the use of a noncanulating electromagnetic flow probe (7). The electromagnetic flowmeter can also be used for measuring aortic flow. Figure 14 shows the typical instantaneous pressures and flow patterns. The sympathetic innervation to the heart can accelerate

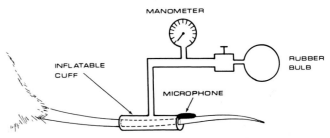

Fig. 13. Blood pressure measurement (indirect).

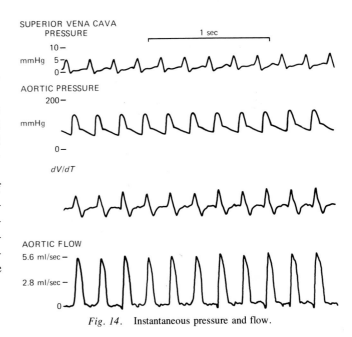

Fig. 14. Instantaneous pressure and flow.

the heart rate from a resting rate of 300 to over 500 beats/min. Parasympathetic stimulation mainly affects the heart rate by producing a bradycardia. It is capable of decreasing the heart rate from a normal 300 beats/min down to a ventricular escape rate of approximately 70–80 beats/min (119).

Blood volume is measured by the dilution technique using a dye or radioactive isotope as the serum protein tag (2).

Blood pressure control in the rat is composed of the same systems as in other mammals as shown in Fig. 15. Many techniques are available for altering this control system. Prolonged deoxycorticosterone acetate (DOCA) treatment with 1% sodium chloride drinking water when given to an uninephrectomized rat will produce malignant hypertension (142). This is probably due to the salt-retaining properties of DOCA and not the properties of angiotensin II (53).

Table III

Cardiovascular Parameters for the Rat

Arterial blood pressure	
Mean systolic (mm Hg)	116
Mean diastolic (mm Hg)	90
Heart rate average (beats/min)	300
Cardiac output (ml/min)	50
Fractional distribution of cardiac output	
Heart (%)	5
Kidney (%)	19.0
Brain (%)	1.5
Adrenal (%)	0.33
Testes (%)	0.71
Splanchnic (%)	19.0
Stomach (%)	1.4
Intestine (%)	13.0
Liver (%)	1.6
Pancreas (%)	1.7
Spleen (%)	1.7
Muscle, skin, bone (%)	47
Other (%)	7.0
Blood volume (ml/kg body weight)	54.3

If the blood flow to the kidneys is limited by placing a constrictor clamp around one of the renal arteries and either removing the other kidney or leaving it in, hypertension will develop. The more severe form of hypertension is produced when the contralateral unclipped kidney is removed. This technique produces "Goldblatt hypertension" (59). The two-kidney Goldblatt hypertensive rat is probably a result of the excess renin–angiotensin produced by the normal kidney, along with the positive sodium and water balance (105,113). The one-kidney Goldblatt hypertensive rat is probably produced by abnormal sodium and water metabolism, and not so much by altered renin-angiotensin production, although the reversal of this form of hypertension by removal of the constrictor does not seem to be related to sodium or water loss (93,110). Rats subjected to removal of a kidney and a high sodium intake develop adrenal regeneration hypertension following enucleation of the adrenal glands (52). It is proposed that the hypertension results from a hypersecretion of a yet to be identified sodium-retaining hormone by the regenerating adrenal (144). Selective inbreeding of susceptible rats has produced a strain of rat that is spontaneously hypertensive (114). Again, abnormal sodium and water metabolism appear to be implicated but no definitive mechanisms have been described.

Accurate knowledge of lymphoid anatomy and lymphatic routes in the laboratory rat has become increasingly important, as this animal is used more frequently in biological investigations. An excellent description of the lymphatic system, along with several well illustrated figures, is given by Tilney (159).

VI. URINARY AND REPRODUCTIVE SYSTEMS

A. Kidneys

The kidneys can be palpated in young animals, but their outlines are lost in fat in mature specimens. The right kidney and adrenal gland lie at a more cranial level than the left. A corresponding difference is noted in the branching patterns of the renal vessel and the relation of these vessels to the aorta and caudal vena cava (51).

The kidney of the rat is unipapillate, that is, it has one papilla and one calyx, and enters the ureter directly. Most animals are multipapillate, but rodents, lagomorphs, and insectivores are unipapillate. The single papilla and calyx lend themselves to cannulization techniques. The rat is one of those animals that have specialized fornices. These are long evaginations of the renal pelvis with epithelium similar to collecting duct epithelium, and in close association with the thin loops of Henle. The fornices probably help build up the urea concentration in the papilla; the papillary urea is much greater in mammals with specialized fornices (rats, rabbit, dog, sheep, opossum) than it is in other animals. The urea in animals with specialized fornices increases the osmotic ceiling of the urine, and the osmotic ceiling is thus affected by the protein ingestion. Other animals have a constant osmotic ceiling (120).

The medullary renal pyramid is well developed with a strong zonation of vascular and tubular elements. Table IV illustrates

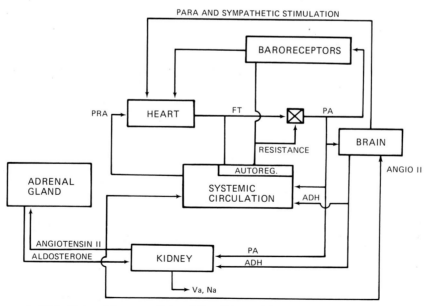

Fig. 15. Blood pressure controls. PRA, Right atrial pressure; FT, total flow; PA, arterial pressure; ADH, antidiuretic hormone; Va, urinary volume; Na, urinary sodium; and Angio II, Angiotensin II hormone.

Table IV

Relation of Body Weight, Kidney Dimensions, and Thickness of Kidney Regions[a,b]

Body weight (gm)	185.5
Weight of one kidney (gm)	1.738
Dimensions (mm)	19×12.5×9
Thickness of cortex (mm%)	4
Outer zone of medulla (mm%)	2.2
Inner zone of medulla (mm%)	6.0

[a] From Abdallah (1).

[b] Average of five animals.

the relative thickness of kidney regions compared to body weight and kidney weight.

In the laboratory rat, there are both short- and long-looped nephrons. Short-looped nephrons arise from the tufts in the middle and outer zone of the cortex (high and middle nephrons). The thin segment of short loops changes to the broad thick segment while still descending, so that the bend of the loop is formed by the latter and the bend occurs in the outmost zone of the medulla or in the medullary ray of the cortex. Long-loop nephrons constitute 28.5% of the total number (Table V). In some parts of the cortex, usually opposite medullary rays, groups of short nephrons occur which sometimes possess no thin segment. The straight descending parts of the proximal tubules join the thick segments directly.

The glomerulus is lobulated and there are a limited number of connective tissue cells between the vessels near the vascular pole (Fig. 16). The visceral layer of Bowman's capsule has flattened cells with slightly projecting oval nuclei. As a rule, the capsular free space is pronounced. The Bowmen's capsule has a thin wall lined with flattened epithelium that may become a little thicker at the urinary pole.

The layer covering the glomerulus is the visceral layer, whereas, the layer covering Bowman's capsule is the parietal layer. The visceral cells are characterized by their foot processes which are sometimes called podocytes. The glomerular endothelium, basement membrane and visceral epithelial cell layer constitute the barrier for glomerular ultrafiltration.

Table V

Percentage Distribution of Different Types of Nephrons[a]

Species	Laboratory rat
No. of nephrons counted	846
Cortical (%)	8.2
Short (%)	44.7
Medullary (%)	18.6
Long (%)	25.8

[a] From Abdallah (1).

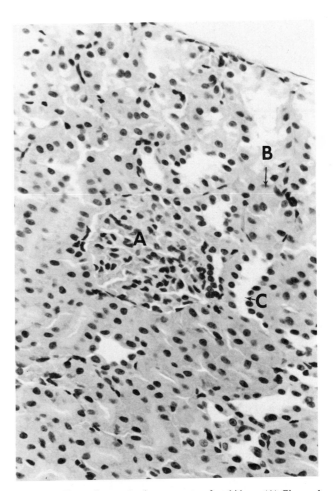

Fig. 16. Photomicrograph of outer cortex of rat kidney. (A) Glomerulus; (B) proximal tubule; (C) distal tubule. ×40.

Normal proteinurea is more marked in rats (0.4 to 1.0 mg/ml urine) (58) than it is in man (0.03 to 0.08 mg/ml) (130). Normal proteinurea is also high in dogs (0.1 to 2.0 mg/ml). Among the mechanisms that increase the amount of protein in the urine, at least in the rat, is sodium depletion (160). Figure 17 depicts the pressure gradients responsible for this fluid movement from blood to urine as measured by direct puncture (79).

When the plasma oncotic pressure is considered to be 20 mm Hg, then the net driving force in the proximal tubular end of the glomerulus is calculated to be 26 mm Hg. This agrees with the pressure difference calculated from stop-flow pressure and the proximal tubular pressure (56).

Other workers using the same micropipette techniques for measuring the pressures in the cortical structure of the kidney, but using a different strain of rat, have found much lower pressures (19).

The glomerular filtration rate (GFR) varies according to the renal location of the glomerulus. The GFR for the high nephrons is 37 ml/min per 10^6 nephrons, whereas, the juxtamedul-

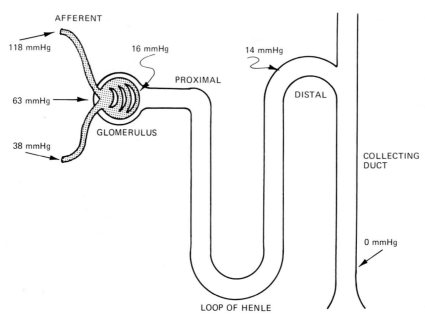

Fig. 17. Nephron pressure gradients.

lary glomeruli have a GFR of 52 ml/min per 10^6 nephrons. The surface area of the juxtamedullary glomeruli is proportionately greater, thus, making it appear that the driving pressure for ultrafiltration is the same in all types of nephrons (10,69).

In the rat (and the dog), the GFR is more variable than it is in man. In man, GFR is reasonably stable. The number of glomeruli in rats increases 2.5-fold in the first 30 days after birth, at which time the adult level is reached. But the size of the glomeruli (and the filtration surface) increases until the rat is 3 or 4 months old (6). Table VI shows other renal excretory parameters.

Because of the presence of superficial nephrons in the cortex of the rat kidney, it has been the model for most micropuncture work directed at the individual transport process along the nephron. This approach offers the most direct method for evaluating tubular function *in vivo*. One method is to collect by micropipette from single nephrons under conditions of free flow. The collected fluid is then analyzed and the tubular collection site is identified by microdissection (167).

The surgical exposure of the renal papilla has made it possible to puncture papillary-collecting ducts, vasa recta, and loops of Henle of juxtamedullary nephrons (168). Therefore, conclusions made in terms of the reabsorptive behavior of electrolytes must take into consideration the nephrons' location (70).

The proximal tubule is characteristic for most rodents. Lining cells are broad and of considerable height, each with a rounded central nucleus (Fig. 17). The cytoplasm is basophilic and contains granular or irregular mitochondria around the nucleus. Regularly arranged parallel rods occupy the basal part of

the cell and are perpendicular to the basement membrane. The next convoluted part of the proximal tubule, as well as the straight part entering the medullary ray and forming the beginning of Henle's loop, stains less darkly, but the cells still contain numerous parallel rods. The brush border is distinct, and a small intensively stained granule can often be observed at the base of each line in the brush border.

Fluid in the proximal tubule in mammals has a chloride

Table VI

Renal Excretory Parameters

Blood urea nitrogen (mg%)[a]	21
Urine volume (ml/24 hr/100 gm body weight)[a]	5.5
Na$^+$ excretion (mEq/24 hr/100 gm body weight)[a]	1.63
K$^+$ excretion (mEq/24 hr/100 gm body weight)[a]	0.83
Urine osmolarity (mOsm/kg of H$_2$O)[a]	1659
GFR (ml/min/100 gm body weight)[a]	1.01
U/P inulin (mg/ml)[a]	431
C_{IN}[b]	
Left	1.23
Right	1.24
C_{PAH}[b]	
Left	2.79
Right	2.68
Filtration fraction[b]	
Left	0.46
Right	0.47

[a] Values from Flamenbaum (48).

[b] Values from Fine (46) in milliliters per minute. C_{IN} = clearance of inulin; C_{PAH} = clearance of PAH (pramino hippuric acid).

concentration higher than in plasma. This is associated with preferential bicarbonate reabsorption, a fall in transtubular pH and diffusion trapping of ammonia. These events are more pronounced in the rat than in the dog (57). For the concept of diffusion trapping of ammonia, the reader is referred to Pitts (121).

At a somewhat variable position along the loop of Henle, the thick cells of the proximal segment are replaced by a flattened epithelium with a pale staining cytoplasm having slightly flattened nuclei projecting toward the lumen. There is no brush border on these cells and mitochondria are very scarce.

In the outer medulla of the rat kidney, all thin limbs of the loop of Henle are descending (115). There is a difference between the rat and the dog in the concentration of sodium along the distal tubule after mannitol loading. The urinary sodium rises after mannitol loading in the dog (166); in the rat, a sharp fall takes place (168).

The thick segment of the loop of Henle has about the same diameter as the proximal tubule. A brush border is still absent.

At the macula densa, where the thick segment joins the diatal convoluted tubule, the structure of the nephrons changes markedly and abruptly, the distal tubule being distinguished by the width of its lumen and light staining of its cells.

The collecting tubules are lined by a cuboidal epithelium having round nuceli. The cytoplasm is slightly basophilic and cell borders are distinct.

The collecting tubules empty into the renal pelvis which is the expanded portion of the ureter within the renal sinus. From this point, the ureter carries the urinary waste products to the bladder, where they are stored for excretion via the urethra. In the male, the urethra extends through the penis and receives the male sex gland products. In the female, this duct opens to the exterior separately of the reproductive orifice.

In the normal rat, even when it is allowed an adequate supply of drinking water, the urine osmolarity is in the region of 1000–1500 mOsm comapred with the normal tissue osmolarity of about 300 mOsm. There is a marked osmotic gradient in the medulla, rising from the corticomedullary junction to the tip of the papilla where it approximates that of the urine. It is suggested that the medullary osmotic gradient, induced by the countercurrent system in the loops of Henle, is augmented by a recirculation of sodium and urea from urine to papillary tissue via the descending vasa recta and perhaps by the osmotic absorption of water from the urine by the medullary interstitial tissue and the vascular bundles. These changes apparently take place through the thin epithelium which covers the outer medulla (80).

Scanning electron microscopic studies of the rats renal papilla have not demonstrated any valvelike arrangement which would prevent urine backflow into the ducts of Bellini (25).

The ability of the laboratory rat to concentrate its urine is about twice that of man. The osmotic ratio (urine/plasma) for man is 4.2; for the rat it is 8.9. The maximum concentrating ability of man for urea, electrolytes (as sodium concentration equivalent), and the total osmotic concentration is 792 mM/liter urea, 460 mEq/liter electrolytes, and 2.9 Osm/liter osmotic concentration (137). The relative medullary thickness of the human kidney is 3.0; of the rat kidney, 5.8. Man has 14% long-looped nephrons; the rat has 28% long-looped nephrons (1).

The rat is the only known animal whose kidney contains significant amounts of L-amino acid oxidase, an enzyme that catalyzes the oxidation of thirteen amino acids. The enzyme is absent in the kidney of the dog, cat, guinea pig, rabbit, pig, ox, and sheep (16).

The kidney of the rat (and rabbit, guinea pig, and sheep) also contains glutamine synthetase, an enzyme that converts ammonium glutamate to glutamine (85). The enzyme is not found in the kidneys of the cat, dog, pig, or pigeon, although it is found in the brain of all vertebrates. Glutamine would, therefore, be found in the renal vein of the rat, but not of the dog or cat.

B. Male Reproductive System

The rat is perhaps one of the most widely used research animals in reproductive physiology.

The testes of the male, which descend between the thirtieth and fortieth day of life, lie in two separate thin-walled scrotal sacs. The inguinal canal remains open throughout life. The testicular artery and the pampiniform venous plexus become surrounded by masses of fat as they enter the inguinal canal. Within the scrotum, several epididymal branches are given off from the internal spermatic artery.

The epididymis is divisible into three portions: the enlarged caput epididymis applied to the proximal end of the testis, and almost completely embedded in fat; the corpus epididymis, a more slender portion lying along the dorsomedial aspect of the testis; and cauda epididymis at the distal pole, which doubles upon itself before giving rise to the ductus deferens (43).

The ductus deferens, accompanied by deferential blood vessels, runs proximally through the inguinal canal and crosses the ureter to enter the urethra. Surrounding the ductus deferens close to its opening into the urethra, is the gland of the ductus deferens.

Within the pelvis, and surrounding the bladder are five pairs of organs: the glands of the ductus deferens; two pairs of prostate glands, one ventral and one dorsal to the ductus deferens; two large hood-shaped and convoluted seminal vesicles; and within the same capsule are the coagulating glands (Fig. 18).

The penis lies within a loose prepuce and has a single cartilaginous or bony process, the os penis, on its ventral wall.

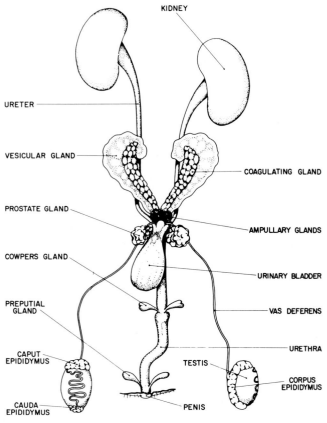

KIDNEY

URETER

VESICULAR GLAND

COAGULATING GLAND

PROSTATE GLAND

AMPULLARY GLANDS

COWPERS GLAND

URINARY BLADDER

PREPUTIAL GLAND

VAS DEFERENS

URETHRA

CAPUT EPIDIDYMUS

TESTIS

CORPUS EPIDIDYMUS

CAUDA EPIDIDYMUS

PENIS

Fig. 18. Male urogenital system, ventral view.

Lying just beneath the skin of the prepuce and opening into its cavity is a pair of rather slender and flattened preputial glands (60).

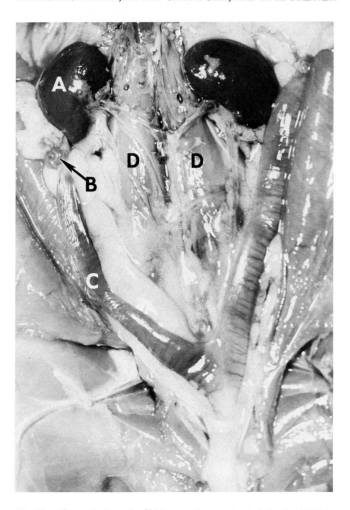

Fig. 19. Gross photograph of kidneys and uterus (ventral view). (A) Right kidney; (B) right ovary; (C) right uterine horn; (D) Psoas muscle.

C. Female Reproductive System

In the female, the ovaries lie close to the kidneys along the lateral border of the psoas muscle and are embedded in fat (Fig. 19). In the adult, the ovary appears as a mass of follicles. The oviduct, a much convoluted tube, has its open end applied to the ovary and its distal end opens into the bicornuate uterus. Surrounding the oviduct is the oviducal mesentery (mesosalpinx) forming a saclike structure, the ovarian bursa. Although the uterine horns appear to be fused distally, there remain two distinct ossa uteri opening into the vagina. Each opening has its own ostium interum and externum, as well as cervical canal (Fig. 20).

The only genital structure of the female that is connected with the urinary system is the clitoris located in a prepuce with bulbi vestibuli glands similar to preputial glands of the male. The bladder empties into the urethra which opens to the exterior at the urethral orifice located at the base of the clitoris.

The vagina is located dorsal to the urethra and is completely separate. The urethra does not enter the vagina or vestibule as is seen in domestic animals.

The approximate onset and duration of the events of the rat estrous cycle are shown in Fig. 21. The arrows show possible causal relationships. A small quantity of luteinizing hormone LH is secreted during the late morning of day 3 which causes estrogen secretion. Luteolysis of the most recent crop of corpora lutea takes place at this time also, probably due to the LH secretion. The estrogen secretion continues until the LH surge, under CNS control, is released on the day of proestrus. The estrogen is responsible for the uterine and vaginal changes, the mating behavior, and luteinizing hormone–follicle-stimulating hormone (LH-FSH) release on the following 2 days. The surge of LH necessary for the triggering of ovulation is released after 14 hr on day 4. This LH release is closely tied to the time of day as defined by the light–dark cycle (light on 5 AM, off 7 PM). There is daily facilitation of this mechanism, but

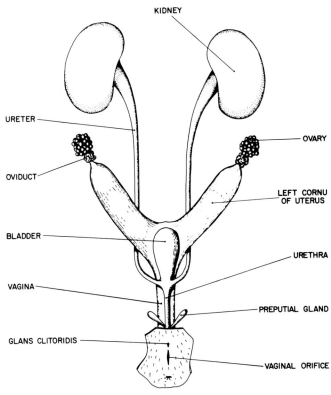

KIDNEY

URETER

OVIDUCT

BLADDER

VAGINA

GLANS CLITORIDIS

OVARY

LEFT CORNU
OF UTERUS

URETHRA

PREPUTIAL GLAND

VAGINAL ORIFICE

Fig. 20. Female urogenital system.

release of LH only occurs on one day of the cycle, which is the day estrogen levels are high. Estrogen levels also promote FSH secretion and possibly prolactin. This LH surge, in addition to causing ovulation, also terminates the ongoing estrogen secretion and starts progesterone secretion by the ovary. The progesterone then acts through the nervous system and helps induce mating behavior. It also causes release of the uterine water. As a result of the lowering of estrogen secretion and its negative feedback effect on LH and FSH, the LH and FSH secretion levels are freed to begin increasing, thus initiating the next crop of follicles.

The 5-day cycle is similar to the 4-day cycle in causal relationships. The LH and estrogen secretions both appear to be delayed. The luteolysis and uterine changes are delayed approximately 24 hr. The "time of day" signal necessary to trigger LH release, in addition to estrogen and LH-FSH release, is delayed approximately 24 hr also, thus prolonging the uterine ballooning and vaginal cornification (138).

VII. ENDOCRINE SYSTEM

The endocrine system of the rat has been studied extensively. It is not appropriate in this chapter to describe the many

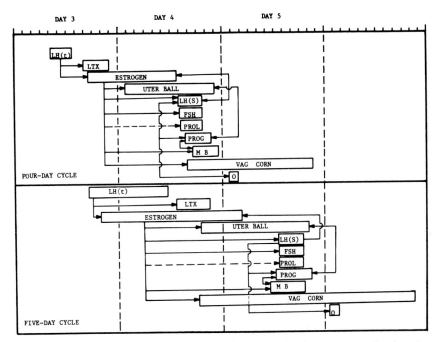

Fig. 21. Rat estrous cycle. LH(t), tonic release of luteinizing hormone; LTX, luteolysis; estrogen, estrogen secretion from the ovary; UTER. BALL., uterine accumulation of ovulatory intraluminal fluid; LH(s), ovulatory surge of LH; FSH, ovulatory surge of follicle-stimulating hormone; PROL., prolactin release on day of proestrus; PROG., progesterone secretion after critical period; M.B., mating behavior; VAG. CORN., vaginal cornification; O, ovulation.

methods which have been used. Therefore, the reader is referred to the following papers for further information (36,38,71,76,153,154,163). The endocrine organs include the pineal body, hypophysis, thyroid, parathyroid, thymus, adrenals, testis, and ovaries.

A. Pineal Body

The pineal body, as in all mammals, is identified by spreading apart the cerebral hemispheres and cerebellum. It appears as a dark, bulb-like protrusion from the caudal portion of the roof of the diencephalon. It would appear that the pineal gland is a senso-neuroendocrine organ or a neuroendocrine transducer which converts neural input into endocrine output. This property is probably due to the presence of adrenergic nerve terminals in its precapillary spaces and to numerous synaptic ribbons (86).

B. Hypophysis

The hypophysis, or pituitary gland, is situated within a bony cavitation, the sella turcica, and is surrounded by a meningeal fold, both of which make it difficult to remove the gland intact with the brain. At about 40–50 days of age, a difference in the weight of the hypophysis occurs between sexes, being greater in the female. With age, this difference tends to increase (153).

Recent investigations have described a pharyngeal hypophysis to occur with high frequency in the rat. This pharyngeal hypophysis presents itself superficially as a median diverticulum in the pharyngeal mucosa underlying the postsphenoid. Also described is a well-defined craniopharyngeal canal which is in continuity with the pharyngeal hypophysis and is similar to it in histological appearance (99).

C. Thyroid

The two thyroid glands lie on either side of the trachea just below the larynx and extend over four to five tracheal rings. These glands are characterized histologically by the presence of colloid in the lumen of the thyroid follicles and by "C" cells which have their origin from the ultimobranchial body (154).

In mammals, the ultimobranchial bodies have been said to disappear or to participate in the formation of thyroid follicles before birth. A recent study has shown that the ultimobranchial bodies do not disappear in Holtzman rats, but remain distinct from the thyroid parenchyma. In addition, it was shown that these bodies give rise to large follicles and an intricate network of ducts which would dispute the theory that the ultimobranchial bodies remain as dormant cysts within the thyroid paren-

chyma. The function of this group of bodies in the rat remains to be shown. They are not involved in the synthesis of thyroid hormones, and they do not synthesize calcitonin.

One thought is that, because of their alkaline phosphate activity and the presence of positive thyroid follicles adjacent to these ultimobranchial bodies, the basal cells of the ultimobranchial bodies may give rise to the new thyroid follicles after birth (24).

D. Parathyroid

The parathyroid glands are located on the anterolateral surface of each lobe of the thyroid. Accessory parathyroids are rare in the rat (71).

E. Thymus

The thymus gland is located just cranial to the heart at the thoracic inlet. It consists of two distinct masses which lie on the ventral aspect of the trachea and is found throughout the life of the rat.

Thymic weight loss occurs during pregnancy and lactation; however, increased cellular activity is observed in the large lymphocytes and lymphoblasts during this time period. It is generally assumed that the thymic regression observed is produced by hormonal factors. The increased activity observed is paradoxical unless its cause is unrelated to that of involution (101).

F. Adrenals

Adrenal glands in the domesticated laboratory rat are much smaller than those found in wild rats. Also, the adrenal gland of the female has been shown to be consistently larger than those of males (36). The adrenal glands are located at the craniomedial pole of each kidney. No lymphatic vessels have been found in the adrenal gland of the rat (161).

The stimulus for glucocorticoid secretion by the adrenal comes in the form of the polypeptide hormone adrenalcorticotropin (ACTH) which, when presented to the adrenal gland by increased release from the anterior pituitary or by increased adrenal blood flow, increases the production of the glucocorticoids (123).

Elevated levels of ACTH are probably responsible for the negative feedback inhibition of the input stimulus, either in the hypothalamus or anterior pituitary. Figure 22 summarizes the feedback control of glucocorticoid secretion.

Peroxisomes are cellular organelles which have been identified in close proximity to mitochondria in cells of the adrenal

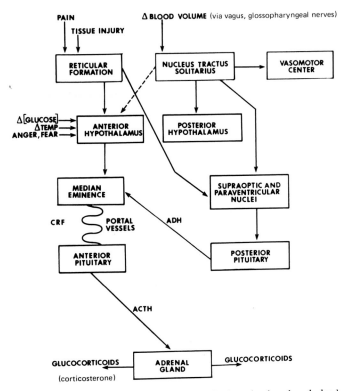

Fig. 22. Control schema for glucocorticord release by the adrenal gland. CRF, corticotropin releasing factor; ADH, antidiuretic hormone; ACTH, adrenal corticotropin hormone.

medulla. It is suggested that these structures may be associated with catacholamine metabolism in sympathetic nervous tissue (8).

VIII. NERVOUS SYSTEM

For almost a century, the rat has been a useful experimental animal model for studies involving the nervous system. As was the case for some other systems of the rat, this chapter does not provide sufficient descriptions for the invetigator who is deeply involved in neurological research. A rather voluminous textbook would be required to do justice to all the areas of the nervous system. One of the most detailed anatomical descriptions of the rat nervous system is that by Greene (60). Another excellent text is the one by Konig and Klippel (83). The intent of this discussion is to point out the major anatomical features of the nervous system and not to provide in depth anatomical descriptions.

A. Peripheral Nervous System

The rat is a mammal and as such has a central nervous system and a peripheral nervous system. The latter is com-

posed of 34 pair of spinal nerves; 8 cervical, 13 thoracic, 6 lumbar, 4 sacral, and 3 caudal. These paired spinal nerves emerge from the spinal cord through the intervertebral foramina. Each nerve is formed of dorsal and ventral roots. The dorsal root is sensory and carries impulses into the spinal cord. The ventral root carries most fibers away from the spinal cord. The first cervical nerves lack a ganglion, whereas, the ganglion of the second cervical is exposed in the angle between the atlas and epistropheus. All other spinal ganglia are enclosed within the vertrebral canal. From the second or third thoracic to the first sacral nerves, pairs of rami communicantes connect the spinal nerves with the sympathetic ganglia. In the lumbar region, the rami are long and slightly irregular in their distribution.

Typical plexuses are formed in the cervical, brachial, and lumbosacral regions by the ventral divisions of the nerves. A detailed description of the distribution of the terminal branches has been described in a previous report (60,87).

Five major nerves arise from the brachial plexus and proceed into the forelimb. These include the ulnar nerve serving the caudal portion of the forearm, the median nerve to the medial surface of the forelimb, the axillary nerve to the shoulder and lateral surface of the forelimb, the musculocutaneous nerve to the shoulder and upper brachium, and the radial nerve to the upper caudal muscles of the forelimb and the lower cranial surface of the forelimb.

The lumbar plexus is formed by the thirteenth thoracic and the first four or five lumbar nerves. Major nerves arising from this plexus include the femoral nerve to the medial surface of the thigh and the saphenous nerve to the medial surface of the thigh.

The spinal cord terminates in a slender filum terminale. The caudal spinal nerves and the filum terminale are collectively called the cauda equina.

The cranial nerves are also considered a part of the peripheral nervous system. Their origins and relations to the brain are illustrated in Fig. 23 and 24. A brief description of each of these nerves is given in Table VII.

The autonomic nervous system is composed of two sympathetic trunks bearing 24 paired ganglia which parallel either side of the vertebral column. Each trunk communicates to the ventral root of the spinal nerves. From the eighth, ninth, and tenth thoracic and first lumbar sympathetic ganglia, four branches arise which unite to form the greater spanchnic nerves of either side. These two nerves pierce the diaphragm to reach the unpaired coeliac ganglion which supplies branches to the diaphragm, cranial mesenteric artery and adjacent viscera.

The lesser splanchnic arises at the level of the third lumbar ganglion and supplies additional branches to the coeliac ganglion. The least splanchnics, or most caudal branches, arise from the third and fourth lumbar ganglia and supply the aortic and caudal mesenteric plexuses.

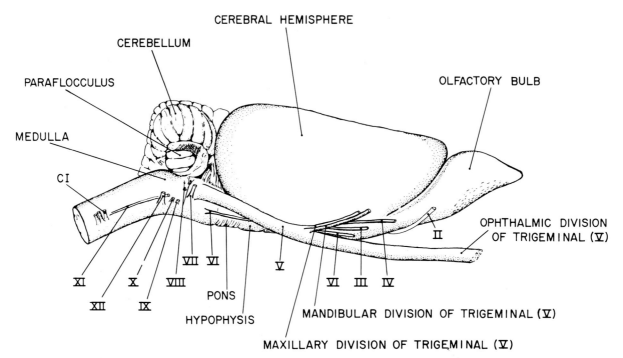

Fig. 23. Lateral view of brain.

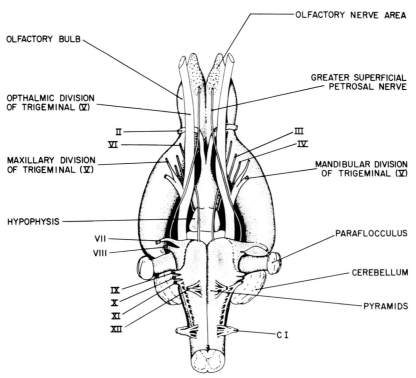

Fig. 24. Ventral view of brain.

Table VII

Chart of the Cranial Nerves

Name	Superficial origin on brain	Foramen of exit from, or entrance to, cranial cavity	Action	Distribution
I. Olfactory	Olfactory bulb	Cribiform plate of ethmoid	Sensory	Nasal epithelium
II. Optic	Anterior corpora bigemina and thalamus	Optic foramen	Sensory	Retina of eye
III. Oculomotor	Pedunculi cerebri	Anterior lacerated foramen	Motor	Dorsal, ventral and medical recti and ventral oblique muscles
IV. Trochlear	Posterior to posterior corpora bigemina	Anterior lacerated foramen	Motor	Dorsal oblique muscle
V. Trigeminal	Pons	Anterior lacerated foramen (ophthalmic and dorsal maxillary branches) and foramen ovale (ventral maxillary branch)	Motor and sensory	Skin, vibrissae, tongue, teeth, muscles of mastication
VI. Abducens	Anterior medulla oblongata	Anterior lacerated foramen	Motor	Lateral rectus muscle
VII. Facial	Medulla oblongata near V	Facial canal to stylomastoid foramen	Sensory and motor	Facial muscles, tongue, salivary and lacrimal glands
VIII. Vestibulochlear	Medulla oblongata, posterior to VII	Internal acoustic meatus	Sensory	Organs of equilibrium and hearing in inner ear
IX. Glossopharyngeal	Medulla oblongata near X	Posterior lacerated foramen	Sensory and motor	Pharynx, tongue, carotid sinus and parotid gland
X. Vagus	Medulla oblongata posterior to VIII	Posterior lacerated foramen	Sensory and motor	Pharynx, larynx, heart, lungs, diaphragm and stomach
XI. Spinal accessory	Medulla oblongata and anterior end of spinal cord	Posterior lacerated foramen	Motor	Muscles of neck and pharyngeal viscera with vagus
XII. Hypoglossal	Medulla oblongata posterior to X	Hypoglossal canal	Motor	Intrinsic, extrinsic tongue muscles

B. Central Nervous System

The central nervous system is composed of the brain and spinal cord. These two structures are covered by three layers of membranes, called the meninges. The outermost of the three membranes is the dura mater. On the brain, the dura mater is folded between the cerebral hemispheres as the falx cerebri and between the cerebrum and cerebellum as the tentorium cerebelli. Along the spinal cord, the dura mater is the outer tough fibrous coat which closely invests or surrounds each spinal nerve for a distance as it branches off from the spinal cord. The innermost layer of the meninges is the very thin and delicate pia mater which adheres to the brain and spinal cord. Between the dura mater and pia mater is a network of thin fibers called the arachnoid.

The brain is composed of two large lessencephalic hemispheres separated by a median fissure. The frontal lobes are not clearly differentiated from the larger and more caudal temporal lobes. Two rather large olfactory bulbs project from the rostral end of the brain. The olfactory nerves terminate here after passing through the fenestrated ethmoid plate from the nasal epithelium. The olfactory lobes are connected to the hippocampal lobes via the olfactory tracts (60).

The convoluted cerebellum occupies the major portion of the caudal end of the brain. It consists of the median vermis and two later hemispheres.

There are four rounded bodies, the corpora quadrigemina, which lie between the cerebral hemispheres and the vermis of the cerebellum. The caudal pair are associated with acute hearing, whereas, the rostral pair are associated with vision.

The most caudal portion of the brain is composed of the medulla oblongata as it tapers back into the spinal cord.

On the ventral aspect of the brain, the origins of the cranial nerves can be seen (Fig. 24). Other structures of importance include the pituitary gland consisting of the stalk, the neurohypophysis, and adenohypophysis. On either side of the infundibulum are fiber tracts, the pendunculi cerebri, which serve to connect the cerebrum and the medulla oblongata. Caudal to the hypophysis can be found the fiber tracts and nuclei of the pons which serve as relay centers to connect the cerebellar hemispheres and the cerebral cortex. Extending caudally from the pons on the ventral aspect of the medulla oblongata are the pyramids. These are fiber tracts carrying information from the cerebral cortex to centers within the spinal cord (27).

The brain, when viewed on a median sagittal plane (Fig. 25) shows a prominent corpus callosum or middle commissure. This is a white band of fibers linking the two cerebral hemi-

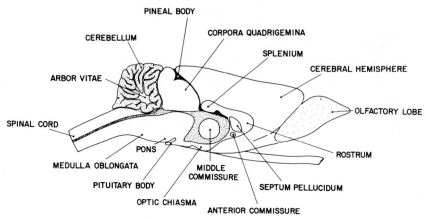

Fig. 25. Brain, median sagittal plane.

spheres. A circular structure lying in the center of the third ventricle is termed the intermediate mass of the thalamus. A very small white fiber tract lies rostral to the intermediate mass and is the rostral commissure. It is closely associated with another band of white fibers connecting the two cerebral hemispheres, the fornix. This structure runs ventral to the corpus callosum, dorsal to the fornix and rostral to the third ventricle. Also identifiable are the crura cerebri, thickened masses of fibers which lie on the floor of the brain and link the fore- and hindbrains. Other structures of importance include the corpora quadrigemina discussed previously, the pineal body, the optic chiasm, the medulla oblongata, the arbor vitae of the cerebellum and the ventricles of the brain. The ventricular system of the rat consists of one central ventricle (the third) connected with two lateral ones (first and second), and a caudal one (the fourth). Each lateral ventricle communicates with the third ventricle through the interventricular foramen (foramen of Monroe) which is on the ventral side of the anterior horn about 1 mm from its caudal end. The cerebral aqueduct is about 3.0 mm long and connects the third and fourth ventricle. The ventricular system communicates with the cerebral subrachnoid space by two lateral aperatures (the foramen of Luschka) only. There does not appear to be a median aperature (foramen of Magendie) in the rat (89).

IX. SPECIAL SENSE ORGANS

The Eye

The eye of the rat shows the main characteristics of all mammalian eyes. Congenital anomalies of the cornea in the rat are few. The normal cornea consists of five layers and is transparent due to regularity of its structural composition and to other factors of a chemical nature which are still incompletely understood. It is approximately 6.80 ± 0.2 mm in diameter in adult males and approximately 6.40 ± 0.1 mm in adult females. The refractive power of the cornea, which is a function of the refraction of its tissue and of the radius of curvature of its surface, is almost twice as high as the lens (68).

The iris arises from the anterior portion of the ciliary body and is a rostral continuation of the choroid. It is completely devoid of pigment in the albino rat and consists of a loose, highly vascular connective tissue. Few anomalies of the iris have been recorded (112).

The pupil is the opening through the center of the iris, and it dilates and contracts to regulate the amount of light entering the eye. Persistence of the pupillary membrane has been seen and is recognized as multiple strands of tissue crossing the pupillary membrane. The incidence of this defect is estimated to be 0.5% (68).

The anterior chamber of the eye is bordered externally by the cornea and internally by the iris and pupil. The posterior chamber of the eye is bordered rostrally by the iris and caudally by the lens. Both the anterior and posterior chambers are filled with aqueous humor. Synechia is frequently found in both chambers (68).

The lens is a transparent, biconvex spherical body which occupies about two-thirds of the intraocular cavity. It is attached to the ciliary body by the suspensory ligament and changes its shape during the process of accommodation. The rate of growth of the lens as investigated by Norrby (112) has shown an increase in weight that follows an asymptotic curve. With advancing age, the rate of growth of the lens and the density of the lens nucleus increases. Typically, the lens of the rat is transparent, and posterior suture lines are poorly discernible in the adult. Occasionally, lens opacities and posterior capsular cataracts are observed. The cataracts are usually unilateral and only rarely is a total cataract observed; a high incidence of spontaneous lenticular changes have been demonstrated biomicroscopically, particularly of the anterior and posterior cortex, the anterior sutures, and the nucleus in Sprague-Dawley rats, but rarely did the changes lead by opacity (13). A similar high incidence of change in the lens of Sherman rats was demonstrated by Balazs, Ohtake, and Noble (12).

The vitreous body fills the vitral cavity between the lens and the retina. The fresh vitreous body is a colorless, structureless, gelatinous mass with a glasslike transparency. Almost 99% of the vitreous body consists of water, the remainder being hydrophilic polysaccharide, especially hyaluronic acid.

In the rat, the hyaloid vessels form two distinct entities. There are the central hyaloid vessels, which consist of three to five arteries running from the disc to the caudal pole of the lens. It is this central leash of vessels which can cause posterior capsular damage which cannot be differentiated ophthalmoscopically from a cataract. Additional hyaloid vessels run from the disc to the equator of the lens and lie close to the surface of the retina (68). Closure of these vessels normally occurs between the seventh and twenty-second day of age (23). Remnants of the hyaloid vessels are seen in 60% of the rats at 6 weeks of age, 5–6% of the rats at 26 weeks of age, and in less than 17% of the rats 1 year of age and older.

The retina is the innermost of the three coats of the eyeball and is responsible for the reception and transduction of light stimuli and the transmission of these signals in the form of nerve impulses to the appropriate portions of the brain.

The retina is divisible into ten distinct layers: pigmented epithelium, photoreceptor cell layer, external limiting membrane, outer nuclear layer, outer plexiform layer, inner nuclear layer, inner plexiform layer, ganglionic cell layer, optic nerve fiber layer, and inner limiting membrane. Several layers are identified in Fig. 26. Because of its developmental origin from the prosencephalon, it is usually considered to be a specially differentiated part of the brain.

In the rat, the fundus shows a pink reflex, and there is no disc or macula. The hyaloid vessels persist for the first 10 days of life and after 12 days of age, the retinal vasculature can be examined. The position of the optic disc is defined as the point from which the retinal vessels radiate outward. There are usually five to eight straight and even-calibered arteries and an equal number of slightly broader and darker veins. Studies have shown that normal light intensities in the laboratory for 24 hr a day for 10 days can easily damage the retina of the rat (68).

Recent work has called attention to a new cell-to-cell contact in the rat retina which share some of the properties of the "basal junctions" in the turtle retina and "flat, superficial contacts" in the frog and primate retina. These contacts re-

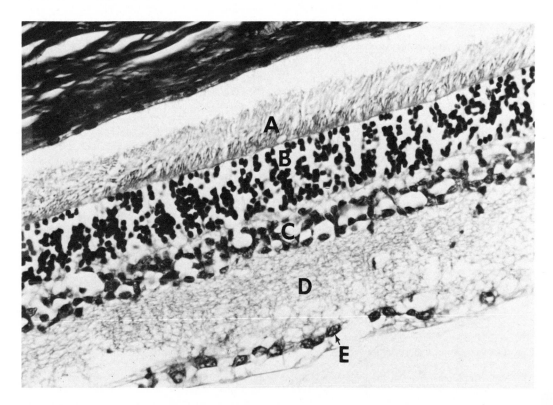

Fig. 26. Photomicrograph of the layers of the retina. (A) Rod and cone layer, (B) outer nuclear layer, (C) inner nuclear layer, (D) inner plexiform, (E) ganglion cell layer. ×40.

semble septatelike contacts in the vertebrate nervous system and between adrenocortical cells. They are not considered to be synapses, but may function in giving rigidity to the receptor terminal (139).

For a more in depth understanding of this area, the reader is referred to the works of Dubin (40) and Sosula and Glow (148). Their findings regarding the fine structural organization of the inner plexiform layer have made a significant contribution to our understanding of the synaptic relationships in the rat retina. It is generally accepted that in most mammals, especially the rat, light influences a large number of autonomic functions and that this influence must be mediated by nervous connections between the retina and the hypothalmus. A nervous pathway which connects the visual system with the hypothalmus was described (40).

Also, differences in organization of the ipsilateral retinal projections have been demonstrated between pigmented and albino strains of rats. The genetic explanation of the attendant albinism and misrouted retinogeniculate axons remains unknown (34).

REFERENCES

1. Abdallah, A., and Tawfik, T. (1969). The anatomy and histology of the kidney of sand rats *(Psammonys obesus). Z. Versuchstierkd.* **11**, 261–275.
2. Ackermann, U. (1975). An approach to the measurement of body fluid compartment volumes in non-steady state conditions in the rat. *Pfluegers Arch.* **355**, 141–150.
3. Adolph, E. F. (1947). Tolerance to heat and dehydration in several species of mammals. *Am. J. Physiol.* **151**, 564–575.
4. Albritton, E. C. (1954). "Standard Values in Nutrition and Metabolism." Saunders, Philadelphia, Pennsylvania.
5. Anand, C., and Han, S. S. (1975). Effects of 5-fluorouracil on exocrine glands. III. Fine structure of Brunner's glands of rats. *J. Anat.* **119**, 1–17.
6. Arataki, M. (1926). On the postnatal growth of the kidney, with special reference to the number and size of the glomeruli (albino rat). *Am. J. Anat.* **36**, 399–436.
7. Arendshorst, W. J., Finn, W. F., and Gottschalk, C. W. (1975). Autoregulation of blood flow in the rat kidney. *Am. J. Physiol.* **288**, 127–133.
8. Arnold, G., and Holtzman, E. (1975). Peroxisomes in rat sympathetic ganglia and adrenal medulla. *Brain Res.* **83**, 509–515.
9. Babero, B. B., Yousef, M. K., Wawerna, J. C., and Bradley, W. G. (1973). Comparative histology of the respiratory appartus of three desert rodents and the albino rat. A view of morphological adaptations. *Comp. Biochem. Physiol.* **44**, 585–597.
10. Baines, A. D., and de Rouffignac, C. (1969). Functional heterogeneity of nephrons. II. Filtration rates, intraluminal flow velocities and fractional water reabsorption. *Pfluegers Arch.* **308**, 260–276.
11. Baker, R. D., Searle, G. W., and Nunn, A. S. (1961). Glucose and sorbose absorption at various levels of rat small intestine. *Am. J. Physiol.* **200**, 301–304.
12. Balazs, T., Ohtake, S., and Noble, J. F. (1970). Spontaneous lenticular changes in the rat. *Lab. Anim. Care* **20**, 215.
13. Balazs, T., and Rubin, L. (1971). A note on the lens in aging Sprague-Dawley rats. *Lab. Anim. Sci* **21**, 267.
14. Berry, R. J. (1901). The true caecal apex, or the vermiform appendix: Its minute and comparative anatomy. *J. Anat. Physiol. Norm. Pathol. Homme Anim.* **35**, 92.
15. Bertalanffy, F. D., and Leblond, C. P. (1953). The continuous renewal of the two types of alveolar cells in the lung of the rat. *Anat. Rec.* **115**, 515–539.
16. Blanchard, M., Green, D. E., Nocito, V., and Ratner, S. (1944). L-Amino acid oxidase of animal tissue. *J. Biol. Chem.* **155**, 421–440.
17. Booth, C. C., and Brain, M. C. (1962). The absorption of tritium-labelled pyridoxine hydrochloride in the rat. *J. Physiol. (London)* **164**, 282–294.
18. Boyne, R., Fell, B. F., and Robb, I. (1966). The surface of the intestinal mucosa in the lactating rat. *J. Physiol. (London)* **183**, 570–575.
19. Brenner, B. M., Troy, J. L., and Daugharty, T. M. (1972). Pressures in cortical structures of the rat kidney. *Am. J. Physiol.* **222**, 246–251.
20. Brunser, O., and Luft, H. J. (1970). Fine structure of the apex of absorptive cells from rat small intestine. *J. Ultrastruct. Res.* **31**, 291–311.
21. Burri, P. H. (1974). The postnatal growth of the rat lung. III. Morphology. *Anat. Rec.* **180**, 77–98.
22. Butcher, E. O. (1934). The hair cycles in the albino rat. *Anat. Rec.* **61**, 5–19.
23. Cairns, J. E. (1959). Normal development of the hyaloid and retinal vessels in the rat. *Br. J. Ophthalmol.* **43**, 385–393.
24. Calvert, R. (1975). Structure of rat ultimobranchial bodies after birth. *Anat. Rec.* **181**, 561–580.
25. Carroll, N., Crock, G. W., Funder, C. C., Green, C. R., Ham, K. N., and Tange, J. D. (1974). Scanning electron microscopy of the rat renal papilla. *J. Anat.* **117**, 447–452.
26. Chew, R. M. (1965). Water metabolism in mammals. *Physiol. Mammol.* **2**, 143–178.
27. Chiasson, R. B. (1958). "Laboratory Anatomy of the White Rat," pp. 1–36. W. C. Brown, Dubuque, Iowa.
28. Clark, S. L., Jr. (1959). The ingestion of proteins and colloidal materials by columnar absorptive cells of the small intestine in suckling rats and mice. *J. Biophys. Biochem. Cytol.* **5**, 41–50.
29. Clarke, R. M. (1970). Mucosal architecture and epithelial cell production rate in the small intestine of the albino rat. *J. Anat.* **107**, 519–529.
30. Clarke, R. M. (1977). The effects of age on mucosal morphology and epithelial cell production in rat small intestine. *J. Anat.* **123**, 805–811.
31. Cleaton-Jones, P. (1972). Anatomical observations in the soft palate of the albino rat. *Anat. Anz.* **131**, 419–424.
32. Cotchin, E., and Roe, F. J. C. (1967). "Pathology of Laboratory Rats and Mice." Blackwell, Oxford.
33. Cowan, M. J., and Crystal, R. G. (1975). Lung growth after unilateral pneumonectomy: quantitation of collagen synthesis and content. *Am. J. Respir. Dis.* **111**, 267–277.
34. Cunningham, T. J., and Lund, R. D. (1971). Laminar patterns in the dorsal division of the latera- geniculate nucleus of the rat. *Brain Res.* **34**, 394–398.
35. Davis, T. R. A., Johnston, D. R., Bell, F. C., and Cremer, B. J. (1960). Regulation of shivering and nonshivering heat production during acclimation of rats. *Am. J. Physiol.* **198**, 471–475.
36. Dhom, G., von Seebach, H. V., and Stephan, G. (1971). Der Geschlechtsdimorphismus der Nebennierenrinde der Ratte; Lichtmikroskopische und histometrische Untersuchungen. *Z. Zellforsch. Mikrosk. Anat.* **116**, 119–135.
37. Domenech-Mateu, J. M., and Boya-Vegué, J. (1975). An ultrastructural study of sinoatrial node cells in the embryonic rat heart. *J. Anat.* **119**, 77–83.

38. Donaldson, H. H. (1924). "The Rat: Data and Reference Tables," 2nd ed., revised and enlarged, Mem. Wist. Inst. Anat. Biol., No. 6. Wistar Inst. Anat. Biol., Philadelphia, Pennsylvania.

39. Dowdle, E. B., Schachter, D., and Shenker, H. (1960). Active transport of Fe^{59} by everted sacs of rat duodenum. *Am. J. Physiol.* **198,** 609–613.

40. Dubin, M. W. (1970). The inner plexiform layer of the vertebrate retina: A quantitative and comparative electron microscopic analysis. *J. Comp. Neurol.* **140,** 479–505.

41. Dugal, L. P., and Thérien, M. (1947). Ascorbic acid and acclimatization to cold environment. *Can. J. Res., Sect. E* **25,** 111–136.

42. Edwards, V. D., and Simon, G. T. (1970). Ultrastructural aspects of red cell destruction in normal rat spleen. *J. Ultrastruct. Res.* **33,** 187–201.

43. Farris, E. J., and Griffith, J. Q. (1949). "The Rat in Laboratory Investigation," 2nd ed. Lippincott, Philadelphia, Pennsylvania.

44. Fawcett, D. W. (1952). A comparison of the histological organization and cytochemical reactions of brown and white adipose tissues. *J. Morphol.* **90,** 363–405.

45. Fawcett, D. W., and Jones, I. C. (1949). The effects of hypophysectomy, adrenalectomy and of thiouracil feeding on the cytology of brown adipose tissue. *Endocrinology* **45,** 609–621.

46. Fine, L. G., Lee, H., Goldsmith, D., Weber, H., and Blaufox, M. D. (1974). Effects of catheterization of renal artery on renal function in the rat. *J. Appl. Physiol.* **37,** 930–933.

47. Fisher, R. B., and Parsons, D. S. (1953). Galactose absorption from the surviving small intestine of the rat. *J. Physiol. (London)* **119,** 224–232.

48. Flamenbaum, W., Huddleston, M. L., McNeil, J. S., and Hamburger, R. J. (1974). Uranyl nitrate-induced acute renal failure in the rat: Micropuncture and renal hemodynamic studies. *Kidney Int.* **6,** 408–418.

49. Fraser, D. A. (1931). The winter pelage of the adult albino rat. *Am. J. Anat.* **47,** 55–87.

50. Friedman, M., and Feed, S. C. (1949). Microphonic manometer for indirect determination of systolic blood pressure in the rat. *Proc. Soc. Exp. Biol. Med.* **70,** 670–672.

51. Fuller, P. M., and Huelke, D. F. (1973). Kidney vascular supply in the rat, cat and dog. *Acta Anat.* **84,** 516–522.

52. Gaunt, R. (1969). The functions of the enucleate adrenal. *Trans. N.Y. Acad. Sci.* [2] **31,** 256–260.

53. Gavras, H., Brunner, H. R., Laragh, J. H., Vaughan, E. D., Jr., Koss, M., Cote, L. J., and Gavras, I. (1975). Malignant hypertension resulting from dexoycorticosterone acetate and salt excess. *Circ. Res.* **36,** 300–309.

54. Gay, W. I., ed. (1965). "Methods of Animal Experimentation," Vol. 1. Academic Press, New York.

55. Gelineo, S. (1964). Organ systems in adaptation. *In* "Handbook of Physiology" (D. B. Dill *et al.,* eds.), Sect. 4, pp. 259–282. Williams & Wilkins, Baltimore, Maryland.

56. Gertz, K. H., Mangos, J. A., Braun, G., and Pagel, H. D. (1966). Pressure in the glomerular capillaries of the rat kidney and its relation to arterial blood pressure. *Pfluegers Arch. Gesamte Physiol. Menschen Tiere* **288,** 369–374.

57. Giebisch, G., and Windhager, E. E. (1973). Electrolyte transport across renal tubular membranes. *In* "Handbook of Physiology," (J. Field, ed.), Sect. 8, Renal Physiology, pp. 315–376. Williams & Wilkins, Baltimore, Maryland.

58. Gilson, S. B. (1949). Studies on proteinuria in the rat. *Proc. Soc. Exp. Biol. Med.* **72,** 608–613.

59. Goldblatt, H., Lynch, J., Hanzal, R. G., and Summerville, W. W. (1934). Studies on experimental hypertension. I. The production of persistent elevation of systolic blood pressure by means of renal ischemia. *J. Exp. Med.* **59,** 347–379.

60. Greene, E. C. (1935). "Anatomy of the Rat." Hafner, New York.

61. Hall, H. D., and Schneyer, C. A. (1964). Paper electrophoresis of rat salivary secretions. *Proc. Soc. Exp. Biol. Med.* **115,** 1001–1005.

62. Halpern, M. H. (1955). The sinoatrial node of the rat heart. *Anat. Rec.* **123,** 425–435.

63. Hamilton, A. I., and Blackwood, H. J. J. (1974). Cell renewal of oral mucosal epithelium of the rat. *J. Anat.* **117,** 313–327.

64. Harrison, G. A. (1974). Ultrastructural response of rat lung to 90 days' exposure to oxygen at 450 mm Hg. *Aerosp. Med.* **45,** 1041–1045.

65. Heap, B. J., and Kiernan, J. A. (1973). Histological, histochemical and pharmacological observations on mast cells in the stomach of the rat. *J. Anat.* **115,** 315–325.

66. Hebb, C. (1969). Motor innervation of the pulmonary blood vessels of mammals. *In* "The Pulmonary Circulation and Interstitial Space" (A. P. Fishman and H. H. Hecht, eds.), pp. 195–222. Univ. of Chicago Press, Chicago, Illinois.

67. Hensel, H., and Hildebrandt, G. (1964). Organ systems in adaptation: The muscular system. *In* "Handbook of Physiology" (D. B. Dill *et al.,* eds.), Sect. 4, pp. 73–90. Williams & Wilkins, Baltimore, Maryland.

68. Heywood, R. (1973). Some clinical observations on the eyes of Sprague-Dawley rats. *Lab. Anim. Sci.* **7,** 19–27.

69. Hierholzer, K., Müller-Suur, R., Gutsche, H. U., Butz, M., and Lichtenstein, I. (1974). Filtration in surface glomeruli as regulated by flow rate through the loop of Henle. *Pfluegers Arch.* **352,** 315–337.

70. Horster, M., and Thurau, K. (1968). Micropuncture studies on the filtration rate of single superficial and juxtamedullary glomeruli in the rat kidney. *Pfluegers Arch.* **301,** 162–181.

71. Hoskins, M. M., and Chandler, S. B. (1925). Accessory parathyroids in the rat. *Anat. Rec.* **30,** 95–98.

72. Howell, A. B. (1926). "Anatomy of the Wood Rat. Comparative Anatomy of the Subgenera of the American Wood Rat (Genus Neotoma)." Williams & Wilkins, Baltimore, Maryland.

73. Hughes, P. C. R., and Tanner, J. M. (1970). A longitudinal study of the growth of the black-hooded rat; methods of measurement and rates of growth for skull, limbs, pelvis, nose-rump and tail lengths. *J. Anat.* **106,** 349–370.

74. Hughes, P. C. R., and Tanner, J. M. (1970). The assessment of skeletal maturity in the growing rat. *J. Anat.* **106,** 371–402.

75. Huntington, G. S., and Schulte, H. von W. (1913). "Studies in Cancer and Allied Subjects," Vol. IV, pp. 315–324. Columbia Univ. Press, New York.

76. Ingle, D. J., and Baker, B. O. (1953). A consideration of the relationship of experimentally produced and naturally occurring pathologic changes in the rat to the adaptation diseases. *Recent Prog. Horm. Res.* **8,** 143–169.

77. Jayaraj, A. P. (1972). Hair growth in oral cavity. *Acta Anat.* **83,** 367–371.

78. Jeffery, P. K., and Redi, L. (1975). New observations of rat airway epithelium: A quantitative and electron microscopic study. *J. Anat.* **120,** 295–320.

79. Källskog, O., Lindbom, L. O., Ulfendahl, H. R., and Wolgast, M. (1975). Kinetics of the glomerular ultrafiltration in the rat kidney. An experimental study. *Acta Physiol. Scand.* **95,** 293–300.

80. Khorshid, M. R., and Moffat, D. B. (1974). The epithelia lining the renal pelvis in the rat. *J. Anat.* **118,** 561–569.

81. Kikkawa, Y. (1970). Morphology of alveolar lining layer. *Anat. Rec.* **167,** 389–400.

82. Kim, S. K., Han, S. S., and Nasjleti, C. E. (1970). The fine structure of secretory glanules in submandibular glands of the rat during early postnatal development. *Anat. Rec.* **168,** 463–476.

83. Kohn, P., Dawes, E. D., and Duke, J. W. (1965). Absorption of

carbohydrates from the intestine of the rat. *Biochim. Biophys. Acta* **107**, 358–362.

83a. König, J. F., and Klippel, R. A. (1963). "The Rat Brian, A Stereotoxic Atlas of the Forebrain and Lower Parts of the Brain Stem." Williams & Wilkins, Baltimore, Maryland.

84. Krahl, V. E. (1964). Anatomy of the mammalian lung. *In* "Handbook of Physiology" (J. Field, ed.), Sect. 3, Vol. I, pp. 213–284. Williams & Wilkins, Baltimore, Maryland.

85. Krebs, :H. A. (1935). Metabolism of amino acids. IV. The synthesis of glutamine from glutamic acid and ammonia, and the enzymic hydrolysis of glutamine in renal tissues. *Biochem. J.* **29**, 1951–1969.

86. Krstic, R. (1974). Ultrastructure of rat pineal gland after preparation by freeze-etching technique. *Cell Tissue Res.* **148**, 371–379.

87. Langley, L. L., Telford, I. R., and Christensen, J. B. (1969). "Dynamic Anatomy and Physiology," 3rd ed. McGraw-Hill (Blakiston), New York.

88. Lee, M. M. C., Chu, P. C., and Chan, H. C. (1969). Effects of cold on the skeletal growth of albino rats. *Am. J. Anat.* **124**, 239–249.

89. Levinger, I. M. (1971). The cerebral ventricles of the rat. *J. Anat.* **108**, 447–451.

90. Liljestrand, G., and Zotterman, Y. (1954). The water taste in mammals. *Acta Physiol. Scand.* **32**, 291–303.

91. Low, F. N. (1952). Electron microscopy of the rat lung. *Anat. Rec.* **113**, 437–449.

92. Low, F. N. (1963). The pulmonary alveolar epithelium of laboratory mammals and man. *Anat. Rec.* **117**, 241–263.

93. Lucas, J., and Floyer, M. A. (1974). Changes in body fluid distribution and interstitial tissue compliance during the development and reversal of experimental renal hypertension in the rat. *Clin. Sci. Mol. Med.* **47**, 1–11.

94. Luciano, L., Real, E., and Ruska, H. (1968). Uber eine glykogenhaltige Bürstenzelle im Rectum der Ratte. *Z. Zellforsch. Mikrosk. Anat.* **91**, 153–158.

95. Ludatscher, R. M. (1968). Fine structure of the muscular wall of rat pulmonary veins. *J. Anat.* **103**, 345–357.

96. Macintosh, S. R. (1975). Observations on the structure and innervation of the rat snout. *J. Anat.* **119**, 537–546.

97. Mann, F. C. (1920). A comparative study of the anatomy of the sphincter at the duodenal end of the common bile-duct with special reference to species of animals without a gall-bladder. *Anat. Rec.* **18**, 355–360.

98. Matheson, D. F. (1968). Studies of the amino acid incorporation into peripheral nerves. Ph.D. Thesis, pp. 1–62. University of London.

99. McGrath, P. (1976). Further observations on the pharyngeal hypophysis and the postsphenoid in the mature male rat. *J. Anat.* **121**, 193–201.

100. McLaughlin, R. F., Tyler, W. S., and Canada, R. O. (1961). A study of the subgross pulmonary anatomy in various mammals. *Am. J. Anat.* **108**, 149–165.

101. McLean, J. M., Mosley, J. G., and Gibbs, A. C. C. (1974). Changes in the thymus, spleen and lymph nodes during pregnancy and lactation in the rat. *J. Anat.* **118**, 223–229.

102. McMaster, P. D. (1922). Do species lacking a gall bladder possess its functional equivalent? *J. Exp. Med.* **35**, 127–140.

103. Miller, I. J., and Preslar, J. A. (1975). Spatial distribution of rat fungiform papillae. *Anat. Rec.* **181**, 679–684.

104. Moe, H., and Bojsen-Moller, F. (1971). The fine structure of the lateral nasal gland (Steno's gland) of the rat. *J. Ultrastruct. Res.* **36**, 127–148.

105. Mohring, J., Mohring, B., Naumann, N. J., Philippi, A., Homsy, E., Orth, H., Dauda, G., Kazda, S., and Gross, F. (1975). Salt and water balance and renin activity in renal hypertension of rats. *Am. J. Physiol.* **228**, 1847–1855.

106. Moore, R. E., and Underwood, M. C. (1963). The thermogenic effects of noradrenaline in newborn and infant kittens and other small animals. *J. Physiol. (London)* **168**, 290–317.

107. Muir, A. R. (1955). The semiatrial node of the rat heart. *Q. J. Exp. Physiol. Cogn. Med. Sci.* **40**, 378–386.

108. Muller, H. B. (1969). The postnatal development of the rat Harderian gland. I. Light microscopic observations. *Z. Zellforsch. Mikrosk. Anat.* **100**, 421–438.

109. Neil, M. W. (1959). The absorption of cystine and cysteine from rat small intestine. *Biochem. J.* **71**, 118–124.

110. Neubig, R. R., and Hoobler, S. W. (1975). Reversal of chronic renal hypertension: Role of salt and water excretion. *Proc. Soc. Exp. Biol. Med.* **150**, 254–256.

111. Norman, A. F., and DeLuca, F. H. (1963). The preparation of H^3-vitamins D_2 and D_3 and their localization in the rat. *Biochemistry* **2**, 1160–1168.

112. Norrby, A. (1958). On the growth of the crystalline lens, the eyeball and cornea in the rat. *Acta Ophthalmol., Supp.* **49**, 5.

113. Oates, H. F., Stokes, G. S., and Storey, B. G. (1975). Plasma renin concentration in hypertension produced by unilateral renal artery constriction in the rat. *Clin. Exp. Pharmacol. Physiol.* **2**, 289–296.

114. Okamoto, K., and Aoki, K. (1963). Development of a strain of spontaneously hypertensive rats. *Jpn. Circ. J.* **27**, 282–293.

115. Osvaldo-Decima, L. (1973). Ultrastructure of the lower nephron. *In* "Handbook of Physiology" (J. Field, ed.), Sect. 8, Renal Physiology, pp. 81–102. Williams & Wilkins, Baltimore, Maryland.

116. Parkins, R. A., Dimitriadou, A., and Both, C. C. (1960). The rates and sites of absorption of ^{131}I-labelled albumin and sodium ^{131}I in the rat. *Clin. Sci.* **19**, 595–604.

117. Parratt, J. R., and West, G. B. (1957). 5-Hydroxytryptamine and tissue mast cells. *J. Physiol. (London)* **137**, 169–192.

118. Permezel, N. C., and Webling, D. D'A. (1971). The length and mucosal surface area of the small and large gut in young rats. *J. Anat.* **108**, 295–296.

119. Pfeffer, M. A., and Frohlich, E. D. (1972). Electromagnetic flowmetry in anesthetized rats. *J. Appl. Physiol.* **33**, 137–140.

120. Pfeiffer, E. W. (1968). Comparative anatomic observations of the mammalian renal pelvis and medulla. *J. Anat.* **102**, 321–331.

121. Pitts, R. F. (1973). Production and secretion of ammonia in relation to acid–base regulation. *In* "Handbook of Physiology" (J. Field, ed.), Sect. 8, Renal Physiology, pp. 455–496. Williams & Wilkins, Baltimore, Maryland.

122. Popper, H., and Volk, B. W. (1944). Absorption of vitamin A in the rat. *Arch. Pathol.* **38**, 71–75.

123. Porter, J. C., and Klaiker, M. S. (1965). Corticosterone secretion as a function of ACTH input and adrenal blood flow. *Am. J. Physiol.* **209**, 811–814.

124. Pullen, A. H. (1975). Studies on the histochemical and physiological maturation of skeletal muscle in the rat. Ph.D. Thesis, pp. 26–75. University of London.

125. Pullen, A. H. (1976). A parametric analysis of the growing CFHB (Wistar) rat. *J. Anat.* **121**, 371–383.

126. Rakshit, P. (1949). Communicating blood vessels between bronchial and pulmonary circulations in the guinea pig and rat. *Q. J. Exp. Physiol. Cogn. Med. Sci.* **35**, 47–53.

127. Rasmussen, A. T. (1916). Theories of hiberation. *Am. Natl.* **50**, 609–625.

128. Rasmussen, A. T. (1922). The glandular status of brown multilocular adipose tissue. *Endocrinology* **6**, 760–770.

129. Rayne, J., and Crawford, G. N. C. (1972). The growth of the muscles of mastication in the rat. *J. Anat.* **113**, 391–408.

130. Rigas, D. A., and Heller, C. G. (1951). The amount and nature of

urinary proteins in normal human subjects. *J. Clin. Invest.* **30,** 853–861.

131. Robert, A. (1971). Proposed terminology for the anatomy of the rat stomach. *Gastroenterology* **60,** 344–345.

132. Rolleston, G. A. (1870). "Forms of Animal Life; being outlines of zoological classification based upon anatomical investigation, and illustrated by descriptions of specimens and of figures." Oxford Univ. Press (Clarendon), London and New York.

133. Ross, D. B. (1962). *In vitro* studies on the transport of magnesium across the intestinal wall of the rat. *J. Physiol. (London)* **160,** 417–428.

134. Rowett, H. G. Q. (1957). "Dissection Guides, the Rat with Notes on the Mouse," pp. 1–28. Holt, New York.

135. Saski, Y., and Wagner, H. N., Jr. (1971). Measurement of the distribution of cardiac output in unanesthetized rats. *J. Appl. Physiol.* **30,** 879–884.

136. Schacter, D., Dowdle, E. B., and Schenker, H. (1960). Active transport of calcium by the small intestine of the rat. *Am. J. Physiol.* **198,** 263–268.

137. Schmidt-Nielsen, B., and O'Dell, R. (1961). Structure and concentrating mechanism in the mammalian kidney. *Am. J. Physiol.* **200,** 1119–1124.

138. Schwartz, N. B., and Waltz, P. (1973). Role of ovulation in the regulation of the estrous cycle. *In* "Engineering Principles in Physiology" (J. H. G. Brown and D. S. Gann, eds.), Vol. 1, pp. 249–260. Academic Press, New York.

139. Schwartz, W. J. (1973). A septate-like contact in the rat retina. *J. Neurocytol.* **2,** 85–89.

140. Sealander, J. A., and Hoffman, C. E. (1967). "Laboratory Manual of Elementary Mammalian Anatomy, with Emphasis on the Rat," 2nd ed., pp. 8–38. Burgess, Minneapolis, Minnesota.

141. Sellers, E. A., Scott, J. W., and Thomas, N. (1954). Electrical activity of skeletal muscle of normal and acclimatized rats on exposure to cold. *Am. J. Physiol.* **177,** 372–376.

142. Selye, H., Hall, C. E., and Rowley, E. M. (1943). Malignant hypertension produced by treatment with desoxycorticosterone acetate and sodium chloride. *Can. Med. Assoc. J.* **49,** 88–92.

143. Shelley, H. J. (1961). Glycogen reserves and their changes at birth and in anoxia. *Br. Med. Bull,* **17,** 137–143.

144. Skelton, F. R. (1955). Development of hypertension and cardiovascular-renal lesions during adrenal regeneration in the rat. *Proc. Soc. Exp. Biol. Med.* **90,** 342–346.

145. Slavin, B. G., and Bernick, S. (1974) Morphological studies on denervated brown adipose tissue. *Anat. Rec.* **179,** 497–506.

146. Smith, D. S., Smith, U., and Ryan, J. W. (1972). Freeze-fractured lamellar body membranes of the rat lung great alveolar cell. *Tissue & Cell* **4,** 457–468.

147. Smith, E. M., and Calhoun, M. L. (1968). "The Microscopic Anatomy of the White Rat." Iowa State Univ. Press, Ames.

148. Sosula, L. and Glow, P. H. (1970). A quantitative ultrastructural study of the inner plexiform layer of the rat retina. *J. Comp. Neurol.* **140,** 439–477.

149. Spector, W. S. (1956). "Handbook of Biological Data," Chapter VIII, p. 437. Saunders, Philadelphia, Pennsylvania.

150. Spencer, R. P., Purdy, S., Hoeldtke, R., Bow, T. M., and Markulis,

M. A. (1963). Studies on intestinal absorption of L-ascorbic acid-1-C^{14}. *Gastroenterology* **44,** 768–773.

151. Spencer, R. P., and Samiy, A. H. (1960). Intestinal transport of L-tryptophan *in vitro:* Inhibition by high concentrations. *Am. J. Physiol.* **199,** 1033–1036.

152. State, F. A., el-Eishi, H. I., and Naga, I. A. (1974). The development of taste buds in the foliate papilla of the albino rat. *Acta Anat.* **89,** 452–460.

153. Stendell, W. (1913). Zur vergleichenden Anatomie und Histologie der Hypophysis cerebri. *Arch. Mikrosk. Anat. Entwicklungsmech.* **82,** 289–332.

154. Stoeckel, M. E., and Porte, A. (1970). Embryonic origin and secretory differentiation of the C cells in foetal rat thyroid. *Z. Zellforsch. Mikrosk. Anat.* **106,** 251–268.

155. Strauss, E. W., and Wilson, T. H. (1958). Effect of intrinsic factor in vitamin B$_{12}$ uptake by rat intestine *in vitro. Proc. Soc. Exp. Biol. Med.* **99,** 224–226.

156. Strong, R. M. (1926). The order, time and rate of ossification of the albino rat skeleton. *Am. J. Anat.* **36,** 313–355.

157. Takino, M., and Watanabe, S. (1937). Uber das Vovkommen der Ganglienzellen von unipolaren Typus in der Lunge des Mehschen und Schweiner. *Acta Sch. Med. Univ. Imp. Kioto* **19,** 317–320.

158. Tenney, S. M., and Remmers, J. E. (1963). Comparative quantitative morphology of mammalian lungs. *Nature (London)* **197,** 54–56.

159. Tilney, N. L. (1971). Patterns of lymphatic drainage in the adult laboratory rat. *J. Anat.* **109,** 369–383.

160. Tobian, L., and Nason, P. (1966). The augmentation of proteinuria after acute sodium depletion in the rat. *J. Lab. Clin. Med.* **67,** 224–228.

161. Verhofstad, A. A. J., and Lensen, W. F. J. (1973). On the occurrence of lymphatic vessels in the adrenal gland of the white rat. *Acta Anat.* **84,** 475–483.

162. Verloop, M. C. (1948). The arterial bronchioles and their anastomoses with the arteria pulmonalis in the human lung. *Acta Anat.* **5,** 171–205.

163. Wassermann, D., and Wassermann, M. (1974). The fine structure of adrenal zona glomerulosa in the adult rat. *Cell Tissue Res.* **149,** 235–243.

164. Weibel, E. R. (1969). The ultrastructure of the alveolar-capillary membrane or barrier. *In* "The Pulmonary Circulation and Interstitial Space" (A. P. Fishman and H. H. Hect, eds.), pp. 9–27. Univ. of Chicago Press, Chicago, Illinois.

165. Wells, T. A. G. (1964). "The Rat: A Practical Guide." Heinemann, London.

166. Wesson, L. G., Jr., and Anslow, W. P., Jr. (1948). Excretion of sodium and water during osmotic diuresis in the dog. *Am. J. Physiol.* **153,** 465–474.

167. Windhager, E. E. (1968). "Micropuncture Techniques and Nephron Function." Appleton, New York.

168. Windhager, E. E., and Giebisch, G. (1961). Micropuncture study of renal tubular transfer of sodium chloride in the rat. *Am. J. Physiol.* **200,** 581–590.

169. Zotterman, Y. (1956). Species differences in the water taste. *Acta Physiol. Scand.* **37,** 60–70.

Chapter 5

Hematology and Clinical Biochemistry

Daniel H. Ringler and Lyubica Dabich

I. INTRODUCTION

This chapter provides normal data on the hematology and clinical biochemistry of the rat. It is emphasized that environmental, sampling, and analytical variables affect specific clinical laboratory values. In addition, attention is called to research manipulations and disease entities that can affect these values. For additional information on normal values and analytical techniques, the references and reference texts such as "Handbook of Laboratory Animal Science" (74) and "Blood and Other Body Fluids" (2) should be consulted.

II. SOURCES OF VARIATION

The hematologic and clinical chemical normal values reported in this chapter were derived largely from a United States National Library of Medicine computer search (MEDLINE) of papers published between 1971 and 1975. Most of these papers reported experimental results and only incidentally reported values for control groups. They were not primary investigations of hematologic or clinical chemical parameters in the general population of normal, untreated rats. A total of 677 relevant papers were published during the 5-year period, of which

100 contained data suitable for inclusion in this chapter.

The problem of establishing the validity and reliability of reported values is highlighted by the following considerations. The stock or strain and the source of the rats used in the above-mentioned studies often were not specified or were identified incompletely. Often the disease status, diet, sex, age, or housing of the rats were not defined. Similarly, sample collection methods including bleeding site, time of day, and restraint or anesthetic used were not well described. Sample handling methods such as anticoagulants used and sample storage methods were often not mentioned. Finally, assay and quality control methods for the various measurements frequently were omitted from the papers. All of these factors have been shown to influence specific hematologic or clinical chemical parameters (Table I). Inasmuch as they are not likely to be controlled in precisely the same way in all laboratories or, indeed, may not be controlled at all, *there can be no "universal" value for any parameter.* Readers should keep this cautionary note in mind in using the information in this chapter.

The establishment of baseline data for the rat is further complicated by the large number of strains, stocks, and substocks recognized within the species. Considerable variation in baseline clinical laboratory parameters can exist among these various populations. For this reason, it is important that rat stocks be identified accurately. Throughout this chapter, strain, stock and source designations (Table II) will be expressed in the nomenclature found in the "International Index of Laboratory Animals" (31) and in "Animals for Research" (3).

Table II

Rat Strain, Stock, and Source Designations

Stocks or Strains

F344	= Fischer 344
LE	= Long-Evans
SD	= Sprague-Dawley
WI	= Wistar
THOM	= Karl Thomae

Microbial Status

BR	= Barrier reared
GF	= Germ-free

If no symbol is designated, the stock is conventionally reared.

Breeder or Supplier

Bir	University of Birmingham Medical School, Birmingham, England
Blu	Blue Spruce Farms, Altamont, New York
Can	Canadian Breeding Farm and Labs., Quebec, Canada
Chbb	Karl Thomae GmbH, Department of Lab Animal Science, 795 Biberach/Riss, Federal Republic of Germany
Crl	Charles River Breeding Labs., Wilmington, Massachusetts
Har	Harlan Industries, Indianapolis, Indiana
Lob	Lobund Laboratory, University of Notre Dame, Notre Dame, Indiana
Nr	Not reported
Osu	Department of Agricultural Chemistry, Oregon State University, Corvallis, Oregon
Sam	Research Labs., State Alcohol Monopoly, Helsinki, Finland
Sch	ARS/Sprague-Dawley, Madison, Wisconsin
Tul	Tulane University School of Medicine, New Orleans, Louisiana

Table I

Variables Affecting Clinical Laboratory Results

Variable	Parameter affected[a]	References
Strain	Glucose, total protein, albumin, differential, hemoglobin, WBC	17,19,39,43,63,104
Source	Total protein, globulin, albumin	104
Age	Differential, glucose, creatinine, hemoglobin, WBC, MCV	19,43,63,101,105
Sex	Total protein, creatinine, albumin, hemoglobin, WBC, MCV	43,62,63,101,105
Diet	Electrolytes, albumin, GOT, GPT, RBC, PCV, Hb, glucose, BUN, Alk. Phos., WBC, differential	11,49,89,111
Disease status	All parameters	39,43,50
Housing method	PCV, WBC, differential total protein	17,27,39,62,100
Bleeding site	Hb, PCV, WBC, pH, differential, glucose	10,30,43,97
Time of sampling	Coagulation, plasma free fatty acids	12,87
Restraint method	pH, protein, calcium, magnesium, glucose	10,51,97,106
Type of sample (serum versus plasma)	Total protein, enzymes, calcium, glucose, inorganic phosphorus, potassium, pH	60,65
Sample storage	GOT, Alk. Phos., Uric acid, cholesterol, total protein, bilirubin, albumin, CPK	7,13,107
Analytic method	Most parameters	26,46,94

[a] WBC, white blood cell count; MCV, mean corpuscular volume; GOT, glutamic oxalacetic transaminase; GPT, glutamic pyruvic transaminase; RBC, red blood cell count; PCV, packed cell volume; Hb, hemoglobin; BUN, blood urea nitrogen; Alk. phos., alkaline phosphatase; CPK, creatinine phosphokinase.

III. MICROTECHNIQUES

Due to inherent difficulties in obtaining large blood volumes from the rat and other small laboratory animals, micromethods are desirable in the experimental animal clinical laboratory. Micromethods or microliter techniques require sample sizes of only 0.02 to 0.1 ml. Ordinary glassware and apparatus except for pipettes are suitable for micromethods. Micropipettes are readily available commerically (94). The vast majority of routine clinical laboratory procedures are available in microtechnique form on semiautomated or automated apparatus (26,94). Even smaller sample size can be achieved with microtechniques that utilize microtest tubes, microcentrifuges and photometers with microcuvettes (72a). Instruments, reagents, and apparatus for microanalysis have been combined into sets by several manufacturers. These include Beckman* and Brinkman (Eppendorf).† Several manual, semiautomated and fully automated microtechnique systems are available from Scientific Products.‡

For many years most individual routine hematologic tests were performed manually with sample sizes of 0.02 ml. Automated equipment has generally required larger initial sample sizes because of sample loading difficulties. However, with use of the Unopette* micropipette diluting system it is possible to determine hemoglobin, hematocrit, erythrocyte count, leukocyte count, and red cell indices with a total blood sample size of 0.0447 ml (44.7 μl). The Unopette system is used in conjunction with Coulter† electronic counters.

IV. QUANTITIES AND UNITS

The results of quantitative measurement in hematology and clinical chemistry are always expressed in some type of unit, generally as amount per unit volume. Unfortunately the units used for reporting these concentrations often vary between countries and localities. In this chapter results are expressed in units currently accepted in most laboratories in the United States (26,46,85,94).

The basic unit of volume is the liter; the basic unit of mass is the gram. Some of the prefixes used for multiples and submultiples of the basic units are listed in Table III. In general, multiples and submultiples should be used so that the numerical value is between 1 and 1000. To conserve space in the text and tables in this chapter, the unit of measurement for each

*Beckman Instruments Inc., 2500 Harbor Blvd., Fullerton, California.
†Brinkman Instruments Inc., Westbury, New York.
‡Scientific Products Division, American Hospital Supply Corp., 1430 Waukegan Rd., McGaw Park, Illinois.
*Becton-Dickinson Inc., Rutherford, New Jersey.
†Coulter Electronic Inc., Hialeah, Florida.

entity is listed in Table IV and commonly used conversions in Table V.

Table III

Prefixes for Units Used in Hematology and Clinical Chemistry

Prefix	Abbreviation	Factor
mega-	M	10^6
kilo-	k	10^3
deci	d	10^{-1}
centi-	c	10^{-2}
milli-	m	10^{-3}
micro-	μ	10^{-6}
nano-	n	10^{-9}
pico-	p	10^{-12}
femto-	f	10^{-15}

Table IV

Units Used in Hematology and Clinical Chemistry

Entity (abbreviation)	Expressed in (abbreviation)
Erythrocytes (RBC)	Millions per microliter (μl) of blood
Hemoglobin (Hb)	Grams per deciliter of blood (gm/dl)
Packed cell volume (PCV)	Percent of unit blood volume (%)
Mean corpuscular volume (MCV)	Femtoliters per RBC (fl)
Mean corpuscular hemoglobin (MCH)	Picograms per RBC (pg)
Mean corpuscular hemoglobin concentration (MCHC)	Grams per deciliter of packed erythrocytes (gm/dl) or percent (%)
Erythrocyte sedimentation rate (ESR)	Millimeters fall in one hour (mm)
Reticulocytes	Number per 100 erythrocytes (%)
Erythrocyte diameter	Micrometers (μm)
Total leukocyte count (WBC)	Thousands per microliter (μl) of blood
Differential leukocyte count (Diff.)	Number per 100 leukocytes (%)
Platelets	Number per microliter (μl) of blood
Viscosity	Centipoise at various shear rates
Osmolality	Milliosmoles per kilogram (mOsm/kg)
Glucose	Milligrams per deciliter (mg/dl)
Serum urea nitrogen (BUN)	Milligrams per deciliter (mg/dl)
Creatinine	Milligrams per deciliter (mg/dl)
Bilirubin	Milligrams per deciliter (mg/dl)
Total serum protein	Grams per deciliter (gm/dl)
Serum protein fractions	Percent of total serum protein or grams per deciliter (gm/dl)
Sodium	Milliequivalents per liter (mEq/liter)
Potassium	Milliequivalents per liter (mEq/liter)
Chloride	Milliequivalents per liter (mEq/liter)
Calcium	Miligrams per deciliter (mg/dl)
Inorganic phosphorus	Milligrams per deciliter (mg/dl)
Magnesium	Milligrams per deciliter (mg/dl)
Partial pressure of carbon dioxide (pCO$_2$)	Millimeters of mercury (mm Hg)
Bicarbonate (HCO$_3^-$)	Milliequivalents per liter (mEq/liter)

(continued)

Table IV (*Continued*)

Units Used in Hematology and Clinical Chemistry

Entity (abbreviation)	Expressed in (abbreviation)
Total carbon dioxide (CO_2)	Milliequivalents per liter (mEq/liter)
Alkaline phosphatase (AP)	King-Armstrong units per deciliter (U/dl)
Creatinine phosphokinase (CPK)	International units per liter (IU/liter)
Lactic dehydrogenase (LDH)	International units per liter (IU/liter)
Glutamic pyruvic transaminase (GPT)	International units per liter (IU/liter)
Glutamic oxalacetic transaminase (GOT)	International units per liter (IU/liter)

Table V

Metric Conversions

1 microliter (μl)	= 1 cubic millimeter $(mm^3)^a$ = 1 lambda[a]
1 deciliter (dl)	= 100 ml[a]
1 micrometer (μm)	= 1 micron[a]

[a]These terms are obsolete according to the Système International d'Unites (S.I.) Convention.

V. HEMATOLOGY

A. Introduction

In the past several years there has been continued progress in delineating the hematology of the rat, although much of the information remains fragmentary and incomplete. The scope of the data has expanded as newer techniques have been applied. Reliability has improved with the use of larger sample sizes and increased attention to control of environmental, sampling, and analytical variables.

B. General Characteristics of the Blood

1. Volume

Blood volume has been determined by means of dye (5,109) and radioisotope (14) dilution techniques and reports range from 5.6 to 7.1 ml per 100 gm body weight (Table VI).

2. Specific Gravity of Whole Blood

Schalm (85) reports a value of 1.053–1.060, whereas Altman (2) cites both a broader and a narrower range: 1.046–1.061 and 1.054–1.058.

3. Blood pH

The reported mean blood pH in the unanesthetized rat is 7.40 (1,67) (also see Table XVI).

4. Whole Blood Viscosity

Zingg *et al.* (112) analyzed blood viscosity as various shear rates and report a mean of 4.3 centipoise at 230 sec^{-1} and 10.2 at 11.5 sec^{-1}.

5. Plasma Viscosity

Unlike whole blood, plasma behaves as a Newtonian fluid with the viscosity remaining constant at various shear rates. The mean viscosity for shear rates of 230 and 115 sec^{-1} is 1.2 centipoise with a range of 1.1 to 1.3 (112).

Table VI

Blood and Plasma Volume

Animal stock[a]	Sex	Body weight (gm)	PCV (%)	Plasma (ml/100 gm BW[b])	RBC (ml/100 gm/BW[b])	Blood (ml/100 gm/BW[b])	Reference
Lob:(WI)GF	Male	260[c] (250–278)	48.1 ± 0.5[d]	3.67 ± 0.1[d]	3.41 ± 0.1[d]	7.08 ± .19[d]	15
Lob:(WI)	Male	277[c] (251–300)	48.3 ± .07	3.88 ± 0.08	3.63 ± 0.1	7.15 ± 0.15	15
Sch:(SD)GF	Female	248 (200–292)	43.2 ± 0.3	3.30 ± 0.07	N.D.[e]	5.80 ± 0.11	109
Sch:(SD)GF	Male	334 (280–421)	46.6 ± 0.4	3.08 ± 0.08	N.D.	5.75 ± 0.14	109
Nr:(WI)	Male and female	262 (182–301)	N.D.	N.D.	N.D.	5.59 ± 0.44	14

[a]See Table II for key to rat stock designations.
[b]BW; body weight in grams.
[c]Total body weight minus cecal weight.
[d]Standard error of the mean.
[e]N.D. = no data.

6. Plasma Osmolality

This value, obtained by means of freezing point depression, is 321 mOsm/kg with a range of 288–336 mOsm/kg (112).

7. Erythrocyte Sedimentation Rate (ESR)

The commonly quoted values of 0.7 mm/hr for males and 1.8 mm/hr for females (2) are somewhat at variance with the more recent data of Zingg, *et al.* (112) who found an ESR of 1.5 mm/hr (1.0–2.5) for males.

C. Peripheral Blood

1. Erythrocytes

a. Populations. Vacet *et al.* (98) have detected the presence of five RBC populations based on RBC size. I, II, and III, present at birth, are superceded by IV and V which appear at 20–24 days. Population IV disappears on day 84, and V then constitutes the final erythrocyte species.

b. Counts. Reports of erythrocyte counts fluctuate widely. Schermer (86) reported a range of 7×10^6 to $9.7 \times 10^6/\mu l$. Creskoff *et al.* (24) found a combined average of $9.35 \times 10^6/\mu l$ for 28 male and 28 female adults. Hulse (50) determined that the value for an inbred albino strain was $8.2 \times 10^6/\mu l$, whereas Vondruska and Greco (101) reported values of 7.22×10^6 to $7.63 \times 10^6/\mu l$ for Crl:(SD)BR females and 7.41×10^6 to $8.45 \times 10^6/\mu l$ for Crl:(SD)BR males.

c. Reticulocyte Counts. According to Creskoff *et al.* (24) the average lies between 3 and 4%. In Hardy's series (43), the range extends from 1.5 to 4.3%. Hulse (50) has reported a lower average reticulocyte value of 2.3%. Godwin *et al.* (39) detected a difference due to stock, housing method, and disease status.

d. Hemoglobin Concentration. Schermer (86) summarized the early work which reflected a range of hemoglobin of 11.4 to 19.2 gm/dl. More recent studies confirm the previous findings (6,44,63,52,92,102,108). Hemoglobin levels have been shown to vary with stock or strain, age, sex, and health status. Table VII summarizes data illustrating this variation.

e. Hemoglobin Composition. Rat hemoglobin has been analyzed recently, and several molecular species have been detected. Shaw and MacLean (91), who examined the hemoglobin of Wistar rats during development and adulthood by means of starch and polyacrylamide electrophoresis, separated adult blood into five components with starch gel electrophoretograms and into six with the polyacrylamide technique. They found that two hemoglobins appear during yolk sac erythropoiesis and two additional hemoglobins are produced during the period of liver hematopoiesis. All four hemoglobins persist in the adult. Garrick *et al.* (37) found six components by DEAE-cellulose chromatography and cellulose acetate electrophoresis. There are five globin chains. Two are nonallelic alpha chains and three are beta chains.

f. Hematocrit (Packed Cell Volume). Until the development of micromethods, the hematocrit was performed infrequently. Now it is apparent that reported values for the hematocrit, like the erythrocyte count and hemoglobin, differ widely (Table VII). As in man, the hematocrit is approximately three times the hemoglobin value with reported means ranging from 40.5 to 53.9%.

Table VII

Erythrocyte Parameters in the Rat

Age (weeks)	Stock[a]	Erythrocytes ($\times 10^6/\mu l$)		PCV (%)		Hemoglobin gm/dl		MCV (fl)		MCHC gm/dl		Reticulocytes (%)		Reference
		M	F	M	F	M	F	M	F	M	F	M	F	
6	Nr:(WI)	N.D.[b]	N.D.	42.4	43.7	13.6	13.8	N.D.	N.D.	N.D.	N.D.	9.0	9.9	43
6	Chbb:(THOM)BR	6.54	6.71	45.3	47.2	14.7	15.4	69.3	70.5	32.6	32.6	3.2	3.0	105
8	Crl:(SD)BR	7.69	7.25	41.7	40.7	15.6	14.9	54.5	55.6	40.1	40.8	N.D.	N.D.	101
14	Chbb:(THOM)BR	8.39	7.35	53.2	52.0	16.9	16.8	62.2	67.3	32.4	32.4	2.1	2.2	105
16	Crl:(SD)BR	8.27	7.62	40.5	39.4	16.0	15.7	49.8	51.9	39.5	40.7	N.D.	N.D.	101
18	Nr:(WI)	N.D.	N.D.	53.9	54.6	17.8	18.1	N.D.	N.D.	N.D.	N.D.	2.5	1.5	43
34	Chbb:(THOM)BR	9.04	8.54	51.9	50.6	17.0	16.7	57.6	59.3	32.8	33.0	2.0	2.1	105
39	Nr:(WI)	N.D.	N.D.	52.5	50.6	16.8	16.8	N.D.	N.D.	N.D.	N.D.	2.3	1.7	43
40	Crl:(SD)BR	8.45	7.49	41.1	38.5	16.1	15.4	48.4	51.2	38.9	39.9	N.D.	N.D.	101

[a] See Table II for key to rat stock designations.
[b] N.D. = no data.

g. Mean Corpuscular Volume (MCV). Several investigators (39,50,101) report that the rat MCV can be affected by age, stock, sex, housing conditions, and health status. Godwin *et al.* (39) have determined the MCV as 61 and 60 fl for barrier reared Long-Evans and Sprague-Dawley rats, respectively, and 54 and 60 fl for conventionally reared Long-Evans and Sprague-Dawley rats, respectively. Vondruska and Greco (101) evaluated MCV in Crl:(SD)BR rats sequentially from 2–25 months of age. The results were characterized by a high initial value, a nadir at 19 months for both sexes, and values at 25 months of 60.2 and 51.8 fl for males and females, respectively. Burns and his colleagues (17) have confirmed the presence of a sex difference in axenic animals, 58.17 fl for females and 53.85 fl for males. Additional data can be found in Table VII.

h. Mean Corpuscular Hemoglobin Concentration. The mean corpuscular hemoglobin concentration (MCHC) determined by Hulse (50) is 32 and 29 pg in two series. Other values are reported in Table VII.

i. Cell Diameter. The rat erythrocyte is a biconcave disk with a diameter of 6.3 μm according to Creskoff *et al.* (24), and 6.8 μm according to Hardy (43). Schermer (86) cites papers confirming both figures.

j. Anisocytosis, Poikilocytosis, and Polychromasia. The rat erythrocytes form a fairly uniform population with more variation in size than shape. Polychromasia or polychromatophilia refers to cells with a gray appearance. They lack the full complement of hemoglobin. These cells are less mature and are equivalent in age to reticulocytes. An increased percentage is often an index of increased erythropoiesis. Godwin *et al.* (39) report that diseased animals have a greater degree of anisocytosis and polychromatophilia than healthy animals.

2. Leukocytes

a. Leukocyte Count. Creskoff *et al.* (24) found no important difference in leukocyte number of differential count between sexes. Their animals averaged 9000 leukocytes/μl with a range of 6000–18,000. In a survey of the literature, Schermer (86) found tremendous differences in reported mean leukocyte counts. Vondruska and Greco (101) determined leukocyte counts in Crl:(SD)BR rats from 2 to 25 months of age. The males averaged 10,000 leukocytes/μl at 2½ months. The nadir of 10,900 occurred at 19 months, and at 25 months the count was 14300. The females averaged 14,140 leukocytes/μl at 2½ months. The minimum value of 7140 occurred at 7 months, and at 25 months the count was 9050. Table VIII illustrates the variation in total leukocyte and differential counts with age, sex, strain, and microbial status.

b. Differential. The predominant white blood cell in the rat is the lymphocyte which may constitute 86% of the leukocytes. Polymorphonuclear neutrophils (PMNs) average 14–20%. Monocytes constitute a modest percentage (up to 6%) of the usual differential count; eosinophils range from 1–4%, and basophils are rare (43.86). (See also Table VIII.)

Table VIII

Leukocyte Parameters in the Rat

Age (weeks)	Stock[a]	WBC (× 10³/μl) M	F	Neutrophil (%) M	F	Lymphocyte (4) M	F	Monocyte (%) M	F	Eosinophilo (%) M	F	Basophil (4) M	F	Reference
6	Nr:(WI)	6.7	7.0	18.8	21.5	76.5	73.5	4.0	3.4	0.2	0.5	N.R.[c]		43
6	Chbb:(THOM)BR	5.0	4.3	26.7	25.5	69.2	69.4	3.0	3.1	0.8	1.0	0.0	0.0	105
8	Crl:(SD)BR	17.2	10.0	15.7	19.3	81.2	77.7	2.2	1.9	0.7	1.0	0.2	0.3	101
14	Chbb:(THOM)BR	5.0	3.2	19.4	23.1	75.2	72.3	2.8	2.2	2.4	2.4	0.0	0.0	105
16	Crl:(SD)BR	14.9	10.3	13.9	14.1	82.8	83.1	2.4	2.0	0.8	0.8	0.0	0.0	101
18	Nr:(WI)	7.3	7.2	16.5	14.7	79.5	81.3	3.1	3.2	0.4	0.5	N.R.		43
34	Chbb:(THOM)BR	5.3	3.9	22.5	26.4	71.3	65.3	3.6	3.6	2.6	2.9	0.0	0.0	105
39	Nr:(WI)	10.5	8.4	20.6	21.2	73.0	74.3	3.8	3.0	3.8	3.0	N.R.		43
40	Crl:(SD)BR	11.1	7.2	20.0	24.3	77.0	73.2	2.2	1.3	0.9	1.1	0.0	0.0	101
6	Nr:(WI)BR	6.0[b]		8.0		89.5		2.0		0.1		N.R.		43
6	Nr:(WI)	6.1		21.0		74.0		4.0		0.4		N.R.		43

[a] See Table II for key to rat stock designations.

[b] Not divided by sex.

[c] N.R., not reported.

i. Polymorphonuclear neutrophil. These cells are approximately 11 μm in diameter. The nucleus is coiled or twisted, and nuclear lobulation is not pronounced (Fig. 1). The cytoplasmic granules stain characteristically, although they are less dense than the specific granules of human neutrophils. They are both alkaline phosphatase (43) and peroxidase (86) positive.

ii Monocytes. The largest of the leukocytes, the monocyte has a convoluted nucleus, abundant cytoplasm, and some azurophilic or reddish-purple granulation (Fig. 2). There may be some difficulty in distinguishing monocytes from large lymphocytes. Schermer (86) points out, however, that rat monocytes have the general characteristics of monocytes of other species.

iii. Eosinophils. These cells, approximately the size of neutrophils, contain a nucleus that may be completely annular (Fig. 3). Large round granules usually fill the cytoplasm.

iv. Basophils. Although classic or circulating basophils are so rare in the rat that some authorities ignore their occurrence, tissue basophils or mast cells can be found on examination of buffy coats from both cardiac and tail blood. These cells contain a round or oval nucleus and numerous small darkly staining granules in contradistinction to the classic basophil with its characteristically lobulated nucleus, vacuoles and large dark staining granules. Basophils also approximate the neutrophil in size (43).

v. Lymphocytes. Like the human equivalent, the rat lymphocytes may range in size from approximately 6 to 15 μm. Unlike the human in whom most of the lymphocytes are large, the rat has a greater portion of small cells. The nucleus of the lymphocyte is made up of chunky-appearing chromatin (Fig. 4). Cytoplasm may be scant to abundant and range from dark to pale blue. Like the monocyte, lymphocytes contain azurophilic granules.

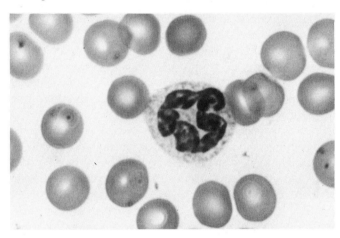

Fig. 1. Rat polymorphonuclear neutrophil surrounded by erythrocytes. Note the simple narrowing between lobes and lack of filamentous bridging in the neutrophil nucleus.

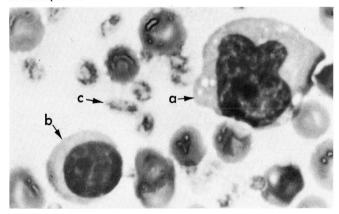

Fig. 2. Rat monocyte (a) with lymphocyte (b). Note the ameboid nucleus and vacuoles in the cytoplasm of the monocyte. Platelets (c) are numerous in this buffy coat preparation.

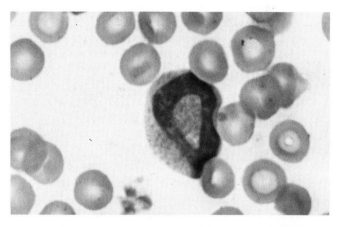

Fig. 3. Rat eosinophil. The nucleus is annular or ropelike rather than lobulated, and cytoplasmic granules are prominent.

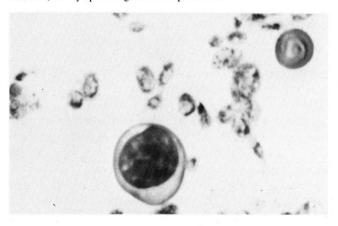

Fig. 4. Rat large lymphocyte surrounded by platelets. The lymphocyte nucleus is eccentric, and the chromatin is characteristically clumped.

3. Platelets

Creskoff *et al.* (24) reported an average value of 800,000 platelets/µl with a range of 500,000 to 1,000,000. Schermer (86), who comments that platelets fuse rendering counts unreliable, reports a similar range: 430,000 to 840,000. Weisse *et al.* (105) found a rise in values in both male and female Chbb:(THOM)BR rats observed from 4 to 36 weeks. Four-week-old males averaged 646,000 and females 645,000 while 36-week-old males averaged 734,000 and females 713,000. Experimental *Mycoplasma arthritidis* infection in rats has been associated with a modest thrombocytosis (6).

D. Bone Marrow

1. Composition

The bone marrow represents 3% of the adult body weight (86). Red marrow fills most of the bony cavities except for the distal two-thirds of the tail where the marrow is yellow (43). Hulse (50) reported the nucleated cell counts from different sites were quite similar. Corso (23) noted little sex difference. Hulse (50) found no evidence that exsanguination reduced the amount of peripheral blood in the bone marrow. His value of 22×10^6 to 3.4×10^6 nucleated cells per microliter of marrow agrees well with the results of Ramsell and Yoffey (82) but is less than that reported by Fruham (34). Corso (23) noted a significant reduction in numbers (3.45×10^6 to 2.68×10^6) between 30 and 40 days of age.

In two series of animals Hulse (50) found a myeloid-erythroid ratio of 1.36 : 1 and 1.16 : 1 which correlates well with other reports. Ramsell and Yoffey (82) listed the differential counts reported by various investigators and found a fairly wide range of values. Hulse also found that erythropoietic cells constituted 39%, myeleopoietic 34%, lymphocytic 24%, and reticulum cells 3% of the cells in the marrow. Note (79) reported that broadest range for lymphocytes: 17–63%.

In fetal animals, the major site of megakaryopoiesis is the liver. A few days after birth, the spleen assumes this function, and it is not until approximately 40 days after birth that the bone marrow becomes the major site of platelet production (71).

Keene and Jandl (55) investigated reticuloendothelial (RE) function in the marrow. They found that RE cells constitute approximately 12% of the nucleated marrow population. Following intravenous injection of antibody-coated radioactively labeled erythrocytes, bone RE cells cleared 7% of the labeled cells. In comparison, liver RE cells cleared 41%, spleen 14%, and lungs 5%. Blockade of the RE system results in inhibition of uptake by liver and spleen and more sequestration in the marrow.

2. Control of Granulopoiesis

There has been intensive effort to find factors equivalent to erythropoietin for controlling granulopoiesis and thrombopoiesis. This goal has not been realized, but other relevant substances have been detected. Graham and McMahon (40) demonstrated that an antiserum to the low molecular weight fraction of rat serum proteins produces specific depression of the maturation of granulocytes in the bone marrow. Rytomaa and Kiviniemi (83) have suggested that a chalone or antimitogenic agent produced by mature granulocytes served as a feedback mechanism. They isolated such a substance and used it to successfully treat rat chronic granulocytic leukemia.

E. Lymphoid Tissue

Hematopoietic stem cells, progenitor cells for erythrocyte, platelets, and the various leukocytes develop in the bone marrow throughout life. The lymphocyte precursors are dispersed via the bloodstream to the primary lymphoreticular organs or maturation cetners where under the influence of the microenvironment they differentiate as lymphocytes. The thymus and the bursa of Fabricus are the primary lymphoid organs in birds. There is now data suggesting that the liver is the mammalian equivalent of the avian bursa (57). From the thymus, the T or thymus-derived lymphocytes travel to the secondary to peripheral lymphoid tissues where they function in cell-mediated immunity. The B or bursal lymphocytes migrate to the germinal centers and medullary cords of the lymph nodes and lymphatic follicles of the spleen where they function in humoral immunity. Lymphocytes in the mouse have been separated into T and B cells, and each population has distinctive cell surface antigens and functions. There are relatively few similar studies in the rat (4,69). The lymphocyte content of different lymphoid organs and tissues of the rat has been reported (57,70,73,76,95). These studies are summarized in Table IX.

F. Coagulation

1. Introduction

Scheving and Pauly (87) have called attention to daily rhythmic variation in blood coagulation times of rats. Some strains are prone to hemorrhagic diseases and hereditary defects of not only the procoagulants but of platelets as well (33).

2. Bleeding Time

According to Creskoff (24) the bleeding time may be determined by snipping small wedges from the periphery of the ear. By this method the bleeding time averages 2 min.

Table IX

Lymphoid Tissue in the Rat

	Albino rats[a] 200 gm (ml/100g BW)	Wistar, male[b] 200 gm		Sprague-Dawley, female[c] 150–200 gm		Mixed stocks[d,e]
		mg	(cells × 10⁹)/organ	mg/100 gm BW	(cells × 10⁹)/100 mg tissue	(cells × 10⁹)/organ
Lymph nodes	0.40 ± 0.044[f]	1410 ± 45[f]	1.715 ± 0.121[f]	—	—	1.9
Spleen	0.21 ± 0.01	940 ± 93	1.100 ± 0.169	277	0.037	0.6
Peyer's Patches	0.005 ± 0.001	280 ± 25	0.382 ± 0.034	—	—	0.22
Thymus	0.13 ± 0.02	417 ± 18	1.071 ± 0.145	161	0.090	1.3
Total	—	3049 ± 103	4.823 ± 0.234	—	—	4.0

[a] From Kendred (57).
[b] From Monden (76).
[c] From McLean et al. (73).
[d] From Trepel (95).
[e] Means calculated from a literature survey.
[f] Standard error of the mean.

3. Clotting Time

Schermer (86) summarized the results of various investigators who reported normal clotting times ranging from 2 to 5 min. There was good correlation between the results obtained with a Bogg coagulometer and those determined in glass capillary tubes.

4. Prothrombin Time

Colvin and Wang (20) determined that the control prothrombin time was 14.4 ± 0.36 sec. Hardy (43), who also used the Quick one-stage method with either rat or human brain thromboplastin, found a mean of 10.5 sec with a range of 8–14, his results agree with the 10.8 ± 0.4 sec cited by Schermer (86). The values reported for individual procoagulants varied widely depending on the method used (86).

5. Partial Thomboplastin Time

A partial thromboplastin time of 21.1 ± 3.7 sec was determined by Tschopp and Zucker (96).

6. Fibrinogen

Beller and Theiss (8) assayed fibrogen in citrated rat plasma according to the Ratnof and Menzie method. They reported a mean value of 184 mg/dl with a range of 158–214 mg/dl. Colvin and Wang (20) found that the control value for oat-fed rats was 232 mg/100 ml with a standard deviation of 14.4 and for chow-fed rats, 202 mg/100 ml with a standard deviation of 6.8.

7. Platelets

Rat platelets adhere to tendon suspensions and glass beads less readily than human platelets (9). DeGaetano (28) found that rat platelets are less sensitive than human platelets to collagen, ristocetin, and some vonWillebrand factor preparations. Platelets of normal rats contain 4.49 μmoles ATP, 0.98 μmoles ADP, and 0.88 μmoles serotonin/10¹¹ platelets (96).

VI. CLINICAL BIOCHEMISTRY

A. Biochemical Constituents of Serum

1. Glucose

Besch and Chou (10) have shown that rat blood collection methods can have a significant effect on the plasma glucose concentration. Under their experimental conditions, the glucose concentration varied significantly depending on the duration of animal handling prior to removal of the sample, method of anesthesia, method and duration of restraint, fasting 24 hr, and method of blood withdrawal. Following blood withdrawal, sample handling is especially important, since glucose in blood at room temperature undergoes glycolysis at the rate of 5% per hour and may disappear faster in the presence of leukocytosis or bacterial contamination (26). Preservation of glucose in a blood sample for longer than 0.5 hr can be accomplished by adding 10 mg sodium fluoride/ml of blood to prevent glycolysis and 0.5 mg of thymol to prevent bacterial growth. Since plasma or serum usually is used for glucose analysis, it may be best to separate the serum or plasma from the cells immediately in order to prevent glycolysis. Plasma glucose levels normally average 10–15 mg/dl higher than whole blood (107).

Many methods of glucose determination have been developed. Earlier methods were based on measurements of re-

Table X

Fasting Blood Glucose[a]

Mean ± SD	N	Stock[b]	Specimen	Reference
152 ± 15.8	34	Blu:(SD)BR	Plasma	19
145 ± 33.3	15	Blu:(LE)	Plasma	63
161 ± 22.3	18	Tul:(F344)GF	Serum	17
134 ± 14.3	110	Crl:(SD)BR	Serum	101
98 ± 8.5	10	Crl:(SD)BR	Serum	45

[a] Values in mg/dl.

[b] See Table II for key to rat stock designations.

Table XII

Serum Creatinine[a]

Mean ± SD	N	Stock[b]	Specimen	Reference
0.44 ± 0.05	15	Blu:(LE)	Serum	63
0.71 ± 0.16	15	Nr:(SD)	Serum	38
0.57 ± 0.04	10	Crl:(SD)BR	Serum	101
1.50 ± 0.79	151	Nr:(WI)	Serum	16

[a] Values in mg/dl.

[b] See Table II for key to rat stock designations.

ducing substances in the blood. In order to overcome nonspecificity, subsequent methods have employed filtrates which are free of non-glucose-reducing substances. Today, enzymatic methods provide the greatest specificity in estimating true blood glucose.

In the rat, fasting blood glucose tends to be higher than the human normal range, and there is considerable variation in reported means of 98 to 152 mg/dl (Table X). Much of this variation is undoubtedly due to environmental, sampling, and analytical differences among the various laboratories.

2. Urea Nitrogen

The early determination of circulating urea nitrogen was made in whole blood, and the result was reported as blood urea nitrogen (BUN). Today most methods determine urea nitrogen in serum, and the result should be termed serum urea nitrogen (SUN). However, by convention and usage the result is still reported as BUN. The preferred specimen for BUN determination is serum but plasma is acceptable. In the rat BUN does not vary with age or sex (45,63,101) and reported normal means vary from 15 to 22 mg/dl (Table XI).

3. Creatinine

Creatinine is the major waste product of creatine metabolism by muscle. Thus, free creatinine appears in the blood. In the

Table XI

Circulatory Urea Nitrogen[a]

Mean ± SD	N	Stock[b]	Specimen	Reference
15.4 ± 2.7	215	Crl:(SD)BR	Serum	101
18 ± 1.9	10	Crl:(SD)BR	Serum	45
17.1 ± 3.0	15	Blu:(LE)BR	Plasma	63
15 ± 3.0	6	Har:(WI)	Serum	41
21.8 ± 4.0	33	Blu:(SD)	Serum	19

[a] Values in mg/dl.

[b] See Table II for key to rat stock designations.

kidney, it is filtered by the glomerulus and actively excreted by the tubules. Most modern methods of creatinine determination are based on the Jaffe reaction, in which creatinine is treated with alkaline picrate solution (26). Lloyd's reagent, an aluminum silicate, is frequently used to remove creatinine from other chromogenic substances prior to the Jaffe reaction.

Serum or heparinized plasma is the preferred specimen for creatinine analysis. In collection of specimens and analysis of results, it should be recognized that acetone, ketones, ascorbic acid, barbiturates, glucose, protein, and sulfobromophthalein may interfere with the Jaffe reaction (94).

Kozma *et al.* (63) reported that serum creatinine was higher in young male rats (2–4 months) than in young females, but by 8 months of age this difference was abolished. Mean normal rat serum creatinine (Table XII) ranges from 0.4 to 1.5 mg/dl depending on the analytical method employed.

4. Bilirubin

Total serum bilirubin concentration is a useful measure of liver dysfunction. The determination of conjugated (Van den Bergh direct) and unconjugated (Van den Bergh indirect) bilirubin fractions has been used in the differential diagnosis of jaundice. Total serum bilirubin rises significantly in the rat following liver damage due to carbon tetrachloride, sodium selenate, or cadmium administration (25,66). It also was shown that serum bilirubin elevation is not as sensitive as the hippuric acid test or sulfobromophthalein (BSP) excretion in revealing long-term impairment of liver function (25).

Selection of a laboratory method for determination of conjugated and total bilirubin is difficult because of inherent problems in many of the methods. See Henry *et al.* "Clinical Chemistry" (46) for an extensive discussion of bilirubin methods.

The preferred specimen for determination of circulating bilirubin is serum, although plasma is acceptable. Specimens are stable for 8 hr at room temperature and overnight at 4°C (107). Freezing may significantly increase total bilirubin levels in serum (7). If immediate analysis is not possible, specimens should be protected from the light, since bilirubin is slowly

Table XIII

Total Bilirubin Levels[a]

Mean ± SD	N	Stock[b]	Specimen	Reference
0.35 ± 0.02	18	Tul:(F344)GF	Serum	17
0.40 ± 0.05	23	Bir:(WI)	Plasma	25
0.19 ± 0.01	10	Can:(SD)BR	Plasma	66
0.12 ± 0.04	15	Nr:(SD)	Serum	38

[a] Values in mg/dl.

[b] See Table II for key to rat stock designations.

oxidized by exposure to light (36).

Reported mean normal total bilirubin levels in the rat ranges from 0.12 to 0.40 mg/dl (see Table XIII).

5. Serum Protein

Total protein may be measured either in serum or plasma. Slightly higher values can be expected in plasma, since it contains fibrinogen which constitutes about 5% of the plasma proteins. The simple biurette method is usually the method of choice for determining total protein, although several other methods are in common use (26). These methods also may be used to quantitate albumin following precipitation of the globulin fraction. The difference between measured total protein and measured albumin generally is reported as globulin.

Albumin is the major serum protein in both man and the rat. However, marked species differences occur in the globulin fractions. In man, the globulin fractions ranked in descending concentration are α_1, β, α_2, and γ (26). In contrast, the major globulin in rat serum is either the α_1 or β fraction depending on the rat strain, followed by α_2- and γ-globulins (103). It has been shown that in rats the γ-globulin fraction rises following

infection with the dwarf tapeworm *Hymenolepis nana* (54). Keraan *et al.* (56) found that serum γ-globulin also rises in the rat following portacaval shunt. They suggested that the elevated levels represented an immune response to bacterial lipopolysaccharides released into the systemic circulation as a result of the portacaval shunt. Conversely, trauma due to subcutaneous injection of turpentine causes a decrease in the albumin and γ-globulin fractions and an elevation of the α_1, α_2, and β components (103). Mean total serum protein values for rats as reported by several investigators range from 6.0 to 7.8 gm/dl (see Table XIV).

6. Blood Electrolytes

Except for inorganic phosphorus, the reported normal values for blood electrolytes in rats fall within or at the upper limit of the human normal range. Laboratory methods that are accurate and precise for human sera are also accurate and precise for rat sera. It appears that the clinicopathologic interpretation of abnormal electrolyte results in the rat is much the same as for man. Reported normal values for rat plasma electrolytes are given in Table XV. References in that table should be consulted for additional information on sampling, sample handling, and analytic methods.

7. Blood Acid–Base Parameters

The blood acid–base status should be considered in interpreting experimental results from anesthetized rats. Several investigators have shown that anesthesia may affect blood gas tensions and acid–base status (53,67,97). Most concluded that, in general, anesthesia does not produce marked acid–base changes unless there is clinically apparent respiratory depression. Table XVI summarizes the arterial blood acid–base parameters reported in anesthetized and unanesthetized rats.

Table XIV

Normal Rat Serum Proteins

Total protein (gm/dl)	Albumin (%)	Globulin				N	Reference
		α_1 (%)	α_2 (4)	β (%)	γ (%)		
6.3 ± 0.23[a]	48 ± 0.7	17 ± 0.4	10 ± 0.5	19 ± 0.3	6 ± 0.2	12	104
6.0 ± 0.10[a]	47 ± 0.5	17 ± 0.5	10 ± 0.3	18 ± 0.3	8 ± 0.4	12	104
7.17 ± 0.39[b]	51.7 ± 5.4	16 ± 4.7	10 ± 2.4	12 ± 2.6	10 ± 2.4	15	63
7.8 ± 0.6[b]	40.3 ± 2.5	13.5 ± 1.1	9.3 ± 0.5	23.1 ± 0.5	13.8 ± 0.5	18	75
NR[c]	42 ± 1	10 ± 1	12 ± 1	22 ± 1	13 ± 1	32	25

[a] Mean ± standard error of the mean.

[b] Mean ± standard deviation of the mean.

[c] NR, Not reported.

Table XV

Blood Electrolytes

Mean ± *SD*	N	Stock[a]	Sex	Age (months)	Specimen	Reference
			Sodium (mEq/liter)			
150.7 ± 1.8	15	Crl:(SD)BR	F	4	Serum	101
151.3 ± 7.6	15	Crl:(SD)BR	M	4	Serum	101
140.4 ± 5.5	20	Tul:(F344)BR	M and F	3.5	Serum	17
140 ± 16	8	Sam:(WI)	M	6	Serum	47
145 ± 3	6	Har:(WI)	F	6[b]	Plasma	41
			Potassium (mEq/liter)			
5.56 ± 0.56	15	Crl:(SD)BR	F	4	Serum	101
5.59 ± 0.68	15	Crl:(SD)BR	M	4	Serum	101
1.76 ± 1.06	18	Tul:(F344)GF	F	3	Serum	17
4.3 ± 0.8	9	Sam:(WI)	M	6	Serum	47
4.8 ± 0.4	6	Har:(WI)	F	6[b]	Plasma	41
			Chloride (mEq/liter)			
112.9 ± 2.5	15	Crl:(SD)BR	F	4	Serum	101
111.5 ± 4.8	15	Crl:(SD)BR	M	4	Serum	101
100 ± 6	8	Sam:(WI)	M	6	Serum	47
			Calcium (mg/dl)			
13.6 ± 1.8	15	Crl:(SD)BR	F	4	Serum	101
14.1 ± 1.2	15	Crl:(SD)BR	M	4	Serum	101
10.6 ± 0.9	18	Tul:(F344)BR	M and F	3.5	Serum	17
10.5 ± 0.8	15	Nr:(SD)	M	2	Serum	32
8.3 ± 1.4	18	Tul:(F344)BR	M and F	3.5	Serum	17
9.1 ± 1.2	15	Nr:(SD)	M	2	Serum	32
			Magnesium (mg/dl)			
2.60 ± 0.16	10	Crl:(SD)BR	F	4	Serum	101
2.34 ± 0.14	10	Crl:(SD)BR	M	4	Serum	101

[a] See Table II for key to rat stock designations.

[b] Age in weeks.

Table XVI

Arterial Blood Acid–Base Parameters in Normal Rats

pH	pCO$_2$ (mm Hg)	HCO$_3$ (mEq/liter)	CO$_2$ (mEq/liter)	N	Experimental conditions[b]	Reference
7.44 ± 0.01[a]	32.7 ± 2.4	21.5 ± 2.4	22.5 ± 1.4	10	Unanesthetized	67
7.46 ± 0.01	32.6 ± 0.8	22.3 ± 0.4	23.3 ± 0.4	30	ip pb 35 mg/kg	67
7.40 ± 0.005	38.0 ± 1.0	23.1 ± 0.6	24.2	9	Unanesthetized	1
7.41 ± 0.009	32.6 ± 2.7	19.3 ± 1.1	20.8	10	ip pb 35 mg/kg	51
7.40 ± 0.03	39.9 ± 1.8	23.8 ± 0.8	NR[c]	8	Unanesthetized	53
7.26	33.7	14.4	15.4	10	Unanesthetized	97
7.36	41.0	21.9	23.2	10	ip pb 60 mg/kg	97

[a] Values are mean ± standard error of the mean.

[b] ip, intraperitoneal injection; pb, pentobarbital sodium.

[c] NR, not reported.

8. Enzymatic Activity in Serum

a. Introduction. All serum enzymes have their origin in cells. The clinically useful enzymes are normally found in serum in relatively low concentrations. Increased serum concentration of a particular enzyme usually signifies damage to the cells from which the enzyme was released. Some enzymes are found in many tissues, while others are uniquely concentrated in only one or two tissues or organs. Increased serum concentration of a ubiquitous enzyme is a less powerful clinical diagnostic tool than is an increase of an enzyme with only a very limited distribution, since the latter tends to indicate more clearly the site of tissue injury. Recent addition has been directed toward identification of different molecular forms (isoenzymes) of widely distributed enzymes in order to enhance the power of these enzymes as clinical diagnostic tools. In the rat, diagnostic serum enzymology has been mainly directly toward early and reliable identification of liver injury due to toxicants (22).

b. Alkaline Phosphatase. Alkaline phosphatase (AP) is found in high levels in osteoblasts, and elevated serum levels are an indication of increased osteoblastic activity. Increased levels of this enzyme are found in the serum of growing children as well as young growing rats (78). In rats, depressed serum alkaline phosphatase activity has been noted during chronic administration of hexachlorophene which also causes significant growth rate depression (78). Schwartz (89) found elevated serum alkaline phosphastase activity in rats fed a restricted diet. Reported normal values for serum alkaline phosphatase are 16 to 48 King-Armstrong units per deciliter (see Table XVII). Note that males have significantly higher normal values than females.

c. Creatinine Phosphokinase. Creatinine phosphokinase (CPK) is found in high concentration in skeletal muscle, myocardium, and brain. Yasin *et al.* (110) have reported that plasma CPK increases in conjunction with the development of skeletal muscle degenerative changes following the administration of vincristine. Plasma CPK also rises following exercise in humans, dogs, and rats (68). The normal mean CPK in the rat is reported as 50 IU/liter of plasma (68).

d. Lactic Dehydrogenase. Lactic dehydrogenase (LDH) is found in high concentration in myocardium, kidney, liver, and skeletal muscle. With appropriate electrophoretic techniques.

LDH is separable into five isoenzymes in varying proportions; thus, the LDH activity of each tissue has a characteristic isoenzyme composition. In the rat, the serum cardiac LDH isoenzyme has been shown to increase during cardiac hypertrophy (80). Rat serum LDH also increases following bacterial infection (75), administration of adrenergic stimulant drugs (72), and exercise (68). Normal rat LDH values vary widely with each of the numerous methods. Even though the result is expressed in international units each method yields a different result, and there is no reliable method for converting one to another. One investigator utilizing General Diagnostics reagents in the Babson and Phillips colorimetric method reports the normal mean LDH as 15 IU/liter of plasma (68).

e. Glutamic Pyruvic Transaminase. Glutamic pyruvic transaminase (GPT) is found in high concentration in the liver and in relatively low concentration in other tissues. This has led to the application of serum GPT determination for the detection of hepatic disease. Since rats are widely used for evaluating the hepatic toxicity of various compounds, serum GPT is one of the most commonly determined clinical biochemical values in the rat (see Section VI, B of this chapter for additional information). Serum GPT in the rat does not vary significantly with respect to age or sex (63). The actual units used in reporting serum GPT concentrations are still in a state of flux. Thus, comparison of results obtained with different methods is difficult. However, normal rat serum GPT usually falls at the upper limit of the human normal range for the particular method employed (63,66,90).

f. Glutamic Oxalacetic Transaminase. Glutamic oxalacetic transaminase (GOT) has been found in the tissues and serum of all mammals studies. It usually is found in high concentration in myocardial, hepatic, skeletal muscle, renal, and cerebral tissue. Because of this wide distribution GOT has found less utility as a diagnostic enzyme in rats. Normal rat serum GOT values do not differ significantly with respect to age or sex, but are usually higher than the human normal range for the particular method employed (21,63).

B. Tests of Hepatic Integrity

The rat has been widely used as an experimental animal for evaluating the hepatic toxicity of various compounds. The most widely used criteria of the toxic action of a substance in rats are reduction in the rate of body weight gain, detection of gross and histologic abnormalities in the organs, changes in organ weights, and increase in mortality rate (5). Cutler *et al.* (25) compared a wide variety of liver function tests including glutamic pyruvic transaminase (GPT), glucose-6-phosphatase (G6P), colloidal red, zinc sulfate tubidity, total bilirubin,

Table XVII

Serum Alkaline Phosphatase[a]

Mean ± *SD*	N	Stock[b]	Age	Sex	Reference
48 ± 6	5	Osu:(WI)	8 weeks	M	78
36 ± 10	6	Osu:(WI)	8 weeks	F	78
21 ± 6	5	Osu:(WI)	16 weeks	M	78
23 ± 7	4	Osu:(WI)	16 weeks	F	78
22 ± 10	110	Crl:(SD)BR	7 months	M	101
16 ± 7	110	Crl:(SD)BR	7 months	F	101

[a] Values in King Armstrong units/dl.
[b] See Table II for key to rat stock designations.

plasma protein electrophoresis, sulfobromophthalein (BSP) clearance and hippuric acid test with the more traditional tests. She concluded that only four tests (plasma albumin, hippuric acid test, GPT, and BSP clearance) discriminated effectively between treated and control groups. Histopathologic examination was the most sensitive test in the detection of long-term liver damage, although several rats showed abnormal function tests and no histologic lesions.

Ghys *et al.* (38) compared hepatic function tests with ultrastructural examination in detecting liver damage due to an experimental antileukemic drug. They found that γ-glutamyl-transpeptidase (GMT), ornithine carbamyltransferase (OCT), and GPT activity in serum was increased and that ultrastructural changes in hepatocytes were extensive. Korsrud *et al.* (59) compared the sensitivity of several serum enzymes for the detection of induced liver damage in the rat. They concluded that histologic damage was detectable at lower doses of each toxicant than were serum enzyme changes and that serum sorbitol dehydrogenase (SDH) activity was the most sensitive enzyme indicator of minimal liver damage. Korsrud *et al.* (61) also reported that serum SDH was more sensitive to carbon tetrachloride-induced liver damage than several other serum enzymes. Clary and Groth (18) compared serum changes in isocitric dehydrogenase (ICD), GOT, GPT, and LDH in detecting liver necrosis due to beryllium and carbon tetrachloride. They concluded that ICD, GOT, and GPT were excellent serum indicators of early lesions. The assessment of liver function and the use of serum enzyme changes in toxicity studies has been reviewed by Cornish (22) and Grice (42).

The sulfobromophthalein (BSP) excretion test is widely used in many species as a measure of hepatic function. The dye is administered intravenously and its disappearance from the blood is determined. Excretion of BSP by the liver involves three steps: uptake of unconjugated dye by hepatic parenchymal cells, conjugation with glutathione, and excretion by active transport into the bile. In the rat BSP conjugates account for most of the BSP appearing in bile (64). It has been shown that phenobarbital pretreatment increases the maximal rate of biliary excretion of BSP (106). The findings of Priestly *et al.* (81) and Schulze (88) indicate that BSP depresses bile flow in the rat. Methods for the BSP excretion test in the rat vary widely (25,48,58,93).

C. Urine Analysis

Laboratory analytic methods that are accurate and precise for human urine are also suitable for rat urine. It appears that the clinicopathologic interpretation of abnormal results in rats is much the same for humans. Reported normal values for urine related parameters are given in Table XVIII. References in that

Table XVIII
Urine Values in Normal Rats

Parameter	Values	Reference
Urine volume (ml/100 gm body weight/24 hr)	3.3	99
	4.2	29
Osmolality (mOsm/kg)	2442	99
	1500	35
Specific gravity	1.022–1.050	45
Protein (mg/dl)	<30	77
pH	7.00	77
	5–7	45
Creatinine (mg/100 gm body weight/24 hr)	5.5	29
Potassium (mEq/100 gm body weight/24 hr)	2.1	29
17-ketosteroid (μg/100 gm body weight/24 hr)	16.4	84

table should be consulted for additional information on sample handling and analytic methods.

REFERENCES

1. Altland, P. D., Brubach, H. F., Parker, M. G., and Highman, B. (1967). Blood gases and acid–base values of unanesthetized rats exposed to hypoxia. *Am. J. Physiol.* **212,** 142–148.
2. Altman, P. L. (1961). "Blood and Other Body Fluids." Fed. Am. Soc. Exp. Biol., Washington, D. C.
3. "Animals for Research" (1975). 9th ed. Inst. Lab. Anim. Resour., Natl. Res. Counc.—Natl. Acad. Sci., Washington, D. C.
4. Balch, C. M., and Feldman, J. D. (1971). Thymus-dependent (T) lymphocytes in the rat. *J. Immunol.* **112,** 79–86.
5. Barnes, J. M., and Denz, F. A. (1954). Experimental methods used in determining chronic toxicity. A critical review. *Pharmacol. Rev.* **6,** 191–242.
6. Baumgartner, R., and Laber, G. (1975). Hämatologische untersuchungen bei ratten mit experimenteller mykoplasma-polyarthritis. *Zentralbl. Bakeriol., Parasitenkd., Infektionskr. Hyg., Abt. I: Reihe A* **232,** 105–112.
7. Bayard, S. P. (1974). Another look at the statistical analysis of changes during storage of serum specimens. *Health Lab. Sci.* **11,** 45–49.
8. Beller, F. K., and Theiss, W. (1973). Fibrin derivatives, plasma hemoglobin and glomerular fibrin deposition in experimental intravascular coagulation. *Thromb. Diath. Haemorr.* **29,** 363–374.
9. Bern, M. M., and Tullis, J. L. (1974). Platelet adhesion from non-anticoagulated rat arterial blood. *Thomb. Diath. Haemorr.* **31,** 493–504.
10. Besch, E. L., and Chou, B. J. (1971). Physiological responses to blood collection methods in rats. *Proc. Soc. Exp. Biol. Med.* **138,** 1019–1021.
11. Bolter, C. P., and Critz, J. B. (1974). Plasma enzyme activities in rats with diet-induced alterations in liver enzyme activities. *Experientia* **30,** 1241–1243.
12. Bortz, W. M., and Steele, L. A. (1973). Synchronization of hepatic cholesterol synthesis, cholesterol and bile acid content, fatty acid synthesis and plasma free fatty acid levels in the fed and fasted rat. *Biochim. Biophys. Acta* **306,** 85–94.
13. Bramanti, G., Ravina, A., and DeFina, V. (1972). Preservability of rat

serum in delayed determinations of some hematochemical parameters. *Boll. Chim. Farm.* **111**, 694–699.

14. Brooks, S. A. (1971). Method of measuring the normal blood volume of the rat using [^{125}I]-labeled albumin. *Aust. J. Exp. Biol. Med. Sci.* **49**, 241–243.

15. Bruckner-Kardosse, E., and Wostmann, B. S. (1974). Blood volume of adult germ-free and conventional rats. *Lab. Anim. Sci.* **24**, 633–635.

16. Burns, K. F., and DeLannoy, C. W. (1966). Compendium of normal blood values of laboratory animals, with indication of variations. 1. Random-sexed populations of small animals. *Toxicol. Appl. Pharmacol.* **8**, 429–437.

17. Burns, K. F., Timmons, E. H., and Poiley, S. M. (1971). Serum chemistry and hematological values for axenic (germfree) and environmentally associated inbred rats. *Lab. Anim. Sci.* **21**, 415–419.

18. Clary, J. J., and Groth, D. H. (1973). Comparative changes in serum enzyme levels in beryllium- or carbon tetrachloride-induced liver necrosis. *Proc. Soc. Exp. Biol. Med.* **143**, 1207–1210.

19. Coleman, E. J. Zbijewska, J. R., and Smith, N. L. (1971). Hematologic values in chronic murine pneumonia. *Lab. Anim. Sci.* **21**, 721–726.

20. Colvin, H. W., Jr., and Wang, W. L. (1974). Toxic effects of warfarin in rats fed different diets. *Toxicol. Appl. Pharmacol.* **28**, 337–348.

21. Cook, J. A., Marconi, E. A., and DiLuzio, N. R. (1974). Lead, cadmium, endotoxin interaction: Effect on mortality and heptatic function. *Toxicol. Appl. Pharmacol.* **28**, 292–302.

22. Cornish, H. H. (1971). Problems posed by observations of serum enzyme changes in toxicology. *Crit. Rev. Toxicol.* **1**, 1–32.

23. Corso, F. (1959). Ricerche sulla valutazione quantatina dell cellule nucleate nel midolo osseo dell'animale esperimento (Ratto albino). *Haematologica* **44**, 1019–1031.

24. Creskoff, A. J., Fitz-Hugh, T., Jr., and Farris, E. J. (1949). Hematology of the rat—methods and standards. *In* "The Rat in Laboratory Investigation" (E. J. Farris and J. Q. Griffiths, eds.), pp. 406–420. Lippincott, Philadelphia, Pennsylvania.

25. Cutler, M. G. (1974). The sensitivity of function tests in detecting liver damage in the rat. *Toxicol. Appl. Pharmacol.* **28**, 349–357.

26. Davidson, I., and Henry, J. B. (1974). "Clinical Diagnosis by Laboratory Methods," 15th ed. Saunders, Philadelphia, Pennsylvania.

27. Deb, C., and Hart, J. S. (1956). Hematological and body fluid adjustments during acclimation to cold environment. *Can. J. Biochem. Physiol.* **34**, 959–966.

28. DeGaetano, G., Donati, M. B., Innocent, I. R. D., and Roncaglioni, M. C. (1974). Platelet adhesion-aggregation reaction. *Thromb. Diath. Haemorr.* **32**, 242.

29. Dilley, J. V., Carter, V. L., and Harris, E. S. (1974). Fluoride ion excretion by male rats after inhalation of one of several fluoroethylenes or hexfluoropropene. *Toxicol. Appl. Pharmacol.* **27**, 582–590.

30. Doell, B. H., and Hegarty, P. V. J. (1970). The haemoglobin concentration of peripheral and central blood of the laboratory rat. *Br. J. Haematol.* **18**, 503–509.

31. Festing, M. F. W. (1975). "International Index of Laboratory Animals," Med. Res. Counc., Lab. Anim. Cent., Carshalton, United Kingdom.

32. Frankel, H. M., Yousef, M. K., Bayer, R., and Dill, D. B. (1972). Blood composition in normothermic and hyperthermic kangaroo rats *Dipodomys merriami* and laboratory rats *Rattus norvegicus*. *Comp. Biochem. Physiol. A* **43**, 733–738.

33. Fritz, T. E., Tolle, D. V., and Flynn, R. J. (1968). Hemorrhagic diathesis in laboratory rodents. *Proc. Soc. Exp. Biol. Med.* **128**, 228–234.

34. Fruham, G. S., and Gordon, A. S. (1953). Quantitative evaluation of total cellular numbers of bone marrow. *Anat. Rec.* **117**, 603.

35. Fusco, M., Malvin, R. L., and Churchill, P. (1966). Alterations in fluid, electrolyte and energy-balance in rats with median eminence lesions. *Endocrinology* **79**, 301–308.

36. Gambino, S. R., and DiRe, J. (1968). "Bilirubin Assay" (revised). Am. Soc. Clin. Pathol., Chicago, Illinois.

37. Garrick, L. M., Sharma, V. S., and Ranney, H. M. (1974). Structural studies of rat hemoglobins. *Ann. N.Y. Acad. Sci.* **241**, 434–435.

38. Ghys, A., Thys, O., Hildebrand, J., and Georges, A. (1975). Relation between hepatic and renal function tests and ultrastructural changes induced by 2-*N*-methylpiperazinomethyl-1,3-diazafluoranthen-1-oxide (AC-3579), a new experimental antileukemic drug. *Toxicol. Appl. Pharmacol.* **31**, 13–20.

39. Godwin, K. O., Fraser, F. J., and Ibbotson, R. N. (1964). Hematological observations on healthy (SPF) rats. *Br. J. Exp. Pathol.* **45**, 514–524.

40. Graham, J. D., and McMahon, C. J. (1971). Control of granulopoiesis in bone marrow of normal albino rats by a maturation factor isolated from rat serum. *Trans. Am. Microse. Soc.* **90**, 238–242.

41. Greenberger, N. H., Perkins, R. L., Cuppage, F. E., and Ruppert, R. D. (1967). Severe metabolic acidosis in the rat induced by toxic doses of tetracycline. *Proc. Soc. Exp. Biol. Med.* **125**, 1194–1197.

42. Grice, H. C. (1972). The changing role of pathology in modern safety evaluation. *Crit. Rev. Toxicol.* **1**, 119–152.

43. Hardy, J. (1967). Hematology of rats and mice. *In* "Pathology of Laboratory Rats and Mice" (E. Cotchin and F. J. C. Roe, eds.), pp. 501–536. Davis, Philadelphia, Pennsylvania.

44. Hebold, V. G., and Bleuel, H. (1971). Hämatologische standard-werte bei der weiblichen und mannlichen ratte (Sprague Dawley). *Z. Versuchstierkd.* **13**, 316–320.

45. Henderson, J. D., Jr. (1976). Riker Laboratories, 3-M Company, St. Paul, Minnesota (personal communication).

46. Henry, R. J., Cannon, D. C., and Winkelman, J. W. (1974). "Clinical Chemistry," 2nd ed. Harper, New York.

47. Hillbom, M. E., and Poso, A. R. (1975). Effects of ethanol on serum electrolytes and respiration in euthyroid and hyperthyroid rats. *Toxicol. Appl. Pharmacol.* **32**, 168–176.

48. Hollander, C. F., DeLefeuw-Israel, F. R., and Arp-Neefjes, J. M. (1968). Bromsulphalein (BSP) clearance in ageing rats. *Exp. Gerontol.* **3**, 147–153.

49. Horowitz, M., and Borut, A. (1973). Blood volume regulation in dehydrated rodents: Plasma colloid osmotic pressure, total osmotic pressure and electrolytes. *Comp. Biochem. Physiol. A* **44**, 1261–1265.

50. Hulse, E. V. (1964). Quantitative cell counts of the bone marrow and blood and their secular variations in the normal adult rat. *Acta Haematol.* **31**, 50–63.

51. Irvine, R. O. H., Dow, J., and Fong, J. (1966). Effect of anesthesia on the collection of blood for acid–base studies in the rat. *Med. Pharmacol. Exp.* **14**, 557–562.

52. Jones, R. A., Jenkins, L. J., Jr., Coon, R. A., and Siegel, J. (1970). Effect of long-term continuous inhalation of ozone on experimental animals. *Toxicol. Appl. Pharmacol.* **17**, 189–202.

53. Kaczmarczyk, G., and Reinhardt, H. W. (1975). Arterial blood gas tension and acid–base status of Wistar rats during thiopental and halothane anesthesia. *Lab. Anim. Sci.* **25**, 184–190.

54. Katiyar, J. C., Tangri, A. N., Ghatak, S., and Sen, A. B. (1973). Serum protein pattern of rats during infection with *Hymenolepis nana*. *Indian J. Exp. Biol.* **11**, 188–190.

55. Keene, W. R., and Jandl, J. H. (1965). Studies of the reticulo-endothelial mass and sequestering function of rat bone marrow. *Blood* **26**, 157–175.

56. Keraan, M., Meyers, O. L., Engelbrecht, G. H. C., Hickman, R., Saunders, S. J., and Terblanche, J. (1974). Increased serum immunog-

lobulin levels following portacaval shunt in the normal rat. *Gut* **15**, 468–472.

57. Kincade, B. W., Moore, M. A. S., Schleger, R. A., and Pye, J. (1975). Lymphocate differentiation from fetal stem cells in ^{89}Sr treated mice. *J. Immunol.* **115**, 1217–1222.

57a. Kindred, J. E. (1942). A quantitative study of the hematopoietic organs of young albino rats. *Am. J. Anat.* **71**, 207–243.

58. Klaassen, C. D. (1974). Comparison of the effects of two-thirds hepatectomy and bile duct ligation on hepatic excretory function. *J. Pharmacol. Exp. Ther.* **191**, 25–31.

59. Korsrud, G. O., Grice, H. G., Goodman, T. K., Knipfel, J. E., and McLaughlan, J. M. (1973). Sensitivity of several serum enzymes for the detection of thioacetamide-, dimethylnitrosamine-, and diethanolamine-induced liver damage in rats. *Toxicol. Appl. Pharmacol.* **26**, 299–313.

60. Korsrud, G. O., and Trick, K. D. (1973). Activities of several enzymes in serum and heparinized plasma from rats. *Clin. Chim. Acta* **48**, 311–315.

61. Korsrud, S. D., Grice, H. C., and McLaughlin, J. M. (1972). Sensitivity of several enzymes in detecting carbon tetrachloride-induced liver damage. *Toxicol. Appl. Pharmacol.* **22**, 474–483.

62. Kozma, C. K. (1967). Electrophoretic determination of serum proteins of laboratory animals. *J. Am. Vet. Med. Assoc.* **151**, 865–869.

63. Kozma, C. K., Weisbroth, S. H., Stratman, S. L., and Conejeros, M. (1969). Normal biological values for Long-Evans rats. *Lab. Anim. Care* **19**, 746–755.

64. Krebs, J. S., and Brauer, R. W. (1958). Metabolism of sulfobromophthalein in the rat. *Am. J. Physiol.* **194**, 37–43.

65. Ladenson, J. H., Tsai, L. B., Michael, J. M., Kessler, G., and Joist J. H. (1974). Selrum versus heparinized plasma for eighteen common chemistry tests. Is serum the appropriate specimen? *Am. J. Clin. Pathol.* **62**, 545–552.

66. Lazar, G., Serra, D., and Tuchweber, B. (1974). Effect on cadmium toxicity of substances influencing reticuloendothelial activity. *Toxicol. Appl. Pharmacol.* **29**, 367–376.

67. Libermann, I. M., Capano, A., Gonzalez, F., Brazzuna, H., Garcia, H., and DeGelabert, A. G. (1973). Blood acid–base status in normal albino rats. *Lab. Anim. Sci.* **23**, 862–865.

68. Loegering, D. J. (1974). Effect of swimming and treadmill exercise on plasma enzyme levels in rats. *Proc. Soc. Exp. Biol. Med.* **147**, 177–180.

69. Lubaroff, D. M. (1973). An alloantigenic marker on rat thymus and thymus-derived cells. *Transplant. Proc.* **5**, 115–118.

70. Lundin, P. M., and Angervall, L. (1970). Effect of insulin on rat lymphoid tissue. *Pathol. Eur.* **5**, 273–278.

71. Marien, G. J., and McFadden, K. D. (1968). A study of megakaryocytes in albino rats. *Can. J. Biol.* **46**, 1053–1058.

72. Marmo, E., Caputi, A. P., and Saini, R. K. (1973). Cardiac histological and serum biochemical changes produced by various adrenergic stimulant drugs in the rat. *Basic Res. Cardiol.* **68**, 318–334.

72a. Mattenheimer, H. (1970). "Micromethods for the Clinical and Biochemical Laboratory." Ann Arbor Sci. Publ., Ann Arbor, Michigan.

73. McLean, J. M., Mosley, J. G., and Gibbs, A. C. C. (1974). Changes in the thymus, spleen and lymph nodes during pregnancy and lactation in the rat. *J. Anat.* **118**, 223–229.

74. Melby, E. C., and Altman, N. H. (1974). "Handbook of Laboratory Animal Science." CRC Press, Cleveland, Ohio.

75. Mitruka, B. M., and Jonas, A. M. (1969). Biochemical changes in serum of pure and mixed *Streptococcus faecalis* and *Bacillus subtilis* infections in rats. *Appl. Microbiol.* **18**, 1072–1076.

76. Monden, Y. (1959). Quantative evaluation of total cellular number and cellular density of the thymolymphatic organs of young albino rats by means of DNA determinations. *Okajimas Folia Anat. Jpn.* **32**, 193–206.

77. Munro, I. C., Moodie, C. A., Kuiper-Goodman, T., Scott, P. N., and Grice, H. C. (1974). Toxicologic changes in rats fed graded dietary levels of ochratoxin A. *Toxicol. Appl. Pharmacol.* **28**, 180–188.

78. Nakaue, H. S., Dost, F. N., and Buhler, D. R. (1973). Studies on the toxicity of hexachlorophene in the rat. *Toxicol. Appl. Pharmacol.* **24**, 239–249.

79. Note, S., Tsunematsu, T., Lieno, K., Tanaka, A., Iwagaki, H., Yokoyaman, I. T., Shinobe, N., Wazizaka, N., and Kumurac, C. (1961). Studies of rat bone marrow by means of a new puncture method. *Acta Haematol. Jpn.* **24**, 16–21.

80. Penney, D. (1974). Short communication: Chronic carbon monoxide exposure: Time course of hemoglobin, heart weight, and lactate dehydrogenase isozyme changes. *Toxicol. Appl. Pharmacol.* **28**, 493–497.

81. Priestly, B. G., and Plaa, G. L. (1970). Reduced bile flow after sulfobromophthalein administration in the rat. *Proc. Soc. Exp. Biol. Med.* **135**, 373–376.

82. Ramsell, T. G., and Yoffey, J. M. (1961). The bone marrow of the adult male rat. *Acta Anat.* **47**, 55–65.

83. Rytomaa, T., and Kiviniemi, K. (1973). Control of granulocyte production. Chalone and anti-chalone. *Eur. J. Cancer* **9**, 515–519.

84. Sackler, A. M., Weltman, A. S., Pandhi, V., and Viadya, R. (1974). Biochemical and endocrine differences between normotensive and spontaneously hypertensive rats. *Lab. Anim. Sci.* **24**, 788–792.

85. Schalm, O. W., Jain, N. C., and Carroll, E. J. (1975). "Veterinary Hematology," 3rd ed. Lea & Febiger, Philadelphia, Pennsylvania.

86. Schermer, S. (1967). "The Blood Morphology of Laboratory Animals," 3rd ed., pp. 43–49. Davis, Philadelphia, Pennsylvania.

87. Scheving, L. E., and Pauly, J. E. (1967). Daily rhythmic variations in blood coagulation times in rats. *Anat. Rec.* **157**, 657–666.

88. Schulze, P., and Czok, G. (1974). Studies on the decrease in bile flow produced by sulfobromophthalein. *Toxicol. Appl. Pharmacol.* **28**, 406–417.

89. Schwartz, E., Tornaben, J. A., and Boxill, G. C. (1973). The effects of food restriction on hematology, clinical chemistry and pathology in the albino rat. *Toxicol. Appl. Pharmacol.* **25**, 515–524.

90. Schwetz, B. A., Leong, B. K. J., and Gehring, P. J. (1974). Embryo- and fetotoxicity of inhaled chloroform in rats. *Toxicol. Appl. Pharmacol.* **28**, 442–451.

91. Shaw, A. R. E., and Maclean, N. (1971). A re-evaluation of the haemoglobin electrophoretogram of the Wistar rat. *Comp. Biochem. Physiol. B* **40**, 155–163.

92. Sherman, H. (1963). Comparative profiles of various strains of rats used in long-term feeding studies. *Lab. Anim. Care* **13**, 793–807.

93. Tonkelaar, E. M., and van Logten, M. J. (1974). Protective action of dithiocarbamates on experimental liver damage produced by carbon tetrachloride. *Toxicol. Appl. Pharmacol.* **30**, 96–106.

94. Toro, G., and Ackerman, P. G. (1975). "Practical Clinical Chemistry." Little, Brown, Boston, Massachusetts.

95. Trepel, F. (1974). Number of lymphocytes in normal man. *Klin. Wochenschr.* **52**, 511–515.

96. Tschopp, T. B., and Zucker, M. B. (1972). Hereditary defect in platelet function in rats. *Blood* **40**, 217–226.

97. Upton, P. K., and Morgan, D. J. (1975). The effect of sampling technique on some blood parameters in the rat. *Lab. Anim.* **9**, 85–91.

98. Vacet, V. G., Meteger, H., Kachel, J., and Ruhenstrut-Bauer, G. (1972) Der Nachweis verscheidener erythrozyten populationen bei der ratte. *Blut* **24**, 42–53.

99. Valtin, H., Sawyer, W. H., and Sokol, H. W. (1965). Neurohypophy-

sial principles in rat homozygous and heterozygous for hypothalamic diabetes insipidus (Brattleboro strain). *Endocrinology* **77,** 701–776.

100. Venturi, V. M., and Rovati, A. L. (1956). Uber blut-und-kochen-mark befunde bei albino ratten. *Arch. Ital. Sci. Farmacol.* **6,** 32–40.

101. Vondruska, J. F., and Greco, R. A. (1973). Certain hematologic and blood chemical values in Charles River CD albino rats. *Bull. Am. Soc. Vet. Clin. Pathol.* **2,** 3–17.

102. Walker, A. I. T., Stevenson, D. E., Robinson, J., Thorpe, E., and Roberts, M. (1969). The toxicity and pharmaco-dynamics of dieldrin (Hedd). *Toxicol. Appl. Pharmacol.* **15,** 345–373.

103. Weimer, H. E., Benjamin, D. C., and Darcy, D. A. (1965). Synthesis of the alpha-glycoprotein (Darcy) of rat serum by the liver. *Nature (London)* **208,** 1221–1222.

104. Weimer, H. E., Roberts, D. M., Villanuevo, P., and Porter, H. G. (1972). Genetic differences in electrophoretic patterns during the phlogistic response in the albino rat. *Comp. Biochem. Physiol. B* **43,** 965–973.

105. Weisse, V. I., Knappen, F., Frölke, W., Guénard, J., Köllmer, H., and Stötzer, H. (1974). Blutwerte der ratte in abhängigkeit van alter und geschlecht. *Arzneim.-Forsch.* **24,** 1221–1225.

106. Whelan, G., and Combes, B. (1975). Phenobarbital-enhanced biliary excretion of administered unconjugated and conjugated sulfobromophthalein (BSP) in the rat. *Biochem. Pharmacol.* **24,** 1283–1286.

107. Winsten, S. (1968). Collection and preservation of specimens. *In* "CRC Handbook of Clinical Laboratory Data" (W. R. Faulkner, J. W. King, and H. C. Damm, eds.), 2nd ed., pp. 80–86. Chem. Rubber Publ. Co., Cleveland, Ohio.

108. Worden, A. N., Noel, P. R. B., Mawdesley-Thomas, L. E., Palmer, A. K., and Fletcher, M. A. (1974). Feeding studies on lenacil in the rat and dog. *Toxicol. Appl. Pharmacol.* **27,** 215–224.

109. Yale, C. E., and Torhorst, J. B. (1972). Critical bleeding and plasma volumes of adult germ-free rats. *Lab. Anim. Sci.* **22,** 497–502.

110. Yasin, R. (1975). Altered plasma creatinine phosphokinase activity in vincristine-treated rats. *Biochem. Pharmacol.* **24,** 745–746.

111. Zbinden, G. (1963). Experimental and clinical aspects of drug toxicity. *Adv. Pharmacol.* **2,** 1–112.

112. Zingg, W., Morgan, C. D., and Anderson, D. E. (1971). Blood viscosity, erythrocyte sedimentation rate, packed cell volume, osmolality, and plasma viscosity of the Wistar rat. *Lab. Anim. Sci.* **21,** 740–742.

Chapter 6

Nutrition

Adrianne E. Rogers

I. INTRODUCTION

Experimental studies of nutrient requirements and the effects of nutrient deficiency or excess have made greater use of rats than of any other animal species. Classic descriptions of experimentally induced deficiency diseases, definition of metabolic or structural functions of nutrients, measurement of nutrient requirements, and their alteration by physiologic or pathologic influences have, more often than not, been made using rats.

The enormous amount of data available on the rat allows firm conclusions as to requirements for many nutrients under specified laboratory conditions. Nevertheless, questions still arise with respect to some nutrients; others previously thought not to be required have been found to enhance growth or improve health by some other criterion when examined under special environmental conditions. Diets composed of purified ingredients which contain all recognized essential nutrients at the required concentration can support normal growth and reproduction in rats, but an investigator may find that increasing the content of one or more nutrients improves performance; thus, the question of nutritional requirements arises again. As biochemical or other assays of performance become more sophisticated, the criteria for dietary adequacy change and may alter the basis for setting a requirement.

Initial studies were made using growth, usually for 4–8 weeks from weaning, and female reproductive performance, i.e., percent conceptions; number, weight, and survival of off-spring; and weight gain of sucklings, as major criteria for dietary adequacy. Diets composed of natural ingredients, developed empirically to support maximum performance, were the standard against which purified or chemically defined diets were measured. The term "purified" used in this chapter describes diets composed of ingredients such as casein, corn starch or sugars and vegetable oils, i.e., components extracted from natural products but not necessarily chemically pure, to which are added mineral salts and chemically pure vitamins. The older term for such diets is "semisynthetic." "Chemically defined" is the term used to describe diets composed of amino acids, sucrose or other mono- or disaccharides, triglycerides or fatty acids, mineral salts, and vitamins of high degrees of purity (3).

Several other criteria have been developed and used in assessing nutrient adequacy. They include (1) nutrient content in storage tissues (e.g., vitamin A in liver) or tissues of which the nutrient is a structural component (e.g., calcium in bone), (2) nutrient content in blood ur urine, (3) activity of enzymes for which nutrient is or is part of a coenzyme, (4) nutrient balance (e.g., net gain or loss of amino acid nitrogen), (5) biochemical alterations highly specific for a given nutrient (e.g., prothrombin time for vitamin K), (6) protein or nucleic acid content of tissues, and (7) behavioral changes. The required dietary concentration of a nutrient indicated by one method may differ from that derived from studies using another method.

The greatest nutritional demands under normal physiologic conditions arise during gestation, lactation, and early post-weaning growth. The dietary concentrations suggested in this chapter are the requirements indicated in studies cited for maintenance of normal rats during these periods and may be greater than the requirements of adult male, or nonpregnant female rats.

Signs of deficiency have been listed briefly since they are extensively described and illustrated in many other reviews (60,162,163,223,80a). Emphasis has been placed on recent studies, when available, in which nutrient needs or effects have been examined in rats maintained under standardized conditions and fed well-defined diets. In many cases, the criteria have been biochemical changes rather than clinically evident signs of deficiency. The ideal level of performance or tissue content, etc., in many cases is not established and therefore must be assumed to be the maximum attained. This is true also for the apparently simple criterion of growth; the assumption is not always borne out, since most rapid growth rate in rats often does not correlate with longest life span or least evidence of disease.

Since data on which requirements are based were obtained in many laboratories under varying conditions of diet mixing, storing, and feeding; rat maintenance; etc., no safety factors have been incorporated to allow for such variations, and one may assume the dietary recommendations are adequate under many different laboratory conditions. Experimental procedures and other alterations of environment may influence nutrient requirements. Significant alteration of requirements for vitamins which are supplied by intestinal microbial synthesis via coprophagy can be induced by prevention of coprophagy or gnotobiotic conditions (Volume II, Chapter 2) (100).

If data have been given in references in terms of nutrient intake per rat per day, they have been converted to give nutrient content in the diet by assuming a daily food intake of 15 gm. Nutrient requirements have been expressed per unit of diet weight and assume a caloric density of approximately 4 kcal/gm.

II. DIETARY CONSIDERATIONS IN EXPERIMENTAL STUDIES

In experimental studies, the composition of the diet fed is of major importance in maintaining animals in good health and yielding consistent results. Effects of dietary intake or composition on experimental studies may be either desired or inadvertent. Review of studies published in 1968–1969 revealed many examples of inadequate dietary intakes by rats included in many different kinds of studies (77). Mechanisms by which diet acts range from simple alteration of intake, which alters intake of test compounds incorporated into the diet, to complex metabolic effects. Dietary effects on hepatic microsomal oxidases are extensive and can alter responses to drugs and other chemicals, including carcinogens, but the direction of change is not reliably predictable (27,73,241). For example, the activity of aflatoxin B_1 (AFB_1), a hepatotoxic and hepatocarcinogenic mold metabolite, is sensitive to changes in hepatic drug metabolism since it requires enzymatic activation and can be deactivated by several different pathways. Rats deficient in lipotropes and amino acids are resistant to single doses of AFB_1 which are lethal to normal rats but are much more susceptible than normal rats to toxicity and carcinogenicity of repeated small doses (205). Protein-deficient rats are unusually susceptible to the toxicity of AFB_1 but relatively resistant to its carcinogenicity (139). Yet both dietary regimens depress hepatic microsomal oxidases. Diet can alter supply of enzyme substrates and thereby increase or decrease enzyme activity and product synthesis. A striking example is the effect of dietary amino acids on brain content of neurotransmitters (59). Tumor development is directly correlated with caloric intake in many instances (30,237). The source of calories also may influence tumor development; isocaloric substitution of fat for carbohydrate enhances induction of mammary tumors (30). Specific nutrients, e.g., vitamin A, choline, methionine, and other amino acids, influence tumor development in certain organs (37,150,171,173,201,202,207,208).

Many diets composed of natural ingredients have been developed for laboratory rats and are available commercially. They support good growth and reproduction and are satisfactory for many experimental purposes. They cannot be used for nutritional studies or studies in which nutrition may exert a significant influence because they vary widely in nutrient and nonnutrient content; in addition they may contain undesirable contaminants. For example, analyses of nutrient content of commercial rat diets showed that variation in nutrient content by a factor of 10 occurred commonly both in different lots from one manufacturer and in different brands; variation by a factor of 100 was not unusual (168). These differences arise in part from variability of composition of natural products and in part from the amounts of vitamins and minerals added. Nitrates, which can be converted to nitrite and form carcinogenic nitrosamines, are present in variable amounts. Lead is a frequent contaminant of such diets. Mycotoxins and insecticides may be found on occasion (65,168).

Natural product diets are available as *open-formula diets* which are composed of natural ingredients but with specification of the exact ingredients (123). This reduces the variability introduced by utilization of different sources of protein, carbohydrate, and fat, as is done in production of the standard commercial products in which the choice depends on supply and cost of ingredients at different times.

For performance of experiments in which nutrient content is a critical factor, i.e., experiments to determine nutrient requirement or the effect of several different concentrations of nutrient on a variable such as tumor development or infectious disease, the most satisfactory solution is to feed a purified diet. Many *purified diets* for rats have been published and will be found in references cited. A committee of the American Institute of Nutrition has published initial studies in rats fed a purified diet which will be evaluated further in long-term studies (3). Examples of diets we have used are presented in Table I. Diet 1 was comparable to a natural product diet in maintaining growth and reproduction (170). It was fed to five generations of Sprague-Dawley rats and their growth, reproduction, and survival were compared to rats fed a natural product diet and maintained under the same conditions. Growth rate was the same in the two dietary groups in all generations (Fig. 1). Organ weights were the same except for a significant increase in liver weight in rats fed the natural ingredient diet (Table II). Reproduction was normal in females fed the purified diet, although there was a small reduction in the number of females mated that produced litters (Table III). Life span was equivalent in the two diet groups.

Diet 2 is a simpler diet which supports growth and survival well. We have used it in many long-term studies. It can be adapted to study deficiency or excess of any of its component vitamins or minerals as well as protein, fat, or carbohydrate. An example of its usefulness is given by a study in which we

Table I

Examples of Composition of Purified Diets for Rats

Component	Diet 1 (%)	Diet 2 (%)
Vitamin-free casein	20.0	22.0
Gelatin	4.0	—
D,L-Methionine	0.2	—
Sucrose	12.0	16.0
Dextrose	—	20.0
Dextrin	—	20.0
Cornstarch	48.8	—
Cellulose (celluflour)	5.0	—
Vegetable oil (Wesson)	5.0	16.0
Salts mixture[a]	4.0	4.0
Vitamin mixture[a]	1.0	2.0

[a] For composition in Diet 1 see Newberne *et al.* (170); for composition in Diet 2 see Newberne and Wogan (174).

examined the influence of vitamin A on chemical induction of colon carcinoma (203). Rats were fed diet 2 with 30,000 IU vitamin A/kg (control), 500,000 IU/kg (excessive), or no

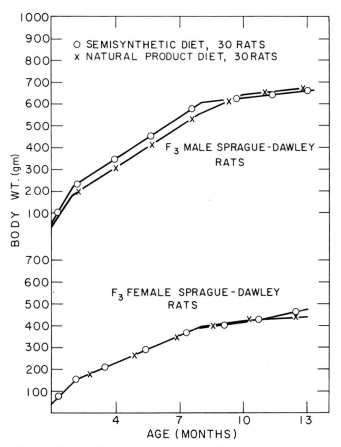

Fig. 1. Growth of male and female Sprague-Dawley rats fed a natural ingredient or a purified (semisynthetic) diet.

Table II

Body and Organ Weights at 1 Year in F_3 Sprague-Dawley Rats Fed Purified or Natural Ingredient Diets[a]

Diet	Sex	Body weight (gm)	Weight (% of body weight)				
			Liver	Kidney	Heart	Gonads	Pituitary
Purified	M	658	2.5[b]	0.56	0.31	0.52	0.002
	F	474	2.6[b]	0.58	0.28	0.02	0.004
Natural ingredient	M	640	3.2	0.67	0.28	0.57	0.002
	F	437	3.3	0.64	0.32	0.03	0.004

[a] From Newberne et al. (170).
[b] Difference from rats fed natural ingredient diet significant $P < 0.01$.

vitamin A. The deficient rats were given diet containing 1000 IU/kg periodically to prevent death from vitamin A deficiency. The rats maintained widely differing serum and hepatic vitamin A content in accord with their dietary treatment (Table IV), and deficient rats grew slowly compared to the other two groups (Fig. 2). Serum and hepatic vitamin A content should be measured because of the buffering capacity of the liver for vitamin A (38).

In the preparation of diets the dry ingredients can be mixed in large batches, stored in a refrigerator or cool room for periods of at least 1 month and used to make the complete diets which can be stored in a refrigerator for 1 week or frozen for longer periods. The complete diets can be fed as mixed or solidified by incorporation into 3% agar solution, 1 : 1, which decreases dissemination of toxic or carcinogenic chemicals incorporated in the diet. Agar diets are easily cut into blocks for weighing and feeding; the remainder can be collected and weighed when intake is to be measured. Agar contains minerals and should be assayed for them if exact levels are important in the particular experiment. Diets can be made in gel form also by utilization of gelatin, but one must take account of amino acids contributed to nutrient intake. Purified diets are available commercially and are satisfactory for many experimental purposes, but they may vary in nutrient content more than can be tolerated in a given experiment.

Since purified diets are composed of specified nutrient sources, levels can be adjusted easily to meet experimental requirements. Many examples will be found in the references. Vitamins and minerals can be omitted from or added to the respective mixes; protein and fat levels can be adjusted, or the sources can be changed to give different amino acid or triglyceride components. Specific amino acids and triglycerides can, of course, be added. Since altering fat content alters caloric content, ratios of calories supplied by protein, fat, and carbohydrate and ratios of vitamins or minerals to calories, adjustments of other nutrients may be required to keep the ratios constant. For example, addition of fat to a complete diet without removing an isocaloric amount of carbohydrate dilutes all the nutrients and changes the balance between them. The same problem arises in studies in which fat or other components are mixed into natural product diets.

Protein sources used for purified diets may contain vitamins or minerals, and starches such as cornstarch may contain amino acids (142,211,249). Chemically defined diets may be required in some cases and must be used for studies of specific amino acid requirements or interactions. They are discussed in Section III,D.

Either antagonistic or synergistic (sparing) interactions between nutrients may occur. One example is the area of lipotropic nutrients. The major lipotropic nutrients, which maintain

Table III

Reproduction in Sprague-Dawley Rats Fed Purified or Natural Ingredient Diet[a]

Diet	Number of litters[b]			Number of pups/litter[c]			Weight of pups (gm)		
	F_0	F_1	F_3	F_0	F_1	F_3	F_0	F_1	F_3
Purified	20/30	21/30	24/29	11.1 ± 0.8	9.4 ± 0.3	10.3 ± 0.8	6.4 ± 0.4	6.0 ± 0.4	6.0 ± 0.4
Natural ingredient	27/28	26/30	29/29	10.6 ± 0.9	11.2 ± 0.8	10.4 ± 1.0	6.1 ± 0.3	6.1 ± 0.4	6.1 ± 0.3

[a] From Newberne et al. (170).
[b] Number litters/number of females bred.
[c] Average ± S.D.

Table IV

Serum and Liver Content of Vitamin A in Rats Fed Different Dietary Concentrations of Vitamin A

Diet (IU/kg)	Vitamin A		Duration of feeding (weeks)	Signs of deficiency	Reference
	Serum (μg/100 ml)	Liver (μg/gm)			
0	6	—	5	+	38
40	6	—	5	+	
160	10	—	5	+	
2,560	44	—	5	−	
10,240	47	—	5	−	
0	4 ± 2	1 ± 1	6	+	
10,240	63 ± 10	147 ± 10	6	−	
0[a]	2 ± 1	0	10–14	+	203
30,000	38 ± 5	64 ± 7	10–15	−	
0[a]	9 ± 2	0.4 ± 0.3	34	+	
30,000	38 ± 5	67 ± 10	24–34	−	
500,000	97 ± 25	683 ± 188	23–26	−	

[a]Diet supplemented with 1000 IU/kg periodically, see Fig. 2.

normal lipid metabolism in liver and other tissues, are methionine, choline, folic acid, and vitamin B_{12}. Combined deficiency of some or all of these in rats fed a high fat diet has been a useful model for human alcoholic liver disease and its attendant pathologic and metabolic abnormalities. Design of satisfactory deficient diets has required a great deal of experimentation because of interactions between the lipotropes. Dietary content of choline or methionine determines the requirement for the other since one spares the other. Folate and vitamin B_{12} also spare choline and methionine but are less effective. Deficiency of choline or methionine induces deficiencies of folate and vitamin B_{12} even if they are present in the diet (18,19,78,82,88,136,242). Protein sources must be low in methionine, therefore, low dietary content of a complete animal protein or higher content of plant proteins (peanut meal, soybean protein) or incomplete animal protein (gelatin) have been used. Increasing dietary lipid to 15–30% by weight enhances the deficiency; chain length and degree of saturation of the lipid also influence its severity. Other examples are the sparing interactions between vitamins A and E and antagonistic interactions between certain amino acids (85,155). Data from which rats' nutrient requirements have been determined are discussed below. Recommended content in diets is summarized in Table V.

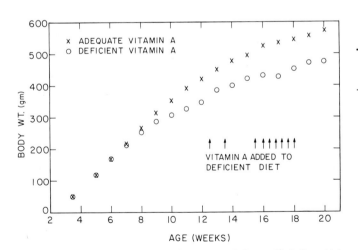

Fig. 2. Growth of male Sprague-Dawley rats fed a purified diet which contained vitamin A (30,000 IU/kg) or was deficient in vitamin A. Deficient rats were fed the diet supplemented with 1000 IU/kg of vitamin A (arrows) when growth ceased, in order to maintain them for a prolonged period.

Table V

Recommended Dietary Content of Nutrients for Rats

Nutrient	Amount in Diet (90% dry matter)
Protein (%)	12.0[a]
Fat (%)[b]	5.0
Digestible energy (kcal/kg)	3800
L-Amino acids	
Arginine (%)	0.6
Asparagine (%)	0.4
Glutamic acid (%)	4.0
Histidine (%)	0.3
Isoleucine (%)	0.5
Leucine (%)	0.75
Lysine (%)	0.7
Methionine[c] (%)	0.6
Phenylalanine[d] (%) (and Tyrosine)	0.8

(*continued*)

Table V (*Continued*)

Recommended Dietary Content of Nutrients for Rats

Nutrient	Amount in Diet (90% dry matter)
Proline (%)	0.4
Threonine (%)	0.5
Tryptophan (%)	0.15
Valine (%)	0.6
Nonessential[e] (%)	0.59
Minerals	
Calcium (%)	0.5
Chloride (%)	0.05
Chromium (mg/kg)	0.3
Copper (mg/kg)	5.0
Fluoride (mg/kg)	1.0
Iodine (mg/kg)	0.15
Iron (mg/kg)	35.0
Magnesium (mg/kg)	400.0
Manganese (mg/kg)	50.0
Phosphorous (%)	0.4
Potassium (%)	0.36
Selenium (mg/kg)	0.10
Sodium (%)	0.05
Sulfur (%)	0.03
Zinc (mg/kg)	12.0
Vitamins	
A (IU/kg)	4000.0
D (IU/kg)	1000.0
E (IU/kg)	30.0
K_1 (μg/kg)	50.0
Choline (mg/kg)	1000
Folic acid (mg/kg)	1.0
Niacin (mg/kg)	20.0
Pantothenate (calcium) (mg/kg)	8.0
Riboflavin (mg/kg)	3.0
Thiamin (mg/kg)	4.0
Vitamin B_6 (mg/kg)	6.0
Vitamin B_{12} (μg/kg)	50.0

[a] Ideal protein adequate to support growth, gestation and lactation. Adult, nonpregnant rats require lower concentrations of protein and amino acids. [See text and Morrison (158).]

[b] Linoleic acid is required at 0.3%.

[c] One-third to one-half can be supplied by L-cystine.

[d] One-third to one-half can be supplied by L-tyrosine.

[e] Mixture of glycine, L-alanine, and L-serine.

III. NUTRIENT REQUIREMENTS

A. Energy

Energy required by rats is the sum of requirements for maintenance, growth, gestation, lactation, physical activity, heat production, and other physiologic or pathologic conditions. Basal metabolic or maintenance requirements can be calculated on the basis of metabolic body size, body weight to the 0.75 power. Diets fed to adult rats can vary in energy density between 2.5 and 5.0 kcal of digestible energy per gram without affecting energy intake, but rapidly growing weanlings require energy density of at least 3 kcal/gm. The requirement during gestation may be 10 to 30% greater than in nonpregnant mature female rats, and lactating rats have an energy requirement two to four times that of nonlactating females (15,148,156,190, 191,198).

Most purified or chemically defined diets fed to rats in the laboratory contain 4 to 4.5 kcals gross energy per gram. Of this total, 90 to 95% is digestible energy, and of that 90 to 95% can be metabolized. In diets made of natural ingredients or in purified diets to which cellulose is added, the digestibility may be somewhat lower. Daily maintenance energy requirement for adult rats is approximately 110 kcal digestible energy per kg body weight$^{0.75}$ (163). Rats adjust food intake to energy requirements within a broad range of dietary caloric content. They can maintain adequate energy intake after dilution of diet by inert materials up to 40 to 50%. At greater dilution energy intake decreases. Lactating females cannot compensate for greater than 10% dilution by inert material because their basal diet intake is nearly maximal. Under conditions of increased protein requirement or decreased protein concentration in the diet, food intake may be increased to meet protein needs, but the response is less well defined than for caloric adjustment (148). Female rats fed 25 or 8% protein were studied to investigate the role of protein in regulation of food intake. Nonpregnant, nonlactating females ate the same amount of each diet and thus responded only to calories. Pregnant rats fed the low protein diet increased food intake to 150% of control intake, but during lactation rats fed 25% protein ate approximately twice as much as rats fed 8% protein. Body weight gain of dams and weight of litters at birth was the same in both diet groups so the extra energy consumed by the low protein rats was not converted into increased body weight but was apparently utilized less efficiently. Therefore, during gestation dietary protein and energy content controlled appetite, but appetite regulation became abnormal during lactation in the face of decreased protein levels (148).

The strain of rat also influences appetite control, growth rate, feed utilization, efficiency, and metabolic rate. In Sprague-Dawley and Buffalo male rats the relationships between energy expenditure, body weight, and energy intake were significantly different as was the energy expenditure fraction consumed in motor activity. Sprague-Dawley rats compensated for dilution of the diet by 25 or 50% nonnutritive fiber but could not compensate fully for 75% dilution. Buffalo rats did not compensate fully at any level (157). Congenitally obese Zucker rats used dietary protein less efficiently for synthesis of body protein than their lean littermates and used dietary energy for fat synthesis rather than for physical activity (45).

Efficiency of utilization of food is influenced by the nutritive value of the diet, the amount of food eaten, the pattern of eating, environmental conditions, and the activity for which energy is being used, i.e., growth, maintenance, milk production, or fat production. Pattern of food intake and diet composition influence absorption of nutrients, the pattern of hepatic and other tissue enzymes responsible for metabolizing the nutrients, particularly carbohydrate and fat, and the rates of utilization of glucose and fat and glycogen storage (14,22,31,33, 45,89,122,133,226).

Food intake is governed by complex metabolic modulators which act either rapidly, i.e. within 1 or 2 days of dietary change, or more slowly, i.e., within 1 or 2 weeks. The modulators correspond to two hypothetical models of control of food intake. The rapidly acting or primary process responds to energy flow from the gut, sensed through receptors in the portal hepatic vascular system, and may initiate and sustain feeding; the secondary process responds to depletion of energy reserves sensed through receptors in the central nervous system. The behavioral response to the rapidly acting modulator is a change in duration of feeding; the response to the second is a change in efficiency of feeding which is the amount of food ingested per unit time (158,191,192).

Studies of control of dietary intake have been applied to tumor-bearing rats in investigations of the etiology of tumor-associated cachexia. The normal hyperphagic response to insulin treatment is not abolished by tumor growth, but dilution of diet with inert material does not induce a normal corrective increase in intake. This suggests that feeding responses to preabsorptive signals, such as gastric distention, and to blood glucose are normal, but response to metabolic signals of energy status is impaired (158).

B. Fat

Dietary fats provide essential fatty acids (EFA) which are required for synthesis of tissue lipids and cellular membranes. In addition, fats are a significant source of calories and are necessary for normal absorption and utilization of the fat-soluble vitamins. Growth, maturation, and reproduction are dependent upon a dietary supply of fat, but a broad range of fat content gives essentially equivalent results. Diets for rats generally contain 5 to 15% fat by weight. A recent extensive review of the literature (163) suggested a minimum level of 5%.

Physiological functions of EFA are (1) promotion of growth, (2) prevention or alleviation of skin abnormalities, (3) maintenance of the normal ratio of phospholipids to triglycerides in tissues, (4) incorporation into phospholipids particularly in the 2-position, (5) prostaglandin formation, and (6) maintenance of a normal ratio of triene to tetraene polyunsaturated fatty acids (47,49,96,97,194,213).

Signs of EFA deficiency in rats are reduced growth rate with a growth plateau 12 to 18 weeks after weaning onto deficient diet, scaling and reduced thickness of the skin, roughening and thinning of hair, necrosis of the tail, fatty liver, renal damage, and electrocardiographic abnormalities. There is failure of reproductive function in both sexes. Basal metabolic rate rises, often before skin or other changes are evident, and abnormalities in mitochondrial function can be demonstrated (96,97,186).

Of the three polyunsaturated fatty acids which have been considered essential (linoleic, linolenic, and arachidonic), linoleic is the most readily available in foods and can be converted by rats to arachidonic acid which is the major EFA in membranes. Linoleic acid also is an important precursor for prostaglandins. There are metabolic interactions between the EFAs and other fatty acids, e.g., oleic, such that the more highly unsaturated fatty acids decrease synthesis of less highly unsaturated acids. Therefore, dietary fatty acids determine the structure of tissue fatty acids and of tissue phospholipids which are rich in unsaturated fatty acids (232).

Polyunsaturated fatty acids are available readily from plants. Green leaves of many plants contain approximately 60% of their fatty acids as linolenic acid, and oils derived from a variety of seeds (corn, cottonseed, soybeans, and peanuts) contain 50% or more linoleic acid. Lard contains approximately 10% linoleic acid, and butter, 2%. Corn oil can contain as much as 63% linoleate, hydrogenated vegetable fat approximately 30%, and beef tallow 3% (97,109). Soybean and cottonseed meals and starches from several sources may contain significant levels of EFAs. Therefore, induction of EFA deficiency requires the use of vitamin-free casein as a protein source and sucrose as a carbohydrate source.

The dietary requirement is approximately 1.3% of calories or 0.3% by weight of linoleic acid in the diet. Diets high in saturated fats, oleic acid, or cholesterol increase the dietary requirement for linoleic acid. Serum prostaglandins were significantly lower in male rats fed a 20% beef fat diet (linoleate approximately 0.6% by weight) than in rats fed a 20% corn oil diet (linoleate 12%) or a diet containing hydrogenated vegetable fat (linoleate 6%) (109). There is an extensive literature on the relative values of different food fats and their utilization by rats (47–49).

It is not known whether both linoleic and linolenic acids are required in the diet of rats. There was no significant difference in growth or reproduction in rats raised to the third generation fed on a diet which contained either 1.25% methyl linoleate or 1.0% methyl linoleate and 0.25% methyl linolenate. The ratios of tissue fatty acids were markedly different in the two dietary groups and reflected the dietary fatty acids. In the third generation, rats changed at weaning from the linoleate diet to the diet which contained both fatty acids showed no increase in weight gain compared to animals maintained on the linoleate diet (238). Many of the fatty acids derived from linolenic acid occur in the brain. In contrast to other tissues, there is little variation in brain content of fatty acids, and there may be a specific need for linolenate in synthesis and function of brain lipids (40). Odd-numbered EFAs occur commonly in bacteria and have been demonstrated in animal tissues. They are present in rat liver in small amounts, 0.5 to 2.9% of total fatty acids. Odd-numbered, polyunsaturated fatty acids corrected almost completely the growth rate of EFA-deficient rats, corrected partially but not completely abnormalities of the skin, and supported normal phospholipid and prostaglandin synthesis (215).

C. Carbohydrates

Carbohydrates are utilized as a source of energy and may be required for normal energy metabolism, although demonstration of requirement has been made only under extreme conditions. The importance of carbohydrate as such in the diet of rats has been indicated by recent studies (2,29,75,125). Rats fed a carbohydrate-free diet which supplied 82% of calories as fatty acids and 18% as protein for 1 week gained weight poorly compared to rats fed a control diet which contained cornstarch. Substitution of triglycerides for fatty acids improved weight gain significantly but not to the level of rats fed control diet. Blood glucose and insulin levels were normal in all groups except rats fed fatty acids in which hyperglycemia followed administration of a glucose load (125). In a subsequent study in the same laboratory, rats were fed diets in which protein contributed 10% of calories, and 90% were contributed by triglycerides, fatty acids, or fatty acids plus glycerol. Control rats were fed a diet in which 80% of the calories were supplied by starch and 10% each by fat and protein. Weight gain was significantly reduced in rats fed any of the high-fat diets; rats fed fatty acids lost weight, and rats fed either triglycerides or fatty acids plus glycerol gained a small amount. Rats fed high fat diets had significant depression of blood glucose, hyperinsulinemia, and a significant increase in body lipid concentration (29).

Increased dietary protein may alleviate metabolic problems induced by high-fat, carbohydrate-free diets. Rats were fed diets which contained either 69% sucrose, 5% corn oil, and 20% casein or 51% lard, 8% corn oil, and 33% casein to provide the same caloric ratios of protein to nonprotein energy sources. Weight gain and protein efficiency ratios were greater in rats fed the high fat diet calculated either on the basis of protein intake or nitrogen retained per kilocalorie intake (55).

If rats are fed diets deficient or marginally deficient in protein or water-soluble vitamins, they grow better if fed insoluble carbohydrates (starches) than if fed soluble carbohydrates (sucrose or glucose). Starches may stimulate growth of intestinal bacteria and increase their synthesis of vitamins which are directly available through absorption or available through cop-

rophagy (86). In addition, the nature of carbohydrates fed to rats can influence lipid metabolism (141). Many carbohydrate sources can be used by rats; they include glucose, sucrose, maltose, fructose, and starches from wheat, corn, and rice. High levels of lactose, galactose, or xylose result in poor growth and cataract formation.

There is probably no absolute requirement for fiber in the rat diet, but generally it is included at about 5%, which is approximately the amount in diets composed of natural ingredients (123).

D. Protein

Protein and amino acid requirements of healthy rats are influenced by physiologic status (e.g., age, growth rate, pregnancy, or lactation), by caloric content of the diet, and by amino acid composition and digestibility of specific proteins utilized. Excessive dietary protein or amino acids can be deleterious; interactions between amino acids determine optimal content in some cases. Protein requirements decline with age after weaning, and requirements of male and nonbreeding female rats above the age of about 3 months are considerably lower then the amount required during the most active growth period. The requirements discussed in this chapter, unless stated otherwise, are based on studies in which the fat content of the diet was approximately 5% by weight. For diets which contain a greater amount of fat, increasing protein to maintain a constant amino acid to calorie ratio should be satisfactory. Mixtures of proteins can be used to supply the necessary essential amino acids or specific supplementation of deficient proteins can be made using L-amino acids. Animals are dependent on an exogenous source of approximately half the 20 amino acids commonly found in their proteins.

Protein deficiency results in decreased food intake and weight gain, anemia, hypoproteinemia, muscle wasting, rough hair coat, irregular estrus, and poor reproduction with fetal resorption or delivery of weak or dead newborn. Protein deficiency in mothers during lactation results in poor growth of offspring. Deficiency of single amino acids may, in some cases, cause specific abnormalities in addition to those cited.

Protein deficiency may have significant effects on experimental results in studies of chemical toxicity and carcinogenesis, immunologic processes, and infection and many other areas of interest. The activity of hepatic and probably other tissue microsomal oxidases (enzymes responsible for metabolism of steroid hormones and many exogenous chemicals) decrease markedly and rapidly in protein deficiency (27). Therefore, the response of rats to toxins and carcinogens may be altered. Rats that are protein deficient may respond abnormally to infectious agents because both cellular and humoral immunologic responses are depressed (167).

As rats grow, the metabolically active visceral organs form an increasingly smaller fraction of total body weight, but skeletal muscle remains a constant fraction of the weight except in obese animals. The most rapid increase in total body protein occurs during the suckling period when both body mass and percentage protein increase. Protein synthesis during that time is not uniform but follows a cyclical pattern which varies from organ to organ. Several techniques are available for study of nutritional requirements in newborn rats. The one used most extensively is alteration of litter size. Individual variations in intake may be large, and dead animals must be replaced since litter size must be maintained throughout the experiment. A second approach is to restrict caloric intake of the dam which reduces the variation in intake among the young to some extent. The primary deficiency induced by these techniques probably is of protein rather than calories. Better control of intake can be achieved by hand-feeding neonates (151).

Several methods are used to evaluate nutritional value of proteins. Protein efficiency ratio (PER) is the grams of weight gain per gram of protein consumed; it may be different at different dietary concentrations of test protein. For example, for casein the highest PER value is found at 7% protein in the diet; for plant proteins the highest PER is at about 15%. Ten percent protein is used in most studies. Strain, age, and sex of rats and duration of feeding influence PER. In final evaluation of proteins it is necessary to take into account quality and quantity of protein and also of the other components of the diet. Protein can be spared by increasing fat or carbohydrate in cases in which part of the protein is being used for energy.

Other methods of protein evaluation include chemical, microbiological, and whole animal bioassays. Net protein utilization (NPU) is a measure of that proportion of food nitrogen retained by test animals and is defined as

Body nitrogen of test group − body nitrogen of nonprotein group/nitrogen consumed by test group

The NPU decreases with increasing protein–calorie ratio because of utilization of protein calories for energy. The relationship between NPU and percentage of protein in the diet depends also on whether protein is being used for growth or maintenance and on the specific protein examined, since amino acids required for maintenance are, in part, different from amino acids required for growth (147,211).

Weanling rats were fed diets which contained 3.6 to 25% protein (lactalbumin) for 3 weeks. There was a steep rise in weight gain as dietary protein increased from 2.6 to 12%; 12% protein and above gave equivalent weight gain. Feed efficiency (grams weight gained per gram of food consumed) was maximum at 12% protein. Protein efficiency ratio was maximal at 8% protein and decreased as protein content of the diet increased above that level. Fecal and urinary nitrogen in-

creased with increasing dietary protein; urinary nitrogen rose markedly at levels greater than 12%, that is, after maximal growth rate was reached. Retention efficiency for most essential amino acids reached a maximum at 6% dietary protein (at which point body weight gain was approximately one-half maximum), plateaued with diets which contained 6 to 10% protein, and then declined (26).

Carcass retention efficiency is the ratio of the increase in content of an amino acid in the carcass to the amount of that amino acid consumed. Dietary amino acids can be (1) incorporated into protein; (2) metabolized to CO_2, water, and urea; or (3) synthesized into nonessential amino acids or other compounds such as creatine. If the major factor operating in utilization of dietary amino acids is requirement for protein synthesis, retention efficiency for each amino acid is inversely related to dietary supply of that amino acid as it approaches the minimal requirement. Maximum retention efficiency is used to determine dietary requirement. In rats fed 6 to 10% protein, 17% of total ingested amino acids were catabolized; in rats fed diets which contained lower concentrations of protein, and were therefore deficient in some amino acids, about 50% of amino acids ingested were catabolized since they could not be used for protein synthesis. At low dietary protein intake there was a change in carcass amino acid composition which indicated alterations in proportions of body proteins (26). If rats were fed casein rather than lactalbumin, maximum weight gain and feed efficiency occurred at 18 to 20% dietary protein and the highest PER at 10 to 12% protein. Fecal nitrogen excretion was the same as in rats fed lactalbumin, and urinary nitrogen rose before maximum weight gain was achieved which indicated that amino acids from casein were not used as efficiently as amino acids from lactalbumin (25).

It has been assumed that nutritive value of protein is determined by the most limiting essential amino acid, and that nutritional quality of proteins decreases linearly as the limiting essential amino acid(s) decrease below the quantity found in an "ideal" protein. Protein utilization should fall regardless of the essential amino acid which is deficient and stop when an essential amino acid is absent from the protein being tested. However, in rats fed diets lacking a single essential amino acid, protein utilization dropped to 20 to 40% of the value for rats fed an adequate diet but not to zero except for diets which lacked threonine, isoleucine, or the sulfur amino acids. Rats may be able to adapt to diets deficient in some essential amino acids by increased coprophagy, alteration of catabolism, or possibly limited synthesis (211).

A recent review of data on protein requirements of rats concluded that growing, pregnant, or lactating rats required 12% ideal protein in the diet or 14% casein supplemented with 0.2% DL-methionine. For maintenance of adult rats the corresponding values were 4.2 and 4.8%. If the requirement is based on energy content of the diet, it can be stated to be 33 mg casein/

kcal gross energy for growth, gestation, and lactation or 12 mg casein for maintenance (163).

If rats are fed diets which contain amino acids in place of protein, mixtures of essential and nonessential amino acids give a greater growth rate than mixtures of only essential amino acids. The range of acceptable ratios of nonessential to essential is wide; satisfactory results have been obtained using ratios of 0.5 to 4.0 in diets which contained 10-15% total amino acids. In addition to amino acids known to be essential from earlier studies, experiments in the last 10-15 years have demonstrated a need for arginine, asparagine, glutamic acid, and probably proline (23,152,163) (Table V). These amino acids may not be required for maintenance of adult rats but are required by growing rats, presumably because they cannot synthesize amino acids at a rate needed for rapid growth. The requirements for amino acids are listed in a pattern which has supported normal growth in rats fed amino acid diets and assume that total dietary protein is 12% and fat 5%. The remainder of the protein requirement can be supplied by a mixture of alanine, glycine, and serine.

In early studies in which dietary protein was replaced by amino acids in amounts equal to or greater than the known requirements of rats, growth rate was not equal to that of rats fed casein, and addition of casein up to 12% to diets which already contained 22% amino acids was required to achieve maximal growth. It was found necessary to define concentrations of essential amino acids and to establish relative or absolute concentrations of nonessential or partially dispensable amino acids, such as arginine, proline, and glutamic acid, since a low intake of one increased the quantity of the others needed for normal growth. Second, rats fed amino acid diets had decreased food intake, but if diets were mixed 1 : 1 with 3% agar, intake and weight gain were normal. Probably water in agar gel diets prevented the osmotic effects of amino acids in the stomach and upper small intestine. Rats fed amino acid diets have been carried through 4 generations with good but not maximal weight gain and with normal reproduction (209,218,219).

The maintenance requirement of young female adult rats was met by lactalbumin at 3.2% of the diet. The rats were then fed amino acid diets deficient in a single amino acid or diets which contained no protein or amino acids. Growth was depressed in all rats fed deficient diets, but only diets which lacked threonine, isoleucine or methionine, and cystine depressed growth as markedly as protein-free diets. Discrepancies were explained in part by the observation that the cornstarch fed contained leucine, methionine, and tyrosine in amounts to provide 15 to 21% of the maintenance requirement, and other essential amino acids were present in amounts that supplied 2 to 10% (211).

Deficiencies of single amino acids have been studied in rats fed *ad libitum* or force-fed. Young rats force-fed amino acid diets deficient in a single essential amino acid were studied at intervals of 3 to 7 days. They had poor hair coats with areas of alopecia and pigmentation from porphyrins around the mouth and nose, were weak and lethargic, and lost weight. Rats fed diets deficient in threonine or histidine became hyperreactive to external stimuli. Threonine, histidine, or methionine deficiency resulted in livers which were enlarged and yellow, and in which periportal hepatocytes were distended with fat. The pancreas was edematous in threonine- or histidine-deficient rats, and the acinar cells had decreased zymogen granules. In methionine-deficient animals the pancreas was normal (85,135,227-229). If rats were force-fed diets deficient in valine, tryptophan, leucine, isoleucine, or lysine there was significant reduction in activities of pancreatic enzymes (135).

Measurement of plasma free amino acids can be used to study amino acid requirements. If rats fed a diet which contains all but one of the essential amino acids in adequate quantities are supplemented with increasing ambients of that amino acid, the blood content of the supplemented amino acid shows an inflection point and then rises sharply when the requirement is reached. This method was used to determine the tryptophan requirement of growing and adult male rats. In young rats the requirement for maximal growth was 0.14%. Older rats (300 gm) required 0.05 to 0.09%. On the basis of plasma amino acid content, the requirement was 0.11% for young rats and 0.04% for older rats (252).

Another method for determination of amino acid requirement is measurement of metabolism or oxidation of specific amino acids. The latter method assumes, as does the carcass retention method, that as dietary supply approaches and exceeds the requirement, amino acids will be utilized increasingly for lipogenesis, gluconeogenesis, excretion, and oxidation. For example, rats were fed a sesame seed protein diet supplemented with lysine. Average daily gain and food efficiency ratio indicated a lysine requirement of approximately 0.7%; study of serum lysine content indicated a requirement of 0.8%. Rats were then injected with radioactive lysine, and their expired CO_2 was collected and measured. At low lysine intakes, little was oxidized, but oxidation rose as diet content increased beyond 0.6% (23).

Interactions between amino acids or between amino acids and other nutrients influence requirements. One-third to one-half the requirement for phenylalanine can be met by tyrosine. The tryptophan requirement (0.15%) assumes an adequate dietary content of niacin. Of the total requirement for methionine and cystine (0.6%), up to one-half can be supplied by L-cystine. Both total sulfur amino acids and the relative proportions supplied by methionine and cystine affect food intake, growth, and body composition of rats. Weight gain and food efficiency ratios were equivalent in rats fed 0.4 to 0.7% methionine in diets which contained 0.6 to 0% cystine, respectively (212). Rats fed 0.2 to 0.8% methionine and no cystine had increased weight gain up to 0.6% with no further improvement at 0.8% methionine. Addition of cystine improved food intake and weight gain at all levels, but the effect was not marked at 0.6% or 0.8% methionine (224).

Methionine supplies methyl groups for synthesis of choline, creatine, carnitine, nucleic acid, histones, etc. It is metabolized more extensively than other essential amino acids; almost twice as much methionine carbon is recovered as CO_2 as carbon from other essential amino acids. Turnover of the methyl carbon is high before methionine is incorporated into protein. Cystine supplementation affects several enzymes of methionine metabolism in rats fed a diet low in methionine (1,13).

Arginine, in addition to being a constituent of protein, is required for transport, storage, and excretion of nitrogen. It is required for growth of young rats but may not be required for maintenance of adult rats. Weanling rats fed amino acid diets complete except for arginine had depressed food intake and growth, although a small positive nitrogen balance was maintained. Urinary excretion of amino acids and urea was increased while excretion of ammonia and creatinine was decreased. Older male rats (150 to 175 gm) fed the arginine-deficient diet had slightly decreased food intake and no significant alteration of weight gain but increased excretion of urea, orotic acid, and citric acid. Activities of hepatic urea cycle enzymes were increased in arginine deficiency, presumably because of increased catabolism of other amino acids. Increased urinary orotic acid may be derived from shunting of ammonia into carbamyl phosphate and pyrimidine synthesis (152).

Specific effects of tryptophan deficiency on eye development have been reported, but may have resulted from combined deficiencies. Female rats fed diets deficient in several amino acids bore young with a 30% incidence of cataracts. Supplementation with either L-tryptophan or DL-α-tocopherol prevented the abnormality. In subsequent studies rats were fed an amino acid diet deficient only in tryptophan. Dams had decreased weight gain and feed efficiency, and offspring had decreased weight gain but did not have cataracts unless the diet was deficient also in α-tocopherol. Unilateral or bilateral lenticular opacities occurred in 33% of offspring of mothers deficient in both nutrients, in 6% of offspring of mothers deficient only in α-tocopherol, but not in offspring of mothers deficient only in tryptophan (24).

Amino acids may have specific functions other than those related directly to protein synthesis. For example, intake and plasma content of tryptophan are closely related to brain content of serotonin and therefore to neurotransmission in the brain. This relationship may be responsible for integration of information about metabolic state and food intake. Administration or endogenous secretion of insulin lowers blood concentration of glucose and most amino acids, but plasma tryptophan is

increased as are brain tryptophan and serotonin (59). The rise of brain tryptophan and serotonin depends not only on plasma tryptophan but also on the relative content of other neutral amino acids which share the brain transport system with tryptophan: tyrosine, phenylalanine, leucine, isoleucine, and valine. With protein intake there is no increase in brain tryptophan or serotonin since most proteins contain the neutral amino acids in addition to tryptophan, and all compete for entry into the brain (59).

Amino acids fed at concentrations greater than required may be toxic (85). Methionine, which is required at slightly less than 1% of the diet, can be toxic at concentrations as low as 2%; toxicity is manifested by growth depression and tissue damage. Adaptation occurs with prolonged intake; expiration of labeled CO_2 from labeled methionine is decreased in rats fed 3% methionine for several days compared to rats fed the diet for the first time. Supplementation of the high methionine diet with glycine or serine is protective; the amino acids may increase metabolism of the methyl carbon. Incorporation of the methyl carbon into phospholipid choline is most important at low methionine concentration, but at higher concentrations conversion to S-methyl-L-cysteine may occur. S-Methyl-L-cysteine is toxic and results in growth depression and anemia, signs associated with methionine toxicity (13).

E. Minerals

In a recent extensive review of dietary mineral requirements of laboratory rats, fifteen elements were listed as required and four others as possibly required (163). Minerals have many functions which range from the structural roles of calcium and phosphate in bone and the osmotic and ionic effects of sodium and potassium to the function of several elements as integral parts of enzymes. In some cases specific functions for essential elements have not been identified but deficiency can be demonstrated by the usual criteria of abnormal growth and reproduction. A few minerals occur in relatively high concentration in one or more tissues and are required in the diet at concentrations of 0.05-0.5%. They are calcium, chloride, phosphorous, potassium, sodium, and sulfur. The trace or micronutrient elements are elements which occur more or less constantly in plants and animals in low concentrations, usually in the ranges from 10^{-6} to less than 10^{-12} gm per gram of wet tissue and are required in the diet at concentrations of 0.1-400 mg/kg.

The elements can be toxic in large amounts, and for several of them, notably selenium, the safety margin is narrow. They can be stored in the body, and stores accumulate in fetuses derived from maternal stores. The absorption and utilization of many trace elements are influenced by other elements or nutrients in the diet. For these reasons, design of experimental diets to test requirements of the trace elements can be complex, and induction of deficiencies may require prolonged depletion or even depletion through more than one generation as well as careful monitoring of the environment for sources of the element other than diet. The importance of iron, copper, and zinc to plants and animals was recognized about 100 years ago; the essentiality and functions of most other trace elements have been determined in the past 50 years. As experimental diets for rats have been increasingly purified and laboratory environments have been protected, recognition of essentiality of other trace elements has been made. In these cases demonstration of essentiality rests upon growth effects, and no specific functions have been recognized from biochemical or pathological material.

Abnormal growth, failure of reproduction, anemia, abnormal neurological function, and deformities of bone and skin commonly are seen in mineral deficiencies as in other nutrient deficiencies. The elements may have actions antagonistic to each other and induce conditioned deficiencies of one another. For example, rats fed excessive levels of zinc grow poorly, are anemic and have poor reproductive function, all of which are due to induction of copper deficiency by interference of zinc with absorption of copper (98). Many elements are required for activity of metalloenzymes. They may be components of the active site on the enzyme or be required to maintain specific protein structure. In a second class of association between trace elements and enzymes, the element occurs as part of a metal–enzyme complex but is less tightly bound to the protein and may not be required for activity of the enzyme.

1. Calcium and Phosphorus

a. Requirement. Calcium and phosphorus are present in high concentrations in bones and teeth as hydroxyapatite. Calcium is required for normal membrane function, blood clotting, and neuromuscular transmission. Phosphorus is a constituent of nucleic acids, proteins, lipids, carbohydrates, and high energy compounds. Absorption and excretion of both elements are regulated by complex interactions of hormones and vitamin D. The dietary requirements are generally expressed in absolute terms and as the ratio of calcium to phosphorus. The requirement of growing rats based on achievement of maximum bone calcification is approximately 0.5% of the diet for calcium and 0.4% for phosphorus (16,32,58). Lower dietary concentrations support maximum weight gain. A ratio of calcium to phosphorus of 1 to 1.5 is recommended during growth, but an increased ratio of 2 : 1 appears preferable after the period of maximum growth in order to maintain normal bone (53). The ideal calcium–phosphorus ratio for growth may be somewhat lower than for complete bone mineralization because of the greater requirement for phosphorus than for calcium in growth of tissues other than bone. Feeding diets adequate in calcium and phosphorus for growth but not for

bone mineralization results in no evidence of abnormality other than decreased bone mineralization. As with many other nutrients, the requirement for calcium and phosphorus decreases if growth decreases due to other inadequacies in the diet (15).

b. Deficiency. Osteoporosis develops in aging rats fed diets excessive in phosphorus relative to calcium and cannot be fully reversed by increasing the dietary calcium. In 8- to 14-month-old rats fed diets which contained 0.6% calcium, cumulative excretion of calcium was low when the diet contained 0.2% phosphorus and increased at 0.6% or more phosphorus. Bone ash was depressed in rats fed 0.6% or more phosphorus. There was no effect of the experimental diets on composition of teeth or plasma calcium concentration. Rats fed 1.8% phosphorus had increased renal calcium and phosphorus and grossly visible renal calcification, but at lower levels of phosphorus intake there was no renal calcification (53). Imbalance or mild deficiencies of calcium and phosphorus impair bone growth and may result in fractures. Severe deficiency of calcium results in growth retardation, increased basal metabolic rate, osteoporosis, paralysis, and hemorrhage.

2. Chloride

a. Requirement. The requirement for chloride in the diet has been set at 0.05% on the basis of a study in which no effect on body weight gain of young rats was found if dietary chloride was raised from 0.05 to 0.2% and on earlier studies which indicated that 0.05% was adequate (163).

b. Deficiency. Rats fed diets deficient in chloride exhibit no specific signs of deficiency, although there is one report of renal fibrosis in chronic deficiency (41).

3. Chromium

a. Requirement. Trivalent chromium is required by rats for normal glucose metabolism, growth, and longevity (216,217). If chromium is rigidly excluded from the diet and environment, 2 mg/kg is required in the drinking water or diet to permit normal weight gain. Under standard laboratory conditions, 0.3 mg/kg is adequate (163,234).

b. Deficiency. Hyperglycemia and glycosuria occur in chromium-deficient rats, but severe diabetes does not develop. Deficient rats have a significant decrease in weight gain and life span.

Addition of trivalent chromium to a natural ingredient diet lowered fasting serum cholesterol and glucose. Since increased serum cholesterol in rats had been associated with dietary sucrose, rats were fed diets which contained refined white sugar, sugar plus chromium acetate (5 mg/kg chromium in drinking water), or unrefined brown sugar. White sugar contained 0.02 to 0.03 mg/kg of chromium and brown sugar, 0.12 to 0.24. Serum cholesterol and glucose were lower in rats fed white sugar plus chromium or brown sugar than in rats fed only white sugar. Brown sugar was more effective than chromium in decreasing blood glucose; it contained several minerals in addition to chromium of which only nickel was known to have a hypoglycemic effect (217).

4. Copper

a. Requirement. Copper is required by rats for hemoglobin synthesis, maintenance of normal hair color, development of bone and elastic tissue, development and function of the central nervous system, and activity of several enzymes which include cytochrome oxidase, tyrosinase, and amine oxidase. The dietary requirement is 5 mg/kg (163).

b. Deficiency. Copper-deficient rats have achromotrichia and anemia. When deficiency becomes extreme there is loss of oxidative capacity in mitochondria and reduced synthesis of ATP and phospholipids. Copper depletion in rats is most easily achieved by depleting female rats of copper and then continuing depletion of their offspring. Rats born to copper-deficient mothers have neurological abnormalities, decreased brain copper content, and decreased myelin formation in the brain (52). Rats are relatively resistant to copper intoxication but gain weight poorly if fed diets markedly excessive in copper.

5. Fluoride

With the development of purified sources of nutrients and careful environmental control, the essentiality of fluoride and several other trace elements has been demonstrated. A basic requirement for fluoride of 2.5 mg/kg for growth and normal tooth development can be shown in rats kept under controlled environmental conditions, but under usual laboratory conditions the requirement is approximately 1 mg/kg (143,163,221). Addition to natural ingredient diets may not be necessary since fluoride occurs widely in natural foods and drinking water. The greatest concentration of fluoride is found in bones and teeth.

6. Iodine

Iodine is required for normal thyroid function, growth, and reproduction. The dietary requirement is 0.15 mg/kg (163).

7. Iron

a. Requirement. Iron is required for hemoglobin synthesis, growth, reproduction, and normal response to infection; it is an integral part of the cytochromes and is a cofactor for many enzymes. The requirement is 35 mg/kg, but much higher concentrations can be fed without evidence of toxicity (145,163).

There is an extensive literature on dietary effects on iron absorption. It is influenced by total body iron stores, interaction with other components of the diet and bacteria in the gastrointestinal tract, and physiologic and pathologic alterations of intestinal mucosal cells. The availability of iron in foods varies with the food and the form in which iron is present (95,119,178,189). Studies of iron absorption and retention can be carried out in animals depleted of iron or, conversely in animals given parenterally a large amount of iron to saturate, insofar as possible, body iron stores. The test dose of iron is usually given after 1 to 2 days of fasting or adaptation to a test diet. In one study rats fed a natural ingredient diet had significantly increased iron absorption after 48 hr of fasting, but fasting did not affect absorption in rats fed purified or cornmeal diets. Greatest iron absorption was measured in rats fed cornmeal, followed in decreasing order by rats fed high protein and low protein diets. Iron absorption changed rapidly when animals were changed from one diet to another, and a change from any diet to the cornmeal diet resulted in marked increase in iron absorption (95). Female rats were fed a diet which supplied less than 10 mg/kg iron until hemoglobin was reduced 3 to 5 gm % and then fed radioactive iron in test diets. Iron-deficient animals absorbed and retained more iron than iron-sufficient animals (5). Iron absorption was reduced by removal of calcium or phosphate from the salt mix or by addition of calcium cabonate. Absorption was less from a diet which contained starch than from diets which contained glucose or disaccharides. Addition of lactose overcame the inhibition of iron absorption induced by starch. Iron absorbed from a corn diet was retained and utilized to a greater extent than iron absorbed from hemoglobin, liver, or meat (5). Marked increases in hepatic iron were induced by feeding rats iron-enriched corn diets; rats fed a complete natural ingredient diet ingested more iron than rats fed the corn diet, but corn-fed rats absorbed a much greater percentage of ingested iron which was deposited in the liver (74).

The corn diets fed in these studies were deficient in lipotropes, amino acids, niacin, and other vitamins and were high in fat. They supported little or no weight gain. Other lipotrope and amino acid-deficient diets enhanced iron absorption (137). It has been postulated that dietary effects on iron absorption may be due in part to chelating activity of dietary phosphates, amino acids, and sugars which increase diffusion of iron across the intestinal mucosal cell. Chicks, mice, and guinea pigs also absorbed greater amounts of iron from corn than from complete natural ingredient diets (93). In rats fed corn, addition of vitamins or casein had little effect on hepatic iron accumulation, but addition of a complete salt mixture decreased iron absorption (92). Addition of phosphate alone may decrease excessive absorption (Table VI).

When rats were fed diets which contained 10% casein and 11% fat, iron absorption was decreased compared to animals fed 15 to 25% casein. A 10% protein, 60% fat diet significantly increased iron absorption compared to an 18% protein, 10% fat diet (118). Thus, 10% protein either increased or decreased iron absorption depending on the fat content of the diet (121).

Many L-amino acids increase the absorption of iron from isolated intestinal loops (127). Germfree rats absorb significantly more iron than conventional rats (72). Iron absorption decreases with increasing age; if absorption was increased by decreasing atmospheric pressure or decreased by administration of iron, age differences still were evident (251). Absorption of iron decreased markedly between 13 and 89 days of age. Up to weaning, rats absorbed 75 to 100% of the administered dose of iron but on weaning absorption dropped to 32%. The effect of weaning was related to age and not to weaning itself (61).

b. Deficiency. In iron-deficient rats, hemoglobin, tissue cytochromes, and myoglobin were decreased; rate of return to normal with iron supplementation varied from tissue to tissue. Repletion led to rapid increase in intestinal cytochrome *c* which was restored to normal in 2 days; hemoglobin was normal in 8 days, but cytochrome *c* and myoglobin deficiency in skeletal muscle did not return to normal until 40 days (44). Enlarged or increased numbers of mitochondria in hepatocytes, bone marrow erythroblasts, and heart muscle have been described in iron deficiency anemia. Young rats fed an iron-deficient diet had depressed weight gain and hemoglobin after 3 weeks. Mitochondrial cross-sectional area was significantly increased after 3 weeks and more markedly increased after 8 weeks. Mitochondrial cytochrome *c* was significantly de-

creased, but microsomal cytochromes *P450* and *B5* were unchanged and responded normally to phenobarbital. With iron repletion the mitochondria returned to normal in 2–5 days. (43).

Severe iron deficiency induced lipemia with significantly increased serum triglycerides, chylomicrons and pre-β-lipoproteins. Cholesterol and phospholipid were elevated in the serum of young born to iron-deficient dams. In further studies in rats that were only moderately anemic, no elevation of serum triglycerides and cholesterol was found (4,5).

8. Magnesium

a. Requirement. Magnesium is required for normal vascular, central nervous system, and cardiac function and normal bone mineralization and cartilage formation. It interacts with both calcium and potassium and is required for maintenance of normal serum and tissue content of these elements. Magnesium is a cofactor for enzymes which catalyze the transfer of phosphorus from ATP to acceptor molecules or from phosphorylated compounds to ADP. It is required in enzymes for which thiamin pyrophosphate is a coenzyme, for certain proteolytic enzymes, and for protein and glycopeptide synthesis.

A concentration of 100 mg/kg in the diet supports normal growth of rats, but approximately 400 mg/kg is required for maintenance of normal blood concentration of magnesium (144). Diets which contain excessive calcium and phosphorus may require additional supplementation with magnesium.

b. Deficiency. Deficiency of magnesium is manifested first by vasodilatation, erythema and then breakdown of skin. Within a few weeks cardiac arhythmias, muscular spasticity, and convulsions develop. The kidneys may be extensively calcified (62,112,197). Muscle concentrations of magnesium and potassium are decreased; tissue sodium and calcium are increased. Reduction of magnesium intake during pregnancy and lactation severely affects both dam and offspring; weight gain is reduced; stillbirths and congenital anomalies are increased. and survival of live offspring to weaning is decreased (106,243).

9. Manganese

a. Requirement. Manganese activates enzyme complexes which catalyze transferase, hydrolase, isomerase, and other reactions and is required for glycoprotein synthesis. In many instances magnesium can substitute for manganese in these enzymes; both elements activate nucleic acid polymerase. The requirement is approximately 50 mg/kg (163).

b. Deficiency. Manganese deficiency results in defective growth, abnormal bone formation, reproductive failure or de-

Table VI

Effect of Dietary Components on Iron Absorption in Rats

Diet	Fe in diet (%)	Hepatic iron content (mg)	Reference
Corn grits, lard	0.3	159	92
Corn grits, lard + K$_2$HPO$_4$ 0.4%	0.3	115	
Corn grits, lard + K$_2$HPO$_4$ 0.8%	0.3	74	
Corn grits, lard + K$_2$HPO$_4$ 2.0%	0.3	40	
Natural product complete	0.3	31	
	0.9	70	
	1.9	172	
Corn grits, lard	0.04	31	
	0.19	87	
	0.3	148	

velopmental abnormalities of the central nervous system, otoliths of the inner ear and skeleton in the offspring, and failure of lactation (105,107).

10. Potassium

a. Requirement. Potassium is the major intracellular cation; it regulates cellular pH and functions in maintenance of normal cell membranes and neuromuscular and cardiac transmission. The requirement for growth is 0.36%; 0.5% may be required for gestation and lactation (79,165).

b. Deficiency. Rats fed diets deficient in potassium rapidly become lethargic and show hair loss, ascites, hydrothorax, atrophy and necrosis of cardiac and skeletal muscle, and degeneration of renal tubular epithelium (163,166).

11. Selenium

a. Requirement. Selenium is an antioxidant and is the cofactor for glutathione peroxidase. Its activity and dietary requirement are influenced by dietary content of vitamin E and sulfur amino acids (164). Both inorganic and organic selenium compounds can be utilized by rats, but the different chemical forms of selenium vary in activity and toxicity. Rats fed a selenium-deficient diet grew normally when selenium was fed at 0.05 mg/kg diet, but significant depression of hepatic and red cell glutathione peroxidase were demonstrated. The hepatic enzyme was maximal in rats fed 0.1 mg/kg. Erythrocyte glutathione perioxidase continued to increase in rats fed selenium up to a toxic concentration of 5 mg/kg (81). A concentration of 0.1 mg/kg was judged adequate. Diets containing more than 1 mg/kg should not be fed to rats because of toxicity.

b. Deficiency. Rats fed diets deficient only in selenium have a gradual decline in serum and muscle selenium content but no other obvious clinical or histologic abnormalities. However, their offspring may have little or no hair growth, poor weight gain and feed efficiency, and atrophy of the skin and its appendages. Organs other than skin are normal (108).

If weaning rats are fed a diet deficient in selenium, vitamin E, and/or sulfur amino acids, they develop fatal hepatic necrosis within approximately 4 weeks, possibly because of damage by lipid peroxides formed in the absence of glutathione peroxidase. The interaction of sulfur amino acids in this system may be through elevation of hepatic glutathione which increases peroxide breakdown (81).

12. Sodium

a. Requirement. Sodium is the major extracellular cation and is required for normal fluid and ionic homeostasis. A diet-ary content of 0.05% is adequate for growth and reproduction. Reproduction in female rats was evaluated at dietary sodium concentrations from 0.01 to 0.09%; body weight gain, maturation, estrous cycles, and reproduction were normal in rats fed 0.03% and above. Maternal weight gain and litter size and weight were significantly decreased in rats fed 0.01%, and litter weight was somewhat reduced in rats fed 0.03%. There was no increase in weight in rats fed amounts greater than 0.03%. Milk content of sodium was decreased at all dietary concentrations below 0.07% (67,68).

b. Deficiency. Hemoconcentration occurred in rats fed 0.01 and 0.03% sodium; the zona glomerulosa of the adrenal cortex was increased in width and had enlarged cells. The renal juxtaglomerular index was increased (67). Severe sodium deficiency results in growth retardation, corneal lesions, infertility, and abnormal bone formation (183).

Induction of hypertension by feeding excessive sodium to newborn rats has been reported but cannot be consistently demonstrated (42,126). In one study increased sodium intake by newborns resulted in dry scaly skin, gastric dilatation, and elevation of serum sodium, but not in hypertension. Cardiac and renal weights were increased, and the brains were histologically abnormal with capillary congestion, focal gliosis, hemorrhage, and neuronal degeneration. The juxtaglomerular index and the width of the zona glomerulosa were decreased (248).

13. Sulfur

Until recently sulfur has not been listed as a required nutrient for rats, but several studies over the last 10 years have demonstrated improved weight gain when sulfur is added to the diet, particularly if the diets are only minimally adequate in methionine. Dietary sulfate is incorporated into cartilage and decreases catabolism of methionine (149). The dietary and metabolic interactions between inorganic and organic sulfur compounds have not been completely defined. A recent review concluded that 0.03% sulfur as sulfate was adequate (112,163).

14. Zinc

a. Requirement. Zinc is a cofactor for many enzymes which include carbonic anhydrase, alcohol dehydrogenase, several pancreatic enzymes, and alkaline phosphatase and is present in high concentration in bone. Rats housed in galvanized cages require 2 to 4 mg/kg diet, but housed in stainless steel cages they require approximate 12 mg/kg for maximum weight gain. If soybean protein is fed, the requirement is 18 mg/kg (63,163).

Absorption of zinc, as of certain other minerals, from the

gastrointestinal tract is decreased by phytic acid, a component of plants, and can be further decreased by dietary calcium because of formation of zinc, calcium, and phytate complexes (83).

Zinc absorption and retention were measured in young adult rats fed 5 or 15% casein in diets which contained 9 or 33 mg/kg of zinc. Zinc retention was positively correlated with both dietary protein and zinc (240). In a study of interactions between zinc and copper, rats were fed the elements in drinking water (2.5 to 40 mg/kg of zinc and 0.25 to 2 mg/kg of copper). Maximal growth was achieved at intakes of 10 to 20 mg/kg of zinc. The lowest zinc content was measured in skin and the highest in testis. There was an inverse correlation between serum, tissue, and hair, zinc and copper. Hair but not hepatic or testicular zinc content varied directly with zinc intake (161,245).

b. Deficiency. In addition to growth retardation, zinc deficiency is manifested by thickening and hyperkeratosis of the skin and esophageal epithelium; hair loss; testicular atrophy; hypoplasia of the coagulating glands, prostate, and seminal vesicles; reproductive failure; and teratogenesis. Alcohol dehydrogenase is responsible for normal metabolism of vitamin A; effects of zinc deficiency in depressing testicular function and decreasing release of vitamin A from liver may be mediated through the deficiency of this enzyme (230).

Many of the signs of zinc deficiency, particularly those in skin and esophagus are readily reversed by treatment with zinc, but testicular damage may not be (11,50). Zinc deficiency depressed carbonic anhydrase, lactic dehydrogenase, and alkaline phosphatase; depression of the enzymes in some tissues was related in part to decreased food intake (103).

15. Other Elements Which May be Required

a. Nickel. A requirement for nickel was demonstrated in rats in an ultraclean environment. Liver homogenates from rats fed no nickel had decreased mitochondrial oxidation, decreased polysomes and increased monosomes compared to homogenates from rats fed the diet supplemented with 3 mg/kg nickle. Deficient females had an increase in stillborn pups (176). The dietary requirement under specified conditions may be considered to be 3 mg/kg, but the figure may be altered by further studies. Nickel induced tumors of the respiratory tract in rats exposed by inhalation (184).

b. Silicon. A requirement for silicon for growth and normal bone formation and mineralization has been demonstrated. Silicon occurs in blood and parenchymal tissues and in high concentrations in bone and connective tissue. It is an integral part of several acid mucopolysaccharides. A growth response of deficient rats was obtained by feeding a dietary supplement

of 500 mg/kg, but lower concentrations were not tested (221). In calcium-deficient rats fed 10, 25, or 250 mg/kg silicon, bone ash increased initially, but with prolonged calcium deficiency bone ash decreased (28).

c. Tin. Under carefully controlled environmental conditions, a significant increase in growth was demonstrated with addition of tin in the form of stannous sulfate at 1 mg/kg diet (221). No specific function for tin is known, but it has been postulated that it contributes to the tertiary structure of proteins and other macromolecules and functions with metalloenzymes. It is found in small amounts in tissues.

d. Vanadium. The evidence for essentiality of vanadium rests on the growth response of rats in a rigidly controlled environment. The dietary content used was 0.1 mg/kg (163).

F. Fat-Soluble Vitamins

1. Vitamin A

a. Requirement. Vitamin A is required by rats for retinal functions; maintenance of normal skin and epithelia of respiratory, urinary, and gastrointestinal tracts; and reproductive function. An influence of the vitamin can be demonstrated in metabolism of carbohydrates, protein, nucleic acids, steroids, and phospholipids. Except for its function in retinal pigment, the mechanism of action of vitamin A is unknown. The functions of water-soluble vitamins are well defined and can be demonstrated in basic reactions in intermediary metabolism which occur in all organisms including bacteria, whereas fat-soluble vitamins are required only by complex multicellular organisms which have highly differentiated and specific functions. The known functions of vitamin A are to some extent separable since, for example, retinoic acid, which is a metabolite of vitamin A, supports growth and epithelial differentiation but not vision or reproduction; it may be the active form of vitamin A for differentiation and maintenance of bone and epithelia. The visual function of vitamin A requires vitamin A aldehyde, retinal, which is formed by alcohol dehydrogenase and NADP from vitamin A alcohol, retinol, in the retina. The role of vitamin A in differentiation is not known, although mechanisms have been proposed (46).

Vitamin A occurs in nature as carotenoids of which the most important is β-carotene. It is cleaved and converted to vitamin A alcohol (retinol) in the mucosa of the small intestine. Retinol is then esterified to retinyl palmitate or stearate, and the esters are transported to the liver where they are stored. Large amounts can be stored in liver and smaller amounts in kidney. When needed, the esters are hydrolyzed to retinol and carried in the blood bound to retinol-binding protein.

Standardization of vitamin A has been established such that 1 IU is equivalent to 0.344 mg all-*trans*-retinyl acetate, 0.55 μg retinyl palmitate, or 0.3 mg all-*trans*-retinol. One milligram of all-*trans* beta-carotene is equal to 0.167 mg of all-*trans*-retinol; other naturally occurring carotenoids are considered to have approximately half the activity of β-carotene (200). Absorption of vitamin A from food varies with the food and its method of preparation. The vitamin is readily destroyed by oxidation and light. Availability of vitamin A from carrots was recently reexamined. Comparison was made of fresh and frozen carrots and carrots which had been sliced, blended, or sonicated and freeze-dried. Sixty-seven to 72% of carotene was present as β-carotene. Rats were fed vitamin A-deficient diets for 3 weeks from weaning and then fed one of the carrot diets for 10 days. Carotene in carrots, as measured by liver storage of vitamin A, was as available to rats as β-carotene administered in cottonseed oil; the physical form of the carrots made no difference (235).

Dietary protein, fat, vitamins D and E, zinc, and unsaturated fatty acids all can affect the absorption and utilization of vitamin A. Because of variation in absorption and utilization of β-carotene, the intake should be two to three times the minimal vitamin A requirement (163,200).

Protein deficiency may result in decreased serum and tissue content of vitamin A, since retinol-binding proteins cannot be synthesized in normal amounts; liver stores of the vitamin under these conditions are normal or even greater than normal. Decreased dietary protein reduces the enzymes of the intestinal mucosa which hydrolyze and synthesize vitamin A esters for transport in serum (69,192). Protein in the diet is related to vitamin A requirement within the range in which protein is increasing growth and therefore increasing the requirement for vitamin A. Development of vitamin A deficiency can be delayed in rats fed protein-deficient diets (196,200).

Although utilization of dietary carotene is influenced by the amount and composition of dietary protein, the response to protein is less consistent with vitamin A itself. Weanling male rats were fed a vitamin A-free diet for 3 weeks and then fed diets which contained 10, 20 or 40% casein and given carotene or retinyl acetate orally. Rats fed 20 or 40% casein gained weight more rapidly than rats fed 10%; after 4 weeks rats fed 40% casein were slightly heavier than rats fed 20%. Hepatic vitamin A storage was significantly increased at each higher concentration of protein in rats given carotene; in rats given retinyl acetate, hepatic storage was increased by increasing protein from 10 to 20%, but increasing from 20 to 40% had no effect. Therefore, marginally protein-deficient rats consistently deposited less vitamin A in the liver than normal rats despite their slower growth rate (117).

Zinc is required for release of vitamin A from liver. Zinc-deficient rats have low serum vitamin A; in children a direct relationship between serum vitamin A and serum zinc has been reported. Rats were fed either 4 or 20 mg/kg zinc in a diet which contained approximately 20,000 IU of vitamin A per kilogram. Hepatic concentration of vitamin A was the same in the two groups, but in rats fed the low zinc diet there was a significant depression of plasma vitamin A. In rats fed diets deficient in both zinc and vitamin A and then repleted with vitamin A alone or with both zinc and vitamin A, hepatic vitamin A increased rapidly in rats given vitamin A alone but serum vitamin A remained significantly depressed unless zinc was added (230).

In recent studies adequacy of vitamin A in the diet has been judged by three criteria: growth rate, cerebrospinal fluid pressure, and absence of squamous metaplasia of the nasolacrimal duct. The requirement was found to be between 1100 and 2500 IU/kg fed as retinyl acetate in beadlet form, i.e., coated with gelatin (38,39,66). Because of the instability of vitamin A, which is susceptible to oxidation, 4000 IU/kg of diet is recommended for growth and maintenance (163).

Rats exposed to infection or treated with toxic or carcinogenic chemicals may require higher maintenance levels of vitamin A because such treatments can deplete hepatic reserves.

Toxicity of vitamin A was studied in female rats given 25,000 to 75,000 IU per rat per day for 16 consecutive days. They had hair loss, abnormal posture and gait, and weight loss. By X ray there was increased radiolucency and cortical thinning of bones, and at autopsy osteoporosis was evident. Microscopically thinning of the epiphyses, irregular ossification, and decreased mineralization were seen. Renal and cardiac calcification were present in rats given 75,000 IU but not 50,000 IU per day. There was a mortality of 60% in rats given 75,000 IU/day, but no mortality in rats given the lower doses (132). Prevention or decreased severity of vitamin A toxicity was reported in rats treated with inducers of hepatic microsomal oxidases (239). Male rats fed 500,000 IU/kg diet, which corresponds to a daily intake of approximately 7500 IU, gained weight normally for about 15 weeks but then growth rate decreased. Serum and hepatic vitamin A were greatly increased (203).

b. Deficiency. Vitamin A deficiency is manifested by cessation of growth, hemoconcentration, elevated cerebrospinal fluid pressure, increased susceptibility to infection, excessive keratinization of the skin, keratinization and subsequently ulceration of the cornea, squamous metaplasia of epithelium of the respiratory and urinary tracts, reduction of mucopolysaccharide synthesis in the gastrointestinal tract, retinal degeneration, and reproductive failure. There may be overgrowth and abnormal mottling of bone. Rats fed deficient diet from weaning stop growing after 6-8 weeks, lose weight, and die at 10-16 weeks. Serum and hepatic vitamin A content decreases; this decrease and development of signs of vitamin A deficiency progress at differing rates depending on the age and rate of growth of rats, their initial stores of vitamin A, and the intervention of infectious diseases. Gnotobiotic rats survive at a plateau of moderate deficiency for months (163).

Induction of vitamin A deficiency can be accomplished by feeding weanling rats a diet devoid of vitamin A until the growth plateau is reached at which time growth essentially ceases; weight loss and inanition do not begin for approximately 1 week after this period. Problems with this approach to the study of vitamin A deficiency are (1) time of onset of deficiency varies from one animal to another, (2) appetite depression occurs so starvation effects are superimposed, and (3) deficient animals are highly susceptible to infection, particularly of skin and lung. One can also study young born to dams fed vitamin A-deficient diet which contains retinoic acid. Retinoic acid maintains weight and condition but not vision or reproduction, and is not stored in tissues. Offspring can then be fed retinoic acid in vitamin A-deficient diet and be made deficient at the desired time by removal of the retinoic acid. Young male rats derived from dams fed natural ingredient diet had liver concentration of vitamin A of 23 to 26 mg/kg. After weaning they were fed a purified diet which contained no vitamin A; 3 weeks later they had no measurable vitamin A in the liver. They were then fed cycles of deficient diet and diet which contained retinoic acid (2 mg/kg). Weight gain was less than in controls; cutaneous hyperkeratinization and periocular porphyrin deposits, signs of deficiency, appeared within 10 days of removing retinoic acid from the diet. Females fed retinoic acid in cycles with deficient diet did not mate, but females supplemented continuously with 5 mg/kg of retinoic acid mated and bore normal litters if given retinyl acetate from day 9 of pregnancy until term; however, the pups did not survive beyond 3 days (130).

Male weanling rats fed a vitamin A-deficient diet from weaning ceased weight gain between 4 and 5 weeks and had enlarged salivary glands with squamous metaplasia and keratinization of the ducts; the testes were atrophic. Supplementation with retinoic acid maintained weight but the animals had testicular atrophy and were blind as measured by electroretinography. Oral administration of retinyl acetate, 5 μg/rat/day, maintained testicular function, and 1 μg/rat/day maintained retinal function; 5 μg/day corresponds to a dietary intake of 1000 IU/kg of diet (39).

2. Vitamin D

a. Requirement. Vitamin D may be considered the precursor of a steroid hormone or hormones active in calcium and phosphorus metabolism. Rats have played a major role in elucidation of its mechanism of action. After absorption it is hydroxylated in the liver and then dihydroxylated in the kidney under regulation by interactions of calcium, phosphorus and parathyroid hormone. The dihydroxy form is thought to be the metabolically active form for all functions of vitamin D. These include absorption of calcium and phosphate from the gastrointestinal tract, maintenance of the two ions in the blood, reabsorption of calcium by the kidney, and mineralization of bone (46). In liver disease activation of vitamin D may be impaired; in rats treated with microsomal oxidase inducers such as phenobarbital, breakdown may be increased which results in decreased vitamin D activity (46,124).

The dietary requirement for vitamin D is 1000 IU/kg when calcium and phosphorus are fed in adequate amounts (163). One international unit is equivalent to 0.025 μg vitamin D_3 (cholecalciferol).

b. Deficiency. Deficient rats have poor bone mineralization and widened, uncalcified epiphyseal cartilage plates (116). Severity of the deficiency is governed by dietary content of calcium and phosphorus.

3. Vitamin E

a. Requirement. Vitamin E is an antioxidant and is required for maintenance of normal red blood cells, muscle, liver, and reproduction. Its function is closely associated with selenium and sulfur-containing amino acids and is apparently carried out through maintenance of normal tissue levels and activity of glutathione and glutathione peroxidase. Glutathione is a free-radical scavenger which blocks toxicity of many compounds. Through the activity of glutathione peroxidase, peroxides formed from unsaturated fatty acids in tissues can be absorbed and tissue destruction can be prevented. Therefore dietary levels of unsaturated fatty acids influence the requirement for vitamin E (114).

Standardization of vitamin E has defined 1 IU as 1 mg DL-α-tocopheryl acetate or 0.74 mg D-α-tocopheryl acetate. The dietary requirement, based on prevention of hemolysis, is 30 IU/kg diet (30 mg DL-α-tocopheryl acetate) for diets which contain up to 5% linoleic acid and adequate selenium and sulfur amino acids. Increasing dietary linoleate 400% increased vitamin E requirement, based on red cell hemolysis, by 40% (20,113,163).

b. Deficiency. In addition to hemolysis, deficient rats have hyaline degeneration of skeletal muscle, ceroid accumulation in muscle and liver, testicular degeneration, and reproductive failure. Hemolysis is the most sensitive indicator of deficiency (20,113). Hepatic necrosis occurs rapidly and is often the cause of death, but livers appear normal until shortly before death. Ceroid deposition is seen in chronic deficiency. Catalase activity was irregularly depressed after 6 weeks of vitamin E deficiency and consistently depressed after 12 weeks; there was some depression of microsomal cytochromes (90).

4. Vitamin K

Deficiency of vitamin K results in failure of the normal blood clotting mechanism, since in its absence thrombin is not generated in plasma and fibrinogen not converted to fibrin. The vitamin is required for hepatic synthesis of prothrombin and clotting factors VII, IX, and X. The activity of vitamin K in prothrombin synthesis is at the point of conversion of the precursor protein to prothrombin by addition of calcium-binding groups; this is the region which demonstrates calcium-dependent binding to phospholipids during activation and is cleaved off when thrombin is generated (46).

Both concentration and type of dietary protein influence the requirement for vitamin K in rats. Rats were fed diets which contained 21% protein from several different sources and were adequate in all nutrients except vitamin K. Casein and lactalbumin supported normal or nearly normal prothrombin levels, but soy or beef protein or egg albumin did not, although rats gained weight at approximately the same rate as rats fed casein or lactalbumin. Increasing casein in the diet increased prothrombin content without affecting weight gain; increasing soy protein to 35% decreased prothrombin and increased weight gain. Extraction of casein with hot ethanol reduced its ability to support normal prothrombin concentration. Unextracted casein contained approximately 0.15–0.6 μg of vitamin K per gram (142). Rats did not maintain normal prothrombin content if they were fed casein and sulfadiazine with no vitamin K or if coprophagy was prevented. Feces collected from untreated rats fed casein contained approximately 16 μg of vitamin K activity per gram (dry weight). The vitamin K requirement of rats fed 21% casein was approximately 0.02 mg/kg of diet. In rats fed 21% soy assay protein, the vitamin K_1 requirement was 0.125 mg/kg of diet. The relationship between dietary protein and serum prothrombin may be due not only to protein content of vitamin K but also to effects on coprophagy or intestinal synthesis of vitamin K (115,142). The vitamin K_1 requirement under most laboratory conditions can be met by a dietary content of 0.05 mg/kg.

G. Water-Soluble Vitamins

Water-soluble vitamins are coenzymes for an extensive group of metabolic reactions. They are absorbed in the upper small intestine, except for vitamin B_{12} and possibly riboflavin which are absorbed in the ileum. They may be available to rats through coprophagy since they are synthesized by intestinal bacteria (100). Most are reasonably stable, but folic acid, vitamin B_{12}, thiamin, and riboflavin can be destroyed by heat, light, and exposure to air.

1. Ascorbic Acid

Rats do not require a dietary source of ascorbic acid, but it can spare certain B vitamins. Five percent ascorbic acid added to a thiamin-deficient diet supported weight gain and increased fecal thiamin in rats. Its efficacy may be explained by the increased fecal vitamin content which is restored to the rat via coprophagy (160,222). Diets severely deficient in nutrients can depress hepatic synthesis and tissue levels of ascorbic acid (140).

2. Biotin

a. Requirement. Biotin, the coenzyme for transcarboxylase enzymes required in synthesis of long-chain fatty acids and purines, is supplied to rats by intestinal bacterial synthesis (7,177).

b. Deficiency. Induction of biotin deficiency requires feeding raw egg white which binds and prevents absorption of the vitamin. Deficient rats develop exfoliative dermatitis, alopecia, achromotrichia, and an abnormal, spastic gait. Biotin-dependent enzymes are decreased and can be restored or maintained by parental administration of 200 μg of biotin once a week, which corresponds approximately to 2 mg/kg of diet (36).

3. Choline

a. Requirement. Choline, a component of acetylcholine and of lecithin in phospholipids, is required by rats, but it can be replaced in chemically defined diets by methionine, and it can be partially replaced or spared by folate or vitamin B_{12} (18,19,137,172,242) (Fig. 3). Interactions of choline, methionine, folate, and vitamin B_{12} (lipotropic agents) are complex. They were first demonstrated with respect to their lipotropic action, i.e., ability to maintain normal hepatic lipid metabolism and prevent accumulation of triglyceride in liver cells (18,19,78,80b,82,88). They interact in hepatic uptake and storage of folates and in control of enzymes which synthesize or degrade methionine. The transmethylation of homocysteine to methionine by methyl groups derived from choline via betaine or other methyl donors is carried out through the folate coenzymes and utilizes vitamin B_{12}; it is central to lipotropic interactions.

The dietary requirement for choline is influenced by dietary lipid content and by chain length and degree of unsaturation of dietary lipids. The total caloric content of the diet also is a factor (18,19,188,212,253,154). In diets which contain 4 to

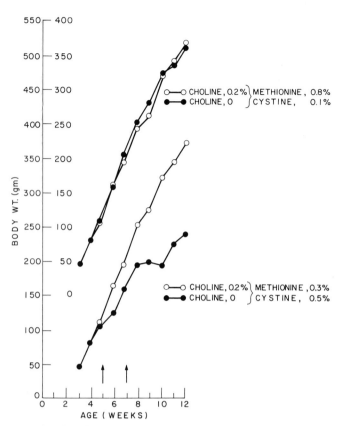

Fig. 3. Growth rate of male Fischer rats fed chemically defined diets which contained different concentrations of methionine, cystine, and choline.

Table VII

Composition of Lipotrope-Deficient Diets[a]

Dietary component	Diet 1 (LD)	Diet 2 (LD)	Diet 3 (LD)
Vitamin-free casein	3	6	—
Methanol-extracted peanut meal	12	25	—
Gelatin	6	—	—
Methanol-extracted fibrin	1	—	—
Amino acid mix	—	—	22.2
Methionine	—	—	0.3
Cystine	0.5	—	0.5
Sucrose, dextrose, dextrin	36.5	42	43.4
Cellulose flour	2	—	—
Vegetable oil	2	20	5
Tallow, lard	30	—	20
Vitamins[b]	2	2	2
Salts	5	5	5
Sodium acetate	—	—	1.6

[a]From Rogers and Newberne (206).

[b]Choline, folic acid, and vitamin B_{12} are omitted from the vitamin mixture to induce severe deficiency. Supplementation with 0.2% choline will induce marginal deficiency.

4.5 kcal/gm and 5 to 10% fat, the requirement is approximately 0.1%. Increasing choline to as much as 0.4% may be required in diets which contain more fat or are low in methionine.

b. Deficiency. Induction of severe deficiency requires feeding young, male rats diets deficient in choline, methionine, folate, and vitamin B_{12}. Examples of diets are given in Table VII. Deficiencies of amino acids other than methionine may be induced because of the use of plant proteins [diets 1(LD) and 2 (LD)] (Table VII); amino acid supplementation or chemically defined diets can be fed to prevent amino acid deficiency [diet 3(LD); Table VII]. A dietary lipid content of 15–30% by weight is used to increase severity of the deficiency (18,19,206). Male rats are more susceptible to the deficiency than female rats (187).

Short-term deficiency induces fatty liver and, in weanling rats, hemorrhagic necrosis of renal proximal tubules (35,57,88,153). Long-term deficiency induces fatty liver, in which the liver cells are markedly distended with fat vacuoles, and then cirrhosis (88,206).

Within 24 hr of first feeding of choline-deficient diet to male weanling rats, droplets of neutral triglyceride and abnormalities

of intracellular membranes are demonstrable in liver cells (35,57). The Golgi apparatus is decreased in size; the smooth and rough endoplasmic reticulum (ER) is dilated and filled with material of low electron density which may be partially synthesized lipoprotein. Lecithin is decreased in organelle membranes, and serum lipoproteins, triglycerides, and cholesterol are decreased (35). With prolonged deficiency, hepatic cholesterol and triglycerides continue to increase; lecithin decreases; serum very low density lipoproteins (VLDL), esterified cholesterol, and triglycerides remain low, in part because of failure of hepatic secretion of lipids and also because esterification of cholesterol in plasma requires lecithin as donor of the fatty acid (80b,254). After 6 to 10 weeks, rats have markedly fatty livers in which cell turnover is increased and triglycerides compose as much as 50% of the total wet weight; after 6 or more months they have cirrhosis. The liver is nodular and contains bands of proliferating fibrovascular tissue in which shunting of blood occurs. Hepatocytes remain fatty, although the fat content may decrease somewhat, and ultimately hepatic failure develops (88,204,254).

Induction of chronic deficiency and cirrhosis requires gradual rather than abrupt removal of choline from the diet of weanlings; if male rats are placed on a choline-deficient diet at weaning, 50 to 90% die within 10 days to 2 weeks of hemorrhagic renal necrosis. Rats killed within the first 4–5 days have no histologic evidence of renal abnormality, but hemorrhagic necrosis of the proximal convoluted tubules develops rapidly thereafter (153). Myocardial necrosis and arterial atheromas and medial necrosis may occur in chronic deficiency (212).

Protection against renal damage is given by diets which con-

Table VIII

Effect of Diet on Hepatic Microsomal Oxidases[a]

Diet	Number of rats	Aminopyrine[b] demethylase	p-Nitroanisole[c] demethylase	Benzopyrene[d] hydroxylase
Control (Diet 2, Table I)	12	481 ± 50	1132 ± 74	184 ± 16
Lipotrope Deficient [Diet 1 (LD), Table VII]	12	305 ± 23	536 ± 116	99 ± 11
P value		0.05	0.01	0.01

[a] From Rogers and Newberne (205). Values are per gram fat-free liver ± S.E.
[b] Micrograms of aminoantipyrene per hour.
[c] Micrograms of p-nitrophenol per hour.
[d] Quinine units.

tain less choline than is required to prevent fatty liver. For induction of long-term deficiency, weanlings are fed a deficient diet supplemented with 0.1–0.2% choline for 2 weeks, 0.05% for 2–4 weeks, and then no choline for the remainder of the experiment. Partial supplementation may be prolonged to prevent or retard development of cirrhosis while inducing fatty liver.

Deficiency of lipotropic agents can affect iron absorption and function of the liver, kidney, and immune system. Iron absorption is increased in lipotrope deficiency; the excess iron is deposited both in normal storage sites (liver, spleen, and bone marrow) and in tissues in which normal rats do not deposit iron, such as pancreas and myocardium. Marginally deficient rats have increased susceptibility to bacterial infection and to toxic and carcinogenic chemicals. They have abnormalities of cellular immunity. Marginal deficiency in pregnant rats induces persistent metabolic and immunologic abnormalities in the offspring (169,247). The kidneys contain abnormal, megalocytic tuble cells with bizarre nuclei (175). Enhanced susceptibility to many different toxic or carcinogenic chemicals may be the result of depression of microsomal oxidases (205).

Correction of growth rate depression and of the hepatic, renal, or absorptive abnormalities can be accomplished most effectively by addition of choline or methionine to the diet; folate and vitamin B_{12} are only partially effective. Supplementation of a lipotrope-deficient diet with choline or methionine is calculated by adding to the total choline content (in mg/kcal) the methionine content using methionine × ⅓ as equivalent to choline. In rats fed a chemically defined diet which contained methionine and cystine in greater amounts than required, addition of choline did not improve weight gain, but choline was required at lower methionine concentrations (Fig. 3) (172).

Supplementation judged effective by the criterion of hepatic lipid content may not correct other abnormalities induced by deficient diets measured against diets which contain casein. For example, we have conducted a series of experiments in

Table IX

Effect of Diet on Acute Toxicity of Aflatoxin B_1 (7 mg/kg)[a]

Route of Administration	Diet[b]	2-Week mortality
Sprague-Dawley Rats		
Intragastric	Marginally lipotrope deficient	0/5
	Control	3/5
Intraperitoneal	Marginally lipotrope deficient	0/5
	Control	5/5
Fischer Rats		
Intragastric	Marginally lipotrope deficient	0/10
	Control	10/10

[a] From Rogers and Newberne (205).
[b] See Table VIII.

which hepatic microsomal oxidase activity and toxicity or carcinogenicity of chemicals for the liver are compared in rats fed an adequate purified diet [diet 2 (Table I)], or a marginally lipotrope-deficient diet [diet 1 (LD) (Table VII) supplemented with 0.2% choline]. Rats fed the marginally deficient diet have increased hepatic lipid, decreased activity of hepatic microsomal oxidases, decreased sensitivity to the toxic effects of a single large dose of aflatoxin B_1 (AFB$_1$) but increased sensitivity to tumor induction by AFB$_1$ and other carcinogens (Tables VIII–X). If the deficient diet is supplemented with lipotropes or with amino acids, which include the lipotrope methionine, to make it equivalent to casein, depression of oxidase activity and enhancement of carcinogenesis are partially corrected. Addition of fat to the control diet does not enhance hepatocarcinogenicity (201,202,207).

4. Folates

Folates are pterines and pteridines which occur naturally conjugated with one or more glutamic acid residues and are stored in rat liver as the pentaglutamate (101,129). Folate coenzymes

Table X

Effects of Lipotrope Deficiency on Chemical Carcinogenesis in Rats

Carcinogen	Target organ	Percentage of rats with tumor fed		Reference
		Control diet[a]	Marginally lipotrope-deficient diet[a]	
Aflatoxin B₁	Liver	6	22	205
Aflatoxin B₁	Liver	11	87	202
Dibutylnitrosamine	Liver	24	64	208
Diethylnitrasamine	Liver	24	60	201
1,2-Dimethylhydrazine	Colon	56	85	207

[a] See Table VIII.

are required for transfer of 1-carbon units in the synthesis of thymidine, purines, choline, methionine, and other compounds.

a. Requirement. The dietary requirement is met in rats by intestinal microbial synthesis of the vitamin, but rats fed diets inadequate in choline, methionine, and vitamin B₁₂ and treated with anticonvulsants or certain chemical carcinogens may become folate-deficient (94,193); pregnancy and lactation may increase the requirement. A range of dietary content of 0.5–10 mg/kg has been used in purified diets; 1 mg/kg is recommended (163).

b. Deficiency. Induction of deficiency in rats requires prolonged feeding of diets which are deficient in lipotropes or contain antibiotics or folate antagonists. In addition to measurement of tissue levels of folate, urinary excretion of formiminoglutamic acid (FIGLU) is used to assess folate nutrition (7). Decreased growth, leukopenia, and anemia are induced in deficient rats; megaloblastic bone marrow or intestinal cells are occasionally reported (102,242,247). The deficiency can also induce defects in thiamin absorption, cell-mediated immunity, and protein synthesis (91,102,247).

5. Niacin

a. Requirement. Niacin forms the dehydrogenase coenzymes NAD and NADP. Simple deficiency of niacin does not occur because it can be synthesized from tryptophan; approximately 40 mg of tryptophan yields 1 mg of niacin. The niacin requirement of the rat is 20 mg/kg diet if tryptophan is fed at a concentration of 0.15% (84,87). Comparison of growth rate or hepatic niacin content showed no advantage of increasing dietary content beyond 20 mg/kg (71).

b. Deficiency. Force-feeding diets deficient in niacin and

tryptophan induces behavioral abnormalities, convulsions, diarrhea, weight loss, and development of rough hair coat and alopecia (23). Corn diets have been used to induce deficiency, but interpretation of the results is complicated by induction of imbalances of amino acids in addition to deficiencies. The relatively high content of leucine in corn may enhance deficiency of tryptophan and niacin, but the results of leucine dietary supplementation have not been consistent (181).

6. Pantothenic Acid

a. Requirement. Pantothenic acid is a constituent of coenzyme A which functions in synthesis of lipids and other acetylation reactions. The dietary requirement is 8 mg/kg as calcium pantothenate (9,10).

b. Deficiency. Pantothenic acid-deficient rats grow and reproduce poorly and have achromotrichia, exfoliative dermatitis, oral hyperkeratosis, necrosis, and ulceration of the gastrointestinal tract. Hemorrhagic adrenal necrosis may occur and cause death after 4–6 weeks of deficiency (195). Deficiency results also in impaired albumin and antibody synthesis, decreased serum globulins, and decreased response of antibody forming cells to antigens (131,210).

7. Riboflavin

a. Requirement. Riboflavin is the precursor of the flavin coenzymes; it is stored in the liver primarily as FAD (7). The coenzymes function with cytochrome *c* reductase, xanthine oxidase, diaphorase, and other oxidation–reduction enzymes. They are required for conversion of tryptophan to nicotinic acid, of pyridoxine to pyridoxal, and in metabolism of folate enzymes (199). The requirement is influenced by the type and level of carbohydrate in the diet; starch decreases the requirement, probably by increasing intestinal synthesis of the vitamin, and sucrose does the reverse. The absolute requirement may be as low as 3.2 μg/rat/day, the equivalent of 0.2 mg/kg of diet, but to allow for excretion an intake of 1.2 mg/kg in the diet was suggested (17,179). Maximum hepatic storage was found in rats fed diets which contained 1–3 mg/kg and maximum growth at 2 mg/kg (71,163). Dams fed 1 mg/kg produced underweight offspring which grew poorly and had decreased brain weight and DNA content. Correction of weight gain and brain size was achieved by feeding dams 2.7 mg/kg during lactation but not by supplementation of the young after weaning (64). Dams fed 4 mg/kg riboflavin in diet produced young with growth rates and hepatic stores equal to the young of dams fed 100 mg/kg (220).

The dietary requirement base on growth and hepatic storage is 2–3 mg/kg, but at least 3 mg/kg is required for normal reproduction; an increase to 4 mg/kg is suggested during gestation. On a caloric basis the requirement is 0.8 mg/1000 kcal.

b. Deficiency. The signs of deficiency are dermatitis, alopecia, weakness, and poor growth. Corneal vascularization and ulceration, cataract formation, anemia, and myelin degeneration may occur (99). Deficient rats may have fatty liver and abnormal hepatic mitochondria. There are complex metabolic effects of riboflavin deficiency: decreased flavoproteins, increased protein turnover, increased enzymes of amino acid metabolism, and decreased mitochondrial respiration and synthesis of ATP (70). Reproductive performance is decreased, with production of abnormal offspring.

In rats fed riboflavin-deficient diet, hepatic and red cell riboflavin and glutathione reductase were measured for 4 to 6 weeks. Riboflavin levels were decreased after 4 weeks and continued to decrease thereafter. Glutathione reductase was significantly decreased at 4 weeks in red cells and at 5 weeks in liver (8). The enzyme may be decreased also in the lens which has increased susceptibility to galactose induction of cataracts (233). Rats deficient in riboflavin had decreased hepatic folate storage despite adequate intake (236).

8. Thiamin

a. Requirement. Thiamin is converted to thiamin pyrophosphate (TPP) which is the storage form and the coenzyme for oxidative reactions which include oxidative decarboxylation.

The dietary requirement for thiamin depends in part on caloric content; it is increased by increasing carbohydrate but may be decreased by dietary fat (222). The sparing effect of fat has not been found in all studies (160). Male weanlings fed either 1.25 or 12.5 mg/kg of thiamin had equivalent food efficiency ratios, but the higher concentration supported a more

rapid growth rate (138). There was no significant effect on growth of increasing dietary thiamin from 5 to 50 mg/kg (111).

Pregnant rats were fed either 4 or 100 mg/kg of thiamin. Fetal content of thiamin was the same in the 2 groups, and there was no significant effect of the higher concentration on growth of offspring (220). The requirement is 4 mg/kg.

b. Deficiency. Thiamin deficiency induces abnormalities of the nervous system and the heart and poor reproduction. Anorexia and weight loss are prominent clinical signs; blood pyruvate may be increased.

Thiamin-deficient rats developed encephalopathy and cardiac hypertrophy at 5 weeks. They were ataxic and had impaired righting reflex and drowsiness which could be corrected by injection of thiamin. Cardiac weight increased an average of 18%, but there were no histologic or ultrastructural abnormalities of the heart. Cardiac and renal ATP and pyruvate carboxylase and brain thiamin were decreased (146).

Blood and tissue transketolase correlate with thiamin status and may be restored *in vitro* by the addition of TPP. The response to TPP may depend on the duration of the deficiency and the stability of the apoenzyme (8,21).

9. Vitamin B$_6$

a. Requirement. The vitamin B$_6$ compounds (pyridoxine, pyridoxal, and pyridoxamine) are phosphorylated and serve as coenzymes for amino acid decarboxylases, transaminases, and other enzymes of amino acid, glycogen, and fatty acid metabolism (7). Approximately 50% of body stores of pyridoxal phosphate is found in muscle, where it serves as coenzyme for glycogen phosphorylase (34,180). The dietary requirement can be based on enzyme activity, weight gain, reproductive performance, or tissue content of pyridoxal phosphate. Red cell transaminase is more sensitive than hepatic transaminase or maximal hepatic B$_6$ storage to dietary content of the vitamin and indicates a requirement of 6–7 mg/kg diet (12).

Male weanling rats fed 1, 2, 4, or 8 mg/kg diet grew at equivalent rates, but liver, serum, and red cell glutamic-pyruvic transaminase (GPT) was maintained only by dietary content of 4 mg/kg and above, and activity was increased by addition of B$_6$ *in vitro* to tissues from all groups (34).

Vitamin B$_6$ is required for normal reproduction and development (154). Offspring of deficient dams have retarded renal differentiation, abnormal cerebral lipids, and increased tissue and urinary cystathionine (51,120,128). In dams fed 3 or 6 mg/kg, maternal weight gain and body and brain weight of offspring were normal; they were slightly lower at 2 mg/kg; 1 mg/kg was clearly inadequate (54). Maximum tissue vitamin content in dams was achieved at dietary concentrations between 9.6 and 19.2 mg/kg, but enzyme activities were normal at 2.4 mg/kg (120). Brain content of B_6, protein, and cerebrosides, thymus weight and DNA content in offspring, and milk content of pyridoxine were decreased at maternal intakes of 1.2 or 2.4 mg/kg but normal at higher intakes. Brain protein in offspring continued to increase up to 19.6 mg/kg. The minimum adequate intake was 4.8 mg/kg (185). The dietary requirement for growth and reproduction is 6 mg/kg.

b. Deficiency. Vitamin B_6-deficient rats have a symmetrical scaling dermatitis on the tail, paws, face, and ears and microcytic anemia; as deficiency increases they become hyperexcitable and convulse (225). Reproduction is poor. There may be deficient production of insulin (104). In deficient rats fed a high protein diet, urinary and blood urea decreased; urinary excretion of free ammonia decreased and excretion of free amino acids increased. Cystathionine and citrulline were excreted in large amounts and can be used as markers for B_6 deficiency. Hepatic serine and threonine dehydratases and cystathionase decreased (182).

10. Vitamin B_{12}

a. Requirement. Vitamin B_{12} is a coenzyme for transmethylation of homocysteine to methionine, and for conversion of methylmalonyl coenzyme A to succinyl coenzyme A (244). It may act also in catabolism of toxic amounts of methionine, since toxicity was blocked by feeding large amounts of B_{12} in the diet (6). The dietary requirement of the rat is not firmly established; it varies with dietary content of choline, methionine, and folic acid, and like folic acid, it is supplied by intestinal microbial synthesis. Induction of isolated vitamin B_{12} deficiency in rats is difficult.

The requirement for vitamin B_{12} is met adequately by a dietary content of 50 μg/kg. Lower concentrations may be adequate when animal protein is fed but probably not in diets which contain only vegetable protein.

b. Deficiency. Induction of deficiency requires feeding vegetable rather than animal protein. Females fed soybean protein supplemented with methionine and choline but not vitamin B_{12} grew at a normal or slightly decreased rate and bred and littered normally. However, the average weight of offspring was decreased, and 10% of litters produced had at least one hydrocephalic member. Hepatic vitamin B_{12} was decreased in dams and offspring. The incidence of abnormalities in the neonates was increased by deletion of choline from the diet. Supplementation with 50 μg/kg diet of vitamin B_{12} supported normal growth and prevented hydrocephalus (250). Vitamin B_{12} deficiency can be induced by feeding unheated soybean flour, but there are associated amino acid deficiencies and evidence of toxicity (56,246).

REFERENCES

1. Aguilar, T. S., Benevenga, N. J., and Harper, A. E. (1974). Effect of dietary methionine level on its metabolism in rats. *J. Nutr.* **104,** 761–771.
2. Akrabawi, S. S., and Salji, J. P. (1973). Influence of meal-feeding on some of the effects of dietary carbohydrate deficiency in rats. *Br. J. Nutr.* **30,** 37–43.
3. American Institute on Nutrition. Ad hoc committee on Standards for Nutritional Studies (1977). Report of one committee. *J. Nutr.* **107,** 1340–1348.
4. Amine, E. K., Desilets, E. J., and Hegsted, D. M. (1976). Effect of dietary fats on lipogenesis in iron deficiency anemic chicks and rats. *J. Nutr.* **106,** 405–411.
5. Amine, E. K., and Hegsted, D. M. (1971). Effect of diet on iron absorption in iron-deficient rats. *J. Nutr.* **101,** 927–935.
6. Areshkina, L. Y., Skorobogatova, E. P., Erofeeva, N. N., Filipovich, E. G., and Annenkov, B. A. (1974). Role of vitamin B_{12} in methionine catabolism. *Appl. Biochem. Microbiol.* **6,** 339–346.
7. Baker, H., and Frank, O. (1968). "Clinical Vitaminology: Methods and Interpretation." Wiley (Interscience), New York.
8. Bamji, M. S., and Sharada, D. (1972). Hepatic gluthathione reductase and riboflavin concentrations in experimental deficiency of thiamin and riboflavin in rats. *J. Nutr.* **102,** 443–447.
9. Barboriak, J. J., Krehl, W. A., and Cowgill, G. R. (1957). Pantothenic acid requirement of the growing and adult rat. *J. Nutr.* **61,** 13–21.
10. Barboriak, J. J., Krehl, W. A., Cowgill, G. R., and Whedon, A. D. (1957). Effect of partial pantothenic acid deficiency on reproductive performance of the rat. *J. Nutr.* **63,** 591–599.
11. Barney, G. H., Orgebin-Crist, M. C., and Macapinlac, M. P. (1968). Genesis of esophageal parakeratosis and histologic changes in the testes of the zinc-deficient rat and their reversal by zinc repletion. *J. Nutr.* **95,** 526–534.
12. Beaton, G. H., and Cheney, M. C. (1966). Vitamin B_6 requirement of the male albino rat. *J. Nutr.* **87,** 125–132.
13. Benevenga, N. J. (1974). Toxicities of methionine and other amino acids. *Agric. Food Chem.* **22,** 2–9.
14. Berdanier, C. D. (1975). Rat strain differences in response to meal feeding. *Nutri. Rep. Int.* **11,** 517–524.
15. Berg, B. N. (1966). Dietary restriction and reproduction in the rat. *J. Nutr.* **87,** 344–8.
16. Bernhart, F. W., Savini, S., and Tomarelli, R. M. (1969). Calcium and phosphorus requirements for maximal growth and mineralization of the rat. *J. Nutr.* **98,** 443–448.
17. Bessey, O. A., Lowry, O. H., Davis, E. B., and Dorn, J. L. (1958). The riboflavin economy of the rat. *J. Nutr.* **64,** 185–202.
18. Best, C. H., Lucas, C. C., and Ridout, J. H. (1954). The lipotropic factors. *Ann. N. Y. Acad. Sci.* **57,** 646–653.
19. Best, C. H., Ridout, J. :H., and Lucas, C. C. (1969). Alleviation of dietary cirrhosis with betaine and other lipotropic agents. *Can. J. Physiol. Pharmacol.* **47,** 73–79.
20. Bieri, J. G. (1972). Aspects of vitamin E metabolism relating to the dietary requirement. Kinetics of tissue α-tocopherol depletion and repletion. *Ann. N. Y. Acad. Sci.* **203,** 181–191.
21. Brin, M. (1966). Transketolase: Clinical aspects. *In* "Methods in Enzymology" (W. A. Wood, ed.), Vol. 9, pp. 506–514. Academic Press, New York.
22. Brody, S. (1945). "Bioenergetics and Growth, with Special Reference to the Efficiency Complex in Domestic Animals." Van Nostrand: Reinhold, Princeton, New Jersey.
23. Brookes, I. M., Owens, F. N., and Garrigus, U. S. (1972). Influence

of amino acid level in the diet upon amino acid oxidation by the rat. *J. Nutr.* **102,** 27–35.

24. Bunce, G. E., and Hess, J. L. (1976). Lenticular opacities in young rats as a consequence of maternal diets low in tryptophan and/or vitamin E. *J. Nutr.* **106,** 222–229.

25. Bunce, G. E., and King, K. W. (1969). Amino acid retention and balance in the young rat fed varying levels of casein. *J. Nutr.* **98,** 168–176.

26. Bunce, G. E., and King, K. W. (1969). Amino acid retention and balance in young rat fed varying levels of lactalbumin. *J. Nutr.* **98,** 159–167.

27. Campbell, T. C., and Hayes, J. R. (1975). Role of nutrition in the drug-metabolizing enzyme system. *Pharmacol. Rev.* **26,** 171–197.

28. Carlisle, E. M. (1974). Silicon as an essential element. *Fed. Proc., Fed. Am. Soc. Exp. Biol.* **33,** 1758–1766.

29. Carmel, N., Konijn, A. M., Kaufmann, N. A., and Guggenheim, K. (1976). Effects of carbohydrate-free diets on the insulin–carbohydrate relationships in rats. *J. Nutr.* **105,** 1141–1149.

30. Carroll, K. K. (1975). Experimental evidence of dietary factors and hormone–dependent cancers. *Cancer Res.* **35,** 3374–3384.

31. Chakrabarty, K., and Leveille, G. A. (1968). Influence of periodicity of eating on the activity of various enzymes in adipose tissue, liver and muscle of the rat. *J. Nutr.* **96,** 76–82.

32. Chandler, P. T., and Cragle, R. G. (1962). Investigation of calcium, phosphorus and vitamin D_3 relationships in rats by multiple regression techniques. *J. Nutr.* **78,** 28–36.

33. Chang, M. L., Schuster, E. M., Lee, J. A., Snodgrass, C., and Ben ton, D. A. (1968). Effect of diet, dietary regimens and strain differences on some enzyme activities in rat tissues. *J. Nutr.* **96,** 368–374.

34. Chen, L. H., and Marlatt, A. L. (1975). Effects of dietary vitamin B_6 levels and exercise on glutamic-pyruvic transaminase activity in rat tissues. *J. Nutr.* **105,** 401–407.

35. Chen, S. H., Estes, L. W., and Lombardi, B. (1973). Lecithin depletion in hepatic microsomal membranes of rats fed on a choline-deficient diet. *Exp. Mole. Pathol.* **17,** 176–186.

36. Chiang, G. S., and Mistry, S. P. (1974). Activities of pyruvate carboxylase and propionyl-CoA carboxylase in rat tissues during biotin deficiency and restoration of the activities after biotin administration. *Proc. Soc. Exp. Biol. Med.* **146,** 21–24.

37. Copeland, D. H., and Salmon, W. D. (1946). The occurrence of neoplasms in the liver, lungs and other tissues of rats as a result of prolonged choline deficiency. *Am. J. Pathol.* **22,** 1059–1067.

38. Corey, J. E., and Hayes, K. C. (1972). Cerebrospinal fluid pressure, growth, and hematology in relation to retinol status of the rat in acute vitamin A deficiency. *J. Nutr.* **102,** 1585–1594.

39. Coward, W. A., Howell, J. McC., Thompson, J. N., and Pitt, G. A. (1969). The retinol requirements of rats for spermatogenesis and vision. *Br. J. Nutr.* **23,** 619–626.

40. Crawford, M. A., and Sinclair, A. J. (1973). The limitations of whole tissue analysis to define linolenic acid deficiency. *J. Nutr.* **102,** 1315–1321.

41. Cuthbertson, E. M., and Greenberg, D. M. (1945). Chemical and pathological changes in dietary chloride deficiency in the rat. *J. Biol. Chem.* **160,** 83–94.

42. Dahl, L. K., Knudsen, K. D., and Heine, M. A. (1968). Effects of chronic excess salt ingestion. Modification of experimental hypertension in the rat by variations in the diet. *Circ. Res.* **22,** 11–18.

43. Dallman, P. R., and Goodman, J. R. (1971). The effects of iron deficiency on the hepatocyte: A biochemical and ultra-structural study. *J. Cell Biol.* **48,** 79–90.

44. Dallman, P. R., and Schwartz, H. C. (1966). Myoglobin and cyto-chrome response during repair of iron deficiency in the rat. *J. Clin. Invest.* **44,** 1631–1638.

45. Deb, S., Martin, R. J., and Hershberger, T. V. (1976). Maintenance requirement and energetic efficiency of lean and obese Zucker rats. *J. Nutr.* **106,** 191–197.

46. DeLuca, H. F. (1975). Function of the fat-soluble vitamins. *Am. J. Clin. Nutr.* **28,** 339–345.

47. Deuel, H. J., Jr. (1957). "The Lipids, their Chemistry and Biochemistry," Vol. III. Wiley (Interscience), New York.

48. Deuel, H. J., Jr., Alfin-Slater, R. B., Well, A. F., Kryder, G. D., and Aftergood, L. (1955). The effect of fat level of the diet on general nutrition. XIV. Further studies of the effect of hydrogenated coconut oil on essential fatty acid deficiency in the rat. *J. Nutr.* **55,** 337–346.

49. Deuel, H. J., Jr., Martin, C. R., and Alfin-Slater, R. B. (1955). The effect of fat level of the diet on general nutrition. XVI. A comparison of linoleate and linolenate in satisfying the essential fatty acid requirement for pregnancy and lactation. *J. Nutr.* **57,** 297–302.

50. Diamond, I., Swenerton, H., and Hurley, L. S. (1971). Testicular and esophageal lesions in zinc-deficient rats and their reversibility. *J. Nutr.* **101,** 77–84.

51. DiPaolo, R. V., Caviness, V. S., Jr., and Kanfer, J. N. (1974). Delayed maturation of the renal cortex in the vitamin B_6 deficient newborn rat. *Pediat. Res.* **8,** 546–552.

52. Dipaolo, R. B., Kanfer, J. N., and Newberne, P. M. (1974). Copper deficiency and the central nervous system. *J. Neuropathol. Exp. Neurol.* **33,** 226–236.

53. Draper, H. H., Sie, T., and Bergan, J. G. (1972). Osteoporosis in aging rats induced by high phosphorus diets. *J. Nutr.* **102,** 1133–1141.

54. Driskell, J. A., Strickland, L. A., Poon, C. H., and Foshee, D. P. (1973). The vitamin B_6 requirement of the male rat as determined by behavioral patterns, brain pyridoxal phosphate and nucleic acid composition and erythrocyte alanine aminotransferase activity. *J. Nutr.* **103,** 670–680.

55. Dror, Y., Sassoon, H. F., Watson, J. J., Mack, D. O., and Johnson, B. C. (1973). Fat versus sucrose as the nonprotein calorie portion of the diet of rats. *J. Nutr.* **103,** 342–346.

56. Edelstein, S., and Guggenheim, K. (1972). Effects of sulfur-amino acids and choline on vitamin B_{12}-deficient rats. *Nutr. Metab.* **13,** 339–343.

57. Estes, L., and Lombardi, B. (1970). Effect of choline deficiency on the Golgi apparatus of rat hepatocytes. *Lab. Invest.* **21,** 374–385.

58. Evans, J. L., and Ali, R. (1968). Calcium utilization and feed efficiency in the growing rat as affected by dietary calcium, buffering capacity, lactose and EDTA. *J. Nutr.* **92,** 417–424.

59. Fernstrom, J. D. (1976). The effect of nutritional factors on brain amino acid levels and monoamine synthesis. *Fed. Proc., Fed. Am. Soc. Exp. Biol.* **35,** 1151–1156.

60. Follis, R. H. (1958). "Deficiency Disease." Thomas, Springfield, Illinois.

61. Forbes, G. B., and Reina, J. C. (1972). Effect of age on gastrointestinal absorption (Fe, Sr, Pb) in the rat. *J. Nutr.* **102,** 647–652.

62. Forbes, R. M. (1966). Effects of magnesium, potassium and sodium nutriture on mineral composition of selected tissues of the albino rat. *J. Nutr.* **88,** 403–410.

63. Forbes, R. M., and Yohe, M. (1960). Zinc requirement and balance studies with the rat. *J. Nutr.* **70,** 53–57.

64. Fordyce, M. K., and Driskell, J. A. (1975). Effects of riboflavin repletion during different developmental phases on behavioral patterns, brain nucleic acid and protein contents, and erythrocyte glutathione reductase activity of male rats. *J. Nutr.* **105,** 1150–1156.

65. Fox, J., Aldrich, F. D., and Boylen, G. W. (1976). Lead in animal foods. *J. Toxicol. Environ. Health* **1,** 461–467.

66. Frier, H. I., Hall, R. C., Jr., Rousseau, J. E., Jr., Eaton, H. D., and Nielson, S. W. (1975). Cerebrospinal fluid pressure and squamous metaplasia in chronic hypovitaminosis A of the male weanling rat. *Conn., Storrs Agric. Exp. Stn., Res. Rep.,* **46,** 1–55.

67. Ganguli, M. C., Smith, J. D., and Hanson, L. E. (1970). Sodium metabolism and its requirement during reproduction in female rats. *J. Nutr.* **99,** 225–234.

68. Ganguli, M. C., Smith, J. D., and Hanson, L. E. (1970). Sodium metabolism and requirements in lactating rats. *J. Nutr.* **99,** 395–400.

69. Ganguly, J. (1969). Absorption of vitamin A. *Am. J. Clin. Nutr.* **22,** 923–933.

70. Garthoff, L. H., Garthoff, S. K., Tobin, R. B., and Mehlman, M. A. (1973). The effect of riboflavin deficiency on key gluconeogenic enzyme activities in rat liver. *Proc. Soc. Exp. Biol. Med.* **143,** 693–697.

71. Gaudin-Harding, F., Griglio, S., Bois-Joyeux, B., de Gasquet, P., and Karlin, R. (1971). Reserves en vitamines du groupe B chex des rats Wistar soumis a des regimes temoins ou hyperlipidiques a deux niveaux vitaminiques. *J. Int. Vitaminol. Nutr.* **42,** 25–32.

72. Geever, E. F., Kan, D., and Levenson, S. M. (1969). Effect of bacterial flora on iron absorption in the rat. *Gastroenterology* **55,** 690–694.

73. Gillette, J. R. (1976). Environmental factors in drug metabolism. *Fed. Proc., Fed. Am. Soc. Exp. Biol.,* **35,** 1142–1147.

74. Gillman, T. (1958). Cell enzymes and iron metabolism in anemias and siderosis. *Nutr. Rev.* **16,** 353–355.

75. Goldberg, A. (1971). Carbohydrate metabolism in rats fed carbohydrate free diets. *J. Nutr.* **101,** 693–697.

76. Gopalan, C., and Rao, B. S. N. (1972). Experimental niacin deficiency. *Methods Achiev. Exp. Pathol.* **6,** 49–80.

77. Greenfield, H., and Briggs, G. M. (1971). Nutritional methodology in metabolic research with rats. *Annu. Rev. Biochem.* **40,** 549–572.

78. Griffith, W. H., and Nyc, J. F. (1954). Choline. *In* "The Vitamins," (W. H. Sebrell, Jr. and R. S. Harris, eds.), Vol. 2, pp. 2–103. Academic Press, New York.

79. Grunert, R. R., Meyer, J. H., and Phillips, P. H. (1951). The sodium and potassium requirements of the rat for growth. *J. Nutr.* **42,** 609–618.

80a. György, P., and Pearson, W. N., eds. (1967). "The Vitamins," 2nd ed., Vols. VI and VII. Academic Press, New York.

80b. György, P., Cardi, E., Rose, C. S., Hirooka, M., and Langer, B. W., Jr. (1967). Lipid transport in experimental dietary hepatic injury in rats. *J. Nutr.* **93,** 568–578.

81. Hafeman, D. G., Sunde, R. A., and Hoekstra, W. G. (1974). Effect of dietary selenium on erythrocyte and liver glutathione peroxidase in the rat. *J. Nutr.* **104,** 580–587.

82. Hale, O. M., and Schaefer, A. E. (1951). Choline requirement of rats as influenced by age, strain, vitamin B_{12} and folacin. *Pro. Soc. Exp. Biol. Med.* **77,** 633–636.

83. Halsted, J. A., Smith, J. C., and Irwin, M. I. (1974). A conspectus of research on zinc requirements of man. *J. Nutr.* **104,** 345–378.

84. Hankes, L. V., Henderson, L. M., Brickson, W. L., and Elvehjem, C. A. (1948). Effect of amino acids on the growth of rats on niacin-tryptophan deficient ratios. *J. Biol. Chem.* **174,** 873–881.

85. Harper, A. E., Benevenga, N. J., and Wohlheuter, R. M. (1970). Effects of ingestion of disproportionate amounts of amino acids. *Physiol. Rev.* **50,** 428–558.

86. Harper, A. E., and Elvehjem, C. A. (1957). Dietary carbohydrates. A review of the effects of different carbohydrates on vitamin and amino acid requirements. *Agric. Food Chem.* **5,** 754–758.

87. Harris, L. J., and Kodicek, E. (1950). Quantitative studies and dose response curves in nicotinamide deficiency in rats. *Br. J. Nutr.* **4,** 13–14.

88. Hartroft, W. S. (1963). Experimental hepatic injury. *In* "Diseases of the Liver" (L. Schiff, ed.), 2nd ed., pp. 101–141. Lippincott, Philadelphia, Pennsylvania.

89. Hartsook, E. W., Hershberger, T. V., and Nee, J. C. M. (1973). Effects of dietary protein content and ratio of fat to carbohydrate calories on energy metabolism and body composition of growing rats. *J. Nutr.* **103,** 167–178.

90. Hauswirth, J. W., and Nair, P. P. (1975). Effects of different vitamin E-deficient basal diets on hepatic catalase and microsomal cytochromes *P-450* and b_5 in rats. *Am. J. Clin. Nutr.* **28,** 1087–1094.

91. Hautvast, J. G. A. J., and Barnes, M. J. (1975). Collagen metabolism in folic acid deficiency. *Br. J. Nutr.* **32,** 457–469.

92. Hegsted, D. M., Finch, C. A., and Kinney, T. D. (1949). The influence of diet on iron absorption. *J. Exp. Med.* **90,** 147–156.

93. Hegsted, D. M., Finch, C. A., and Kinney, T. D. (1952). The influence of diet on iron absorption. III. Comparative studies with rats, mice, guinea pigs, and chickens. *J. Exp. Med.* **96,** 115–119.

94. Herbert, V. (1973). The five possible causes of all nutrient deficiency: Illustrated by deficiencies of vitamin B_{12} and folic acid. *Am. J. Clin. Nutr.* **26,** 77–88.

95. Higginson, J., Grady, H., and Huntley, C. (1963). The effects of different diets on iron absorption. *Lab. Invest.* **12,** 1260–1269.

96. Holman, R. T. (1970). Biological activities of and requirements for polyunsaturated acids. *Prog. Chem. Fats Other Lipids* **9,** 611–682.

97. Holman, R. T. (1968). Essential fatty acid deficiency. *Prog. Chem. Fats Other Lipids* **9,** 279–348.

98. Horvath, D. J. (1976). Trace elements and health. *In* "Trace Substances and Health" (P. M. Newberne, ed.), Chapter 5, pp. 319–331. Dekker, New York.

99. Horwitt, M. K. (1954). Riboflavin. *In* "The Vitamins" (W. H. Sebrell, Jr. and R. S. Harris, eds.), Vol. 3, pp. 380–391. Academic Press, New York.

100. Hotzel, D. (1967). Contributions of the intestinal microflora to the nutrition of the host. *Vitam. Horm. (N.Y.)* **24,** 115–171.

101. Houlihan, C. M., and Scott, J. M. (1972). The identification of pteroylpentaglutamate as the major folate derivative in the rat liver and the demonstration of its biosynthesis from exogenous H^3 pteroyl glutamate. *Biochem. Biophys. Res. Commun.* **48,** 1675–1681.

102. Howard, L., Wagner, C., and Schenker, S. (1974). Malabsorption of thiamin in folate-deficient rats. *J. Nutr.* **104,** 1024–1032.

103. Huber, A. M., and Gershoff, S. N. (1973). Effects of dietary zinc on zinc enzymes in the rat. *J. Nutr.* **103,** 1175–1181.

104. Huber, A. M., Gershoff, S. N., and Hegsted, D. M. (1964). Carbohydrate and fat metabolism and response to insulin in vitamin B_6-deficient rats. *J. Nutr.* **82,** 371–378.

105. Hurley, L. S. (1968). Approaches to the study of nutrition in mammalian development. *Fed. Proc., Fed. Am. Soc. Exp. Biol.* **29,** 193–198.

106. Hurley, L. S., Cosens, G., and Theriault, L. L. (1976). Teratogenic effects of magnesium deficiency in rats. *J. Nutr.* **106,** 1254–1260.

107. Hurley, L. S., Wooten, E., and Everson, G. J. (1961). Disproportionate growth in offspring of manganese-deficient rats. 2. Skull, brain and cerebrospinal fluid. *J. Nutr.* **74,** 282–288.

108. Hurt, H. D. Cary, E. E., and Visek, W. J. (1971). Growth, reproduction, and tissue concentrations of selenium in the selenium-depleted rat. *J. Nutr.* **101,** 761–766.

109. Hwang, D. H., Mathias, M. M., Dupont, J., and Meyer, D. L. (1975). Linoleate enrichment of diet and prostaglandin metabolism in rats. *J. Nutr.* **105,** 995–1002.

111. Itokawa, Y., and Fujiwara, M. (1973). Changes in tissue magnesium, calcium and phosphorus levels in magnesium-deficient rats in relation to thiamin excess or deficiency. *J. Nutr.* **103,** 438–443.

112. Jacob, M., and Forbes, R. M. (1969). Effects of magnesium defi-

ciency, dietary sulfate and thyroxine treatment on kidney calcification and tissue protein-bound carbohydrate in the rat. *J. Nutr.* **99,** 51–57.

113. Jager, F. C. (1972). Long-term dose-response effects of vitamin E in rats. *Nutr. Metab.* **14,** 1–7.

114. Jager, F. C., and Houtsmuller, U. M. T. (1971). Effect of dietary linoleic acid acid on vitamin E requirement and fatty acid composition of erythrocyte lipids in rats. *Nutr. Metab.* **12,** 3–12.

115. Johnson, B. C., Mameesh, M. S., Metta, V. V., and Rama Rao, P. B. (1960). Vitamin K nutrition and irradiation sterilization. *Fed. Proc., Fed. Am. Soc. Exp. Biol.* **19,** 1038–1044.

116. Jones, J. H. (1971). Vitamin D requirement of animals, *In* "The Vitamins" (W. H. Sebrell, Jr. and R. S. Harris, eds.), 2nd ed., Vol. 3, pp. 285–289. Academic Press, New York.

117. Kamath, S. K., MacMillan, J. B., and Arnrich, L. (1973). Dietary protein and utilization of carotene or retinyl acetate in rats. *J. Nutr.* **102,** 1579–1584.

118. Kaufman, N., Klavins, J. V., and Kinney, T. D. (1958). Excessive iron absorption in rats fed low protein, high fat diets. *Lab. Invest.* **7,** 369–376.

119. Kinney, T. D., Hegsted, D. M., and Finch, C. A. (1949). The influence of diet on iron absorption. *J. Exp. Med.* **90,** 137–145.

120. Kirksey, A., Pang, R. L., and Lin, W. J. (1975). Effects of different levels of pyridoxine fed during pregnancy superimposed upon growth in the rat. *J. Nutr.* **105,** 607–615.

121. Klavins, J. V., Kinney, T. D., and Kaufman, N. (1962). The influence of dietary protein on iron absorption. *Br. J. Exp. Pathol.* **113,** 172–180.

122. Kleiber, M. (1975). "The Fire of Life: An Introduction to Animal Energetics." R. F. Krieger Publ. Co., New York.

123. Knapka, J. J., Smith, K. P., and Judge, F. J. (1974). Effect of open and closed formula rations on the performance of three strains of laboratory mice. *Lab. Anim. Sci.* **24,** 480–487.

124. Kolata, G. B. (1975). Vitamin D: Investigations of a new steroid hormone. *Sciences* **187,** 635–636.

125. Konijn, N. M., Muogbo, D. N. C., and Guggenheim, K. (1971). Metabolic effects of carbohydrate-free diets. *Isr. J. Med. Sci.* **6,** 498–505.

126. Koretsky, S. (1959). Role of salt and renal mass in experimental hypertension. *Arch. Pathol.* **68,** 11–22.

127. Kroe, D., Kinney, T. D., Kaufman, N., and Klavins, J. V. (1963). The influence of amino acids on iron absorption. *Blood* **21,** 546–552.

128. Kurtz, D. J., Levy, H., and Kanfer, J. N. (1972). Cerebral lipids and amino acids in the vitamin B_6-deficient suckling rat. *J. Nutr.* **102,** 291–298.

129. Kutzbach, C., Galloway, E., and Stokstad, E. L. R. (1969). Influence of vitamin B_{12} and methionine on levels of folic acid compounds and folate enzymes in rat liver. *Proc. Soc. Exp. Biol. Med.* **124,** 801–805.

130. Lamb, A. J., Apiwatanaporn, P., and Olson, J. A. (1974). Induction of rapid synchronous vitamin A deficiency in the rat. *J. Nutr.* **104,** 1140–1148.

131. Lederer, W. H., Kumar, M., and Axelrod, A. E. (1975). Effects of pantothenic acid deficiency on cellular antibody synthesis in rats. *J. Nutr.* **105,** 17–25.

132. Leelaprute, V., Boonpucknavig, V., Bhamarapravati, N., and Weerapradist, W. (1973). Hypervitaminosis A in rats. *Arch. Pathol.* **96,** 5–9.

133. Leveille, G. A., and Chakrabarty, K. (1968). Absorption and utilization of glucose by meal-fed and nibbling rats. *J. Nutr.* **96,** 69–75.

134. Loosli, J. K., Lingenfelter, J. F., Thomas, J. W., and Maynard, L. A. (1944). The role of dietary fat and linoleic acid in the lactation of the rat. *J. Nutr.* **28,** 81–88.

135. Lyman, R. L., and Wilcox, S. S. (1963). Effect of acute amino acid deficiencies on carcass composition and pancreatic function in the force-fed rat. II. Deficiencies of valine, lysine, tryptophan, leucine and isoleucine. *J. Nutr.* **79,** 37–44.

136. MacDonald, R. A., Jones, R. S., and Pechet, G. S. (1965). Folic acid deficiency and hemochromatosis. *Arch. Pathol.* **80,** 153–160.

137. MacDonald, R. A., and Pechet, G. S. (1965). Experimental hemochromatosis in rats. *Am. J. Pathol.* **46,** 85–103.

138. Mackerer, C. R., Mehlman, M. A., and Tobin, R. B. (1973). Effects of chronic acetylsalicylate administration on several nutrition and biochemical parameters in rats fed diets of varied thiamin content. *Biochem. Med.* **8,** 51–60.

139. Madhaven, T. V., and Gopalan, C. (1968). The effect of dietary protein on carcinogenesis of oflatoxin. *Arch. Pathol.* **85,** 133–137.

140. Majumder, A. K., Nandi, B. K., Subramanian, N., and Chatterjee, I. B. (1975). Growth and ascorbic acid metabolism in rats and guinea pigs fed cereal diets. *J. Nutr.* **105,** 233–239.

141. Marshall, M. W., Durand, A. M. A., and Adams, M. (1971). Different characteristics of rat strains: Lipid metabolism and response to diet. *In* "Defining the Laboratory Animal," pp. 381–412. Natl. Acad. Sci., Washington, D. C.

142. Matschiner, J. T., and Doisy, E. A., Jr. (1965). Effect of dietary protein on the development of vitamin K deficiency in the rat. *J. Nutr.* **86,** 93–99.

143. Maurer, R. L., and Day, H. G. (1957). The non-essentiality of fluorine in nutrition. *J. Nutr.* **62,** 561–573.

144. McAleese, D. M., and Forbes, R. M. (1961). The requirement and tissue distribution of magnesium in the rat as influenced by environmental temperature and dietary calcium. *J. Nutr.* **73,** 94–106.

145. McCall, M. G., Newman, G. E., Obrien, J. R. P., Valberg, L. S., and Witts, L. J. (1962). Studies in iron metabolism. 1. The experimental production of iron deficiency 2. The Effects of experimental iron deficiency in the growing rat. *Br. J. Nutr.* **16,** 305–323.

146. McCandless, D. W., Hanson, C., Speeg, K. V., Jr., and Schenker, S. (1970). Cardiac metabolism in thiamin deficiency in rats. *J. Nutr.* **100,** 991–1002.

147. McLaughlan, J. M., and Campbell, J. A. (1969). Methodology of protein evaluation. *In* "Mammalian Protein Metabolism" (H. N. Munro and J. B. Allison, eds.), Chapter 29, pp. 391–418. Academic Press, New York.

148. Menaker, L., and Navia, J. M. (1973). Appetite regulation in the rat under various physiological conditions: The role of dietary protein and calories. *J. Nutr.* **103,** 347–352.

149. Michels, F. G., and Smith, J. T. (1966). A comparison of the utilization of organic and inorganic sulfur by the rat. *J. Nutr.* **87,** 217–220.

150. Miller, E. C., and Miller, J. A. (1972). Approaches to the mechanisms and control of chemical carcinogenesis. Bertner Foundation Award Lecture. *In* "Environment and Cancer," pp. 5–39. Williams & Wilkins, Baltimore, Maryland.

151. Miller, S. A. (1969). Protein metabolism during growth and development. *In* "Mammalian Protein Metabolism" (H. N. Munro and J. B. Allison, eds.). Chapter 26, pp. 183–227. Academic Press, New York.

152. Milner, J. A., Wakeling, A. E., and Visek, W. J. (1974). Effect of arginine deficiency on growth and intermediary metabolism in rats. *J. Nutr.* **104,** 1681–1689.

153. Monserrat, A. J., Porta, E. A., Ghoshal, A. K., and Hartman, S. B. (1974). Sequential renal lipid changes in weanling rats fed a choline-deficient diet. *J. Nutr.* **104,** 1496–1502.

154. Moon, W. H. Y., and Kirksey, A. (1973). Cellular growth during prenatal and early postnatal periods in progeny of pyridoxine-deficient rats. *J. Nutr.* **103,** 123–133.

155. Moore, T. (1957). "Vitamin A." Am. Elsevier, New York.

156. Morrison, S. D. (1956). The total energy and water metabolism during pregnancy in the rat. *J. Physiol. (London)* **134**, 650-654.

157. Morrison, S. D. (1973). Differences between rat strains in metabolic activity and in control systems. *Am. J. Physiol.* **224**, 1305-1308.

158. Morrison, S. D. (1976). Generation and compensation of the cancer cachectic process by spontaneous modification of feeding behavior. *Cancer Res.* **36**, 228-233.

159. Munro, H. N. (1969). Evolution of protein metabolism in mammals. *In* "Mammalian Protein Metabolism" (H. N. Munro and J. B. Allisson, eds.), Chapter 25, pp. 133-176. Academic Press, New York.

160. Murdock, D. S., Donaldson, M. L., and Gubler, C. J. (1974). Studies on the mechanism of the "thiamin-sparing" effect on ascorbic acid in rats. *Am. J. Clin. Nutr.* **27**, 696-699.

161. Murthy, L., Klevay, L. M., and Petering, H. G. (1974). Interrelationships of zinc and copper nutriture in the rat. *J. Nutr.* **104**, 1458-1465.

162. National Research Council (1972). "Nutrient Requirements of Laboratory Animals," 2nd rev. ed., Vol. 10. NRC, Washington, D. C.

163. National Research Council (1978). "Nutrient Requirements of Laboratory Animals," 3rd rev. ed., Vol. 10. NRC, Washington, D. C.

164. National Research Council (1971). "Selenium in Nutrition." NRC, Washington, D. C.

165. Nelson, M. M., and Evans, H. M. (1961). Dietary requirements for lactation in the rat and other laboratory animals. *In* "Milk: The Mammary Gland and its Secretions" (S. K. Kon and A. T. Cowie, eds.), Vol. 2, pp. 137-191. Academic Press, New York.

166. Newberne, P. M. (1963). Cardiorenal lesions of potassium depletion or steroid therapy in the rat. *Am. J. Vet. Res.* **25**, 1256-1266.

167. Newberne, P. M. (1974). The influence of nutrition on response to infectious disease. *Adv. Vet. Sci. Comp. Med.* **17**, 265-289.

168. Newberne, P. M. (1975). Influence on pharmacological experiments of chemicals and other factors in diets of laboratory animals. *Fed. Proc., Fed. Am. Soc. Exp. Biol.* **34**, 209-218.

169. Newberne, P. M., and Gebhardt, B. M. (1974). Pre- and postnatal malnutrition and responses to infection. *Nutr. Rep. Int.* **7**, 407-420.

170. Newberne, P. M., Glaser, O., Friedman, L., and Stillings, B. (1973). Safety evaluation of fish protein concentrate over five generations of rats. *Toxicol. Appl. Pharmacol.* **24**, 133-141.

171. Newberne, P. M., and Rogers, A. E. (1973). Rat colon carcinomas associated with aflatoxin and marginal vitamin A. *J. Natl. Cancer Inst.* **50**, 439-448.

172. Newberne, P. M., Rogers, A. E., Bailey, C., and Young, V. R. (1969). Induction of liver cirrhosis in rats by purified amino acid diets. *Cancer Res.* **29**, 230-235.

173. Newberne, P. M., Rogers, A. E., and Wogan, G. N. (1968). Hepatorenal lesions in rats fed a low lipotrope diet and exposed to aflatoxin. *J. Nutr.* **94**, 331-343.

174. Newberne, P. M., and Wogan, G. N. (1968). Sequential morphologic changes in aflatoxin B₁ carcinogenesis in the rat. *Cancer Res.* **28**, 770-781.

175. Newberne, P. M., and Young, V. R. (1966). Effect of diets marginal in methionine ;and choline with and without vitamin B₁₂ on rat liver and kidney. *J. Nutr.* **89**, 69-79.

176. Nielsen, F. H., and Ollerich, D. A. (1974). Nickel: A new essential trace element. *Fed. Proc., Fed. Am. Soc. Exp. Biol.* **33**, 1767-1772.

177. Numa, S., Hashimoto, T., Iritani, N., and Okazaki, T. (1970). Role of acetyl coenzyme A carboxylase in the control of fatty acid synthesis. *Vitam. Horm. (N.Y.)* **23**, 213-243.

178. Nutrition Reviews (1964). Iron absorption and diet. *Nutr. Rev.* **22**, 306-309.

179. Nutrition Reviews (1972). The fate of riboflavin in the mammal. *Nutr. Rev.* **30**, 75-79.

180. Nutrition Reviews (1975). Regulation of liver metabolism of pyridoxal phosphate. *Nutr. Rev.* **33**, 214-217.

181. Nutrition Reviews (1976). Effect of high intake of leucine on urinary tryptophan and niacin metabolites in humans. *Natl. Rev.* **340**, 105-107.

182. Okada, M., and Suzuki, K. (1974). Amino acid metabolism in rats fed a high protein diet without pyridoxine. *J. Nutr.* **104**, 287-293.

183. Orent-Keiles, E. R., Robinson, A., and McCollum, E. V. (1937). The effects of sodium deprivation on the animal organism. *Am. J. Physiol.* **119**, 651-661.

184. Ottolenghi, A. D., Haseman, J. K., Payne, W. W., Fack, H. L., and MacFarland, H. M. (1975). Inhalation studies of nickel sulfide in pulmonary carcinogenesis of rats. *J. Natl. Cancer Inst.* **54**, 1165-1172.

185. Pang, R. L., and Kirksey, A. (1974). Early postnatal changes in brain composition in progeny of rats fed different levels of dietary pyridoxine. *J. Nutr.* **104**, 111-117.

186. Panos, T. C., Finerty, J. C., and Wall, R. L. (1956). Increased metabolism in fat-deficiency: Relation to dietary fat. *Proc. Soc. Exp. Biol. Med.* **93**, 581-587.

187. Patek, A. J., Jr., Erenoglu, E., O'Brian, N. M., and Hirsch, R. L. (1969). Sex hormones and susceptibility of the rat to dietary cirrhosis. *Arch. Pathol.* **87**, 52-56.

188. Patek, A. J., Jr., Kendall, R. F., de Fritsch, N. M., and Hirsch, R. L. (1966). Cirrhosis-enhancing effect of corn oil. *Arch. Pathol.* **82**, 596-601.

189. Pennell, M. D., Davies, M. I., Rasper, J., and Motzok, I. (1976). Biological availability of iron supplements for rats, chicks, and humans. *J. Nutr.* **106**, 265-274.

190. Peterson, A. D., and Baumgardt, B. R. (1971). Food and energy intake of rats fed diets varying in energy concentration and density. *J. Nutr.* **101**, 1057-1067.

191. Peterson, A. D., and Baumgardt, B. R. (1971). Influence of level of energy demand on the ability of rats to compensate for diet dilution. *J. Nutr.* **101**, 1069-1074.

192. Peterson, P. A., Nilsson, S. F., Ostberg, L., Rask, L., and Vahlquist, A. (1975). Aspects of the metabolism of retinol-binding protein and retinol. *Vitam. Horm. (N.Y.)* **32**, 181-214.

193. Poirier, L. A., and Whitehead, V. M. (1973). Folate deficiency and formiminoglutamic acid excretion during chronic diethylnitrosamine administration to rats. *Cancer Res.* **33**, 383-388.

194. Pudelkewicz, C., Seufart, J., and Holman, R. T. (1968). Requirements of the female rat for linoleic and linolenic acids. *J. Nutr.* **94**, 138-146.

195. Ralli, E. P., and Dumm, M. E. (1953). Relation of pantothenic acid to adrenal cortical function. *Vitam. Horm. (N.Y.)* **11**, 133-158.

196. Rechcigl, M., Jr., Berger, S., Loosli, J. K., and Williams, H. H. (1962). Dietary protein and utilization of vitamin A. *J. Nutr.* **76**, 435-440.

197. Reeves, P. G., and Forbes, R. M. (1972). Prevention by thyroxine of nephrocalcinosis in early magnesium deficiency in the rat. *Am. J. Physiol.* **222**, 220-224.

198. Richardson, L. R., Godwin, J., Wilkes, S., and Cannon, M. (1964). Reproduction performance of rats receiving various levels of dietary protein and fat. *J. Nutr.* **82**, 257-262.

199. Rivlin, R. S. (1970). Riboflavin metabolism. *N. Engl. J. Med.* **283**, 463-472.

200. Rodriguez, M. S., and Irwin, M. I. (1972). A conspectus of research on vitamin A requirements of man. *J. Nutr.* **102**, 909-968.

201. Rogers, A. E. (1977). Reduction of *N*-nitrosodiethylamine carcinogenesis in rats by lipotrope or amino acid supplentation of a marginally deficient diet. *Cancer Res.* **37**, 194-199.

202. Rogers, A. E. (1975). Variable effects of a lipotrope-deficient high-fat diet on chemical carcinogenesis in rats. *Cancer Res.* **35**, 2469-2474.

203. Rogers, A. E., Herndon, B. J., and Newberne, P. M. (1973). Induction by dimethylhydrazine of intestinal carcinoma in normal rats and rats fed high or low levels of vitamin A. *Cancer Res.* **33**, 1003-1009.

204. Rogers, A. E., and MacDonald, R. A. (1965). Hepatic vasculature and cell proliferation in experimental cirrhosis. *Lab. Invest.* **14**, 1710-1726.

205. Rogers, A. E., and Newberne, P. M. (1971). Diet and aflatoxin B₁ toxicity in rats. *Toxicol. Appl. Pharmacol.* **20**, 113-121.

206. Rogers, A. E., and Newberne, P. M. (1973). Alcoholic or nutritional fatty liver and cirrhosis. Animal model of human disease. *Am. J. Pathol.* **73**, 817-820.

207. Rogers, A. E., and Newberne, P. M. (1975). Dietary effects on chemical carcinogenesis in animal models for colon and liver tumors. *Cancer Res.* **35**, 3427-3431.

208. Rogers, A. E., Sanchez, O., Feinsod, F., and Newberne, P. M. (1974). Dietary enhancement of nitrosamine carcinogenesis. *Cancer Res.* **34**, 96-99.

209. Rogers, Q. R., and Harper, A. E. (1965). Amino acid diets and maximal growth in the rat. *J. Nutr.* **87**, 267-273.

210. Roy, A. K., and Axelrod, A. E. (1971). Protein synthesis in liver of pantothenic acid deficient rats. *Proc. Soc. Exp. Biol. Med.* **138**, 804-807.

211. Said, A. K., and Hegsted, D. M. (1970). Response of adult rats to low dietary levels of essential amino acids. *J. Nutr.* **100**, 1363-1376.

212. Salmon, W. D., and Newberne, P. M. (1962). Cardiovascular disease in choline-deficient rats. *Arch. Pathol.* **73**, 190-209.

213. Samuelsson, B. (1972). Biosynthesis of prostaglandins. *Fed. Proc., Fed. Am. Soc. Exp. Biol.* **31**, 1442-1450.

215. Schlenk, H. (1972). Odd numbered and new essential fatty acids. *Fed. Proc., Fed. Am. Soc. Exp. Biol.* **31**, 1430-1435.

216. Schroeder, H. A. (1966). Chromium deficiency in rats: A syndrome simulating diabetes mellitus with retarded growth. *J. Nutr.* **88**, 439-445.

217. Schroeder, H. A. (1969). Serum cholesterol and glucose levels in rats fed refined and less refined sugars and chromium. *J. Nutr.* **97**, 237-242.

218. Schultz, M. O. (1957). Nutrition of rats with compounds of known chemical structure. *J. Nutr.* **61**, 585-596.

219. Schultz, M. O. (1956). Reproduction of rats fed protein-free amino acid rations. *J. Nutr.* **60**, 35-45.

220. Schumacher, M. F., Williams, M. A., and Lyman, R. L. (1965). Effect of high intakes of thiamine, riboflavin and pyridoxine on reproduction in rats and vitamin requirements of the offspring. *J. Nutr.* **86**, 343-349.

221. Schwarz, K. (1974). Recent dietary trace element research, exemplified by tin, fluorine, and silicon. *Fed. Proc., Fed. Am. Soc. Exp. Biol.* **33**, 1748-1757.

222. Scott, E. M., and Griffith, I. V. (1957). A comparative study of thiamine-sparing agents in the rat. *J. Nutr.* **61**, 421-436.

223. Sebrell, W. J., Jr., and Harris, R. S., eds. (1967). "The Vitamins," 2nd ed., Vols. I-V. Academic Press, New York.

224. Shannon, B. M., Howe, J. M., and Clark, H. E. (1972). Interrelationships between dietary methionine and cystine as reflected by growth, certain hepatic enzymes and liver composition of weanling rats. *J. Nutr.* **102**, 557-562.

225. Sherman, H. (1954). Pyridoxine and related compounds. *In* "The Vitamins" (W. H. Sebrell, Jr. and R. S. Harris, eds.), 1st ed., Vol. 3, pp. 265-276. Academic Press, New York.

226. Sibbald, I. R., Bowland, J. P., Robblee, A. R., and Berg, R. T. (1957). Apparent digestible energy and nitrogen in the food of the weanling rat. Influence of food consumption on nitrogen retention and carcass composition. *J. Nutr.* **61**, 71-85.

227. Sidransky, H., and Farber, E. (1958). Chemical pathology of acute amino acid deficiencies. *Arch. Pathol.* **66**, 119-134.

228. Sidransky, H., and Farber, E. (1958). Chemical pathology of acute amino acid deficiencies. II. Biochemical changes in rats fed. threonine- or methionine-devoid diets. *Arch. Pathol.* **66**, 135-149.

229. Sidransky, H., and Verney, E. (1969). Chemical pathology of acute amino acid deficiencies: Morphologic and biochemical changes in young rats force-fed a threonine-deficient diet. *J. Nutr.* **96**, 349-358.

230. Smith, J. C., Jr., McDaniel, E. G., Fan, F. F., and Halsted, J. A. (1973). Zinc: A trace element essential in vitamin A metabolism. *Science* **181**, 954-955.

231. Spector, W. S. (1956). "Handbook of Biological Data." Saunders, Philadelphia, Pennsylvania.

232. Sprecher, H. W. (1972). Regulation of polyunsaturated fatty acid biosynthesis in the rat. *Fed. Proc., Fed. Am. Soc. Exp. Biol.* **31**, 1451-1457.

233. Srivastava, S. K., and Beutler, E. (1970). Increased susceptibility of riboflavin deficient rats to galactose cataract. *Experientia* **26**, 250 (only).

234. Staub, H. W., Reussner, G., and Thiessen, R. (1970). Serum cholesterol reduction by chromium in hypercholesterolemic rats. *Science* **166**, 746-747.

235. Sweeney, J. P., and Marsh, A. C. (1974). Liver storage of vitamin A by rats fed carrots in various forms. *J. Nutr.* **104**, 1115-1120.

236. Tamburro, C., Frank, O., Thomson, A. D., Sorrell, M. F., and Baker, H. (1971). Interactions of folate, nicotinate, and riboflavin deficiencies in rats. *Nutr. Rep. Int.* **4**, 185-189.

237. Tanhenbaum, A. (1959). Nutrition and cancer. *In* "The Physiopathology of Cancer" (F. Hamburger, ed.), 2nd ed., pp. 517-562. Harper (Hoeber), New York.

238. Tinoco, J., Williams, M. A. Hincenbergs, I., and Lyman, R. L. (1971). Evidence for nonessentiality of linolenic acid in the diet on the rat. *J. Nutr.* **101**, 937-945.

239. Tuchweber, B., Garg, B. D., and Salas, M. (1976). Microsomal enzyme inducers and hypervitaminosis A in rats. *Arch. Pathol.* **100**, 100-105.

240. Van Campen, D., and House, W. A. (1974). Effect of a low protein diet on retention of an oral dose of Zn⁶⁵ and on tissue concentrations of zinc, iron, and copper in rats. *J. Nutr.* **104**, 84-90.

241. Vessell, E. S., Lang, C. M., White, W. J., Passananti, G. T., Hill, R. N., Clemens, T. L., Liu, D. K., and Johnson, W. D. (1976). Environmental and genetic factors affecting the response of laboratory animals to drugs. *Fed. Proc., Fed. Am. Soc. Exp. Biol.* **35**, 1125-1132.

242. Vitale, J. J., and Hegsted, D. M. (1970). Effects of dietary methionine and vitamin B₁₂ deficiency on folate metabolism. *Br. J. Haematol.* **17**, 467-475.

243. Wang, F. L., Wang, R., Khairallah, E. A., and Schwartz, R. (1971). Magnesium depletion during gestation and lactation in rats. *J. Nutr.* **101**, 1201-1210.

244. Weissbach, H., and Taylor, R. T. (1970). Roles of vitamin B₁₂ and folic acid in methionine synthesis. *Vitam. Horm. (N.Y.)* **23**, 415-440.

245. Whanger, P. D., and Weswig, P. H. (1971). Effect of supplementary zinc on the intracellular distribution of hepatic copper in rats. *J. Nutr.* **101**, 1093-1098.

246. Williams, D. L., and Spray, G. H. (1973). The effects of diets containing raw soya-bean flour on the vitamin B₄₁₂ status of rats. *Br. J. Nutr.* **29**, 57-63.

247. Williams, E. A. J., Gross, R. L., and Newberne, P. M. (1975). Effect of folate deficiency on the cell-mediated immune response in rats. *Nutr. Rep. Int.* **12**, 137-148.

248. Wilson, R. B., Smith, D. M., and Newberne, P. M. (1974). Excess sodium chloride intake in neonatal rats. *Arch. Pathol.* **96**, 372-376.

249. Witting, L. A., and Horwitt, M. K. (1964). Effects of dietary selenium, methionine, fat level and tocopherol on rat growth. *J. Nutr*. **84,** 351–360.

250. Woodard, J. C., and Newberne, P. M. (1966). Relation of vitamin B_{12} and 1-carbon metabolism to hydrocephalus in the rat. *J. Nutr*. **88,** 375–381.

251. Yeh, S. D. J., Soltz, W., and Chow, B. F. (1965). The effect of age on iron absorption in rats. *J. Gerentol*. **20,** 177–180.

252. Young, V. R., and Munro, H. N. (1973). Plasma and tissue tryptophan levels in relation to tryptophan requirements of weanling and adult rats. *J. Nutr*. **103,** 1756–1763.

253. Zaki, F. G., Bandt, C., and Hoffbauer, F. W. (1963). Fatty cirrhosis in the rat. IV. The influence of different levels of dietary fat. *Arch. Pathol*. **75,** 654–660.

254. Zaki, F. G., Hoffbauer, F. W., and Grande, F. (1966). Fatty cirrhosis in the rat. Effect of dietary fat. *Arch. Pathol*. **80,** 323–331.

Chapter 7

Reproduction and Breeding

Dennis E. J. Baker

I. INTRODUCTION

In attempting to review the advances made during the last thirty or so years in the breeding of the laboratory rat, it is advantageous to use an ecologic approach as a basis for classification and order. This approach, described by Park (66), describes the animal in its environment from the biotic, nutritive, spatial, climatic, and temporal perspectives.

II. BIOTIC ASPECTS

A. Puberty

Differentiation of sex in mature rats by observation of the perineum is easily learned (Fig. 1), but separating the sexes in prepubertal animals is more difficult (Fig. 2). An examination of the anogenital space shows that the distance between the anus and the genital papilla is much greater in males than females (Fig. 3). When learning to identify the sexes of young rats, it is sometimes useful to examine littermates until two markedly different anogenital spaces are found and then to compare each of the remaining littermates to these two.

In both sexes, puberty occurs at about 50–60 days of age. The vagina opens at about 72 days (34–109 days range) and the testes descend during the period of 15–51 days of age. Eckstein *et al.* (25), using descendants of Wistar or Charles River rats, showed that vaginal opening occurs at 35.6 days (S.E. 0.8) and first estrus occurs at 36.4 days (S.E. 1.2).

Traditionally, rats have been mated at an age of 100–120 days and fertility has been regarded as maximum between 100 and 300 days of age. Kali and Amir (41) showed that mating rats at 35 days of age produced smaller litters but pups were of similar body weight to progeny from rats mated at 80 days of age. Dams mated at either 35 or 80 days of age developed to the same body weight at 106 days of age.

Ovulatory estrous cycles usually begin at about 77 days but may start at any time from 45 to 147 days of age.

A method for provoking ejaculation and sperm collection by the ip injection of halothane or penthrane has been described by Kessler (43).

B. Estrous Cycle

Vaginal changes in the rat are closely related to the estrous cycle, to the extent that examination of vaginal fluid and cells provide a reliable method for determining its stages.

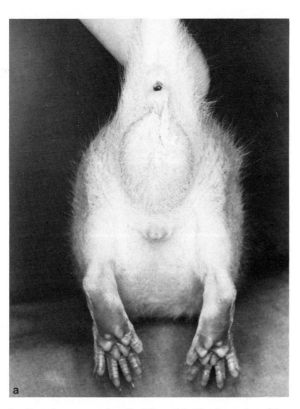 

Fig. 1. Sex of mature rats is easily differentiated by observation of the perineum. (a) The anogenital space is greater in males and the scrotum is prominent. (b) Note the shortened anogenital space and separate urethral papilla at the ventral commissure of the vulva.

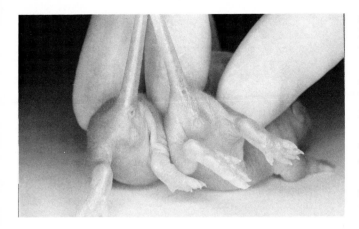

Fig. 2. Sex differentiation of neonatal rat pups is best accomplished by comparison of anogenital space, which is greater in the male pup on the right.

The vaginal wall has a dry, white, and lusterless appearance during estrus but changes to moist and pink during metestrus (53). These changes are associated with cornification of the surface layers during estrus and extensive desquamation at the end of this stage.

Histologic changes in the vaginal epithelium and exfoliated cells associated with various stages of the estrous cycle have been described by Asdell (5). He reports that as proestrus approaches, the cell types in the vaginal fluids consist of nucleated epithelial cells and leukocytes with an occasional cornified cell. During proestrus the leukocytes disappear, nucleated epithelial cells increase in number and are gradually replaced by cornified nonnucleated cells. The beginning of estrus usually occurs when the smear contains 75% nucleated and 25% cornified cells. Cornified cells are eventually the only ones present. Gradually a flat nucleated ("pavement") epithelial cell appears. During the latter part of the cornified stage, the vaginal fluid becomes both abundant and caseous, and

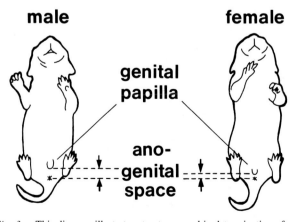

Fig. 3. This diagram illustrates structures used in determination of sex in neonatal rats.

estrous may end during this stage. The end of estrous is followed by the appearance of large number of leukocytes and the virtual disappearance of cornified cells.

Descriptions of the vaginal fluids during the stages of the estrous cycle have been reported by Long and Evans (53) and by Young, Boling, and Blandau (91) and have been summarized by Asdell (5). A comparison of the cycles of common laboratory rodents has been made by Friedmann *et al.* (32). A correlation of the stages of the estrous cycle and the changes occurring in the superficial genitalia, uterus and ovary is presented in Table I. Change in pH of vaginal gluids may also be used as an indicator of estrus (10). A technique for measuring the vaginal impedance of the rat has been described by Bartos (8), who used the elevation of impedance at proestrus to determine the optimum time for mating. The absence of a cyclic peak in impedance also has been used as a reliable indicator of pregnancy. However, using different probes, Wolf and Horn (88) found no correlation between the stage of the estrous cycle and electric conductivity in the rat vagina.

Behavior of the female rat during estrus has been described and is useful in determining the existence of estrus. An early sign of estrus (30) is ear quivering elicited by gently stroking the head or back. The copulatory response test of Blandau (11), in which digital stimulation of the pelvic region produces a characteristic lordosis, in indicative that the rat is in estrus. Farris (29) claims that the most accurate method for determining the onset and length of estrus is by observing the amount of running activity by the turntable method. Regardless of the method used, wide individual behavioral variations must be expected, and these are influenced by such factors as sampling techniques, handling (63), and time of day.

C. Conception

An accurate estimate of the date of conception is often useful, but since the actual time of conception is virtually impossible to determine, indirect indicators are used.

A commercial supplier (3) of Sprague-Dawley descended rats reports that presence of sperm in vaginal smears is a useful indicator of conception and claims that 90% of "sperm-positive" rats undergo successful shipment, gestation, and parturition. Chow and Augustin (16) defined conception as the "first appearance of sperm in the vagina" and reported a 94% success rate (of 34 "sperm-positive" rats, 32 gave birth). A study of the predictability of pregnancy in Sprague-Dawley descended rats by Szabo *et al.* (80) concluded that, contrary to the widely held view, vaginal cytology is not the most accurate predictor of pregnancy in mice and rats. They showed that detection of sperm in the vagina was one of the most reliable criteria for prediction of pregnancy (see Table II). A similar conclusion was reached by Heinecke *et al.* (38).

Table I

Changes Associated with Stages of the Estrous Cycle in the Rat[a]

Stage and duration	Vaginal fluids		Superficial genitalia	Uterus	Ovary	pH of vaginal contents
	Long and Evans (53)	Young, Boling, and Blandau (91)				
Proestrus (ca. 12 hr)	Small, round nucleated cells only	Stage I, small round nucleated cells to 75% cornified cells. Estrus begins when 25% of cells are cornified	Vulva slightly swollen; vagina dry	Vascular engorgement; distention by fluid, lumen to 5 mm diameter	Enlargement of follicle	5.4
Estrus (ca. 12 hr)	Cornified cells only	Stage II, from cornified cells only, to cornified 25%, pavement cells 75%	Vulva swollen; vagina dry	Maximum distention in early part of this stage followed by regression to normal	Large follicles; maturation of egg	4.2
Metestrus I (ca. 15 hr)	Late cornified stage; abundant caseous fluid	Stage III, pavement cells only	Vulva still swollen; caseous mass in vagina	Increase in vacuolar degenerational epithelium which began during estrus	Ovulation	
Metestrus II (ca. 6 hr)	Cornified cells and leukocytes	Stage IV, pavement cells and leukocytes	No swelling; mucosa moist	Epithelium begins to regenerate	Eggs in oviduct	
Diestrus (ca. 57 hr)	Epithelial cells and leukocytes		No swelling; mucosa moist	Epithelium regenerated; lumen to 2.5 mm in diameter	Corpora lutea formed	6.1

[a] Based on data reported by Nicholas (63), Asdell (5), Long and Evans (53), and Young et al. (91).

The production of large numbers of pregnant rats is facilitated by the induction of synchronized estrus. May (55) achieved synchronization by placing polyurethane sponges, containing 0.75 mg medroxyprogesterone in the vagina of 90 to 110-day-old female rats. The sponges were left in place for 7 days after which the females were placed in cages previously occupied by male rats (Whitten effect). When the sponges were removed, the females were injected with 3 IU (international units) of pregnant mare serum. Within 34 hr 93% of the females had mated.

An improvement on this technique was developed by May (56) which avoided the use of intravaginal sponges. Estrus was suppressed by administering 40 mg medroxyprogesterone in 200 ml ethanol per liter of drinking water. This solution was prepared fresh each day and administered for 6 days, after which estrus was induced by the intramuscular injection of 1 IU of pregnant mare's serum.

Table II

Number of Rats to be Mated in Order to Obtain at Least Ten Pregnant Animals[a]

Mating sign(s)	50% of the time	90% of the time
Vaginal plug	13	27
Pan plug	11	13
Sperm	11	12
Vaginal plug, pan plug	11	13
Vaginal plug, sperm	11	12
Pan plug, sperm	11	12
Vaginal plug, pan plug, sperm	11	12

[a] Based on data from Szabo et al. (80).

D. Gestation Period

Gestation averages 22–23 days from copulation to parturition. Abdominal enlargement is usually evident by the thirteenth day of pregnancy. A clear mucoid vaginal discharge is commonly observed 1½ to 4 hr prior to delivery of the first pup.

E. Parturition

The average duration of parturition is 1½ hr within a range of 55 min through 3 hr and 50 min. Failure of the dam to lick her vulva seems to be correlated with difficult delivery. Dystocia appears to be rare even in commonly used strains of rats, possibly due to the selection that has occurred over many generations of laboratory breeding and to the adequacy of the levels of Vitamin A in the diet (74). However, Tuchmann *et al.* (81) have reported that the prenatal administration of acetylsalicylic acid to rats will prolong labor up to 18 hr and also result in dystocia with possible secondary death of fetuses *in utero.*

Rat pups are born hairless with eyes and ear canals closed (Fig. 4). By 7–10 days they are fully haired and have patent ear canals and opened eyes (Fig. 5). Figure 6 shows the appearance of rat pups at weaning age.

F. Menopause

Menopause begins during 450–540 days of age. Except in special cases, such as precious genetic material, efficient breeding life ceases long before the onset of menopause. (See Section VI on temporal aspects for a discussion of optimal breeding life span.)

G. Microbic Environment

Many of the problems of intercurrent infection, particularly murine respiratory mycoplasmosis, have been reduced by the

Fig. 5. These 10-day-old nursing pups are fully haired, have patent ears, and opened eyes.

production and maintenance of specific pathogen-free (SPF) rats. A more complete discussion of bacterial and mycoplasma infections, as they apply to fertility, is presented in this volume, Chapters 9 and 10.

A review of the efficiency of an isolation facility for rat breeding has been given by Van der Waaij and Van Bekkum (82). A system for producing axenic and SPF mice proposed by Baker (6) could be applied to rats also. Kappel, Nelson, and Weisbroth (42) have developed a screening technique for the monitoring of a mycoplasma-free colony of rats.

When it is not possible to use gnotobiotic techniques, methods of screening and selection have been reported to yield

Fig. 4. Rats are born hairless with closed ear canals and fused eyelids. Note the elliptical white spot on the left abdomen of the rat pup in lower right of the photo which represents milk in the pups' stomach.

Fig. 6. Rat pups approximately 15 days of age.

encouraging results in eliminating pathogens from colonies of rats. By the elimination of excretors and the application of rigorous hygenic measures Kunze and Marwitz (45) were able to eliminate *Salmonella enteritidis* from a rat colony.

III. NUTRITION

This section addresses itself to nutritional requirements for optimum reproduction in the rat. For a more detailed discussion of the subject of nutrition of the rat see this volume, Chapter 6. Extensive reviews of this topic have been published by Russell (74), Warner (83), and Cumming and Cumming (21), and form the basis for this section.

A. Food Consumption

Food intake in the mature female is lower during estrus than during diestrus sometimes by as much as 6 gm/day. Table III shows the nutrient requirements for a breeding rat. The average daily consumption by a pregnant female increases from a maintenance rate of 30 gm/day to a level of 19 gm/day by the twentieth day of gestation. Following a drop at parturition, back to approximately the maintenance level, average daily food consumption rises sharply to a level of 40 gm/day at the twentieth day of lactation. It has been estimated that for each

individual pup the female rat consumes 20 gm of food during the 22 days of gestation and 72 gm of food during 28 days of lactation. When fed an adequate diet, the breeding female can be expected to gain approximately 85 gm during gestation, 60% during the last week and to show a net gain or loss of 10 gm during lactation.

An essential quality of any diet is that it support optimum reproduction and development of the young. The definition of optimum will have to be made by each worker taking into consideration such factors as the strain, breeding and mating methods, litter size, body weight of mature rats, environmental temperatures, coprophagy, and the many other sources of influence on the rat's interaction with its diet. A listing of the minimum concentrations of nutrients in a diet suitable for breeding is given in Table IV. However, it is emphasized that

Table III

Daily Nutrient Requirements for Breeding—Female Rats[a]

	Maintenance	Gestation	Lactation
Average total feed (gm)	13	19	33
Gross energy (meal)	52	76	131
Metabolizable energy (kcal)	47	68	118
Essential fatty acids (mg)	Trace	20	80
Crude protein (gm)	0.91	3.8	6.6
Net protein (gm)	0.52	2.3	4.0

[a]Based on data from Warner (83).

Table IV

Nutrient Content of a Breeding Female Rat Diet Containing *Minimum* Amounts of Required Nutrients[a]

Gross energy (kcal/kg)	4000	Minerals (mg/100 gm)	
Metabolizable energy (kcal/kg)	3600	Calcium	600
Fat (%)	5	Phosphorus	500
		Sodium	500
Crude protein (%)	20	Potassium	500
Net protein (%)	12	Chlorine	25
		Magnesium	50
Essential fatty acids (%)	0.3	Manganese	3.3
		Iron	2.5
Net amino acids (%)		Copper	0.5
L-Tryptophan	0.2	Zinc	0.4
L-Histidine	0.54	Iodine	0.015
L-Lysine	1.24	Vitamin A (IU/100 gm)	400
L-Leucine	0.80	Vitamin D (IU/100 gm)	100
L-Isoleucine	0.50	α-Tocopherol (mg/100 gm)	3
L-Phenylalanine	0.90	Menadione	0.1
L-Methionine	1.00	Thiamin-HCl	0.4
L-Threonine	0.50	Riboflavin	0.4
L-Valine	0.70	Pyridoxine-HCl	0.6
L-Argenine	0.75	Niacin	3.0
		Ca pantothenate	1.0
Nonessential amino acids	4.87	Choline chloride (mg/100 gm)	80.0
		Vitamin B$_{12}$ (μg/100 gm)	5.0

[a]Based on data reported by Russell (74), Warner (83), and Cuthbertson (22).

minimum requirements for a nutrient are influenced by the amounts of other nutrients in the diet.

In species such as the rat in which both lactation and gestation may be occurring concurrently, severe nutritional demands are made by the animal. When comparing a breeding diet with a maintenance diet, Weihe (84) showed a 31% difference in weaned young produced.

B.   Proteins

A level of 20% protein from mixed animal and vegetable sources is adequate for reproduction and the rearing of young rats. The feeding of high levels of protein (circa 50%) impairs the efficiency of protein utilization and also may have an adverse effect on the sexual development of the female. The need for 0.90% L-phenylalanine may be reduced by one-third if replaced by L-tyrosine. Similarly L-cystine can replace one-half of the need for L-methionine.

C.   Carbohydrates

There is no specific need for carbohydrates except as an energy source. In marginal diets the complex carbohydrates, starch, and dextrin promote a higher growth rate than do the soluble mono- and disaccharides.

D.   Essential Fatty Acids and Fat

The presence of linoleic acid or other unsaturated fatty acid is essential in the diet of rats for normal estrous cycles, prevention of testicular degeneration and for normal reproduction (74).

In a study of two different levels of fat in the diet, Nolan and Alexander (64) found no significant differences in reproductive performance which could be ascribed to the 5 or 18% fat levels.

E.   Minerals

Both the level and the proportions of calcium and phosphorus are important for reproduction and lactation with 500 mg/100 gm diet of phosphorus and a Ca:P ratio between 1 : 1 and 2 : 1 as the optimum. A marginal magnesium deficiency allows good growth but affects the next generation in which deficiency symptoms develop at an early age. A high intake of either calcium or phosphorus increases the requirement for magnesium. Satisfactory reproduction has been reported on daily intakes of manganese ranging from 0.1 mg/rat to 100

mg/rat. A level of 1 mg/rat/day is optimum for nursing mothers to obtain good growth in their young.

The requirement for potassium in the pregnant rat is 15 mg/day (75 mg/100 gm diet), but this need increases during lactation to approximately 50 mg/day or more. Subnormal levels of either sodium or phosphorus increase the requirement for potassium.

F.   Vitamins

The need for all vitamins, except ascorbic acid, in rat diets is well established. Knowledge of deficiency symptoms (see this volume, Chapter 6) is valuable when the vitamin content of a sterilized diet is under consideration or in the preparation of diets for animals carrying a limited intestinal flora.

Diets low in vitamin A cause testicular degeneration, cornification of the vaginal epithelium, dystocia, and fetal malformation and may influence the estrous cycle adversely. When diet in the form of dried pellets is autoclaved, approximately 40% of the vitamin A is destroyed. When the diet is autoclaved in the form of a wet mash (4 parts dry meal to 5 parts water by weight), then an increased loss to 60% is incurred (20). The requirement for vitamin D is related to calcium and phosphorus metabolism, and if the diet contains adequate amounts of these elements at a favorable ratio, some workers question the need for any additional Vitamin D. However, the addition of 100 IU/100 g diet permits a greater range of variability in the calcium and phosphorus content of the diet and hastens, slightly, both growth and calcification.

Vitamin E deficiency results in resorption of embryos and testicular degeneration in young rats. Vitamin E in the diet may be destroyed easily by the presence of rancid fats or certain metallic salts (21).

Under ordinary circumstances the rat can synthesize in its intestinal tract sufficient vitamin K for its needs. Deaths of SPF rats are reported to have occurred both in a breeding colony and under test conditions. When the basic diet was fortified with 1 ppm of vitamin K_3 these deaths ceased (35). It was suggested that these SPF rats did not carry strains of bacteria capable of synthesizing vitamin K in adequate amounts.

A deficiency of thiamin can result in impaired reproduction. Since losses of thiamin in steam-sterilized diets can approach or even exceed 80%, it is wise to include an assay of thiamin levels as part of a routine quality control program. While it is helpful to fortify diets for sterilization with additional thiamin, care must be taken to reduce the losses from autoclaving, since much of the breakdown product of thiamin is in the form of oxythiamin, an antivitamin (89). Lack of riboflavin upsets the estrous cycle, and lack of pyridoxine causes sterility in males. The rat does not require ascorbic acid.

As pointed out by Lane-Petter and Pearson (51), apart from

specific deficiencies, the adequacy of a diet is only relative to its application. They recommend, therefore, that long-term, large-scale studies be conducted over three or four generations in order to determine the fitness of a diet to a particular application.

IV. SPATIAL ASPECTS

A method for calculating the floor area needed for a given output of rats (and other small laboratory animals) has been provided by Festing and Bleby (31).

$$A = \frac{RG(OW)}{T(1-F)} \left(\frac{1}{P(1-K)} + \frac{W}{D} \right)$$

where A = area of animal room needed
R = number of units (ft^2 or m^2) needed to service a linear unit (ft or m) of wall space
G = width of single cage (ft or m)
T = number of shelves
F = average proportion of animals not used due to fluctuations in demand and supply (all used = 0.0. none used = 1.0)
OW = average number of animals to be used per week (output)
P = productivity (i.e., number of weaned animals produced per cage per week)
K = proportion of animals not suitable for use (e.g., 0.5 if only one sex used)
W = maximum length of growing period
D = number of growing animals maintained per breeding size cage.

Where previous experience is not available or applicable, these authors offer estimates for the rat which is shown in Table V. The number of cages needed can be determined from

$$N = OW \left(\frac{1}{P(1-K)} + \frac{W}{D} \right)$$

Table V

Estimates of Values for Some Variables in Calculation of Floor Area Needed for Breeding and Growing Rats[a]

Variable	Estimated Value
R	3–5 ft^2/linear ft
F	0.25 when all males but only 50% of females used
P	0.6–1.5 inbred line, pairs
	1.5–3.0 noninbred line, pairs
	3.0–5.0 polygamous systems

[a] Adpated from Festing and Bleby (31).

when N = number of cages needed and the other variables are as previously defined.

Cage Type and Bedding

The cage floor area recommended for a female and her litter is 150 in.^{2} Mundy and Porter (61) determined that cage floor area appeared to have no effect on the average estrous cycle.

To assure that rats use their behavioral heat regulatory mechanism, Weihe (85) recommended that they be housed in solid bottom cages with some form of bedding, and this is the practice of most commercial breeders (2). Mulder (60) has reported that pregnant rats will select a bedding, showing a preference for bedding of wood origin. However, Smith *et al.* (78) report death in infant rats due to sawdust bedding. There appears to be a concensus (in the United States) on the desirability of wood shavings or wood chips, although Lane-Petter and Lane-Petter (48) recommend dried sugar beet pulp, a material readily available in England. Nolan and Alexander (64) were able to wean a greater percentage of pups by using shredded paper rather than wood shavings. Smith *et al.* (78) suggest that this difference could be due to loss of pups from ingestion of wood particles from the shavings. A comparison of excelsior, shredded paper, and no nesting material was made by Norris and Adams (65). They weaned 94, 41, and 28% of pups born from the respective nesting treatments.

The effect of frequency of cage cleaning on productivity was studied by Cisar *et al.* (17) who concluded that cleaning twice each week was better than cleaning once per week and that the extra cleaning could be justified economically.

V. CLIMATIC ASPECTS

A. Pheromones

Pheromones are chemicals secreted from the body which affect communication and influence behavior between members of the same species. They are widespread throughout the chordata and other taxa, although their influence on mammalian behavior is still poorly understood. Sex attractants (both male and female), dominance odors, and territorial and home range markers are common in mammals (87). Although more studies have been made on mice, there are some significant reports on rats.

Odors associated with sex, frustration, social dominance, stress (71), and maternal care (39) reliably affect rat behavior. Given the absence of identified specialized external scent glands, chemicals in urine (71), feces (52), or from skin and mammae (39) are the probable sources of pheromones in the rat.

Whitten (86) reported that the estrous cycle of the mouse is apparently modified by the presence of a male or his excreta. A practical application of this finding, in the reproduction of mice, has been described by Ross (73). The modification of the estrous cycle of the mouse described by Whitten (86) consists of the induction of estrus in approximately 50% of the female population 3 nights after exposure to males or male excreta. Induction of estrus by such exposure is known as the Whitten effect. There are conflicting reports in the literature concerning the operation of the Whitten effect in rats. Ritchie (72) suggests that it does not operate in rats. However, in a study of the synchronization of estrus in rats, May (55) first treated female rats with progesterone (see Section II,C on conception) and then injected them with pregnant mare's serum. Those rats exposed to cages previously occupied by males showed a 93% incidence of mating, whereas those placed in clean cages showed only an 83% incidence of mating. May (55) introduced the males to the females, hence it is possible that territorial behavior could account for his findings rather than the Whitten effect. It is common practice among commercial breeders to introduce the female to cage occupied by males. While it is possible that the Whitten effect operates in rats, the definitive evidence is lacking.

Calhoun (15) showed that rats are territorial, and Richards and Stevens (71) have produced evidence that rats mark with urine. Voided urine from intact males has an avoidance effect on the normal adult male. This quality is not present in bladder urine of adult males nor in voided urine of castrated or immature males. Gawienowski (36) suggests that the marking pheromone is produced by an androgen-controlled gland(s) and is released into the urine during micturation.

From a study of fear-conditioned and naive rats, King and Pfister (44) suggested that odors generated by stressed male rats do not have alarm pheromone properties.

The role played by the olfactory stimuli in the regulation of the estrous cycle of two strains of Wistar rats has been described by Aron (4). Sprinkling the floor of the cages with male or female urine caused a shortening of the estrous cycle to 4 days in 50% of females which had previously regularly cycled at 5 days. While olfactory bulb deprivation suppressed the effects of female urine, it did not prevent the action of male urine under the experimental conditions. Licking may substitute for the mechanisms involved in the regulation of estrous rhythm by the pheromone. It has been suggested by Young (90) that in the male rat, odor from a sexually receptive female is capable of inducing the release of neurohypophyseal hormones which possibly influence sexual behavior, sperm transport, and penile sensitivity.

Pup attraction to lactating rats has been demonstrated in both Wistar and Sprague-Dawley rats (40). This attraction is effected by a maternal pheromone produced as a result of an attraction inducing stimulus originating in the pups. Pups from any litter constitute an adequate stimulus, a factor of considerable importance when cross fostering techniques are used. Leon (52) has demonstrated that the maternal attractant pheromone is in the cecotrophe (contents of cecum) fraction of the anal excreta. Using various diets he showed that although a maternal pheromone may be emitted by all members of the species, the individual odor is characteristic of the microbial action specific to each maternal diet.

The effectiveness with which infant rats initiate suckling is influenced by a pheromone from the mothers ventral skin which offers an olfactory or gustatory cue to the infant rat for orientation and attachment to the nipples (39). Weanling rats feed and explore in areas containing residual olfactory cues in preference to clean areas (33), and it is suggested that these cues can play a role in directing weanlings to their first meals of solid food in the natural environment.

It is apparent that many factors bearing on reproduction are influenced by pheromones. When considered together with Calhoun's (15) findings of decreased production in overcrowded populations, then the importance of adequate ventilation, hygiene, and population density is emphasized.

B. Temperature and Humidity

It is recommended by the Institute of Laboratory Animal Resources (ILAR) (62) that the room temperature for rats should not fluctuate more than $\pm 2°F$ around the mean within a range of 70°–80°F. Relative Humidity (RH) should be kept within a range of 40 to 70%.

According to Lane-Petter and Pearson (51), for every square meter of a well-stocked room of rats, between 1.5 and 2.0 kg of water will be generated each 24 hr. This figure assumes that no condensation or absorbtion takes place. To remove this amount of water vapor requires something between ten and twenty air changes per hour.

The climate within the cage is usually warmer and more humid due to decreased ventilation. This is particularly true if cage filters are used (see this volume, Chapter 8).

C. Light

The role of lighting in a breeding unit has been reviewed by Lane-Petter and Pearson (51). They recommend a lighting cycle of 12 hr light and 12 hr darkness. Glare can be avoided by the use of diffusers or fluorescent lighting. Intensity, while avoiding extremes, should consider ease of working for the human staff. The color balance of the light should approach that of sunlight, since this will permit observation of the condition of eyes and extremities without the addition of extraneous color.

D. Sound

Noise is known to reduce maternal care activity in rats (18) with consequent lowering of production. Some commercial breeders use recorded music or radios, to provide a background level of sound, which may effectively reduce the impact of other noises.

A problem in institutions raising rats occurs with the need to test fire alarms. One attempt to circumvent this led to the development of an alarm which, while audible to humans, was not significantly audible to rats. Gamble (34) studied this "silent" alarm and compared it to the conventional one. He found that a significant alteration occurred in the cyclicity of rats exposed to conventional fire alarms, but that no alteration occurred with the use of the "silent" alarm or no alarm.

VI. TEMPORAL ASPECTS

With the advent of SPF animals and the accompanying technology, it soon became apparent that the question of "*will* a colony become contaminated with pathogens was more realistically phrased as *when* will a colony become contaminated?" Experience soon showed that current technology permitted a high degree of confidence in freedom from contamination for only 1–2 years, after which confidence decreased. Therefore, periodic reestablishment of breeding colonies has become common practice.

If one regards a breeding unit as independent (in terms of contamination) from other such units, then the probability of spreading contamination can be decreased by routinely eliminating the colony after a prescribed period. This approach differs significantly from the "immortal" colony in which a proportion of the breeding stock is replaced each week but the colony per se is never eliminated. An effective program would

Table VI

Sequence and Duration of Events for Elimination and Reestablishment of a Rat Colony

Event	Time (weeks)
Clean and disinfect room, cages and equipment	2
Set up 20% of capacity each week	5
Most economical breeding life (*vide infra*)	28
Total time for life of colony	35

be to rigorously monitor the animals at periodic intervals with an effective microbial quality control program. Colonies need not be phased out unless, and until, contamination by pathogens is demonstrated. If a fixed renewal schedule is desired, however, it may be programmed as follows.

For example, if five units exist, one unit would be eliminated and replacement started each 7 weeks. Fewer than 5 rooms available results in unacceptably wide fluctuations in production, and in such cases this system is not recommended. If larger numbers of rooms are available then the set-up of each room can be increased anywhere from the 20% to the full 100% of the room's capacity (see Table VI).

A determination of the most economical life of a breeding unit can be made by assuming that the cost of operation of such a unit is constant for each day of its life. Data from a significant number of breeding units is collected and averaged and the cumulative production divided into the cumulative number of days needed to yield that production.

Table VII shows data from an actual breeding colony and Fig. 7 shows the number of days to produce one young plotted against time in weeks. Table VII shows that both the unselected and the selected breeders produce young at the lowest unit cost at about 17 weeks from initial mating, but the selected breeders produce young at 29 or 34 weeks lower than the lowest cost ever attained by the unselected breeders. What then

Table VII

Production Data for Unselected Breeders and for Breeders Selected for High Reproduction

Days since mating	Unselected breeders		Selected breeders	
	Average cumulative production	Average number of days to produce one young[a]	Average cumulative production	Average number of days to produce one young[a]
42	3.5	11.86	3.3	12.73
84	15.1	5.55	14.5	5.79
119	26.2	4.54	30.2	3.94
182	36.8	4.94	42.8	4.25
203	39.3	5.16	47.5	4.27
238	42.7	5.57	52.9	4.49

[a] Average number of days to produce one young = days since mating/average cumulative production.

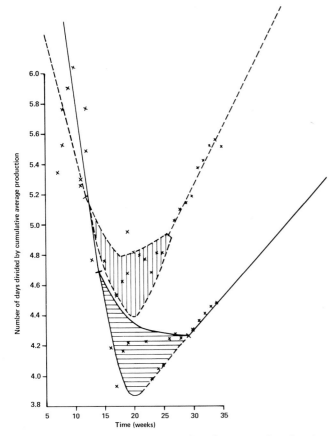

Fig. 7. Number of days for the production of one young for selected and unselected breeders. Dashed line and vertical hatching, unselected breeders; solid line and horizontal hatching, selected breeders.

Table VIII

Values of Q (Index of Productivity)[a]

Days since mating	Unselected breeders	Selected breeders
42	8	8
84	18	17
119	22	26 {Mean = 26.6, S.D. = 6.5}
182	22	23
203	19	23
238	18	22

[a] Q = (cumulative number of young produced × 100)/cumulative number of days to most recent weaning from first mating.

A mathematically simple technique for estimating productivity has been devised by Lane-Petter et al. (46), who propose an index (Q) which is the number of young produced per hundred days. Use of this index in measuring such factors as dietary influence on productivity has been reported by Porter et al. (69). Dinsley (23) also used the Q-index in a study of the influence of female weight on productivity. A useful nomogram for calculating Q in breeding ratios of 1 ♂ × 1 ♀ to 1 ♂ × 6 ♀ has been prepared by Lane-Petter and Pearson (51).

Table VIII shows the values of Q for the two colonies previously presented. From this can be seen that differences greater than 1.0 are apparent by day 119. To approach the production of selected breeders, animals from either colony to be selected as future breeder should come from a female or breeding unit having a value of Q = 32.0 or greater by the seventeenth week (119 days) since first mating. The value of Q = 32.0 was derived by adding the product of the standard deviation and Z* = 0.842 to the mean, to yield the value of Q at the 80th percentile (24). Hence, all future breeders will most likely then be selected from the offspring of that 20% of the population who are most productive. The use of the 80th percentile is arbitrary. If fewer breeders are required, then a higher percentile can be calculated; if more breeders are required then a lower percentile value of Q would have to be calculated. The selection of future breeders from third or subsequent litters has been recommended by Falconer (28).

is the optimum length of the breeding life of a female rat and how can one select for maximum production? Figure 7 shows that in both selected and unselected breeders there is a large area of uncertainty (hatched areas), but this uncertainty changes to a definite increase in unit costs at about the twenty-ninth and twenty-sixth weeks, respectively. The selected breeders would, therefore, be expected to have a useful breeding life approximately 10% longer than the unselected. This increase will, in turn, reduce daily cost of maintaining a breeding unit, since the expense involved in maintaining future breeding animals would be amortized over a long period.

The answer to the question of the optimum breeding life of a female rat is determined by estimating the time at which the curve of number of days to produce one young is committed to an increase beyond the "area of uncertainty." In the selected breeder example this would be 29 weeks. A similar technique in more mathematical detail has been provided by Gridgeman and Taylor (37). Since selection of breeding stock and environmental change will influence the curve, regular reassessment is advised.

VII. MANAGEMENT

A. Mating Systems

A number of excellent reviews of breeding and production techniques are to be found in the literature. The basic elements

*The Z score measures how many standard deviations a raw score (e.g., 80th percentile) is from the mean (24).

are thoroughly covered by Falconer (28) and a more recent discussion of production techniques is provided by Lane-Petter and Pearson (51).

B. Monogamous Mating

Rats may be monogamously mated by permanently placing one male and one female together for their entire breeding life. Young pups are removed from the cage at weaning.

This method is commonly used in the production of inbred strains but less commonly so in outbred strains. Advantages include the ease of record keeping, particularly productivity, and the maximal opportunity for postpartum matings, although this point may have been somewhat overemphasized in the past. In the two colonies previously presented, both were monogamously mated. The unselected breeders showed 43% postpartum conceptions leading to live births and the selected breeders showed only 53% such postpartum conceptions. It must be remembered that in monogamous matings one male must be supported for each female. If productivity be expressed on the basis of each breeding *animal,* then it can be seen that the productivity may be somewhat lower than may be obtained from harem matings.

C. Polygamous Mating

In a study of factors influencing productivity of SPF Wistar descendant rats, Masson and Gomond (54) showed that mating one male to four females and removing the pregnant female so that she might litter in isolation was more efficient than monogamous mating. Similar conclusions were reached by Rapp and Hedrich (70). A commercial breeder reports a ratio of 1 : 4 for random-bred rats (1). Since rats do not usually nurse well in a communal setting, it is best to isolate the visibly pregnant female into an isolation-maternity cage. The use of numbers tattooed on the ears simplifies record keeping and aids in returning the female to the appropriate harem after her litter is weaned. Polygamous matings are commonly used in random-bred strains but may also be used with inbred strains.

D. Production Methods

The production of large numbers of inbred rats is best achieved by the use of multistage breeding (51). The use of gnotobiotic techniques as a means of separating the microflora from the gene pool has also necessitated the use of multistage breeding for the production of large numbers of both inbred and random-bred rats (6). Essentially, multistage breeding uses a small colony to provide breeders for a larger colony, which in turn provides breeders for an even larger colony, and so on

until a colony of sufficient size to produce the desired number of rats is established.

E. Inbred Colonies

A nucleus colony can be established by monogamous matings of the offspring of a single pair. (A male and a female of different litters from a single pair are just as much brother and sister as are littermates.) Replacement breeders in the nucleus colony *must* consist of animals from one monogamous mating only, for each generation. In this manner the pedigree will be maintained via a single line system (28) which is illustrated in Fig. 8.

Animals produced by the nucleus colony are monogamously brother and sister mated to form an expansion colony. The offspring from the expansion colony may be either used by the investigator or used as breeding stock for a production colony. In the production colony animals from the expansion colony may be randomly and polygamously mated without significantly influencing genetic fixation (1). However, all offspring from the production colony should be destined for the investigator and *none* should be used for continued breeding. A system of identification of the nucleus, expansion, and production colony cages by means of colored cage cards has been described by Lane-Petter and Pearson (51). This system is particularly useful when the colonies are maintained in the same room.

F. Random-Bred Colonies

In the random-bred colony the major genetic concern is to maintain the degree of heterozygosity in the colony. In a closed

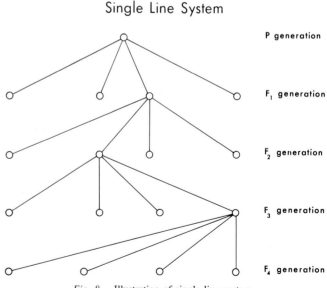

Single Line System

P generation

F_1 generation

F_2 generation

F_3 generation

F_4 generation

Fig. 8.   Illustration of single line system.

colony some degree of inbreeding is unavoidable, but efforts should constantly be exercised to minimize inbreeding. Certainly, brother and sister matings are to be avoided. The rate of inbreeding depends on the number of parents whose offspring will be used for breeding and the rate of inbreeding can be calculated from the formula $100/8M + 100/8F =$ percent inbreeding increase in each generation where M is the number of male parents and F the number of female parents (28). Falconer (28) recommends that there be at least twenty parents of each sex (1.25% inbreeding increase) and 50 parents of each sex (0.50% inbreeding increase) if the total size of the colony permits. Eggenberger (26) has shown that population size and the method of selecting breeder replacements are decisive in maintaining the genetic structure of random-bred colonies.

In order to assure that future breeders are selected from the entire population various rotation systems have been proposed (1,7,68), but it should be borne in mind that Eggenberger (26) has shown that different rotation systems produce different subcolonies. Figure 9 shows a simple rotation method in which ABCD represent four numerically equal segments of the breeding unit. Identification can be made conveniently by using the appropriate letter as a prefix to the cage card number. In use, for example, a pregnant female from group A with a value of $Q = 32$ (*vide supra*) by the seventeenth week, might be selected as a parent of future breeders. The males from her litter would go to be mated with females of group B to constitute a new mating in group C. The females from her litter would be mated with males from group D to form a new mating in group B.

Rotation Method

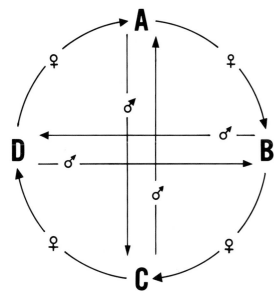

Fig. 9. Simple rotation method of breeding.

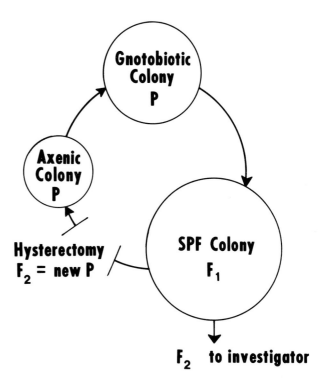

Fig. 10. Multistage colonies for rats produced from axenic ancestors.

Multistage breeding also can be applied if it is desired to separate the rat from its microflora by use of hysterectomy and gnotobiotic techniques. As illustrated in Fig. 10 females are selected from the SPF colony, undergo hysterectomy and their young, the F_2 generation become the P generation of a new cycle. This P generation is raised in the axenic state to allow confirmation of the germfree status. Next, the axenic rats are associated with pure cultures to establish the gnotobiotic P generation. If the true gnotobiotic state is to be maintained the colony must be housed within an isolator system. If, however, this is not necessary then considerable savings can be made by housing the colony within a filtered air, laminar flow system (9). The animals of the P generation are bred to produce the F_1 generation who then become the breeding stock of the SPF colony. Their young, the F_2 generation, are distributed to the investigator and some of the F_1 generation are selected as parents of the new P generation. A technique for the hysterectomy, which reduces the risk of contamination of newly delivered fetuses by infected maternal tissue, has been described by Lane-Petter (50) and developed by Erfle (27). A method for planning hysterectomies for the establishment of SPF colonies of inbred rats has been developed by Parrot *et al.* (67). Foster females were mated on days 1–3, and donor females were mated on day 4. Hysterectomies were then performed on the donor females on day 26 (22 days after mating).

The costs of production incurred by a breeding operation continue to be of importance in both industry and academia. Methods of conducting time and motion studies in breeding colonies and such an analysis at the Laboratory Animals Centre, England, have been published by Bleby and Porter (12).

G. Cross-Fostering

Techniques for producing rats of increased uniformity, increased productivity, and of a sex ratio adjusted to match the anticipated demand have been developed by Lane-Petter *et al.* (47,49) and Bowtell *et al.* (13). Essentially, cross-fostering is done by sexing the young of all litters born in one day, when the young are 2 days old. Unisexual litters of 14 pups are assembled and fostered onto one of the lactating females. Litters are assembled to reflect the anticipated demand for males and females. Usually this results in discarding a number of female pups. Females not selected as lactating mothers are returned to the male where they commonly come into estrus within 3 or 4 days. The time spent on cross-fostering is less than the time otherwise spent on weaning when cross-fostering is not practiced. It has been shown that of rats thus raised, 80% will achieve a predicted body weight within a range of 5% either way of that predicted weight. Meehan *et al.* (58) have shown that cross-fostering is superior to natural litters for the production of uniformity of body weight. Brown *et al.* (14) showed a uniformity of response to follicle-stimulating hormone in females from cross-fostered litters, and Sharpe *et al.* (75) showed that variation in both ovarian response to gonadotropin and in barbiturate sleeping time was greater in natural litters than in cross-fostered litters. However, bisexually reared females produce larger litters than unisexually raised females (76), and bisexually raised females produce litters in which males predominate (77).

ACKNOWLEDGMENTS

The author wishes to thank Joan B. Maxwell and the staff of the Instructional Resources Center of the State University of New York, Dehli, for their assistance in preparation of the manuscript and of the figures, respectively. The support extended by Wilbur M. Farnsworth and Kenneth C. Pyle during the preparation of the manuscript is gratefully acknowledged.

REFERENCES

1. Anonymous (1963). Breeding and mating techniques for laboratory mice and rats. *Charles River Dig.* **2**, 1-4.
2. Anonymous (1965). Bedding for laboratory rodents. *Charles River Dig.* **4**, 1-4.
3. Anonymous (1970). "Growth Curve and Price List." Holtzman Company, Madison, Wisconsin.
4. Aron, C. (1973). Phéromones et régulation de la durée du cycle oestral chez la ratte. *Arch. Anat. Histol. Embryol.* **56**, 209-216.
5. Asdell, S. A. (1964). "Patterns of Mammalian Reproduction," 2nd ed. Cornell Univ. Press, Ithaca, New York.
6. Baker, D. E. J. (1967). Gnotobiotics applied to the standardization of laboratory mice. *In* "International Symposium on Laboratory Animals" (R. H. Regamey *et al.*, eds.), pp. 31-42. Karger, Basel.
7. Barber, B. R. (1972). Experiences in large scale rat and mouse breeding. *J. Inst. Anim. Tech.* **23**, 146-154.
8. Bartos, L. (1977). Vaginal impedance measurement used for mating in the rat. *Lab. Anim.* **11**, 53-55.
9. Beall, J., Tooring, E., and Runkle, R. (1971). A laminar flow system for animal maintenance. *Lab Anim. Sci.* **21**, 206-212.
10. Beily, J. S. (1939). Hydrogen ion concentration changes in the vaginal fluid of the rat during an estrous cycle. *Endocrinology* **25**, 275-277.
11. Blandau, R. J., Boling, J. L., and Young, W. C. (1941). The length of heat in the albino rat as determined by the copulatory response. *Anat. Rec.* **79**, 453-463.
12. Bleby, J., and Porter, G. (1969). Timed assessment of the duties of animal technicians. *J. Inst. Anim. Tech.* **20**, 12-24.
13. Bowtell, C. W., and Lane-Petter, M. E. (1968). Breeding and cross fostering technique of CFE rats. *J. Inst. Anim. Tech.* **19**, 185-191.
14. Brown, P. S., and Wyatt, A. C. (1970). Litter differences in cross fostered rats. *Lab. Anim.* **4**, 241-243.
15. Calhoun, J. B. (1962). Population density and social pathology. *Sci. Am.* **206**, 139-148.
16. Chow, B. F., and Augustin, C. E. (1965). Induction of premature birth in rats by a methionine antagonist. *J. Nutr.* **87**, 293-296.
17. Cisar, C. F., and Nelson, G. (1967). Effects of frequency of cage cleaning on rat litters prior to its weaning. *Lab. Anim. Care* **17**, 215-217.
18. Cumming, C. N. W., and Cumming, E. L. W. (1955). Factors capable of affecting experimental results in laboratory animals. *Carworth Farms Q. Lett.* No. 39, pp. 1-4.
20. Cumming, C. N. W., and Baker, D. E. J. (1963). The use of the flexible film isolator in the large scale production of small laboratory animals. *Lab. Anim. Care* **13**, 624-628.
21. Cumming, E. L. W., and Cumming, C. N. W. (1957). The laboratory rat. *Carworth Farms Q. Lett.* Nos. 47-51 and 76.
22. Cuthbertson, W. F. (1957). Nutrient requirements of rats and mice. *Proc. Nutr. Soc.* **16**, 70.
23. Dinsley, M. (1966). The influence of male weight on productivity in the C57BL/LAC inbred mouse strain. *Lab Anim. Care* **16**, 322-326.
24. Dixon, W. J., and Massey, F. J. (1969). "Introduction to Statistical Analysis," 3rd ed. McGraw-Hill, New York.
25. Eckstein, B., Golan, R., and Shani, J. (1973). Onset of puberty in the immature female rat induced by 5α–Androstane-3B,17B-diol. *Endocrinology* **92**, 941-945.
26. Eggenberger, E. (1973). Model population zur Beurteilung von Rotationssystemen in der Versuchstierzucht. *Z. Versuchstierkd.* **15**, 297-331.
27. Erfle, V. (1976). A method for obtaining germfree rodents by "wet-hysterectomy." *Z. Versuchstierkd.* **18**, 313-317.
28. Falconer, D. S. (1962). Breeding. *In* "Notes for Breeders of Common Laboratory Animals" (G. Porter and W. Lane-Petter, eds.), pp. 111-125. Academic Press, New York.
29. Farris, E. J. (1941). Apparatus for recording cyclical activity in the rat. *Anat. Rec.* **81**, 357-362.
30. Farris, E. J. (1949). Breeding and parturition in the albino rat. *In* "The Rat in Laboratory Investigation" (E. J. Farris and J. Q. Griffith, eds.), 2nd ed., pp. i-iii. Hafner, New York.
31. Festing, M., and Bleby, J. (1968). A method for calculating the area of breeding and growing accommodation required for a given output of small laboratory animals. *Lab. Anim.* **2**, 121-129.

32. Friedmann, J. C., Mahovy, G., and Tuffrau, H. (1968). Quelque caractéristiques de la physiolgie sexuelle chez les rongeurs de laboratoire. *Exp. Anim.* **1**, 111-117.

33. Galef, B. G., and Heiber, L. (1976). Role of residual olfactory cues in the determination of feeding site selection and exploration patterns of domestic rats. *J. Comp. Physiol. Psychol.* **90**, 727-739.

34. Gamble, M. R. (1976). Fire alarms and oestrus in rats. *Lab. Anim.* **10**, 161-163.

35. Gaunt, I. F., and Lane-Petter, W. (1967). Vitamin K deficiency in S. P. F. rats. *Lab. Anim.* **1**, 147-149.

36. Gawienowski, A. M., DeNicola, D. B., and Stacewicz-Sapuntzakis, M. (1976). Androgen dependance of a marking pheromone in rat urine. *Horm. Behav.* **7**, 401-405.

37. Gridgeman, N. T., and Taylor, J. M. (1972). Maximization of long term productivity in a rat colony. *Lab. Anim.* **6**, 203-206.

38. Heinecke, H., Schussling, G., and Fuchs, A. (1961). Zur Bestimmung des Beginns der Trachtigkeit bei der Ratte. *Z. Versuchstierkd.* **1**, 107-109.

39. Hofer, M. A., Shair, H., and Singh, P. (1976). Evidence that maternal ventral skin substances promote suckling in infant rats. *Physiol. Behav.* **17**, 131-136.

40. Holinka, C. F., and Carlson, A. D. (1976). Pup attraction to lactating Sprague-Dawley rats. *Behav. Biol.* **16**, 489-505.

41. Kali, J., and Amir, S. (1973). The effect of mating male rats at 35 or 80 days of age on pup production and maternal weight. *Lab. Anim.* **7**, 61-64.

42. Kappel, H., Nelson, J. B., and Weisbroth, S. H. (1974). Development of a screening technic to monitor a mycoplasmafree Blu:(LE) Long-Evans rat colony. *Lab. Anim. Sci.* **24**, 768-772.

43. Kessler, G. (1974). Eine Methode zur Provokation von Ejakulation bei der Ratte. *Z. Versuchstierkd.* **16**, 73-74.

44. King, M. G., Pfister, H. P., and DiGiusto, E. L. (1975). *Behav. Biol.* **13**, 175-81.

45. Kunze, B., and Marwitz, T. (1967). Tilgung einer *Salmonella enteriditis*-infektion in einem Rattenzuchbestand. *Z. Versuchtierkd.* **9**, 257-259.

46. Lane-Petter, W., Brown, A. M., Cook, M. J., Porter, G., and Tuffery, A. A. (1963). Measuring productivity in breeding small animals. *Nature (London)* 183-339.

47. Lane-Petter, W., Lane-Petter, M. E., and Bowtell, C. W. (1968). Intensive breeding of rats. I. Crossfostering. *Lab. Anim.* **2**, 35-39.

48. Lane-Petter, W., and Lane-Petter, M. E. (1970). Dried sugar beet pulp as bedding. *J. Inst. Anim. Tech.* **21**, 31-32.

49. Lane-Petter, W., and Lane-Petter, M. E. (1971). Toward standardized laboratory rodents: The manipulation of rat and mouse litters. *In* "Defining the Laboratory Animal" (Schneider, H. A., ed.), pp. 3-10. Natl. Acad. Sci., Washington, D. C.

50. Lane-Petter, W. (1971). Dry and wet hysterectomy. *Z. Versuchstierkd.* **13**, 126-127.

51. Lane-Petter, W., and Pearson, A. E. G. (1971). "The Laboratory Animal—Principles and Practice." Academic Press, New York.

52. Leon, M. (1975). Dietary control of maternal pheromone in the lactating rat. *Physiol. Behav.* **14**, 311-319.

53. Long, J. A., and Evans, H. M. (1922). The estrous cycle of the rat and its associated phenomenon. *Mem. Univ. Calif.* **6**, 1-148.

54. Masson, P., and Gomond, P. (1971). Modalités de production du rat Wistar, A. F dams unéunite E. O. P. S. *Exp. Anim.* **4**, 73-79.

55. May, D. (1969). Synchronization of estrus in the rat. *J. Inst. Anim. Tech.* **20**, 155-161.

56. May, D., and Simpson, K. (1971). An improved method for synchronizing estrus in the rat. *J. Inst. Anim. Tech.* **22**, 133-39.

58. Meehan, A. P., Wyat, A. C., Cowley, L. G., and Brown, D. S. (1971). Uniformity of cross-fostered rats: Variability of body weight with size of foster groups. *J. Inst. Anim. Tech.* **22**, 155-160.

60. Mulder, J. B. (1974). Bedding selection by rats. *Lab. Anim. Dig.* 27-30.

61. Mundy, L. A., and Porter, G. (1969). Some effects of physical environment on rats. *J. Inst. Anim. Tech.* **20**, 78-81.

62. National Academy of Sciences (1969). "Rodents: Standards and Guidelines for the Breeding Care and Management of Laboratory Animals." Subcommittee on Rodent Standards, NAS, Washington, D. C.

63. Nicholas, J. S. (1949). Experimental methods and rat embryos. *In* "The Rat in Laboratory Investigation" (E. J. Farris and J. Q. Griffith, eds.), 2nd ed., p. 542. Hafner, New York.

64. Nolan, G. A., and Alexander, J. C. (1966). Effects of diet and type of nesting material on the reproduction and lactation of the rat. *Lab. Anim. Care* **16**, 327-36.

65. Norris, M. L., and Adams, C. E. (1976). Incidence of pup mortality in the rat with particular reference to nesting materials, maternal age and parity. *Lab. Anim.* **10**, 165-169.

66. Park, T. (1962). Beetles, competition and populations. *Science* **138**, 1369-1375.

67. Parrot, R. F., and Eveleight, J. R. (1970). A method of planning hysterectomies and its use in establishing an SPF colony of inbred mice and rats. *J. Inst. Anim. Tech.* **21**, 161-166.

68. Poiley, S. M. (1960). A systematic method of breeder rotation for non-inbred laboratory animal colonies. *Proc. Anim. Care Panel* **10**, 159-166.

69. Porter, G., Lane-Petter, W., and Horne, N. (1963). Assessment of diets for mice. *Z. Versuchstierkd.* **2**, 171-182.

70. Rapp, K., and Hedrich, H. J. (1974). Reproductionsleitungen und deren phaenotypischer Zusammenhaenger bei Han: Wistar Ratten unter verscheidenen Haltungsbedingungen. *Z. Versuchstierkd.* **16**, 85-101.

71. Richards, D. B., and Stevens, A. (1974). Evidence for marking with urine by rats. *Behav. Biol.* **12**, 517-523.

72. Ritchie, D. H., and Humphrey, J. K. (1970). Some observations on the mating of rats. *J. Inst. Anim. Tech.* **21**, 100-105.

73. Ross, M. (1961). Practical applications of the Whitten effect. *J. Anim. Tech. Assoc.* **12**, 21-29.

74. Russell, F. C. (1948). Diet in relation to reproduction and the viability of the young. Part I. Rats and other laboratory animals. *Commonw. Bur. Anim. Nutr., Tech. Commun.* No. 16.

75. Sharpe, R. M., Morris, A., Wyatt, A. C., and Brown, P. S. (1972). Variability of response in cross-fostered rats. *Lab. Anim.* **6**, 225-234.

76. Sharpe, R. M., Morris, A., and Wyatt, A. C. (1973). The effect of the sex of littermates on the subsequent behavior and breeding performance of cross-fostered rats. *Lab. Anim.* **7**, 51-59.

77. Sharpe, R. M., and Wyatt, A. C. (1974). Sex ratio and weaning body weight differences in the offspring of unisexually and bisexually-reared cross-fostered rats. *Lab. Anim.* **8**, 61-69.

78. Smith, P. C., Stanton, J. S., Buchanan, R. D., and Tanticharoenyos, P. (1968). Intestinal obstruction and death in suckling rats due to sawdust bedding. *Lab. Anim. Care* **18**, 224-228.

80. Szabo, K. T., Free, S. M., Birkhead, H. A., and Gay, P. (1969). Predictability of pregnancy from various signs of mating in mice and rats. *Lab. Anim. Care* **19**, 822-825.

81. Tuchmann-Duplessis, H., Hiss, D., Mottot, A., and Rosner, I. (1975). Effects of prenatal administration of acetylsalicylic acid in rats. *Toxicology* **3**, 207-11.

82. Van der Waaij, D., and Van Bekkum, D. W. (1967). Isolation facilities for rat breeding: The efficiency of an isolation unit. *Lab. Anim. Care* **17**, 532-541.

83. Warner, R. G. (1962). Nutrient requirements of the laboratory rat. N.A.S.—N.R.C., Publ. **990**.

84. Weihe, W. H. (1965). Traechtigkeit und Laktation der Ratte usw. *Z. Versuchstierkd.* **7,** 1–22.

85. Weihe, W. H. (1971). The significance of the physical environment for the health and the state of adaptation of laboratory animals. *In* ''Defining the Laboratory Animal'' (Schneider, H. A., ed.), pp. 353–375. Natl. Acad. Sci., Washington, D. C.

86. Whitten, W. K. (1956). Modification of the oestrus cycle of the mouse by external stimuli associated with the male. *J. Endocrinol.* **13,** 399–404.

87. Wilson, E. O. (1975). ''Sociobiology: The New Synthesis.'' Belknap Press, Cambridge, Massachusetts.

88. Wolf, von D., and Horn, J. (1977). Untersuchungen zur elektrischen Leitfaehigkeit der Vaginalschleimhaut der Ratte in Verlauf des Sexualzyklus. *Z. Versuchstierkd.* **19,** 36–39.

89. Wostmann, B. S. (195). Nutrition of the germfree mammal. *Ann. N. Y. Acad. Sci.* **78,** 175–82.

90. Young, P. M. (1976). Effects of olfactory stimuli upon the unit activity of antidromically identified neurosecretory neurones in the paraventricular nuclei of the rat. *J. Reprod. Fertil.* **46,** 493–494.

91. Young, W. C., Boling, J. L., and Blandau, R. J. (1941). *In* ''Patterns of Mammalian Reproduction'' (S. A. Asdell, ed.) 2nd ed., p. 333. Cornell Univ. Press, Ithaca, New York.

Chapter 8

Housing to Control Research Variables

Henry J. Baker, J. Russell Lindsey, and Steven H. Weisbroth

I. INTRODUCTION

Experimental reproducibility has been the accepted axiom of scientific research since the time of Claude Bernard (23). The early workers who established the rat as a laboratory species made serious efforts toward this objective through innovative husbandry practices, facility design, genetic standardization, and control of enzootic infections (see Chapter 1, this volume). Subsequently, the progress toward controlling variables which affect the biologic responses of laboratory rodents was slow until the 1950s.

In the decade following World War II, there was unprecedented expansion of animal research under conditions of rapidly increasing sophistication. While the general idea of subtle variables in laboratory animal experiments had long been recognized, the increasing sophistication in research methodology forced a more realistic appreciation of their total impact. There was pressure from the scientific community to reduce the effects of indigenous diseases (114,204). Major advances resulted through the application of gnotobiotic principles to large-scale rodent production (79,82,83,216). An essential corollary was the development of facilities and husbandry methods capable of maintaining rats free of the common infectious diseases (19,30,44,80,125,144,211). Concurrently, there was gradual accumulation of compelling evidence that numerous other (i.e., noninfectious) environmental variables can have profound effects on biologic response (127,153,232).

Despite the remarkable progress made to date toward refinement of the laboratory rat as a standardized research subject, biological variability in this species is all too frequently a continuing, important reality. The sheer complexity of a mammal such as the rat makes it unlikely that it can be standardized to the degree desired by exacting scientists. Nevertheless, sophisticated inquiry demands that all reasonable measures be taken to control research variables. As a first step toward coping with this complex issue, it is useful to adopt the concept that the biologic response of the laboratory rat is an expression of multiple genetic and environmental effects at each point from zygote to death (153) (Fig. 1). While not entirely new, this concept recognizes the importance of complex interactions between genetic and environmental factors. It emphasizes the need to define laboratory rats in terms of *both* genetics and environment (physical, chemical, and microbial factors) and to report these crucial data in scientific publications.

The purpose of this chapter is to outline those considerations whereby maximum experimental reproducibility can be achieved in research using the rat. Since no single management program can hope to meet the needs of every research project, the approach here will be to emphasize principles and concepts which investigators can apply to their own unique conditions.

II. PHYSICAL FACTORS AFFECTING BIOLOGIC RESPONSE

Contemporary husbandry practices have evolved largely by trial and error in an attempt to provide optimal conditions for the health and well being of laboratory rats. The need for efficient, economical animal care procedures also has influ-

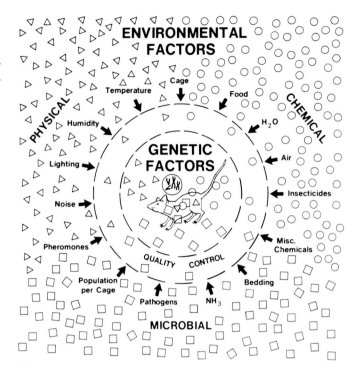

Fig. 1. Conceptual view of the laboratory rat. Biological response results from effects of genetic and multiple environmental factors (physical, chemical, and microbial) at all points from zygote to death. Quality control maintains genetic purity while preventing, minimizing, or standardizing effects of environmental factors. [After Lindsey *et al.* (153.)]

enced this developmental process. However, it is now clear that virtually all factors in the environment where laboratory rats are housed exert potentially important influences on biologic response (153,181,232). Thus, it follows that husbandry practices must be *selected* and *standardized* to give the highest possible degree of biological control.

A. Facilities

The design and proper maintenance of physical facilities greatly influence the level of excellence in husbandry practice that can be achieved. Space for housing rats should be located in a structurally sound, vermin-proof (especially wild rodent-proof) building physically separate from laboratories and areas used primarily for human occupancy. A high degree of security against unnecessary traffic must be achieved. Functions within the animal facility which usually must be accommodated in addition to housing include (1) animal receiving and quarantine; (2) cage washing and sanitation; (3) diet storage and preparation; (4) supply and equipment storage; (5) waste disposal; (6) personnel lockers, showers, sinks, and toilets; (7) specialized laboratories (such as surgery, necropsy, diagnostic procedures, radiography, infectious disease and biohazard containment); and (8) administrative activities. The arrangement of space assigned to these functions should optimize traffic flow to limit cross-contamination between clean and soiled materials. The essential characteristics of various classes of barriers designed to control microbiological contamination have been described recently (127) and are discussed in Section IV.

Room dimensions should accommodate standard racks and caging units such as those described below. Housing areas should be constructed with monolithic wall, floor, and ceiling finishes that minimize accumulation of debris, facilitate microbial decontamination, and discourage harborage for vermin. Rooms should be windowless, with light provided by fluorescent fixtures. Additional principles of facilities design can be found in the "Guide for the Care and Use of Laboratory Animals" (165), and in a number of recent publications (30,44,119,125,159,216).

B. Cage Design

The cage constitutes the immediate environment of the rat (which differs substantially from the room environment) (5,24,32,168,169,252,253). Two basic cage styles are in popular use for housing rats: solid bottom (Fig. 2) and wire bottom cages (Fig. 3). The former, often referred to as "shoe box" cages, are rectangular boxes with solid floor and walls, fabricated of metal or plastic. If metal is used, stainless steel is preferred because of its durability and chemical inertness. Galvanized steel cages are less durable and result in ingestion of zinc (a serious variable in studies of zinc deficiency) by rats that gnaw at cage parts or lick metal particles from their fur while grooming. For most research applications plastic cages

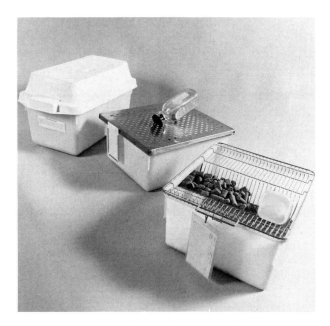

Fig. 2. Solid bottom plastic rat cages showing wire and perforated metal tops, food and water bottle wells, filter cap, and identification cards.

are preferred over metal because of economy, seamless construction, durability, and chemical inertness. Materials such as polycarbonate (clear) or polypropylene (translucent) are very durable, with high impact and heat resistance. Less durable, inexpensive plastics such as polystyrene may be used in instances where contamination by infectious agents or hazardous materials makes disposable cages desirable. Clear plastics facilitate observation of the occupants, but deny them the security of an opaque enclosure.

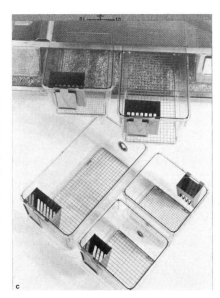

Fig. 3. (a) Rack of wire bottom metal rat cages (b) with food hopper, water bottle, and litter pan. (c) Suspended wire bottom cages are also constructed with plastic sides.

The shoe box-type enclosure is completed by a cover which can be constructed of perforated sheet metal or wire (Fig. 2). In some instances shoe box cages are suspended from a rack so that a shelf forms the top (Fig. 4). In all cases the top should fit tightly to prevent the occupants from gnawing edges of the cage. The top should be locked in place to prevent displacement and escape by larger rats (Fig. 5).

Intracage ventilation of shoe box cages is markedly influenced by design of the top. Wire rod tops provide the greatest freedom of air movement, but even this type can restrict air exchange by as much as 25% (212). This type also may serve to hold pelleted food and a water bottle, further restricting ventilation (Fig. 2). If cages of this type are fitted with fiber canopies, intracage ventilation will be even further compromised (Figs. 2 and 4). The principle of individual cage isolation by use of filter "caps" or "bonnets" was developed by Kraft to control transmission of viral diarrheas in infant mice (141). Cage filters are very effective in controlling these infections, presumably by restricting the exchange of particulate fomites between cages (216). Evidence that cage filters have a beneficial effect in controlling infectious diseases of rats

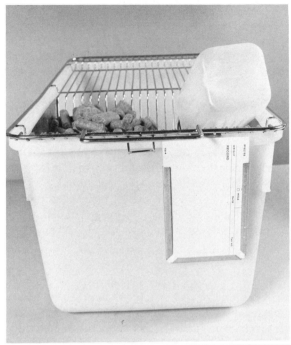

Fig. 5.   Solid bottom plastic cages and wire top with locking clips.

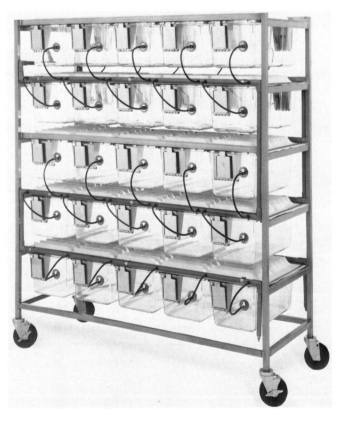

Fig. 4.   Rack of suspended clear plastic solid bottom cages with filter caps and automatic watering.

is lacking (although they probably are advantageous in certain circumstances). The advantages of using filters should be weighed carefully against the possible disadvantages. For example, murine respiratory mycoplasmosis (MRM) is potentiated by ammonia (32). Since cage filters are known to increase the accumulation of intracage ammonia, they may actually have a deleterious net effect.

Absorbent material called "bedding" is normally used in shoe box cages to absorb urine and provide nesting material. Optimal characteristics of bedding include high capacity to absorb moisture without dehydration (107) or other injury (34) of newborn pups, dust-free, nonabrasive, free from noxious chemicals or pathogenic agents, economical, and convenient disposal. A variety of cellulose products, particularly wood particles, are commonly used for this purpose.

Bedding constitutes a potentially rich source of biologic variability because of the close, prolonged contact of rats with such material and the likelihood of their ingesting the material. Wood products also may be contaminated with excreta from wild rodents or birds, and must be sterilized prior to use as bedding. It also must be remembered that such products may be contaminated by exogenous chemicals such as preservatives, defoliants, pesticides, paints, and aflatoxins (193). Avoidance of such contaminants depends largely upon the reliability of the supplier, but product suitability can be assured only by specific testing (i.e., systematic monitoring for contaminants may be necessary).

Endogenous constituents of wood also have been implicated as complicating factors. Cedar (68,189,232) and certain other soft wood (231) beddings are known to induce hepatic microsomal enzymes. The active ingredients in cedar are the volatile hydrocarbons, cedrene, and cedrol (106,236). Such chemicals have not been identified in hardwoods to date, which probably accounts for the increasing popularity of hardwood bedding materials for rodents in recent years. Recent evidence has suggested that aldehydes and lignins emitted by some woods during processing may be carcinogenic (1,207).

Specific research (e.g., radionuclides and toxicants) may require the use of wire grid inserts in shoe box cages to elevate the occupants above the cage floor and minimize direct contact with excreta (Fig. 6).

Fig. 7. Wheeled rack with multiple shelves to support solid bottom cages.

Fig. 6. Wire insert in clear plastic cage used to elevate occupants above the cage floor to minimize direct contact with excreta. Absorbent material may be used below the insert.

Shoe box cages usually are supported on wheeled racks approximately 5 feet long by 5 feet high with multiple shelves. Cages can be arranged on the shelves in one or two rows depending on the width of the shelves (Fig. 7). The position of a cage on a rack and the position of a rack within the room has significant influence on intracage ventilation, temperature, and light intensity (47,244,253,254). Elimination of this variable requires cage position assignment by standard methods of randomization and/or rotation in a systematic manner.

Suspended wire bottom cages usually are constructed entirely of metal, but can have plastic sides with wire floors (Fig. 3). Cages of this design are suspended from racks to permit excrement to fall through the wire floor and collect on appropriate absorbent material. Absorbent paper products usually are preferred over particulate materials because they are convenient to handle and easily disposed. The use of antibiotic impregnated absorbent pads should be discouraged because of the possibility that the antibiotic may constitute a subtle variable for some experiments. Suspended wire bottom cages facilitate frequent removal of excreta without having to manipulate the occupants. Intracage ventilation also is high, so that ammonia levels can be kept low. These apparent advantages may be offset by the free exchange of air and indigenous infections between cages. Such cages are suitable for housing juveniles and adult rats, but not for whelping and raising young.

Standards for cage population density are usually stated in

terms of the area required for occupants to achieve normal postural movements or behavioral patterns. The "Guide for the Care and Use of Laboratory Animals" (165) suggests the following minimum cage sizes for rats.

Weight of rat (gm)	Floor area/rat (in.²)[a]
Up to 100	17 (110 cm²)
100–200	23 (148 cm²)
200–300	29 (187 cm²)
Over 300	40 (258 cm²)

[a] Height should be at least 7 inches (17.8 cm).

It is not widely appreciated that population density has considerable impact on biological characteristics. In addition to dramatic effects on reproduction (43) and behavior (55), population density has important metabolic effects. Barrett and Stockham (18) showed that rats housed singly for 18 hr had plasma cortisone levels about half those of animals housed 20 per cage. In contrast, prolonged individual housing of rats had the opposite effect. After being housed individually for several weeks, rats became irritable and aggressive, had larger adrenals and thyroids, and showed increased hepatic microsomal enzyme activity. This effect, referred to as the *isolation stress phenomenon* (17,53,54), also has been shown to enhance ethanol consumption in the rat (188).

Housing density also has been found to influence immune responsiveness (89,123,190,198,220,234). Joasoo and McKenzie (122) housed rats 1, 2, and 10 per cage and immunized them with thyroglobulin. They found that the *in vitro* response of sensitized splenic lymphocytes to thyroglobulin was increased by crowding and decreased by isolation in female rats, while both the crowded and isolated male rats responded by a decrease in reactivity of lymphocytes to the antigen. It appears that environmental influences on activity of pituitary (95,96), adrenal (155), and sex (49) hormones contribute heavily to immune responsiveness of the individual animal. It is well known that handling and other minimally stressful procedures can cause significant changes in serum levels of several hormones (18,64). Psychological factors are known to modify host response to experimentally induced and spontaneous diseases (8,9).

Few workers have investigated the influence of exercise on biological responses of rats. In one study, Hoffman *et al.* (113) demonstrated that exercise markedly reduced growth of the Walker 256 transplantable tumor in a Wistar stock of rats compared with nonexercised controls.

Identification of cage occupants can be accomplished by water-resistant cage labels (Fig. 5). However, unambiguous identification requires permanent marking of individual rats by ear punch code, ear tagging (Fig. 8), or tattooing (also see Chapter 1, Volume II). A combination of cage and individual

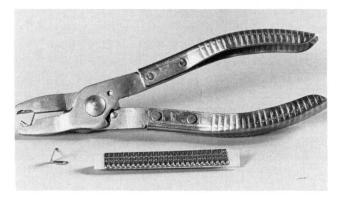

Fig. 8. Numbered ear tags and applicator used for permanent identification of individual rats.

identification may be preferred, depending on the research protocol. Definitive identification of animals should be mandatory anytime two or more strains having the same coat color are housed in the same room.

Maintenance of facilities and caging in a sanitary condition is essential for effective infectious disease control, stabilization of metabolic functions, and provision of comfortable living conditions. The application of sanitation procedures for reducing environmental contamination due to pathogenic microorganisms and metazoan parasites is a well established principle of infectious disease control. The more subtle effects of unsanitary environments on the biologic response of research animals have been appreciated only recently. Vesell *et al.* (233) demonstrated that a dirty environment can cause impairment of hepatic microsomal enzyme function in rats. While the exact mechanism(s) of this effect remain unclear, the effect on research results can be so profound that the inherent principle cannot be ignored.

Odor counteractants should not be used in experimental animal facilities because they merely disguise the evidence of poor sanitation and are themselves potent modifiers of biologic response (see Section III,D). In short, there is no substitute for good sanitation. Regular schedules must be established for rigorous cleaning of room surfaces, cages, and equipment. Frequency of cleaning is determined by cage design, population density, and other factors pertinent to each facility. Normally, primary cage enclosures should be sanitized at least twice weekly. Water bottles and sipper tubes should be sanitized and refilled twice weekly to accommodate for the loss of chlorine and growth of bacteria over this period. It is generally accepted that washing is facilitated by the use of detergents and disinfectants. However, these substances represent subtle variables for some experiments (215) and must be removed by thorough rinsing to avoid potential effect on biologic response. Because of the threat of such chemical variables, it may be necessary for some studies to eliminate the use of detergents altogether and rely on mechanical cleaning exclusively. The operation of mechanical cage washers at a water temperature of 180°F (82°C) is considered an effective method for eliminating ordinary rodent pathogens (165).

C. Food and Water

Nutritionally complete diets, available from commercial sources, are generally adequate for normal growth, reproduction, and maintenance of rats. However, because of specific dietary considerations, standard commercial diets may not be optimal for some studies (146,214). It must be appreciated that animal diets are potentially important sources of variability in many experiments using rats (21,22). There are two main considerations, both having a chemical basis: (a) variations in concentration of essential ingredients and (b) the occurrence of unwanted contaminants. (For a complete discussion of nutrition of the rat, see this volume, Chapter 6.)

Rat diets are usually formulated to one of the following levels of ingredient purity (117,127,165).

1. Purified or chemically defined diets are formulated with pure chemicals including individual amino acids and carbohydrates.
2. Semipurified diets include combinations of pure chemical and natural ingredients.
3. Natural diets consist of ingredients such as oats, corn, soybeans, fish, and meat products. This type is the most economical and widely used.

Although not generally appreciated, it has become increasingly clear that under the usual conditions of animal feed manufacture, concentrations of essential dietary ingredients may vary remarkably in different batches of diet formulated from natural products. The reasons are numerous and complex, including such diverse considerations as geographic and climatic factors influencing the nutritional value of plant products used in diets, the conditions of processing, and the length of time and conditions of storing diets. The resulting wide range of variation which may occur in concentrations of nutrients, such as vitamins, minerals, and proteins, can significantly modify the biologic response of rats to many experimental procedures (38,178,179).

Because of the proprietary nature of commercial diets, detailed information on ingredient composition is not available; these diets are designated *closed formula*. Little stability in composition can be expected from manufacturers that routinely vary the concentration of ingredients to accommodate for availability and/or cost of ingredients. Even products marketed as *constant formula* may have substantial variation in ingredient concentration. The highest degree of consistency in quantitative and qualitative composition of natural diets is achieved

by the use of *open formula* diets in which the concentration of each defined ingredient is specified. With this type of diet, adjustments in composition can be made to meet the requirements of specific research objectives (117,127,137).

The second potential problem with ordinary commercial diets made from natural products is contamination with deleterious chemicals. Most such contaminants in animal diets occur in low concentrations which may cause no clinical signs of toxicity but alter the results of experiments (46,127,153, 178-180). The list of contaminants which have been recognized in animal food and/or drinking water is now quite impressive and includes nitrates, cadmium, selenium, lead, arsenic, aluminum, mercury, nickel, insecticides, mycotoxins, herbicides, chloroform, estrogenic substances, polycyclic hydrocarbons, phenothiazines, phenylthiazoles, polychlorinated biphenyls, and antibiotics (180). The National Center for Toxicological Research has established recommended maximum allowable concentrations for many of these substances in experimental animal diets (127). Regular monitoring of diets for these compounds is essential quality control in some studies (127,180).

Lead provides a good example of the extremely subtle effects which may occur when a chemical contaminates animal diets in low concentration. Lead, in subclinical doses, has been found to suppress resistance of mice to *Salmonella typhimurium* (111), increase susceptibility of rats (210) and chicks (229) to bacterial endotoxin, decrease phagocytosis in rats (227), and decrease antibody production in mice (139). Cadmium reportedly produces similar effects (65,138,209).

Rat diets can be prepared in several physical forms including pelleted, meal, semimoist or gel, and liquid. *Pelleted diet* is the most common form of commercial product. It is low in moisture, dense, and uniform in size and shape. Manufacturing usually involves extrusion through steel dies under high temperature and pressure. This form is convenient for routine use but excludes easy addition of ingredients after manufacture and, depending on method of processing, may be too hard for ready acceptance by rodents (77,78).

Pelleted diets can be fed to rats from hoppers of various designs (Figs. 2,3, and 9). It is beneficial for weaning rats to be offered diet on the cage floor, but this is not considered good management for routine feeding of adults because of the potential contamination of food by excreta (165).

Meal diets consist of dry, coarse powders that facilitate addition of substances, but are difficult to feed. Although special diets can be procured in pelleted form, they usually are available in meal or powdered form which should be fed from wide mouth cups fitted with a lid that minimizes spillage (Fig. 10).

Semimoist or gel diets are prepared from powdered ingredients with the addition of water (50%, v/v) and agar (0.1%) to achieve a semisolid form (173). Additives can be included with ease, feeding is simplified, acceptance is increased, waste is

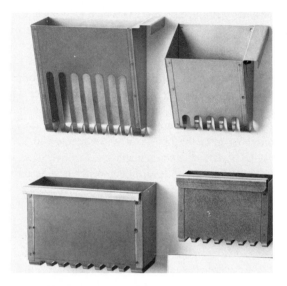

Fig. 9. Food hoppers of various styles to hold pelleted rat diet inside solid or wire bottom cages.

Fig. 10. Wide mouth cups for powdered diet with inverted cave lid to reduce spillage.

reduced, and unwanted contamination of the environment with ingredients (especially toxic or hazardous additives) is reduced.

Liquid diets are used for specialized purposes, such as low antigen diets.

The physical form of diets can have significant influence on acceptance, intake of free water, and exposure of oral surfaces to nutrients (e.g., production of caries with high sucrose diets).

Maintenance of microbiologically defined rats in barrier facilities usually requires diets that are heat sterilized or pasteurized (see Volume II, Chapter 2). The use of diets fortified with heat-labile nutrients is necessary to assure nutritional adequacy following heat treatment (81,116,149,191,249,257).

Diets should be stored in a manner that preserves nutritional quality and prevents contamination by chemicals, infectious agents, or vermin. To achieve optimal preservation, bulk diets and dietary ingredients should be stored in a separate room away from areas of high contamination (e.g., cage washing). Containers of food should be stored in a dry (50% or less relative humidity), cool (72°F or less) environment in a manner that facilitates free circulation of air and prevents contamination by water or other medium. Stacking packages on standard warehouse pallets is a useful practice. Exclusion of insects and wild or feral rodents is essential. After bulk containers are opened, diet should be stored in moisture-proof, vermin-resistant containers. Diets with a high moisture content (that are not canned), such as gelled diet, should be refrigerated or frozen.

Water is a crucial nutrient for which rats have a particularly high requirement (246). The need for free water is accentuated by the use of diets with low moisture content. Malfunctions of a watering system that limit availability of water can have disastrous effects on rats in a very short time. Also, water is an extremely important source of chemical contaminants and pathogens. The mineral content of water as well as additives and contaminants derived from plumbing or utensils may require careful monitoring. Hyperchlorination and/or acidification have been used to control microbial contaminants of water, particularly *Pseudomonas* sp., in immunocompromised rodents (73,74,88,148,157,158). However, complications due to corrosion of equipment and the biologic effects (69) of such additives must be considered.

Water is normally supplied to rats by the water bottle/sipper tube method or by pressure-fed, plumbed systems utilizing lick-activated mechanical valves (so-called "automatic waterers"). Each method has unique advantages and limitations. The water bottle/sipper tube system consists of an airtight reservoir bottle made of glass or plastic. A stainless steel tube with a constricted aperture or ball valve is connected to the bottle with a tight-fitting rubber stopper (Figs. 2,3, and 11). The unit is inverted on the cage with the aperture at suitable height above the bedding or floor. When a rat licks the aperture or displaces the ball, a drop of water is released and an equivalent volume of air is drawn up the tube into the bottle to equalize pressures (and simultaneously inoculating the water with contaminants such as *Pseudomonas* sp. from the oral cavity). This system is inexpensive, effective, relatively resistant to malfunction, and limits the water-borne exchange of pathogens to the occupants of single cages. Limitations include a high labor requirement for frequent (every 2 to 3 days) sanitization, filling and reassembly, plus a high rate of contamination caused by aspiration of oral secretions with displacement air. Failure to provide a small air pocket in the system or a malfunctioning ball valve can prevent water flow. Conversely, an air leak or contact of the aperture with bedding can result in premature loss of water and flooding of shoe box cages.

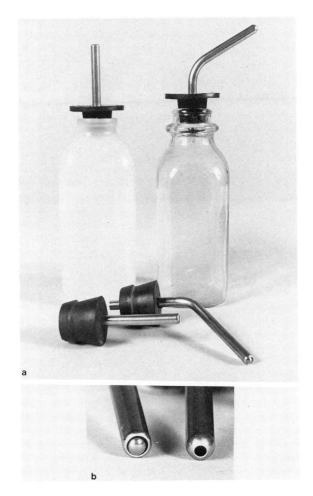

Fig. 11. (a) Plastic and glass quart-size water bottles with rubber stopper and stainless steel sipper tube. (b) Sipper tubes have constructed aperture or ball valve.

Pressure-fed systems require less labor and reduce aspiration of oral secretions because water is delivered at 2 to 4 psi. However, such systems are more expensive, require special plumbing in the facility, enhance the risk of cage-to-cage exchange of pathogens, and are more prone to mechanical failure. Flooding caused by malfunctioning licking valves is a very serious problem associated with this system. The consequences can be minimized by use of cages with perforated bottoms or location of valves outside of the living compartment of shoe box cages (Fig. 4).

D. Shipping

When rats are transported, (e.g., from production to research facilities) behavioral and physiological uniformity is disrupted. Studies of this phenomenon in the laboratory rat have shown that animals lose weight during the transport phase between breeder and using laboratory (60,76,84,100,156,219,246). The weight loss is more severe in hot weather than in cold (219), air shipment is more stressful than surface transport (60,219), indices of dehydration (e.g., the hematocrit) are measurably affected subsequent to shipment (156), and aged rats are more susceptible to these effects than young animals (60,156). One study demonstrated that the observed weight loss and dehydration are substantially related to lack of availability of water during shipment, and that these effects are directly related to duration of shipment (246). As a minimum, rats should be shipped in small compatible groups in adequately sized shipping boxes. They should be protected against extremes of cold, but given adequate ventilation, particularly during hot weather. Spacing devices should be provided on the shipping containers to prevent close stacking of crates that would restrict ventilation.

Food and water or high moisture foods should be provided to rats during shipment. Numerous materials have been investigated for this purpose including fruits and vegetables (e.g., apples and potatoes) (246), canned high moisture carnivore diets, and gelled rodent diets (84). However, of the group, water and pelleted diets have been shown to be the most effective in minimizing weight loss and dehydration during shipment (246).

Adequate time following shipment should be provided at the receiving institution before experimentation to allow rats to return to a state of physiologic stability. This period, the *equilibration time,* has been defined as the time interval for recovery of body weight in recently shipped adolescent rats to within one standard deviation of the growth curve of unshipped rats of the same age and sex of the same stock maintained at the production plant (246). The equilibration time is directly related to the duration of the transport phase and the type of moisture supplement provided, and varies from 1 to 5 days (246).

Shipment via common carriers introduces the additional risk of transmission of infectious disease from shipping container to shipping container when rats from different sources are pooled for shipment or receipt in common areas. Fiber filter covers over ventilation ports of shipping containers are in common use to reduce this risk; however, filtered containers can safely carry only about 50% of the biomass of nonfiltered (thus better ventilated) containers because of inhibited heat dissipation from the container. Such filters can be used only with air-conditioned shipment conditions during hot weather because of the risk of deaths due to hyperthermia.

E. Environment

1. Temperature and Humidity

Temperature, humidity, and ventilation are extremely important environmental variables. Together they determine relative heat loss or retention, and ultimately contribute heavily to metabolic rate. As emphasized previously, intracage environment directly influences many biologic responses of caged rats (but as a practical matter, environmental standards usually refer to room conditions!). Since cage surfaces facilitate rapid equilibration of temperature, differences in and outside of cages are usually small (24,99) but may reach a differential of 3° to 5°C higher within cages with filter tops (212,217,254). In contrast, differences in humidity are influenced by air exchange, and thus intracage humidity is usually higher than room conditions (91,212).

While much data has been compiled on the effects of climatic extremes, few studies have focused on small variations in temperature and humidity (104,112,224). Current recommendations suggest that rats be housed at 65°–75°F (18°–24°C) with 45–55% relative humidity (RH) (70,143,165,240). Although rats can adapt to a wide range of environmental temperatures (10°–30°C) (240,241), the time for adaptation is long (3–6 weeks) (92a); therefore, sudden fluctuations must be avoided.

A recent review (242) of temperature effects on drug action emphasizes the importance of adequately defining the size of animal populations, type of cages (wire mesh or boxes with bedding), and the ambient temperatures used in drug toxicity trials. For example, the LD_{50} for amphetamine at 27°C was found to average 78.9 mg/kg for singly caged mice compared to 11.6 mg/kg for mice housed in groups of 10. Weihe (242) emphasizes the existence of at least two patterns of drug toxicity responses (90) in mice and rats dependent on cage temperature. The first is V- or U-shaped with minimum toxicity around thermal neutrality and increasing toxicity at lower and higher cage temperatures. In the second, there is a linear response with increasing toxicity being directly correlated with increasing cage temperature. Related observations have been reported by others (35,131,205,256).

Baetjer (14) demonstrated that temperature and humidity can influence susceptibility to infectious diseases. For example, she demonstrated that mice were more susceptible to the PR8 strain of influenza virus when maintained at 35.6°C, 22% RH than at 35.6°C, 90% RH.

One of the most visible indicators that temperature and humidity can seriously affect laboratory rats is the condition known as "ringtail." This lesion of young animals is characterized by annular constrictions with or without subsequent sloughing of part or all of the tail (Fig. 12). The precise mechanism responsible is unknown, but it is generally thought to be associated with high temperature and low (below 40%) RH (66,72,182,223,225). Prolonged temperatures of 80°–100°F (26.6°–37.7°C) have been observed to cause deaths and retarded testicular development with infertility in male rats (194).

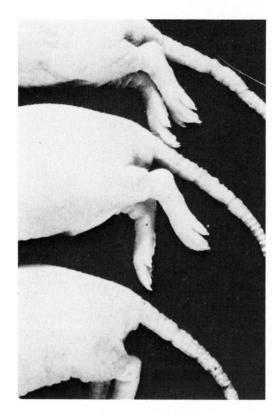

Fig. 12.   Ring tail lesions in young rats associated with low environmental humidity.

2. Room Ventilation

A properly designed system that provides regular exchange of air is essential to control temperature and humidity, dilute chemical pollutants, and control transmission of infectious agents. Current standards recommend a ventilation rate of 0.815 ft³/min/250 gm rat or 10–15 changes of room air per hour (165). Recirculation of air requires adequate systems for removal of particulate matter to reduce transmission of pathogens. Optimal filtration provides 99.97% retention of particles greater than 3 μm. With such a system, addition of at least 10% fresh makeup air is recommended. Mass air displacement of HEPA filtered air has been adapted to racks, room enclosures, and entire facilities to reduce particulate contamination (Fig. 13).

Areas where minimal disease stocks are housed should be ventilated under slight positive static air pressure compared with halls, quarantine areas, cage washing, and other contaminated areas. Since continuous ventilation is crucial for life support of rats caged in a windowless environment, the mechanical equipment must be designed with failsafe characteristics such as redundant parts and emergency electrical generators.

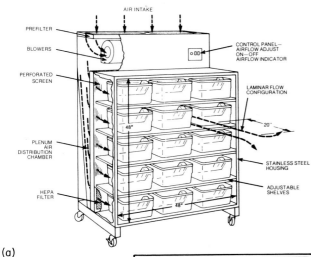

(a)

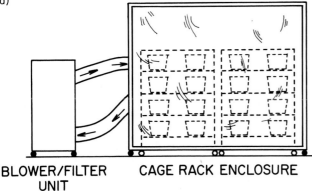

BLOWER/FILTER UNIT   **CAGE RACK ENCLOSURE**

(b)

Fig. 13. (a) High volume, unidirectional, filtered air flow cabinet. (b) Room size enclosures with mass air displacement of filtered air can accommodate multiple racks of cages.

Odors must be controlled by improved sanitization and/or increasing the proportion of fresh makeup air, not by the use of chemical odor counteractants (see Section III,D).

Pheromones are potent chemicals excreted by rodents that modify many behavioral characteristics, particularly those associated with reproduction (42,55,58,63,166,248). Although, these chemical communicators have been shown to be influenced by positive pressure ventilation systems (248), the effects of routine laboratory rat management systems have not been defined.

3. Lighting

Photoperiod (light–dark cycling, usually stated as L : D in hours) has been demonstrated to be one of the most important of the many environmental factors that modify biologic response (57,108,164,170). In addition, light intensity and quality (wavelength) have profound effects on many physiological characteristics. Therefore, intensity, quality, and duration of light should be standardized and carefully controlled in experimental rat rooms.

Housing areas should be windowless so that light characteristics can be critically controlled with artificial lighting. Duration of light per 24-hr period is known to have significant influence on reproduction, and 13 to 14 hr of light daily is thought to be best for breeding of rats (167). Light–dark cycles can be controlled readily by the use of electronic timing devices.

The intensity and quality of artificial light are governed by number, power rating, and emission spectra of the lights being used. Light intensity varies inversely with the square of the distance from the source. Consequently, animals housed on the top shelf of a rack are exposed to far greater light intensity than those on the bottom. Furthermore, the rack and cage design (opaque, translucent, or transparent materials) also modify the intensity and quality of environmental light. Standardization of cage design and randomizing the position of animals within the room will reduce variables imposed by such factors (47,244).

Optimal light intensity (measured in foot-candles as lumen/ft^2 or lux as lumen/m^2) and quality [wavelength spectrum as angstrom units (Å)] are not well understood in terms of the levels most beneficial for or detrimental to laboratory rats (243,244). The current recommendation is for 75–125 ft-c (807–1345 lumens/m^2) at the cage level (165). However, levels of this intensity (184) and lower (239) are known to cause degeneration of photoreceptors in the eyes of rats, and this effect becomes more pronounced with increasing age (146,185,239). Albino rats apparently are more susceptible than pigmented rats (199). Similar effects have been reported in the mouse (202).

Photoperiodicity regulates (but does not control) circadian rhythms (109). The term *circadian rhythm* (meaning "around a day'') was introduced by Halberg *et al.* (101) to describe endogenous rhythms or biological clocks of physiologic functions (108). Much of the pioneering work on this subject using the rat was done by C. P. Richter (26a). Until recent years it was not fully appreciated that these rhythms may profoundly influence experimental data. The number of behavioral, physiological, and biochemical parameters known to have a 24-hr periodicity is impressive (a few examples are cited in references 3,4,13,25,28,40,41,61,67,102,110,121, 130,147,163,171,172,195–197,203,213,251). Biologic responses such as duration of pentobarbital narcosis, time to death after whole body irradiation, and mortality from a standard dose of ethanol deviate from the 24 hr mean by at least 50% during a full 24-hr cycle (12,206).

4. Noise

Rats have an acute sense of hearing. Since noise is known to produce dramatic physiological and behavioral changes, every effort should be made to eliminate loud, high-pitched, and alarming noises (71). It is believed (11) that rats, like man, experience mechanical damage to the ears due to sounds of 160 decibels (db), pain at 140 db, and signs of inner ear disturbance after prolonged exposure to sounds around 100 db. Accordingly, it is recommended that permissable noise levels in animals facilities not exceed 85 db.

Geber *et al.* (92) exposed rats to average sound levels of 83 db (octave band noise level 46–78 db for 20–4800 cps) for periods ranging from minutes up to 6 min/hr for 3 weeks. This resulted in increased eosinophils in peripheral blood, increased serum cholesterol, increased adrenal ascorbic acid levels, and increased adrenal weights. Friedman *et al.* (87) also have shown that noise (continuous sound and intensity of 102 db or intermittent sound of 200-cycle square wave with duration of 1 sec and intensity of 114 db) resulted in elevation of serum lipids in rats. Rats are known to be disturbed by fire alarms in the frequency range usually manfactured for human audibility. Alarms have been developed in lower frequencies (450 Hz) that are audible to humans, but not rats, to avoid the startle reaction to sudden alarms (48).

F. Monitoring

Much of the monitoring essential to assuring a quality physical environment in an animal facility is merely frequent observation with conscientious attention to the needs of the animals, equipment, and building. Preventive maintenance must always be emphasized in order to avoid interruptions in the uniformity of conditions. This may involve the testing of water and each batch of diet and bedding for chemical and/or microbial contaminants prior to acceptance and use (46,127,180). Environ-

mental variables, such as temperature, humidity, lighting, air flow, and noise level, require regular monitoring and control (124,127,237). Numerous other factors may require monitoring dependent upon conditions and specific research protocols.

III. CHEMICAL FACTORS AFFECTING BIOLOGIC RESPONSE

Experimental animals exist in complex environments. All components of these environments are subject to substantial variations in chemical composition and thus are capable of influencing biologic responses. In addition, each of these environmental components has the potential of being contaminated inadvertently by an enormous diversity of man-made and naturally occurring chemicals which may be deleterious to the animal's health or, perhaps even worse, cause subtle aberrations in experimental data. For these reasons, it is imperative that modern scientists be well informed about controlling the chemical environment of their experimental animals.

The number of environmental chemicals which find their way into animal facilities and influence research data is quite impressive (Table I). However, one must remember that the additional possibilities are almost infinite, and, in most facilities, few if any chemicals are monitored at all. Theoretically, it seems probable that some environmental chemical will eventually be found to affect virtually every response which can be measured in an animal. The majority of examples reported to date have involved the hepatic microsomal enzymes

Table I

Partial List of Chemicals Occurring in Laboratory Animal Environments That Alter Biologic Responses[a]

Air pollutants
Dust and bedding particles, trace volatile anesthetics, animal room deodorants (volatile hydrocarbons, eucalyptol), disinfectant sprays (peracetic acid), vinyl chloride, ammonia, insecticides, piperonyl butoxide

Food and water
Nitrates, cadmium, arsenic, lead, aluminum, mercury, nickel, insecticides, mycotoxins, herbicides, chloroform, food additives, estrogenic compounds, polycyclic hydrocarbons, phenothiazines, phenothiazoles, flavones, antibiotics

Bedding, caging, and equipment
Detergents, disinfectant, soaps, ethylene oxide, wood alkaloids, cedrene, cedrol, ammonia, lignin, aldehydes, antibiotics, microbiocides

Test chemicals and drugs
Mutagens, teratogens, carcinogens, toxic agents, drugs, vaccines

Organic solvents
Ethers, alcohols, chloroform, carbon tetrachloride, acetone

[a] Modified from Newberne and Fox (180).

(HME) and immune systems (153).

The HME are a large assortment of enzymes occurring in the microsomal fraction of hepatic cells. Although similar enzymes are found in other organs, such as lung and kidney, the liver contains the body's major pool of such enzymes. Because of their principal role in metabolism and elimination of exogenous substances from the body, the HME are of particular interest in pharmacological and biochemical studies using rats.

Literally hundreds of environmental chemicals and drugs have been shown to alter HME activity. Some such chemicals enhance, while others impair, HME activity. Thus, the possibilities for altering biologic responses involving these enzyme systems are extremely complex. As a simple example, rats exposed to very small concentrations of an insecticide (perhaps, by a commercial pest control service man unknown to the investigator) may be far less susceptible to the induction of cancers by a chemical carcinogen being given experimentally. The problem becomes much more complex, however, in many facilities where multiple environmental chemicals (some HME inducers and other HME suppressors) are present (36,51,52,145).

Although the data are far less voluminous, the evidence is just as convincing that environmental chemicals also can alter other biologic responses. Immune responses may be altered dramatically by gaseous pollutants of many kinds (115,152) and by a variety of chemicals entering the body by other routes (153). Numerous biochemical and pathological responses are known to be influenced by dietary substances (38,162) including chemical contaminants (179,180). Some of the better known examples of these environmental factors will be discussed below.

A. Gaseous Pollutants

Exposure of laboratory rats by inhalation of potent chemical modifiers of biological response ranks along with ingestion and skin–mucous membrane contact of chemicals as leading causes of unwanted research variables. Sources of air pollutants may be classified as (a) those introduced by man into the animal room environment and (b) those originating in the cage environment.

Research laboratories are virtually unrivaled as repositories of vast varieties of potent chemicals. Complete separation of rat housing areas from locations where such chemicals are stored and used is an obvious, but often violated, principle. Frequently, organic solvents, such as acetone, ether, and chloroform, are used carelessly in animal housing areas. The well-known and dramatic nephrotoxic effect of minute concentrations of chloroform on male mice of certain inbred strains should provide sufficient warning against such indiscretion

(62). Volatile (and explosive) chemicals should not be used in animal housing areas. For protection of experimental animals and personnel alike, their use should be limited to chemical fume hoods. The use of volatile chemicals, such as insecticides and odor counteractants, in routine animal husbandry and animal facility maintenance must receive careful consideration and scrupulous control. Whenever possible, nonvolatile substitutes should be used (180).

Only in recent years has there been active investigation of environmental air quality within animal cages. Thus, the various parameters of the gaseous environment in cages is poorly understood. At present it appears that the main gaseous pollutants derived from accumulation of animal wastes in cages are carbon dioxide and ammonia (31). Serrano (212) demonstrated that, depending upon cage design, population density, and animal activity, the intracage level of CO_2 in standard cages housing eight adult mice may increase as much as eight-fold over room air (from 524 to 4517 ppm). Although the biological effects of increased CO_2 have not been defined, differences of this magnitude conceivably could have important effects on biologic response.

Serrano (212) also studied NH_3 in mouse cages, showing that NH_3 usually was not detected until the third to sixth day, but by the seventh day NH_3 levels reached 21 to 177 ppm. Levels of 220 to 350 ppm were obtained in some cages. Similar results were obtained for mice by Murakami (169), who also reported much higher levels at night and in cages housing males compared to females. Gamble and Clough (91) showed that levels of NH_3 of 25 ppm and greater were common in rodent cages, and NH_3 concentration was directly related to humidity. Flynn (75) reported levels of NH_3 exceeding 700 ppm in mouse cages with filters. Lindsey and Conner (152) demonstrated a relationship between cage cleaning frequency and intracage NH_3 for adult rats housed in shoe box cages. Their data also show a dramatic effect of sex on NH_3 concentration presumably because of body size (male rats are usually larger than females). The threshold limit value (maximum allowable level for work environment of 8 hr per day, 5 days per week) for man is 25 ppm (7). Concentrations above this level are common in rat cages. Thus, rats housed according to contemporary practices generally live in environments seriously polluted by NH_3.

The biologic effects of increased environmental NH_3 are poorly understood. Visek (235) believes that as a metabolite NH_3 has severe effects on intermediary metabolism and shortens life span of animal cells, among other deleterious effects. It also has been suggested (232) that NH_3 may have accounted for the impairment of HME by dirty rat cages observed by Vesell *et al.* (233). Lindsey and co-workers (32,152) have demonstrated that levels of NH_3 normally encountered in rat cages significantly influence the pathogenesis of murine respiratory mycoplasmosis (MRM) due to *Mycoplasma pulmonis* (see this volume, Chapter 10). This strongly suggests that intracage NH_3 may influence the results of many studies of the respiratory tract of rats.

B. Insecticides

The use of pesticides for vermin control in animal facilities is a common source of some of the most serious complications of animal research due to environmental contamination. Indeed, the first documented observation of an alteration in drug metabolism caused by an environmental chemical resulted from the use of chlordane insecticide in an experimental animal colony (51). The HME are stimulated by halogenated hydrocarbon insecticides and inhibited by organophosphate insecticides or pesticide synergists of the methylene–dioxyphenyl type (85,105,140). The complex interactions of insecticides and other environmental chemicals with drugs have been reviewed by Conney and Burns (52). Insecticides have been shown to alter immune response (238), induce lymphocytopenia in mice (133), and induce other metabolic disturbances (192). Because of the high risk of unwanted complications caused by pesticides, it is generally recommended that nonchemical means of vermin control be used to their maximum in animal facilities. When absolutely necessary, insecticides may be used, but then only sparingly and with utmost discretion. Investigators should be informed about the intended use of insecticides and only specific approved compounds should be used. It is preferable to use nonvolatile forms of insecticides and limit applications to areas outside of rat housing rooms.

C. Drugs

Drugs may be used for treatment or control of disease, anesthesia or euthanasia, and to alter physiological status for specific research objectives (e.g., induction of diabetes with alloxan). Such drugs are certain to have profound effects on metabolism other than the primary target system (174). The deleterious consequences of these effects are compounded when investigators are not aware that drugs are being used. For example, the results of early attempts to induce caries in rats were inconsistent because some stocks were treated with antibiotics which inhibited cariogenic streptococci. The effects that some drugs might have on specific research procedures often are not well defined, and researchers may be forced to perform additional experiments to evaluate their effects on the research objective. The effects that drugs have on research results depend upon (a) the characteristics of the active ingredient, (b) the characteristics of vehicles or dilutents, (c) route and site of administration, (d) frequency of treatment, and (e) dosage level.

The addition of chemicals to food and water is a useful method for administration of experimental materials, but many factors influence the suitability of this approach for specific agents. The extent to which the material is absorbed via the gastrointestinal tract must be considered. The stability of a drug or chemical in the presence of nutrients or other chemicals (chlorine in water), light, and bacterial degradation also should be considered. Use of distilled water, brown glass receptacles, and frequent replacement of treated drinking water or food reduces the effect of this complication. Direct effects of drugs or chemicals on the gastrointestinal tract (including the gut flora) may affect health as well as alter the rate of drug absorption. These side effects are of special significance with antimicrobial agents. Dosage of material administered in this way varies with the quantity of food and water consumed, and usually cannot be controlled precisely because of wastage and competition among occupants of a single cage. Furthermore, if the test chemical imparts an undesirable flavor, the consumption of food or water may be reduced below expected levels. One method designed to compensate partially for variations in food or water consumption is the paired feeding of test and control animals. It may become necessary to reduce the availability of food and/or water for control rats to match the quantity consumed by the test rats. It should be recognized that the unabsorbed portions of the test material or metabolites of a drug may be present in excreta. Coprophagy may result in repeated exposure to the drug in an unexpected way. The use of wire cage floors reduces, but does not eliminate, this complication.

As indicated previously, environmental factors have significant effects on drug action. For example, sleeping time of rats administered a standard dose of pentobarbital varies as much as 100% depending on time of administration in relationship to light–dark periodicity (206).

D. Cleaners and Deodorizers

A large number of chemical compounds, including soaps, detergents, wetting agents, disinfectant, solvents, and others, often are used in great profusion for "cleaning" equipment, floors, etc., in animal facilities. Many (perhaps most) of these chemicals are highly volatile substances which have pleasant smells, but are notorious for their abilities to alter hepatic microsomal enzyme function (52,145). They also may alter immunoresponsiveness (215).

Animal room deodorizing agents usually are complex mixtures of volatile hydrocarbons or essential oils. These substances are well known inducers and inhibitors of hepatic microsomal enzymes (45,86,129).

E. Chemical Sterilizers

Peracetic acid is an efficient bactericide, fungicide, and virucide (136) which occupies a central role in sterilization of gnotobiotic equipment. Its caustic effects are such that personnel who work with it must wear protective gas masks, clothing, and gloves (191). Rats housed under gnotobiotic conditions do not have such protection and are almost invariably exposed to this chemical each time materials are passed in and out of the isolator. This could pose a serious problem to studies employing such animals, as recent data (27) has shown peracetic acid to be a carcinogen and a cancer promoter.

The practice of sterilizing animal bedding by fumigation with ethylene oxide may have devastating effects on experimental results and thus is no longer recommended. Treatment of bedding with this chemical has been associated with occurrence of bleeding disorders and intracardiac thrombosis in mice and rats (10,89,160). Exposure of animal diets to ethylene oxide also has been shown to reduce the concentrations of vitamins and proteins (16,250).

F. Monitoring

Continuous vigilance is essential to prevent the entry of unnecessary chemicals into animal environments. Specific testing for individual chemicals may be necessary for the successful performance of certain studies. The testing of air quality should differentiate between room air and that of the intracage environment where pollutants such as NH_3 are most concentrated (6,32,47,124,180).

IV. MICROBIAL FACTORS AFFECTING BIOLOGIC RESPONSE

A. The Problem

While research complications due to overt infectious diseases are significant, the effect of clinically silent infections are devastating because they often go undetected until alterations in experimental data herald their presence (103,175, 177,208). In the rat, mycoplasmosis (150,151,176) and hemobartonellosis (18,151) are notorious for their extremely subtle and profound influence on a great variety of studies. Recent reviews (97,127,151,153,221) chronicle the complicating effects of such infections (also see this volume, Chapter 10). Limited data are available on the incidence of infectious diseases in laboratory rodents (39,56,114,187,204).

B. Solving the Problem

The solution to complications caused by infections in the laboratory rat was found in the application of gnotobiotic technology. The science of gnotobiology arose as a consequence of investigations directed principally to the exploration of germfree life. Although a new field where most of the technical innovations are even now less than 20 years old, it was quickly realized that application of the principles of gnotobiology to problems of animal disease offered an ingenious opportunity to break the cycle of intercurrent disease. The basic principle was to prevent disease by exclusion. Full-term fetuses would be free from infectious disease if aseptically delivered into a sterile environment and maintained behind a barrier that excluded all microbial forms except those willfully introduced by the operator. Rapid exploitation of these principles in both commercial and academic sectors resulted in remarkably successful solutions to the further reduction of infectious disease in laboratory animals particularly mice and rats.

C. Definitions

A number of terms have evolved as descriptors for various levels of microbic association in rats (154,200). At one extreme are *axenic* (or *germfree*) and *gnotobiotic* (those associated with a single and known microbic form) rats. Gnotobiotic (GN, from the Greek "of known biology") rats are associated with more than one known microbic form, although this term may also embrace axenic rats. The terms *barrier-reared* (BR) and *defined flora* (DF) have come to describe that class of rats reared from conception as gnotobiotes. Because it was recognized early that germfree animals were sufficiently different morphologically (93,94) and physiologically (2,50) from "normal" animals, the goal was production of laboratory rats with the least microbic burden consistent with a "normal" physiologic state, but devoid of pathogenic microbes. The term *specific pathogen-free* (SPF) was introduced to describe rats freed of certain specific pathogens by the use (principally) of "gnotobiotic" technology. By definition this implies that pathogens known to be absent must be indicated, otherwise the term SPF is unwarranted and even misleading—the situation which actually exists in its usual usage! By consensus, *conventional* (CV) rats include rats from which pathogenic microbes have not been excluded by the application of gnotobiotic technology. Animals of the latter category usually maintain an assortment of pathogens as latent or overt infections.

In common breeding practice, a cycle is established where axenic rats are produced by hysterectomy and reared behind germfree barriers. Axenics, in turn, are given the known minimum microbic flora to produce *gnotobiotes* (59,228). Gnotobiotes are used to produce *barrier-reared production colonies*. These colonies produce rats designated as BR or SPF (92,93).

There are varying levels of security attached to the maintenance of the barriers and implicit microbiologic monitoring protocols attached to each level (154). The major characteristics that determine degree of security of a barrier system include (a) microbial definition of animals housed in the barrier, (b) methods for processing equipment and supplies into the barrier, (c) entry of personnel, (d) animal housing systems, (e) environmental systems, and (f) monitoring practices (127).

A program for prevention of disease by exclusion is justified by the benefits of performing research with disease-free rats. Regardless of the precision with which terms accurately bracket the degree of microbial association, it is difficult at present to translate this ideal consistently into standard biomedical research practices. There is no public regulation of the quality these terms imply, and hence no restriction on their use, whether warranted or not and whether misrepresentation is unwitting or deliberate. Protection of users at present (in the United States) is based on the reputation of individual vendors and the pressure exerted on them by consumers free to buy elsewhere. For these reasons, many institutions have established programs for monitoring animal health status.

In recognition of this problem in England, a voluntary system of designated laboratory animal quality grades based on evaluation of sample animals by an independent laboratory has been initiated (226). Under this scheme, a system of five numbered categories has been established that vary according to decreasing pathogenicity of permitted pathogens. This system purportedly provides the consumer with accurate information on which to base purchase decisions (142).

"Animals for Research," a directory of laboratory animal sources and supplies, contains a listing of the principal commercial sources of laboratory rats in the United States (118). The current directory lists approximately 32 commercial sources for rats. The sources are listed according to what they offer—BR, DF, GF, GN, or nonspecified (presumably CV) animals. Of the 32 suppliers, only four offer GN or DF rats. In all cases, these suppliers also actively produce animals in the BR (a term preferred to SPF) categories. At least 11 suppliers who have no facilities for production of gnotobiotic animals claim to have BR rats for sale. In many cases, such colonies are established by the purchase of GN (or GF, or DF) animals from institutional or commercial suppliers, and these animals are used to foster rear, (behind a barrier) young derived by hysterectomy from dams originating from the breeder's CV colony (132). This arrangement is an economical expedient that gets around the need to have a gnotobiote production plant with expensive and technologically complex facilities.

Among large commercial vendors there seems to be a consensus that no barrier facility can be maintained indefinitely free of unwanted microbial contaminants. In consequence, the practice of fixed and periodic renewal has been generally adopted. Axenics are periodically rederived and hand reared. Barrier facilities are periodically vacated, sanitized, and reestablished. In the commercial sector only the breeders' integrity is available to compel periodic renewal. If periodic renewal is not practiced, the colony is destined for gradual deterioration and loss of quality. The endpoint of such deterioration is animals indistinguishable from ordinary CV rats.

Differences in the quality of BR or SPF rats from different suppliers are recognized. These differences show themselves in the presence of diseases, often unmasked by experimental stress, that should not be present in recently acquired BR or SPF animals. In turn, these diseases occur because of neglect and indifference on the part of breeders who fail to undertake colony renewal, who do not monitor colonies to detect contaminants, or who do not have available the personnel or laboratory facilities for this type of quality control. In the absence of independently evaluated quality grades, the buyer is forced to determine by subjective or experiential means the quality of rats purchased from various sources. The other side of this coin is that uninformed investigators often purchase high quality BR or SPF animals and promptly jeopardize that quality by housing them in communal facilities with CV animals.

D. Reduced Variability of Disease-Free Rats

Having implied that there are differences in BR and SPF rats, as opposed to CV rats, it is important to explore in some detail what they are and particularly how they might bear on the interpretation of experimental results using laboratory rats. As an introduction to this subject, it would be well to point out that in spite of the obvious importance that BR versus CV comparisons be made in parallel from colonies of similar genetic origin, there is not much literature available for this kind of side-by-side comparison, and even less where studies were extended longitudinally into senescence. The comparisons are grouped into three main areas of reference: (a) general physiologic indices, such as growth, longevity, hematologic profiles, pulmonary function, etc.; (b) differences in the incidence of disease agents, diseases, and lesions known to be associated with an infectious etiology; and (c) diseases and lesions not definitely known to be associated with an infectious etiology.

It has been well documented that BR or SPF reared rats grow faster, have large litters, utilize food better (even fats), and mature at heavier weights than do their CV counterparts of both sexes (79,183). The BR rats are more uniform physically, respond to stress more uniformly, and are more resistant to the deleterious consequences of certain stresses, e.g., anesthetization (2).

The hematologic indices of BR and CV rats have been compared with inconsistent results, perhaps because CV rats have had varying degrees of illness. In general WBC counts are ordinarily significantly elevated in CV rats, especially when murine mycoplasmosis (MRM) is accompanied by other concurrent respiratory tract infections (37,50,98). Even when the WBC count is not elevated, however, there is a tendency for a shift in CV rats to higher numbers of neutrophils at the expense of lymphocytes. There is a noted and significant trend toward lower blood glucose levels in BR versus CV rats (50). The differences noted between cell types and numbers in circulating blood in turn reflect local demand for these cells. Lungs from CV rats weigh more than lungs from their BR counterparts and have increased numbers of neutrophils and lymphocytes in lung tissues as detected by enumeration of cell types in lung washings (2,26).

Because of the cardinal importance of MRM as a disease of CV rats, considerable attention has been given to the comparison of cardiopulmonary function in CV versus BR or SPF rats. The pulmonary complications consequent to MRM impose a state of permanent compensated cardiac function on the rat with MRM. It is known that the right ventricle often is hypertrophied in rats with MRM, that higher systemic blood pressure is maintained during pentobarbital anesthesia in SPF rats, and that the left ventricle contracts more efficiently under the same tension load in SPF rats (2). Other cardiopulmonary function indices in which differences have been noted include the increased lower airways resistance of non-SPF rats (134), the lower oxygen saturation of arterial blood in rats with MRM (135), and the elevated acid mucopolysaccharide content of lungs from CV animals (120,230,255).

Among the more dramatic differences to be expected in BR as opposed to CV rats is the shift to the right in the cumulative population mortality curve over the lifetime of the sample group (186). In other words, the life span is increased. By 12 months one can expect 1% mortality in the BR group, compared to 6% in the conventional group; by 24 months a reasonable expectation would be 20% of the BR group versus 45% in the CV group. The median age at death would be 14 months in the CV group compared to 30 months in the SPF group. The oldest 5% would live 30–39 months in CV rats compared to 40–51 months in SPF rats of the same strain.

Considerable information has accummulated on the incidence and pathology of infectious diseases in laboratory rats and how they directly contribute to the difference in physiologic indices discussed previously (15,127,150,151, 153,221). It is, therefore, unnecessary for us to review it here. Less information is available with regard to the interaction and synergistic effects between various infectious agents. The process of derivation by hysterectomy and maintenance behind barriers simultaneously strips off all infectious diseases not vertically transmitted, hence the observer sees cumulative differences. Thus, in CV rats one sees not only the deleterious effects of MRM but also the cumulative burden of experience with arthropod and helminth parasites, protozoan parasites of the tissues, blood and gastrointestinal tract, primary and secondary bacterial pathogens, and viral orphans of unknown pathogenicity to their rat host.

Although there are many lesions that occur in the course of life in rats, certain of them occur with sufficient frequency to assume the status of lesions characteristic of the species, especially in aged rats (33). Some of these have been studied in great detail. They are more fully discussed in Chapter 14 of this volume, but include chronic nephropathy, myocardial degeneration, and various tumors (20,218). For each of these diseases, it has been pointed out that there is an age of maximum probability of lesion onset, a characteristic range of distribution about this maximum value, and a finite limit to the percentage of animals that will acquire each type of lesion. Longevity in the rat, as in other species, appears to be largely determined by those forces (genetic, infectious, nutritional, physical, etc.) that hasten or retard the onset of lesions of the major degenerative diseases (218). Information is available with regard to the comparative incidence between CV and SPF rats for several of these lesions (20,186,218,245).

The incidence of chronic nephropathy and polyarteritis (periarteritis) is known to begin at an earlier age and to involve more animals in the case of CV rats compared to BR rats of the same strain (186). Information is lacking as to whether this implies an infectious cause of these lesions or whether the presence of pathogens in some way lowers the threshold for their onset.

Such evidence as is available indicates that the incidence, age of onset, and numbers involved for each of the various spontaneous tumors is not influenced by gnotobiotic technology and barrier maintenance (186). More BR animals survive into old age. Thus, more tumors appear as a function of the number of animals at risk, but, otherwise, there appear to be no differences between incidences of tumors in BR and CV rats.

E. Limitations

The introduction of laboratory rats produced under circumstances rendering them initially free of all recognized pathogens (or even when contaminated only with organisms of low pathogenic potential) greatly aids efforts toward reducing research variables. However, unless continuously shielded by appropriate barrier or containment programs, such animals become exposed to pathogens and the process of conventionalization is initiated once again. Therein lies the major problem at present. While advances in gnotobiology provided easy methods for eliminating pathogens from rodent stocks, it has proved virtually impossible to maintain them free of pathogens for extended periods of time! The rapidity with which such animals become exposed, infected, and exhibit the clinical signs of infectious disease depends on complex interactions of many factors.

Although there is little published evidence of disease transmission during the shipment phase (between leaving the vendor and introduction to the consumer's facility), the circumstances of closely packed boxes from vendors of BR or SPF rats with those from CV sources at certain pooling points (e.g., airplane cargo holds, delivery vans, receiving areas, "quarantine rooms") provide abundant opportunity if filtered shipping containers are not used.

Once received at the using facility, the process of continuous shielding must be maintained in order to continue the BR or SPF condition through the investigative phases. Although discussed in greater detail in Section II of this chapter, use of such management practices as unidirectional flow of ventilatory air with controlled temperature and humidity, equipment, food and bedding, and personnel from areas of low contamination ("clean") to areas of high contamination ("dirty"); filtered cages, sterilized bedding; sterilized or pasteurized diets; treated drinking water, adequate equipment and surface sanitization; disease monitoring programs; etc., are all links in a chain that retard or prevent exposure of rats to infectious agents. The opportunity for exposure to pathogens is greatly influenced by the health status of other animals resident within the facility. Thus, if a mixture of both CV and SPF animals is present, the opportunities for exposure are greater than in circumstances where vendor selection programs may be used to limit introduction of CV animals.

Finally, many factors may play roles in determining whether, once exposed, animals will develop a latent infection or overt disease that will interfere with their use as laboratory subjects. Although genetic differences in susceptibility of rats to a variety of bacterial pathogens have been the subject of investigation (201,222), little has been accomplished in a practical way to exploit these differences as a means of disease control.

F. Monitoring

Routine monitoring of the animals and environment of a barrier is essential to assure that microbial exclusion is being achieved. Microbiological monitoring includes testing sterilization procedures and culturing room surfaces, equipment, food, bedding, water, air, etc. Monitoring the animal occupants of the barrier is perhaps the most sensitive and relevant indicator of change in microbial environment and presence of unwanted organisms. Testing of BR rat colonies usually requires periodic (quarterly necropsy of a portion of the population (0.25 to 0.5%) (127). Testing should include culture for bacteria (including mycoplasmas), serological testing for

viruses, parasitological tests, and histopathological search for lesions of infectious diseases (29,126–128,161,164a,221,247).

V. ETHICAL AND LEGAL CONSIDERATIONS

Previous discussions in this chapter have emphasized the adoption of housing methods that achieve optimal scientific results. It should be apparent that humane treatment is equally important and entirely consistent with scientific considerations. Scientists and even humane workers tend to consider humane treatment of rats in a context separate from (and below) that given to pet species and primates. Laboratory rats are intelligent, sentient creatures that respond positively to kindness. Therefore, they should be extended the same consideration for their comfort and well being as any other species.

In the United States laws and policies regulating the use of rats in biomedical research include The Animal Welfare Act of 1966 (and its amendments of 1970 and 1976), the National Institutes of Health Policy on Responsibility for Care and Use of Animals (1978), and the Good Laboratory Practices Act of 1978.

Institutions using laboratory rats may participate in a voluntary accreditation program conducted by the American Association for Accreditation of Laboratory Animal Care (AAALAC). Accreditation requires compliance with standards that are appreciably higher than the minimal standards of the Animal Welfare Act. The published standards by which AAALAC judges animal care is the "Guide for the Care and Use of Laboratory Animals" (165). Accreditation by AAALAC is accepted by the Department of Health, Education and Welfare as assurance that an institution's animal care program complies with NIH policy.

ACKNOWLEDGMENTS

The authors gratefully acknowledge the contribution of photographs by Mr. Neil Campbell, Director of Design, Lab Products, Inc.; Dr. Frank M. Loew, Associate Professor, Johns Hopkins School of Medicine; and Mr. E. D. Olfert, Animal Resources Centre, University of Saskatchewan, Canada.

REFERENCES

1. Acheson, E. D., Cowdell, R. H., Hadfield, E., and Macbeth, R. G. (1968). Nasal cancer in woodworkers in the furniture industry. *Br. Med. J.* **2**, 587–596.
2. Albrecht, I., and Souhrada, J. (1971). Defining the laboratory rat for cardiovascular research. *In* "Defining the Laboratory Animal," pp. 616–625. Natl. Acad. Sci., Washington, D. C.

3. Alder, R. (1967). Behavioral and physiological rhythms and the development of gastric erosions in the rat. *Psychosom. Med.* **29,** 345–353.

4. Alder, R., and Friedman, S. B. (1968). Plasma cortisone response to environmental stimulation: Effects of duration of stimulation and the 24-hour adrenocortical rhythm. *Neuroendocrinology* **3,** 378–386.

5. Allander, C., and Abel, E. (1973). Some aspects of the differences of air conditions inside a cage for small laboratory animals and its surroundings. *Z. Versuchstierkd.* **15,** 20–34.

6. American Conference of Governmental Industrial Hygienists (1972). "Air Sampling Instruments for Evaluation of Atmospheric Contaminants," 4th ed. ACGIH, Cincinnati, Ohio.

7. American Conference of Governmental Industrial Hygienists (1976). "Documentation of the Threshold Limit Values for Chemical Substances and Physical Agents in the Workroom Environment with Supplements for those Substances Added or Changed Since 1971," 3rd ed., p. 11. ACGIH, Cincinnati, Ohio.

8. Amkraut, A. A., Solomon, G. F., and Kraemer, H. C. (1971). Stress, early experience and adjuvant-induced arthritis in the rat. *Psychosom. Med.* **33,** 203–214.

9. Ander, R. (1967). The influence of psychological factors on disease susceptibility in animals. *In* "Husbandry of Laboratory Animals" (M. L. Conalty, ed.), pp. 219–238. Academic Press, New York.

10. Angevine, D. M., and Furth, J. (1942). A fatal disease of middle-aged mice characterized by myocarditis associated with hemorrhage in the pleural cavity. *Am. J. Pathol.* **19,** 187–195.

11. Anthony, A. (1962). Criteria for acoustics in animal housing. *Lab. Anim. Care* **13,** 340–347.

12. Aschoff, J. (1967). Desynchronization and resynchronization of human circadian rhythms. AGARD *Conf. Proc.* **25,** 1–11.

13. Austin, C. R., and Braden, A. W. H. (1954). Time relations and their significance in the ovulation and penetration of eggs in rats and rabbits. *Aust. J. Biol. Sci.* **7,** 179–194.

14. Baetjer, A. M. (1968). Role of environmental temperature and humidity in susceptibility to disease. *Arch. Environ. Health* **16,** 565–570.

15. Baker, H. J., Cassell, G. H., and Lindsey, J. R. (1971). Research complications due to *Haemobartonella* and *Eperythrozoon* infections in experimental animals. *Am. J. Pathol.* **64,** 625–656.

16. Bakerman, H., Romine, M., Schricher, I. A., Takahashi, S. M., and Mickelson, O. (1956). Stability of certain B vitamins exposed to ethylene oxide in the presence of diolene chloride. *J. Agric. Food Chem.* **4,** 956–959.

17. Balazs, T., and Dairman, W. (1967). Comparison of microsomal drug-metabolizing enzyme systems in grouped and individually caged rats. *Toxicol. Appl. Pharmacol.* **10,** 409–410.

18. Barrett, A. M., and Stockham, M. A. (1963). The effect of housing conditions and simple experimental procedures upon the corticosterone level in the plasma of rats. *J. Endocrinol.* **26,** 97–105.

19. Beall, J. R., Torning, F. E., and Runkle, R. S. (1971). A laminar flow system for animal maintenance. *Lab. Anim. Sci.* **21,** 206–212.

20. Berg, B. N. (1967). Longevity studies in rats. II. Pathology of aging rats. *In* "Pathology of Laboratory Rats and Mice" (E. Cotchin and F. J. C. Roe, eds.), pp. 749–785. Davis, Philadelphia, Pennsylvania.

21. Berg, B. N., and Simms, H. S. (1960). Nutrition and longevity in the rat. II. Longevity and onset of disease with different levels of food intake. *J. Nutr.* **71,** 255–263.

22. Berg. B. N., and Simms, H. S. (1961). Nutrition and longevity in the rat. III. Food restriction beyond 800 days. *J. Nutr.* **74,** 23–32.

23. Bernard C. (1865). "An Introduction to the Study of Experimental Medicine" (English translation, 1927, by H. C. Guen, Dover, New York; reprinted 1957).

24. Besch, E. L. (1975). Animal cage room dry-bulb and dew-point temperature differentials. *ASHRAE Trans.* **81,** 549–558.

25. Besch, E. L. (1975). Animal cage room dry-bulb and dew-point temperature differentials. *Lab. Anim. Sci.* **27,** 54–59.

26. Binns, R., Clark, G. C., and Healey, P. (1971). Physiological and histopathological measurements on lungs of laboratory rats maintained under barrier and under conventional conditions. *Lab. Anim.* **5,** 57–66.

26a. Blass, E. M., ed. (1976). "The Psychobiology of Curt Richter." New York Press, Baltimore, Maryland.

27. Bock, F. G., Myers, H. K., and Fox, H. N. (1975). Cocarcinogenic activity of peroxy compounds. *J. Natl. Cancer Inst.* **55,** 1359–1361.

28. Bowman, R. E., Wolf, R. C., and Sackett, G. P. (1970). Circadian rhythms of plasma 17-hydrocorticosteroids in the infant monkey. *Proc. Soc. Exp. Biol. Med.* **133,** 342–344.

29. Box, P. G. (1976). Criteria for producing high quality animals for research. *Lab. Anim. Sci.* **26,** 334–338.

30. Brick, J. O., Newell, R. R., and Doherty, D. Go. (1969). A barrier system for a breeding and experimental rodent colony: Description and operation. *Lab. Anim. Care* **19,** 93–97.

31. Briel, J. E., Kruckenberg, S. M., and Besch, E. L. (1972). "Observations of Ammonia Generation in Laboratory Animal Quarters," Publ. No. 72-03. Inst. Environ. Res., Kansas State University, Manhattan.

32. Broderson, J. R., Lindsey, J. R., and Crawford, J. E. (1976). The role of environmental ammonia in respiratory mycoplasmosis of rats. *Am. J. Pathol.* **85,** 115–130.

33. Bullock, B. C., Banks, K. L., and Manning, P. J. (1968). "The Laboratory Animal in Gerontological Research," Publ. No. 1951. Inst. Lab. Anim. Resour., Natl. Acad. Sci., Washington, D. C.

34. Burkhart, C. A., and Robinson, J. L. (1978). High rat pup mortality attributed to the use of cedar-wood shavings as bedding. *Lab. Anim.* **12,** 221–222.

35. Burns, J. H. (1961). The effect of temperature on the response to drugs. *Br. Med. Bull.* **17,** 66–69.

36. Burns, J. J. (1968). Variation of drug metabolism in animals and the prediction of drug action in man. *Ann. N.Y. Acad. Sci.* **151,** 959–967.

37. Burns, R. F., Timmons, E. H., and Poiley, S. M. (1971). Serum chemistry and hematological valves for axenic (germfree) and environmentally associated inbred rats. *Lab. Anim. Sci.* **21,** 415–419.

38. Campbell, T. C., and Hayes, J. R. (1974). Role of nutrition in the drug-metabolizing enzyme system. *Pharmacol. Rev.* **26,** 171–197.

39. Carthew, P., and Verstraete, A. (1978). Serological survey of accredited breeding colonies in the United Kingdom for common rodent viruses. *Lab. Anim.* **12,** 29–32.

40. Cayen, M. N., Givner, M. L., and Kraml, M. (1972). Effect of diurnal rhythm and food withdrawal on serum lipid levels in the rat. *Experientia* **28,** 502–503.

41. Chedid, A., and Nair, V. (1972). Diurnal rhythm in endoplasmic reticulum of rat liver: Electron microscopic study. *Science* **175,** 176–179.

42. Chipman, R. K., and Bronson, F. H. (1968). Pregnancy blocking capacity and inbreeding in laboratory mice. *Experientia* **24,** 199–200.

43. Christian, J. J., and LeMunyan, C. D. (1958). Adverse effects of crowding on lactation and reproduction of mice and two generations of their progeny. *Endocrinology* **63,** 517–529.

44. Christie, R. J., Williams, F. P., Whitney, R. A., Jr., and Johnson, D. J. (1968). Techniques used in the establishment and maintenance of a barrier mouse breeding colony. *Lab. Anim. Care* **18,** 544–549.

45. Cinti, D. L., Lemelin, M. A., and Christian, J. (1976). Induction of liver microsomal mixed-function oxidases by volatile hydrocarbons. *Biochem. Pharmacol.* **25,** 100–103.

46. Clarke, H. E., Coates, M. E., Eva, J. K., Ford, D. J., Milner, C. K., O'Donoghue, P. N., Scott, P. P., and Ward, J. J. (1977). Dietary standards for laboratory animals: Report of the Laboratory Animals Centre Diets Advisory Committee. *Lab. Anim.* **11,** 1–28.

47. Clough, G. (1976). The immediate environment of the laboratory animal. *In* "Control of the Animal House Environment" (T. McSheehy,

ed.), Lab. Anim. Handb. No. 7, pp. 77–94. Lab. Anim. Ltd., London.

48. Clough, G., and Fasham, J. A. L. (1975). A "silent" fire alarm. *Lab. Anim.* **9,** 193–196.

49. Cohn, D. A., and Hamilton, J. B. (1976). Sensitivity to androgen and the immune response: Immunoglobulin levels in two strains of mice, one with high and one with low target organ responses to androgen. *J. Reticuloendothel. Soc.* **20,** 1–10.

50. Coleman, D. J., Zbijewska, J. R., and Smith, N. L. (1971). Hematologic values in chronic murine pneumonia. *Lab. Anim. Sci.* **21,** 721–726.

51. Conney, A. H. (1969). Drug metabolism and therapeutics. *N. Engl. J. Med.* **280,** 653–660.

52. Conney, A. H., and Burns, J. J. (1972). Metabolic interactions among environmental chemicals and drugs. *Science* **178,** 676–586.

53. Consolo, S., Garrattini, S., and Valzelli, L. (1965). Sensitivity of aggressive mice to centrally acting drugs. *J. Pharm. Pharmacol.* **17,** 594–595.

54. Dairman, W., and Balzs, T. (1970). Comparison of liver microsomal systems and barbiturate sleep times in rats caged individually or communally. *Biochem. Pharmacol.* **19,** 951–955.

55. Davis, D. E. (1978). Social behavior in a laboratory environment. *In* "Laboratory Animal Housing," pp. 44–63. Inst. Lab. Anim. Resour., Acad. Sci., Washington, D. C.

56. Descôteaux, J. P., Grignon-Archambault, D., and Lussier, G. (1977). Serologic study on the prevalence of murine viruses in five Canadian mouse colonies. *Lab. Anim. Sci.* **27,** 621–626.

57. DeWitt, G. H., ed. (1976). "Symposium on Biological Effects and Measurement of Light Sources," DHEW (FDA) Publ. No. 77-8002. U.S. Govt. Printing Office, Washington, D. C.

58. Dominic, C. J. (1966). Observations on the reproductive pheromones of mice. *J. Reprod. Fertil.* **11,** 407–414.

59. Dubos, R. J., Schaedler, R. W., Costello, R., and Hoet, P. (1965). Indigenous, normal and autochthonous flora of the gastrointestinal tract. *J. Exp. Med.* **122,** 67–82.

60. Dymsza, H. A., Miller, S. A., Maloney, J. R., and Foster, H. L. (1963). Equilibration of the laboratory rat following exposure to shipping stresses. *Lab. Anim. Care* **13,** 60–65.

61. Ede, M. D. M. (1974). Circadian rhythms of drug effectiveness and toxicity. *Clin. Pharmacol. Ther.* **14,** 925–935.

62. Eschenbrenner, A. B., and Miller, E. (1945). Sex differences in kidney morphology and chloroform necrosis. *Science* **102,** 302–303.

63. Etkin, W., ed. (1965). "Social Behavior and Organization Among Vertebrates." Univ. of Chicago Press, Chicago, Illinois.

64. Euker, J. S., Meites, J., and Riegle, G. D. (1975). Effects of acute stress on serum LH and prolactin in intact, castrate and dexamethasone-treated male rats. *Endocrinology* **96,** 85–92.

65. Exon, J. H., Patton, N. M., and Koller, L. D. (1975). Hexamitiasis in cadmium-exposed mice. *Arch. Environ. Health* **30,** 463–464.

66. Farris, E. J. (1950). The rat as an experimental animal. *In* "The Care and Breeding of Laboratory Animals" (E. J. Farris, ed.), pp. 43–78. Wiley, New York.

67. Feigin, R. D., San Joaquin, V. H., Haymond, M. W., and Wyatt, R. G. (1969). Daily periodicity of susceptibility of mice to pneumococcal infection. *Nature (London)* **224,** 379–380.

68. Ferguson, H. C. (1966). Effect of red cedar chip bedding on hexobarbital and pentobarbital sleep time. *J. Pharm. Sci.* **55,** 1142–1148.

69. Fidler, I. J. (1977). Depression of macrophages in mice drinking hyperchlorinated water. *Nature (London)* **270,** 735–736.

70. Fioretti, M. C., Riccardi, C., Menconi, E., and Martini, L. (1974). Control of the body temperature in the rat. *Life Sci.* **14,** 2111–2119.

71. Fletcher, J. I. (1976). Influence of noise on animals. *In* "Control of the Animal House Environment" (T. McSheehy, ed.), Lab. Anim. Handb. No. 7, pp. 51–62. Lab. Anim. Ltd., London.

72. Flynn, R. J. (1959). Studies on the etiology of ringtail of rats. *Proc. Anim. Care Panel* **9,** 155–160.

73. Flynn, R. J. (1963). *Pseudomonas aeruginosa* infection and radiobiological research at the Argonne National Laboratory: Effects, diagnosis, epizootiology and control. *Lab. Anim. Care* **13,** 25–35.

74. Flynn, R. J. (1963). The diagnosis of *Pseudomonas aeruginosa* infection of mice. *Lab. Anim. Care* **13,** 126–129.

75. Flynn, R. J. (1968). A new cage cover as an aid to laboratory rodent disease control. *Proc. Soc. Exp. Biol. Med.* **129,** 714–717.

76. Flynn, R. J., Poole, C. M., and Tyler, S. A. (1971). Long distance air transport of aged laboratory mice. *J. Gerontol.* **26,** 201–203.

77. Ford, D. J. (1977). Effect of autoclaving and physical structure of diets on their utilization by mice. *Lab. Anim.* **11,** 235–239.

78. Ford, D. J. (1977). Influence of diet pellet hardness and particle size on food utilization by mice, rats and hamsters. *Lab. Anim.* **11,** 241–246.

79. Foster, H. L. (1958). Large scale production of rats free of commonly occurring pathogens and parasites. *Proc. Anim. Care Panel* **8,** 92–100.

80. Foster, H. L. (1962). Establishment and operation of SPF colonies. *In* "Problems of Laboratory Animal Disease" (R. J. C. Harris, ed.), pp. 249–259. Academic Press, New York.

81. Foster, H. L., Black, C. O., and Pfan, E. S. (1964). A pasteurization process for pelleted diets. *Lab. Anim. Care* **14,** 373–381.

82. Foster, H. L., Foster, S. J., and Pfan, E. S. (1963). The large scale production of caesarian originated barrier sustained mice. *Lab. Anim. Care* **13,** 711–718.

83. Foster, H. L., and Pfan, E. S. (1963). Gnotobiotic animal production at the Charles River Breeding Laboratories, Inc. *Lab. Anim. Care* **13,** 609–632.

84. Foster, H. L., Trexler, P. C., and Rumsey, G. (1967). A canned sterile shipping diet for small laboratory rodents. *Lab. Anim. Care* **17,** 400–405.

85. Fouts, J. R. (1970). Some effects of insecticides on hepatic microsomal enzymes in various animal species. *Rev. Can. Biol.* **29,** 377–389.

86. Fouts, J. R. (1976). Overview of the field: Environmental factors affecting chemical or drug effects in animals. *Fed. Proc., Fed. Am. Soc. Exp. Biol.* **35,** 1162–1165.

87. Friedman, M., Byers, S. O., and Brown, A. E. (1967). Plasma lipid responses of rats and rabbits to an auditory stimulus. *Am. J. Physiol.* **212,** 1174–1178.

88. Fritz, T. E., Brennen, P. C., Giolitti, J. A., and Flynn, R. J. (1968). Interrelationships between x-irradiation and the intestinal flora of mice. *In* "Gastro-Intestinal Radiation Injury" (M. F. Sullivan, ed.), pp. 279–290. Excerpta Med. Found., Amsterdam.

89. Fritz, T. E., Tolle, D. V., and Flynn, R. J. (1968). Hemorrhagic diathesis in laboratory rodents. *Proc. Soc. Exp. Biol. Med.* **128,** 228–234.

90. Fuhrman, G. J., and Fuhrman, F. A. (1961). Effects of temperature on the action of drugs. *Annu. Rev. Pharmacol.* **1,** 65–78.

91. Gamble, M. R., and Clough, G. (1976). Ammonia build-up in animal boxes and its effect on rat tracheal epithelium. *Lab. Anim.* **10,** 93–104.

92. Geber, W. F., Anderson, T. A., and Van Dyne, V. (1966). Physiologic responses of the albino rat to chronic noise stress. *Arch. Environ. Health* **12,** 751–754.

92a. Gelineo, S. (1934). Influence du milieu thermique d'adaptation sur la thermogenese des homéothermes. *Ann. Physiol. Physicochim. Biol.* **10,** 1083.

93. Giddens, W. E., Jr., and Whitehair, C. K. (1969). The peribronchial lymphocytic tissue in germfree defined-flora, conventional and chronic murine pneumonia affected rats. *In* "Germfree Biology" (E. Mirando and N. Black, eds.), pp. 75–84. Plenum, New York.

94. Giddens, W. E., Jr., Whitehair, C. K., and Carter, G. (1971). Morphologic and microbiologic features of trachea and lungs in germfree, defined-flora, conventional and chronic respiratory disease-

affected rats. *Am. J. Vet. Res.* **32,** 115–129.

95. Gisler, R. H., Bussard, A. E., Mazie, J. C., and Hess, R. (1971). Hormonal regulation of the immune response. I. Induction of an immune response *in vitro* with lymphoid cells from mice exposed to acute systemic stress. *Cell. Immunol.* **2,** 634–645.

96. Gisler, R. H., and Schenkel-Hulliger, L. (1971). Hormonal regulation of the immune response. II. Influence of regulation of pituitary and adrenal activity on immune responsiveness *in vitro. Cell. Immunol.* **2,** 646–657.

97. Gledhill, A. W. (1965). Significance of pathogen burden. *Food Cosmet. Toxicol.* **3,** 37–46.

98. Godwin, K. O., Fraser, F. J., and Ibbotson, R. N. (1964). Haematological observations on healthy (SPF) rats. *Br. J. Exp. Pathol.* **45,** 514–524.

99. Gorton, R. L., Woods, J. E., and Besch, E. L. (1976). System load characteristics and estimation of animal heat loads for laboratory animal facilities. *ASHRAE Trans.* **82,** 107–111.

100. Grant, L., Hopkinson, P., Jennings, G., and Jenner, F. A. (1971). Period of adjustment of rats used for experimental studies. *Nature (London)* **232,** 135.

101. Halberg, F., Halber, E., Barnum, C. P., and Bittner, J. J. (1959). Physiologic 24-hour periodicity in human beings and mice, the lighting regimen and daily routine. *Publ. Am. Assoc. Adv. Sci.* **55,** 803–879.

102. Halberg, F., Haus, E., Cordoso, S. S., Scheving, L. E., Kuhl, J. F. W., Shiotsuka, R., Rosene, G., Pauly, J. E., Runge, W., Spaulding, J. F., Lee, J. K., and Good, R. A. (1973). Toward a chronotherapy of neoplasic tolerance of treatment depends upon host rhythms. *Experientia* **29,** 909–934.

103. Hanna, M. G., Jr., Nettesheim, P., Richter, C. B., and Tennant, R. W. (1973). The variable influence of host microflora and intercurrent infections on immunological competence and carcinogenesis. *Isr. J. Med. Sci.* **9,** 229–238.

104. Hardy, J. D. (1961). Physiology of temperature regulation. *Physiol. Rev.* **41,** 521–606.

105. Hart, L., and Fouts, J. (1965). Further studies on the stimulation of hepatic microsomal drug metabolizing enzymes by DDT and its analogs. *Arch. Exp. Pathol. Pharmakol.* **249,** 486–500.

106. Hashimoto, M., Davis, D. C., and Gillette, J. R. (1972). Effect of different routes of administration of cedrene on hepatic drug metabolisms. *Biochem. Pharmacol.* **21,** 1514–1517.

107. Hastings, J. S. (1967). Long term use of vermiculite. *J. Inst. Anim. Tech.* **18,** 184–190.

108. Hastings, J. W. (1970). The biology of circadian rhythms from man to microorganisms. *N. Engl J. Med* **282,** 435–441.

109. Hastings, J. W., and Menaker, M. (1976). Physiological and biochemical aspects of circadian rhythms. *Fed. Proc., Fed. Am. Soc. Exp. Biol.* **35,** 2325–2357.

110. Haus, E., Halberg, F., Scheving, L., Pauly, J. E., Cordoso, S., Kuhl, J. F. W., Sothern, R. B., Shiotsuka, R. N., and Hwang, D. W. (1972). Increased tolerance of leukemic mice to arabinosyl cytosine given on schedule adjusted to circadian system. *Science* **177,** 80–82.

111. Hemphill, F. E., Kaeberle, M. L., and Buck, W. B. (1971). Lead suppression of mouse resistance to *Salmonella typhimurium. Science* **172,** 1031–1032.

112. Hensel, H. (1955). Mensch und Warmblutige Tiere. *In* "Temperatur und Leben" (H. Precht, J. Christophersen, and H. Hensel, eds.), pp. 329–466. Springer-Verlag, Berlin and New York.

113. Hoffman, S. A., Paschkis, K. E., Debias, D. A., Cantarow, A., and Williams, T. L. (1962). The influence of exercise on the growth of transplanted rat tumors. *Cancer Res.* **22,** 597–599.

114. Holdenreid, R., ed. (1966). "Viruses of Laboratory Rodents," Natl. Cancer Inst. Monog. No. 20, p. XIII–XIV. Natl. Cancer Inst., Bethesda, Maryland.

115. Holt, P. G., and Keast, D. (1977). Environmentally induced changed in immunological function: Acute and chronic effects of tobacco smoke and other atmospheric contaminants in man and experimental animals. *Bacteriol. Rev.* **41,** 205–216.

116. Institute of Laboratory Animal Resources (1970). "Gnotobiotes. Standards and Guideline for the Breeding, Care, and Management of Laboratory Animals." Natl. Acad. Sci., Washington, D. C.

117. Institute of Laboratory Animal Resources (1972). "Nutrient Requirements of Laboratory Animals," 2nd rev. No. 10. Inst. Lab. Anim. Resour., Natl. Acad. Sci., Washington, D. C.

118. Institute of Laboratory Animal Resources (1975). "Animals for Research," 9th ed. Inst. Lab. Anim. Resour., Natl. Acad. Sci., Washington, D. C.

119. Institute of Laboratory Animal Resources (1978). "Laboratory Animal Housing," Inst. Lab. Anim. Resour., Natl. Acad. Sci., Washington, D. C.

120. Irvani, J., and van As, A. (1972). Mucus transport in the tracheobronchial tree of normal and bronchitic rats. *J. Pathol.* **106,** 81–93.

121. Izquierdo, J. N., and Gibbs, S. J. (1972). Circadian rhythms of DNA synthesis and mitotic activity in hamster cheek pouch epithelium. *Exp. Cell Res.* **71,** 402–408.

122. Joasoo, A., and McKenzie, J. M. (1976). Stress and immune response in rats. *Int. Arch. Allergy Appl. Immunol.* **50,** 659–663.

123. Johnson, T., Lavender, J. F., Hultin, E., and Rasmussen, A. F., Jr. (1963). The influence of avoidance-learning stress on resistance to cocksackie B virus in mice. *J. Immunol.* **91,** 569–575.

124. Johnstone, M. W., and Scholes, P. F. (1976). Measuring the environment. *In* "Control of the Animal House Environment," pp. 113–128. Lab. Anim. Ltd., London.

125. Jonas, A. M. (1965). Laboratory animal facilities. *J. Am. Vet. Med. Assoc.* **146,** 600–606.

126. Jonas, A. M. (1973). The role of veterinary diagnostic support laboratories in a research animal colony. *In* "Research Animals in Medicine" (L. T. Harmison, ed.), DHEW Publ. No. (NIH) 72-333. U.S. Govt. Printing Office, Washington, D. C.

127. Jonas, A. M., chairman (1976). Long-term holding of laboratory rodents. *ILAR News* **19,** 1–25.

128. Jonas, A. M. (1976). The research animal and the significance of a health monitoring program. *Lab. Anim. Sci.* **26,** 339–344.

129. Jori, A., Bianchetti, A., and Prestini, P. E. (1969). Effect of essential oils on drug metabolism. *Biochem. Pharmacol.* **18,** 2081–2085.

130. Jori, A., Di Salle, E., and Santini, V. (1971). Daily rhythmic variation and liver drug metabolism in rats. *Biochem. Pharmacol.* **29,** 2965–2969.

131. Kalser, S. C., and Kunig, R. (1969). Effect of varying periods of cold exposure on the action and metabolism of hexobarbital. *Biochem. Pharmacol.* **18,** 405–412.

132. Kappel, H. K., Kappel, J. P., Weisbroth, S. H., and Kozma, C. K. (1969). Establishment of a hysterectomy-derived pathogen-free breeding nucleus of BLU:(LE) rats. *Lab. Anim. Care* **19,** 738–742.

133. Keast, D., and Coales, M. R. (1967). Lymphocytopenia induced in a strain of laboratory mice by agents commonly used in treatment of ectoparasites. *Aust. J. Exp. Biol. Med. Sci.* **45,** 645–650.

134. King, T. K. C. (1966). Mechanical properties of the lung in the rat. *J. Appl. Physiol.* **21,** 259–264.

135. King, T. K. C., and Bell, D. P. (1966). Arterial blood gases in specific-pathogen-free and bronchitic rats. *J. Appl. Physiol.* **21,** 237–241.

136. Kline, L. B., and Hull, R. N. (1960). The virucidal properties of peracetic acid. *Am. J. Clin. Pathol.* **33,** 30–33.

137. Knapka, S. J., Smith, K. P., and Judge, F. T. (1974). Effect of open and closed formula rations on the performance of three strains of laboratory mice. *Lab. Anim. Sci.* **24,** 480–487.

138. Koller, L. D. (1973). Immunosuppression produced by lead, cadmium, and mercury. *Am. J. Vet. Res.* **34,** 1457–1458.

139. Koller, L. D., and Kovacic, S. (1974). Decreased antibody formation in mice exposed to lead. *Nature (London)* **250,** 148–150.

140. Kolmodin, B., Azarnoff, D. L., and Sjöqvist, F. (1969). Effect of environmental factors in drug metabolism: Decreased plasma half-life of antipyrine in workers exposed to chlorinated hydrocarbon insecticides. *Clin. Pharmacol. Ther.* **10,** 638–642.

141. Kraft, L. M. (1958). Observations on the control and natural history of epidemic diarrhea of infant mice (EDIM). *Yale J. Biol. Med.* **31,** 121–137.

141a. Lalich, J. L., and Allen, J. R. (1971). Protein overload nephropathy in rats with unilateral nephrectomy. II. Ultrastructural study. *Arch. Pathol.* **91,** 372–382.

142. Lamb, D. (1975). Rat lung pathology and quality of laboratory animals: The users view. *Lab. Anim.* **9,** 1–8.

143. Lang, C. M., chairman (1977). Laboratory animal management: Rodents. *ILAR News* **20,** L1–L15.

144. Lang, C., and Harrell, G. T., Jr. (1971). "A Comprehensive Animal Facility for a College of Medicine." USHEW, PHS, NIH, Bethesda, Maryland.

145. Lang, C., and Vesell, E. S. (1976). Environmental and genetic factors affecting laboratory animals: Impact on biomedical research. *Fed. Proc., Fed. Am. Soc. Exp. Biol.* **35,** 1123–1124.

146. LaVail, M. M. (1976). Survival of some photoceptor cells in albino rats following long-term exposure to continuous light. *Invest. Ophthalmol.* **15,** 64–72.

147. LeBouton, A. V., and Handler, S. D. (1971). Persistent circadian rhythmicity of protein synthesis in liver of starved rats. *Experientia* **27,** 1031–1032.

148. Les, E. P. (1968). Effect of acidified-chlorinated water on reproduction in C_3H/HeJ and $C_{57}B1/6J$ mice. *Lab Anim. Care* **18,** 210–213.

149. Ley, F. J., Bleby, J., Coates, M. E., and Patterson, J. S. (1969). Sterilization of laboratory animal diets using gamma radiation. *Lab. Anim.* **3,** 221–254.

150. Lindsey, J. R., Baker, H. J., Overcash, R. G., Cassell, G. H., and Hunt, C. E. (1971). Murine chronic respiratory disease: significance as a research complication and experimental production with *Mycoplasma pulmonis*. *Am. J. Path.* **64,** 675–716.

151. Lindsey, J. R., Cassell, G. H., and Baker, H. J. (1978). Diseases due to mycoplasmas and rickettsias. *In* "Pathology of Laboratory Animals" (K. Benirschke, F. M. Garner, and T. C. Jones, eds.), Vol. II, pp. 1481–1550. Springer-Verlag, Berlin and New York.

152. Lindsey, J. R., and Conner, M. W. (1978). Influence of cage sanitization frequency on intracage ammonia (NH_2) concentration and progression of murine respiratory mycoplasmosis in the rat. *Zentralbl. Bakteriol. Parasitenkd. Infektionskr. Hyg.* **241,** 215.

153. Lindsey, J. R., Conner, M. W., and Baker, H. J. (1978). Physical, chemical and microbial factors affecting biologic response. *In* "Laboratory Animal Housing," pp. 37–43. Inst. Lab. Anim. Resour., Natl. Acad. Sci., Washington, D. C.

154. Luckey, T. D. (1963). "Germfree Life and Gnotobiology," pp. 486–489. Academic Press, New York.

155. MacManus, J. P., Whitfield, J. F., and Youdale, T. (1971). Stimulation by epinephrine of adenyl cyclase activity, cyclic AMP formation, DNA synthesis, and cell proliferation in populations of rat thymic lymphocytes. *J. Cell. Comp. Physiol.* **77,** 103–106.

156. Manning, P. J., and Banks, K. L. (1968). The effects of transportation on the aged rat. *N.A.S.—N.R.C., Publ.* **1591,** 98–103.

157. McDougal, P. T., Wolf, N. S., Stenback, W. A., and Trentin, J. J. (1967). Control of *Pseudomonas aeruginosa* in an experimental mouse colony. *Lab. Anim. Care* **17,** 204–214.

158. McPherson, C. W. (1963). Reduction of *Pseudomonas aeruginosa* and coliform bacteria in mouse drinking water following treatment with hydrochloric acid or chlorine. *Lab. Anim. Care* **13,** 737–744.

159. McSheehy, T., ed. (1976). "Control of the Animal House Environment." Lab. Anim. Ltd., London.

160. Meier, H., and Hoag, W. G. (1966). Blood coagulation. *In* "Biology of the Laboratory Mouse" (E. L. Green, ed.), pp. 373–376. McGraw-Hill, New York.

161. "Microbiological Examination of Laboratory Animals for Purposes of Accreditation." Med. Res. Counc., Lab. Anim. Centre, Woodmansterne Road, Carshalton, Surrey, England.

162. Miltenberger, R., and Oltersdorf, U. (1978). The B-vitamin group and the activity of hepatic microsomal mixed-function oxidases of the growing Wistar rat. *Br. J. Nutr.* **39,** 127–137.

163. Mitropoulos, K. A., Balasubcamaniam, S., Gibbons, G. F., and Reaves, B. E. A. (1972). Diurnal variation in the activity of cholesterol 7-hydroxylase in the livers of fed and fasted rats. *FEBS Lett.* **27,** 203–206.

164. Mock, E. J., Norton, H. W., and Frankel, A. I. (1978). Daily rhythmicity of serum testosterone concentration in the male laboratory rat. *Endocrinology* **103,** 1111–1121.

164a. Mohr, J. W. (1974). Health screening of incoming animals. *Lab Animal (U.S.A.)* **3,** 22–57.

165. Moreland, A. F., chairman (1978). "Guide for the Care and Use of Laboratory Animals," DHEW Publ. No. (NIH) 78-23. U.S. Govt. Printing Office, Washington, D. C.

166. Mugford, R. A., and Newell, N. W. (1971). The preputial glands as a source of aggression-promoting odors in mice. *Physiol. Behav.* **6,** 247–249.

167. Mulder, J. B. (1971). Animal behavior and electromagnetic energy waves. *Lab. Anim. Sci.* **21,** 389–393.

168. Multiple authors (1971). Proceedings of the Symposium on "Environmental Requirements for Laboratory Animals," Publ. 71-02. Dept. Physiol. Sci., Coll. Vet. Med. and Inst. Environ. Res. Coll. Eng., Kansas State University, Manhattan.

169. Murakami, H. (1971). Differences between internal and external environments of the mouse cage. *Lab. Anim. Sci.* **21,** 680–684.

170. Murakami, H., and Watanabe, Y. (1973). Rhythm of water intake of mice in the daytime under continuous darkness. *J. Comp. Physiol. Psychol.* **85,** 272–276.

171. Nair, V., and Casper, R. (1969). The influence of light on daily rhythm in hepatic drug metabolizing enzymes in rat. *Life Sci.* **8,** 1291–1298.

172. Nash, R. E., and Llanos, J. M. E. (1971). Twenty-four hour variations in DNA synthesis of a fast-growing and a slow-growing hepatoma: DNA synthesis rhythm in hepatoma. *J. Natl. Cancer Inst.* **47,** 1007–1012.

173. Navia, J. M. (1977). Preparation of diets used in dental research. "Animal Models in Dental Research," pp. 151–167. Univ. of Alabama Press, University, Alabama.

174. Nebert, D. W., and Felton, J. S. (1976). Importance of genetic factors influencing the metabolism of foreign compounds. *Fed. Proc., Fed. Am. Soc. Exp. Biol.* **35,** 1133–1141.

175. Nelson, J. B. (1960). The problems of disease and quality in laboratory animals. *J. Med. Educ.* **35,** 34–43.

176. Nelson, J. B. (1967). Respiratory infections of rats and mice with emphasis on indigenous mycoplasms. *In* "Pathology of Laboratory Rats and Mice" (E. Cotchin and F. J. C. Roe, eds.), pp. 259–289. Davis, Philadelphia, Pennsylvania.

177. Nettesheim, P., Schrieber, H., Creasia, D. A., and Richter, C. B. (1974). Respiratory infections and the pathogenesis of lung cancer. *Recent Results Cancer Res.* **44,** 138–157.

178. Newberne, P. M. (1975). Influence on pharmacological experiments of

chemicals and other factors in diets of laboratory animals. *Fed. Proc., Fed. Am. Soc. Exp. Biol.* **34,** 209-218.

179. Newberne, P. M., chairman (1978). Control of diets in laboratory animal experimentation. *ILAR News* **21,** A3-A11.

180. Newberne, P. M., and Fox, J. G. (1978). Chemicals and toxins in the animal facility. *In* "Laboratory Animal Housing," pp. 118-138. *Inst. Lab. Anim. Resour.,* Natl. Acad. Sci., Washington, D. C.

181. Newton, W. M. (1978). Environmental impact on laboratory animals. *Adv. Vet. Sci. Comp. Med.* **22,** 1-28.

182. Njaa, L. R., Utne, F., and Bracklan, O. R. (1957). Effect of relative humidity on rat breeding and ringtail. *Nature (London)* **180,** 290-291.

183. Nolen, G. A., and Alexander, J. C. (1965). A comparison of the growth and fat utilization of caesarian derived and conventional albino rats. *Lab. Anim. Care* **15,** 295-304.

184. O'Steen, W. K., and Anderson, K. V. (1972). Photoreceptor degeneration after exposure of rats to incandescent illumination. *Z. Zellforsch. Mikrosk. Anat.* **127,** 306-313.

185. O'Steen, W. K., Anderson, K. V., and Shear, C. R. (1974). Photoreceptor degeneration in albino rats: Dependency on age. *Invest. Ophthalmol.* **13,** 334-339.

186. Paget, G. E., and Lemon, P. G. (1965). The interpretation of pathology data an analysis of the pathological lesions encountered in specific pathogen free rats. *In* "Pathology of Laboratory Animals" (W. E. Ribelin and J. R. McCoy, eds.), pp. 382-405. Thomas, Springfield, Illinois.

187. Parker, J. C., Tennant, R. W., and Ward, T. G. (1966). Prevalence of viruses in mouse colonies. *Natl. Cancer Inst. Monogr* **20,** 25-36.

188. Parker, L. F., and Radow, B. L. (1974). Isolation stress and volitional ethanol consumption in the rat. *Physiol. Behav.* **12,** 1-3.

189. Pick, J. R., and Little, J. M. (1965). Effect of type of bedding material on thresholds of pentylenetetrazol convulsions in mice. *Lab. Anim. Care* **15,** 29-33.

190. Plaut, S. M., Alder, R., Friedman, S. B., and Ritterson, A. L. (1969). Social factors and resistance to malaria in the mouse: Effects of group vs. individual housing on resistance to *Plasmodium berghei* infection. *Psychosom. Med.* **31,** 536-540.

191. Pleasants, J. R. (1974). Gnotobiotics. *In* "Handbook of Laboratory Animal Science," (E. C. Melby, Jr. and N. H. Altman, eds.), Vol. I, pp. 119-174. CRC Press, Inc., Cleveland, Ohio.

192. Poland, A., Smith, D., Kuntzman, R., Jacobson, M., and Conney, A. H. (1970). Effect of intensive occupational exposure to DDT on phenylbutazone and cortisol metabolism in human subjects. *Clin. Pharmacol. Ther.* **11,** 724-732.

193. Port, C. D., and Kaltenbach, J. P. (1969). The effect of corncob bedding on reproductivity and leucine incorporation in mice. *Lab. Anim. Care* **19,** 46-49.

194. Pucak, G. J., Lee, C. S., and Zaino, A. S. (1977). Effects of prolonged high temperature on testicular development and fertility in the male rat. *Lab. Anim. Sci.* **27,** 76-77.

195. Radzialowski, F. M., and Bousquet, W. F. (1967). Circadian rhythm in hepatic drug metabolizing activity in the rat. *Life Sci.* **6,** 2545-2548.

196. Radzialowski, F. M., and Bousquet, W. F. (1968). Daily rhythmic variation in hepatic drug metabolism in the rat and mouse. *J. Pharmacol. Exp. Ther.* **163,** 229-238.

197. Ramaley, J. A. (1972). Changes in daily serum corticosterone values in maturing male and female rats. *Steroids* **20,** 185-197.

198. Rasmussen, A. F., Jr. (1969). Emotions and immunity. *Ann. N.Y. Acad. Sci.* **164,** 458-461.

199. Reiter, R. J. (1973). Comparative effects of continual lighting and pinealectomy on the eyes, the harderian glands and reproduction in pigmented and albino rats. *Comp. Biochem. Physiol. A* **44,** 503-509.

200. Report of the Committee on Standardized Nomenclature of the Association for Gnotobiotics (1969). *Newsl. Assoc. Gnotobiotics* **5,** 3.

201. Robinson, R. (1965). "Genetics of the Norway Rat," pp. 354-365. Pergamon, Oxford.

202. Robinson, W. G., and Kuwabara, T. (1976). Light-induced alterations of retinal pigment epithelium in black, albino and beige mice. *Exp. Res.* **22,** 549-557.

203. Romero, J. A. (1976). Influence of diurnal cycles on biochemical parameters of drug sensitivity: The pineal gland as a model. *Fed. Proc., Fed. Am. Soc. Exp. Biol.* **35,** 1157-1161.

204. Rowe, W. P., Hartley, J. W., and Huebner, R. J. (1962). Polyoma and other indigenous mouse viruses. *In* "The Problems of Laboratory Animal Diseases" (R. J. C. Harris, ed.), p. 131. Academic Press, New York.

205. Sanvordeker, D. R., and Lambert, H. J. (1974). Environmental modification of mammalian drug metabolism and biological response. *Drug Metab. Rev.* **3,** 201-229.

206. Scheving, L. E., Vedral, D. F., and Pauly, J. E. (1968). A circadian susceptibility rhythm in rats to pentobarbital sodium. *Anat. Rec.* **160,** 741-749.

207. Schoenatal, R. (1973). Carcinogenicity of wood shavings. *Lab. Anim.* **7,** 47-49.

208. Schrieber, H., Nettesheim, P., Lijinsky, W., Richter, C. B., and Walburg, H. E., Jr. (1972). Induction of lung cancer in germfree, specific-pathogen-free, and infected rates by N-nitrosoheptamine: Enhancement by respiratory infection. *J. Natl. Cancer Inst.* **49,** 1107-1114.

209. Schroeder, H. A., Balassa, J. J., and Vinton, W. H., Jr. (1965). Chromium, cadmium, and lead in rats: Effects on life span, tumors, and tissue levels. *J. Nutr.* **86,** 51-66.

210. Selye, H., Tuchweber, B., and Bertok, L. (1966). Effect of lead acetate on susceptibility of rats to bacterial endotoxins. *J. Bacteriol.* **91,** 884-890.

211. Serrano, L. J. (1971). Defined mice in a radiobiological experiment. *In* "Defining the Laboratory Animal," pp. 13-41. Inst. Lab. Anim. Resour.—Natl. Acad. Sci., Washington, D. C.

212. Serrano, L. J. (1971). Carbon dioxide and ammonia in mouse cages: Effect of cage covers, population, and activity. *Lab. Anim. Sci.* **21,** 75-85.

213. Shackelford, P. G., and Feigin, R. D. (1973). Periodicity of susceptibility to pneumococcal infection: Influence of light and adrenocortical secretions. *Science* **182,** 285-287.

214. Silberberg, M., and Silberberg, R. (1955). Diet and life span. *Physiol. Rev.* **35,** 347-362.

215. Silverman, M. S., and La Via, M. F. (1976). Letter to the editor. *Immunology* **117,** No. 6.

216. Simmons, M. L., and Brick, J. O. (1970). "The Laboratory Mouse Selection and Management," pp. 153-156. Prentice-Hall, Englewood Cliffs, New Jersey.

217. Simmons, M. L., Robie, D. M., Jones, J. B., and Serrano, L. J. (1968). Effect of a filter cover on temperature and humidity in a mouse cage. *Lab. Anim.* **2,** 113-120.

218. Simms, H. S. (1967). Longevity studies in rats. I. Relation between life span and age of onset of specific lesions. *In* "Pathology of Laboratory Rats and Mice" (E. Cotchin and F. J. C. Roe, eds.), pp. 733-797. Davis, Philadelphia, Pennsylvania.

219. Slanets, C. A., Fratta, I., Crouse, C. W., and Jones, S. C. (1956). Stress and transportation of animals. *Proc. Anim. Care Panel* **7,** 278-289.

220. Solomon, G. F. (1969). Stress and antibody response in rats. *Int. Arch. Allergy Appl. Immunol.* **35,** 97-104.

221. Squire, R. A., chairman (1971). "A Guide to Infectious Diseases of Mice and Rats." Inst. Lab. Anim. Resour., Natl. Acad. Sci., Washington, D. C.

222. Stratman, S. L., and Conejeros, M. (1969). Resistance to experimental bacterial infection among different stocks of rats. *Lab. Anim. Care* **19**, 742-745.

223. Stuhlman, R. A., and Wagner, J. E. (1971). Ringtail in *Mystromys albicaudatis:* A case report. *Lab. Anim. Sci.* **21**, 585-587.

224. Swift, R. W. (1944). The effect of feed on the critical temperature for the albino rat. *J. Nutr.* **28**, 359-364.

225. Totton, M. (1958). Ringtail in newborn Norway rats—A study of the environmental temperature and humidity on incidence. *J. Hyg.* **56**, 190-196.

226. Townsend, G. H. (1969). The grading of commercially-bred laboratory animals. *Vet. Rec.* **85**, 225-226.

227. Trejo, R. A., Di Luzio, N. R., Loose, L. D., and Hoffman, E. (1972). Reticuloendothelial and hepatic functional alterations following lead acetate administration. *Exp. Mol. Pathol.* **17**, 145-158.

228. Trexler, P. C., and Skelly, B. J. (1963). A microbic flora for the gnotobiotic animal. *Lab. Anim. Care* **13**, 609-615.

229. Truscott, R. B. (1970). Endotoxin studies in chicks: Effects of lead acetate. *Can. J. Comp. Med.* **34**, 134-137.

230. Ventura, J., and Domaradzki, M. (1967). Role of mycoplasma infection in the development of experimental bronchiectasis in the rat. *J. Pathol. Bacteriol.* **93**, 342-348.

231. Vesell, E. S. (1967). Induction of drug-metabolizing enzymes in liver microsomes of rats and mice by softwood bedding. *Science* **157**, 1057-1058.

232. Vesell, E. S., Lang, C. M., White, W. J., Passananti, G. T., Hill, R. N., Clemens, T. L., Liu, D. K., and Johnson, W. D. (1976). Environmental and genetic factors affecting the response of laboratory animals to drugs. *Fed. Proc., Fed. Am. Soc. Exp. Biol.* **35**, 1125-1132.

233. Vesell, E. S., Lang, C. M., White, W. J., Passananti, G. T., and Tripp, S. L. (1973). Hepatic drug metabolism in rats: Impairment in a dirty environment. *Science* **179**, 896-897.

234. Vessey, S. H. (1964). Effects of grouping on levels of circulating antibodies in mice. *Proc. Soc. Exp. Biol. Med.* **115**, 225-255.

235. Visek, W. J. (1974). Some biochemical considerations in utilization of nonspecific nitrogen. *J. Agric. Food Chem.* **22**, 174-184.

236. Wade, A. E., Holl, J. E., Hilliard, C. C., Molton, E., and Greene, F. E. (1968). Alteration of drug metabolism in rats and mice by an environment of cedarwood. *Pharmacology* **1**, 317-328.

237. Walker, P. H. (1976). Monitoring, control and alarm systems. *In* "Control of the Animal House Environment," pp. 277-290. Lab. Anim. Ltd., London.

238. Wassermann, M., Wassermann, D., Gershon, Z., and Zellermayer, L. (1969). Effects of organochlorine insecticides on body defenses. *Ann. N.Y. Acad. Sci.* **160**, 393-401.

239. Weihe, I., and Stotzer, H. (1974). Age- and light-dependent changes in the rat eye. *Virchows Arch. Pathol. Anat. Histol.* **362**, 145-156.

240. Weihe, W. H. (1965). Temperature and humidity climatograms for rats and mice. *Lab. Anim. Care* **15**, 18-28.

241. Weihe, W. H. (1971). The significance of the physical environment for the health and state of adaptation of laboratory animals. *In* "Defining the Laboratory Animal," pp. 353-378. Inst. Lab. Anim. Resour., Natl. Acad. Sci., Washington, D. C.

242. Weihe, W. H. (1973). The effect of temperature on the action of drugs. *Annu. Rev. Pharmacol.* **13**, 409-425.

243. Weihe, W. H. (1976). Influence of light on animals. *In* "Control of the Animal House Environment" (T. McSheehy, ed.), Lab. Anim. Handb. No. 7, pp. 63-76. Lab. Anim., Ltd., London.

244. Weihe, W. H., Schidlow, J., and Strittmatter, J. (1969). The effect of light intensity on the breeding and development of rats and golden hamsters. *Int. J. Biometeorol.* **13**, 69-79.

245. Weisbroth, S. H. (1972). Pathogen-free substrates for gerontologic research: Review, sources and comparison of barrier-sustained vs. conventional laboratory rats. *Exp. Gerontol.* **7**, 417-426.

246. Weisbroth, S. H., Paganelli, R. G., and Salvia, M. (1977). Evaluation of a disposable water system during shipment of laboratory rats and mice. *Lab. Anim. Sci.* **27**, 186-194.

247. White, W. J., Newbauer, F., and Lang, C. M. (1977). Diagnostic lab for an ARF. *Lab Animal (U.S.A.)* **6**, 22-28.

248. Whitten, W. K., Bronson, F. H., and Greenstein, J. A. (1968). Estrus-inducing pheromone of male mice. *Science* **161**, 584-585.

249. Williams, F. P., Christie, R. J., Johnson, D. J., and Whitney, R. A. (1968). A new autoclave system for sterilizing vitamin-fortified commercial rodent diets with lower nutrient loss. *Lab. Anim. Care* **18**, 195-199.

250. Windmueller, H. G., Ackerman, C. J., and Engel, R. W. (1959). Reaction of ethylene oxide with histidine, methionine and cysteine. *J. Biol. Chem.* **234**, 395-402.

251. Wongwiwat, M., Sukapanit, S., Triyanond, C., and Sawyer, W. D. (1972). Circadian rhythm of the resistance of mice to acute pneumococcal infection. *Infect. Immun.* **5**, 442-448.

252. Woods, J. E. (1975). Influence of room air distribution on animal cage environments. *ASHRAE Trans.* **81**, 559-571.

253. Woods, J. E. (1978). Interactions between primary (cage) and secondary (room) enclosures. *In* "Laboratory Animal Housing," pp. 65-83. Inst. Lab. Anim. Resour., Natl. Acad. Sci., Washington, D. C.

254. Woods, J. E., Besch, E. L., and Nevins, R. G. (1974). Heat and moisture transfer in filter-top rodent cages. *AALAS Publ.* **74-3**, Abstr. 77.

255. Wusterman, F. S., Johnson, D. B., Dodgson, K. S., and Bell, D. P. (1968). The use of "normal" rats in studies on the acid mucopolysaccharides of lung. *Life Sci.* **7**, 1281-1287.

256. Yamauchi, C., Takahashi, H., and Ando, A. (1965). Effects of environmental temperature on physiological events in mice. I. Relationship between environmental temperature and number of caged mice. *Jpn. J. Vet. Sci.* **27**, 471-478.

257. Zimmerman, D. R., and Wostmann, B. S. (1963). Vitamin stability in diets sterilized for germfree animals. *J. Nutr.* **79**, 318-322.

Chapter 9

Bacterial and Mycotic Diseases

Steven H. Weisbroth

I. INTRODUCTION

The intent of this chapter will be to review the pertinent microbiological, pathological, diagnostic, and medical features of naturally occurring bacterial and mycotic infections of the laboratory rat. The emphasis will, of course, be on those pathogenic agents of greatest current significance. It is important to recognize, however, that viewed in perspective, an evolutionary process has been at work just as surely on the microbial flora as on the host itself since domestication of the rat approximately one century ago (379). As the ecological niche of the rat has been sequentially modified by refinements in laboratory environments, the dominant microbial species accompanying these changes have shifted to forms unrecognized as rat pathogens at the beginning of this process. As a general statement it may be said that primary bacterial pathogens posing major epizootic hazard were among the first to be eliminated by these ecological changes. Examples include *Salmonella* sp. and *Streptobacillus moniliformis*, which figure prominently as agents encountered in the decades just prior to and after the turn of the twentieth century. These are organisms believed still common in wild rats, and the concept that laboratory rats differed little in a microbial sense from their wild counterparts in those earlier times is a valid one. Under the generally improved circumstances of laboratory environments that obtained between the turn of the century and the mid-1950s, the main accomplishments were to reduce the quantitative incidence of primary pathogens, but not to effect a qualitative change in the distribution of bacterial pathogens carried by rat hosts.

As this pattern applies to the laboratory rat, by the early 1950s disease as a practical matter had been apparently reduced to a small number of extremely important diseases of high morbidity. The bacterial components of this spectrum were now seen to include *Streptococcus pneumoniae*, *Pasteurella pneumotropica*, and *Corynebacterium kutscheri*, entities of minor or unrecognized significance at the turn of the century. Moreover, although exposure alone is generally sufficient to produce infection with clinical signs and lesions in the case of such frank primary pathogens as *Salmonella* sp. and *Streptococcus pneumoniae*, in modern times the increasing significance of latent infection with bacterial forms of lower pathogenicity has been recognized. By latent infection is meant the carriage of infectious organisms by the host, whose presence is unrecognized or masked until a suitable stress sufficiently impairs the defensive competence of the host in a direction favoring multiplication and invasion of the microbe accompanied by clinical signs and lesions. Such bacteria as *Corynebacterium kutscheri*, *Pasteurella pneumotropica*, *Bacillus piliformis*, and *Pseudomonas aeruginosa* characteristically occur in the rat as latent infections. Stressors impairing immunocompetence, such as shipment, other intercurrent infectious diseases, vitamin deficiency, X irradiation, administration of adrenal corticosteroids, immunosuppressive drugs, thymectomy, and others, are documented examples of the types of incitant stimuli that often are known to precede or trigger clinical signs and/or microscopic lesions associated with particular bacterial diseases. It is to these diseases that we will now turn our attention.

II. BACTERIAL DISEASES

A. Streptobacillosis (*Streptobacillus moniliformis*)

Considered only within the context of its status as a pathogen of laboratory rats, *Streptobacillus moniliformis* occupies a somewhat ambiguous role. It has been associated with various aspects of respiratory disease in laboratory rats since early times; however, two other features of the organism have received intense scrutiny, tending to obscure or deemphasize the limited pathogenicity of the organism in its natural reservoir host. These other features include the etiological significance of *S. moniliformis* in rat-bite and Haverhill fever of man, and also, the cultural instability of the organism leading to L-phase variants that were for a certain period of time confused with *Mycoplasma* of rat origin.

The binomial name *Streptobacillus moniliformis*, first suggested by Levaditi *et al.* (264), will be used throughout this text in accordance with the latest (eighth) edition of "Bergey's Manual" (66), although based on relationship to the actinobacilli, Wilson and Miles have retained the name *Actinobacillus moniliformis* (496), which is here considered to be a synonym. Both of these texts (66,496) and Nelson (322) describe the evolution of synonymy with this organism and the following list of names may be found there and in the older literature: *Streptothrix muris ratti*, *Nocardia muris*, *Actinomyces muris ratti*, *Haverhilia multiformis*, *Actinomyces muris*, *Asterococcus muris*, *Proactinomyces muris*, *Haverhilia moniliformis*, *Actinobacillus muris*, and *Bacillus actinoides* var. *muris*. Current taxonomy includes this organism with the family Bacteroidaceae.

Although a streptothrix-like organism had been recognized in the blood of some human patients with recurrent fever following rat bites, the organism had not been isolated in pure culture and characterized until 1914 (406). The first etiologically confirmed case of human streptobacillosis in America was described in 1916 (39). The association of the organism with human disease was clear from the beginning, but many years were to elapse before an understanding of the relationships between *S. moniliformis* and its natural reservoir host, the rat, were established. The earliest associations of *S. moniliformis* and its rat host were in the context of respiratory disease, particularly pneumonia (221,468,469,470) and otitis media (321,322), and it was this context that was contributory to the confusion discussed ahead between *S. moniliformis* L-phase variants and *Mycoplasma pulmonis*. Although Tunicliff's original isolates were from pneumonic tissues, and in addition, she was to claim that intraperitoneal injection of *S. moniliformis* caused pneumonias (468,469), critical examination of this work casts doubt on whether the organism caused pneumonia or merely was reisolated from (preexisting) pneumonic tissues. Moreover, the weight of experimental and circumstantial evidence since has, to the contrary, established the inability of this organism to cause respiratory disease in rats (322,435). It was not until 1933 that Strangeways (435) reported the asymptomatic occurrence of *S. moniliformis* in both wild and laboratory rats as a commensal of the nasopharynx. The disease had inadvertently been caused in mice receiving *Trypanosoma equiperdum*-infected blood from rat donors. Investigation revealed that the manner of collecting blood from the rat donors, i.e., by "a blow on the head of sufficient violence to cause blood to flow freely from the head and nose"

(435) was instrumental in directing her attention to the nasopharynx because heart blood collected from the rats did not (except on two occasions) transfer *S. moniliformis* to the mice. Subsequent investigations established the nasopharynx as the usual location of this organism in the rat and provided the ecological link to explain the injection of *S. moniliformis* into humans following rat bites, at the time a puzzle of some significance.

At the present time *S. moniliformis* is believed to be a commensal of low pathogenicity for the rat. It has not been etiologically associated with naturally occurring disease processes in this species except as a secondary invader. Experimental infections in rats with certain isolates ("strain C") given intravenously, have been used as a model of the infective arthritides (258,259), but these bone and joint lesions have not been seen in the rat under natural circumstances. Indeed, it suggests that *S. moniliformis* occurs as a septicemia rarely, if ever, in naturally infected rats. While it was widely distributed in laboratory rats during the first half of this century, variably as high as 50% (321,329,435), the incidence of *S. moniliformis* has markedly decreased in modern times under the influence of generally improved sanitation in animal care facilities and the production of laboratory rats by gnotobiotic derivation. It continues, however, to be occasionally isolated from both wild and conventional laboratory rats (283), and rat-bite fever remains an occupational hazard of laboratory personnel (10,33,53,95,158,160,175,199,249,288,362,387,421,454) (see also this volume, Chapter 15).

Laboratory rats may also act as zoonotic reservoirs for *S. moniliformis* infections in laboratory mice (129,131). Unlike the rat, mouse infections with *S. moniliformis* are not known to be lesionless and asymptomatic, and the mouse is not thought to be a carrier of commensal streptobacilli. Infections in mice may be acute with generalized septicemia (erythema multiforme), or more chronic and characterized by polyarthritis, especially of the lower hind extremity and tail vertebrae (129–130,131,274,478). Natural infection of pregnant mice may also result in abortions (274,401). The mode of transmission from rat to mouse has been investigated but remains unknown (265, 478). Infections do not persist in mouse colonies rigidly separated from rats (160). In this connection, the mouse appears to mimic human infections in terms of the ecology of *S. moniliformis*. In the mouse, as in man, septicemia and polyarthritis may follow rat bites (265), or be unrelated to this mode of transmission (274,478). Similarly, Haverhill fever in man (erythema arthriticum epidemicum) occurred as epidemics in Haverhill, MA in 1926 and in Chester, PA in 1925, in which the source of infection was eventually traced to infected milk to which rats had access (102,342,346,347). Even in more modern times, human *S. moniliformis* infections may be unrelated to rat bites (184,247,336,429), although direct inoculation by rat bite appears to be the most common mode of transmission to man. There appear to be very few human cases on record of streptobacillary rat-bite fever following mouse bite (15,158).

The morphology of the organism is variable and pleomorphic depending on environmental factors and age of the culture (185,323). Smears from blood or other clinical materials most commonly show small (less than 1 μm wide and 1–5 μm long) gram-negative rods and filaments (125). Blood, serum, or ascitic fluid is required for growth; however, growth on Loeffler's serum agar slants is poor. Growth does not occur in nutrient or infusion broths or agars unless supplemented with one of these body fluids. Following subculture on 20% horse serum infusion agar, slender interwoven masses of filaments 0.4–0.6 μm develop. Occasional filaments show spherical, oval, fusiform, or club-shaped swellings that may occur terminally, subterminally, or irregularly (hence the name "moniliformis") by 6–12 hr after subculture (496). After 12–18 hr, the filaments fragment into chains of bacillary, coccoid, or fusiform bodies. Some 24–30 hr after subculture on agar, tiny coccoid or ring forms are seen, as well, accompanied by masses of bubbles and irregularly shaped cholesterol-containing droplets. Gram negativity is variable, and staining is more intense in the monilia-like swellings than in the filaments (Fig. 1). In 20% horse serum infusion broth cultures the morphology is similar, but aggregate filamentous masses are characteristically seen giving the appearance of a coarsely granular sediment. The supernatant is typically clear, without turbidity, and contains small flocculent masses of organisms giving the tube a snow flake or cotton ball appearance when the tube is agitated (Fig. 2). Although described as a facultative anaerobe, ordinary aerobic incubator conditions suffice for isolation and propagation. Growth may be enhanced in 8–10% CO_2 environments, which are recommended (421).

Colonies on serum or ascitic fluid infusion agar are 1–3 mm in diameter after 48–72 hr of growth (Fig. 3), are circular, low convex, may vary from colorless to grayish white, be translucent or opaque, and have a smooth, glistening appearance with butyrous consistency (186). Best results have been obtained using trypticase soy agar with 20% horse or rabbit serum (421). There is little or no increase in size of colonies with prolonged incubation. On 5% sheep blood agar colonies are barely visible at 24 hr, but 1–3 mm by 48–72 hrs, and are smooth, gray, translucent, and essentially nonhemolytic. Cystine trypticase broth, 10% horse serum, and 1% of various carbohydrates may be used for study of fermentative reactions (421). Other cultural and biochemical characteristics are summarized in Table I. The organism is viable in serum broth and agar cultures for 2–4 days at 37°C, 7–10 days at 3°–4°C, 14–15 days at 3°–4°C in sealed cultures, and for several years in infected tissues, body fluids, or cultures maintained at from −25° to −70°C (66).

It is believed that all isolates of *S. moniliformis* may undergo

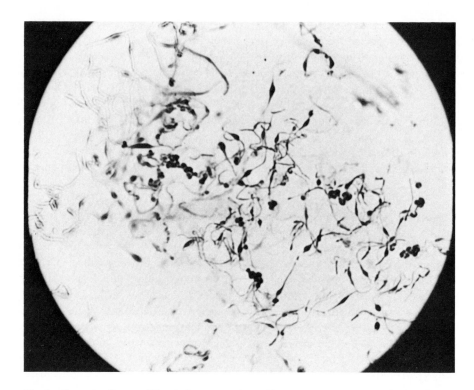

Fig. 1. Streptobacillus moniliformis from agar culture. Giemsa. ×1000. (Courtesy of J. B. Nelson.)

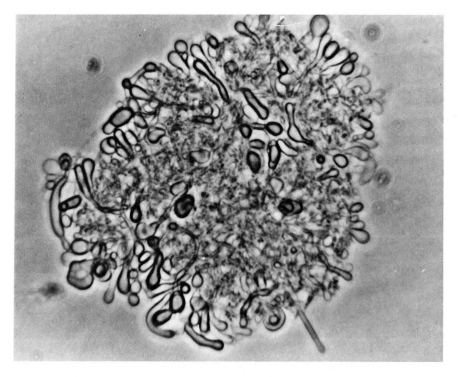

Fig. 2. Streptobacillus moniliformis colony from broth culture. Unstained. ×1200. (Courtesy of J. B. Nelson.)

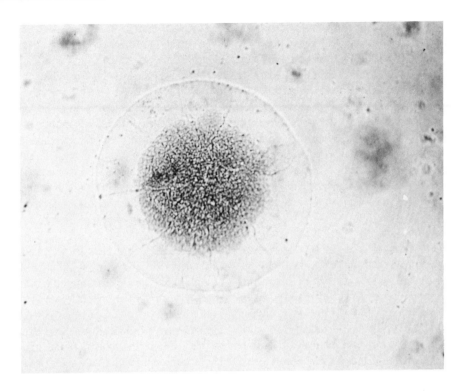

Fig. 3. *Streptobacillus moniliformis* colony on agar culture. Note "fried egg" appearance. Unstained. ×100. (Courtesy of J. B. Nelson.)

Table I

Characteristic Biochemical Reactions and Other Properties of *Streptobacillus moniliformis*[a]

Characteristic	Reaction or property
Morphology	Gram-negative, pleomorphic, filaments or fragmented filaments (bacillary) with moniliform swellings
CO_2 requirement	Enhances growth but not required
Catalase	(−)
Oxidase	(−)
Growth on nutrient agar or broth	(−), requires serum or blood
Indole	(−)
MR/VP	(−)/(−)
Glucose fermentation	(+) acid
Growth in 10% serum nutrient broth	(+) whitish flocculent aggregates, clear supernate
Growth on blood agar	1–2 mm colonies in 48–72 hours, smooth, gray, nonhemolytic
Gelatin liquifaction	(−)
Nitrate reduction	(−)
H_2S production	(−)
Arginine hydrolysis	(+)

[a]From Buchanan and Gibbons (66), Cowan (87), and Wilson and Miles (496).

reversible conversion to transitional phase variants. The term L1 was first used by Kleineberger (235,236,240) to designate the occurrence in *S. moniliformis* cultures of small, filterable (Berkfeld V) variants morphologically resembling the bovine pleuropneumonia organism. Indeed, a sensation was caused when Kleineberger affirmed that the L1 variant of *S. moniliformis* belonged to the pleuropneumonia-like organisms (PPLO). In this designation the "L" (for Lister Institute) was later applied to any bacterial variant reproducing in the form of very small cells lacking rigid cell walls. Individual isolates of *S. moniliformis* vary in degree of stability and the L phase is typically unstable. L-phase variants differ from the bacillus in cellular and colonial morphology and in susceptibility to antibiotics, e.g., penicillin, that act primarily on the cell wall. Occasionally, stable L variants may be achieved by manipulation, which do not undergo reversion to the parental type. One such, "LI Rat 30," was stabilized by Kleineberger in 1938 and reversion has never since occurred (237). The individual cells of LI Rat 30 vary from 0.3 to 3 μm with a mean of 1.0 ± 0.5 μm. On serum or ascitic fluid infusion agar, colonies of the LI variant average about 300–500 μm in diameter (compared to 1000–3000 μm in the parental form), will show oil-like droplets, and the central portion will penetrate the agar to a depth of 30–50 μm. Bacterial L variants may not be distinguished from true *Mycoplasma* colonies on the basis of size, "fried egg" appearance, or deep brown color in the central area (130,287).

However, the usual reversion of L variants to parental forms on supportive media without inhibitors will ordinarily serve to identify bacteria. The L-phase variant with incomplete cell wall is resistant to penicillin concentrations as much as 10,000 times higher than is the normal bacillus with cell wall (287,368), and is ordinarily only encountered in the laboratory following subculture, although it has been isolated directly from pneumonic rats (237) and from the blood of penicillin-treated rat-bite fever patients (100). When inoculated into mice, the L-phase variant is nonpathogenic; however, it frequently reverts *in vivo* to the bacillary form with full recovery of the pathogenic properties of the original *S. moniliformis* isolate (130).

Great confusion was inadvertently introduced into the gradually emerging understanding of the etiology of respiratory disease of laboratory rats when in 1937 Kleineberger and Steabben (238,239) isolated hitherto undescribed organisms from pneumonic rats that they termed L3 and believed to be L-phase variants of *S. moniliformis*. In that same year, Nelson (326) described "coccobacilliform bodies" isolated from mice with respiratory disease, and several years later also from rats (325). Accumulated findings in these and other laboratories led to the conclusion that the L3 and coccobacilliform bodies were identical, and they have since been classified within the Mycoplasmatales as *Mycoplasma pulmonis* (128). It is now generally understood that L1 and L3 are not associated, that the L1 variant (and other L variants) of *S. moniliformis* are unrelated to the mycoplasmas (PPLO), and that *S. moniliformis* and/or its L variants play no necessary role in murine respiratory mycoplasmosis (see this volume, Chapter 10).

Comparatively little is known about the immunology of *Streptobacillus moniliformis*, although opsonins, agglutinins, and complement fixing (CF) antibodies have been demonstrated in the postinfective sera of both experimentally and naturally infected rats (185,186,469), naturally and experimentally infected mice (324,396,478), and in human *S. moniliforms* infections (63,158,175,247'378). Use of Tween 80 in media (1.5%) used to propagate *S. moniliformis* for agglutination antigen appears to prevent the organism's tendency to clump (247). Agglutinins may not appear in promptly treated human cases (199). Both direct (247) and indirect (200,247) immunofluorescence have been used to confirm *S. moniliformis* isolates and antibodies from rat-bite fever patients. The L-phase variant shares at least one common antigen with the bacillus but lacks at least one antigen present in the bacterial form (240). Antigenic differences in *S. moniliformis* isolates from various species have not been reported (or even investigated), although such isoaltes may have varying biological characteristics. Guinea pig isoaltes, for instance, are strict anaerobes, biochemically inert, and when injected intraperitoneally into mice, are pyogenic with abscess formation rather than septicemic and polyarthritic as is the case with rat,

mouse, and human isolates (7,423). Some doubt exists, however, as to whether or not guinea pig isolates are truly *S. moniliformis*, or rather, taxonomically removed; "Bergey's Manual" classifies such organisms as *Sphaerophorus caviae* (66).

Diagnosis is established by isolation and cultural characterization of *S. moniliformis*. Important confirmatory information may be obtained by intraperitoneal and foot pad inoculation of test mice with causation of septicemia and embolically distributed lesions including polyarthritis from which *S. moniliformis* may be reisolated. Sufficient evidence is not available as to whether, or not, antibodies invariably occur in asymptomatic rats so as to make serological screening useful in monitoring rat colonies.

With regard to treatment, little is known about the efficacy of antibiotic treatments in abolishing latent infections of the nasopharynx in rats. Considerable information is available, of course, with regard to antibiotic treatment of systemic *S. moniliformis* infections in man (247,387). However, this information is of little utility in predicting treatment methods for the rat, in which the ecology of the organism is clearly different. There is no evidence that the organism may hematogenously traverse the rat uterus, or reside as a commensal in rat reproductive tract. Infections have not been reported to occur in rats derived by gnotobiotic means and shielded from proximity to conventional rats.

B. Spirochetosis (*Spirillum minus*)

The status of *Spirillum minus* as a bacterial pathogen of rats, as with *Streptobacillus moniliformis*, is somewhat uncertain. Like *S. moniliformis*, the pathogenicity of *S. minus* for its natural hosts has been overshadowed by the interest attending this organism as a cause of rat-bite fever in man. Although illness following rat bites had been known since ancient times in India, where it is believed to have originated (388), and for centuries in Japan as sodoku (*so* =rat; *doku* =poisoning) it was not until 1902 that the disease was described by Japanese workers in European journals (309) and given the name "Rattenbisskrankheit," or rat-bite fever. The latter paper led to the recognition of rat-bite fever as a clinical entity and to the realization that a number of previous accounts of "poisoning" following rat bites had been made from America and the European continent [see McDermott (285) and Robertson (375,376) for early reviews]. Previous to Miyake's report, it had been recognized that wild *Rattus norvegicus* (=*Mus decumanus*) could carry spirillar organisms in the blood (76), and slightly later (1906) in mice (57,493). The association was left, however, until 1915, when Futaki and his co-workers discovered this cause of rat-bite fever in man (150). Readers will recall that only 1 year previously *S. moniliformis* also had been associated with rat-bite fever in man (406) and a controversy was initiated that raged for almost three decades as to the "true" cause of rat-bite fever. As so often happens in science, the passage of time and the careful documentation of clinical cases and laboratory data was to reveal that both of these organisms would be recognized as causing similar, but nonetheless different, syndromes of "fever" following rat bites (63). The evidence was summarized at about the same time by Allbritten *et al.* (8) and Farrell *et al.* (106), who carefully distinguished three diseases: sodoku and streptobacillosis following rat bite, and Haverhill fever following (presumed) ingestion of *S. moniliformis* [see also reviews by Beeson (3) and Watkins (484)]. Sodoku and streptobacillosis were differentiated on the basis of longer incubation period (14 days); the formation of an indurated "chancre" with regional lymphadenitis related to the bite site, more regularly periodic recurrent fever; maculopapular, erythematous rash spreading from the initial lesion and absence or rare occurrence of petechiae; and polyarthritis in the case of sodoku (*S. minus* infection). Streptobacillosis, on the other hand, was usually shorter in onset (within 10 days), septicemic, with morbilliform or petechial hemorrhages, endocarditis, irregularly recurrent fever, and polyarthritis. All known cases of rat-bite fever contracted from laboratory rats and mice have been of the streptobacillary type (160). Human rat-bite fever of the *S. minus* type, in India, has a seasonal epidemiology (79).

The organism has been identified in a number of other hosts including dog, cat, weasel, ferret, and still others, although it is believed that carnivore infections represent zoonoses contracted from rodent carriers in the course of predation (183). At present only one species, *S. minus*, is taxonomically recognized (66). The older literature, however, contains references to *Spirochaeta morsus muris*, *Spirillum minor*, *Spirochaeta laverani*, *Spironema minor*, *Leptospira morsus minor*, *Spirochaeta muris*, and *Spirochaeta petit*, which are regarded as synonyms. The frequent early references to this organism as a spirochete were based not only on morphologic criteria. Indeed, it is a matter of some historic interest that Ehrlich used the mouse–*S. minus* system in experiments leading to development of salvarsan for chemotherapy of human syphilis (285), that early cases of *S. minus* rat-bite fever in man were successfully treated with salvarsan (91,183,425,438), and that the organism has been advocated as a model system for the study of syphilitic infections (436).

Spirillum minus is classified by "Bergey's Manual" within the Spirillaceae, and it appears to be the only species in the genus that is both parasitic and, at present, incultivable on artificial media. As identified in fixed and stained blood smears from infected humans or laboratory rodents, the cells are gram negative, appear short and thick but are approximately 0.5 μm in width and 1.7–5 μm in length, and have two to six spirals [see Bayne-Jones (28) for illustrations]. The organism has tufts of flagellae (2,275) at each end and is actively motile.

The little that is known about the pathogenicity of *S. minus* in its natural hosts (rats and mice) is related to the diagnostic inoculation of laboratory rats and mice with blood (or strains of *S. minus* initially derived from) rat-bite fever patients. Although some surveys of both wild and laboratory rats and mice have indicated the prevalence of the organism in the decades surrounding the World War II (14,63,100,126,204,215,241), these surveys have not reported the occurrence of lesions referrable to *S. minus*. At the same time, the presence of unrecognized intercurrent disease was so prevalent at the time studies were made in experimental laboratory hosts, the interpretation of clinical signs and lesions said to be caused by *S. minus* must be viewed with considerable uncertainty. It appears to have been neither reported, isolated, or otherwise studied since demonstration in a clinical case of rat-bite fever in 1969 (82).

The experimental disease has been studied by the intraperitoneal injection of blood into laboratory rats, mice, guinea pigs, rabbits, cats, ferrets, and rhesus monkey. These studies achieve a fair consensus on placing the incubation period in experimental hosts at about 3–5 days or longer (91,251,252,285,436,493). Indeed, it has been pointed out that the observation of *S. minus* in the blood of experimental hosts prior to 3 days postinoculation should be regarded as intercurrent (preexisting) infection of the host with *S. minus* (93,125,126,241,376). The height of infection in terms of demonstrable organisms by dark-field examination of thick films occurs 2–3 weeks after inoculation of experimental rats, mice, and guinea pigs. Experimental animals may remain infective 6–12 months after inoculation (252). Few, if any, signs other than an initial bacteremia have been reported in mouse hosts (285,375). In the rat and guinea pig, however, when given subcutaneously, *S. minus* infection may typically also cause indurated or ulcered chancroid lesions at the inoculation site, with regional lymphadenopathy and fever reminiscent of human sodoku-type rat-bite fever (285). Spirilla frequently are demonstrable in the serous discharge from primary lesions (285).

In considering the essential features of the experimental disease in rats, McDermott (285) described the following successive phases: incubation, development of primary inflammatory lesion, and lymphadenopathy with organisms initially limited to these tissues; a secondary septicemic stage with organisms demonstrable in the blood; a latent stage in which the blood was free of (demonstrable) spirilla and there were no obvious lesions; and finally, after many months, a tertiary stage with gummatoid lesions (abscesses and granulomas) of the lungs, spleen, lymph nodes, and liver. Gummas have been described as long-term lesions by others as well (252,436). The gummatoid lesions particularly are suspect of being otherwise induced and merit reexamination. It is worth restating, however, that none of these lesions has been reported as occurring in naturally infected rat hosts.

Diagnosis of *S. minus* infection may be made by demonstration of the actively motile organisms in wet mounts made of primary lesion exudates or peripheral blood by dark-field or phase-contrast microscopy. Blood films stained with Wright's, Giemsa, or silver impregnation stains also should be examined. Since it is recognized that septicemic phases may be brief, failure to demonstrate the organism in peripheral blood should always be supported by culture (to isolate *S. moniliformis*, if present) and animal inoculation. It is recommended that four mice and two guinea pigs be used to screen each suspect blood sample and that preinoculation blood samples be examined from the test animals to preclude use of intercurrently infected mice or rats (63,381). Suspect blood should be inoculated subcutaneously, intraperitoneally, and intradermally (496). Blood and peritoneal fluid from the test animals should be examined microscopically at weekly intervals for 4 weeks. Peritoneal fluid is usually richer in organisms than blood (215). An equal number of animals should be identically inoculated and examined using an aliquot of suspect blood heat treated to 52°C for 1 hr (381). The heart and tongue of test animals at the conclusion of the test period (4 weeks) should be used to make Giemsa-stained impression smears. Organisms often are present in fair numbers in tongue and cardiac tissue when scarce (as in latent phases) in blood or peritoneal fluid (215,381,496).

Live organisms in fresh preparations have been immobilized by immune sera (285), but serologic tests have not been advocated for diagnosis because of the technical problems encountered in making antigenic suspensions. Approximately one-half of human cases are positive by *Treponema pallidum* serodiagnostic methods and the importance of taking a serum sample early in the course for this purpose has been stressed (504). Rabbits infected with *S. minus* blood suspensions have been reported to develop positive Weil–Felix reactions with *Proteus* OXK strains (400).

The mechanism of zoonotic transfer of *S. minus* from infected rats via bite lesions is unknown. The organism has not been demonstrated in saliva, although this question has been investigated (252,285). It has been observed in urine of human rat-bite fever cases (252), and isolated (by inoculation) from urine of naturally and experimentally infected rats (204). The current hypothesis suggests that bleeding gum abrasions at the time of bite may result in bite inoculation. There have been only two human cases of record following mouse bite despite the documented widespread distribution of *S. minus* in both wild and laboratory mouse populations (215).

The organism has not been reported in laboratory animals in modern times, and it is unknown, therefore, whether it can be vertically transmitted to gnotobiotically derived animals. That

it may not be is suggested by McDermott's experiment in which *S. minus* could not be demonstrated in the progeny of infected female mice (285). Chemotherapy of naturally acquired or experimental *S. minus* infection has not not been reported for the rat in recent times, although human cases have been reported as successfully treated with penicillin or streptomycin (215,387). A battery of antibiotics has been assessed for efficacy in prevention of transmission of *S. minus* from infected donor mice to treated mouse recipients (449).

C. Streptococcosis (*Streptococcus pneumoniae*)

It was pointed out in the introduction to this chapter that *Streptococcus pneumoniae* was unrecognized as a pathogen of laboratory rats during the first half of this century. The organism appears not to have been encountered even in microbial surveys of respiratory flora from laboratory rats during this period (41,321,322), unlike similar surveys undertaken later (283,487). As will be described in more detail later, *S. pneumoniae* has been observed in recent times in the context of acute respiratory episodes most commonly, but not limited to, adolescent and young adult age groups. The evidence does not suggest that these infections occurred and were unrecognized earlier, but rather that fundamental changes in the ecology of the host occurred rendering it more susceptible to this organism.

The first report of respiratory epizootics in laboratory rats due to *S. pneumoniae* was published by Mirick *et al.* in 1950 (306). However, as late as 1963 it was written that acute respiratory episodes in rats of any age were rare (467), and as late as 1969 that bacterial pneumonias, including those caused by *S. pneumoniae*, were unlikely to be encountered in properly managed facilities (60). Accumulating evidence and experience with respiratory disease in laboratory rats since the 1950s have documented the increasing frequency of *S. pneumoniae* infections. It is currently regarded as a widespread and major primary bacterial pathogen of laboratory rats. The pathogenicity of the organism is sufficient to serve as a primary cause of disease in the absence of physiologic stress or other infectious incitants, although it has been recognized as frequently associated with *Mycoplasma* infections of rats.

In conformity with the eighth edition of "Bergey's Manual" (66), the name *Streptococcus pneumoniae* has been adopted for this organism, and the former names *Diplococcus* and *Pneumococcus* regarded as synonyms.

That the organism is widespread among commercial rat breeding stocks was documented in 1969 (487). In the latter study, 19 of 22 commercial sources were found to be offering for sale adolescent rat carriers of *S. pneumoniae* with varying degrees of clinical signs. The majority of animals from which pneumococci were isolated were asymptomatic carriers of the

Fig. 4. Streptococcus pneumoniae infection of upper respiratory tract with sanguinopurulent exudate from nares.

organism in the nasoturbinate mucosa. It was concluded that the organism is generally carried in the upper respiratory tract, particularly the nasoturbinates and middle ear in the absence of clinical signs. Of 20 rats without clinical signs, no pneumococci were isolated from pulmonary tissues, while *S. pneumoniae* was isolated from 17 in nasoturbinate washings.

The disease is not ordinarily latent, however, and clinical signs of respiratory disease with varying degrees of mortality are more typical (3,21,22,122,208,230,308,465). Serosanguinous to mucopurulent nasal exudates are frequently the first signs of clinical illness (Fig. 4 and 5) and ordinarily precede pulmonary involvement. Rhinitis, sinusitis, conjunctivitis, and otitis media are common gross lesions of upper respiratory infection. Outward bulging of the tympanic membrane may be noted as a sign of pus under pressure in the middle ear. Histologically, mucopurulent exudates are seen to overlay the respiratory mucosa of the turbinates, sinuses, eustachian tube, nasolacrimal duct, and tympanic cavity. Acute inflammatory cells, primarily neutrophils, and also more chronic inflammatory cells including plasma cells and lymphocytes infiltrate the mucosa and underlying submucosa of these tissues. Abundant numbers of organisms may be observed in the exudates and superficial levels of the mucosa by means of tissue gram stains or smears. Concomitant clinical signs include postural changes (e.g., hunching) abdominal breathing as pneumonia supervenes, dyspnea, conjunctival exudation, anorexia with loss of weight, depression, and gurgling or snuffling respiratory sounds (rales). The onset of clinical signs and lesions is more often acute or subchronic than chronic and affects rats of all ages, but particularly younger age groups.

The ordinary route of progression of upper respiratory infections is by descent to pulmonary tissues. Frequently the right intermediate lobe of the lung is the first to be affected. Initially fibrinous bronchopneumonias in certain lobes develop rapidly into classical confluent fibrinous lobar pneumonia (Fig. 6).

Fig. 5. Streptococcus pneumoniae infection of upper respiratory tract. Note evidence of nasal wiping on medial aspect of forepaw (arrow).

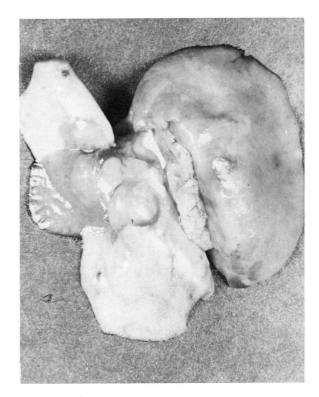

Fig. 6. Streptococcus pneumoniae infection. Fibrinous consolidation of right intermediate and left lobes of lung (lobar pneumonia).

Frothy, serosanguinous fluid exudes from the cut trachea. Microscopically, the mucosa of the trachea, bronchi, and bronchioles is necrotic and eroded, with fluid and purulent exudates, often with blood, in the lumina (Fig. 7). The alveolar capillaries are congested, and alveoli themselves are filled with blood, proteinaceous fluids, neutrophils, or variable combinations of these. Tissue gram stains reveal abundant gram-positive cocci in affected tissues, in microcolonies, as freely distributed single cells, and intracellularly within phagocytic cells, including neutrophils (Fig. 8). Death may supervene at this point depending on the degree of pulmonary involvement. Frequently, however, the organism escapes to the thoracic cavity and the lesions of fibrinous pleuritis, pleural effusion, and fibrinous pericarditis are common (Fig. 9). Organisms may escape from the thoracic organs to become septicemic and septicemia is a frequent terminating event. Less commonly, the organism is embolically distributed to any of a variety of organs and locations, and such complications as purulent arthritis; focal necrosis and/or infarction of the liver, spleen and kidneys, and fibrinopurulent peritonitis, orchitis, (Figs. 10,11), and meningitis have been observed (22,308,487). It is known that the serum biochemistry of naturally infected rats is disturbed with increases in several serum enzymes (including glutamic-pyruvic transaminase, glutamic-oxalecetic transaminase, and lactate dehydrogenase) and α- and β-globulins (308).

Diagnosis is established by retrieval of *S. pneumoniae* from affected tissues and supported by characteristic gross and microscopic lesions. Rats may be screened for *S. pneumoniae* by the plating of nasopharyngeal swabs (1,306) or saline washings of the tympanic cavity and nasoturbinates (487), as these upper respiratory locations are more likely to yield the organism in carriers than pulmonary tissues. The nares should be prepared for examination by snipping off the tip of the nose with sterile scissors; and the middle ear by reflecting the skin and ear from the external auditory meatus. Sterile saline may be introduced and withdrawn from either location by means of sterile plugged Pasteur pipettes. Both techniques follow the method of Nelson (327). The pipette should puncture the tympanic membrane to enter the bulla. Primary clinical samples (swabs, washings, exudates, heart blood, etc.) are plated directly onto 5% blood agar. Growth is facilitated by 10% CO_2 and some isolates require microaerophilic environments (496). Colonies of capsulated pneumococci on blood agar are raised, circular, and 1–2 mm in diameter with steeply shelving sides and an entire edge. The colonies are dome shaped and glistening (especially type 3) early, but by 24–48 hr collapse in the center due to autolysis, giving a typical concave umbilication to the top of the colony. *Streptococcus pneumoniae* colonies are surrounded by a small (alpha) zone of hemolysis with greenish (viridans) discoloration in the media, and the organism is grouped with the α-hemolytic streptococci.

The morphology of *S. pneumoniae* is consistent from both

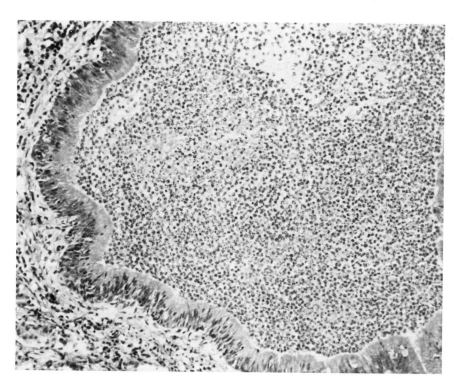

Fig. 7. Streptococcus pneumoniae infection. Bronchus with purulent exudate composed mainly of acute inflammatory cells. Note integrity of epithelium. Hematoxylin and eosin. ×1250.

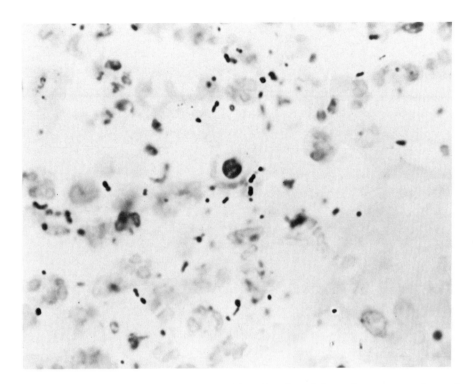

Fig. 8.   *Streptococcus pneumoniae* infection. Smear of bronchial exudate. Note diplococcal arrangement. Gram stain. ×1200.

clinical materials, e.g., smears of exudates or tissues, and from subcultivation on artificial media. The individual cells are ovoid or lanceolate and typically occur as pairs, with occasional short chains, surrounded by polysaccharide capsules. The capsule may be demonstrated by Gin's method (22). Gram-staining reactions may be inconsistent from clinical ma-

terials; however, the majority of smears and isolates are distinctly gram positive with clear nonstaining capsular material surrounding the cells and are of value in presumptive identification.

Suspicious colonies from primary isolates should be individually picked and established in isolation by subculture on

Fig. 9.   *Streptococcus pneumoniae* infection with fibrinous pleuritis and pericarditis. Note liver at bottom for orientation.

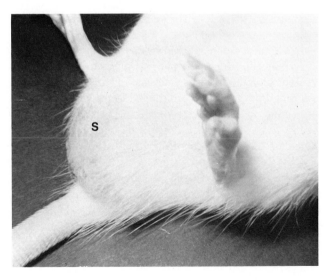

Fig. 10.   *Streptococcus pneumoniae* orchitis. Note swollen scrotum (S).

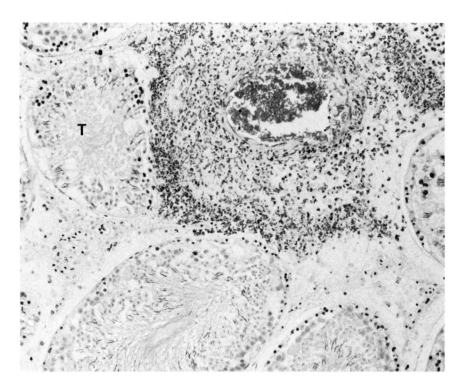

Fig. 11. Streptococcus pneumoniae orchitis. Note coagulative necrosis of tubule (T) to left of arteriole and emigration of leukocytes. Hematoxylin and eosin. × 1250.

blood agar. Identity of the isolate should be confirmed by characteristic reactions in differential media, sensitivity to optochin and bile solubility, and serological typing. Twenty-four hour glucose enriched (Todd–Hewitt) broth subcultures may be used for the serological typing. Characteristic biochemical reactions are summarized in Table II. Optochin (hydrocuprein hydrochloride) sensitivity is established by a zone of inhibited growth surrounding optochin (5 μg) impregnated paper disks placed on the surface of evenly streaked blood agar plates (54). Sensitive isolates are inhibited for a zone of about 5 mm from the edge of the disk. Occasional viridans streptococci are inhibited by optochin; however, pneumococci are rarely resistant (1). Bile solubility may be ascertained by resuspension of the organisms from 5 ml of a centrifuged overnight Todd–Hewitt broth culture in 0.5 ml of the supernate. To each of two small tubes (3 × 0.5 inches) are added 2 ml phosphate buffer, pH 7.8, and one drop of concentrated organism resuspension. Two drops of 10% sodium deoxycholate are added to one of the tubes, and both are mixed and incubated for 1 hr. Isolates that are bile soluble will clear (i.e., autolyze) during this period, and it is generally understood that bile solubility simply acts to accelerate the natural autolytic process that causes central collapse of colonies on blood agar (496). Bile solubility of *S. pneumoniae* isolates is more variable than optochin sensitivity (1,496).

Biochemically confirmed isolates of *S. pneumoniae* should

Table II

Characteristic Biochemical Reactions and Other Properties of *Streptococcus pneumoniae*[a]

Characteristic	Reaction or property
Growth on blood agar	α-Hemolysis with viridans discoloration, colony surface frequently concave after 24–48 hr
Morphology	Gram-positive diplococci
Carbohydrate acid from:	
Glucose	(+)
Glycerol	(−)
Lactose	(+)
Maltose	(+)
Mannitol	(−)
Raffinose	(+)
Salicin	(−)
Sorbitol	(−)
Sucrose	(+)
Trehalose	(+)
Voges-Proskauer (VP)	(−)
Litmus milk	Acid
Gelatin liquifaction	(−)
Aesculin hydrolysis	(−)
Arginine hydrolysis	(−)
Hippurate hydrolysis	(−)
Bile solubility	(+)
Optochin sensitivity	(+)
Growth in 6% NaCl	(−)

[a] From Buchanan and Gibbons (66), Cowan (87), and Wilson and Miles (496).

be typed by the use of type-specific antisera in the Neufeld–Quellung reaction (17,18). Quellung reactions are based on swelling of the capsules surrounding individual organisms in the presence of specific antisera (Fig. 12) and may be performed by the microscopic examination of cover-slipped slides with a mixture of one loopful each of an overnight Todd–Hewitt broth culture, specific antiserum, and methylene blue (487). Typing of the isolate may be helpful in determination of the source of the infection. Weisbroth and Freimer found that rat isolates from various colonies were usually monotypic and that the same type appeared to prevail in distinct geographical patterns (487). Most commonly, types 2, 3, and 19 have been encountered in rats on the American continent (3,22, 208,306, 308,487) and less commonly types 8 and 16 (21,122).

Although it is recognized that rats respond with antibodies to natural infections with *S. pneumoniae,* this property has not received a great deal of attention as a diagnostic method, perhaps because of the ease with which the organism may be culturally retrieved. It has been established that certain rat stocks are more resistant to certain pneumococcal types, that young animals are more susceptible than mature rats (386), and that natural resistance to experimental infection with pneumococcal types commonly found in rats (e.g., types 2 and 3) has been experienced (385,386). Similarly, although both direct and indirect immunofluorescence have been used to lo-

cate *S. pneumoniae* in infected tissues (85,228), and pneumococcal antigens have been detected by counter electrophoresis of clinical specimens (428), they have not received attention as routine diagnostic methods in rats.

It is recognized that laboratory mice are very sensitive to the intraperitoneal injection of clinical samples or broth cultures (87,257a,312,362) containing *S. pneumoniae,* with death often occurring in 24–72 hr (97) with dilutions containing as few as five microorganisms (122). While animal inoculation has been advocated as a biological sieve for human diagnostic specimens, especially where speed may be of the essence (365), it has not generally been advocated as a diagnostic method in the rat (1).

Antibiograms of isolates of *S. pneumoniae* from the rat have generally established the susceptibility of the organism to antibiotics effective against streptococci (339). Antibiotic sensitivity should be established prior to treatment, since penicillin-resistant isolates have been retrieved from naturally infected rats (306). Most isolates are sensitive to penicillin, however, and benzathine based penicillin has been used for effective treatments in the course of epizootics (3,21,487). Rats also have been effectively treated by intraperitoneal injection of type-specific (rabbit) antiserum (306). It has been generally agreed that antibiotics are useful in the amelioration or prevention of systemic infection (septicemia, pneumonia); however, antibiotics are thought to be ineffective in eradication

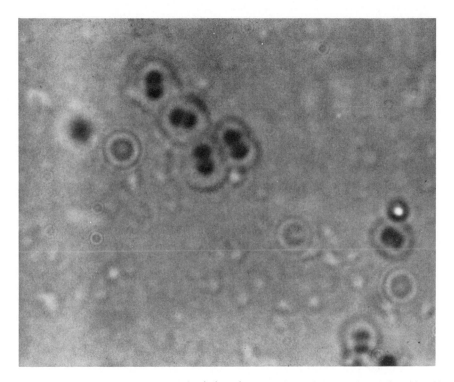

Fig. 12.   Neufeld-Quellung reaction with *Streptococcus pneumoniae* in homologous typing antiserum and methylene blue. Note swollen capsule around diplococci. × 1200.

of organism from infected rat colonies, with remission following withdrawal of treatment. It was concluded (487) that the bacteria in and on the mucosa of the upper respiratory tract are shielded from antibiotic access. The organism may be excluded by establishment of barrier colonies of rats derived gnotobiotically. No pneumococci were isolated from commercial sources of rats thus derived in the course of a microbiologic survey (487).

Rats have been used as the host in model systems employing *S. pneumoniae* in experimental infections. Such studies have established a number of factors bearing on susceptibility, including splenectomy (35,49,263), age (385,386), genetic stock (385), iron deficiency (415), and pulmonary edema (217). Other reports have explored the effect of *S. pneumoniae* infections on such host systems as nitrogen metabolism and protein synthesis (356), serum proteins and enzymes (308), blood pH and electrolytes (105), hepatic enzymes (71,96), thyroid physiology (414), and the effect of antihyperlipidemic agents, e.g., clofibrate (355).

D. Pseudomoniasis (*Pseudomonas aeruginosa*)

The association of *Pseudomonas aeruginosa* with disease states of laboratory rats and mice has only rarely been encountered as a naturally occurring phenomenon in immunologically competent animals. As a fair generality, it may be stated that *Pseudomonas* is a normally occurring, but not necessary, commensal of conventional laboratory rodents. It assumes the status of a significant and serious pathogen in the course of experimental treatments that impair normal host defensive responses. The importance of pseudomoniasis as a consequence of such treatments is intensified by the ubiquitous and almost universal distribution of the organism in conventional stocks of rodents, other laboratory species, and man, and the ability of the organism to thrive in a variety of inanimate materials outside of mammalian systems that include the bedding, feces, water bottles, disinfectant and detergent solutions, etc. In the discussion that follows, information derived from experience with mouse infections will be borrowed, where appropriate, because of the similarity of epizootiology in these two species and comparatively greater availability of information.

The notoriety of pseudomoniasis has been primarily established in connection with "early mortality" (3–8 days postirradiation) consequent to lethal whole body X irradiation in the 750–1000 rad range. Rats and mice infected with *Pseudomonas* die earlier and in larger numbers than uninfected animals (4,115,116,162,170,176,177–180, 191, 300, 301,441,492, 502), although experience has indicated that lower dose ranges of X irradiation court infection as well. Similarly, supervening pseudomoniasis has complicated burn research in laboratory animals (293,302,303,305,384) and has

indeed, been used as a model for burn wound contamination in man (114,119,124,167,257,302,309,319,434,452,453). *Pseudomonas* infections in laboratory rodents have complicated or been provoked by other experimental stressors, e.g., cortisone injections (304), infections with other microorganisms (119), anti-lymphocyte serum (167), tumors (114), cold stress (172), neonatal thymectomy (441), and injection abscesses or surgical procedures (115,441,506). With the exception of a single report of otitis media in a closed mouse colony from which *P. aeruginosa* was incriminated (101,333), there is no convincing evidence that clinically overt disease syndromes may be induced by *Pseudomonas* in unstressed laboratory rats and mice.

The ecology of *P. aeruginosa* within animal care facilities is complex and independent of animal colonization. The organism grows well at ordinary room temperatures and may be introduced by a variety of nonsterile supply fomites that include food, bedding, inadequately sanitized cages, food hoppers, water bottles, bottle stoppers, and sipper tubes (29,195,476,492). Considerable evidence has accumulated that the organism may be introduced by human contacts (animal care technicians, investigators) even within barrier facilities, primarily by contact with ungloved hands (476,492). Human carriers appear to be relatively frequent [10–15% of fecal samples and 90% of sewage cultures (206,267,374, 497)]. The organism has been cultivated from the floor, water faucets, water pipe lines, tap water, cages, water bottles, and sipper tubes within animal rooms (29,195, 464,476,492,502). *Pseudomonas aeruginosa* thrives in quarternary ammonium disinfectant solutions (266,348,361) and has been cultured from a number of presumably sterile parenteral materials [see review by Flynn (115)].

It is generally believed that the other floral components of the oropharynx and gastrointestinal tract of untreated conventional animals are effective in regulating the population dynamics of *Pseudomonas in vivo*. Repeated animal colony survey data indicates that the ordinary incidence of *Pseudomonas* carriers is no higher than 5–20% (29,191,195), in comparison to the near 100% incidence in lethally irradiated rats and mice from the same colonies. Ironically, reduction of these other microbial forms by antibiotic treatments or by establishment of the gnotobiotic state, in effect, removes their inhibitory effect and facilitates colonization by *Pseudomonas* to a much higher degree when the animals are exposed to it (115,195,476). *Pseudomonas* appears to localize in the oropharynx, upper respiratory, and rectocolonic tissues in colonized normal animals. It is thought that isolations of *P. aeruginosa* from deeper reaches of the gastrointestinal tract, i.e., duodenum or ileum, represent organisms being transported through the tract, rather than localization or colonization of these sites (191).

The organism ascends from the normal areas of localization

in vivo to become septicemic following treatments that deprive the host of competence to inhibit invasion. Septicemia and hematogenous distribution to such organs as the spleen, lungs, and liver generally describe the pathogenesis of pseudomoniasis, and there is some evidence that even peripheral localizations, e.g., abscesses of middle and inner ears, pyelonephritis, and skin burn wound sepsis may be hematogenously initiated (452,453). The gross pathology, therefore, is generally that of a septicemia, i.e., congestion of splanchnic and thoracic viscera with occasional abscesses of organs with capillary nets. The usual circumstance, however, is that of fulminating infection with onset of death prior to pyogenic lesion development.

Diagnosis is established by the characteristic case history that usually involves immunosuppressive pretreatment or treatments which induce a shocklike state and by isolation of *P. aeruginosa*. It may be isolated from heart blood or spleen of septicemic cadavers. The organism is aerobic and grows readily on blood agar, usually producing pigments that darken the agar initially and also a powerful hemolysin that may clear the entire plate. The colonial morphology is irregularly round, 2–3 mm in diameter, and usually with a smooth matt surface, butyrous consistency, and floccular internal structure. Colonial variants that include mucoid, rough umbonate, rugose, or smaller coliform-like forms are not uncommon. The organism is a typical gram-negative rod, 1.5–3 μm in length, by 0.5 μm wide. Suspect colonies should be isolated in pure culture and confirmed by a number of tests that characterize the pigment producing and biochemical properties of *P. aeruginosa* (see Table III). Although taxonomy of *Pseudomonas aeruginosa* has not been in recent flux, it may be found in the older literature under the names *P. pyocyanea, Bacterium aeruginosum, B. pyocyaneum, B. aerugineum, Bacillus aeruginosus,* and *B. pyocyaneus.*

More typically, however, the organism is suspected in dying animals or monitoring programs are employed to screen laboratory animals or component equipment. A number of specialized isolation or detection systems for presumptive cultural identification of *P. aeruginosa* oriented to its biochemical properties have been developed for such purposes. Such studies have emphasized the desirability of enrichment broths for primary isolation followed by subculture to selective or differential semisolid media. Direct inoculation of agar plates for primary isolation from test materials, e.g., water, swabs, feces, or tissue samples, even when using such recommended media as brilliant green agar, *Pseudomonas* agar P or F (King's medium B-Flo agar), glycerol–peptone agar (King's medium A), or Wensinck's glycerol agar have been shown to yield fewer isolations than when the same test materials are selectively enriched in broths prior to subculture on agar (195). A number of broths for primary culture have been explored, including Wensinck's glycerol enrichment broth (492), tet-

Table III

Characteristic Biochemical Reactions and Other Properties of *Pseudomonas aeruginosa*[a]

Characteristic	Reaction or property
Morphology	Gram-negative rods
Motility	Motile
Pigments	Pyocyanin (blue-green), fluorescein (yellow)
Growth at 42°C	(+)
Growth at 5°C	(−)
Oxidase/catalase	(+/+)
Growth on blood agar	2–3 mm colonies, early discoloration, later hemolysis
Nitrate reduction	(+)
Growth on brilliant green	(+), reddish colonies, alkaline (red) agar
Growth on glycerol-peptone	(+), 2–3 mm colonies, blue-green pigment in medium
Gelatin liquifaction	(+)
Urease	(+)
Arginine hydrolase	(+)
Lysine decarboxylase	(−)
Carbohydrates, acid from:	
Glucose	(+)
Lactose	(−)
Maltose	(−)
Mannitol	(+)
Salicin	(−)
Sucrose	(−)
Xylose	(+)
Dulcitol	(−)
Inulin	(−)
Raffinose	(−)

[a] From Buchanan and Gibbons (66), Cowan (87), Gilardi (157), Lennette *et al.* (257a), and Wilson and Miles (496).

rathionate broth, and Koser's citrate medium, which may be enriched by addition (10% by volume) of brain–heart infusion broth (29,502). There is some evidence favoring Koser's citrate medium as the broth of choice for this purpose (195).

Presumptive determination of the presence of *P. aeruginosa* may be made by the observation of water-soluble greenish-blue pigments (pyocyanin) in the medium, which may be extracted in chloroform. The latter method without subculture has been used for routine screening of water and swabs of water bottles and bottle parts (29,118,292). Other test materials, including oropharyngeal swabs, rectal swabs, feces, and tissue samples, should be subcultured from 24-hr broth primary cultures onto agar plates, e.g., brilliant green or glycerol–peptone agars. On brilliant green, *P. aeruginosa* colonies are reddish and a pronounced red color (non-lactose-fermenting, basic pH) develops in the agar. On glycerol–peptone agar, the greenish-blue pigment may be seen diffusing through the medium. Colonies should be confirmed as oxidase positive by the development of a purple color within 15 sec when inoculating loop samples are applied to filter paper disks soaked in Kovac's reagent (191).

Final confirmation of individual isolates should be made by establishment of a biochemical profile consistent with the summary in Table V (see Section II,G). Several systems have been established for the serological and pyocin typing of *P. aeruginosa* (see following reviews 66,87,257a,496); however, little correlation between type and pathogenicity or type and origin of animal isolate has emerged from these studies (277,284).

Investigations of the method of choice for screening rodent populations have generally established a higher rate of success from culture of oropharyngeal swabs than fecal samples, and the former is recommended for monitoring programs designed to detect shedder or carrier animals (464). Similarly, screening of water bottle contents after 2–4 days and water bottle parts have generally yielded higher percentages or indications of carrier animals than fecal cultures (29,116,117,195,502), although it has been shown that water bottles may frequently fail to yield *Pseudomonas* from cages with mice having positive oropharyngeal cultures, particularly when the bottles are well sanitized (464).

A variety of means have been explored to cope with the problems of pseudomoniasis in immunosuppressed or traumatized animals. Although there are some antibiotics [e.g., gentamacin (286,437)] and bacteriocidins [e.g., pyocin (296)] that have been used to extend the life of X-irradiated animals (500), the approach of mass treatments using antibiotics for control of *Pseudomonas* within rodent colonies has been generally concluded to be ineffective (177,195,286). Similarly, the use of bacterins for active immunization and passively administered anti-*P. aeruginosa* hyperimmune sera have been explored and found to have limited benefit in control of pseudomoniasis in immunosuppressed animals (268,286).

The treatment of drinking water with sodium hypochloride to provide 10–12 ppm chlorine, or water adjusted to pH 2.5–2.8 by the addition of concentrated hydrochloric acid, or preferably, both has been found to suppress *Pseudomonas* sufficiently to prevent its effects in irradiated or burned animals (29,195,292,502). Potable tap water chlorinated to 2 ppm at the source has been found to drop to less than 1 ppm at the tap and to have little inhibitory effect on *Pseudomonas* (262,464). Similarly, chlorine at the 10 ppm level initially is effective in sanitizing water bottles to eliminate *Pseudomonas*, but drops to 6 ppm by 24 hr, 2.5 ppm by 48 hr, and 0.25 ppm by 72 hr and resembles tap water (292). Such bottles, with sodium hypochlorite alone, may become recontaminated by 48–72 hr if the mice are oropharyngeal carriers of *Pseudomonas*.

It was found earlier that *Pseudomonas* is quite sensitive to low pH (2.5–2.8) and that chlorinated–acidified drinking water was effective in eliminating the *Pseudomonas* carrier state (and substantially suppressing coliform and *Proteus* populations) in laboratory mice (402). The combination of acidified and chlorinated water is recommended for mass treatment as a

standard method for control of pseudomoniasis in immunocompromised animals. Palability studies have indicated little, if any, preference for untreated tap water compared to chlorine–acid–treated drinking water (292,456). However, at least one study has indicated retarded growth on the basis of reduced intake in some mouse strains (261). Several studies through reproductive life time cycles have failed to demonstrate deleterious effects of such treatments (315); indeed, most parameters of reproductive performance have increased (260,292,402). The low pH has no demonstrable effect on rat tooth enamel (457). It is known that a small proportion of rats and mice on treated water may continue to shed *Pseudomonas* in the feces (195,464). If it is desirable to do so, such animals may be detected by fecal or oropharyngeal screening and removed from the colony. The decision to recommend initiation of acidification–chlorination treatment of drinking water in rat colonies resolves itself to an assessment of the risk of pseudomoniasis in the course of experimentation and should be determined by the nature of the research program. No specific disease ameliorating or preventive effects with other defined rat pathogens have been shown for chlorinated–acidified drinking water.

E. Tyzzer's Disease (*Bacillus piliformis*)

Tyzzer's disease occurs as an acute (or latent) fatal epizootic with cardinal features of focal necrosis of the liver, myocarditis, enteritis, and diarrhea. It has been observed preeminently in the laboratory mouse (72,127,132,153,227,229,297,343, 352,373,395,466), but naturally occurring disease has been described in the laboratory rat (147,218,431,432,445,507), gerbil (75,353,495), hamster (141,444,495), rabbit (9,90,151, 477), primate (331), cat (243), and horse (174,455). The condition is named after Ernest Tyzzer of Harvard, who initially reported the disease as a fatal epizootic of Japanese waltzing mice, *Mus bactrianus* (232). Tyzzer's original report of the disease, with descriptions of its histopathology, still stands as the fundamental study. The disease has emerged as one of the major infectious diseases of the mouse, and to a much lesser extent of the rat, with world-wide distribution.

Ganaway *et al.* (152), in a recent review, have pointed out that only two articles had been reported on the disease between 1917 and 1933, only about six more between 1933 and 1959, and of 35 articles between 1959 and 1971, 23 had appeared since 1965 alone. In that same context, the disease was recognized only in *Mus musculus, Mus bactrianus,* and their crosses prior to 1965, while naturally occurring disease syndromes have since been recognized in a variety of mammalian species. It is uncertain if the disease is emerging or, rather, being subjected to increasingly more accurate surveillance. The degree to which this disease has interfered with the interpretation of experimental results is difficult to assess for several reasons; however, it is thought to be substantial. On the one hand, it is likely that the disease frequently has gone unrecognized by investigators, and, on the other, it is a general fact that ruined experiments are seldom published. In England, Tyzzer's disease is claimed to have ruined more carcinogenesis experiments than any other single disease (496).

Little about antecedent factors predisposing the pathogenesis and history of natural infections is known. The disease may appear in epizootic proportions with seemingly little in the way of predisposing stress, but more commonly appears to be precipitated into overt, clinically apparent disease by any of a number of factors that impair host immunocompetence. Precedent experimental treatments that include tumor transplantation (89,448,471), administration of steroid or other immunosuppressive drugs (145,331,441,446,507), leukocyte injections (441), and X irradiation (441,446,447) have either been observed in the course of outbreaks or have been used to facilitate transmission studies with the Tyzzer's agent. Other factors bearing on susceptibility are known (or believed) to include poor sanitation and overcrowding (153,343,373), transportation stress (343), nutritional status (276), dietary form (153), and carbon tetrachloride liver injury (448). In assessing the issue of genetic susceptibility and resistance, Gowen and Schott (164,447) were unable to relate resistance in *M. musculus, M. bactrianus,* and their crosses to the waltzing factor,

dominant white, or sex, and concluded that resistance was best interpreted as depending on a major dominant factor. In the light of present knowledge, it appears that the dominant factor was more likely precedant latent infection and immunity than chromosomal in nature.

The taxonomic status of *Bacillus piliformis,* the Tyzzer's disease agent, has not been established. The agent is not listed in the current (eighth) edition of "Bergey's Manual" (66). Wilson and Miles (496) have "quite tentatively" proposed the name *Actinobacillus piliformis* based on the morphologic analogy of monilial thickenings at irregular locations in the vegetative rods seen in *Streptobacillus (Actinobacillus) moniliformis (muris).* However, the analogy appears to end there, and the name originally proposed by Tyzzer has been retained—*Bacillus piliformis.* The organism stains poorly with hematoxylin, although it is generally recognizable, with some experience, in hematoxylin and eosin stained tissue sections. Methylene blue or Giemsa stains are satisfactory for demonstration of the organism in smears and tissue sections. The silver impregnation stains (e.g., Warthin–Starry or Levaditti) similarly are satisfactory, and may be the stains of choice, for demonstration in tissue sections (Fig. 13). The organism is periodic acid-Schiff (PAS) positive and gram negative (the latter determined by the Brown and Brenn method preferably). Excellent color illustrations may be found in the publications by Jonas *et al.* (218) and Fujiwara *et al.* (147).

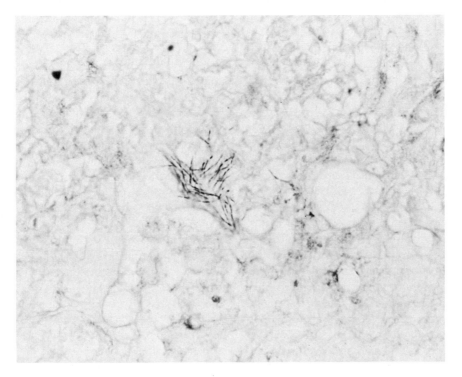

Fig. 13. *Bacillus piliformis* infection of rat liver. Note silver-stained, slender, beaded vegetative rods. Gomori. × 800.

The morphology of the organism is pleomorphic, but general agreement has been reached that the vegetative phase of long slender rods approximately $0.5 \times 8{-}10$ μm (or more) are the obligately intracellular forms seen in large numbers (sheaves or bundles) within infected cells. Sporelike monilial thickenings are seen less frequently in tissue sections from liver and intestine, but more frequently in yolk sac preparations from infected embryonating chicken eggs (88). Although Tyzzer believed the organism to be nonmotile, observation of motility by phase-contrast microscopy (9,88), supported by demonstration of peritrichous flagellae using electron microscopy (141,218,245), have led to its being accepted as motile.

It is generally agreed that *B. piliformis* may not be successfully cultivated in cell-free media. The two reports to the contrary by Kanazawa and Imai (226) and Simon (418) have not been supported by other workers. The organism may be propagated in embryonating eggs by yolk sac inoculations of 10- to 12-day chick embryos (9,88,133) without loss of pathogenicity. The vegetative phase, i.e., the bundles of long slender rods seen in infected cells, is extremely unstable. Infective preparations made from heavily infected livers, even when injected into cortisone-prepared substrate hosts, are likely to fail if the preparation has been made from a donor animal dead more than 2 hr. Preparations made from infected yolk sacs lose substantial activity after 15–20 min at room temperature and even more at 37°C. Infectivity is completely lost after 24 hr at 4°C. No method has been found to stabilize the vegetative form; preparations frozen at −70°C, centrifuged for any reason, or lyophilized have failed (in some hands) to sustain activity. The vegetative phase is killed by heating to 45°C for 15 min.

The spore form is considerably more stable, however, and will survive 56°C temperatures for 1 hr (but not 80°C for 30 min). Infectivity of spore forms may be retained in dead yolk sac-inoculated chicken eggs for 1 year at room temperature (88) and for similar periods in infected mouse and rabbit bedding at room temperature (9,80,471). The latter observation has provided support for the consensual view that natural infections are initiated by peroral ingestion of spore forms from the bedding contaminated by infected animals shedding spores in the feces. Both Tyzzer (471) and Gard (153) used contaminated cages for transmission studies. Transmission studies directly exploring the oral route have produced varying results. Peroral infections in the young rat (218,445), rabbit (9), and gerbil (75) have generally been successful with characteristic production of both intestinal and liver lesions. In the mouse, Gorer used dried liver blocks as a means of stabilizing the sporulated form, which were ground up and placed in cooked cereal for oral transmission studies (163). Interestingly, Fujiwara *et al.* (144) have recently reported higher rates of success with raw liver blocks than with ground up liver suspensions and proposed that cannibalism might serve for transmission and maintenance of the cycle in breeder colonies. The latter

studies routinely induced Tyzzer's disease by the peroral route (gastric intubation) in untreated suckling mice up to 6 to 7 days old, enhanced transmission in cortisone-treated weanlings, and in adults starved 24–48 hr prior to gastric intubation (144). It has been suggested that variability in transmission studies may relate to fragility of the vegetative form *in vitro* and/or failure to administer sufficient spores for peroral transmission (152). Support for this view was demonstrated in the higher infectivity rate obtained in nonfasted adult mice inoculated with 10^7 organisms directly into the duodenum or cecum (38 and 50%, respectively) than intragastrically (13%), and also, demonstration that the infectivity rate was dose related (144,442).

A number of lines of evidence now suggest that the pathogenesis of natural infections follows the sequence: establishment of primary infection in tissues of the jejunum, ileum, and cecum; followed by ascension of organisms via the portal vein to the liver; and bacteremic embolization to other tissues, e.g., the myocardium. Natural and experimental (peroral) infections in the rabbit, rat, gerbil, and hamster characteristically include intestinal localization of the organism with lesion production in the intestine as well as liver and (more variably) my;ocardial lesions (75,151,218,255,318,477,495,507). In natural and experimental peroral infections in the mouse, intestinal lesions are more variable, and liver lesions do not appear dependent on prior development of intestinal lesions (144). Bacteremia occurs routinely subsequent to induction of liver lesions (442) and elevation of serum enzymes (GPT, GOT) indicative of liver destruction (195,317). Experimental infections of any host species given by intravenous or intraperitoneal routes typically induce liver and myocardial lesions but not intestinal lesions. It was originally shown by Rights *et al.* (373) that the organism could be propagated in brain tissue inoculated intracerebrally, and several studies have extended this observation (148,334,335). Similarly, the intracerebral route does not induce intestinal lesions.

Few clinical signs specific to naturally occurring Tyzzer's disease in the mouse and rat have been reported. Even in the presence of enteritic lesions, signs of diarrhea have been variable and not reported for the rat, although diarrhea is more frequent in other species. Affected rats are depressed, with ruffled hair coat and short period of illness. The few outbreaks that have been studied have indicated low morbidity but high mortality in rat groups. It has been mentioned previously that the serum transaminases are elevated in sick animals (145,317). In the mouse both morbidity and mortality may be high; Tyzzer lost his entire colony of 79 animals (471).

Gross lesions in the rat are suggestive (Figs. 14–16) and include flaccid segmental dilatation of affected portions of the intestines (up to three to four times normal diameter) with an edematous, atonic ileum (218). Ileal lesions may extend to adjacent cecum and jejunum. On the basis of these lesions Jonas *et al.* (218) have suggested that several earlier reports of

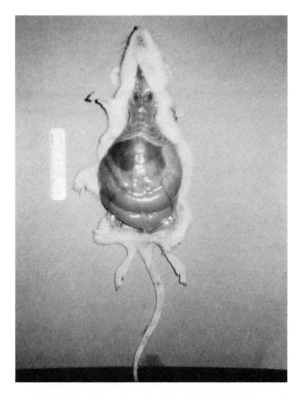

Fig. 14. Tyzzer's disease of adolescent rat with skin reflected. Note enlargement of ileal loops *in situ* and pot-bellied appearance.

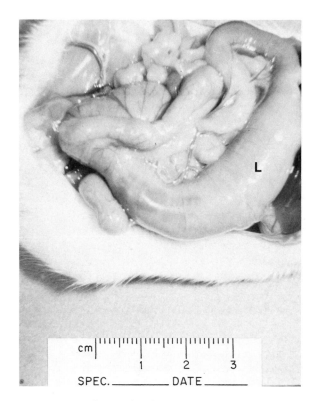

Fig. 15. Tyzzer's disease of adolescent rat in Fig. 14 with peritoneum reflected and ileum *in situ*. Note flaccid dilatation of ileum (L).

megaloileitis or segmental ileitis in the rat (154,203) may have actually have unrecognized instances of Tyzzer's disease [although it has recently been pointed out (113) that the lesion is not specific and may be induced by chloral hydrate alone]. The liver (Fig. 17) typically has few to many disseminated pale foci up to several millimeters in diameter scattered throughout the parencyma (218,507). Circumscribed, grayish foci may be seen in the myocardium in some cases. Excellent color illustrations of gross lesions may be seen in the report by Jonas *et al.* (218). The mesenteric lymph nodes are usually swollen. No other lesions are seen at necropsy.

Definitive diagnosis of Tyzzer's disease in the rat continues to rest on histopathologic criteria, although it has been pointed out that *B. piliformis* may be isolated from infected tissues, cultivated via yolk sac inoculations in chick embryo, and these isolates used to satisfy Koch's postulates (9,88). Microscopic lesions in the liver are initiated by foci of coagulative necrosis varying in size from several micrometers several millimeters, with a tendency to centrolobular orientation in association with central veins and distinct demarcation between necrotic and adjacent normal hepatic parenchyma. *Bacillus piliformis* typically may be found in viable hepatocytes on the periphery of necrotic foci. Necrotic foci thought to be early developing are relatively acellular and barely eosinophilic, with hepatic cell

ghosts and karyolytic debris (218). Lesions interpreted to be later in development (Fig. 18) demonstrate moderate to marked infiltration with neutrophils, Kupfer cells, macrophages, and

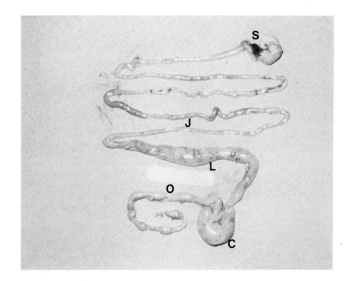

Fig. 16. Tyzzer's disease of adolescent rat. Entire gastrointestinal tract to show proportions of unaffected jejunum (J) and megaloileitis (L). Stomach (S), cecum (C), and colon (O) are normal.

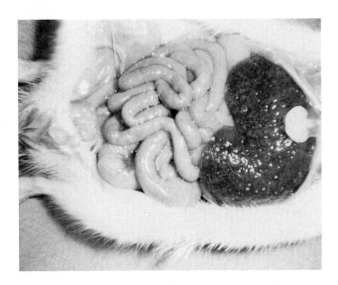

Fig. 17. Tyzzer's disease of adolescent rat. Note focal process in liver. Small, whitish foci are mineralized "older" lesions.

tively acellular "early" foci may be seen, along with infiltrated "later" foci, and still others with fibroplasia and giant cells. These observations are suggestive of successive showers of organisms from the intestines. The ultrastructure of hepatic lesions has been described (140,141,218,245).

Intestinal lesions are characterized microscopically by transmural involvement of the intestinal wall, particularly the ileum, in a process varying from segmental sharply demarcated necrosis to acute and chronic inflammatory infiltration of all levels that include the mucosa, submucosa, and muscular and serosal layers (Fig. 20). Edema and hemorrhage are common in the subserosal and muscular layers. *Bacillus piliformis* may be found, often profusely, within epithelial cells of the villi and their crypts (Fig. 21). The organism has not been seen deep to the mucosa (218). Although Jonas *et al.* (218) described villi as intact but blunted, Japanese workers have (without detailed microscopic descriptions) described the intestinal process in the rat as an ulcerative enterocolitis (507). Myocardial lesions, when present, vary in size from only several myofibers to complete transmural involvement (Fig. 22). Inflammatory infiltrates are variable, and the myocardial process was described as degenerative rather than necrotic. *Bacillus piliformis* may be demonstrated within affected myocardial muscle cells (Fig. 23).

The occurrence of subclinical infection and the latent carrier state may be assumed from several lines of evidence; chief

fibroblasts, which may almost obliterate previously acellular foci of necrosis (151). Mineralized material (Fig. 19) similar to that seen in rabbit lesions has been seen in necrotic foci of rat livers (218). Giant cells and histiocytes may aggregate peripheral to infiltrated necrotic foci (218). In any given liver relationship.

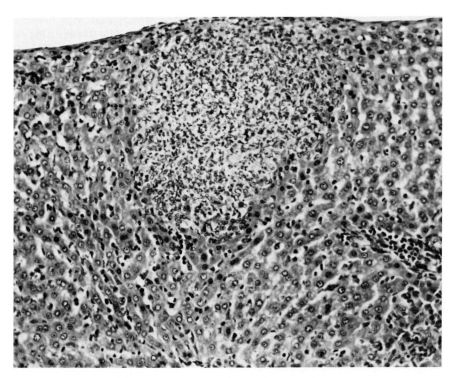

Fig. 18. Tyzzer's disease in rat liver. Infiltration of necrotic focus with both acute and chronic inflammatory cells. Note sharp demarcation of lesion margin from adjacent normal parenchyma. Hematoxylin and eosin. ×200.

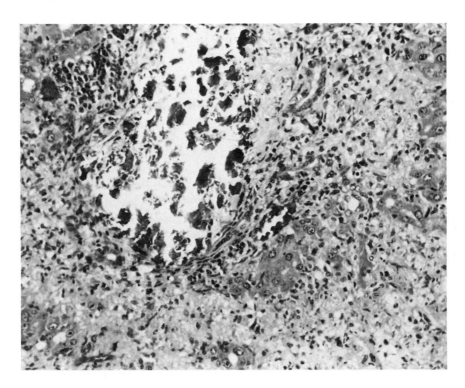

Fig. 19. Tyzzer's disease in rat liver. Mineralized focus representing "older" lesion. Note diminished intensity of inflammatory reaction. Hematoxylin and eosin. ×200.

among them are the frequent reports of intercurrent epizootics subsequent to immunosuppressive treatments and the documented results of immunosuppressive and immunodiagnostic screening techniques. Using antigens derived from infected liver suspensions, it has been established by fluorescent antibody techniques, complement fixation, and agar gel immunodiffusion that the organism is immunogenic and that the anitibodies thus induced are protective and generally are naturally present in younger age groups than in retired breeders (138,142,149,334). Fluorescent antibody techniques, both direct and indirect, have been used to locate *B. piliformis* antigen in tissues (134,141,142,397). Cortisone provocation using 2.5 mg cortisone or hydrocortisone daily (for 2 days) or single 5-mg injections given subcutaneously has been adopted as a routine method for clinical expression of the latent carrier state and is recommended for diagnostic surveillance programs (229,443,507).

Similarly, the latent state may be determined in naturally infected adult animals which frequently carry antibody levels too low to be detected by ordinary serologic assay. The basis of the test is the anamnestic antibody response to a defined challenge with killed bacterial or viral antigens (139). The principle has been shown to be equally useful for serologic monitoring of rodent colonies for other pathogens, especially mouse hepatitis virus, but also including Sendai virus, *Corynebacterium kutscheri,* and *Salmonella enteritidis.* As a rule, antibody levels are determined on sera from the sixth day after antigen injection. The (controlled) assumption has been made that primary antibody responses occur subsequent to the ninth day, whereas those that occur prior to it are truly anamnestic and reflect sensitization from previous experience with the antigen (139). The principle offers great promise for serologic surveillance of rodent colonies for latent and opportunistic pathogens.

The degree of host specificity of Tyzzer's isolates and the hazards of interspecific transmission are of great interest and have been examined to some degree (143,146). These studies have shown that heterologous transmission (i.e., rat isolate to mouse, and vice versa) is less virulent than homologous transmission, particularly by the peroral and intravenous routes, that immunity to infective challenge was higher in homologous systems, and that while rat and mouse isolates had certain antigens in common, they also had other unshared antigens. Based on the greater stability of vegetative suspensions from rat liver, Fujiwara *et al.* (143) have suggested that *B. piliformis* may originally have been a pathogen of the rat, being forced to mutate to a form with surface changes in the heterologous (mouse) host. The general view at the present time is that regardless of host origin, all isolates must be considered as causative agents of Tyzzer's disease in heterologous hosts.

The sensitivity of *B. piliformis* to various antibiotics has

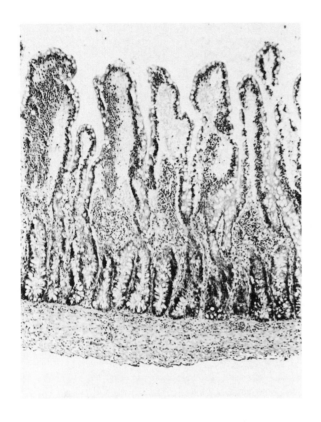

Fig. 20. Tyzzer's disease in rat ileum. Transmural involvement in inflammatory reaction. Note intense infiltration below villous epithelium. Hematoxylin and eosin. × 80.

been investigated primarily from the standpoint of treatment and also to shed, by antibiogram, some light on the taxonomic position of the organism. These studies have established the effectiveness, for protection against experimental infection, of cephaloridine, tetracycline, and chlorampenicol (88,89,447, 508). Penicillin and its analogues were less effective; kanamycin, streptomycin, and sulfamethizole are ineffective (508).

The emphasis has been on prevention of infection, however, rather than treatment. The importance of good sanitary standards were emphasized previously in the discussion on transmission and factors known to predispose infection. The disease may be excluded by gnotobiotic principles and maintenance within barriers (205). There is no evidence that *B. piliformis* may traverse the placental barrier.

F. Pseudotuberculosis (*Corynebacterium kutscheri*)

Pseudotuberculosis in laboratory rats and mice is caused by *Corynebacterium kutscheri,* first isolated and described from the mouse in 1894 (246). Synonymy with this organism is confusing, both because of the descriptive name for gross lesion morphology (pseudotuberculosis) and also because of other similarly named but distinct bacterial entities, e.g., *Corynebacterium pseudotuberculosis* (the Preisz–Nocard bacterium) and *Pasteurella* (now *Yersinia*) *pseudotuberculosis.* The synonymy of *Y. pseudotuberculosis* is additionally confusing, since it includes *Bacterium pseudotuberculosis rodentium* (182,407,408). Both of these latter organisms are culturally

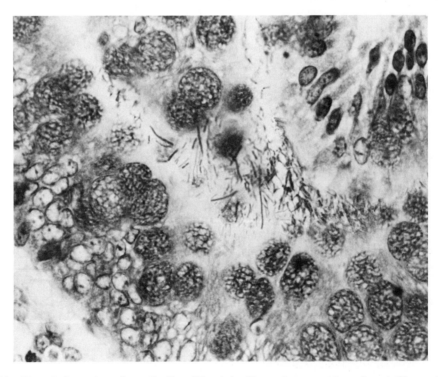

Fig. 21. Tyzzer's disease in rat ileum. *Bacillus piliformis* in villous submucosa at base of crypt. Giemsa. × 800.

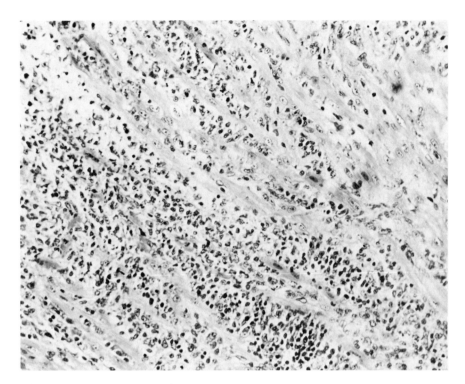

Fig. 22. Tyzzer's disease of rat myocardium. Note myocarditis with degeneration of myocardial bundles and infiltration of chronic inflammatory cells. Hematoxylin and eosin. ×200.

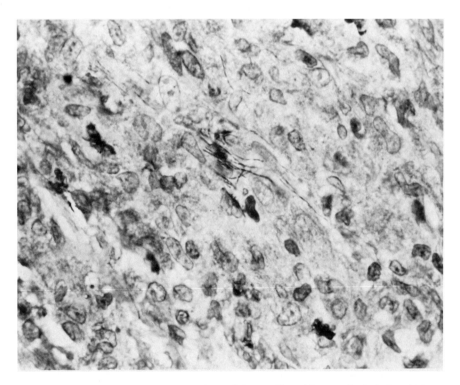

Fig. 23. Tyzzer's disease of rat myocardium. Note intracellular *Bacillus piliformis* rods. Periodic acid-Schiff. ×800.

and biochemically distinct from *C. kutscheri,* and there is no convincing evidence that they may cause natural or intercurrent disease syndromes in laboratory rats and mice. The older literature contains references to *Bacillus pseudotuberculosis murium* (123,198,370), *Corynethrix pseudotuberculosis murium* (404), and *Corynebacterium pseudotuberculosis murium* (404), which are regarded as synonyms of *C. kutscheri.* Additionally, as pointed out by Giddens *et al.* (156), older references to *Bacillus muris* (234,307) have been cited in reviews on murine pneumonia that describe isolation of gram-positive diphtheroids from rat pulmonary abscesses consistent with those associated with *C. kutscheri.* Wilson and Miles (496) have continued *C. murium* as the current name for this organism; however, "Bergey's Manual" (66) retains the name *C. kutscheri,* and it is adopted here as the standard practice for this volume. "Bergey's Manual" lists *Bacterium kutscheri* and *Mycobacterium pseudotuberculosis* as additional synonyms.

As with many of the bacterial pathogens of laboratory rats and mice, *C. kutscheri* has been encountered as a primary pathogen in unprovoked epizootics (or single cases) in the rat (32,123,156,328,344,474) as well as the mouse (34,51,198, 246,349,396,488,489,501). More commonly, however, in both species, inapparent or latent infections have been unmasked by experimental treatments lowering host resistance or impairing immunocompetence. Such treatments have included cortisone (12,73,108,345,377,11,13,31,107,256,427,507), X irradiation (403,404), infection with other organisms [e.g., ectromelia (250) and *Salmonella* (459)], and vitamin deficiency [particularly biotin (156,170) and pantothenic acid (410–412, 510–512)]. Indeed, the rat was long considered relatively insusceptible and frequently refractory to unconditioned transmission attempts (51,123,246,410) and was thought to require immunosuppressive or nutritionally deficient conditioning (51,246,410). It is known that heritably conditioned resistance (and susceptibility) occurs among various mouse (345,488) and rat (411,412) strains. Pathogenicity for other host species appears quite limited, although *C. kutscheri* has been isolated from guinea pig (328,473).

The concept of latency has been investigated in some detail by the research group of Pierce-Chase and Fauve in the mouse (107–109,345). Their initial observation was that randomly chosen mice could be placed into one of two categories based on their suceptibility or resistance to a defined intravenous challenge with *C. kutscheri.* They found that resistant mice were actually resistant to superinfection because a state of premunition (i.e., infection immunity) existed in these stocks. In short, the resistant mice were immunologically sensitized (and protected) by virtue of preexisting latent infections. They found further that the latent organism was an avirulent variant of *C. kutscheri* that was carried in the same organs as the virulent form and that shortly following a suitable stress, e.g.,

cortisone administration, these avirulent "A" bacterial populations shifted to the virulent "K" form. While these results have not been entirely supported by others (64), they substantially clarify the nature of the latent state with *C. kutscheri* and the requirement for immunocompetence for continuation of the resistant state. It is assumed that the frequent isolations of *C. kutscheri* from the middle ear, lungs, and abscessed preputial glands in the course of routine diagnostic monitoring of essentially normal conventional rats (283) actually represent cultural confirmation of latent infection in resistant hosts.

Clinical signs in unprovoked rats typically are those associated with subacute respiratory disease. Variable degrees of upper respiratory signs are seen, including reddish to mucopurulent discharges from the nares and medial canthus. Affected rats are depressed and anorexic with humped posture and ruffled hair coat. Respiratory rales are common. Untreated animals typically die 1–7 days after the onset of signs. All age groups may be affected.

Gross lesions in the rat may be confined to pulmonary tissues and consist of numerous pale foci 1–5 mm in size scattered throughout the parenchyma of the lung (Fig. 24). Typically the lesions are liquified centrally but may be caseous. Occasional areas of coalescence of foci may be quite large, up to 1.5 cm and involve substantial portions of a given lobe. The tissue surrounding foci is congested, and entire lobes may be consolidated. Fibrinous adhesions to the thoracic wall are not uncommon and occasionally make retraction of the lung difficult. Such adhesions become fibrous in more chronic cases. Similar foci may infrequently be seen in the liver and kidney of infected rats, unlike the situation in the mouse where

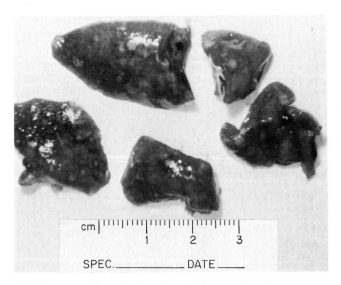

Fig. 24. Corynebacterium kutscheri infection of rat lung. Note numerous small foci of infection in parenchyma.

hematogenous septic embolization and abscessation of organs, with capillary nets, e.g., the liver and kidney, is the rule (488). Infrequently, arthritic lesions of the pedal extremities (Fig. 25) and subcuticular abscesses may be seen (111,328). It was pointed out earlier that abscesses of the preputial glands (Fig. 26) and middle ears frequently yield *C. kutscheri* in the absence of other more serious lesions. Under conditions of cortisone preconditioning the appearance of hematogenous distribution may be more apparent. In discussing the issue of how organisms get to the lungs, Giddens *et al.* reasoned that inasmuch as the pulmonary lesions were interstitial rather than related to the bronchi, hematogenous embolization of the lung was more consistent with the observations than inhalation, and pointed out that pulmonary infections in their own experiments followed oral or intraperitoneal inocula, and those induced by LeMaistre and Thompsett (256) were provoked by the intracardiac injection of *C. kutscheri* (156). Similarly, Fauve *et al.* recovered latent *C. kutscheri* from the lungs of asymptomatic lesionless mice (108).

Microscopic lesions in any of the affected tissues consist of abscesses that vary in size from several micrometers up to the size seen grossly at necropsy. The centers of such lesions are necrotized and, depending on the stage, may be coagulative and relatively acellular early, liquifactive (later) as acute inflammatory cells infiltrate, and mature as inspissated or caseous abscesses. The inflammatory infiltrate surrounding and invading necrotized areas consists of neutrophils early in the course and includes macrophages, lymphocytes and plasma cells as the lesions become subacute. Fibroplasia is not pronounced in pulmonary lesions; epitheloid cells, calcification, or giant cells are not seen. Thus, although the lesions are spoken of in the literature as being "granulomatous," Giddens

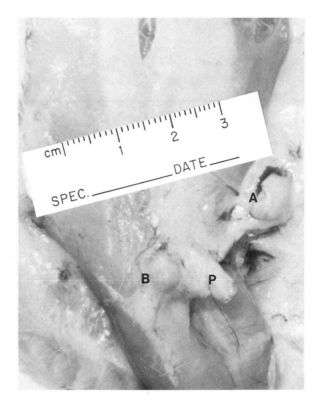

Fig. 26. Corynebacterium kutscheri abscesses of preputial glands. Skin reflected to show opened abscess (A), intact abscess of contralateral gland (B), and penis (P).

et al. (256) have pointed out that there is little histopathologic support for this description and thus, the term "pseudotuberculosis." Although grayish-blue amorphous bacterial colonies may be seen, typical organisms are not readily discernible in hematoxylin and eosin stained material. They may easily be demonstrated by Giemsa or tissue gram (Brown and Brenn or Gram-Weigert) stains as irregular palisades or "Chinese letter" arrangements of gram-positive rods. In the lung, suppurative exudates are seen in the brochioles and bronchi. Interstitial tissues peripheral to pulmonary abscesses are congested, alveoli are filled with proteinaceous fluids, and leukocytes may be seen as perivascular cuffs around blood vessels. Foci are believed to enlarge and coalesce by radial expansion (488). Microscopic lesions are illustrated in Figs. 27 and 28.

Presumptive diagnosis may be made at necropsy by smears of affected tissues (or in sectioned material), in which typical aggregates of *C. kutscheri* are demonstrated by Giemsa or gram stains (Fig. 29). Gram reactions may be variable in clinical material. Definitive diagnosis requires cultural isolation and biochemical characterization of the organism. Aseptically collected suspect tissues, washings of the nasal turbinates or middle ears, or organ homogenates should be cultured directly on 5% blood agar plates incubated aerobically at 37°C. Colonies of *C. kutscheri* on blood agar are 1–2 mm in size

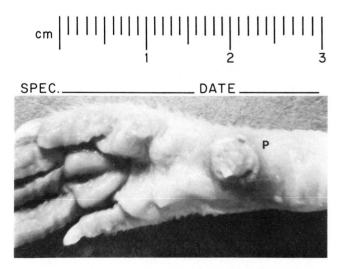

Fig. 25. Corynebacterium kutscheri pododermatitis of rat forepaw (P).

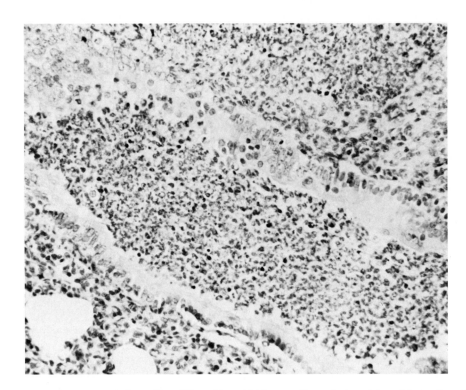

Fig. 27. *Corynebacterium kutscheri* pneumonia in rat. Note filling of bronchial lumen with purulent exudate and destruction of epithelium in upper left. Hematoxylin and eosin. × 160.

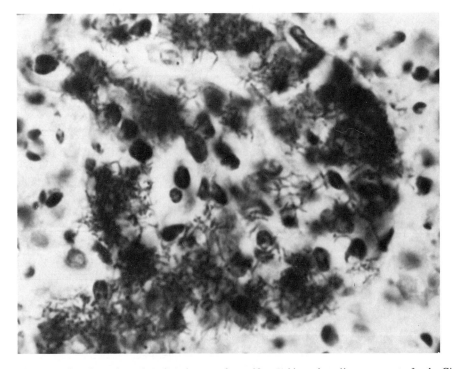

Fig. 28. *Corynebacterium kutscheri* microcolony in pulmonary focus. Note ''chinese letter'' arrangement of rods. Giemsa. × 1200.

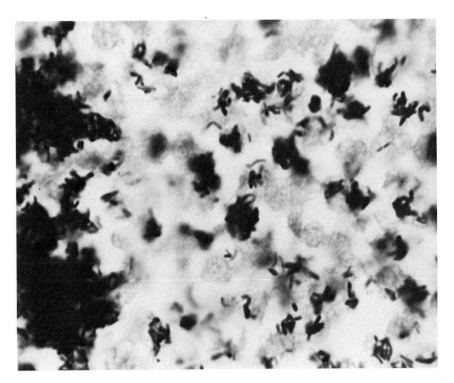

Fig. 29. *Corynebacterium kutscheri* infection of preputial gland. Tissue imprint to illustrate "chinese letter" arrangement of gram-positive rods. Gram stain. ×1200.

after 24 hr, circular, entire, dome-shaped, grayish-white in color, smooth, and nonhemolytic. Growth and biochemical characteristics are summarized in Table IV. It is pointed out that *C. pseudotuberculosis* is hemolytic on blood agar (66, 496).

Deliberate cortisone provocation (5–10 mg subcutaneously) has been used as a means of expressing latent infections and is advocated for diagnostic surveillance of rat and mouse colonies (229,472,507). Overtly infected animals develop antibody titers that may be detected by tube or microplate agglutination and indirect immunofluorescence (139,489). These methods are suitable for diagnostic surveillance of colonies in which the organism occasionally surfaces in clinically apparent forms. As was pointed out in the discussion of Tyzzer's disease, however, latently infected animals typically have antibody responses too low to be serologically detectable. Weisbroth and Scher (489) were unable to detect antibodies in lesionless mice. Fujiwara and his colleagues have explored the use of the anamnestic response test with this organism and greatly increased serologic detection of the latent state (139). As with Tyzzer's disease, killed suspensions of organisms are injected and the test animals bled on the sixth day. Antibody responses, if any, by the sixth day are held to be anamnestic and reflective of previous experience (sensitization) with *C. kutscheri*.

Visibly ill animals should be treated by medication of the water or individual injections of suitable antibiotics following a determination of antibiotic sensitivity *in vitro*. The organism has been shown to be sensitive *in vitro* to a variety of narrow-range and broad-spectrum antibiotics (489), and antibiograms have been developed for *C. kutscheri* rat isolates (339). A number of unprovoked epizootics have been brought under control by the expedient of simply removing visibly ill animals from the colony when the morbidity is low (32,156,489). It is emphasized that such procedures, i.e., antibiotic treatment and rigid culling, serve to remove the visibly ill tip of the iceberg but leave behind the great bulk of inapparently ill reservoir hosts.

The disease may be prevented or excluded from rat colonies by maintenance of microbiologic barrier systems. Although *C. kutscheri* has been excluded from breeding colonies on many occasions by derivation or rederivation from conventional animals without (to the authors' knowledge) transplacental contamination of term fetuses, a recent report has documented vertical transmission of *C. kutscheri* to neonatal animals (223).

G. Salmonellosis (*Salmonella* sp.)

Salmonella infections were among the first to be recognized as naturally occurring enzootic or epizootic diseases of labora-

Table IV

Characteristic Biochemical Reactions and Other Properties of *Corynebacterium kutscheri*[a]

Characteristic	Reaction or property
Growth on blood agar	Grayish or yellowish-white 1–3 mm colonies; no hemolysis
Morphology	Gram-positive rods with metachromasia in Chinese letter arrangement
Carbohydrate, acid from:	
Glucose	(+)
Xylose	(±)
Lactose	(−)
Sucrose	(+)
Maltose	(+)
Salicin	(+)
Mannitol	(−)
Dulcitol	(−)
Fructose	(+)
Mannose	(+)
Aesculin hydrolysis	(+)
Growth on MacConkey	(−)
H_2S production	(±)
Indol	(−)
Urease	(+)
Nitrate reduction	(+)
Gelatin liquifaction	(−)
Litmus milk	No change
Catalase	(+)
Oxidase	(−)
MR/VP	(−) (−)

[a]References from Buchanan and Gibbons (66), Weisbroth and Scher (489), and Wilson and Miles (496).

tory rats and mice. Indeed, the first isolation of *S. typhimurium* was made by Loeffler in the course of an epizootic of laboratory mice in 1892 (340). The scientific literature of the next three decades (1895–1925) was particularly rich on the subject of rat salmonellosis and dealt with perhaps three major themes: incidence and significance of salmonellosis in wild rats as vectors of human food contamination, the use of "rat viruses" (*Salmonella* cultures) for biological control of wild rat populations, and the documentation and description of intercurrent salmonellosis in laboratory rat colonies.

These studies have established that although a large number of *Salmonella* species have been sporadically isolated from rats (and mice), only two have been recovered with sufficient frequency as to be regarded as typical of murine salmonellosis: *S. typhimurium* and *S. enteritidis*. The older literature contains a confusing synonomic array for both species, as would be expected for a large bacterial group undergoing taxonomic study for almost 100 years. The synonymy for *S. typhimurium* includes *Bacillus typhimurium*, *Bacterium aertrycke*, *S. pestis caviae*, *S. aertrycke*, and *S. psittacosis*. In like manner, the following are regarded as synonymous with *S. enteritidis: Bac-*

terium enteritidis, *Bacillus enteritidis*, Danysz bacillus, *S. enteriditis* var. *danysz*, and Gaertner's (or Gärtner's) bacillus. Both *Salmonella* species have essentially identical epizootiology in rats, and the discussion that follows applies to both, except where noted.

As mentioned previously, surveys of wild rat populations established that these animals were, and remain, enzootic vectors of human *Salmonella* food poisoning, particularly of dairy, bakery, and meat products (27,222,232,233,270,299, 316,393,398,399,430,480,491). Although the incidence of survey reports has declined in recent decades, such studies indicate the continuing association of wild rats with *Salmonella* (61,64,253,254,291,479). For this reason, wild rats must still be regarded as significant hazards for the introduction of salmonellas to laboratory rat colonies.

Considerable early interest attended the practical use of *Salmonella* cultures as rodenticides. They were prepared for this purpose on a large scale by commercial and public health organizations. The idea was first suggested in 1893 by the success observed by Loeffler with the use of *S. typhimurium* in combatting a plague of field mice in Thessaly (340). Another of the early isolates was the bacillus (*S. enteritidis*) isolated by Danysz from an epizootic of field mice in Charny-en-Seine, France (92). The isolate was early observed to have considerable pathogenicity for rats. These agents were known as "rat viruses" and enjoyed popularity for about 15 years (1895–1910) for employment as rodenticides. Such cultures as the Liverpool virus, the Azoa virus of Wherry, the Danysz and Issatschenko viruses, and Ratin I and II (all isolates of *S. enteritidis*) were widely used on the European continent, England, and in the United States as rat "poisons" (23, 24,222,279,340). They are still used for this purpose in eastern Europe (462). The predictable curtailment of these agents came about when it was recognized that they could not be contained in the field, and they were, in fact, indicted in numerous human epidemics (27,222,399).

Salmonella infections have traditionally been internationally regarded as among the most important of laboratory animal diseases. The renown was not misplaced since it was a serious and common disease of laboratory rats for about four decades (1890–1930), as the frequent reports during that period indicate (25,26,43,55,70,135,340,360). Much of the earlier literature, prior to 1914, has been summarized by Pappenheimer and von Wedel (340). The naturally occurring disease has been reported on several occasions subsequently (67,219,244, 353,364,367,390,451), but has not been reported from laboratory rats in the United States since 1939 (219). The epizootiology of this disease has undergone a radical transition in American rodent stocks in recent years, attributable perhaps in equal measure to the concepts of gnotobiotic technology with the principle of disease exclusion, to the elevation of standards of animal care generally, and as a consequence of

Federal laws (91:579 and amendments) and the accreditation scheme of the American Association for Accreditation of Laboratory Animal Care. The situation now in most countries with similar standards is that the disease is comparatively rare.

The clinical form of the disease in rats may vary from that of an acute epizootic with substantial mortality to that of a clinically latent disease that may be silent until the animals are placed under investigatory stress. There are many factors known to influence the clinical expression of *Salmonella* infections in laboratory rats and mice. They include the level of acquired immunity, genetic resistance, virulence and dosage of the organism, age of the host, nutritional balance, and any of a variety of stressors that may act to lessen the immunologic competence or general vitality of the host (40,50,169,168, 171,225,271,280,289,330,380,450,459,485). It also is worth mentioning that the pathologic effects of *Salmonella* infection as a monocontaminant in rat (505), as in mouse (281,389), are considerably different (asymptomatic) than in microbially associated animals. Natural infections in laboratory rats tend to be enzootic after the acute phase with only modest increases in mortality or overt clinical signs.

Such signs have been observed to include reduced activity, thirst, ruffled fur, hunched posture, anorexia, weight loss, and softer lighter stools (67,278,340,364). Diarrhea has been an inconstant sign under conditions of natural infection, seen more commonly as an accompaniment to acute episodes, and less frequently in enzootically infected colonies (67). Experimental infections of rats with *S. typhimurium* by Maenza and his colleagues (278) have established the brief incubation period of 2–3 days prior to development of clinical signs including diarrhea. Diarrhea incidence averaged 15–20% of infected rats but was increased in rats fasted prior to infection and by overcrowding. Maenza *et al.* confirmed the results of others that the diarrheal sign was dose dependent (44,278). The incidence of diarrhea was maximal 4 days after infection and ranged in persistence from 1 to 21 days. The character of the stools in rats with diarrhea ranged from formless but normally colored, to mucoid and occasionally bloody bowel movements. The latter have been seen by others as well (340). Experimental *Salmonella* diarrhea has been extensively investigated as a diarrhea model in rat (358,359,505).

The development of methemoglobinemia during septicemic phases was first observed by Boycott (55). This transitory sign causes browning of the normally pink eyes and an anemic, bluish discoloration of the tail, feet, and ears in albino rats. These signs, confirmed by others, disappear shortly after death (26,55,340,360). Several reports have described sanguinous discharges accumulating about the conjunctiva and nares as terminal events (340). The clinical signs of gastrointestinal disturbance and methemoglobinemia may be suggestive, but neither is sufficiently unique or constant to be of diagnostic value.

The gross findings in fulminating and acute cases are limited to those consistent with septicemia. Most abdominal organs, particularly the liver and spleen, are congested, and lesions may be thus limited. Subchronic and chronic cases, with longer duration prior to death, generally have specific lesions. The disease is preeminently a classic enteric infection of the ileum and cecum; however, intestinal lesions may extend on either side of this location. Bacteremic or septicemic ascension to the liver and spleen occurs typically. It is doubtful if the cardiac and pulmonary lesions described in the older literature (340) are *Salmonella* related, since they have not been described by more recent investigators (67).

Gross lesions of the intestines include thickening of the ileum and cecum with intraluminal accumulations of semi-liquid feces, gas, and flecks of blood (Fig. 30). The mesenteric and serosal vessels are congested, and the mesenteries edematous (67). In animals with diarrhea the intestinal lesions typically extend into the jejunum and include as key features large and small ulcers of the cecal epithelium and, less commonly, of the terminal ileum. The ulcers are usually covered with fibrinous exudate. Microscopically, the ileal and cecal lesions are seen to consist of diffuse enteritis with infiltration of the lamina propria with macrophages, neutrophils, and, later, lymphocytes (Fig. 31). The crypt epithelium is typically hyperplastic; however, the villous epithelium shows degenerative changes. Ulcers of the ileal mucosa, infrequently present, tend to be microscopic in size, whereas those of the cecum are larger (up to 20 mm with irregular outline) and grossly apparent (Fig. 32). Cecal ulcers were not always accepted as occurring as direct consequences of salmonellosis in rats, and the

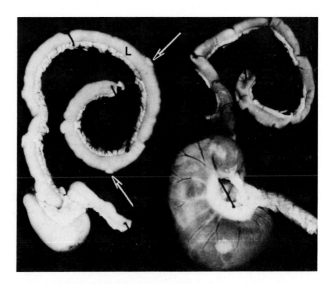

Fig. 30. Salmonella typhimurium infection of rat gastrointestinal tract (left) and normal gastrointestinal tract (right). In tract at left note thickened ileum (L) and swollen Peyer's patches (arrows). (Courtesy of R. M. Maenza.)

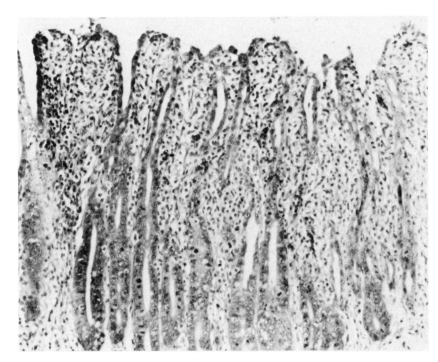

Fig. 31.   *Salmonella typhimurium* infection of rat ileum. Note infiltration of lamina proprin with inflammatory cells. Hematoxylin and eosin. × 160. (Courtesy of R. M. Maenza.)

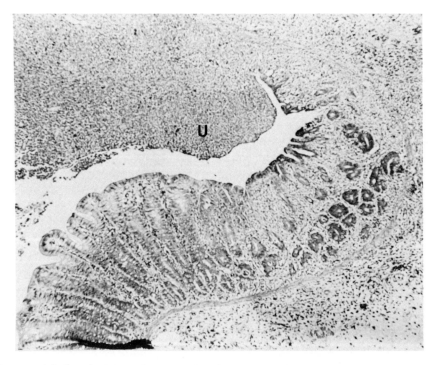

Fig. 32.   *Salmonella typhimurium* infection of rat cecum. Note ulceration of cecal epithelium with fibrinonecrotic overlay toward top of photograph (U). (Courtesy of R. M. Maenza.)

relationship had been seriously questioned (42–48,219,433). They are now recognized, however, as lesions attributable to *Salmonella* infection (67,278). It must be emphasized that subclinically infected animals without diarrhea have qualitatively similar, but quantitatively less extensive, lesions on both gross and microscopic levels of observation. A positive correlation has been shown to exist between the signs of diarrhea and the severity of ileal lesions (278). Similarly, the persistence of diarrhea is related to the severity of ileal lesions, with signs subsiding as the tissues heal. The intestines of surviving animals were histologically healed by 4–5 weeks after infection (278).

Grossly apparent indications of reticuloendothelial involvement are initiated early in the course and include enlargement of the Peyer's patches of the ileum, the mesenteric lymph nodes, and spleen. Microscopically, the Peyer's patches and lymph nodes are edematous, have inflammatory infiltrates, and may show focal necrosis or granuloma formation. Splenomegaly is a common feature. Such spleens are dark red or with a grayish capsule and may approximate 4 cm in length by 1 cm width (67,340). Microscopically, the splenomegaly is seen to be due to intense congestion of splenic sinuses with spilling over of erythrocytes into the white pulp. Extensive inflammatory infiltrates, particularly of phagocytic cells, are seen in splenic tissues. Focal lesions varying from necrosis to granulomata are commonly seen.

The liver is variably congested and dark red or pale and friable. Focal lesions (1–2 mm) may be seen grossly under the capsule and on cut surface but more commonly are not. Upon microscopic examination, however, small areas of coagulative, acellular necrosis are seen (Fig. 33) as has been described in mice (467). More recently, granuloma formation in the liver has been described (270). It is uncertain whether the granulomas are truly different or merely represent different stages of the same basic process. In the original histopathologic description provided by Pappenheimer and von Wedel, small microthrombi ("cell fragment thrombi"), composed of mononuclear phagocytes containing cell fragments and bacteria, degenerating phagocytes, cellular debris, and fibrin, were observed in organs with focal necrosis, e.g., spleen, liver, and lymph nodes (340). The authors concluded that the focal necroses were actually sequella of embolically occluded (thrombosed) microvasculature. One is tempted to recall, in support of this concept, the pulmonary thrombi in hamster salmonellosis described as pathognomonic by Innes *et al.* (209). Certain lesions seen experimentally using the rat *Salmonella* model, e.g., cholangitis (16), placentitis (173), and polyarthritis (481), have not been experienced with natural infections.

Diagnosis of salmonellosis in the rat is based on a consistent case history, isolation of the organism in pure culture, and cultural and immunologic confirmation of the identified isolate as one of the *Salmonella* serotypes. Although methemoglobinemia was mentioned above as a clinical observation in rat

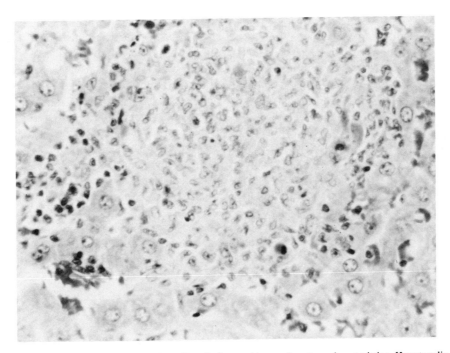

Fig. 33. Salmonella typhimurium infection of rat liver. Hepatic focus with granulomatous characteristics. Hematoxylin and eosin. ×375.

salmonellosis, and research results have documented changes in serum enzyme concentrations (503), these changes are not specific and are of little diagnostic value.

It has been pointed out, in the mouse, that difficulty should not be experienced in the retrieval of *Salmonella* species encountered during acute epizootics (313). Large numbers of animals die of the infection, and the many ill animals shed high ratios of *Salmonella* organisms into the feces. More problems have been encountered in the diagnosis of salmonellosis in asymptomatic carrier animals from enzootically infected colonies. The asymptomatic carrier animal (mouse) has been investigated in some detail (216), and it is understood from such studies that the relationship is one of intracellular multiplication by the organism within a mileu of humoral antibody. Exacerbation of the disease in recovered carriers is prevented by specific antibody, but complete elimination is inhibited by smoldering replication of the bacteria within macrophages. Such animals may shed *Salmonella* intermittently in the feces over long periods of time and represent infective reservoirs of the disease. That essentially the same factors are involved in the rat is generally accepted, and the asymptomatic carrier rat has been amply documented (67).

Although clearly of great diagnostic importance, the focus of infection in recovered carriers is not well understood and probably is not the same in each carrier. Several studies have attempted to localize the most likely organs or other locations in carrier animals for more efficient diagnostic retrieval (67,193,194,282,313,314,420). The consensus is that while a greater number of isolates may be retrieved from intestines, compared to liver, spleen, heart, or other organs, none of these tissue locations results in higher percentage of isolations than from feces alone. For this reason, diagnostic monitoring programs for murine salmonellosis have been oriented to fecal culture.

The factors relating to efficacy of retrieval of *Salmonella* from rat and mouse feces under varying conditions have been studied. It is known that recovery from individual animal samples is higher than from pooled samples (282,420), that the highest number of isolates will be obtained from the first six fecal pellets in any given animal (282), and that storage up to 8 days in fluid media does not result in greater degree of isolations than dry feces stored for 8 days (282). Similarly, storage temperatures [whether frozen, at room temperature (20°C), or at 40°C] did not affect percent isolations (282). It is recommended that feces shipped for diagnostic culture be kept cool and dry in order to minimize growth of species other than *Salmonella* which may interfere with isolation.

In like manner, primary culture media used for isolation of the organism prior to biochemical characterization have been explored in great detail. Consensus has been established that the highest numbers of isolations follow the sequence of over-night enrichment in a broth medium to permit preferential multiplication of *Salmonella* (compared to coliforms) followed by plating on differential solid agar media. The media have been thoroughly reviewed by Margard *et al.* (282), but they, and others (36,171,292,313), recommend the sequence of incubating fecal pellets or tissue macerates in selenite-F plus cystine broth overnight followed by streaking onto brilliant green agar as the method of choice. Suspicious colonies are picked from the brilliant green plates and inoculated on triple sugar iron (TSI). Isolates consistent with *Salmonella* sp. are then confirmed by agglutination in polyvalent antisera and characterized biochemically according to the profile in Table V. The serotyping of isolates is beyond the competence of most laboratories, and characterized isolates should be forwarded to a recognized diagnostic center for complete identification.

Although it is recognized that both rats and mice respond immunologically to infective challenge by *Salmonella* (26,56,67,83,84,197,231,420,458), it is generally agreed that postrecovery titers, especially in immune asymptomatic carriers, are too low and inconsistent to be of great value in diagnosis. Numerous studies have documented the two- to threefold higher number of successful isolations of *Salmonella* from feces compared to detection of antibodies in the same animals (197,390,391,420). For this reason immunodiagnosis has not been generally recommended, other than as an adjunct to cultural methods for diagnosis. Of the methods explored, passive hemagglutination in the presence of complement (hemolysis) was found to be much more sensitive than tube agglutination (313) and may offer promise as a means of detecting very low postrecovery titers in carriers recorded as negative by standard agglutination tests. Similarly, vaccination has not been endorsed as a method of prevention, although the concept has been seriously explored. Vaccination with bacterins or bacteriophage has not been completely effective in either preventing infection in vaccinated individuals or in preventing spread when vaccinated and nonvaccinated mice were placed in close proximity (20,69,207,366,459–461,483).

The concept of treatment of infected individuals has been discouraged because of the incomplete effectiveness of various treatment regimens (171). The problem has been that while the majority of animals may be apparently freed of *Salmonella,* a residue of infection always remains to act as a reservoir for susceptible animals (65). Similarly, while fecal testing programs oriented to detection and elimination of infected animals may be statistically effective, it is almost impossible, with certainty, to eradicate the infection from a given colony by these means. Current recommendations favor destruction of colonies in which the disease has been detected, followed by replacement with gnotobiotically derived stock known to be free of *Salmonella*. There is no evidence that the organism may traverse the placental barrier.

Table V

Characteristic Biochemical Reactions and Other Differential Properties of Rodent Enterobacteriaceae[a,b]

Characteristic	EC	ET	YE	CF	S	EH	P	H	SM	EA	KA
Catalase	+	+	+	+	+	+	+	+	+	+	+
Oxidase	−	−	−	−	−	−	−	−	−	−	−
Motility	+	+	+	+	+	+	+	+	+	+	−
Growth in KCN	−	−	−	+	−	−	+	+	+	+	+
Simmons citrate	−	−	−	+	+	+	+	+	+	+	+
MR	+	+	+	+	+	−	+	−	+	−	−
VP	−	−	−	−	−	+	−	+	+	+	+
Indole	+	+	−	−	−	−	+	−	−	−	−
Gelatin liquifaction	−	−	−	−	−	+	+	−	+	+	−
Urease	−	−	+	−	−	−	+	−	−	−	+
Phenylalanine	−	−	−	−	−	+	+	−	−	−	−
H$_2$S on triple sugar iron	−	+	−	+	+	−	+	−	−	−	−
Lysine decarboxylase	+	+	−	−	+	−	−	+	+	+	+
Ornithine decarboxylase	+	+	+	+	+	−	−	+	+	+	−
Carbohydrates, acid from:											
Adonitol	−	−	−	−	−	−	−	−	v	+	+
Arabinose	+	−	+	+	+	+	−	+	−	v	v
Dulcitol	v	−	−	v	+	−	−	−	−	+	+
Glycerol					v	v	+	+	+	+	+
Inositol	−			−	v	v	−	−	v	+	+
Lactose	+	−	−	v	−	v	−	−	−	+	+
Maltose	+	+	+	+	+	+	+	+	+	+	+
Mannitol	+	−	+	+	+	+	−	+	+	+	+
Raffinose					+				−	+	+
Rhamnose	v	−	−	+	+	+	−	+	−	+	+
Salicin	v	−	−	v	−	v	v	−	+	+	+
Sorbitol	v	−	+	+				v	+	+	+
Sucrose	v	−	+	v	−	+	+	v	+	+	+
Trehalose	+	−	+	+	+	+	+	+	+	+	+
Xylose	v	−	+	+	+	+	v	v	v	+	+

[a] From Cowan (87), and Lennette *et al.* (257a).

[b] Key to abbreviations: EC, *Escherichia coli;* ET, *Edwardsiella tarda;* YE, *Yersinia enterocolitica;* CF, *Citrobacter freundii;* S, *Salmonella,* Arizona; EH, *Erwinia herbicola;* P, *Proteus;* H, *Hafnia alvei;* SM, *Serratia marcesens;* EA, *Enterobacter aerogenes;* KA, *Klebsiella aerogenes.*

H. Pasteurellosis (*Pasteurella pneumotropica*)

Since its original description (212), *Pasteurella pneumotropica* has been associated mainly with the respiratory tract of laboratory rodents, a predilection emphasized in the species name by Jawetz in 1950 (213,214). It is uncertain if the organism has emerged since the late 1940s or has only been recognized as a distinct entity since that time. Sporadic reports of pasteurellosis in rats occur prior to that time, notably those of Schipper (405), and Meyer and Batchelder (298), but these reports describe organisms culturally more similar to *P. multocida.* Similarly, the older literature contains references to respiratory infection in rats with organisms that on close examination resemble what is now recognized as *P. pneumotropica* (181,202,283).

In any event, *P. pneumotropica* has emerged since the 1950s as an infectious disease of laboratory rats of great importance. It has been described as a clinical entity in laboratory mice on a number of occasions (36,58,166,190,213,320,490,494,498). *Pasteurella pneumotropica* is carried as a latent organism in respiratory and other tissues of a number of host species beyond laboratory rats and mice, including wild rats (475), carnivores (48,332,475), hamsters (58,290,426), guinea pigs (426), rabbits (426), and man (187–189,220). These other species should be regarded as reservoirs of potential *P. pneumotropica* infectivity for laboratory rats and strengthen the accepted principle of separation of species within animal facilities.

A number of surveys since the original (188) have established the widespread distribution of *P. pneumotropica* in breeding stocks of rats and mice, (77,120,192,196,201,371, 426,463,494,509), frequently in the absence of clinically overt disease. Indeed, ubiquitous latency has remained one of characteristics of the organism. Even under conditions of experimental infection, the organism may be entirely latent and clinically silent (68,192,211,475). Alternatively, it may occupy the role of secondary opportunist, potentiating or synergistically enhancing the pathogenicity of Sendai virus (210,211) or *Mycoplasma pulmonis* (59,269). Finally, it may act as a primary pathogen associated with production of disease and otherwise meeting Koch's postulates for cause and effect (59,60,475). Little is known regarding the triggering mechanisms promoting progression from latency to pathogenicity with this organism, although such factors as impairment of pulmonary phagocytic defense mechanisms are known to favor multiplication of *P. pneumotropica* (161,210,211). This factor, at least, may be used to interpret the synergism of infections coupled with other pulmonary pathogens.

Under conditions of natural infection, *P. pneumotropica* has been encountered frequently as a component of the latent microflora of the nasoturbinates, conjunctiva, nasolacrimal duct, trachea, lungs, uterus, and, less commonly, deeper organs (i.e., liver, spleen, kidney) of laboratory rats (320,494). Recent reports involving latent infection in axenic rats indicate that the organism readily colonizes the intestines where it may be carried asymptomatically for long periods of time (311). Because of the usual association of *P. pneumotropica* with respiratory disease, it is possible that intestinal colonization has not been adequately investigated. Although the difficulties of isolating *P. pneumotropica* from the profusion of vigorous enteric forms encountered in the feces of conventional rats are appreciated, it is not certain that enteric colonization has significance beyond that of a peculiarity of gnotobiosis. In defense of a more common enteric occurrence, however, it has been pointed out that intranasal inoculation of fecal extracts have been sufficient to result in respiratory infection by *P. pneumotropica* (494). The organism is poorly transmitted by aerosol and thought to be naturally transmitted by direct contact and fecal contamination (494).

Disease associated with *P. pneumotropica* in the rat has been encountered in three general categories; (a) upper respiratory-associated infections including rhinitis, sinusitis, otitis media, conjunctivitis, and ophthalmitis (231,494,509); (b) pulmonary infections including bronchopneumonia, most commonly in the presence of other infectious agents (494); and (c) pyogenic infections (abscesses) of superficial tissues, e.g., skin (furunculosis) and mammary gland (mastitis) (475,494).

The pathology is not distinctive and resembles that caused by any of a variety of infectious forms in similar sites. Latent infections in the lungs, upper respiratory tract, uterus, and intestines frequently occur without histopathologic indication

of epithelial inflammation. In the lungs of mice areas of consolidation with perivascular and peribronchial infiltration by acute and chronic inflammatory cells may be produced (59), although this lesion is not distinctive. It has been pointed out that the pathologic architecture of experimental pulmonary disease caused by *P. pneumotropica* and *Mycoplasma pulmonis* in combination more resembles naturally occurring chronic respiratory conditions in rodents than either organism used singly (59). In this connection it is worth emphasizing that disease conditions with *P. pneumotropica* have been more commonly encountered in conventional rats and mice than they have been in gnotobiotes. Isolator or barrier breaks involving gnotobiotes have more frequently tended to be clinically silent, perhaps reinforcing the status of the organism as an organism of ordinarily low pathogenicity requiring the resistance-lowering presence of a primary pathogen.

Diagnosis of *P. pneumotropica* infection is established by retrieval of the organism from infected tissues and isolation in pure culture. The difficulties of fulfilling Koch's postulates with isolates from clinically overt infections have been previously reviewed. Serologic assessment may be helpful in establishing the pathogenicity of isolates from clinically overt infections as discussed below.

Primary isolation of the organism is accomplished readily on blood agar using swabs from epithelial surfaces; aspirates from nasoturbinates, trachea, or uterus; and organ homogenates. Isolation from feces is facilitated by selective enrichment in gram-negative (GN) broth (311). Many isolates fail to grow on MacConkey agar. On blood agar the colonies are 1 mm in diameter by 24 hr and 2 mm by 48 hr. The organism is nonhemolytic, although greening of the agar deep to the surface may occur. Colonies are dome shaped, entire, and usually grayish-yellow. The organism is a gram-negative, nonmotile rod or coccobacillus. Biochemical characterization of *P. pneumotropica* is summarized in Table VI, which lists the major reactive profile for this organism. The most constant differential features include oxidase, catalase, urease, and indole positivity; fermentation of glucose, sucrose, and maltose with acid (but not gas); and failure to utilize citrate or to ferment mannitol or sorbitol. Biochemical variants are common (104,190,201), however, and have been recently summarized (417). Two major groups of *P. pneumotropica* are now recognized that accommodate much of the reported variation; (a) isolates that are pale yellow on blood agar and indole positive (the majority of American isolates) and (b) isolates that are white on blood agar and indole negative (417).

Diagnostic serology with *P. pneumotropica* using agglutination or complement-fixation methodology has been investigated sufficiently to establish that latent infections are usually immunologically silent (196,490). Clinically overt infections, e.g., pneumonia or abscesses, usually induce substantial antibody titers (490). Since most immunologic studies have

Table VI

Characteristic Biochemical Reactions and Other Properties of *Pasteurella pneumotropica* and Related Forms[a]

Characteristic	P. pneumotropica	P. multocida	P. hemolytica	P. ureae
Catalase	(+)	(+)	(+)	(+)
Oxidase	(+)	(+)	(+)	(+)
Growth on MacConkey	(−)	(−)	(+)	(−)
Carbohydrate, acid from:				
Glucose	(+)	(+)	(+)	(+)
Sucrose	(+)	(+)	(+)	(+)
Lactose	(+)	(−)	(+)	(−)
Maltose	(+)	(+)	(+)	(+)
Mannitol	(−)	(+)	(+)	(+)
Sorbitol	(−)	(+)	(+)	(+)
Xylose	(+)	(+)	(+)	(−)
Salicin	(−)	(−)	(−)	(−)
Raffinose	(+)	(−)	(+)	(−)
Citrate utilization	(−)	(−)	(−)	(−)
Nitrate reduction	(+)	(+)	(+)	(+)
Nitrite reduction	(−)	(+)	(+)	(+)
Indole	(+)	(+)	(−)	(−)
Gelatin hydrolysis	(−)	(−)	(−)	(−)
Urease	(+)	(−)	(−)	(+)
Methyl Red/Voges-Proskauer (MR/VP)	(−/−)	(−/−)	(−/−)	(−/−)

[a]From Buchanan and Gibbons (66), Cowan (87), Lennette *et al.* (257a), and Wilson and Miles (496).

utilized as antigen organisms isolated from the same animal colony, the occurrence of serotypes is uncertain, although the issue has not been seriously investigated. Immunofluorescence has been used to study tissue localization of *P. pneumotropica* (59).

Attempts to control the clinical disease with antibiotics have had mixed but generally disappointing results (196,494). These results may have been due in part to the frequent (but unrecognized) association of *P. pneumotropica* with antibiotic-refractory primary pathogens. Clinical manifestations of infection may be controlled by individual treatments with antibiotics, especially chloramphenicol, although such treatments should be recognized as ineffective in eliminating the organism from a rat colony (166,196,428).

The organism may be excluded from rat stocks by gnotobiotic derivation (or rederivation), with the proviso that only culturally negative uteri be used. Lack of appreciation of the latent colonization of rat (and mouse) uterus by *P. pneumotropica* more than any other single factor has been responsible for the reintroduction of this organism to isolators or barriers of gnotobiotic rodents, and this mechanism is a frequent cause of barrier "breaks" (37,77). Several experimental reports have established the tropism of *P. pneumotropica* for rodent uterus (38,68), although some controversy exists as to whether such

infections ascend from the vagina or are hematogenously distributed. Studies have established that *P. pneumotropica* colonization of rat uterus may approach 60–70% under conditions of natural infection (77,165,248). Rodents with infected uteri may conceive and produce full-term fetuses as effectively as with uteri devoid of *P. pneumotropica* (121).

I. Potential Pathogens

There are several organisms occasionally isolated in the course of microbial diagnosis with rats that are difficult to categorize as to pathogenicity and which do not recur in periodic association with disease states in this species. Such organisms as *Bordetella (Alcaligenes) bronchiseptica* and *Klebsiella pneumoniae* fall within this group. While not definitely associated with disease states in rat, the latter bacteria are commonly associated with disease in other species. When isolated, the question becomes one of significance as to whether one has isolated a "pathogen" or a "commensal," and whether one is dealing with an "infection" or a "floral resident." The position taken in this chapter is reductionist in principle; if there is no evidence from the literature that an organism is naturally involved in disease states, it is not here regarded as a natural pathogen for rat. It may reinforce, however, the principle of separation of species within animal care facilities.

1. *Klebsiella pneumoniae*

Klebsiella pneumoniae, the Friedlander bacillus, has long been associated as a commensal of low incidence in rat respiratory tract. The organism is readily isolated on blood agar from aspirates, swabs, or organ homogenates plated on blood agar and incubated aerobically. The diagnostic features of *K. pneumoniae,* including the distinctive capsule, are outlined in standard works in clinical microbiology (66,463,496). The early European literature on this organism is reviewed by Cohrs, Jaffe, and Meessen (81). Surveys taken in the 1930s have reported the low incidence (1–5%) of *Klebsiella* among the bacteria isolated from rat middle ear (320,471), and similar low incidence has been recently reported among the isolates from rats monitored in the English animal accreditation program (426). Despite limited evidence of pathogenicity for mice (103,486), there is no convincing evidence that *Klebsiella* has ever been indicted in a natural disease process in rats, nor has it emerged among the opportunists associated with disease in immunocompromised laboratory rats.

2. *Bordetella (Alcaligenes) bronchiseptica*

Bordetella bronchiseptica is the name retained for this organism in the eighth edition of "Bergey's manual" (66), although the English have reclassified the organism as a species of the more closely related *Alcaligenes* (87). The organism is readily isolated from respiratory tissues onto blood agar and may be identified according to criteria in the standard works cited above. *Bordetella bronchiseptica* is occasionally isolated during diagnostic microbiology in rats, from those clinically normal as well as those showing signs of disease. It has been reported as an isolate of low incidence from rats submitted for microbiologic examination in the English accreditation program (426). Although the organism is recognized as a frank pathogen in guinea pig [see review in Wagner and Manning (482)], there is no critical evidence that *B. bronchiseptica* is involved in naturally occurring disease processes in the laboratory rat. In his review of this organism, Winsser described lesions in rats from which *B. bronchiseptica* was recovered (499); however, the rats in that study were conventional, and the role assumed for *B. bronchiseptica* cannot with confidence be attributed solely to that organism. In recognition of the ambiguity of the Winsser study, Burek *et al.* (68) have recently explored the experimental pathogenicity of *B. bronchiseptica* for laboratory rats. There is no question that the organism induced acute to subchronic bronchopneumonias in both germfree and conventional rats. The current situation is

that while the pathogenicity of the organism for rats has been demonstrated (68), there is no published evidence suggestive of natural infection resulting in disease.

III. MYCOTIC DISEASES

A. Dermatomycosis (*Trichophyton mentagrophytes*)

Dermatomycosis (dermatophytosis, ringworm, or favus) has long been associated with wild and laboratory rodents. Although the literature, particularly older reports, is rife with synonymy, it is now generally concluded that rodent ringworm is confined as a practical matter to a single fungal species, *Trichophyton mentagrophytes.* This species, however, is one of the most polymorphic of the dermatophytes, and failure to recognize its range of forms has led to confusion in taxonomy. Two principal forms of the organism are recognized; (a) a zoophilic variant with granular colonial surface and red pigmentation named *T. mentagrophytes* var. *mentagrophytes* and (b) an anthropophilic type with white, fluffy colonial surface and no pigmentation designated *T. mentagrophytes* var. *interdigitalis* (5). Many reports, particularly in the mouse, refer to *T. quinckeanum* as a separate and distinct species, but most modern mycologists consider *T. quinckeanum* to be synonymous with *T. mentagrophytes* var. *mentagrophytes* (6), and it is here so regarded. Both variants may commonly infect laboratory rodents.

Dermatomycosis is more common as a disease of laboratory mice (52,61,78,94,99,272,295,341,372), and guinea pigs (224,294,310,337,338,350) than rats (98,99,155,354), as the volume of literature indicates. It has not been reported in laboratory rats since 1965 (354). Periodic survey work indicates that *T. mentagrophytes* is not uncommon in wild rats (155,424) and mice 62), although, as will be pointed out below, the asymptomatic carrier state may be more common than is realized. The disease in rats may assume an epizootic form with many of the animals showing lesions or may be insidious and characterized by lesionless carriers. In both infective modes, substantial hazard to human contacts is present; indeed, human infection by persons handling the animals is frequently the first indication that the infection is in the colony. Most human infections occur on the exposed, relatively hairless parts of the body, especially the hands and arm (see also this volume, Chapter 15).

As mentioned above, the infection assumes variable form in rats and is thought to be influenced by a number of factors that include those directly bearing on susceptibility or resistance, e.g., age, genetic constitution, immunologic competence, and phase of hair growth cycle, as well as other less understood factors. Cortisone injections to test this hypothesis experimentally failed to influence degree of infection when compared to untreated guinea pigs (112). The few reported epizootics with lesions (in rats) all occurred in nonutilized animals prior to experimentation (99,354). Lesions, when present, may occur in the skin of any area but are most common on the neck, back, and base of the tail. Lesions are not uniformly discoid with alopecia and raised margins as is classically described, but rather may have a scurfy or erythematous papular–pustular appearance with irregular, patchy hair loss. Lesions on the tail (typically seen in the mouse), were not seen in rats by Povar in the outbreak he described (354).

Diagnosis of dermatophytosis is established by demonstration of fungal elements in skin scrapings and isolation of the causative organism by culture. Histopathology of affected skin is supportive in the attribution of lesion development to isolated dermatophytes if invasion of epidermal structures can be demonstrated. Histologic sections stained with Gridley fungus stain reveal fungal elements in the superficial epithelium and invasion of hair follicles. Secondary invasion of fungal lesions by bacteria with suppurative inflammation is commonly observed and is the cause of kerionlike lesions in both animals and man.

Skin scrapings should be carefully taken from the lesion periphery, mounted in 10% KOH under a vaseline-ringed cover slip, and examined under the microscope immediately and again in 30 min. When present, septate mycelia are observed in squamous cells. Small spore (2–3 μm) ectothrix invasion of hairs, especially near the base, are seen in *T. mentagrophytes* infections. Scrapings, similarly collected, should be inoculated onto the surface of a suitable agar medium and cultivated aerobically at room temperature for at least 10 days before discarding as negative. Suitable media include DTM (dermatophyte test medium with color indicators) (74) or Sabouraud's medium with cycloheximide and chloramphenical to inhibit nondermatophytic contaminants (369,382). Typical microscopic feature of *T. mentagrophytes* (macroconidia, spiral coils) should be demonstrated (409).

Variations of MacKenzie's hair brush technique may be used to evaluate the incidence of lesionless, asymptomatic carrier rats (273). There is some evidence to indicate that this state may occur as frequently in the rat (98,99) as is more commonly recognized in the guinea pig, mouse, and cat (99,110,136, 137,159,295,383). In this technique, the animal to be screened is held over an opened petri-dish of suitable agar medium and the hair brushed with a sterile surgical scrub brush so that hairs, flakes, and desquamated cellular debris fall directly down on the medium surface. The plate is incubated as mentioned above.

The ideal (and proven) method for control in research colonies entails destruction of the affected groups and sterilization or disinfection of equipment and environmental surfaces. When this is not immediately possible because of research

investment in particular animals, treatment may be considered. Although the clinical efficacy of feeding griseofulvin to treat dermatomycosis has had mixed results in guinea pigs (78,350), its effectiveness for this purpose has not been evaluated in rats. The organism is not known to traverse the placenta and has not been recovered from barrier-reared laboratory rats.

B. Deep Mycoses

The deep or systemic mycoses (fungal infections other than dermatomycosis) are regarded as extremely rare in rats; indeed, two comprehensive reviews covering the period up to 1954 (409) and another to 1967 (422) failed to document a single episode in unimpaired laboratory rats. Wild *R. norvegicus,* and other rat species have been encountered as both asymptomatic and lesioned carriers of several fungal species and are regarded as playing some role as natural reservoir hosts for certain organisms. The review cited above may be consulted for information related to wild rats.

Aspergillosis has been reported recently as a naturally occurring outbreak of pulmonary infections in laboratory rats in India (419). This report appears to be the first case study documenting *Aspergillus* sp. infection in rats not rendered susceptible by experimental treatments. In their review of systemic mycoses as experimental diseases in rat hosts, Smith and Austwick (422) concluded that rats were quite refractory to most fungal infections; however, a number of reports have documented aspergillosis in cortisone-treated laboratory rats and mice as unanticipated outbreaks (392,394,413,416). That natural immunity is important in resistance is suggested both by the lethal aspergillosis induced experimentally in mice with anti-lymphocyte serum (440), and by the protection induced by immunization (86).

In these reports, as in the natural outbreak above (419), aspergillosis presented as pulmonary infections with milliary distribution in that organ primarily, although hematogenous distribution to other organs with capillary nets, e.g., liver and kidney, was observed to lesser degree. Diagnosis is established by cultural retrieval of the organism and supportive histopathologic description. Sabouraud's agar with bacterial inhibitors (chloramphenicol, penicillin, and streptomycin) is satisfactory for this purpose, but care must be used to avoid modern media designed for isolation of dermatophytes, since the cycloheximide (Actidione) used as inhibitor will prevent growth by "contaminants" such as *Aspergillus*. Isolated *Aspergillus* cultures should be speciated according to the criteria outlined by Austwick (19), which emphasizes subculture on Czapek-Dox medium for optimal growth of differential features useful in speciation.

Granuloma formation is typical of pulmonary aspergillosis in all species, including rat. The lesions are seen microscopically to consist of macrophage and epitheloid cells centrally, and peripheral aggregation with both acute and chronic inflammatory cells, eosinophils, and giant cells of the Langhans type (419). Congestion, microscopic hemorrhage, and septal thickening occurs in surrounding parenchyma. *Aspergillus* is seen quite distinctly in hematoxylin and eosin stained material, but Gridley's fungus stain or Gomori–Grocott should be used for careful study. The *Aspergillus* species may be differentiated in tissue from those causing phycomycosis by the more uniform and smaller hyphae (2.5 μm in diameter), regular branching at 45° angles, and prominent septae. In contrast, phycomycotic fungi in tissue have thicker hyphae (10 μm in diameter), irregular branches, and are nonseptate.

Phycomycosis (or mucormycosis) is the general term used to describe tissue infection by nonseptate fungi of the genera *Mucor, Absidia,* and *Rhizopus.* The infection is quite rare in all species and has most oftenly been reported in association with primary conditions lowering resistance or otherwise predisposing to phycomycosis, e.g., diabetis mellitus, cortisone treatment, and immunosuppressive therapy. Mycotic encephalitis may be induced experimentally in mice by cortisone treatment (439,440). Phycomycotic encephalitis has been reported recently in rats as a naturally occurring disease prior to experimentation (363). The lesions presented as purulent, necrotizing foci with acute inflammatory responses. Phycomycotic nonseptate hyphae up to 10–15 μm in diameter were demonstrated with various stains. The authors emphasized that the immaturity of these rats (2–4 weeks) was predisposing, since neither their dams nor older rats in the same room developed signs or lesions.

REFERENCES

1. Accreditation Microbiological Advisory Committee (1971). "Microbiological Examination of Laboratory Animals for Purposes of Accreditation." Med. Res. Counc. Lab. Anim. Cent., Carshalton, Surrey, U.K.
2. Adachi, K. (1921). Flagellum of the microorganism of rat-bite fever. *J. Exp. Med.* **33,** 647–652.
3. Adams, L. E., Yamauchi, Y., Carleton, J., Townsend, L., and Kim, O. J. (1972). An epizootic of respiratory tract disease in Sprague-Dawley rats. *J. Am. Vet. Med. Assoc.* **161,** 656–660.
4. Ainsworth, E. J., and Forbes, P. D. (1961). The effect of *Pseudomonas* pyrogen on survival of irradiated mice. *Radiat. Res.* **14,** 767–774.
5. Ajello, L. (1974). Natural history of the dermatophytes and related fungi. *Mycopathol. Mycol. Appl.* **53,** 93–110.
6. Ajello, L., Bostick, L., and Cheng, S. L. (1968). The relationship of *Trichophyton quinckeanum* to *Trichophyton* mentagrophytes. *Mycologia* **60,** 1185–1189.
7. Aldred, P., Hill, A. C., and Young, C. (1974). The isolation of *Streptobacillus moniliformis* from cervical abscesses of guinea-pigs. *Lab. Anim.* **8,** 275–277.
8. Allbritten, F. F., Sheely, R. F., and Jeffers, W. A. (1940). *Haverhilia multiformis* septicemia. *J. Am. Med. Assoc.* **114,** 2360–2363.
9. Allen, A. M., Ganaway, J. R., Moore, T. D., and Kinard, R. F. (1965). Tyzzer's disease syndrome in laboratory rabbits. *Am. J. Pathol.* **4,** 859–882.
10. Anonymous (1975). Rat-bite fever a lab job risk. *Hosp. Pract.* **10,** 53.
11. Antopol, W. (1950). Anatomic changes in mice treated with excessive doses of cortisone. *Proc. Soc. Exp. Biol. Med.* **73,** 262–265.
12. Antopol, W., Glaubach, S., and Quittner, H. (1951). Experimental observations with massive doses of cortisone. *Rheumatism* **7,** 187–197.
13. Antopol, W., Quittner, H., and Saphra, I. (1953). "Spontaneous" infections after the administration of cortisone and ACTH. *Am. J. Pathol.* **29,** 599–600.
14. Araujo, E. de (1931). Diagnostico experimenta de sodoco. *Bras.-Med.* **45,** 924–928.
15. Arkless, H. A. (1970). Rat-bite fever at Albert Einstein Medical Center. *Pa. Med.* **73,** 49.
16. Arora, H. L., and Sangal, B. C. (1973). *Salmonella typhi* cholangitis in rats—an experimental study. *Indian J. Pathol. Bacteriol.* **16,** 28–34.
17. Austrian, R. (1974). *Streptococcus pneumoniae (Pneumococcus).* In "Manual of Clinical Microbiology" (E. H. Lennette, E. H. Spaulding, and J. P. Truant, eds.), 2nd ed., pp. 109–115. Am. Soc. Microbiol., Washington, D. C.
18. Austrian, R. (1976). The Quellung reaction, a neglected microbiologic technique. *Mt. Sinai J. Med.* **43,** 699–709.
19. Austwick, P. K. C. (1974). Medically important *Aspergillus* species. In "Manual of Medical Microbiology" (E. H. Lennette, E. H. Spaulding, and J. P. Truant, eds.), 2nd ed., pp. 550–556. Am. Soc. Microbiol., Washington, D. C.
20. Badakhsh, F. F., and Herzberg, M. (1969). Deoxycholate-treated, nontoxic, whole-cell vaccine protective against experimental salmonellosis of mice. *J. Bacteriol.* **100,** 738–827.
21. Baer, H. (1967). *Diplococcus pneumoniae* type 16 in laboratory rats. *Can. J. Comp. Med. Vet. Sci.* **31,** 216–218.
22. Baer, H., and Preiser, A. (1969). Type 3 *Diplococcus pneumonia* in laboratory rats. *Can. J. Comp. Med.* **33,** 113–117.
23. Bahr, L. (1905). Uber die zur Vertilgung von Ratten und Mausen benutzten Bakterien. *Zentralbl. Bakteriol., Parasitenkd., Infektionskr. Hyg., Abt. I: Orig.* **39,** 263–274.
24. Bahr, L., Raebiger, M., and Grosso, G. (1910). Ratin I und II, sowie uber die Stellung der Ratinbazillus zur Gartnergruppe. *Zentralbl. Bakteriol., Parasitenkd., Infektionskr. Hyg., Abt. I: Orig.* **54,** 231–234.
25. Bainbridge, F. A., and Boycott, A. E. (1909). The occurrence of spontaneous rat epidemics due to Gaertner's bacillus. *J. Pathol. Bacteriol.* **13,** 342.
26. Ball, N. D., and Price-Jones, C. (1926). A laboratory epidemic in rats due to Gaertner's bacillus. *J. Pathol. Bacteriol.* **29,** 27–30.
27. Bartram, J. T., Welch, H., and Ostrolenk, M. (1940). Incidence of members of the *Salmonella* group in rats. *J. Infect. Dis.* **67,** 222–226.
28. Bayne-Jones, S. (1931). Rat-bite fever in the United States. *Int. Clin.* **3,** 235–253.
29. Beck, R. W. (1963). A study in the control of *Pseudomonas aeruginosa* in a mouse breeding colony by the use of chlorine in the drinking water. *Lab. Anim. Care* **13,** 41–46.
30. Beeson, P. B. (1943). Problem of etiology of rat-bite fever: Report of two cases due to *Spirillum minus. J. Am. Med. Assoc.* **123,** 332–334.
31. Berlin, B. S., Johnson, G., Hawke, W. D., and Lawrence, A. G. (1952). The occurrence of bacteremia and death in cortisone treated mice. *J. Lab. Clin. Med.* **40,** 82–89.
32. Bhandari, J. C. (1976). A spontaneous outbreak of pseudotuberculosis in rats. *27th Annu. Sess., Am. Assoc. Lab. Anim. Sci.* Abstract No. 73.
33. Biberstein, E. L. (1975). Rat bite fever. In "Diseases Transmitted from

Animals to Man'' (W. T. Hubbert, W. F. McCulloch, and P. R. Schnurrenberger, eds.), 6th ed., pp. 186–188. Thomas, Springfield, Illinois.

34. Bicks, V. A. (1957). Infection of laboratory mice with *Corynebacterium murium*. *Aust. J. Sci.* **20,** 20–23.

35. Biggar, W. D., Bogart, D., Holmes, B., and Good, R. A. (1972). Impaired phagocytosis of *Pneumococcus* type 3 in splenectomized rats. *Proc. Soc. Exp. Biol. Med.* **139,** 903–908.

36. Black, D. J. (1975). *Pasteurella pneumotropica* isolated from a Harderian gland abscess of a mouse. *Lab. Anim. Dig.* **9,** 74–77.

37. Blackmore, D. K. (1972). Accidental contamination of a specified-pathogen-free unit. *Lab. Anim.* **6,** 257–271.

38. Blackmore, D. L., and Casillo, S. (1972). Experimental investigation of uterine infections of mice due to *Pasteurella pneumotropica*. *J. Comp. Pathol.* **82,** 471–475.

39. Blake, F. G. (1916). The etiology of rat-bite fever. *J. Exp. Med.* **23,** 39–60.

40. Blanden, R. V., Mackaness, G. B., and Collins, F. M. (1966). Mechanisms of acquired resistance in mouse typhoid. *J. Exp. Med.* **124,** 585–599.

41. Block, O., Jr., and Baldock, H. (1937). A case of rat-bite fever with demonstration of *Spirillum minus*. *J. Pediatr.* **10,** 358–360.

42. Bloomfield, A. L., and Lew, W. (1941). Protective effect of sulfaguinidine against ulcerative cecitis in rats. *Proc. Soc. Exp. Biol. Med.* **48,** 363–368.

43. Bloomfield, A. L., and Lew, W. (1943). Prevention by succinylsulfathiazole of ulcerative cecitis in rats. *Proc. Soc. Exp. Biol. Med.* **51,** 28–29.

44. Bloomfield, A. L., and Lew, W. (1942). Significance of *Salmonella* in ulcerative cecitis of rats. *Proc. Soc. Exp. Biol. Med.* **51,** 179–182.

45. Bloomfield, A. L., and Lew, W. (1943). Increased resistance to ulcerative cecitis of rats on a diet deficient in the vitamin B complex. *J. Nutr.* **25,** 427–431.

46. Bloomfield, A. L., and Lew, W. (1943). Prevention of infectious ulcerative cecitis in the young of rats by chemotherapy of the mother. *Am. J. Med. Sci.* **205,** 383–388.

47. Bloomfield, A. L., and Lew, W. (1948). Cure of ulcerative cecitis of rats by streptomycin. *Proc. Soc. Exp. Biol. Med.* **69,** 11–14.

48. Bloomfield, A. L., Rantz, L. A., Lew, W., and Zuckerman, A. (1949). Relation of a specific strain of *Salmonella* to ulcerative cecitis of rats. *Proc. Soc. Exp. Biol. Med.* **71,** 457–461.

49. Boart, D., Biggar, W. D., and Good, R. A. (1972). Impaired intravascular clearance of *Pneumococcus* type-3 following splenectomy. *Res. J. Reticuloendothel. Soc.* **11,** 77–87.

50. Bohme, D. H., Schneider, H. A., and Lee, J. M. (1954). Some pathophysiological parameters of natural resistance to infection in murine salmonellosis. *J. Exp. Med.* **110,** 9–26.

51. Bonger, J. (1901). *Corynethrix pseudotuberculosis murium*, ein neuer pathogener Bacillus fur Mause. *Z. Hyg. Infektionskr.* **37,** 449–475.

52. Booth, B. H. (1952). Mouse ringworm. *Arch. Dermatol. Syphilol.* **66,** 65–69.

53. Borgen, L. O., and Gaustad, V. (1948). Infection with *Actinomyces muris ratti (Streptobacillus moniliformis)* after bite of a laboratory rat. *Acta Med. Scand.* **130,** 189–198.

54. Bowers, E. F., and Jeffries, L. R. (1955). Optochin in the identification of *Str. pneumoniae*. *J. Clin. Pathol.* **8,** 58–60.

55. Boycott, A. E. (1911). Infective methaemoglobinaemia in rats caused by Gaertner's bacillus. *J. Hyg.* **11,** 443–472.

56. Boycott, A. E. (1919). Note on the production of agglutinins by mice. *J. Pathol. Bacteriol.* **23,** 126.

57. Breinl, A., and Kinghorn, A. (1906). A preliminary note on a new Spirochaeta found in a mouse. *Lancet* **2,** 651–652.

58. Brennan, P. C., Fritz, T. E., and Flynn, R. J. (1965). *Pasteurella pneumotropica:* Cultural and biochemical characteristics and its association with disease in laboratory animals. *Lab. Anim. Care* **15,** 307–312.

59. Brennan, P. C., Fritz, T. E., and Flynn, R. J. (1969). The role of *Pasteurella pneumotropica* and *Mycoplasma pulmonis* in murine pneumonia. *J. Bacteriol.* **97,** 337–349.

60. Brennan, P. C., Fritz, T. E., and Flynn, R. J. (1969). Murine pneumonia: A review of the etiologic agents. *Lab. Anim. Care* **19,** 360–371.

61. Brown, C. M., ;and Parker, M. T. (1957). *Salmonella* infections in rodents in Manchester with special reference to *Salmonella enteritidis* var *danysz*. *Lancet* **2,** 1277–1279.

62. Brown, G. W., and Suter, I. I. (1969). Human infections with mouse favus in a rural area of South Australia. *Med. J. Aust.* **2,** 541–542.

63. Brown, T. M., and Nunemaker, J. C. (1942). Rat-bite fever: A review of the American cases with reevaluation of etiology: Report of cases. *Bull. Johns Hopkins Hosp.* **70,** 201–328.

64. Bruce, D. L., Bismanis, J. E., and Vickerstaff, J. M. (1969). Comparative examinations of virulent *Corynebacterium kutscheri* and its presumed avirulent variant. *Can. J. Microbiol.* **15,** 817–818.

65. Bruner, D. W., and Moran, A. (1949). *Salmonella* infections of domestic animals. *Cornell Vet.* **39,** 53–63.

66. Buchanan, R. E., and Gibbons, N. E., eds. (1974). ''Bergey's Manual of Determinative Bacteriology,'' 8th ed. Williams & Wilkins, Baltimore, Maryland.

67. Buchbinder, L., Hall, L., Wilens, S. L., and Slanetz, C. A. (1935). Observations on enzootic paratyphoid infection in a rat colony. *Am. J. Hyg.* **22,** 199–213.

68. Burek, J. D., Jersey, G. C., Whitehair, C. K., and Carter, G. R. (1972). The pathology and pathogenesis of *Bordetella bronchiseptica* and *Pasteurella pneumotropica* in conventional and germfree rats. *Lab. Anim. Sci.* **22,** 844–849.

69. Cameron, C. M., and Fuls, W. J. P. (1974). A comparative study on the immunogenicity of live and inactivated *Salmonella typhimurium* vaccines in mice. *Onderstepoort J. Vet. Res.* **41,** 81–92.

70. Cannon, T. R. (1920). Bacillus enteritidis infection in laboratory rats. *J. Infect. Dis.* **26,** 402–404.

71. Canonico, P. G., White, J. D., and Powanda, M. C. (1975). Peroxisome depletion in rat liver during pneumococcal sepsis. *Lab. Invest.* **33,** 147–150.

72. Carda, P., Barros, C., and Salamanka, M. E. (1959). Epidemic of *Bacillus piliformis* infection in a colony of white mice. *Ann. Inst. Invest. Vet. (Madrid)* **9,** 153–160.

73. Caren, L. D., and Rosenberg, L. T. (1966). The role of complement in resistance to endogenous and exogenous infection with a common mouse pathogen *Corynebacterium kutscheri*. *J. Exp. Med.* **124,** 689–699.

74. Carroll, H. W. (1974). Evaluation of dermatophyte-test medium for diagnosis of dermatophytosis. *J. Am. Vet. Med. Assoc.* **165,** 192–195.

75. Carter, G. R., Whitenack, D. L., and Julius, L. A. (1969). Natural Tyzzer's disease in Mongolian gerbils (*Meriones unguiculatus*). *Lab. Anim. Care* **19,** 648–651.

76. Carter, H. V. (1888). Note on the occurrence of a minute bloodspirillum in an Indian rat. *Sci. Mem. Med. Officers Army India* **3,** 45–48.

77. Casillo, S., and Blackmore, D. K. (1972). Uterine infection caused by bacteria and mycoplasma in mice and rats. *J. Comp. Pathol.* **82,** 477–482.

78. Cetin, E. T., Tashinoglu, M., and Volkan, S. (1965). Epizootic of *Trichophyton* mentagrophytes in white mice. *Pathol. Microbiol.* **28,** 839–846.

79. Chopra, R. N., Basu, B. C., and Sen, J. (1939). Rat-bite fever in Calcutta. *Indian Med. Gaz.* **74,** 449-451.

80. Christensen, L. R. (1968). Commentary following presentation by Niven, J. S. N. *Z. Versuchstierkd.* **10,** 174.

81. Cohrs, P., Jaffe, R., and Meessen, H. (1958). "Pathologie der Laboratoriumstiere." Springer-Verlag, Berlin and New York.

82. Cole, J. S., Stole, R. W., and Bulger, R. J. (1969). Rat-bite fever. Report of 3 cases. *Ann. Intern. Med.* **71,** 979-981.

83. Collins, F. M. (1968). Recall of immunity in mice vaccinated with *Salmonella enteritidis* or *Salmonella typhimurium. J. Bacteriol.* **95,** 2014-2021.

84. Collins, F. M. (1970). Immunity to enteric infection in mice. *Infect. Immun.* **1,** 243-250.

85. Coons, A. H., Creech, H. S., Jones, R. N., and Berliner, E. (1942). The demonstration of penumococcal antigens in tissues by the use of fluorescent antibody. *J. Immunol.* **45,** 159-170.

86. Corbel, M. J., and Eades, S. M. (1977). Examination of the effect of age and acquired immunity on the susceptibility of mice to infection with *Aspergillus fumigatus. Mycologia* **60,** 79-85.

87. Cowan, S. T. (1974). "Cowan and Steel's Manual for the Identification of Medical Bacteria." Cambridge Univ. Press, London and New York.

88. Craigie, J. (1966). *Bacillus piliformis* (Tyzzer) and Tyzzer's disease of the laboratory mouse. I. Propagation of the organism in embryonated eggs. *Proc. R. Soc. London, Ser. B* **165,** 35-61.

89. Craigie, J. (1966). *Bacillus piliforms* (Tyzzer) and Tyzzer's disease of the laboratory mouse. II. Mouse pathogenicity of *B. piliformis* grown in embryonated eggs. *Proc. R. Soc. London, Ser. B* **165,** 61-78.

90. Cutlip, R. C., Amtower, W. C., Beall, C. W., and Matthews, P. J. (1971). An epizootic of Tyzzer's disease in rabbits. *Lab. Anim. Sci.* **21,** 356-361.

91. Dalal, A. K. (1914). Case of rat-bite fever, treated with intravenous injection of neo-salvarsan. *Practitioner* **92,** 449.

92. Danysz, J. (1914). Un microbe pathogène pour les rats (Mus decumanus et mus ratus) et son application à la destruction de ces animaux. *Ann Inst. Pasteur, Paris* **28,** 193-201.

93. Das Gupta, B. M. (1938). Spontaneous infection of guinea-pigs with a spirillum, presumably *Spirillum minus.* Carter, 1887. *Indian Med. Gaz.* **73,** 140-141.

94. Davies, R. R., and Shewell, J. (1965). Ringworm carriage and its control in mice. *J. Hyg.* **63,** 507-515.

95. Dawson, M. H., and Hobby, G. L. (1939). Rat-bite fever. *Trans. Assoc. Am. Physicians* **54,** 329-332.

96. De Rubertis, F. R., and Woeber, K. A. (1972). The effect of acute infection with *Diplococcus pneumoniae* on hepatic mitochondrial alpha-glycerophosphate dehydrogenase activity. *Endocrinology* **90,** 1384-1387.

97. "Diagnostic Procedures and Reagents" (1963). 4th ed., p. 225. Public Health Assoc., New York.

98. Dolan, M. M., and Fendrick, A. J. (1959). Incidence of *Trichophyton mentagrophytes* in laboratory rats. *Proc. Anim. Care Panel* **9,** 161-164.

99. Dolan, M. M., Kligman, A. M., Kobylinski, P. G., and Motsavage, M. A. (1958). Ringworm epizootics in laboratory mice and rats: Experimental and accidental transmission of infection. *J. Invest. Dermatol.* **30,** 23-35.

100. Dolman, C. E., Kerr, D. E., Chang, H., and Shearer, A. R. (1951). Two cases of rat-bite fever due to *Streptobacillus moniliformis. Can. J. Public Health* **42,** 228-241.

101. Ediger, R. D., Rabstein, M. M., and Olson, L. D. (1971). Circling in mice caused by *Pseudomonas aeruginosa. Lab. Anim. Sci.* **21,** 845-848.

102. Editorial (1939). Haverhill fever (erythema arthriticum epidemicum). *J. Am. Med. Assoc.* **113,** 941.

103. Ehrenworth, L., and Baer, H. (1956). The pathogenicity of *Klebsiella pneumoniae* for mice: The relationship to the quantity and rate of production of type-specific capsular polysaccharide. *J. Bacteriol.* **72,** 713-717.

104. Eisenberg, G. H. G., Jr., and Cavanaugh, D. C. (1974). *Pasteurella. In* "Manual of Clinical Microbiology" (E. H. Lennette, E. H. Spaulding, and J. P. Truant, eds.), 2nd ed., pp. 246-249. Am. Soc. Microbiol., Washington, D. C.

105. Elwell, M. R., Sammons, M. L., Liu, C. T., and Beisel, W. R. (1975). Changes in blood pH in rats after infection with *Streptococcus pneumoniae. Infect. Immun.* **11,** 724-726.

106. Farrell, E., Lord, G. H., and Vogel, J. (1939). Haverhill fever: Report of a case with review of the literature. *Arch. Intern. Med.* **64,** 1.

107. Fauve, R. M., and Pierce-Chase, C. H. (1967). Comparative effects of corticosteroids on host resistance to infection in relation to chemical structure. *J. Exp. Med.* **125,** 807-821.

108. Fauve, R. M., Pierce-Chase, C. H., and Dubos, R. (1964). Corynebacterial pseudotuberculosis in mice. II. Activation of natural and experimental latent infections. *J. Exp. Med.* **120,** 283-304.

109. Fauve, R. M., Bouanchaud, D., and Delaunay, A. (1966). Resistance cellulaire a l'infection bacterienne. *Ann. Inst. Pasteur, Paris* **110,** 106-117.

110. Feuerman, E., Alteras, I., Honig, E., and Lehrer, N. (1975). Saprophytic occurrence of *Trichophyton mentagrophytes* and *Microsporum gypseum* in the coats of healthy laboratory animals (Preliminary report). *Mycopathologia* **55,** 13-16.

111. Fischl, V., Koech, M., and Kussat, E. (1931). Infektarthritis bei muriden. *Z. Hyg. Infektionskr.* **112,** 421-425.

112. Fisher, M., and Sher, A. M. (1972). Virulence of *Trichophyton mentagrophytes* infecting steroid-treated guinea pigs. *Mycopathol. Mycol. Appl.* **47,** 121-127.

113. Fleischman, R. W., McCracken, D., and Forbes, W. (1977). Adynamic ileus in the rat induced by chloral hydrate. *Lab. Anim. Sci.* **27,** 238-243.

114. Flynn, R. J. (1960). "Progress Report: Diseases of Laboratory Animals. *Pseudomonas* Infection of Mice," Semi-Annu. Rep., ANL-6264, pp. 155-157. Biol. Med. Res. Div., Argonne Natl. Lab., Lemont, Illinois.

115. Flynn, R. J. (1963). Introduction: *Pseudomonas aeruginosa* infection and its effects on biological and medical research. *Lab. Anim. Care* **13,** 1-6.

116. Flynn, R. J. (1963). *Pseudomonas aeruginosa* infection and radiobiological research at the Argonne National Laboratory: Effects, diagnosis, epizootiology, and control. *Lab. Anim. Care* **13,** 25-35.

117. Flynn, R. J. (1963). The diagnosis of *Pseudomonas aeruginosa* infection in mice. *Lab. Anim. Care* **13,** 126-130.

118. Flynn, R. J., and Greco, I. (1962). "Progress Report: Disease and Care of Laboratory Animals. Further Studies on the Diagnosis of *Pseudomonas aeruginosa* Infection of Mice," Semi-ann. Rep. ANL 6535. Biol. Med. Res. Div., Argonne Natl. Lab., Lemont, Illinois.

119. Flynn, R. J., Ainsworth, E. J., and Greco, I. (1961). Progress Report: Diseases and Care of Laboratory Animals. I. Effects of *Pseudomonas* Infection of Mice, Summ. Rep., ANL-6368, pp. 35-37. Biol. Med. Res. Div., Argonne Natl. Lab., Lemont, Illinois.

120. Flynn, R. J., Brennan, P. C., and Fritz, T. E. (1965). Pathogen status of commercially produced laboratory mice. *Lab. Anim. Care* **15,** 440-448.

121. Flynn, R. J., Simkins, R. C., Brennan, P. C., and Fritz, T. E. (1968). Uterine infection in mice. *Z. Versuchstierkd.* **10,** 131-136.

122. Ford, T. M. (1965). An outbreak of pneumonia in laboratory rats associated with *Diplococcus pneumoniae* type 8. *Lab. Anim. Care* **15,** 448-451.

123. Ford, T. M., and Joiner, G. N. (1968). Pneumonia in a rat associated

with *Corynebacterium pseudotuberculosis,* a case report and literature survey. *Lab. Anim. Care* **18,** 220–223.

124. Fox, C. L., Jr., Sampath, A. C., and Stanford, J. W. (1970). Virulence of *Pseudomonas* infection in burned rats and mice. Comparative efficacy of silver sulfadiazine and mafenide. *Arch. Surg.* **101,** 508–512.

125. Francis, E. (1932). Rat-bite fever and relapsing fever in the United States. *Trans. Assoc. Am. Physicians.* **47,** 143–151.

126. Francis, E. (1936). Rat-bite fever spirochetes in naturally infected white mice. *Mus musculus. Public Health Rep.* **51,** 976–977.

127. Francis, R. A. (1970). Tyzzer's disease in laboratory animals. *J. Inst. Anim. Tech.* **21,** 167–171.

128. Freundt, D. E. (1957). Mycoplasmatales. *In* "Bergey's Manual of Determinative Bacteriology," (R. S. Breed, E. G. D. Murray, and N. R. Smith, eds.), 7th ed., pp. 914–926. Williams & Wilkins, Baltimore, Maryland.

129. Freundt, E. A. (1956). *Streptobacillus moniliformis* infection in mice. *Acta Pathol. Microbiol. Scand.* **38,** 231–245.

130. Freundt, E. A. (1956). Experimental investigations into the pathogenicity of the L phase variant of *Streptobacillus moniliformis. Acta Pathol. Microbiol. Scand.* **38,** 246–248.

131. Freundt, E. A. (1959). Arthritis caused by *Streptobacillus moniliformis* and pleuropneumonia-like organisms in small rodents. *Lab. Invest.* **8,** 1358–1366.

132. Friedmann, J. C., Guillon, J. C., Vassor, M. J., and Mahouy, G. (1969). La maladie de Tyzzer de la souris. *Exp. Anim.* **2,** 195–213.

133. Fries, A. S. (1977). Studies on Tyzzer's disease: Isolations and propagation of *Bacillus piliformis. Lab. Anim.* **11,** 75–78.

134. Fries, A. S. (1977). Studies on Tyzzer's disease: Application of immunofluorescence for detection of *Bacillus piliformis* and for demonstration and determination of antibodies to it in sera from mice and rabbits. *Lab. Anim.* **11,** 69–73.

135. Friesleben, M. (1927). Das Vorkommen von Bazillen der Paratyphys-B-Gruppe in gesunden Schlachttieren sowie in Ratten und Mausen. *Dtsch. Med. Wochenschr.* **53,** 1589–1591.

136. Fuentes, C. A., and Aboulafia, R. (1955). *Trichophyton mentagrophytes* from apparently healthy guinea pigs. *AMA Arch. Dermatol.* **71,** 478–480.

137. Fuentes, C. A., Bosch, Z. E., and Boudet, C. C. (1956). Occurrence of *Trichophyton mentagrophytes* and *Microsporum gypseum* on hairs of healthy cats. *J. Invest. Dermatol.* **23,** 311–313.

138. Fujiwara, K. (1967). Complement fixation reaction and agar-gel double diffusion test in Tyzzer's disease of mice. *Jpn. J. Microbiol.* **11,** 103–117.

139. Fujiwara, K. (1971). Problems in checking inapparent infections in laboratory mouse colonies: An attempt at serological checking by anamnestic response. *In* "Defining the Laboratory Animal," pp. 77–92. Intl. Comm. Lab. Anim., Natl. Acad. Sci., Washington, D. C.

140. Fujiwara, K., Fukuda, S., Takagaki, Y., and Tajima, Y. (1963). Tyzzer's disease in mice. Electron microscopy of the liver lesions. *Jpn. J. Exp. Med.* **33,** 203–212.

141. Fujiwara, K., Kurashina, H., Matsunuma, N., and Takahashi, R. (1968). Demonstration of peritrichous flagella of Tyzzer's disease organism. *Jpn. J. Microbiol.* **12,** 361–363.

142. Fujiwara, K., Takahashi, R., Kurashina, H., and Matsunuma, N. (1969). Protective serum antibodies in Tyzzer's disease of mice. *Jpn. J. Exp. Med.* **39,** 491–504.

143. Fujiwara, K., Yamada, H., Ogawa, H., and Oshima, Y. (1971). Comparative studies on the Tyzzer's organisms from rats and mice. *Jpn. J. Exp. Med.* **41,** 125–134.

144. Fujiwara, K., Hirano, N., Takenaka, S., and Sato, K. (1973). Peroral infection in Tyzzer's disease of mice. *Jpn. J. Exp. Med.* **43,** 33–42.

145. Fujiwara, K., Takagaki, Y., Naiki, M., Maejima, K., and Tajima, Y. (1964). Tyzzer's disease in mice. Effects of corticosteroids on the for-

mation of liver lesions and the level of blood transaminases in experimentally infected animals. *Jpn. J. Exp. Med.* **34,** 59–75.

146. Fujiwara, K., Kurashina, H., Magaribuchi, T., Takenaka, S., and Yokoiyama, S. (1973). *Jpn. J. Exp. Med.* **43,** 307–316.

147. Fujiwara, K., Takagaki, Y., Maejima, K., Kato, K., Naiki, M., and Tajima, Y. (1963). Tyzzer's disease in mice: Pathologic studies on experimentally infected animals. *Jpn. J. Exp. Med.* **33,** 183–202.

148. Fujiwara, K., Maejima, K., Takagaki, Y., Maiki, M., Tajima, Y., and Takahashi, R. (1964). Multiplication des organismes de Tyzzer dans les tissus cérébraux de la souris expérimentalement infectée. *C. R. Seances Soc. Biol.* **158,** 407–413.

149. Fujiwara, K., Kurashina, H., Maejima, K., Tajima, Y., Takagaki, Y., and Naiki, M. (1965). Actively induced immune resistance to experimental Tyzzer's disease of mice. *Jpn. J. Exp. Med.* **35,** 259–275.

150. Futaki, K., Takaki, S., Taniguchi, T., and Osumi, S. (1916). The cause of rat-bite fever. *J. Exp. Med.* **23,** 249–250.

151. Ganaway, J. R., Allen, A. M., and Moore, T. D. (1971). Tyzzer's disease of rabbits: Isolation and propagation of *Bacillus piliformis* (Tyzzer) in embryonated eggs. *Infect. Immun.* **3,** 429–437.

152. Ganaway, J. R., Allen, A. M., and Moore, T. D. (1971). Tyzzer's disease. *Am. J. Pathol.* **64,** 717–732.

153. Gard, S. (1944). *Bacillus piliformis* infection in mice, and its prevention. *Acta Pathol. Microbiol. Scand., Suppl.* **54,** 123–134.

154. Geil, R. G., Davis, D. L., and Thomson, S. W. (1961). Spontaneous ileitis in rats: A report of 64 cases. *Am. J. Vet. Res.* **22,** 932–936.

155. Georg, L. K. (1960). "Animal Ringworm in Public Health," Public Health Serv. Publ. No. 727, pp. 9–17. US Govt. Printing Office, Washington, D. C.

156. Giddens, W. E., Keahey, K. K., Carter, G. R., and Whitehair, C. K. (1968). Pneumonia in rats due to infection with *Corynebacterium kutscheri. Pathol. Vet.* **5,** 227–237.

157. Gilardi, G. L. (1976). *Pseudomonas* species in clinical mircobiology. *Mt. Sinai J. Med. N.Y.* **43,** 710–726.

158. Gilbert, G. L., Cassidy, J. F., and Bennett, N. M. (1971). Rat-bite fever. *Med. J. Aust.* **2,** 1131–1134.

159. Gip, L., and Martin, B. (1964). Occurrence of *Trichophyton mentagrophytes asteroid* on hairs of guinea pigs without ringworm lesions. *Acta Derm.-Venereol.* **44,** 208–210.

160. Gledhill, A. W. (1967). Rat-bite fever in laboratory personnel. *Lab. Anim.* **1,** 73–76.

161. Goldstein, E., and Green, G. M. (1967). Alteration of the pathogenicity of *Pasteurella pneumotropica* for the murine lung caused by changes in pulmonary antibacterial activity. *J. Bacteriol.* **93,** 1651–1656.

162. Gonsherry, L., Marston, R. Q., and Smith, W. W. (1953). Naturally occurring infections in untreated and streptomycin-treated X-irradiated mice. *Am. J. Physiol.* **172,** 359–364.

163. Gorer, P. A. (1947). Some observations on the diseases of mice. *In* "The Care and Management of Laboratory Animals" (A. N. Worden, ed.), Williams & Wilkins, Baltimore, Maryland.

164. Gowen, J. W., and Schott, R. G. (1933). Genetic predisposition to *Bacillus piliformis* among mice. *J. Hyg.* **33,** 370–378.

165. Graham, W. R. (1963). Recovery of a pleuropneumonia-like organism (PPLO) from the genitalia of the female albino rat. *Lab. Anim. Care* **13,** 719–723.

166. Gray, D. F., and Campbell, A. L. (1953). The use of chloramphenicol and foster mothers in the control of natural pasteurellosis in experimental mice. *Aust. J. Exp. Biol.* **31,** 161–166.

167. Grogan, J. B. (1969). Effect of antilymphocyte serum on mortality of *Pseudomonas aeruginosa*-infected rats. *Arch. Surg. (Chicago)* **99,** 382–384.

168. Groschel, D. (1970). Genetic resistance of inbred mice to *Salmonella.*

Zentralbl. Bakteriol., Parasitenkd., Infektionskr. Hyg., Abt. 1: Orig. **215,** 441–444.

169. Guggenheim, K., and Buechler, E. (1947). Nutritional deficiency and resistance to infection. The effect of caloric and protein deficiency on the susceptibility of rats and mice to infection with *Salmonella typhimurium. J. Hyg.* **45,** 103–109.

170. Gundel, J., Gyorgy, P., and Pager, W. (1932). Experimentelle Beobachtungen zu der Frage der Resistenzverminderung und Infektion *Z. Hyg. Infektionskrankh.* **113,** 629–644.

171. Habermann, R. T., and Williams, F. P. (1958). Salmonellosis in laboratory animals. *J. Natl. Cancer Inst.* **20,** 933–947.

172. Halkett, J. A. E., Davis, A. J., and Natsios, G. A. (1968). The effect of cold stress and *Pseudomonas aeruginosa* gavage on the survival of three-week-old swiss mice. *Lab. Anim. Care* **18,** 94–96.

173. Hall, G. A. (1974). An investigation into the mechanism of placental damage in rats inoculated with *Salmonella dublin. Am. J. Pathol.* **77,** 299–312.

174. Hall, W. C., and Van Kruiningen, H. T. (1974). Tyzzer's disease in a horse. *J. Am. Vet. Med. Assoc.* **164,** 1187–1189.

175. Hamburger, M., and Knowles, H. C., Jr. (1953). *Streptobacillus moniliformis* infection complicated by acute bacterial endocarditis. *Arch. Intern. Med.* **92,** 216–220.

176. Hammond, C. W. (1954). The treatment of post-irradiation infection. *Radiat. Res.* **1,** 448–458.

177. Hammond, W. (1963). *Pseudomonas aeruginosa* infection and its effects on radiobiological research. *Lab. Anim. Care* **13,** 6–11.

178. Hammond, C. W., Tompkins, M., and Miller, C. P. (1954). Studies on susceptibility to infection following ionizing radiation. I. The time of onset and duration of the endogenous bacteremias in mice. *J. Exp. Med.* **99,** 405–410.

179. Hammond, C. W., Colling, M., Cooper, D. B., and Miller, C. P. (1954). Studies on susceptibility to infection following ionizing radiation. II. Its estimation by oral inoculation at different times postirradiation. *J. Exp. Med.* **99,** 411–418.

180. Hammond, C. W., Ruml, D., Cooper, C. B., and Miller, C. P. (1955). Studies on susceptibility to infection following ionizing radiation. III. Susceptibility of the intestinal tract to oral inoculation with *Pseudomonas aeroginosa. J. Exp. Med.* **102,** 403–411.

181. Harr, J. R., Tinsley, I. J., and Weswig, P. H. (1969). *Haemophilus* isolated from a rat respiratory epizootic. *J. Am. Vet. Med. Assoc.* **155,** 1126–1130.

182. Hass, V. A. (1938). A study of pseudotuberculosis rodentium recovered from a rat. *U.S. Public Health Rep.* **53,** 1033–1038.

183. Hata, S. (1912). Salvarsantherapie der Rattenbisskrankheit in Japan. *Muench. Med. Wochenschr.* **59,** 854.

184. Hazard, J. B., and Goodkind, R. (1932). Haverhill fever (erythema arthriticum epidemicum) a case report and bacteriologic study. *J. Am. Med. Assoc.* **99,** 534–538.

185. Heilman, F. R. (1941). A study of *Asterococcus muris (Streptobacillus moniliformis).* I. Morphologic aspects and nomenclature. *J. Infect. Dis.* **69,** 32–44.

186. Heilman, F. R. (1951). A study of *Asterococcus muris (Streptobacillus moniliformis).* II. Cultivation and biochemical activities. *J. Infect. Dis.* **69,** 45–51.

187. Henriksen, S. D. (1962). Some *Pasteurella* strains from the human respiratory tract. A correction and supplement. *Acta Pathol. Microbiol. Scand.* **55,** 355–356.

188. Henriksen, S. D., and Jyssum, K. (1960). A new variety of *Pasteurella hemolytica* from the human respiratory tract. *Acta Pathol. Microbiol. Scand.* **50,** 443.

189. Henriksen, S. D., and Jyssum, K. (1961). A study of some *Pasteurella* strains from the human respiratory tract. *Acta Pathol. Microbiol. Scand.* **51,** 354–368.

190. Heyl, J. G. (1963). A study of *Pasteurella strains from animal sources. Antonie van Leeuwenhoek* **29,** 79–83.

191. Hightower, D., Uhrig, H. T., and Davis, J. I. (1966). *Pseudomonas aeruginosa* infection in rats used in radiobiology research. *Lab. Anim. Care* **16,** 85–93.

192. Hill, A. (1974). Experimental and natural infection of the conjunctiva of rats. *Lab. Anim.* **8,** 305–310.

193. Hoag, W. G., and Rogers, J. (1961). Techniques for the isolation of *Salmonella typhimurium* from laboratory mice. *J. Bacteriol.* **82,** 153–154.

194. Hoag, W. G., Strout, J., and Meier, H. (1964). Isolation of *Salmonella* spp. from laboratory mice and from diet supplements. *J. Bacteriol.* **88,** 534–536.

195. Hoag, W. G., Strout, J., and Meier, H. (1965). Epidemiological aspects of the control of *Pseudomonas* infection in mouse colonies. *Lab. Anim. Care* **15,** 217–225.

196. Hoag, W. G., Wetmore, P. W., Rogers, J., and Meier, H. (1962). A study of latent *Pasteurella* infection in a mouse colony. *J. Infect. Dis.* **11,** 135–140.

197. Hobson, D. (1957). Chronic bacterial carriage in survivors of mouse typhoid. *J. Pathol. Bacteriol.* **73,** 399–410.

198. Hojo, E. (1939). On the *Bacillus pseudo-tuberculosis murium* prevalent in mouse. *Jpn. J. Exp. Med.* **10,** 113.

199. Holden, F. A., and MacKay, J. C. (1964). Rat-bite fever—an occupational hazard. *Can. Med. Assoc. J.* **91,** 78.

200. Holmgren, E. B., and Tunevall, G. (1970). Rat-bite fever. *Scand. J. Infect. Dis.* **2,** 71–74.

201. Hooper, A., and Sebesteny, A. (1974). Variation in *Pasteurella pneumotropica. J. Med. Microbiol.* **7,** 137–140.

202. Hoskins, H. P., and Stout, A. L. (1920). *Bacillus bronchisepticus* as the cause of an infectious respiratory disease of the white rat. *J. Lab. Clin. Med.* **5,** 307–310.

203. Hottendorf, G. H., Hirth, R. S., and Peer, R. L. (1969). Megaloileitis in rats. *J. Am. Vet. Med. Assoc.* **155,** 1131–1135.

204. Humphreys, F. A., Campbell, A. G., Driver, M. W., and Hatton, G. N. (1950). Rat-bite fever. *Can. J. Public Health* **41,** 66–71.

205. Hunter, B. (1971). Eradication of Tyzzer's disease in a colony of barrier-maintained mice. *Lab. Anim.* **5,** 271–276.

206. Hunter, C. A., and Ensign, P. R. (1947). An epidemic of diarrhea in a newborn nursery caused by *Pseudomonas aeruginosa. Am. J. Public Health* **37,** 1166–1169.

207. Ibraham, H. M., and Schutze, H. (1928). A comparison of the prophylactic value of the H, O and R antigens of *Salmonella aertrycke,* together with some observations on the toxicity of its smooth and rough variants. *Br. J. Exp. Pathol.* **9,** 353–360.

208. Innes, J. R. M., McAdams, A. J., Yevich, P. (1956). Pulmonary disease in rats. A survey with comments on "chronic murine pneumonia." *Am. J. Pathol.* **32,** 141–159.

209. Innes, J. R. M., Wilson, C., and Ross, M. A. (1956). Epizootic *Salmonella enteritidis* infection causing septic pulmonary phlebothrombosis in hamsters. *J. Infect. Dis.* **98,** 133–141.

210. Jakab, G. J. (1974). Effect of sequential innoculation of Sendai virus and *Pasteurella pneumotropica* in mice. *J. Am. Vet. Med. Assoc.* **164,** 723–728.

211. Jakob, G. J., and Dick, E. C. (1973). Synergistic effect in viralbacterial infection: Combined infection of the murine respiratory tract with Sendai virus and *Pasteurella pneumotropica. Infect. Immun.* **8,** 762–768.

212. Jawetz, E. (1948). A latent pneumotropic *Pasteurella* of laboratory animals. *Proc. Soc. Exp. Biol. Med.* **68,** 46–48.

213. Jawetz, E. (1950). A pneumotropic *Pasteurella* of laboratory animals 1. Bacteriological and serological characteristics of the organism. *J. Infect. Dis.* **86,** 172–183.

214. Jawetz, E., and Baker, W. H. (1950). A pneumotropic *Pasteurella* of laboratory animals. II. Pathological and Immunological studies with the organism. *J. Infect. Dis.* **86**, 184–196.

215. Jellison, W. L., Eneboe, P. L., Parker, R. R., and Hughes, L. E. (1949). Rat-bite fever in Montana. *Public Health Rep.* **64**, 1661–1665.

216. Jenkin, C. R., Rowley, D., and Auzins, I. (1964). The basis for immunity in mouse typhoid. I. The carrier state. *Aust. J. Exp. Biol. Med. Sci.* **42**, 215–228.

217. Johanson, W. G., Jr., Jay, S. J., and Pierce, A. K. (1974). Bacterial growth *in vivo*. An important determinant of the pulmonary clearance of *Diplococcus pneumoniae* in rats. *J. Clin. Invest.* **53**, 1320–1325.

218. Jonas, A. M., Percey, D., and Craft, J. (1970). Tyzzer's disease in the rat. *Arch. Pathol.* **90**, 561–566.

219. Jones, B. F., and Stewart, H. L. (1939). Chronic ulcerative cecitis in the rat. *Public Health Rep.* **54**, 172–175.

220. Jones, D. M. (1962). A *Pasteurella*-like organism from the human respiratory tract. *J. Pathol. Bacteriol.* **83**, 143–151.

221. Jones, F. S. (1922). An organism resembling *Bacillus actinoides* isolated from pneumonic lungs of white rats. *J. Exp. Med.* **35**, 361–366.

222. Jordan, E. O. (1925). The differentiation of the paratyphoid enteritidis group. IX. Strains from various mammalian hosts. *J. Infect. Dis.* **36**, 309–329.

223. Juhr, N. C., and Horn, J. (1975). Modellinfektion mit Corynebacterium kutscheri bei der Maus. *Z. Versuchstierkd.* **17**, 129–141.

224. Kaffka, A., and Reith, H. (1960). *Trichophyton mentagrophytes*—Varianten bei Laboratoriumstieren. *Zentralbl. Bakteriol., Parasitenkd., Infektionskr. Hyg., Abt. 1: Orig.* **177**, 96–106.

225. Kampelmacher, E. H., Guinee, P. A. M., and van Noorle Jansen, L. M. (1969). Artificial *Salmonella* infections in rats. *Zentralbl. Veterinaermed., Reihe B* **16**, 173–182.

226. Kanazawa, K., and Imai, A. (1959). Pure culture of the pathogenic agent of Tyzzer's disease of mice. *Nature (London)* **184**, 1810.

227. Kaneko, J., Fujita, H., Matsuyama, S., Kojima, H., Asakura, H., Nakamura, Y., and Kodama, T. (1969). An outbreak of Tyzzer's disease among the colonies of mice. *Bull. Exp. Anim.* **9**, 148–156.

228. Kaplan, M. H., Coons, A. H., and Deanne, H. W. (1950). Localization of antigen in tissue cells. III. Cellular distribution of penumpococcal polysaccharides types II and III in the mouse. *J. Exp. Med.* **91**, 15–30.

229. Karasek, E. (1970). Der Kortisontest zum Nachweis latenter infektionen bei Versuchsmäusen. *Z. Versuchstierkd.* **12**, 155–161.

230. Kelemen, G., and Sargent, F. (1946). Nonexperimental pathologic nasal findings in laboratory rats. *Arch. Otolaryngol.* **44**, 24–42.

231. Kent, R. L., Lutzner, M. A., and Hansen, C. T. (1976). The masked rat: An x-ray-induced mutant with chronic blepharitis, alopecia and pasteurellosis. *J. Hered.* **67**, 3–5.

232. Kerrin, J. C. (1928). *Bacillus enteriditis* infection in wild rats. *J. Pathol. Bacteriol.* **31**, 588–589.

233. Khalil, A. M. (1934). The incidence of organisms of *Salmonella* group in wild rats and mice in Liverpool. *J. Hyg.* **38**, 75–78.

234. Klein, E. (1903). Discussion of the Pathologic Society of London. *Lancet* **1**, 238–239.

235. Kleineberger, E. (1935). The natural occurrence of pleuropneumonia-like organismsin apparent symbiosis with *Streptobacillus moniliformis* and other bacteria. *J. Pathol. Bacteriol.* **40**, 93–105.

236. Kleineberger, E. (1936). further studies on *Streptobacillus moniliformis* and its symbiont. *J. Pathol. Bacteriol.* **42**, 587–598.

237. Kleineberger, E. (1938). Pleuropneumonia-like organisms of diverse provenance: Some results of an enquiry into methods of differentiation. *J. Hyg.* **38**, 458–476.

238. Kleineberger, E., and Steabben, D. B. (1937). On a pleuropneumonia-like organism in lung lesions or rats with notes on the clinical and pathological features of the underlying condition. *J. Hyg.* **37**, 143–152.

239. Kleineberger, E., and Steabben, D. B. (1940). On the association of the pleuropneumonia-like organisms L3 with bronchiectatic lesions in rats. *J. Hyg.* **40**, 223–227.

240. Kleineberger, E. (1942). Some new observations bearing on the nature of the pleuropneumonia-like organism known as L1 associated with *Streptobacillus moniliformis*. *J. Hyg.* **42**, 485–497.

241. Knowles, R., Das Gupta, B. M., and Sen, S. (1936). Natural *Spirillum minus* infection in white mice. *Indian Med. Gaz.* **71**, 210–212.

242. Koopman, J. P., and Janssen, F. G. J. (1975). The occurrence of salmonellas in laboratory animals and a comparison of three enrichment methods used in their isolation. *Z. Versuchstierkd.* **17**, 155–158.

243. Kovatch, R. M., and Zebaith, G. (1973). Naturally occurring Tyzzer's disease in a cat. *J. Am. Vet. Med. Assoc.* **162**, 136–138.

244. Kunze, B., and Marwitz, T. (1967). Tilgung einer *Salmonella enteritidis*—Infektionen in einem Rattenzuchtbestand. *Z. Versuchstierkd.* **9**, 257–259.

245. Kurashina, H., and Fujiwara, K. (1972). Fine structure of the mouse liver infected with the Tyzzer's organism. *Jpn. J. Exp. Med.* **42**, 139–154.

246. Kutscher, D. (1894). Ein Beitrag zur Kentniss det bacillaren Pseudotuberculose der Nagethiere. *Z. Hyg. Infektionskr.* **18**, 327–342.

247. Lambe, D. W., Jr., McPhedran, A. M., Mertz, J. A., and Stewart, S. (1973). *Streptobacillus moniliformis* isolated from a case of Haverhill fever: Biochemical characterization and inhibitory effect of sodium polyanethol sulfonate. *Am. J. Clin. Pathol.* **60**, 854–860.

248. Larsen, B., Markovetz, A. J., and Galask, R. P. (1976). The bacterial flora of the female rat genital tract. *Proc. Soc. Exp. Biol. Med.* **151**, 571–574.

249. Larson, C. L. (1941). Rat-bite fever in Washington D.C. due to *Spirillum minus* and *Streptobacillus moniliformis*. *Public Health Rep.* **56**, 1961–1969.

250. Lawrence, J. J. (1957). Infection of laboratory mice with *Corynebacterium murium*. *Aust. J. Sci.* **20**, 147.

251. Leadingham, R. S. (1928). Rat-bite fever. *J. Med. Assoc., Ga.* **17**, 16–19.

252. Leadingham, R. S. (1938). Rat-bite fever (sodoku). Report of 5 cases. *Am. J. Clin. Pathol.* **8**, 333–344.

253. Lee, P. E. (1955). *Salmonella* infections of urban rats in Brisbane, Queensland. *Aust. J. Exp. Biol. Med. Sci.* **33**, 113–115.

254. Lee, P. E., and Mackerras, I. M. (1955). *Salmonella* infections of Australian native animals. *Aust. J. Exp. Biol. Med. Sci.* **33**, 117–125.

255. Lee, Y. S., Hirose, H., Ogisho, Y., Goto, N., Takahashi, R., and Fujiwara, K. (1976). Myocardiopathy in rabbits experimentally infected with the Tyzzer's organism. *Jpn. J. Exp. Med.* **46**, 371–382.

256. LeMaistre, C., and Thompsett, R. (1952). The emergence of pseudotuberculosis in rats given cortisone. *J. Exp. Med.* **95**, 393–407.

257. Lemperle, G. (1967). Stimulation of the reticuloendothelial system in burned rats infected with *Pseudomonas aeruginosa*. *J. Infect. Dis.* **117**, 7–14.

257a. Lennette, E. H., Spaulding, E. H., and Truant, J. P., eds. (1974). ''Manual of Clinical Microbiology,'' 2nd ed. Am. Soc. Microbiol., Washington, D. C.

258. Lerner, E. M., and Silverstein, E. (1957). Experimental infection of rats with *Streptobacillus moniliformis*. *Science* **126**, 208–209.

259. Lerner, E. M., and Sokoloff, L. (1959). The pathogenesis of bone and joint infection produced in rats by *Streptobacillus moniliformis*. *Arch. Pathol.* **67**, 364–372.

260. Les, E. P. (1968). Effect of acidified-chlorinated water on reproduction in C3H/HeJ and C57BL/6J mice. *Lab. Anim. Care* **18**, 210–213.

261. Les, E. P. (1968). Environmental factors influencing body weight of C57BL/6J and DBA/2J mice. *Lab. Anim. Care* **18**, 623–625.

262. Les, E. P. (1973). ''Acidified-chlorinated Drinking Water for Mice,'' Jax Notes, No. 415 (August). Jackson Laboratory, Bar Harbor, Maine.

263. Leung, L. S. E., Szal, G. J., and Drachman, R. H. (1972). Increased

susceptibility of splenectomized rats to infection with *Diplococcus pneumoniae*. *J. Infect. Dis.* **126,** 507-513.

264. Levaditi, C., Nicolau, S., and Poincloux, P. (1925). Sur le role étiologique de *Streptobacillus moniliformis* (nov. spec.) dans l'érythème polymorphe aigu septicemique. *C. R. Hebd. Seances Acad. Sci.* **180,** 1188-1190.

265. Levaditi, C., Selbie, R. F., and Schoen, R. (1932). Le rheumatisme infectieux spontane de la souris provogue par le *Streptobacillus moniliformis. Ann. Inst. Pasteur, Paris* **48,** 308-343.

266. Lowbury, E. J. L. (1951). Contamination of cetrimide and other fluids with *Pseudomonas pyocyanea. Br. J. Ind. Med.* **8,** 22-25.

267. Lowbury, E. J. L., and Fox, J. (1954). The epidemiology of infection with *Pseudomonas pyocyanea* in a burns unit. *J. Hyg.* **52,** 403-416.

268. Lusis, P. I., and Soltys, M. A. (1971). Immunization of mice and chinchillas against *Pseudomonas aeruginosa. Can. J. Comp. Med.* **35,** 60-66.

269. Lutsky, I., and Organick, A. B. (1966). Pneumonia due to *Mycoplasma* in gnotobiotic mice. I. Pathogenicity of *Mycoplasma penumoniae, Mycoplasma salivarium,* and *Mycoplasma pulmonis* for the lungs of conventional and gnotobiotic mice. *J. Bacteriol.* **92,** 1154-1163.

270. Macalister, G. H., and Brooks, R. St. J. (1914). Report upon the post-mortem examination of rats at Ipswich. *J. Hyg.* **14,** 316-330.

271. Mackaness, G. B., Blanden, R. V., and Collins, F. M. (1966). Host-parasite relations in mouse typhoid. *J. Exp. Med.* **124,** 573.

272. Mackenzie, D. W. R. (1961). *Trichophyton mentagrophytes* in mice; infections of humans and incidence amongst laboratory animals. *Sabouraudia* **1,** 178-182.

273. Mackenzie, D. W. R. (1963). "Hairbrush diagnosis" in the detection and eradication of non-fluorescent scalp ringworm. *Br. Med. J.* **2,** 363-365.

274. Mackie, T. J., Van Rooyen, C. E., and Gilroy, E. (1933). An epizootic disease occurring in a breeding stock of mice: Bacteriological and experimental observations. *Br. J. Exp. Pathol.* **14,** 132-136.

275. MacNeal, W. J. (1907). A spirochaete found in the blood of a wild rat. *Proc. Soc. Exp. Biol. Med.* **4,** 125.

276. Maejima, K., Fujiwara, K., Takagaki, Y., Naiki, M., Kucashin, H., and Tajima, Y. (1965). Dietetic effects on experimental Tyzzer's disease of mice. *Jpn. J. Exp. Med.* **35,** 1-10.

277. Maejima, K., Urano, T., Itoh, K., Fujiwara, K., Homma, J. Y., Suzuki, K., and Yanabe, M. (1973). Serological typing *Pseudomonas aeruginosa* from mouse breeding colonies. *Jpn. J. Exp. Med.* **43,** 179-184.

278. Maenza, R. M., Powell, D. W., Plotkin, G. R., Focmal, S. B., Jervis, H. R., and Sprinz, H. (1970). Experimental diarrhea: *Salmonella* enterocolitis in the rat. *J. Infect. Dis.* **121,** 475-485.

279. Mallory, F. B., and Ordway, T. (1909). Lesions produced in the rat by a typhoid-like organism (Danysz virus). *J. Am. Med. Assoc.* **52,** 1455.

280. Marecki, N. M., Hsu, H. S., and Mayo, D. R. (1975). Cellular and humoral aspects of host resistance in murine salmonellosis. *Br. J. Exp. Pathol.* **56,** 231-243.

281. Margard, W. L., and Peters, A. C. (1964). A study of gnotobiotic mice monocontaminated with *Salmonella typhimurium. Lab. Anim. Care* **14,** 200-207.

282. Margard, W. L., Peters, A. C., Dorko, N., Litchfield, J. H., Davidson, R. S., and Rheins, M. S. (1963). Salmonellosis in mice—Diagnostic procedures. *Lab. Anim. Care* **13,** 144-165.

283. Matheson, B. H., Grice, H. C., and Connell, M. R. E. (1955). Studies of middle ear disease in rats. I. Age of infection and infecting organisms. *Can. J. Comp. Med. Vet. Sci.* **19,** 91-97.

284. Matsumoto, H., Tazaki, T., and Kato, T. (1968). Serological and pyocine types of *Pseudomonas aeruginosa* from various sources. *Jpn. J. Microbiol.* **12,** 111-119.

285. McDermott, E. N. (1928). Rat-bite fever; study of the experimental disease, with a critical review of the literature. *Q. J. Med.* **21,** 433-458.

286. McDougall, P. T., Wolf, N. S., Steinbach, W. A., and Trentin, J. J. (1967). Control of *Pseudomonas aeruginosa* in an experimental mouse colony. *Lab. Anim. Care* **17,** 204-215.

287. McGee, A. A., and Wittler, R. G. (1969). The role of L phase and other wall-defective microbial variants in disease. *In* "The Mycoplasmatoles and the L-phase of Bacteria" (L. Hayflick, ed.), pp. 697-720. Appleton, New York.

288. McGill, R. C., Martin, A. M., and Edmunds, P. N. (1966) Rat-bite fever due to *Streptobacillus moniliformis. Br. Med. J.* **1,** 1213-1214.

289. McGuire, E. A., Young, V. R., Newberne, P. M., and Payne, B. J. (1968). Effects of *Salmonella typhimurium* infection in rats fed varying protein intakes. *Arch. Pathol.* **86,** 60-68.

290. McKenna, J. M., South, F. E., and Mussachia, X. J. (1970). *Pasteurella* infection in irradiated hamsters. *Lab. Anim. Care* **20,** 443-446.

291. McKeil, J. A., Rappay, D. E., Cousineau, J. G., Hall, R. R., and McKenna, H. E. (1970). Domestic rats as carriers of leptospires and salmonellae in Eastern Canada. *Can. J. Public Health* **61,** 336-340.

292. McPherson, C. W. (1963). Reduction of *Pseudomonas aeruginosa* and coliform bacteria in mouse drinking water following treatment with hydrochloric acid or chlorine. *Lab. Anim. Care* **13,** 737-745.

293. McRipley, R. J., and Garrison, D. W. (1964). Increased susceptibility of burned rats to *Pseudomonas aeruginosa. Proc. Soc. Exp. Biol. Med.* **115,** 336-338.

294. Menges, R. W., and Georg, L. K. (1956). An epizootic of ringworm among guinea pigs caused by *Trichophyton mentagrophytes. J. Am. Vet. Med. Assoc.* **128,** 395-398.

295. Menges, R. W., Georg, L. K., and Habermann, R. T. (1957). Therapeutic studies on ringworm-infected guinea pigs. *J. Invest. Dermatol.* **28,** 233-237.

296. Merrikin, D. J., and Terry, C. S. (1972). Use of pyocin 78-C2 in the treatment of *Pseudomonas aeruginosa* infection in mice. *Appl. Microbiol.* **23,** 164-169.

297. Meshorer, A. (1974). Tyzzer's disease in laboratory mice. *Refuah Vet.* **31,** 41-42.

298. Meyer, K. F., and Batchelder, A. P. (1926). A disease in wild rats caused by *Pasteurella muricida* n. sp. *J. Infect. Dis.* **39,** 386-412.

299. Meyer, K. F., and Matsumura, K. (1927). The incidence of carriers of *B. aertrycke (B. pestiscaviae)* and *B. enteritidis* in wild rats of San Francisco. *J. Infect. Dis.* **41,** 395-404.

300. Miller, C. P., Hammond, C. W., and Tompkins, M. A. (1950). The incidence of bacteremia in mice subjected to total body X-irradiation. *Science* **11,** 540-541.

301. Miller, C. P., Hammond, C. W., and Tompkins, M. (1951). The role of infection in radiation injury. *J. Lab. Clin. Med.* **38,** 331-343.

302. Millican, R. C. (1963). *Pseudomonas aeruginosa* infection and its effects in non-radiation stress. *Lab. Anim. Care* **13,** 11-19.

303. Millican, R. C., and Rust, J. D. (1960). Efficacy of rabbit *Pseudomonas* antiserum in experimental *Pseudomonas aeruginosa* infection. *J. Infect. Dis.* **107,** 389-396.

304. Millican, R. C., Rust, J. D., Verder and Rosenthal, S. M. (1957). Experimental chemotherapy of *Pseudomonas* infections. I. Production of fatal infections in cortisone-infected mice. *Antibiot. Annu.* pp. 486-493.

305. Millican, R. C., Evans, G., and Markley, K. (1966). Susceptibility of burned mice to *Pseudomonas aeruginosa* and protection by vaccination. *Ann. Surg.* **163,** 603-610.

306. Mirick, G. S., Richter, C. P., Schaub, I. G., Franklin, R., MacCleary, R., Schipper, G., and Spitznager, J. (1950). An epizootic due to *Pneumococcies* type 11 in laboratory rats. *Am. J. Hyg.* **52,** 48-53.

307. Mitchell, D. W. H. (1912). Bacillus muris as the etiologic agent of pneumonitis in white rats and its pathogenicity for laboratory animals. *J. Infect. Dis.* **10,** 17-23.

308. Mitruka, B. M. (1971). Biochemical aspects of *Diplococcus pneumoniae* infections in laboratory rats. *Yale J. Biol. Med.* **44,** 253-264.

309. Miyake, H. (1902). Ueber die Rattenbissenkrankheit. *Mitt. Grenzgeb. Med. Chir.* **5,** 231-262.

310. Mohapatra, L. N., Gugnani, H. C., and Shivrajan, K. (1964). Natural infection in laboratory animals due to *Trichophyton mentagrophytes* in India. *Mycopathol. Mycol. Appl.* **24,** 275-280.

311. Moore, T. D., Allen, A. M., and Ganaway, J. R. (1973). Latent *Pasteurella Pneumotropica* infection of the intestine of gnotobiotic and barrier-held rats. *Lab. Anim. Sci.* **5,** 657-661.

312. Morch, E. (1947). Mechanism of *Pneumococcus* infection in mice. *Acta Pathol. Microbiol. Scand.* **24,** 169-180.

313. Morello, J. A., Digenio, T. A., and Baker, E. E. (1964). Evaluation of serological and cultural methods for the diagnosis of chronic salmonellosis in mice. *J. Bacteriol.* **88,** 1277-1282.

314. Morello, J. A., Digenio, T. A., and Baker, E. E. (1965). Significance of salmonellae isolated from apparently healthy mice. *J. Bacteriol.* **89,** 1460-1464.

315. Mullink, J. W. M. A., and Rumke, C. L. (1971). Reaction on hexobarbital and pathological control of mice given acidified drinking water. *Z. Versuchstierkd.* **13,** 196-200.

316. Nafiz, M. (1935). Untersuchungen an wilden Ratten in München auf das Vorkommen von Typhus und Paratyphuskeimen. *Arch. Hyg.* **113,** 245-246.

317. Naiki, M., Takagaki, Y., and Fujiwara, K. (1965). Note on the change of transaminases in the liver and the significance of the transaminase ratio in experimental Tyzzer's disease of mice. *Jpn. J. Exp. Med.* **35,** 305-309.

318. Nakayama, M., Saegusa, J., Itoh, K., Kiuch, Y., Tamura, T., Veda, K., and Fujiwara, K. (1975). Transmissible enterocolitis in hamsters caused by Tyzzer's organisms. *Jpn. J. Exp. Med.* **45,** 33-41.

319. Nance, F. C., Lewis, V. L., Jr., and Hines, J. L. (1970). *Pseudomonas* sepsis in gnotobiotic rats. *Surg. Forum.* **21,** 234-235.

320. Needham, J. R., and Cooper, J. E. (1976). An eye infection in laboratory mice associated with *Pasteurella pneumotropica*. *Lab. Anim.* **9,** 197-200.

321. Nelson, J. B. (1930). The bacteria of the infected middle ear in adult and young albino rats. *J. Infect. Dis.* **46,** 64-75.

322. Nelson, J. B. (1930). The reaction of the albino rat to the intraaural administration of certain bacteria associated with middle ear disease. *J. Exp. Med.* **52,** 873-883.

323. Nelson, J. B. (1931). The biological characters of *B. actinoides* variety *muris*. *J. Bacteriol.* **21,** 183-195.

324. Nelson, J. B. (1933). The reaction of antisera for *B. actinoides*. *J. Bacteriol.* **26,** 321-327.

325. Nelson, J. B. (1937). Infectious catarrh of mice. *J. Exp. Med.* **65,** 833-860.

326. Nelson, J. B. (1940). Infectious catarrh of the albino rat. I. Experimental transmission in relation to the role of Actinobacillus muris. *J. Exp. Med.* **72,** 645-654.

327. Nelson, J. B. (1963). Chronic respiratory disease in mice and rats. *Lab. Anim. Care* **13,** 137-143.

328. Nelson, J. B. (1973). Response of mice to *Corynebacterium kutscheri* on footpad injection. *Lab. Anim. Sci.* **23,** 370-372.

329. Nelson, J. B., and Gowen, J. W. (1930). The incidence of middle ear infection and pneumonia in albino rats at different ages. *J. Infect. Dis.* **46,** 53-63.

330. Newberne, P. M., Hunt, C. E., and Young, V. R. (1968). The role of diet and the veticuloendothelial system in the response of rats to *Salmonella typhimurium* infection. *Br. J. Exp. Pathol.* **49,** 448-457.

331. Niven, J. S. F. (1968). Tyzzer's disease in laboratory animals. *Z. Versuchstierkd.* **10,** 168-174.

332. Olson, J. R., and Meadows, T. R. (1969). *Pasteurella pneumotropica* infection resulting from a cat bite. *Am. J. Clin. Pathol.* **51,** 709-710.

333. Olson, L. D., and Ediger, R. D. (1972). Histopathologic study of the heads of circling mice infected with *Pseudomonas aeruginosa*. *Lab. Anim. Sci.* **22,** 522-527.

334. Onodera, T., and Fujiwara, K. (1970). Experimental encephalopathy in Tyzzer's disease of mice. *Jpn. J. Exp. Med.* **40,** 295-323.

335. Onodera, T., and Fujiwara, K. (1972). Neuropathology of intraspinal infection in mice with the Tyzzer's organism. *Jpn. J. Exp. Med.* **42,** 263-282.

336. Osimani, J. R., Nairac, R., Hormaeche, C., Fonseca, D., de Hormaeche, R. D., and Portillo, M. (1972). Fiebre estreptobacilar por mordedura de rata. *Rev. Latinoam. Microbiol.* **14,** 197-201.

337. Otcenasek, M., Stros, K., Krivanek, K., and Komarek, J. (1974). Epizoocie dermatofytozy ve velkochovu morcat. *Vet. Med. (Prague)* **19,** 277-281.

338. Owens, D. R., and Wagner, J. E. (1975). *Trichophyton mentagrophytes* infection in guinea pigs. *Lab. Anim. Dig.* **10,** 14-18.

339. Owens, D. R., Wagner, J. E., and Addison, J. B. (1975). Antibiograms of pathogenic bacteria isolated from laboratory animals. *J. Am. Vet. Med. Assoc.* **167,** 605-609.

340. Pappenheimer, A. M., and von Wedel, H. (1914). Observations on a spontaneous typhoid-like epidemic of white rats. *J. Infect. Dis.* **14,** 180-215.

341. Parish, H. J., and Craddock, J. (1931). Ringworm epizootic in mice. *Br. J. Exp. Pathol.* **12,** 209-213.

342. Parker, F., Jr., and Hudson, N. P. (1926). The etiology of Haverhill fever (erythema arthriticum epidemicum). *Am. J. Pathol.* **2,** 357-379.

343. Peace, T., and Soave, O. A. (1969). Tyzzer's disease in a group of newly purchased mice. *Lab. Anim. Dig.* **5,** 8-10.

344. Pestana de Castro, A. F., Giorgi, W., and Ribeiro, W. B. (1964). Estudo de umna amostra de *Corynebacterium kutscheri* isolada de ratos e camundongos. *Arq. Inst. Biol., (Sao Paulo)* **31,** 91-99.

345. Pierce-Chase, C. H., Fauve, M., and Dubos, R. (1964). Corynebacterial pseudotuberculosis in mice. I. Comparative susceptibility of mouse strains to experimental infection with *Corynebacterium kutscheri*. *J. Exp. Med.* **120,** 267-281.

346. Place, E. H., and Sutton, L. E. (1934). Erythema arthriticum epidemicum. *Arch. Intern. Med.* **54,** 659-684.

347. Place, E. H., Sutton, L. E., and Willner, O. (1926). Erythema arthriticum. Preliminary report. *Boston Med. Surg. J.* **194,** 285-287.

348. Plotkin, S. A., and Austrian, R. (1958). Bacteremia caused by *Pseudomonas* sp. following the use of materials stored in solutions of a cationic surface-active agent. *Am. J. Med. Sci.* **235,** 621-627.

349. Polak, M. F. (1944). Epidémie survenue parmi des souris blanches à la suite d'une infection par le *Corynebacterium pseudotuberculosis murium*. *Antonie van Leeuwenhoek* **10,** 23-27.

350. Pombier, E. C., and Kim, J. C. S. (1975). An epizootic outbreak of ringworm in a guinea pig colony caused by *Trichophyton mentagrophytes*. *Lab. Anim.* **9,** 215-221.

351. Port, C. D., Richter, W. R., and Moise, S. M. (1970). Tyzzer's disease in the gerbil (*Meriones unguiculatus*). *Lab. Anim. Care* **20,** 109-111.

352. Porter, G. (1952). An experience of Tyzzer's disease in mice. *J. Inst. Anim. Tech.* **2,** 17-18.

353. Potel, J. (1967). (Latent Salmonella infection in laboratory animals.) *Zentralbl. Bakteriol., Parasitenkd., Infektionskr. Hyg., Abt. 1: Orig.* **203,** 292-295.

354. Povar, M. L. (1965). Ringworm (*Trichophyton mentagrophytes*). Infection in a colony of albino Norway rats. *Lab. Anim. Care* **15,** 264-265.

355. Powanda, M. C., and Canonico, P. G. (1976). Protective effect of

clofibrate against *S. pneumoniae* infection in rats. *Proc. Soc. Exp. Biol. Med.* **152,** 437-430.

356. Powanda, M. C., Wannemacher, R. W., Jr., and Cockerell, G. L. (1972). Nitrogen metabolism and protein synthesis during pneumococcal sepsis in rats. *Infect. Immun.* **6,** 266-271.

357. Powell, D. W., Plotkin, G. R., Maenza, R. M., Solberg, L. I., Catlin, D. H., and Formal, Σ. B. (1971). Experimental diarrhea. 1. Intestinal water and electrolyte transport in rat *Salmonella* enterocolitis. *Gastroenterology* **60,** 1053-1064.

358. Powell, D. W., Plotkin, G. R., Solberg, L. I., Catlin, D. H., Maenza, R. M., and Formal, S. B. (1971). Experimental diarrhea 2. Glucose-stimulated sodium and water transport in rat *Salmonella* enterocolitis. *Gastroenterology* **60,** 1065-1075.

359. Powell, D. W., Solberg, L. I., Plotkin, G. R., Catlin, D. H., Maenza, R. M., and Formal, S. B. (1971). Experimental diarrhea. 3. Bicarbonate transport in rat *Salmonella* enterocolitis. *Gastroenterology* **60,** 1076-1086.

360. Price-Jones, C. (1926-1927). Infection of rats by Gartner's bacillus. *J. Pathol. Bacteriol.* **30,** 45-54.

361. Pyrah, L. N., Goldie, W., Parsons, F. M., and Raper, F. P. (1955). Control of *Pseudomonas pyocanea* infection in a urological ward. *Lancet* **2,** 314-317.

362. Rake, G. (1936). Pathology of *Pneumococcus* infection in mice following intranasal installation. *J. Exp. Med.* **63,** 17-31.

363. Rapp, J. P. and McGrath, J. T. (1975). Mycotic encephalitis in weanling rats. *Lab. Anim. Sci.* **25,** 477-480.

364. Ratcliffe, H. L. (1949). Spontaneous diseases of laboratory rats. *In* "The Rat in Laboratory Investigation" (E. J. Farris and J. Q. Griffith, eds.), 2nd ed., pp. 515-530. Lippincott, Philadelphia, Pennsylvania.

365. Rathbun, H. K., and Govani, I. (1967). Mouse inoculation as a means of identifying pneumococci in the sputum. *Johns Hopkins Med. J.* **120,** 46-48.

366. Rauss, K., and Kalovich, B. (1971). *Salmonella* immunity in mice. *Zentralbl. Bakteriol., Parasitenkd., Infektionskr. Hyg., Abt. 1: Orig.* **216,** 32-53.

367. Ray, J. P., and Mallick, B. B. (1970). Public Health significance of *Salmonella* infections in laboratory animals. *Indian Vet. J.* **47,** 1033-1037.

368. Razin, S., and Boschwitz, C. (1968). The membrane of the *Streptobacillus moniliformis* L phase. *J. Gen. Microbiol.* **54,** 21-32.

369. Rebell, G., and Taplin, D. (1970). "Dermatophytes. Their Recognition and Identification," 2nd. rev., pp. 10-19. Univ. of Miami Press, Coral Gables, Florida.

370. Reed, D. M. (1902). The *Bacillus pseudotuberculosis murium* its streptothrix forms and pathogenic action. *Johns Hopkins Hosp. Rep.* **9,** 525-541.

371. Rehbinder, C., and Tschappat, V. (1974). *Pasteurella pneumotropica,* isoliert von der Konjunktivalschleimhaut gesunder Laboratoriumsmause. *Z. Versuchstierkd.* **16,** 359-365.

372. Reith, von H. (1968). Spontane und experimentelle Pilzerkrankungen bei Mausen. *Z. Versuchstierkd.* **10,** 75-81.

373. Rights, F. L., Jackson, E. B., and Smadel, F. E. (1947). Observations on Tyzzer's disease in mice. *Am. J. Pathol.* **23,** 627-635.

374. Ringen, L. M., and Drake, C. H. (1952). A study of the incidence of *Pseudomonas aeruginosa* from various natural sources. *J. Bacteriol.* **64,** 841-845.

375. Robertson, A. (1924). Causal organism of rat-bite fever in man. *Ann. Trop. Med. Parasitol.* **18,** 157-175.

376. Robertson, A. (1930). *Spirillum minus* Carter 1887, the etiological agent of rat-bite fever: A review. *Ann Trop. Med. Parasitol.* **24,** 367-410.

377. Robinson, H. J., Phares, H. F., and Graessle, O. E. (1968). Effects of indomethacin on acute, subacute, and latent infections in mice and rats. *J. Bacteriol.* **96,** 6-13.

378. Robinson, L. B. (1963). *Streptobacillus moniliformis* infections. *In* "Diagnostic Procedures and Reagents" (A. H. Harris and M. B. Coleman, eds.). 4th ed., pp. 642-651. Am. Public Health Assoc., New York.

379. Robinson, R. (1965). "Genetics of the Norway Rat," pp. 5-6. Pergamon, Oxford.

380. Robson, H. G., and Vas, S. I. (1972). Resistance of inbred mice to *Salmonella typhimurium. J. Infect. Dis.* **126,** 378-386.

381. Rogosa, M. (1974). *Streptobacillus moniliformis* and *Spirillum minus. In* "Manual of Clinical Microbiology" (E. H. Lennette, E. H. Spaulding, and J. P. Truant, eds.), 2nd ed., pp. 326-332. Am. Soc. Microbiol., Washington, D. C.

382. Rosenthal, S. A., and Furnari, D. (1957). The use of a cyclohesimide-chloramphenicol medium in routine culture for fungi. *J. Invest. Dermatol.* **28,** 367-371.

383. Rosenthal, S. A., and Wapnick, H. (1963). The value of MacKenzie's hair brush technic in the isolation of *Trichophyton mentagrophytes* from clinically normal guinea pigs. *J. Invest. Dermatol.* **41,** 5-6.

384. Rosenthal, S. M., Millican, R. C., and Rust, J. (1957). A factor in human gamma globulin preparations active against *Pseudomonas aeruginosa* infections. *Proc. Soc. Exp. Biol. Med.* **94,** 214-217.

385. Ross, V. (1931). Oral immunization against *Pneumococcus* types II and III and the normal variation in resistance to these types among rats. *J. Exp. Med.* **54,** 875-898.

386. Ross, V. (1934). Protective antibodies following oral administration of *Pneumococcus* types 2 and 3 to rats, with some data for types 4, 5 and 6. *J. Immunol.* **27,** 273-306.

387. Roughgarden, J. W. (1965). Antimicrobial therapy of rat-bite fever; a reivew. *Arch. Intern. Med.* **116,** 39-54.

388. Row, R. (1918). Cutaneous spirochetosis produced by rat-bite in India. *Bull. Soc. Pathol. Exot.* **11,** 88-195.

389. Ruitenberg, E. J., Guinee, P. A. M., Kruyt, B. C., and Berkvens, J. M. (1971). *Salmonella* pathogenesis in germ-free mice. *Br. J. Exp. Pathol.* **52,** 192-197.

390. Sacquet, E. (1958). La salmonellose du rat de laboratoire: Detection et traitment des porteurs sains. *Rev. Fr. Etud. Clin. Biol.* **3,** 1075-1078.

391. Sacquet, E. (1959). Valeur de la sero-agglutination dans le diagnostic des salmonellosis du rat et de la souris. *Rev. Fr. Etud. Clin. Biol.* **4,** 930-932.

392. Sagi, T., and Lapis, L. (1956). Unter dem Enfluss det Cortison behandlung entstehende lungenaspergillose bei Ratten. *Acta Microbiol. Acad. Sci. Hung.* **3,** 337-340.

393. Salthe, O., and Krumweide, C. (1924). Studies on the paratyphoid-enteriditis group. VIII. An epidemic of food infection due to a paratyphoid bacillus of rodent origin. *Am. J. Hyg.* **4,** 23-32.

394. Sandhu, K. K., Sandhu, R. S., Damodaran, V. N., and Randhawa, H. S. (1970). Effect of cortisone on brochopulmonary aspergillosis in mice exposed to spores of various *Aspergillus* species. *Sabouraudia* **8,** 32-38.

395. Saunders, L. Z. (1958). Tyzzer's disease. *J. Natl. Cancer Inst.* **20,** 893-897.

396. Savage, N. L. (1972). Host-parasite relationships in experimental *Streptobacillus moniliformis* arthritis in mice. *Infect. Immun.* **5,** 183-190.

397. Savage, N. L., and Lewis, D. H. (1972). Application of immunofluorescence to detection of Tyzzer's disease agent (*Bacillus piliformis*) in experimentally infected mice. *Am. J. Vet. Res.* **33,** 1007-1012.

398. Savage, W. D., and Read, W. J. (1914). Gaertner group bacilli in rats and mice. *J. Hyg.* **13,** 343-352.

399. Savage, W. G., and White, P. B. (1923). Rats and *Salmonella* group

bacilli. *J. Hyg.* **21,** 258–261.

400. Savoor, S. R., and Lewthwaite, R. (1941). The Weil-Felix reaction in experimental rat-bite fever. *Br. J. Exp. Pathol.* **22,** 274–292.

401. Sawicki, L., Bruce, H. M., and Andrewes, C. H. (1962). *Streptobacillus moniliformis* infection as a probable cause of arrested pregnancy and abortion in laboratory mice. *Br. J. Exp. Pathol.* **43,** 194–197.

402. Schaedler, R. W., and Dubos, R. J. (1962). The fecal flora of various strains of mice. Its bearing on their susceptibility to endotoxin. *J. Exp. Med.* **115,** 1149–1160.

403. Schechmeister, J. L. (1956). Pseudotuberculosis in experimental animals. *Science* **123,** 463–464.

404. Schechmeister, J. L., and Adler, F. L. (1953). Activation of pseudotuberculosis in mice exposed to sublethal total body radiation. *J. Infect. Dis.* **92,** 228–239.

405. Schipper, G. J. (1947). Unusual pathogenicity of *Pasteurella multocida* isolated from the throats of common wild rats. *Bull. Johns Hopkins Hosp.* **81,** 333–356.

406. Schottmuller, H. (1914). Zur Atiologie und Klinik der Bisskrankheit Rattin, Katzen-, Eichhornchen-Bisskrankheit). *Dermatol. Wochenschr.* **58,** Suppl. 77–103.

407. Schutze, H. (1928). *Bacterium pseudotuberculosis rodentium.* *Arch. Hyg.* **100,** 181.

408. Schutze, H. (1932). Studies in *B. pestis* antigens. II. The antigenic relationship of *B. pestis* and *B. pseudotuberculosis rodentium.* *Br. J. Exp. Pathol.* **13,** 289–292.

409. Schwarz, J. (1954). The deep mycoses in laboratory animals. *Proc. Anim. Care Panel* **5,** 37–70.

410. Seronde, J. (1954). Resistance of rats to inoculation with *Corynebacterium* pathogenic in pantothenate deficiency. *Proc. Soc. Exp. Biol. Med.* **85,** 521–524.

411. Seronde, J., Zucker, L. M., and Zucker, T. F. (1955). The influence of duration of pantothenate deprivation upon natural resistance of rats to a *Corynebacterium.* *J. Infect. Dis.* **97,** 35–38.

412. Seronde, J., Zucker, T. F., and Zucker, L. M. (1956). Thiamine, pyridozine and pantothenic acid in the natural resistance of the rat to a *Corynebacterium* infection. *J. Nutr.* **59,** 287–298.

413. Sethi, K. K., Salfelder, K., and Schwarz, J. (1964). Pulmonary fungal flora in experimental pulmocystosis of cortisone treated rats. *Mycopathol. Mycol. Appl.* **24,** 121–129.

414. Shambaugh, G. E., 3rd, and Beisel, W. R. (1966). Alterations in thyroid physiology during pneumococcal septicemia in the rat. *Endocrinology* **79,** 511–523.

415. Shu-heh, W., Chu, Welch, K. J., Murray, E. S., and Hagsted, D. M. (1976). Effect of iron deficiency on the susceptibility to *Streptococcus pneumoniae* infection in the rat. *Nutr. Rep. Int.* **14,** 605–609.

416. Sidransky, H., and Friedman, L. (1959). The effect of cortisone and antibiotic agents on experimental pulmonary asperigillosis. *Am. J. Pathol.* **35,** 169–184.

417. Simmons, D. J. C., and Simpson, W. (1977). The biochemical and cultural characteristics of *Pasteurella pneumotropica.* *Med. Lab. Sci.* **34,** 145–148.

418. Simon, P. C. (1977). Isolation of *Bacillus piliformis* from rabbits. *Can. Vet. J.* **18,** 46–48.

419. Singh, B., and Chawla, R. S. (1974). A note on an outbreak of pulmonary aspergillosis in albino rat colony. *Indian J. Anim. Sci.* **44,** 804–807.

420. Slanetz, C. A. (1948). The control of *Salmonella* infections in colonies of mice. *J. Bacteriol.* **56,** 771–775.

421. Smith, C. D., and Sampson, C. C. (1960). Studies of *Streptobacillus moniliformis* from case of human rat-bite fever. *Am. J. Med. Technol.* **26,** 47–50.

422. Smith, J. M. B., and Austwick, P. K. C. (1967). Fungal diseases of

rats and mice. *In* "Pathology of Laboratory Rats and Mice" (E. Cotchin and F. J. C. Roe, eds.), pp. 681–719. Davis, Philadelphia, Pennsylvania.

423. Smith, W. (1941). Cervical abscesses of guinea pigs. *J. Pathol. Bacteriol.* **53,** 29–37.

424. Smith, W. W., Menges, R. W., and Georg, L. I. (1957). Ecology of ringworm fungi on commensal rats from rural premises in southwestern Georgia. *Am. J. Trop. Med. Hyg.* **6,** 81–85.

425. Spaar, R. C. (1923). Two cases of rat-bite fever. Rapid cure by the intravenous injection of neo-salvarsan. *J. Trop. Med. Hyg.* **26,** 239.

426. Sparrow, S. (1976). The microbiological and parasitological status of laboratory animals from accredited breeders in the United Kingdom. *Lab. Anim.* **10,** 365–373.

427. Speirs, R. S. (1956). Effect of oxytetracycline upon cortisone induced pseudotuberculosis in mice. *Antibiot. Chemother. (Washington, D. C.)* **6,** 395–399.

428. Spencer, R. C., and Savage, M. A. (1976). Use of counter and rocket immunoelectrophoresis in acute respiratory infections due to *Streptococcus pneumoniae.* *J. Clin. Pathol.* **29,** 187–190.

429. Sprecher, M. W., and Copeland, J. R. (1947). Haverhill fever due to *Streptobacillus moniliformis.* *J. Am. Med. Assoc.* **134,** 1014–1016.

430. Staff, E. J., and Grover, M. L. (1936). An outbreak of *Salmonella* food infection caused by filled bakery products. *Food Res.* **1,** 465–479.

431. Stedham, M. A., and Bucci, T. J. (1969). Tyzzer's disease in the rat. *Lab. Invest.* **20,** 604.

432. Stedham, M. A., and Bucci, T. J. (1970). Spontaneous Tyzzer's disease in a rat. *Lab. Anim. Care* **20,** 743–746.

433. Stewart, H. L., and Jones, B. F. (1941). Pathologic anatomy of chronic ulcerative cecitis: A spontaneous disease of the rat. *Arch. Pathol.* **31,** 37–54.

434. Stone, H. H., Given, K. S., and Martin, J. D., Jr. (1967). Delayed rejection of skin homografts in *Pseudomonas* sepis. *Surg., Gynecol. Obstet.* **124,** 1067–1070.

435. Strageways, W. L. (1933). Rats as carriers of *Streptobacillus.* *J. Pathol. Bacteriol.* **37,** 45–51.

436. Stuhmer, A. (1929). Die Rattenbiszerkrankung (sodoku) als Modell-infektion für Syphilisstudien. *Arch. Dermatol. Syph.* **158,** 98–110.

437. Summerlin, W. T., and Artz, C. P. (1966). Gentamicin sulfate therapy of experimentally induced *Pseudomonas* septicemia. *J. Trauma* **6,** 233–238.

438. Surveyor, N. F. (1913). A case of rat-bite fever treated with neosalvarsan. *Lancet* **2,** 1764.

439. Swenberg, J. S. Koestner, A., and Tewari, R. P. (1969). Experimental mycotic encephalitis. *Actual Neuropathol.* **13,** 75–90.

440. Swenberg, J. A. Koestner, A., and Tewari, R. P. (1969). The pathogenesis of experimental mycotic encephalitis. *Lab. Invest.* **21,** 365–373.

441. Taffs, L. F. (1974). Some diseases in normal and immunosuppressed experimental animals. *Lab. Anim.* **8,** 149–154.

442. Takagaki, Y., and Fujiwara, K. (1968). Bacteremia in experimental Tyzzer's disease of mice. *Jpn. J. Microbiol.* **12,** 129–143.

443. Takagaki, Y., Naiki, M., Ito, M., Noguchi, and G., and Fujiwara, K. (1967). Checking of infections due to *Corynebacterium* and Tyzzer's organism among mouse breeding colonies by cortisone injection. *Bull. Exp. Anim.* **16,** 12–19.

444. Takagaki, Y., Ogisho, Y., Sato, K., and Fujiwara, K. (1974). Tyzzer's disease in hamsters. *Jpn. J. Exp. Med.* **44,** 267–270.

445. Takagaki, Y., Tsuji, K., and Fujiwara, K. (1968). Tyzzer's disease-like liver lesions observed in young rats. *Exp. Anim.* **17,** 67–69.

446. Takagaki, Y., Naiki, M., Fujiwara, K., and Tajima, Y. (1963) Maladie

de Tyzzer expérimentale de la souris traitee avec la cortisone. *C. R. Seances Soc. Biol. Ses Fil.* **157,** 438–441.

447. Takagaki, Y., Ito, M., Naiki, W., Fujiwara, K., Okugi, M., Maejima, K., and Tajima, Y. (1966). Experimental Tyzzer's disease in different species of laboratory animals. *Jpn. Exp. Med.* **36,** 519–534.

448. Takenaka, J., and Fujiwara, K. (1975). Effect of carbon tetrachloride on experimental Tyzzer's disease of mice. *Jpn. J. Exp. Med.* **45,** 393–402.

449. Tani, T., and Takano, S. (1958). Prevention of *Borrelia duttoni, Trypanosoma gambiense, Spirillum minus* and *Treponema pallidum* infections conveyable through transmission. *Jpn. J. Med. Sci. Biol.* **11,** 407–413.

450. Tannock, G. W., and Smith, J. M. B. (1972). The effect of food and water deprivation (stress) on *Salmonella*-carrier mice. *J. Med. Microbiol.* **5,** 283–289.

451. Taylor, J., and Atkinson, J. D. (1956). *Salmonella* in laboratory animals. *Lab. Anim. Bur. Collect. Pap. (Carshalton)* **4,** 57–66.

452. Teplitz, C., Davis, D., Mason, A. D., Jr., and Moncrief, J. A. (1964). *Pseudomonas* burn wound sepsis. I. Pathogenesis of experimental *Pseudomonas* burn wound sepsis. *J. Surg. Res.* **4,** 200–216.

453. Teplitz, C., Davis, D., Walker, H. L., Raulston, G. L., Mason, A. D., Jr., and Moncrief, J. A. (1964). *Pseudomonas* burn wound sepsis. II. Hematogenous infection at the junction of the burn wound and the unburned hypodermis. *J. Surg. Res.* **5,** 217–222.

454. Thjotta, T., and Jonsen, J. (1947). *Streptothrix (Actinomyces) muris ratti (Streptobacillus moniliformis)* isolated from a human infection and studied as to its relation to Emmy Kleineberger's LI. *Acta Pathol. Microbiol. Scand.* **24,** 336–351.

455. Thomson, G. W., Wilson, R. W., Hall, E. A., and Physick-Sheard, P. (1977). Tyzzer's disease in the foal: Case reports and review. *Can. Vet. J.* **18,** 41–43.

456. Thunert, A., and Heine, W. (1975). Zur Trinkwasserversorgung von SPF-Tieranlager. III. Erhitzung und Ansauerung von Trinkwasser. *Z. Versuchstierkd.* **17,** 50–52.

457. Tolo, K. J., and Erichsen, S. (1969). Acidified drinking water and dental enamel in rats. *Z. Versuchstierkd.* **11,** 229–233.

458. Topley, W. W. C., and Ayrton, J. (1924). Biologic characteristics of *B. enteritidis (aertrycke)*. *J. Hyg.* **23,** 198–222.

459. Topley, W. W. C., and Wilson, G. S. (1922–1923). The spread of bacterial infection. The problem of herd immunity. *J. Hyg.* **21,** 243–249.

460. Topley, W. W. C., and Wilson, J. (1925). Further observations on the role of the Twort-d'Hérelle phenomenon in the epidemic spread of mouse typhoid. *J. Hyg.* **24,** 295–300.

461. Topley, W. W. C., Wilson, J., and Lewis, E. R. (1925). Immunization and selection as factors in herd-resistance. *J. Hyg.* **23,** 421–436.

462. Trakhanov, D. F., and Kadirov, A. F. (1972). Izuchenie vospriimchivosti porosyat k bakteriyam Isachenko. *Probl. Vet. Sanit.* **42,** 248–253.

463. Tregier, A., and Homburger, F. (1961). Bacterial flora of the mouse uterus. *Proc. Soc. Exp. Biol. Med.* **108,** 152–154.

464. Trentin, J. J., Van Hoosier, G. L., Jr., Shields, J., Stepens, K., and Stenback, W. A. (1966). Establishment of a caesarean-derived, gnotobiote foster nursed inbred mouse colony with observations on the control of *Pseudomonas*. *Lab. Anim. Care* **16,** 109–118.

465. Tucek, P. C. (1971). *Diplococcal pneumonia* in the laboratory rat. *Lab. Anim. Dig.* **7,** 32–35.

466. Tuffery, A. A. (1956). The laboratory mouse in Great Britain. IV. Intercurrent infection (Tyzzer's disease). *Vet. Rec.* **68,** 511–515.

467. Tuffery, A. A., and Innes, J. R. M. (1963). Diseases of laboratory mice and rats. *In* "Animals for Research: Principles of Breeding and Management" (W. Lane-Petter, ed.), pp. 47–107. Academic Press, New York.

468. Tunnicliff, R. (1916). *Streptothrix* in bronchopneumonia of rats similar to that in rat-bite fever. A preliminary report. *J. Am. Med. Assoc.* **66,** 1606.

469. Tunnicliff, R. (1916). *Streptothrix* in bronchopneumonia of rats similar to that in rat-bite fever. *J. Infect. Dis.* **19,** 767–771.

470. Turner, R. G. (1929). Bacteria isolated from infections of the nasal cavities and middle ear of rats deprived of vitamin A. *J. Infect. Dis.* **45,** 208–213.

471. Tyzzer, E. E. (1917). A fatal disease of the Japanese waltzing mouse caused by a sport-bearing bacillus (*Bacillus piliformis* N. sp.). *J. Med. Res.* **37,** 307–338.

472. Utsumi, K., Matsui, Y., Ishikawa, T., Fukagawa, S., Tatsumi, H., Fujimoto, K., and Fujiwara, K. (1969). Checking of corynebacterial infection in rats by cortisone treatment. *Bull. Exp. Anim.* **18,** 59–67.

473. Vallee, A., Guillon, J. C., and Cayeux, R. (1969). Isolement d'une souche de *Corynebacterium kutscheri* chez un cobaye. *Bull. Acad. Vet. Fr.* **42,** 797–800.

474. Vallee, A., and Levaditi, J. C. (1957). Abcès miliaires des reins observes chex le rat blanc et provoque par un corynebacterium aerobié voisin du type *C. kutscheri*. *Ann. Inst. Pasteur, Paris* **93,** 468–474.

475. Van der Schaff, A., Mullink, J. W. M. A., Nikkels, R. J., Goudswaard, J. (1970). *Pasteurella pneumotropica* as a causal microorganism of multiple subcutaneous abscesses in a colony of Wistar rats. *Z. Versuchstierkd.* **12,** 356–362.

476. Van Der Waay, D., Zimmerman, W. M. T., and Van Bekkum, D. W. (1963). An outbreak of *Pseudomonas aeruginosa* infection in a colony previously free of this infection. *Lab. Anim. Care* **13,** 46–53.

477. Van Kruiningen, H. J., and Blodgett, S. B. (1971). Tyzzer's disease in a Connecticut rabbitry. *J. Am. Vet. Med. Assoc.* **158,** 1205–1212.

478. Van Rooyen, C. R. (1936). The biology pathogenesis and classification of *Streptobacillus moniliformis*. *J. Pathol. Bacteriol.* **43,** 455–472.

479. Varela, G., Olarte, J., and Mata, F. (1948). Salmonellas en las ratas de la Ciudad de Mexico estudio de 1927 *Ratas norvegicus*. *Rev. Inst. Salubr. Enferm. Trop., Mexico City* **9,** 239–243.

480. Verder, E. (1927). The wild rat as a carrier of organisms of the paratyphoid-enteriditis group. *Am. J. Public Health* **17,** 1007.

481. Volkman, A., and Collins, F. M. (1973). Polyarthritis associated with *Salmonella*. *Infect. Immun.* **8,** 814–827.

482. Wagner, J. E., and Manning, P. J., eds. (1976). "The Biology of the Guinea Pig." Adacemic Press, New York.

483. Walker, H. L., Mason, A. D., Jr., and Raulston, G. L. (1964). Surface infection with *Pseudomonas aeruginosa*. *Ann. Surg.* **160,** 297–305.

484. Watkins, C. G. (1946). Rat-bite fever. *J. Pediatr.* **28,** 429–448.

485. Webster, L. T. (1924). Microbic virulence and host susceptibility in paratyphoid-enteriditis infection of white mice. *J. Exp. Med.* **39,** 129–135.

486. Webster, L. T. (1930). The role of microbic virulence, dosage, and host resistance in determining the spread of bacterial infections among mice. II. *B. Friedlaenderi*- like infection. *J. Exp. Med.* **52,** 909–929.

487. Weisbroth, S. H., and Freimer, E. H. (1969). Laboratory rats from commerical breeders as carriers of pathogenic pneumococci. *Lab. Anim. Care* **19,** 473–478.

488. Weisbroth, S. H., and Scher, S. (1968). *Corynebacterium kutscheri* infection in the mouse. I. Report of an outbreak, bacteriology, and pathology of spontaneous infections. *Lab. Anim. Care* **18,** 451–458.

489. Weisbroth, S. H., and Scher, S. (1968). *Corynebacterium kutscheri* infection in the mouse. II. Diagnostic serology. *Lab. Anim. Care* **18,** 459–468.

490. Weisbroth, S. H., Scher, S., and Boman, I. (1969). *Pasteurella*

pneumotropica abscess syndrome in a mouse colony. *J. Am. Vet. Med. Assoc.* **155**, 1206–1210.

491. Welch, H., Ostrolenk, M., and Bartram, M. T. (1941). Role of rats in the spread of food poisoning bacteria of the Salmonella group. *Am. J. Public Health* **31**, 332–340.

492. Wensinck, F. D., Van Bekkum, D. W., and Renaud, H. (1957). The prevention of *Pseudomonas aeruginosa* infections in irradiated mice and rats. *Radiat. Res.* **7**, 491–499.

493. Wenyon, C. M. (1906). Spirochetosis of mice. N. sp. in the blood. *J. Hyg.* **6**, 580–585.

494. Wheater, D. F. W. (1967). The bacterial flora of an S.P.F. colony of mice, rats, and guinea pigs. *In* ''Husbandry of Laboratory Animals'' (M. L. Conalty, ed.), pp. 343–360. Academic Press, New York.

495. White, D. J., and Waldron, M. M. (1969). Naturally-occurring Tyzzer's disease in the gerbil. *Vet. Rec.* **85**, 111–114.

496. Wilson, G. S., and Miles, A. A., eds. (1975). ''Topley and Wilson's Principles of Bacteriology, Virology, and Immunity,'' 6th ed., Vols. I and II. Williams & Wilkins, Baltimore, Maryland.

497. Wilson, M. G., Nelson, R. C., Phillips, L. H., and Boak, R. A. (1961). New source of *Pseudomonas aeruginosa* in a nursery. *J. Am. Med. Assoc.* **175**, 1146–1148.

498. Wilson, P. (1976). *Pasteurella pneumotropica* as the causal organism of abscesses in the masseter muscle of mice. *Lab. Anim.* **10**, 171–172.

499. Winsser, J. (1960). A study of *Bordetella bronchiseptica*. *Proc. Anim. Care Panel* **10**, 87–101.

500. Wolf, N., Stenback, W., Taylor, P., Graber, C., and Trentin, J. (1965). Antibiotic control of post-irradiation deaths in mice due to *Pseudomonas aeruginosa*. *Transplantation* **3**, 585–589.

501. Wolff, H. L. (1950). On some spontaneous infections observed in mice. I. *C. kutscheri* and *C. pseudotuberculosis*. *Antonie van Leeuwenhoek* **16**, 105–110.

502. Woodward, J. M. (1963). *Pseudomonas aeruginosa* infection and its control in the radiobiological research program at Oak Ridge National Laboratory. *Lab. Anim. Care* **13**, 20–25.

503. Woodward, J. M., Camblin, M. L., and Jobe, M. H. (1969). Influence of bacterial infection on serum enzymes of white rats. *Appl. Microbiol.* **17**, 145–149.

504. Woolley, P. V., Jr. (1936). Rat-bite fever. Report of a case with serologic observations. *J. Pediatr.* **8**, 693–696.

505. Wostman, B. S. (1970). Antimicrobial defense mechanisms in the *Salmonella typhimurium* associated ex-gremfree rat. *Proc. Soc. Exp. Biol. Med.* **134**, 294–299.

506. Wyand, D. S., and Jonas, A. M. (1967). *Pseudomonas aeruginosa* infection in rats following implantation of an indwelling jugular catheter. *Lab. Anim. Care* **17**, 261–267.

507. Yamada, A., Osada, Y., Takayama, S., Akimoto, T., Ogawa, H., Oshima, Y., and Fujiwara, K. (1970). Tyzzer's disease syndrome in laboratory rats treated with adrenocorticotropic hormone. *Jpn. J. Exp. Med.* **39**, 505–518.

508. Yokoiyama, S., and Fujiwara, K. (1971). Effects of antibiotics on Tyzzer's disease. *Jpn. J. Exp. Med.* **41**, 49–58.

509. Young, C., and Hill, A. (1974). Conjunctivitis in a colony of rats. *Lab. Anim.* **8**, 301–304.

510. Zucker, T. F. (1957). Pantothenate deficiency in rats. *Proc. Anim. Care Panel* **7**, 193–202.

511. Zucker, T. F., and Zucker, L. M. (1954). Pantothenic acid deficiency and loss of natural resistance to a bacterial infection in the rat. *Proc. Soc. Exp. Biol. Med.* **85**, 517–521.

512. Zucker, T. F., Zucker, L. M., and Seronde, J. (1956). Antibody formation and natural resistance in nutritional deficiencies. *J. Nutr.* **59**, 299–308.

Chapter 10

Mycoplasmal and Rickettsial Diseases

Gail H. Cassell, J. Russell Lindsey, Henry J. Baker, and Jerry K. Davis

I. INTRODUCTION

Infectious diseases undermine the validity of many experiments that utilize laboratory animals. Clinically inapparent infections are particularly troublesome since they are often completely unsuspected by the investigator. In the laboratory rat, mycoplasmal and rickettsial diseases are prime examples of this type of "silent infection." They not only present diagnostic problems, but also interact with host physiology to induce subtle changes in biologic response adversely influencing studies in many different fields (4,112).

In this chapter we will describe the essential features of these infections, emphasizing mechanisms by which they interfere with research and also summarizing the recent advances in methods for their detection and control.

II. MYCOPLASMAL DISEASES

For many years mycoplasmas have been known to cause respiratory, joint, and genital diseases of great economic importance in farm animals (73,149). However, only in recent years have the mycoplasmas of rodents been recognized and appreciated as having a major impact on biomedical research.

It seems appropriate to describe briefly the biology of mycoplasmas and differentiate them from other microbes discussed elsewhere in this text. Mycoplasmas are the smallest free living microorganisms. They differ from viruses by containing both RNA and DNA, and most can be cultivated on artificial media. In contrast to bacteria, they lack a cell wall and are surrounded only by a cell membrane which is similar to the plasma membrane of mammalian cells (73). Owing to their unique characteristics, mycoplasmas are assigned to a separate class Mollicutes, order Mycoplasmatales, which is subdivided into a family requiring sterols, Mycoplasmataceae, and another family capable of growth without sterols, Acholeplasmataceae. The former family contains two genera *Mycoplasma* and *Ureaplasma* and the latter contains only one genus *Acholeplasma*. Spiroplasmataceae is a new family which contains mostly plant pathogens.

The mycoplasmal flora of the rat is poorly understood. While the use of improved methods for isolating and characterizing these agents recently has led to recognition of many new species in other animals (73,115), no intensive effort has been made to survey the mycoplasmal flora of rats. Nevertheless, it is apparent that several species occur in rats and recognition of further species probably can be anticipated in the future (Table I).

Mycoplasma pulmonis and *Mycoplasma arthritidis* are the established mycoplasmal pathogens of rats. Therefore, the dis-

cussion which follows is limited to these two. Major emphasis is given to *M. pulmonis* since it apparently is the most common. Additional features of murine mycoplasmas have recently been reviewed (14a,113).

A. *Mycoplasma pulmonis*

1. Respiratory Disease

The respiratory disease caused by *M. pulmonis* in rats and mice has been designated by many different terms over the years (Table II). This resulted from several lines of reasoning generally dominated by a single common error—separating as distinct diseases the lesions in the upper respiratory tract (rhinitis and otitis media) from those in the lower tract (laryngitis, tracheitis, and pneumonia). Consequently, different etiologic agents and factors often were postulated for the two groups of lesions.

Mycoplasmas were recognized as a possible cause of respiratory tract disease in rats around 1940 in the pioneering studies of Klieneberger (100) and Nelson (123). Yet only within the last few years have the works of Kohn and Kirk (102), Lindsey *et al.* (112), Whittlestone *et al.* (174), and Jersey *et al.* (83) conclusively shown that *M. pulmonis* alone will reproduce all of the characteristic clinical and pathological features of the natural disease when inoculated into rats reared and maintained so as to exclude all other pathogens. The lesions of both the natural and experimental diseases also now have been shown to be suppressed by administration of tetracycline (2,38, 53,68,112,129,166), and *M. pulmonis* can consistently be demonstrated within the lesions throughout the respiratory tract by cultural, immunofluorescent, and ultrastructural methods (112,129). Based upon experimental studies, it is clear that *M. pulmonis* respiratory disease is slow in onset and of long duration. Consequently, there are various stages of pathological lesions and a lack of uniformity of lesions, in part due to the large number of variables which can affect the development of lower respiratory tract disease. It appears almost certain then that the more than twenty descriptive terms advanced over the years (Table II) designate a single clinicopathologic entity which is due primarily to *M. pulmonis*. Thus, the term murine respiratory mycoplasmosis (MRM) now seems to be preferred.

a. Clinical Manifestations. Except for the terminal stage of the disease when weight loss, roughened hair coat, nasal and ocular discharges, and dyspnea are seen, MRM is most typically a clincally silent infection. Even in the later stages of disease, it is generally correct to say that the disease can be heard better than seen, since snuffling and rales are often present in the absence of other clinical manifestations. Other signs

Table I

Mycoplasmas Isolated from Laboratory Rats

Species	Isolation site	Implicated in disease	Reference(s)
M. pulmonis	Conjunctiva	+	(102, 112, 174)
	Respiratory tract	+	(13, 53, 63, 107, 110)
	Middle ear		
	Genital tract	+	(14, 86, 87)
M. arthritidis	Nasopharynx and	−	(154)
	Middle ear		
	Joint	+	(28, 81, 100)
Gough (58B)[a]	Conjunctiva	−	(75, 179)
	Nasopharynx	−	(75)
Noncultivable[b]	Lung	+	(56)

[a]This strain has been isolated only recently. It appears to be serologically distinct from all other mycoplasmas, but has not yet been speciated (75).

[b]A mycoplasma-like organism, noncultivable on artitifical media, has been demonstrated by electron microscopy in bronchiectatic rat lungs. Although this probably was *M. pulmonis* which for unknown reasons did not grow on the media in use, there remains the possibility that noncultivable strains or species exist.

Table II

Chronological List of Terms Used in the Literature and Probably Concerned with Murine Respiratory
Mycoplasmosis[a]

Year	Term and reference	Year	Term and reference
1916	Bronchopneumonia (163)	1956	Chronic murine pneumonia (52, 80, 126)
1923	Nasal sinusitis (34)	1957	Chronic pneumonia of rats (64, 133)
1924	Suppurative otitis (117)	1957	Enzootic bronchiectasis (65)
1928	Pulmonary suppuration (118)	1957	Virus pneumonia (168)
1939	Labyrinthitis (98, 166)	1957	Snuffling disease (168)
1940	Infectious catarrh (40, 123)	1961	Otitis media (65, 128)
1945	Gray lung pneumonia (1)	1967	Chronic bronchitis (55)
1946	Endemic murine pneumonia (85, 124)	1968	Chronic respiratory disease of BALB/c mice (39)
1955	Middle ear disease (66)	1970	Respiratory disease of rats (107)

[a]Reproduced from *Ann. N.Y. Acad. Sci.* **225,** 409 (1973).

are frequent rubbing of the eyes and, less frequently, head tilt due to labyrinthitis. When held in a vertical position by the tail, rats with labyrinthitis characteristically rotate their bodies rapidly. In contrast, rats with normal inner ears remain rigidly suspended. Occasionally, in infected rats a red-stained zone along the margin of the eyelids and external nares may be seen due to excessive secretion of porphyrin by the lacrimal glands (not serosanguinous exudate!). This is not specific for MRM, since it also occurs in other diseases of the rat.

Even though morbidity due to MRM is usually low, a high incidence can occur in epizootics. Mortality is cumulative over many months; thus the disease has special significance in studies of chronic disease states (e.g., gerontology and cancer research).

b. Gross and Microscopic Lesions. There is little correlation between clinical signs and pathological alterations, especially in the lungs. Lungs which appear grossly normal may show extensive involvement microscopically. In addition, upper respiratory tract disease can exist in the absence of lower respiratory tract disease. One of the most noteworthy features of MRM is the lack of uniformity of lesions observed in rats of the same age, from the same colony, or even from the same cage (83). Therefore, as emphasized later in this chapter (Section C), the pathologic diagnosis of this disease is dependent upon microscopic evaluation of the entire respiratory tract (113).

The main gross lung lesions are atelectasis, consolidation, and bronchiectasis (Fig. 1). In advanced cases, entire lobes can be affected. Consolidated areas are characterized by firm, dark red, slightly depressed areas interspersed with areas of yellow to gray coloration. Bronchiectasis (and bronchiolectasis) is characterized by distended airways containing mucopurulent exudate and often appearing grossly as yellow spots on the lung. Variable amounts of mucopurulent exudate also are found in the nasal passages, trachea, and tympanic cavities.

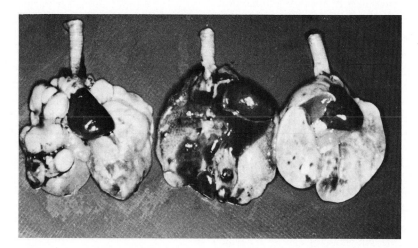

Fig. 1. Gross view of lungs from naturally infected rats 4 months of age showing advanced lesions of *M. pulmonis* infection. From left to right showing multiple abscesses, consolidation, and atelectasis. (Courtesy of Dr. P. B. Carter, Trudeau Institute, Saranac Lake, New York.)

The lesions in all levels of the respiratory tract are characterized microscopically by neutrophilic exudate, epithelial hyperplasia of respiratory mucosa, and submucosal infiltration and/or hyperplasia of lymphoid cells. The principal lesions are acute and chronic rhinitis, otitis media, laryngotracheitis, and bronchopneumonia. Recently, the normal anatomy of rat nasal passages (58,95), tympanic cavity (58), and lung (59) and the alterations seen in naturally and experimentally infected rats have been described (58,102,112,113,174).

Purulent rhinitis and otitis media seem to be the most common microscopic lesions observed in *M. pulmonis*-infected rats and are often seen in the absence of other lesions. Accumulation of mucopurulent exudate along the nasal turbinates is characterized by an increase in subepithelial lymphocytes and goblet cells with a decrease or absence of cilia in many areas. Otitis media usually accompanies rhinitis and is frequently bilateral. Thickening of the lining membrane is a consistent finding. Submucosal lymphocyte accumulation is pronounced in the tympanic cavity, with the eustachian tube and oropharynx being frequently involved. In more severe cases, the cavity may be virtually filled by neutrophils, which are later replaced by collagen with small glandular spaces around the periphery.

Severe lymphocytic infiltration of the submucosa with encroachment upon the epithelial layer and mucopurulent exudate in the lumen are the most common findings in MRM laryngitis and tracheitis. The laryngeal glands also are frequently involved.

The lung lesions of MRM have been described by many investigators (58,102,112,113,179). Normal rat lungs, like those of many mammalian species have small numbers of lymphocytes in the bronchial submucosa, particularly at bifurcations or between bronchi and adjacent blood vessels (Fig. 2). This "bronchial-associated lymphoid tissue" (BALT) is present in the lungs of all rats regardless of age or microbial flora (57). Val *et al.* (165) recently have characterized rat BALT by light and electron microscopy and found it to consist almost entirely of large and small lymphocytes. Using immunofluorescence, we (35a) recently have characterized these lymphocytes in F344 rats as to their major classes. Twenty-six percent of these lymphocytes reacted with rabbit anti-rat thymocyte sera. An additional 25% reacted with anti-rat FAB sera and, thus, are B cells (of these, 9.5% carry IgA on their surface, 7% carry IgM, and 5.4% carry IgG). Half (49%) do not carry either immunoglobulin or T surface markers and are thus "null" cells.

Bronchial-associated lymphoid tissue proliferation is one of the earliest and most pronounced microscopic changes following infection of the lung with *M. pulmonis* (Fig. 3). Approximately twice as many viable lymphocytes can be recovered from the infected lung as from the normal rat lung. The largest increases are in the T lymphocyte and the IgA-bearing B lym-

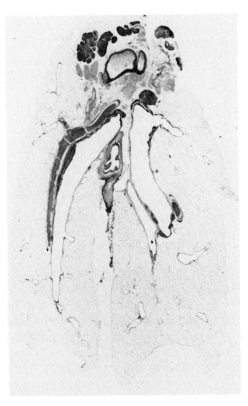

Fig. 2. Horizontal section of lung from uninfected disease-free rat. Three lobes are shown well: Left, right cranial and right caudal. Note the delicate bronchial wall in the left lobe with minimal lymphoid tissue at each bifurcation off the primary bronchus. These aggregates are entirely normal along with a very narrow zone of lymphocytes in the wall of primary bronchi elsewhere but not discernible at this magnification. Hematoxylin and eosin. ×4.

phocyte. In both cases 2.6 times as many cells carrying these markers are present in the infected lung. There also is an increase (1.6 times normal) in the number of "null" cells (35a).

Associated with BALT proliferation in MRM is increased mucin production and accumulation of polymorphonuclear leukocytes in the bronchial lumen. The neutrophilic exudate may increase until the bronchi become greatly distended (Fig. 4). In addition, the epithelium may become squamoid, and glandlike spaces often form around the bronchi. Although most of the involvement is in the bronchi and bronchioles, lung parenchyma also may be affected. Compensatory emphysema has been reported in more chronic cases of MRM (132). The disease pattern may vary remarkably between lobes, or between different areas of the same lobe.

2. Genital Tract Disease

Mycoplasma pulmonis also causes chronic genital disease in rats (13,14,53,63,86,87,107). The frequency of genital infec-

Lesions seen in the uteri range from mild metritis to marked pyometra and endometritis (Fig. 6). Hyperplasia and dilatation of the endometrial glands often gives a "Swiss cheese" pattern. Epithelial changes include hyperplasia, squamous metaplasia, and polyp formation. Organisms can be shown by immunofluorescence to be present on the mucosa and in affected glands (Fig. 7). The incidence of infection in the male genital tract is not known.

The entire spectrum of lesions seen in the natural female genital disease can be reproduced experimentally by intravenous or intravaginal inoculation of pathogen-free rats with viable *M. pulmonis* (14). This results in gross lesions in the genital tract of 20% of female rats, microscopic lesions in 80%, and mycoplasmal colonization of the entire tract in all inoculated animals. Again, the most common lesions were salpingitis with distention of the ovarian bursae (Figs. 8 and 9). Although the ovary remains uninvolved in most cases, focal areas of injury do occur. At 49 days postinoculation, one of the most notable findings was that uteri from animals with salpingitis and peri-oophoritis often appeared histologically normal, although organisms could be shown by immunofluorescence to

Fig. 3. Section of lungs from a rat 28 days following intranasal inoculation of 10^6 CFU *M. pulmonis*. Note the peribronchial lymphoid hyperplasia. Indentations between lymphoid nodules on the lateral walls of the large primary bronchi represent emergence points of second-order bronchi. Hematoxylin and eosin. ×4.

tion has not been surveyed widely, but a recent study of a few colonies has demonstrated the infection to be as high as 30 to 40% of females in a given colony (14,63). As in the respiratory disease, little correlation exists between clinical signs and pathological alterations. Genital tracts that appear normal grossly sometimes show extensive involvement microscopically. Gross pathology is limited to approximately 30% of naturally infected females. In most instances, unilateral purulent oophoritis and salpingitis are the major lesions. The ovarian capsule may be thickened and distended with clear fluid or, occasionally, purulent exudate (Fig. 5). The Fallopian tubes may likewise be distended. Histological examination reveals neutrophilic exudate in the oviduct lumen, hyperplasia of oviductal epithelium, and lymphoid infiltration in the submucosa. The ovarian bursa of such animals is distended by pink fluid and an admixture of inflammatory cells.

Gross observation of the uteri usually reveals no change, except that undeveloped or partially resorbed feti are seen occasionally. Histologically, a large number of the uteri show evidence of mild metritis characterized by a few polymorphonuclear leukocytes in the lumen and in the uterine glands.

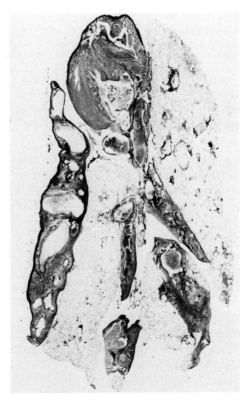

Fig. 4. Section of lungs from a rat 6 months following intranasal inoculation of 10^6 CFU *M. pulmonis*. Note the advanced bronchiectasis involving bronchi and bronchioles. Neutrophils are present in the lumen of some of the airways. The surrounding lung parenchyma is atelectatic with a mixed population of infiltrating cells. Hematoxylin and eosin. ×4.

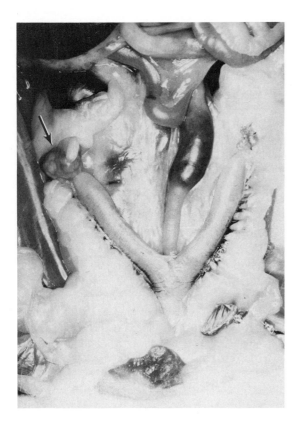

Fig. 5. Gross view of naturally infected rat uterus 5 months of age showing distension of the ovarian bursa (arrow) and enlarged oviducts containing purulent exudate.

is clear, however, that both are chronic progressive entities characterized by a very delicate equilibrium between the mycoplasma and the rat which slowly shifts to favor the mycoplasma. The slow evolution of the respiratory disease is somewhat analogous to that seen in "slow virus" diseases.

There is evidence to suggest that offspring of affected mothers may acquire the organism *in utero* (89,162) and/or as a result of aerosol transmission in the first few weeks of life (101,111). Although the relative importance of these two modes of transmission has not been delineated, it is clear that vertical transmission at an early age is the rule in infected colonies. A slowly progressing respiratory disease is initiated early and appears to persist throughout the animal's life. Such animals probably serve as reservoirs of infection, readily transmitting the organism via airborne transmission to cagemates or rats in the same shipping container (6,83), or to adjacent rats housed in wire mesh cages (6,112). Separation of rats by the use of shoebox-type cages (J. R. Lindsey, unpublished data) or by distance (6) greatly reduces the rate of transmission and retards the development of disease. Horizontal transmission, even between cagemates, generally appears slow, as development of

line the epithelial surface. By 8 months postinoculation, severe disease was still present in ovaries, oviducts, and uteri.

Among males inoculated intravenously, less than 1% of vas deferens cultures are positive. When they are positive, organisms usually can also be demonstrated in the epididymis, sometimes associated with epididymitis. Infrequently, infected males have testicular atrophy (14).

Preliminary studies (14) indicate that 20% of experimentally infected females are infertile. An additional 60% give birth to less than half as many pups as uninfected animals. At autopsy many of those with reduced fertility are pregnant in only one horn of the uterus. These findings correlate well with the greater than 50% reduction in birth rate of naturally infected rats displaying *M. pulmonis* genital infections (14). Similar findings have been obtained in mice experimentally infected with *M. pulmonis* (62).

3. Epizootiology and Pathogenesis of *M. pulmonis* (Respiratory and Genital Diseases)

Both the pathogenesis and epizootiology of MRM and *M. pulmonis* genital tract disease are incompletely understood. It

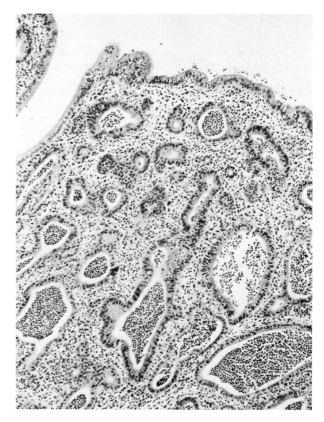

Fig. 6. Uterus of a rat with natural *M. pulmonis* infection, showing marked endometritis. The hyperplastic glands demonstrate the "Swiss cheese" pattern seen in endometrial hyperplasia. Hematoxylin and eosin. ×150.

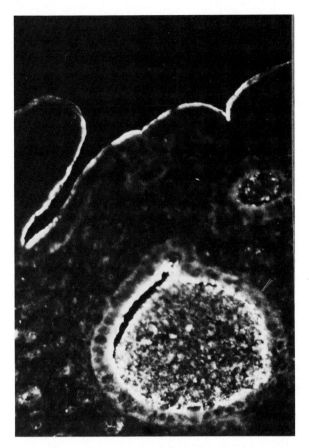

Fig. 7. Endometrium of rat shown in Fig. 5. Immunofluorescence staining of organisms shows heavy concentration lining the endometrial and glandular epithelium. ×600.

severe pulmonary disease in the recipient(s) requires a few weeks or even months. However, once clinical infections begin appearing, the disease may seem to spread throughout the colony rapidly (74).

There is contradictory evidence which suggests that mycoplasmas remain viable for several days in drinking water, food, feces, and on other material (87,167) which might serve as vehicles of infection. Other animal species also may serve as reservoirs of infection. *Mycoplasma pulmonis* has been isolated from wild rats (75), rabbits (36), and guinea pigs (88).

The nasal passages and middle ears are the preferred sites of natural *M. pulmonis* infection in the rat. These are the sites most commonly found infected and diseased due to *M. pulmonis* in conventional rat stocks. Furthermore, these organs are by far the most easily and consistently infected upon intranasal inoculation of broth cultures of the organism. Thus, the upper respiratory tract appears to serve as the major nidus (or reservoir) of infection from which the organism may be transmitted to other rats (parent to offspring and horizontally) and to other organs within the same host, i.e., the lower respiratory tract and/or the genital tract.

Experimentalists over the years have had major difficulties in consistently reproducing the lung disease of MRM by the intranasal inoculation of *M. pulmonis*—undoubtedly one of the greatest enigmas surrounding this disease to date! If infection and disease are so readily established in the upper respiratory tract, why is the subsequent occurrence of pneumonia so erratic and unpredictable? And why are the disease spectra within and among infected colonies so different? The answers involve complex interactions among host, organism, and environment,

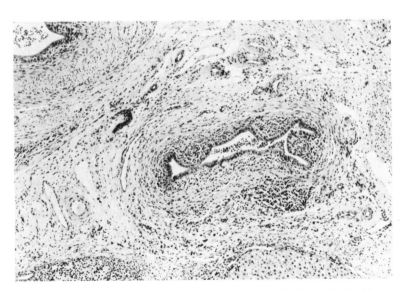

Fig. 8. Oviduct of rat 49 days postintravenous inoculation of 10^7 CFU *M. pulmonis,* showing chronic salpingitis with neutrophils in lumen and lymphocytes in submucosa. Hematoxylin and eosin. ×100.

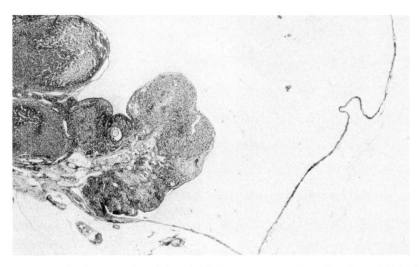

Fig. 9. Ovarian bursa from rat 49 days after intravenous inoculation of 10^7 CFU *M. pulmonis*. Note distension and thickening of the bursa. Hematoxylin and eosin. ×50.

with environmental factors more likely serving as the critical determinants in most infected colonies.

Recently, it has become apparent that intracage environments for housing laboratory rats are often seriously polluted by ammonia (NH_3) due to bacterial action on soiled bedding. Broderson *et al.* (11) have shown that concentrations of NH_3 from 25 to 250 ppm significantly increase severity of the rhinitis, otitis media, tracheitis, and pneumonia (including bronchiectasis) characteristic of MRM. The prevalence of pneumonia but not other respiratory lesions of MRM showed a strong tendency to increase directly with environmental NH_3 concentration. These findings have been confirmed by Lindsey and Conner (113a) in a large series of studies comparing the effects of frequency of cage sanitization (cyclic fluctuations in intracage NH_3) on progression of MRM. The results indicate that even in the face of cyclic fluctuations in environmental NH_3, progression of pulmonary lesions of MRM correlate directly with mean daily NH_3 concentration in the rat's environment. The mechanisms by which NH_3 potentiates MRM are unknown.

Although supporting data are scanty at present, it seems likely that other environmental factors may influence the pathogenesis of MRM. Lane-Petter *et al.* (107) suggested that increased sulfur dioxide and carbon dioxide from a defective ventilation system may have precipitated one outbreak. Systematic studies are needed to assess the potential effects of other factors of the intracage environment on pathogenesis of MRM, including carbon dioxide, temperature, humidity, and types of bedding.

Certainly there are other microbial agents which when present might alter the progression of MRM. *Mycoplasma pneumoniae* has been shown to enhance, to a great extent, *Streptococcus pneumoniae* pneumonia in hamsters (114). *Streptococcus pneumoniae* also has been isolated in conjunction with *M. pulmonis* from animals dying of MRM (59). Many viral agents have been shown to alter the phagocytic potential of rodent alveolar macrophages (145). Thus, without actually causing overt disease, they could predispose the host to *M. pulmonis* infection. Sendai virus, which alters the phagocytic response of mouse macrophages toward *Pasturella pneumotropica* (82) has, in fact, been incriminated in epizootics of MRM in which a high fatality rate was seen (106). Ganaway *et al.* (53) have discussed the importance of other bacteria in perpetuating *M. pulmonis* disease in rats, but, for the most part, these proposed interactions between *M. pulmonis* and other agents in the rat remain poorly defined.

Age is also known to influence the development of MRM; older rats are more susceptible (83,111). Furthermore, the disease progresses much more rapidly in older animals, probably due to the declining immune system. Recent work by Bilder (9) suggests that maximal immune competence in the rat is reached at the onset of puberty (4 weeks), and immune competence declines significantly with age (4 to 96 weeks). This may explain why weanling rats appear to be more resistant to MRM than older animals. In a recent experiment in our laboratory 3-month-old animals were found to be consistently less susceptible to lower lung disease than 5- and 8-month-old animals (35). Eight-month-old females also were more susceptible to mycoplasmal genital disease. In addition, 3-month-old animals were significantly more protected by immunization than were 5- and 8-month-old animals. Susceptibility was not affected by sex.

Although different mouse strains have been shown to vary in their response to *M. pulmonis* (70,71,158), data which suggest differences in rat strain susceptibility are only circumstantial and sketchy (49,50). Guenther *et al.* (67) recently have shown that the immune response genes are linked to the major histocompatibility system in the rat as in the mouse. Therefore, different rat strains probably do differ in their response to *M. pulmonis*.

The female rat can acquire uterine infection early in life, in some cases even prior to coitus. This conceivably could be due to transfer of the organism from the respiratory passages with subsequent ascending spread to the uterus. Juhr (87), however, has shown that there is a septicemic phase in *M. pulmonis* respiratory disease so that organisms may reach the urogenital tract via the hematogenous route.

Little is known about enhancing factors for *M. pulmonis* genital infections. However, testosterone propionate given at 3 days of age has been shown to increase the incidence of salpingitis and oophoritis in infected conventional rats (110).

Inherent characteristics of *M. pulmonis* almost certainly contribute to the pattern of genital and respiratory disease. Successful parasitism for this organism and most other pathogenic mycoplasmas requires intimate contact with host cell surfaces. *Mycoplasma pulmonis* colonizes the respiratory epithelium throughout the respiratory tract. When found in the genital tract, it preferentially localizes on the surface of genital epithelial cells. Following attachment, a very intimate association is formed between the *M. pulmonis* membrane and that of the host cell (Figs. 10 and 11). The wide variety of cellular changes reported in mycoplasmal infection, i.e., epithelial destruction, hyperplasia, and metaplasia, are probably due to mycoplasmal utilization of host cell components and/or release of toxic metabolic wastes such as hydrogen peroxide (10). Like many other mycoplasmas (27), *M. pulmonis* induces ciliostasis in tracheal organ cultures (171a). If this phenomenon occurs *in vivo*, it might help to explain the large accumulation of exudates in major airways.

Mycoplasma pulmonis genital infection may cause *in utero* or, at least, congenital transmission of the organism and infertility. The effect of *M. pulmonis* on fertilization and preimplantation development has been studied *in vitro* (47). Mouse sperm preincubated with *M. pulmonis* were not as efficient in fertilizing ova as noninfected sperm. There also was a highly significant and consistent reduction in embryonic development when infected sperm did manage to fertilize the ova, with relatively few embryos reaching the blastocyst stage. In addition, *M. pulmonis*-induced ciliostasis might influence ova transport.

4. Immune Response

Most of the information concerning the immune response to *M. pulmonis* has been derived from studies using mice. Mice respond vigorously to either natural or experimental infection

(18,111). The characteristics of the response seem to be determined by route of infection. When mice are infected by the intranasal route, the immune response is predominately local in nature. The submucosa of the respiratory tract and the lung parenchyma are rapidly infiltrated by lymphoid cells, many of which differentiate into plasma cells. Local production of antibody of all the major classes has been demonstrated by immunofluorescence both in the infected lung and in hilar lymph nodes (18). T lymphocytes have been demonstrated indirectly in local infiltrates. Pulmonary infiltrates are reduced both in athymic (nude) mice and in those treated with anti-thymocyte. serum. T cells definitely function as helper cells, as antibody production is minimal in their absence (20). Other functions are possible but as yet unproved.

Both athymic mice and T cell depressed mice survive *M. pulmonis* infection in the absence of T cells or significant antibody production (16,20). Lymphoid cells collected from lungs of infected nude mice show neither thymic markers nor immunoglobulins on their surface. These "null" cells may be responsible for the survival of athymic animals.

When the initial infection is established by less than 10^5 colony-forming units (CFU) of organisms, mice usually recover with few signs of clinical disease. Even when larger doses are given, the development of the local response in surviving animals eventually limits the mycoplasmas to the respiratory epithelium and clears the organisms from the alveoli (18). This is probably accomplished by cooperation with either the local alveolar macrophage or recruited circulating phagocytes. The inability to end the infection seems to be linked to the inability of the immune response to attack organisms already established on the respiratory epithelium.

When *M. pulmonis* is injected intravenously into mice, the resulting immune response is systemic. This does not rule out a concomitant local response, however, as shown by passive transfer of immune serum which protects mice from *M. pulmonis*-induced polyarthritis. Immune serum also will protect mice against intranasal challenge while immune cells do not (19).

It would not be appropriate to transfer this information piecemeal from the mouse to the rat, or to any other species. The response definitely differs between the two species. The pulmonary infiltrate in rats is predominantly lymphocytic except in some animals with very active disease where plasma cells occur in large numbers. Rats, unlike mice, develop little or no metabolic inhibiting antibodies (69) and only very low levels of complement fixing (CF) and hemagglutinating (HA) antibodies (102). In addition, the development of the CF response parallels the progress of infection in that antibody titers increase with the development of lung lesions, and titers are consistently higher in rats with more severe lesions.

Rats do appear to mount a local response after infection with *M. pulmonis*. The strongest evidence for this is the massive

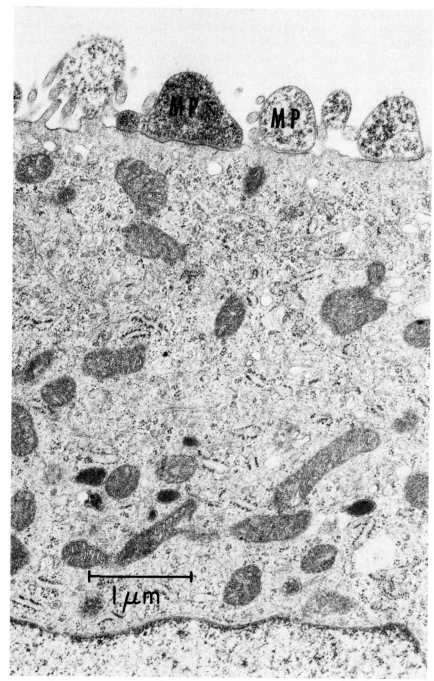

Fig. 10. Mycoplasma pulmonis (MP) attached to endometrium of rat uterus. ×27,500.

proliferation of BALT (not present in normal mice). The role of this tissue in immunity is undefined, but it seems to contain precursor cells destined to produce s-IgA (8a). BALT proliferation is known to be associated with clearing of *Bordetella bronchiseptica* from the respiratory epithelium of the dog (7) and may be involved in the production of local immunity to *M. pulmonis* in the rat. It is not certain, however, if BALT prolif-

eration is directed against *M. pulmonis,* which is known to act as a nonspecific mitogen for rat lymphocytes (60,122). Thus, the increase in BALT may be entirely nonspecific.

When active immunization of rats was attempted using formalin-killed organisms, the results indicated a dichotomy in immune mechanisms operative in the upper versus the lower respiratory tract (15). Intravenous vaccination protected against lung lesions but not upper respiratory tract lesions. Intranasal vaccination provided protection where it was applied. The severity of rhinitis was decreased, but no protective effect was seen at any other level of the tract. These results contrast markedly with those obtained using live *M. pulmonis* given intravenously (iv), which gave marked protection to all levels of the respiratory tract (15). Not only were no lung lesions seen but both the incidence and severity of rhinitis, otitis media, and laryngitis were markedly reduced. This probably is due to establishment of a chronic infection in other tissues with a resulting heightened immune response (15).

The hypothesis that immunity in the rat depends upon cellular rather than humoral mechanisms is supported by passive transfer of immunity using spleen cells from immunized rats.

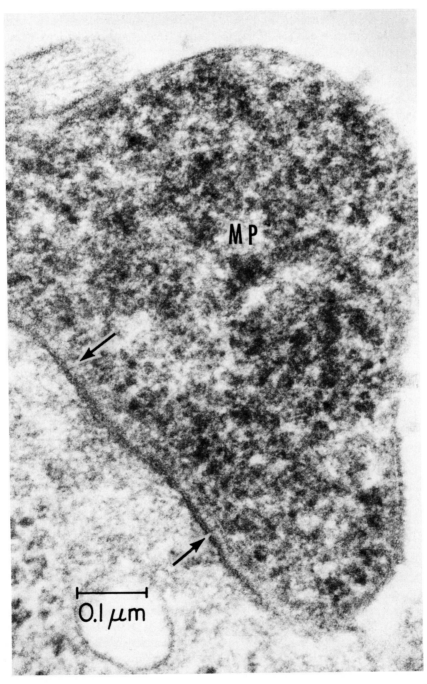

Fig. 11.   Spikelike projections (arrows) extend between membranes of *M. pulmonis* (MP) and surface epithelial cell of rat uterus. ×177,000.

In contrast to the results in mice, preliminary results indicate that cells but not serum are protective in rats.

No data exist concerning the immune response of rats to genital mycoplasmal infection, but the infiltration of lymphoid cells into the genital mucosa of infected females suggests that these rats are mounting an adaptive response. As in the respiratory tract, lymphocytes, not plasma cells, seem to be the predominant cell type in all levels of the genital tract, except the oviduct where the reverse is true (14).

From the above discussion, it is evident that rats do respond immunologically to *M. pulmonis* infection and can in fact be protected against disease by a preexisting immune response elicited by both active and passive immunization. Perhaps the most puzzling facet in this disease is failure of rats to recover from natural infection. Mycoplasmas can be recovered from rats for at least 8–12 months following infection of the lung (174) or genital tract (14). How then do mycoplasmas manage to persist in the presence of an immune response that is at least theoretically capable of destroying them?

A seemingly trite explanation for mycoplasmal persistence is seclusion of mycoplasmas in sites inaccessible to immune effector mechanisms. For example, the tendency of mycoplasmas to reside between the cilia on respiratory epithelial cells might prevent their clearance by the mucociliary escalator. Also their partial envelopment by host cell cytoplasmic processes has been suggested to prevent combination with antibody and/or phagocytosis (152). Even so, these explanations do not seem totally satisfactory.

There are several more potentially effective mechanisms that recently have been proposed (17). The most likely of these with regard to *M. pulmonis* in rats is alteration of lymphocyte responsiveness. Since *M. pulmonis* is a nonspecific mitogen for rat lymphocytes, it obviously could misdirect the immune response by preempting helper cells or could misdirect immune regulatory mechanisms. It is also quite possible that immune mechanisms involved in recovery from infection are different from those involved in immunity.

B. *Mycoplasma arthritidis*

Polyarthritis in laboratory and wild rats is caused most commonly by *Mycoplasma arthritidis* (28,29,140). This organism also can be isolated from tissues other than joint, but its relationship to other diseases is unknown. It has been recovered from subcutaneous abscesses (177), otitis media (138), bronchiectatic lung (154), and paraovarian abscesses (138), in addition to the oropharynx of healthy rats (154).

1. Clinical Findings

Although Ito *et al.* (81) reported natural outbreaks involving many animals showing clinical signs of arthritis, usually the infection is endemic in a colony, with only a few animals showing clinical disease. The joints most commonly affected are the tibiotarsals and the radiocarpals. Less frequently the intertarsal or phalangeal joints are swollen. In the spontaneous disease, three or four limbs may be simultaneously involved (46).

The experimental intravenous inoculation of large numbers of *M. arthritidis* induces visible erythema, swelling, and limited motion of joints in 3 to 5 days (170). These lesions may resolve or progress to massive bony ankylosis of affected structures (151). Much of the long bone deformity appears to be caused by exuberant periosteal new bone formation (134). A varying number of infected rats develop flaccid paralysis of the hindlimbs. This usually occurs in animals with severe arthritis, but has been observed in rats with no peripheral lesions. Although skeletal involvement is the principal manifestation of the infection, conjunctivitis and urethritis also may be present (29). Most clinical manifestations regress by the seventh week.

2. Pathologic Findings

There are several detailed histologic descriptions of rat arthritis due to *M. arthritidis* (28,134,169,170). Under controlled experimental conditions the first microscopic lesions are seen in joints and tendon sheaths 48 to 72 h following the intravenous inoculation of the organism, when polymorphonuclear leukocytes begin accumulating (Fig. 12). Immunofluorescent studies suggest that the organisms are localized in this area for the first week of infection (76). Over a period of several days the joint space becomes distended with exudate, and edema occurs in the adjacent connective tissue and skeletal muscle. Within 10 days, articulating cartilage often shows considerable destruction. Repair usually occurs in the face of continued suppuration. Shortly after the appearance of synovitis, periosteal proliferation and new bone formation appear (Fig. 13). A similar suppurative process involves the intervertebral articulations if they are affected. Perineural inflammation and edema compresses nerve roots within the spinal foramina leading to hindlimb paralysis. Unless the motor nerves are completely destroyed, the entire process is self-limiting as the arthritis regresses. Lesions in other tissues are rare.

3. Epizootiology and Pathogenesis

The natural disease occurs more frequently in young rats (from 8 to 93 days old) with no special relationship between the incidence of the disease and sex of the animal (81). It appears that *M. arthritidis* may remain in a latent state in joints, or may reside normally in the respiratory tract and under certain conditions be introduced into the joint causing severe sporadic or epizootic joint disease (169). Considering the fact that arthritic lesions are mostly seen in the forelimbs and that

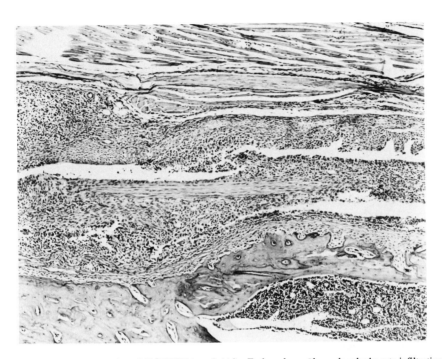

Fig. 12. Microscopic appearance 7 days after injection of 10^8 CFU *M. arthritidis*. Early polymorphonuclear leukocyte infiltration of the tendon and tendon sheath can be seen. Hematoxylin and eosin. ×600.

rats lick and gnaw forelimbs habitually, mouth to foot inoculation has been proposed (81,169). However, attempts to reproduce this experimentally have been unsuccessful (81).

The disease cannot be transmitted readily by direct contact of healthy with infected animals. In view of this and the large number of organisms required to induce experimental arthritis, subsidiary factors may be involved in the initiation of the disease (169). Ward and Cole (169) have observed that mycoplasmal arthritis seems most prevalent when rats are housed in unsuitable cages which injure the feet. They also suggest that a synergistic effect may occur between mycoplasmas and other infectious agents or foreign material. Most early reports of the natural disease involve such a possibility. The incidence and severity of arthritis can be reduced by exposure of the animals to cold (70°F) and ultraviolet light (54).

Mycoplasma arthritidis factors involved in the pathogenesis of arthritis are undefined. However, the organism recently has been shown to produce gelatinases (178). There also seems to be a correlation between infection and inhibition of DNA repair (104). Infected animals also have been shown to produce anti-DNA antibodies (99). In addition, *M. arthritidis* possesses antigens which cross-react with certain rat antigens (12). All of these may contribute to the pathogenesis of mycoplasmal arthritis, but, as Ward and Cole have concluded, the mechanism of pathogenicity of rat arthritis is a complex one and will require much effort to unravel (169).

4. Immune Response

Recovery from infection is followed by resistance to reinfection (22,28), and immunity can be transferred passively to normal rats by convalescent serum (22). The protective properties of the serum, however, cannot be adsorbed with washed *M. arthritidis*. This observation is consistent with the theory that immunity is related to antitoxin, although no conclusive data exist for a soluble toxin produced by *M. arthritidis*.

Rats produce complement-fixing antibody and a defective mycoplasmacidal antibody, i.e., antibody capable of destroying only resting mycoplasmas (26). These defective antibodies probably do not play a role in immunity to *M. arthritidis*, but may contribute to disease pathogenesis by giving rise to infective antigen–antibody complexes. Rats apparently do not produce metabolic inhibiting and hemagglutinating antibodies against *M. arthritidis* (22).

Although it is not possible to transfer immunity passively with cells, the involvement of cellular immunity to *M. arthritidis* has not been explored fully. However, preliminary studies suggest that sensitized lymphocytes do develop in response to *M. arthritidis*, but it is not known whether these are B or T cells (22a). Rat peritoneal macrophages fail to kill *M. arthritidis* even in the presence of convalescent rat serum (25). The production of defective antibodies coupled with this defective macrophage response suggest that *M. arthritidis* infection

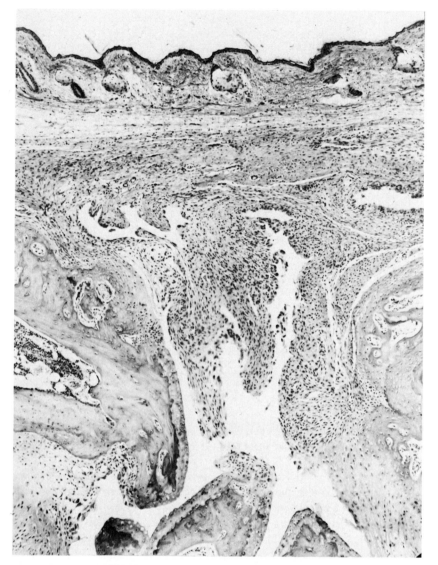

Fig. 13.   Microscopic appearance of joint 30 days after injection of 10^8 CFU *M. arthritidus* showing distension of all synovial spaces and marked proliferation and fibroplasia of the synovial membrane. Hematoxylin and eosin. $\times 180$.

in rats should be chronic. It is, therefore, surprising that the disease is self-limiting.

C.   Detection and Control

1.   Detection

Presently there is a great need for a rapid, simple screening test for diagnosis of mycoplasmal diseases in rodents. Unfortunately, it is currently necessary to use several procedures in conjunction. These include (1) cultural isolation, (2) detection of mycoplasmas in infected tissue by fluorescent or peroxidase-labeled specific antibody, (3) histopathology, and (4) detection of specific antibodies by serology.

a. Culture.   Ganaway *et al.* (53) and Kappel *et al.* (94) have demonstrated the effectiveness of culture as a screening technique for monitoring the mycoplasma status of rat colonies. It is possible to use oropharyngeal swabs and/or nasal washings collected from anesthetized rats. Both *M. pulmonis* (75) and *M. arthritidis* are most frequently isolated from these sites, even in the absence of overt disease. Since the technique does not require that the animal be killed to establish the diagnosis, all of the animals at risk may be evaluated efficiently and economically (53,94). However, it must be remembered that negative results from a few rats in a colony by any technique do not prove that a colony is free of *M. pulmonis* (94).

Careful studies have not been performed to determine the incidence of *M. pulmonis* carriage in the genital tract. We know that as many as 30% of females in a given colony may carry the organism in the uterus. Unfortunately, we do not know how frequently genital infection occurs in the absence of respiratory infection. These data must be obtained before one can use culture of the respiratory tract alone as an indication of the mycoplasma status of a colony. Uterus, oviduct, and vas deferans can be cultured by gentle washings with a maximum of 0.3 ml of sterile broth.

When animals are killed for isolation procedures, tracheo-bronchial lavage (0.6 ml sterile broth) is superior to culture of infected lung or tracheal tissue (159). As already mentioned, in cases in which there are no obvious clinical manifestations, the nasopharynx seems to be the best isolation site.

Culture of synovial membrane, as well as other infected tissues, deserves special consideration. Mycoplasmas may not grow as well from concentrated as from dilute tissue suspensions due to the presence of a nonspecific inhibitor, probably lysolecithin (92). Addition of either ammonium reinechate or lysophospholipase appears to abolish the inhibitory activity (91).

All pathogenic mycoplasmas require a complex medium for growth. Hayflick's medium (72), pH 7.8–8.0 [1.8 gm mycoplasma broth base (BBL) without crystal violet, or 3 gm mycoplasma agar base (BBL) with 0.05% glucose, 10% yeast extract (Microbiological Associates), 20% heat-inactivated horse serum (FLOW Laboratories), 500 units/ml penicillin, 1 : 2000 thallium acetate, and 0.05% phenol red] will support the growth of most strains of *M. pulmonis* and *M. arthritidis*. However, the latter grows best in medium containing arginine in lieu of glucose.

Various lots of horse serum differ in their growth promoting effects, and some batches of other components will not support growth at all, e.g., some penicillins delay growth and some agars are inhibitory. Culture vessels, their washing water, and all medium components should ideally be of tissue culture quality (173).

Each new batch of media, either broth or agar, should be pretested for its ability to support growth (173). Without such tests, negative isolations can be meaningless. Complete broth can be stored at $-20°C$ while pretest cultures are performed. Storage up to 2 months is possible with no decrease in growth-supporting activity. Agar plates can be stored for up to 3 weeks in sealed containers at 4°C.

Both *M. pulmonis* and *M. arthritidis* grow well aerobically at 37°C. Some strains are easier to isolate on agar than in broth; for others, the reverse is true. Blind restreaking of plates and/or rapid passes in broth every 48 h for several passes improves the isolation rate. Cultures should not be reported as negative until at least 3 weeks have lapsed.

If a positive culture is obtained, one must prove that mycoplasmas were isolated and preferably establish which species. The minimal standards of identifying an isolate as a mycoplasma have been proposed by the Subcommittee on the Taxonomy of Mycoplasmatales (156). Isolates are usually identified as a mycoplasma by demonstrating characteristic minute colonies that do not revert to bacterial forms in the absence of penicillin. Additional identification characteristics include demonstration by light and electron microscopy of pleomorphic organisms bounded by a single, triple-layered membrane without remnants of a cell wall. For speciating mycoplasmas, immunofluorescence (37), growth inhibition (21), or complement fixation (51) may be used.

Ogata *et al.* (127) recently have examined the biological, growth inhibition, and gel diffusion characteristics of 115 strains of *M. pulmonis* isolated from 73 rats from five different colonies suffering from MRM. They found that antisera to some strains would not inhibit the growth of heterologous strains. However, they did find that antisera to strain "T" would inhibit the growth of 110 different strains, suggesting that one must be careful in selecting the standard strain used in serological studies of *M. pulmonis*. The number of serotypes of *M. arthriditis* is not known.

It should be emphasized that even using appropriate methods, isolation failures may occur, either because different strains of the same species may have different requirements or because of batch variability in media (156). There has been at least one report of noncultivable mycoplasmalike organisms associated with respiratory disease in rats (56). One should be aware of the problems with cultural isolation and, therefore, be persistent in making a definitive diagnosis.

b. Detection of Organisms in Infected Tissue. Isolation, cultivation, and positive identification of mycoplasmas may take as long as 3 weeks and can be very laborious. Indirect identification of mycoplasmas in infected tissues by immunofluorescence is more rapid and, in many cases, more reliable.

We have found indirect immunofluorescence very useful in diagnosis of MRM and genital mycoplasmosis. Organisms can be detected in tracheobronchial or genital lavages collected in small amounts of phosphate-buffered saline and concentrated by centrifugation or in purulent exudates from lung abscesses and infected middle ears. A drop of the sample is placed on a microscope slide, air dried, fixed in 95% cold ethanol, and then stained and examined by standard immunofluorescent procedures. All parts of the lower respiratory tract, genital tract, joints, and other organs can also be examined by immunofluorescence by special tissue processing (144).

The need for special equipment limits general acceptance of

mycoplasmal disease diagnosis by fluorescent microscopy. This can be circumvented by substituting peroxidase-labeled reagents for fluorescein-labeled reagents (76a,121,136). Tissues or specimens stained with the former can be screened by regular light microscopy.

The Giemsa-stained touch method of Whittlestone (172,174) is useful for detection of organisms in heavily infected tissues by regular light microscopy, but only with considerable practice can one become adept at recognizing mycoplasmas by this method. This technique may be used to make a rapid presumptive diagnosis but it must later be confirmed by culture or fluorescence. This technique also may be used for detection of organisms that are not cultivable, but their identity should be substantiated by electron microscopy.

c. Histopathology.   Because of the difficulties often encountered with the above methods, diagnosis of mycoplasmal diseases must be correlated closely with pathological features while eliminating other likely infectious agents. The insidious nature of most mycoplasmal diseases often means that there can be an active disease process without apparent gross lesions. For this reason, pathologic interpretation requires microscopic evaluation. This especially holds true for the diagnosis of MRM (113) in which all levels of the respiratory and genital tracts must be closely examined.

d. Serology.   The opportunity for indirect detection of mycoplasmal infection in rats by serology is somewhat limited. Because most of the infections are local and limited to mucosal or serosal surfaces, serum antibodies are usually low. In the early stages of disease when the number of organisms is small and, more particularly, when the infections are latent, the organisms are confined to the upper respiratory tract, and diagnosis by conventional serology is not possible.

Due to its sensitivity, the recent development of an enzyme-linked immunosorbent assay (ELISA) for detection of *M. pulmonis* antigen and antibodies may circumvent this problem (77). In addition, the assay will allow tail blood, nasal washings, and oropharyngeal swabs to be used as specimens. The assay can easily be adapted for screening of *M. arthritidis* infections as well. Low cost, simplicity, and rapidity make the test extremely promising.

2.   Control

A few investigators have claimed success (albeit, usually temporary) in eliminating MRM from breeding stocks through programs involving rigid selection (79,125) coupled with administration of antibacterial drugs (58,62), or principles of cesarean derivation combined with strict isolation procedures (93,94). In the past decade, much emphasis was placed on the latter technique with the creation of a large family of terms,

such as specific pathogen-free (SPF) and pathogen-free (PF), to describe the stocks for commercial purposes. In actual practice, maintenance of these stocks for long periods of time, with some exceptions, has resulted in disappointment because of reapperance of MRM. One of the most notable studies in this regard is the isolation of *M. pulmonis* from a ''germfree'' colony (53). Although these disappointments generally have been attributed to a break in strict isolation procedures, recent results obtained in our laboratory (14) and others (53) suggest that the true cause may be more insidious. As already discussed, *M. pulmonis*, the etiologic agent of MRM, also colonizes the female genital tract. This would suggest that hysterectomy should not be relied upon to exclude *M. pulmonis* from a barrier protected area. The fact that this organism may be transferred *in utero* also suggests that cesarean derivation alone may not be the answer.

Our recent studies (15) showing that rats can be protected from MRM by immunization are encouraging and give hope that an effective vaccine might be developed. It must be kept in mind, however, that the precise relationship between the respiratory tract disease and the genital tract disease is not known. An effective vaccine may well be required to protect against both, and immunization studies involving genital mycoplasmosis have not yet been conducted. Studies by Broderson *et al.* (11) and Lindsey and Conner (113a) suggest that environmental control will reduce the severity of MRM. Perhaps a combination of immunization and environmental control will be required for optimal protection against disease.

Mycoplasma arthritidis does not appear to be as ubiquitous in contemporary rat stocks as *M. pulmonis*. Arthritis due to this organism is not nearly as common as respiratory or genital mycoplasmosis. Immunoprophylactic methods to control *M. arthritidis* do not appear to be necessary at this time. However, investigators do need to be aware of its presence and the possible influence that it can have upon their research.

D.   Research Complications Imposed by Mycoplasmas

Respiratory and genital mycoplasmoses seriously affect the usefulness of rats for research purposes. The variation in severity of *M. pulmonis* infection (latency to lethality) in older animals indeed makes it a serious problem. Although *M. arthritidis* apparently occurs less frequently, it too can have devastating effects on research results. For the uninformed investigator, these organisms pose a particular threat due to their subtle interactions with their mammalian host.

Previous reviews by Lindsey *et al.* (112) and others (80,126,171) have documented a large number of specific examples in which MRM complicated studies using rats in respiratory disease, gerontological, nutritional, toxicological, and behavioral research. It is clear that MRM has an enormous total impact on long-term studies of all types using rats. In fact, MRM often has been declared as the number one problem in chronic studies using rodents (171).

Many acute studies using rats are also affected by MRM. For example, Schreiber *et al.* (147) observed that the rate of respiratory tract cancers induced by a carcinogen was markedly influenced by *M. pulmonis* infection. In another example, Kimbrough and Gaines (96) and Kimbrough and Sedlak (97) reported that the insect chemosterilant, hexamethylphosphoramide, when administered orally to rats, caused pneumonia. This was shown later to be natural MRM, greatly enhanced by the chemical (131). Vitamin A- and E-deficient diets have also been shown to enhance MRM (164).

Owing to the subtle effects of *M. pulmonis* on the respiratory tract, infected animals are unsuitable for a wide range of respiratory disease investigations. Atmospheric pollution, tobacco smoke, and industrial diseases (coal dust, silica dust, asbestos dust) are areas of research which are particularly vulnerable. This recently has been emphasized by the noted lung pathologist, Lamb (105). Immunological investigations using the rat respiratory tract also would be particularly compromised especially in view of the marked BALT hyperplasia.

Genital mycoplasmosis also must be considered a threat with regard to its potential effects on studies of reproductive biology in the rat. Since much of our knowledge on reproductive physiology has been obtained from studies in conventional rats, one wonders how much of it has been influenced by the presence of *M. pulmonis*. The fact that *M. pulmonis* infection can reduce breeding efficiency by 50% impinges not only upon research but also upon the production of rodents.

Mycoplasma arthritidis also has been recognized as a complicating factor in many research projects using rats. It has been incriminated as a complicating infection in experimental arthritis induced by various experimental methods (23), particularly in adjuvant arthritis and as a contaminant causing polyarthritis following transplantation of certain rat tumors (176,177). In view of this, the participation of preexisting *M. arthritidis* infection in any form of experimental arthritis must be carefully excluded before proper interpretation of pathogenesis can be made.

Mycoplasma arthritidis is known to induce interferon production (24) and to suppress antibody formation and lymphocyte transformation *in vitro* (5) and to some extent *in vivo* (90). These effects also may be manifested when infected animals are used for experimental purposes.

II. RICKETTSIAL DISEASES

Bartonella, Hemobartonella, Eperythrozoon, and *Grahamella* spp. are rickettsial parasites of the family Bartonellaceae. *Bartonella bacilliformis,* the only species in the genus *Bar-*

tonella, is the cause of human bartonellosis (oroya fever, Carrion's disease, Guaitara fever, or Verruga peruana) (120). Species of the remaining genera are host-specific parasites of a wide range of mammalian species including many laboratory animals. It should be emphasized that *Grahamella* spp. have been found in a variety of wild rodents, and, while this genus has not been studied extensively, it contains potentially important parasites of captive wild rats. *Hemobartonella muris* infects both wild and laboratory *Rattus norvegicus,* and this discussion will be limited to that species, except to point out similarities or differences with *Eperythrozoon coccoides,* the analogous parasite of laboratory mice.

Hemobartonella muris is a small (350 nm) (157), pleomorphic gram-negative organism and occurs as an extracellular parasite of erythrocytes. It is not cultivable on artificial media and has been shown to be quite unstable *in vitro*. It is inactivated rapidly by drying or exposure to disinfectants. Whole blood is rendered noninfectious after incubation for 3 h at 37°C. Whole blood or plasma frozen at −3°C remains infectious for 7–10 days, while freezing at −70°C preserves infectivity for several months (4,161). Additional features of the basic microbiology of the Bartonellaceae will be found in recent reviews (103,120).

A. Clinical Manifestations

Natural infections of *H. muris* are almost always inapparent. In contrast, rats with *H. muris* infection activated by splenectomy develop severe hemolytic anemia (hematocrit 20% or less), profound reticulocytosis (to 30% or more), and, frequently, hemoglobinuria and death (44). Hemoglobinuria may be heralded by dark red discoloration of bedding stained by urine. Intact rats given *H. muris* alone rarely show anything more than transient mild parasitemia and reticulocytosis (to approximately 10% of RBC) (141).

B. Lesions

1. Anatomical Lesions

The reticuloendothelial system is primarily responsible for limiting multiplication of these parasites, and splenomegaly is often the only gross evidence of infection.

Microscopically, dramatic changes take place in the spleens of mice following *E. coccoides* infection (4). By the fourth or fifth day, the splenic corpuscles are transformed into massive sheets of blasts and stem cells containing scattered macrophages in whose cytoplasm cellular debris is commonly seen. These cells multiply rapidly and to such an extent that they may temporarily predominate in both the white and red

Table III

Carbon Clearance in Rats Infected with *H. muris*[a]

	Clearance half-time (min)[c]				
Treatment[b]	59-Day-old Nat	128-Day-old Nat	143-Day-old Nat	152-Day-old Exp	160-Day-old Nat
Normal control	10.6	17.0	24.2	45.0	—
H. muris infected	4.3	16.3	—	12.0	26.7
H. muris, Tr	—	—	23.5	—	—
M. muris, SX	—	8.3	9.8	—	8.9
H. muris, Tr, SX	—	—	27.6	—	—
Control, SX	—	—	—	39.0	—

[a] Adapted from Elko and Cantrell (41).

[b] Tr, treated with oxophenarsine; SX, splenectomy.

[c] Nat, naturally infected; Exp, experimentally infected.

pulp. By the sixth or seventh day, many have differentiated into erythroid elements, and the normal pattern of splenic pulp is being reestablished. Large numbers of plasma cells are seen from the tenth day onward, correlating with marked elevation of immunoglobulin (4). Increased numbers of Kupffer cells may be observed in the liver during the first 2 weeks. Although quantitative data are not available, similar changes are thought to occur in rats infected with *H. muris*.

2. Physiological Lesions

Functional capacity of the reticuloendothelial system, as measured by clearance of intravenously injected carbon particles, may be markedly influenced by *H. muris* infection. Elko and Cantrell (41) studied carbon clearance rates in rats infected with *H. muris* (Table III). They found hyperphagocytosis in naturally infected young rats, experimentally infected old rats, and rats with active parasitemias provoked by splenectomy. Phagocytic activity was normal in uninfected rats and in rats cleared of *H. muris* infection by treatment with oxophenarsine. These studies indicate that during periods of active parasitemia (e.g., shortly after initial infection or during disease activation), there is marked potentiation of phagocytosis, while during the latent infection, phagocytic activity declines to normal levels. Also, hyperphagocytosis does not persist after *H. muris* is eliminated by chemotherapy.

C. Epizootiology and Transmission

1. Natural Transmission

Infections due to *H. muris* are characterized by a peculiar detente between parasite and host whereby infection may per-

sist for long periods (probably the life of the host) without causing overt disease. During this inapparent or latent state, blood and other tissues are highly infectious to susceptible members of the host species, even though organisms are not seen in peripheral blood. This unique feature constitutes one of the foremost reasons for perpetuation of these agents in laboratory animal stocks, and favors their accidental dissemination in biologic materials used in research.

Blood-sucking ectoparasites play an important role as vectors in the natural transmission of these agents (31–33). In a study of transmission of *H. muris* by the rat louse, *Polyplax spinulosa*, Crystal (31–33) found that lice which fed on infected rats were able to transmit the infection by biting susceptible rats. Most well-managed contemporary laboratory rat stocks are free of these ectoparasites, but their presence forecasts probable enzootic infection with *H. muris*. Some workers have reported enzootic infection with *H. muris* in ectoparasite-free rat colonies, and experimental evidence suggests possible transplacental transmission of the infection (30,161). These considerations are important because cesarean derivation is widely used in the production of laboratory rats. In our experience, rats from cesarean-derived stocks maintained according to strict barrier techniques have been uniformly free of *H. muris*, although reliable estimates of the incidence of this infection in contemporary rat stocks are not available.

Transmission of this infection is readily accomplished by all parenteral routes of administration, and some workers have even demonstrated transmission by ingestion and inhalation (161). The likelihood of fortuitous transmission is enhanced markedly by the exceptionally high infectivity and small size of these organisms. A single parasitized erythrocyte is capable of establishing infection.

2. Inadvertent Transmission

Accidental transmission resulting from *in vivo* passage of whole blood, plasma, or other blood-laden biologic material contributes significantly to perpetuation of the problem. Inadvertent transmission with transplantable tumor tissue is of particular concern since many widely used experimental tumor lines have been shown to be contaminated with these agents (143,153). Present evidence indicates that transmission in tumor transplants probably represents nothing more than mechanical transfer of infected blood entrapped in tumor tissue, but there is a suggestion that certain types of tumors may activate and enhance the virulence of *H. muris*. Furthermore, the presence of this organism may influence tumor growth patterns and recipient susceptibility to implanted tumors (3,142).

Sacks *et al.* (142,143) reported finding a filterable agent that caused fulminant hemolytic anemia in tumor-bearing rats. They found this agent in a variety of tumor lines and observed uniform resistance to tumors among rats harboring the agent (Table IV). Moore *et al.* (119) conclusively identified Sacks' "filterable hemolytic anemia agent" as *H. muris*. Such experiences clearly reflect the hazards imposed by these agents on studies utilizing tumor transplantation techniques, and indicate the need for further studies of the mechanisms of *H. muris* infection in altering susceptibility to tumors.

3. Activation of Latent Infections

A variety of procedures that compromise the reticuloendothelial system result in activation of latent *Hemobartonella* infections. Familiarity with these mechanisms is a key step toward circumventing invalidation of research data attributable to the infection, since many of the procedures known to precipitate active disease are widely used in biomedical research.

The bulk of phagocytic activity necessary to suppress these infections resides in the spleen. Consequently, splenectomy is a potent activator of latent *H. muris* infection. In rats infected with this agent, splenectomy leads to a rapid increase in numbers of organisms, resulting in overt parasitemias and hemolytic episodes occurring 3–10 days after surgery. Duration and severity of the active disease appear to depend heavily on the capacity of other phagocytes (principally Kupffer cells) to clear the blood of parasites and once again reduce the load of organisms to covert levels. Thus, splenectomy of older rats infected with *H. muris* results in severe hemolytic disease which is frequently fatal. There is ample evidence that phagocytic activity declines markedly with age in rats, which may partially account for this severe response (41). Furthermore, *H. muris* is bound firmly to the plasmalemma of erythrocytes, causing marked distortion of the cells (Fig. 14), which, in turn, enhances erythrophagocytosis. Rudnick and Hollingsworth (141) found substantial reduction in the life span of erythrocytes infected with *H. muris*, indicating that erythrophagocytosis and erythrolysis may be major components of the active disease in rats.

Drugs that inhibit spleen function also have been shown to activate *Hemobartonella* infections. Pomerat *et al.* (137,160) and Thomas *et al.* (160) demonstrated that active *H. muris* infection occurred in latently infected rats treated with anti-rat spleen serum of rabbit origin. The same antiserum prepared against heterologous spleen material failed to activate latent *H. muris* infection.

Certain alkyl esters of fatty acids are well known reticuloendothelial system depressants. Stuart (155) described profound depression of phagocytosis in mice given a single intravenous injection of ethyl palmitate. He also reported massive splenic necrosis after large doses of this compound. Recently, Finch *et al.* (45) observed activation of latent hemobartonellosis in rats after intravenous injection of ethyl palmitate (3 gm/kg).

Polonium-210, a strong alpha-emitting nuclide that selectively injures spleen, has been shown to promptly and consistently activate latent *H. muris* infection in rats (148). This effect of ^{210}Po closely resembles the response to surgical splenectomy and contrasts sharply with the effect of whole-body irradiation.

Cortisone is used occasionally as a diagnostic aid in unmasking other clinically silent infections of laboratory rats and mice

Table IV

Altered Susceptibility to Transplantable Tumors due to Latent *H. muris* Infection[a]

Treatment	Percent of rats rejecting transplanted tumor				
	Murphy-Sturm	Walker 256	Fibrosarcoma	Miller hepatoma	Flexner-Jobling
H. muris	86.7	44.4	60.0	88.9	68.8
Uninfected	12.5	0	30.0	0	12.5

[a]Adapted from Sacks *et al.* (142).

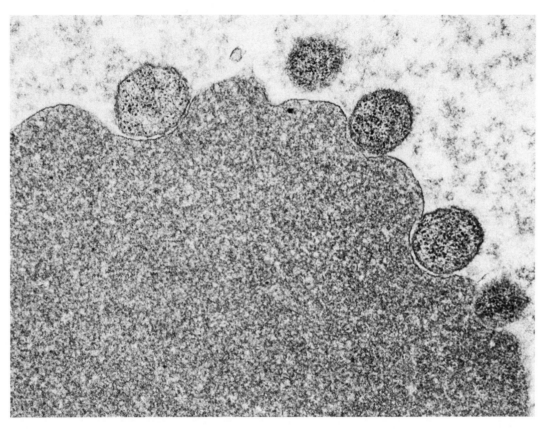

Fig. 14. Transmission electron micrograph of rat erythrocyte infected with *H. muris*, showing concavity of erythrocyte plasmalemma and discrete separation between organism and red cell. ×33,400.

(43,48). Therefore, it is important to recognize that intensive cortisone treatment (5 mg/day for 24 days) is ineffective in activating latent *H. muris* infection in rats and may actually inhibit the activating effect of splenectomy (108,146). Other immunosuppressive drugs, such as anti-lymphocyte serum and cyclophosphamide, also appear to be ineffective in activating *H. muris* infection.

Whole-body γ-irradiation of 700–750 R has been shown to activate latent hemobartonellosis in 150 to 200-gm rats (8,139,148). In contrast to splenectomy, whole-body irradiation does not uniformly activate latent *H. muris* infection in rats. When the infection is activated, the resulting parasitemia and anemia usually are not severe. Furthermore, rats infected with *H. muris* have lower mortality and survive longer than noninfected rats when exposed to high doses of γ radiation. The explanation for this unexpected response is not known.

The following examples of disease activation caused by experimental procedures only indirectly related to reticuloendothelial suppression should serve to emphasize the need for constant vigilance against complications due to these infections. Wills and Mehta (175) attempted to produce pernicious anemia in pregnant rats by feeding a diet deficient in vitamins A and C. They successfully induced severe anemias in these rats but found the cause to be active hemobartonellosis. They subsequently observed activation of latent *H. muris* infection in male and nonpregnant female rats fed the same diet, although in these animals the disease occurred less frequently and was less severe than in pregnant rats. Surgical splenectomy of normal rats used in these experiments revealed a high incidence of latent hemobartonellosis, but the authors failed to offer a satisfactory explanation for the potentiating effect of their experimental diet on *H. muris* infection.

The role of transplantable tumors in the accidental transmission of these agents has been mentioned previously. It has been observed also that some rats and mice receiving tumor transplants or affected with spontaneous tumors experience patent *Hemobartonella* parasitemia with hemolytic anemia (142,143). Sacks and Egdahl (142,143) reported that Sprague-Dawley rats receiving transplants of Walker 256 carcinosarcoma or Miller hepatoma exhibited fulminant hemolytic anemia which frequently terminated fatally. Since these animals received nothing other than transplants of tumor material, it follows that in some instances neoplasia can play a primary role in activation of these infections. At present, the mechanisms responsible are unknown.

D. Diagnosis and Control

There is ample evidence that measures, such as ectoparasite control and modern techniques of obtaining and rearing laboratory rats, have reduced the incidence of these infections to a level far below the near universal infection of laboratory rats prior to the 1930s. However, the continuum of reports and the unpublished experience of many investigators argue strongly that the problem is far from erradicated. In view of current uncertainty regarding the incidence of these infections, effective control measures must be based on systematic testing of laboratory rat stocks. Much of the aversion to routine testing for these infections is attributed to the lack of reliable tests and the inherent difficulty of the few methods available.

In the past, disease activation with subsequent detection of organisms in peripheral blood provided the most reliable procedure for detecting latent infections. Surgical splenectomy is the most consistent and potent activator of these infections, but, as mentioned previously, there is considerable variation in host and parasite response to this stimulus. As one example, there is lack of consistent disease activation in young rats latently infected with *H. muris*. The cause of this apparent resistance is not clear, but it probably depends on effectiveness of extrasplenic phagocytes, such as Kupffer cells, in controlling *H. muris* replication. Technical difficulties surrounding surgical splenectomy can be circumvented by use of ethyl palmitate to induce splenic necrosis, but additional studies are needed to confirm the reliability of this procedure (84). *H. muris* parasitemias occur 5–10 days after splenectomy of infected rats and do not reach high levels. Following splenectomy, infected rats usually undergo a severe hemolytic episode which is often fatal.

Positive identification of organisms in peripheral blood smears is made extremely difficult by their small size and resemblance to erythrocyte inclusions. For example, basophilic stippled erythrocytes, which occasionally reach high concentrations in the peripheral blood of normal young rodents, are easily confused with these organisms (Fig. 15).

In Romanowsky-stained blood smears, *H. muris* appear as solid coccoid elements arranged in clusters, chains, or singly on erythrocytes, are rarely seen free in plasma (Fig. 16), and stain blue to polychromatophilic. Acridine orange staining (Fig. 17) has been recommended to increase the sensitivity of parasite identification, especially when their concentration is low (150). When examined by dark field fluorescence microscopy,

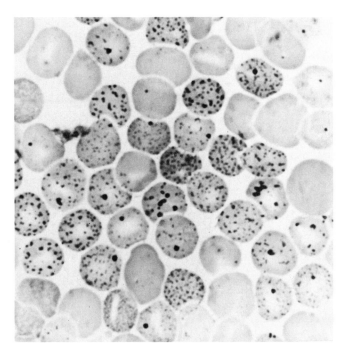

Fig. 15. Punctate basophilic stippling of erythrocytes from peripheral blood of a young, normal gerbil. Giemsa stain. ×1500.

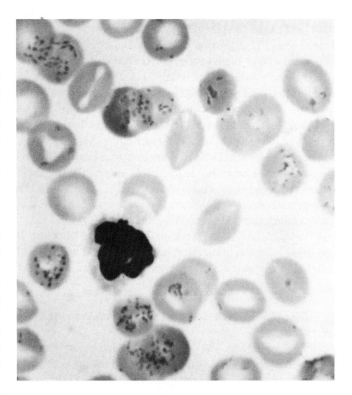

Fig. 16. Rat blood infected with *H. muris,* showing solid coccoid organisms limited to surface of intact erythrocytes or lysed red cell fragments. Giemsa stain. ×2000.

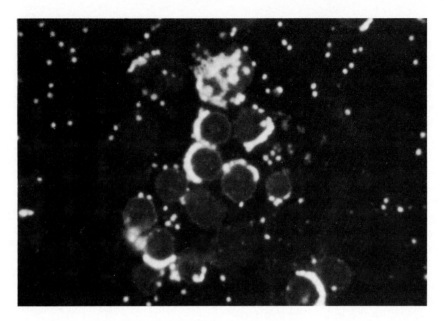

Fig. 17. Acridine orange staining of *H. muris*-infected rat erythrocytes. ×1500.

stained organisms fluoresce yellow-green to red-orange, depending on the procedure of acridine orange staining used. Fluorescing structures such as Howell-Jolly bodies, reticulocytes, and platelet fragments may be confused with organisms. Scanning and transmission electron microscopy studies have shown that these organisms are spherical and intimately attached to the erythrocyte plasmalemma (Fig. 14).

Uninfected, splenectomized animals provide the most sensitive test subjects for detecting these agents in various biological materials. It is important to appreciate that the incubation period and intensity of parasitemias are highly dose dependent, and occasionally serial animal passage is necessary to raise the concentration of organisms to levels that are readily detectable. Intact (with spleen) rats virtually never develop patent parasitemias, even though enormous numbers of *H. muris* organisms are given.

A useful indirect fluorescent antibody test for detecting *E. coccoides* infection of mice has been reported (4) and is adaptable to the diagnosis of hemobartonellosis in rats.

E.   Research Complications due to *Hemobartonella muris*

Clinically silent *H. muris* infections of laboratory rats rank among the most subtle complications of modern biomedical research. As already mentioned, this rickettsia causes profound alterations in the response of infected hosts to a great variety of experimental procedures, especially those utilized in studies of reticuloendothelial function, radiation injury, antibody production, experimental viral and protozoal infections, interferon induction, and tumor transplantation. Aberrations in experimental data after activation of latent infections or accidental transmission of these agents often provide the only clue to their presence. Perhaps the most subtle complications due to *H. muris* result by the interactions of these organisms with other infectious agents. The scientific literature is replete with reports describing interactions between *Hemobartonella* and a wide variety of other microorganisms. Unfortunately, most of these were discovered unwittingly by investigators who found the deviant behavior of some experimental infection to be caused by concurrent hemobartonellosis. It was perhaps prophetic that Mayer's original discovery (116) of *H. muris* resulted from the activation of latent hemobartonellosis in rats he intended to use in studying pathogenesis of experimental *Trypanosoma brucei* infection. The interrelationships between *Hemobartonella* and other microorganisms are complex and difficult to reduce to common denominators. However, it is clear that the profound effect of this parasite on reticuloendothelial function exerts a strong influence on host response to other pathogens.

Experimental plasmodial infections of laboratory rats and mice rank as one of the foremost animal models in malaria research. Early in the development of this model, Hsu *et al.* (78) and others (130,135,161) discovered that latent *H. muris* infections were potentiated by concurrent *Plasmodium* infection, and, furthermore, the course of malaria in such animals was altered markedly by active hemobartonellosis. Despite these early warnings, reports have continued to appear describing the complicating effect of intercurrent *H. muris* infections on experiments employing the rat malaria model. Of even greater concern is the likelihood that some workers have completely overlooked the influence of these agents on malaria studies in rats.

At the present time, it appears that some stock inocula of *P. berghei, P. chabaudi,* and other rodent malaria species are contaminated with *H. muris,* and, to further magnify the dilemma, there is little firm evidence to substantiate the contamination status of most of these inocula. In addition, when an inoculum known to be free of contamination is passed in rats whose carrier status with reference to *H. muris* is not clearly defined, the spector of possible contamination is raised once again.

Hsu and Geiman (78) observed that *P. berghei* induced the appearance of *H. muris* in latently infected rats. In younger rats (100 gm), *Hemobartonella* appeared 3–7 days after inoculation of *P. berghei,* while the appearance of *H. muris* was frequently delayed up to 21 days postinoculation in older rats (100–150 gm). Occasionally, patent *H. muris* parasitemias were not detectable in *P. berghei*-infected rats weighing up to 300 gm. The intensity and duration of *H. muris* parasitemia varied greatly in all age groups and, in most animals, organisms disappeared before the development of severe anemia. Plasmodial parasitemias were erratic, and peak concentration of *P. berghei* appeared to be related to increased release of young erythrocytes resulting from the anemia due to *Hemobartonella.* More than 50% of rats with the dual infection died, usually due to severe anemia. In contrast to this pattern of disease progression, rats infected with *P. berghei* alone showed (a) slowly and regularly progressing plasmodial parasitemias, (b) lower peak parasitemia, (c) few immature erythrocytes, (d) shorter incubation period, (e) lower death rates (14%), and (f) death occurring before the development of severe anemia.

Glasgow *et al.* (61) reported studies of the effect of *E. coccoides* on interferon response in mice. Mice infected with *E. coccoides* and control mice were inoculated with Newcastle disease virus, poly(I:C), and Chicungunya virus. Serum samples were collected and assayed for interferon 2–4 hr after inoculation. Their data showed a 90–99% reduction in interferon response to all of the inducing agents tested in mice infected with *E. coccoides.* These observations have important implications for interferon research, and while similar data are not yet available to define the relationship of *H. muris* infection in rats to interferon response, analogous findings should be expected.

ACKNOWLEDGMENTS

Supported by United States Public Health Grants RR00959, AI/HL 19741, HD-11447, RR00463 and research funds of the Veterans Administration.

G.H.C. is the recipient of United States Public Health Service Research Career Development Award 1K04 HL 00387.

REFERENCES

1. Andrewes, C. H., and Glover, R. W. (1945). Grey lung virus: An agent pathogenic for mice and other rodents. *Br. J. Exp. Path.* **26,** 379–387.
2. Andrewes, C. H., and Niven, J. S. F. (1950). A virus from cotton rats: Its relation to gray lung virus. *Br. J. Exp. Path.* **31,** 773–778.
3. Arison, R. N., Cassaro, J. A., and Shonk, C. E. (1963). Factors which affect plasma lactic dehydrogenase in tumor bearing mice. *Proc. Soc. Exp. Biol. Med.* **113,** 497–50.
4. Baker, H. J., Cassell, G. H., and Lindsey, J. R. (1971). Research complications due to *Haemobartonella* and *Eperythrozoon* infections in experimental animals. *Am. J. Pathol.* **64,** 625–656.
5. Barile, M. F., and Levinthal, B. G. (1968). Possible mechanism for mycoplasma inhibition of lymphocyte transformation induced by phytohaemagglutinin. *Nature (London)* **219,** 751–752.
6. Bell, D., and Wheeler, S. M. (1970). Susceptibility of caesarian-derived rats to natural infection with chronic respiratory disease. *Lab. Anim.* **4,** 45–53.
7. Bemis, D. A., Greisen, H. A., and Appel, M. J. G. (1977). Pathogenesis of canine bordetellosis. *J. Infect. Dis.* **135,** 753–762.
8. Berger, H., and Linkenheimer, W. H. (1962). Activation of *Bartonella muris* infection in x-irradiated rats. *Proc. Soc. Exp. Biol. Med.* **109,** 271–273.
8a. Bienenstock, J., Johnston, N., and Perey, D. Y. E. (1973). Characterization of bronchial associated lymphoid tissue in rabbits. *Lab. Invest.* **28,** 693–698.
9. Bilder, G. E. (1975). Studies on immune competence in the rat: Changes with age, sex, and strain. *J. Gerontol.* **30,** 641–646.
10. Brennan, P. C., and Feinstein, R. N. (1969). Relationship of hydrogen peroxide production by *Mycoplasma pulmonis* to virulence for catalase-deficient mice. *J. Bacteriol.* **98,** 1036–1040.
11. Broderson, J. R., Lindsey, J. R., and Crawford, J. E. (1976). The role of environmental ammonia in respiratory mycoplasmosis of rats. *Am. J. Pathol.* **85,** 115–130.
12. Cahill, J. F., Cole, B. C., Wiley, B. B., and Ward, J. R. (1971). Role of biologic mimicry in the pathogenisis of rat arthritis induced by *Mycoplasma arthritidis. Infect. Immun.* **3,** 24–35.
13. Casillo, S., and Blackmore, D. K. (1972). Uterine infections caused by bacteria and mycoplasma in mice and rats. *J. Comp. Pathol.* **82,** 477–482.
14. Cassell, G. H., Carter, P. B., and Silvers, S. H. (1976). Genital disease in rats due to *Mycoplasma pulmonis:* Development of an experimental model. *Proc. Soc. Gen. Micro.* **3,** 150.
14a. Cassell, G. H., and Hill, A. (1979). Murine and other small animal mycoplasmas. *In* "The Mycoplasmas" (M. F. Barile, S. Razin, J. G. Tully, and R. F. Whitcomb eds.), Vol. 2, Chapter 8, pp. 235–273. Academic Press, New York.
15. Cassell, G. H., and Davis, J. K. (1978). Active immunization of rats against *Mycoplasma pulmonis* respiratory disease. *Infect. Immun.* **21,** 69–75.
16. Cassell, G. H., Davis, J. K., and McGhee, J. R. (1978). Pathogenesis and immunogenesis of *Mycoplasma pulmonis* respiratory disease in nude mice. In preparation.
17. Cassell, G. H., Davis, J. K., Wilborn, W. M., and Wise, K. S. (1978). Pathobiology of mycoplasmas. *In* "Microbiology 1978" (D. Schlessinger, ed.), pp. 339–403. Am. Soc. Microbiol., Washington, D.C.

18. Cassell, G. H., Lindsey, J. R., and Baker, H. J. (1974). Immune response of pathogen-free mice inoculated intranasally with *Mycoplasma pulmonis*. *J. Immunol.* **112**, 124–136.

19. Cassell, G. H., Lindsey, J. R., Overcash, R. G., and Baker, H. J. (1973). Murine *Mycoplasma* respiratory disease. *Ann. N.Y. Acad. Sci.* **225**, 395–412.

20. Cassell, G. H., and McGhee, J. R. (1975). Pathologic and immunologic response of nude mice following intranasal inoculation of *Mycoplasma pulmonis*. *J. Reticuloendothel. Soc.* **18**, 342.

21. Clyde, W. A. (1964). *Mycoplasma* species identification based upon growth inhibition by specific antisera. *J. Immunol.* **92**, 958–965.

22. Cole, B. C., Cahill, J. F., Wiley, B. B., and Ward, J. R. (1969). Immunological responses of the rat to *Mycoplasma arthritidis*. *J. Bacteriol.* **98**, 930–937.

22a. Cole, B. C., Golightly-Rowland, L., and Ward, J. R. (1975). Chronic proliferative arthritis of mice induced by *Mycoplasma arthritidis:* Demonstration of a cell-mediated immunity to *Mycoplasma* antigens *in vitro. Infect. Immun.* **11**, 1159–1161.

23. Cole, B. C., Miller, M. L., and Ward, J. R. (1969). The role of mycoplasma in rat arthritis induced by 6-sulfanilamidoindazole. *Proc. Soc. Exp. Biol. Med.* **130**, 994–1000.

24. Cole, B. C., Overall, J. C., Lombardi, P. S., and Glasgow, L. A. (1975). *Mycoplasma*-mediated hyporeactivity to various interferon inducers. *Infect. Immun.* **12**, 1349–1354.

25. Cole, B. C., and Ward, J. R. (1973). Interaction of *Mycoplasma arthritidis* and other mycoplasmas with murine peritoneal macrophages. *Infect. Immun.* **7**, 691–699.

26. Cole, B. C., and Ward, J. R. (1973). Detection and characterization of defective mycoplasmacidal antibody produced by rodents against *Mycoplasma arthritidis. Infect. Immun.* **8**, 199–207.

27. Collier, A., and Baseman, J. B. (1973). Organ culture techniques with mycoplasmas. *Ann. N.Y. Acad. Sci.* **225**, 277–289.

28. Collier, W. A. (1939). Infectious polyarthritis of rats. *J. Pathol. Bacteriol.* **48**, 579–589.

29. Cotchin, E., and Roe, F. J. (1967). "Pathology of Laboratory Rats and Mice," pp. 373–383. Davis, Philadelphia, Pennsylvania.

30. Crosby, W. H., and Benjamin, N. R. (1961). Frozen spleen reimplanted and challenged with *Bartonella. Am. J. Pathol.* **39**, 119–127.

31. Crystal, M. M. (1958). The mechanism of transmission of *Hemobartonella muris* (Mayer) of rats by the spined rat louse, *Polyplax spinulosa* (Burmeister). *J. Parasitol.* **44**, 603–606.

32. Crystal, M. M. (1959). Extrinsic incubation period of *Hemobartonella muris* in the spined rat louse, *Polyplax spinulosa. J. Bacteriol.* **77**, 511.

33. Crystal, M. M. (1959). The infective index of the spined rat louse, *Polyplax spinulosa* (Burmeister), in the transmission of *Hemobartonella muris* (Mayer) of rats. *J. Econ. Entomol.* **52**, 543–544.

34. Daniels, A. L., Armstrong, M. E., and Hutton, M. K. (1923). Nasal sinusitis produced by diets deficient in fat-soluble A vitamin. *J. Am. Med. Assoc.* **81**, 828–829.

35. Davis, J. K., Carter, P. B., and Cassell, G. H. (1978). The influence of age on the susceptibility of rats to *Mycoplasma* respiratory disease. In preparation.

35a. Davis, J. K., Maddox, P., Thorp, R., and Cassell, G. H. (1979). Immunofluorescent characterization of lymphocytes in lungs of rats infected with *Mycoplasma pulmonis. Infect. Immun.* In press.

36. Deeb, B. J., and Kenny, G. E. (1967). Characterization of *Mycoplasma pulmonis* variants isolated from rabbits. I. Identification and properties of isolates. *J. Bacteriol.* **93**, 1416–1424.

37. Del Guidice, R. A., Robillard, N. F., and Carski, T. R. (1967). Immunofluoroescence identification of *Mycoplasma* on agar by use of indirect incident illumination. *J. Bacteriol.* **93**, 1205–1209.

38. Dolowy, W. C., Hess, A. L., and McDonald, G. O. (1960). Oxytetracycline hydrochloride in the treatment of 590 rats in an outbreak of infectious catarrh. *Ill. Vet.* **3**, 20–24

39. Ebbesen, P. (1968). Chronic respiratory disease in Balb/c mice. I. Pathology and relation to other murine lung infections. II. Characteristics of the disease. *Am. J. Pathol.* **53**, 219–243.

40. Edward, D. G. (1947). Catarrh of the upper respiratory tract in mice and its association with pleuropneumonia-like organisms. *J. Pathol. Bacteriol.* **59**, 209–221.

41. Elko, E. E., and Cantrell, W. (1968). Phagocytosis and anemia in rats infected with *Hemobartonella muris. J. Infect. Dis.* **118**, 324–332.

42. Farris, E. J. (1950). The rat as an experimental animal-with practical notes on the breeding and rearing of the albino rat and the gray Norway rat. *In* "The Care and Breeding of Laboratory Animals" (Edmond J. Farris, ed.), pp. 43–78. Wiley, New York.

43. Fauve, R. M., Pierce-Chase, C. H., and Dubos, R. (1964). *Corynebacterium pseudotuberculosis* in mice. *J. Exp. Med.* **120**, 283–304.

44. Finch, S. C., and Jonas, A. M. (1973). Ethyl palmitate-induced bartonellosis as an index of functional splenic ablation. *RES, J. Recticuloendothel. Soc.* **13**, 20–26.

45. Finch, S. C., Kawaskaki, S., Prosnitz, L., Clemett, A. R., Jonas, A. M., and Perillie, P. E. (1970). Ethyl palmitate induced medical splenectomy in rats. *Proc. Int. Cong. Soc. Hematol., 12th, 1968* pp. 241–252.

46. Findlay, G. M., MacKenzie, R. D., MacCallum, F. O., and Klieneberger, E. (1939). The etiology of polyarthritis in the rat. *Lancet* **2**, 7–10.

47. Fraser, L. R., and Taylor-Robinson, D. (1977). The effect of *Mycoplasma pulmonis* on fertilization and preimplantation development *in vitro* of mouse eggs. *Fertil. Steril.* **28**, 488–498.

48. Frenkel, J. K., Good, J. T., and Schultz, J. A. (1966). Latent pneumocystis infection of rats, relapse, and chemotherapy. *Lab. Invest.* **15**, 1559–1577.

49. Freudenberger, C. B. (1932). A comparison of the Wistar albino and the Long-Evans hybrid strains of the Norway rat. *Am. J. Anat.* **50**, 293–349.

50. Freudenberger, C. B. (1932). The incidence of middle ear infection in the Wistar albino and the Long-Evans hybrid strain of the Norway rat. *Anat. Rec.* **54**, 179–184.

51. Frey, M. L., Thomas, G. B., and Hale, P. A. (1973). Recovery and identification of mycoplasmas from animals. *Ann. N.Y. Acad. Sci.* **225**, 334–346.

52. Ganaway, J. R., and Allen, A. M. (1969). Chronic murine pneumonia of laboratory rats: Production and description of pulmonary-disease-free rats. *Lab. Anim. Care* **19**, 71–79.

53. Ganaway, J. R., Allen, A. M., Moore, T. D., and Bohner, H. J. (1973). Natural infection of germ-free rats with *Mycoplasma pulmonis. J. Infect. Dis.* **127**, 529–537.

54. Gardner, D. L. (1960). The experimental production of arthritis. A review. *Ann. Rheum. Dis.* **19**, 297–317.

55. Gay, F. W. (1967). Fine structure and location of the mycoplasma-like gray lung and rat pneumonia agents in infected mouse lung. *J. Bacteriol.* **94**, 2048–2061.

56. Gay, F. W., Maguire, M. E., and Baskerville, A. (1972). Etiology of chronic pneumonia in rats and a study of the experimental disease in mice. *Infect. Immun.* **6**, 83–91.

57. Giddens, W. E., and Whitehair, C. K. (1969). The peribronchial lymphocytic tissue in germ-free, defined-flora, conventional and chronic murine pneumonia-affected rats. *In* "Germ-Free Biology" (E. A. Mirand and N. Black, eds.), pp. 75–84. Plenum, New York.

58. Giddens, W. E., Whitehair, C. K., and Carter, G. R. (1971). Morphologic and microbiologic features of nasal cavity and middle ear

in germ-free, defined-flora, conventional and chronic respiratory disease-affected rats. *Am. J. Vet. Res.* **32,** 99–114.

59. Giddens, W. E., Whitehair, C. K., and Carter, G. R. (1971). Morphologic and microbiologic features of trachea and lungs in germfree, defined-flora, conventional and chronic respiratory disease-affected rats. *Am. J. Vet. Res.* **32,** 115–129.

60. Ginsberg, H. S., and Nicolet, J. (1973). Extensive transformation of lymphocytes by a *Mycoplasma* organism. *Nature (London), New Biol.* **246,** 143–146.

61. Glasgow, L. A., Odugbemi, T., Dwyer, P., and Ritterxon, A. L. (1971). *Eperythrozoon coccoides.* I. Effect on the interferon response in mice. *Infect. Immun.* **4,** 425–430.

62. Goeth, H., and Appel, K. R. (1974). Experimentelle Unter suchungen uber die Beeintrachtigung der Fertilitat durch Mycoplasmeninfektionen des Genital traktes. *Zentralbl. Bakteriol., Parasitenkd., Infektiansler. Hyg., Abt. 1: Orig., Reihe A* **228,** 282–289.

63. Graham, W. R. (1963). Recovery of a pleuropneumonia-like organism (P.P.L.O.) from the genitalia of the female rat. *Lab. Anim. Care* **13,** 719–724.

64. Gray, J. E. (1963). Naturally occurring and sulfonamide-induced lesions in rats during a one year toxicity study. *Am. J. Vet. Res.* **24,** 1044–1059.

65. Greselin, E. (1961). Detection of otitis media in the rat. *Can. J. Comp. Med.* **25,** 274–276.

66. Grice, H. C., Gregory, E. R. W., and Connell, M. R. E. (1955). Diagnosis of middle and inner ear disease in rats. *J. Am. Vet. Med. Assoc.* **127,** 430–432.

67. Guenther, E., Rude, E., Meyer-Delius, M., and Stark, O. (1973). Immune response genes linked to the major histocompatibility system in the rat. *Transplant. Proc.* **5,** 1467–1469.

68. Habermann, R. T., Williams, F. P., McPherson, C. W., and Every, R. R. (1963). The effect of orally administered sulfamerazine and chlortetracycline on chronic respiratory disease in rats. *Lab. Anim. Care* **13,** 28–40.

69. Haller, G. J., Boiarski, K. W., and Sommerson, N. L. (1973). Comparative serology of *Mycoplasma pulmonis. J. Infect. Dis.* **S127,** 38–45.

70. Hannan, P. C. T. (1971). Observations on the arthritogenic properties on Sabin's type C murine mycoplasma (*Mycoplasma histotropicum*). *J. Gen. Microbiol.* **67,** 363–365.

71. Hannan, P. C. T. (1976). Variations in the arthritogenicity of *Mycoplasma pulmonis* in different strains of mouse. *Proc. Soc. Gen. Microbiol.* **3,** 147.

72. Hayflick, L. (1965). Tissue cultures and mycoplasmas. *Tex. Rep. Biol. Med.* **23,** 285–303.

73. Hayflick, L. (1970). "The Mycoplasmatales and the L-Phase of Bacteria." Appleton, New York.

74. Hill, A. (1972). Transmission of *Mycoplasma pulmonis* between rats. *Lab. Anim.* **6,** 331–336.

75. Hill, A. (1974). Mycoplasmas of small animals. *In* "Mycoplasmas of Man, Animals, Plants and Insects" (J. M. Bové and J. F. Duplan, eds.), pp. 311–316. INSERM, Paris.

76. Hill, A., and Dagnall, G. J. (1975). Experimental polyarthritis in rats produced by *Mycoplasma arthritidis. J. Comp. Pathol.* **85,** 45–52.

76a. Hill, A. C. (1978). Demonstration of *Mycoplasma* in tissue by the immunoperoxidase technique. *J. Infect. Dis.* **137,** 152–154.

77. Horowitz, S. A., and Cassell, G. H. (1978). Enzyme-linked immunosorbent assay for detection of *Mycoplasma pulmonis* antibodies. *Infect. Immun.* **22,** 161–170.

78. Hsu, D., and Geiman, Q. M. (1952). Synergistic effect of *Hemobartonella muris* on *Plasmodium berghei* in white rats. *Am. J. Trop. Med. Hyg.* **1,** 747–760.

79. Innes, J. R., Donati, E. J., Ross, M. A., Stouffer, R. M., Yevich, P. P., Wilson, C. E., Farber, J. R., Pankevicius, J. A., and Downing, T. O. (1957). Establishment of a rat colony free from chronic murine pneumonia. *Cornell Vet.* **47,** 260–280.

80. Innes, J. R. M., McAdams, A. J., and Yevich, P. (1956). Pulmonary disease in rats: A survey with comments on "chronic murine pneumonia." *Am. J. Pathol.* **32,** 141–160.

81. Ito, S., Imaizumi, K., Tajima, Y., Endo, M., and Koyama, R. (1957). Disease of rats caused by a pleuropneumonia-like organism. *Jpn. J. Exp. Med.* **27,** 243–248.

82. Jakab, G. J., and Dick, E. C. (1973). Synergistic effect in viral-bacterial infection: Combined infection of the murine respiratory tract with Sendai virus and *Pasterurella pneumotropica. Infect. Immun.* **8,** 762–768.

83. Jersey, G., Whitehair, C. K., and Carter, G. R. (1973). *Mycoplasma pulmonis* as the primary cause of chronic respiratory disease in rats. *J. Am. Vet. Med. Assoc.* **163,** 599–604.

84. Jonas, A. M., and Finch, S. C. Activation of hemobartonellosis in rats by chemical splenectomy. (Unpublished data.)

85. Joshi, N. N., and Dale, D. G. (1965). Etiology of murine endemic pneumonia. *Rev. Can. Biol.* **24,** 169–178.

86. Juhr, von N. C. (1971). Untersuchungen zur chronischen murinen pneumonie. Vorkommen und resistenz von *Mycoplasma pulmonis* in der umwelt. *Z. Versuchstierkd.* **13,** 210–216.

87. Juhr, von N. C. (1971). Untersuchungen zur chronischen murinen pneumonie. Pathogenese der *Mycoplasma pulmonis* infektion bei der ratte. *Z. Versuchstierkd.* **13,** 217–223.

88. Juhr, von N. C., and Obi, S. (1970). Uterusinfektionen beim Meerschweinchen. *Z. Versuchstierkd.* **12,** 383–387.

89. Juhr, von N. C., Obi, S., Hiller, H. H., and Eichberg, J. (1970). Mycoplasmen bei Keimfreien ratten mausen. *Z. Versuchstierkd.* **12,** 318–320.

90. Kaklamanis, E., and Pavlatos, M. (1972). The immunosuppressive effect of *Mycoplasma* infection. I. Effect on the humoral and cellular response. *Immunology* **22,** 695–702.

91. Kaklamanis, E., Stavropoulos, K., and Thomas, L. (1971). The mycoplasmacidal action of homogenates of normal tissues. *In* "Mycoplasma and L-Forms of Bacteria" (S. Madoff, ed.), pp. 27–35. Gordon & Breach, New York.

92. Kaklamanis, E., Thomas, L., Stavropoulos, K., Borman, I., and Boshwitz, C. (1969). Mycoplasmacidal action of normal tissue extracts. *Nature (London)* **221,** 860–862.

93. Kappel, H. K., Kappel, J. P., and Weisbroth, S. H. (1969). Establishment of a hysterectomy-deprived pathogen-free nucleus of BLU:(LE) rats. *Lab. Anim. Care* **19,** 738–741.

94. Kappel, H. K., Nelson, J. B., and Weisbroth, S. H. (1974). Development of a screening technique to monitor a mycoplasma-free Blu: (LE) Long-Evans rat colony. *Lab. Anim. Sci.* **24,** 768–772.

95. Kelemen, G. (1962). Histology of the nasal and paranasal cavities of germ-free-reared and ex-germ-free rats. *Acta Anat.* **48,** 108–113.

96. Kimbrough, R., and Gaines, T. B. (1966). Toxicity of hexamethylphosphoramide in rats. *Nature (London)* **211,** 146–147.

97. Kimbrough, R. D., and Sedlak, Y. A. (1968). Lung morphology in rats treated with hexamethylphosphoramide. *Toxicol. Appl. Pharmacol.* **12,** 60–67.

98. King, M. D. (1939). Labyrinthitis in the rat and a method for its control. *Anat. Rec.* **74,** 215–222.

99. Klein, G., and Wottama, A. (1975). DNA-repair and DNA-antibodies during experimental *Mycoplasma* arthritis. *Stud. Biophys.* **50,** 27–31.

100. Kleineberger, E. (1940). The pleuropneumonia-like organisms: Bacteriological features and serological relationships of strains from various sources. *J. Hyg.* **40,** 204–222.

101. Klieneberger, E., Nobel, E. (1962). "Pleuropneumonia-like Organisms (PPLO) Mycoplasmataceae," pp. 13-15. Academic Press, New York.

102. Kohn, D. F., and Kirk, B. E. (1969). Pathogenicity of *Mycoplasma pulmonis* in laboratory rats. *Lab. Anim. Care* **19**, 321-330.

103. Kreier, J. P., and Ristic, M. (1968). Haemobartonellosis, eperythrozoonosis, grahamellosis and ehrlichiosis. *In* "Infectious Blood Diseases of Man and Animals" (D. Weinman and M. Ristic, eds.), Vol. 2, pp. 387-472. Academic Press, New York.

104. Laber, G., Schutze, E., Teherani, D. K., Tuschl, H., and Altmann, H. (1974). Correlation between the occurrence of *Mycoplasma* and DNA repair inhibition in spleen cells of rats during experimental mycoplasma arthritis. *Stud. Biophys.* **50**, 21-26.

105. Lamb, D. (1975). Rat lung pathology and quality of laboratory animals: The user's view. *Lab. Anim.* **9**, 1-8.

106. Lane-Petter, W. (1970). A ventilation barrier to the spread of infection in laboratory animal colonies. *Lab. Anim.* **4**, 125-134.

107. Lane-Petter, W., Olds, R. J., Hacking, M. R., and Lane-Petter, M. E. (1970). Respiratory disease in a colony of rats. I. The natural disease. *J. Hyg.* **68**, 655-662.

108. Laskowski, L., Stanton, M. F., and Pinkerton, H. (1954). Chemotherapeutic studies of alloxan, dehydroascorbic acid, and related compounds in murine hemobartonellosis. *J. Infect. Dis.* **95**, 182-190.

109. Lastikka, L., Virsu, M. L., Halkka, O., Erikkson, K., and Estola, T. (1976). Goniomitosis in rats affected by *Mycoplasma* or *Macrolides*. *Med. Biol.* **54**, 146-149.

110. Leader, R. W., Leader, I., and Witschi, E. (1970). Genital mycoplasmosis in rats treated with testosterone propionate to produce constant estrus. *J. Am. Vet. Med. Assoc.* **157**, 1923-1925.

111. Lemcke, R. M. (1961). Association of PPLO infection and antibody response in rats and mice. *J. Hyg.* **59**, 401-412.

112. Lindsey, J. R., Baker, H. J., Overcash, R. G., Cassell, G. H., and Hunt, C. E. (1971). Murine chronic respiratory disease. Significance as a research complication and experimental production with *Mycoplasma pulmonis*. *Am. J. Pathol.* **64**, 675-716.

113. Lindsey, J. R., Cassell, G. H., and Baker, H. J. (1978). Diseases due to mycoplasmas and rickettsias. *In* "Pathology of Laboratory Animals (K. Benirschke, F. Garner, and C. Jones, eds.), Vol. 2, Chapter 15, pp. 1482-1550. Springer-Verlag, Berlin and New York.

113a. Lindsey, J. R., and Conner, M. W. (1978). Influence of cage sanitization frequency on intracage ammonia (NH₃) concentration and progression of murine respiratory mycoplasmosis in the rat. *Zbl. Bakt. Hyg.* **241**, 215.

114. Liu, C., Jayanetra, P., Voth, D. W., Muangmanee, L., and Cho, C. T. (1972). Potentiating effect of *Mycoplasma pneumoniae* infection on the development of pneumococcal septicemia in hamsters. *J. Infect. Dis.* **125**, 603-612.

115. Maramorosch, K. (1973). Mycoplasma and mycoplasma-like agents of human, animal, and plant diseases. *Ann. N.Y. Acad. Sci.* **225**, 1-532.

116. Mayer, M. (1921). Uber einige bakterienahnlicke Parasiten der Erythrozyten bei Menchen und Tieren. *Arch. Schiffs- Trop.-Hyg.* **25**, 150-151.

117. McCordock, H. A., and Congdon, C. C. (1924). Suppurative otitis of the albino rat. *Proc. Soc. Exp. Biol. Med.* **22**, 150-154.

118. Moise, T. S., and Smith, A. H. (1928). Observations on the pathogenesis of pulmonary suppuration in the albino rat. *Proc. Soc. Exp. Biol. Med.* **26**, 723-725.

119. Moore, D. H., Arison, R. H., Tanaka, H., Hall, W. T., and Chanowitz, M. (1965). Identity of the filterable hemolytic anemia agent of Sacks with *Hemobartonella muris*. *J. Bacteriol.* **90**, 1669-1674.

120. Moulder, J. W. (1974). The rickettsias. *In* "Bergey's Manual of Determinative Bacteriology" (R. E. Buchanan and N. E. Gibbons, eds.), 8th ed., pp. 882-925. Williams & Wilkins, Baltimore, Maryland.

121. Nakane, P. K., and Kawaoi, A. (1974). Peroxidase-labeled antibody. A new method of conjugation. *J. Histochem. Cytochem.* **22**, 1084-1091.

122. Naot, Y., Tully, J. G., and Ginsburg, H. (1977). Lymphocyte activation by various *Mycoplasma* strains and species. *Infect. Immun.* **18**, 310-317.

123. Nelson, J. B. (1940). Infectious catarrh of the albino rat. I. Experimental transmission in relation to the role of *Actinobacillus muris*. II. The causal relation of coccobacilliform bodies. *J. Exp. Med.* **72**, 645-662.

124. Nelson, J. B. (1946). Studies on endemic pneumonia of the albino rat. I. The transmission of a communicable disease to mice from naturally infected rats. *J. Exp. Med.* **84**, 7-14.

125. Nelson, J. B. (1951). Studies on endemic pneumonia of the albino rat. IV. Development of a rat colony free from respiratory infection. *J. Exp. Med.* **94**, 377-385.

126. Newberne, P. M., Salmon, W. D., and Hare, V. W. (1961). Chronic murine pneumonia in an experimental laboratory. *Arch. Pathol.* **72**, 224-233.

127. Ogata, M., Ohta, T., and Atobe, H. (1967). Studies on mycoplasmas of rodent origin. Mycoplasmas from the chronic respiratory disease of rats. *Jpn. J. Bacteriol.* **22**, 618-627.

128. Olson, L. D., and McLune, E. L. (1968). Histopathology of chronic otitis media in the rat. *Lab. Anim. Care* **18**, 478-485.

129. Organick, A. B., Siegesmund, K. A., and Lutsky, I. I. (1966). Pneumonia due to *Mycoplasma* in gnotobiotic mice. II. Localization of *Mycoplasma pulmonis* in the lungs of infected gnotobiotic mice by electron microscopy. *J. Bacteriol.* **92**, 1164-1176.

130. Ott, K. J., and Stauber, L. A. (1967). *Eperythrozoon coccoides* influence on course of infection of *Plasmodium chabaudi* in mouse. *Science* **155**, 1546-1548.

131. Overcash, R. G., Lindsey, J. R., Cassell, G. H., and Baker, H. J. (1976). Enhancement of natural and experimental respiratory mycoplasmosis in rats by hexamethylphosphoramide. *Am. J. Pathol.* **82**, 171-189.

132. Palecek, F., and Holusa, R. (1971). Spontaneous occurrence of lung emphysema in laboratory rats. A quantitative functional and morphological study. *Physiol. Bohemoslov.* **20**, 335-344.

133. Pankevicius, J. A., Wilson, C. E., and Farber, J. F. (1957). The debatable role of pleuropneumonia-like organisms in the etiology of chronic pneumonia of rats. *Cornell Vet.* **47**, 317-325.

134. Parkes, M. W., and Wrigley, F. (1951). Arthritis in rats produced by pleuropneumonia-like organisms. *Ann. Rheum. Dis.* **10**, 177-181.

135. Peters, W. (1965). Competitive relationship between *Eperythrozoon coccoides* and *Plasmodium berghei* in the mouse. *Exp. Parasitol.* **16**, 158-166.

136. Polak-Vogelzang, A. A., and Hagenaars, R. (1976). Identification of mycoplasmas and acholeplasmas by means of an indirect immunoperoxidase test. *Proc. Soc. Gen. Microbiol.* **3**, 151.

137. Pomerat, C. M., Frieden, E. H., and Yeager, E. (1947). Reticuloendothelial immune serum (RES). V. An experimental anemia in *Bartonella* infected rats produced by anti-blood immune serum. *J. Infect. Dis.* **80**, 154-163.

138. Preston, W. S. (1942). Arthritis in rats caused by pleuropneumonia-like organisms and the relationship of similar organisms to human rheumatism. *J. Infect. Dis.* **70**, 180-184.

139. Rekers, P. E. (1951). The effect of x-irradiation on rats with and without *Bartonella muris*. *J. Infect. Dis.* **88**, 224-229.

140. Ribelin, W. E., and McCoy, J. R. (1965). "The Pathology of Laboratory Animals," pp. 11-14. Thomas. Springfield, Illinois.

141. Rudnick, P., and Hollingsworth, J. W. (1959). Lifespan of rat erythrocytes parasitized by *Bartonella muris*. *J. Infect. Dis.* **104**, 24-27.

142. Sacks, J. H., Clark, R. F., and Egdahl, R. H. (1960). Induction of

tumor immunity with a new filterable agent. *Surgery* **48**, 244–260 and 270–271.

143. Sacks, J. H., and Egdahl, R. H. (1960). Protective effects of immunity and immune serum on the development of hemolytic anemia and cancer in rats. *Surg. Forum* **10**, 22–25.

144. Sainte-Marie, G. (1962). A paraffin embedding technique for studies employing immunofluorescence. *J. Histochem. Cytochem.* **10**, 250–256.

145. Sawyer, W. D. (1969). Interaction of influenza virus and its effect on phagocytosis. *J. Infect. Dis.* **119**, 541.

146. Scheff, G. J., Scheff, I. M., and Eiseman, G. (1956). Concerning the mechanism of *Bartonella* anemia in the splenectomized rat. *J. Infect. Dis.* **98**, 113–120.

147. Schreiber, H., Nettesheim, P., Lijinsky, W., Richter, C. B., and Walburg, H. E. (1972). Induction of lung cancer in germ-free, specific-pathogen-free, and infected rats by *N*-nitrosoheptamethyleneimine: Enhancement by respiratory infection. *J. Natl. Cancer Inst.* **49**, 1107–1114.

148. Scott, J. K., and Stannard, J. N. (1954). Relationship between *Bartonella muris* infection and acute radiation effects in the rat. *J. Infect. Dis.* **95**, 302–308.

149. Sharp, J. T. (1970). "The Role of Mycoplasmas and L-Forms of Bacteria in Disease." Thomas. Springfield, Illinois.

150. Small, E., and Ristic, M. (1967). Morphologic features of *Hemobartonella felis*. *Am. J. Vet. Res.* **28**, 845–851.

151. Sokoloff, L. (1951). Joint diseases of laboratory animals. *J. Natl. Cancer Inst.* **20**, 965–969.

152. Stanbridge, E. (1971). Mycoplasmas and cell cultures. *Bacteriol. Rev.* **35**, 206–227.

153. Stansly, P. G. (1965). Nononcogenic infectious agents associated with experimental tumors. *Prog. Exp. Tumor Res.* **7**, 224–258.

154. Stewart, D. D., and Buck, G. E. (1975). The occurrence of *Mycoplasma arthritidis* in the throat and middle ear of rats with chronic respiratory disease. *Lab. Anim. Sci.* **25**, 769–775.

155. Stuart, A. E. (1962). Experimental necrosis of spleen. *J. Pathol. Bacteriol.* **84**, 193–200.

156. Subcommittee on the Taxonomy of *Mycoplasmatales* (1972). Proposal for minimal standards for descriptions of new species of the order Mycoplasmatales. *Int. J. Syst. Bacteriol.* **22**, 184–188.

157. Tanaka, H., Hall, W. T., Sheffield, J. B., and Moore, D. H. (1965). Fine structure of *Hemobartonella muris* as compared with *Eperythrozoon coccoides* and *Mycoplasma pulmonis*. *J. Bacteriol.* **90**, 1735–1749.

158. Taylor, G., Taylor-Robinson, D., and Slavin, G. (1974). Effect of immunosuppression on arhtritis in mice induced by *Mycoplasma pulmonis*. *Ann. Rheum. Dis.* **33**, 376–382.

159. Taylor-Robinson, D., Denny, F. W., Thompson, G. W., Allison, A. C., and Mardh, P. A. (1972). Isolation of mycoplasmas from lungs by a perfusion technique. *Med. Microbiol. Immunol.* **158**, 9–15.

160. Thomas, T. B., Pomerat, C. M., and Frieden, E. H. (1949). Cellular reactions in *Hemobartonella* infected rats with anemia produced by anti-blood immune serum. *J. Infect. Dis.* **84**, 169–186.

161. Thurston, J. P. (1955). Observations on the course of *Eperythrozoon coccoides* infections in mice, and the sensitivity of the parasite to external agents. *Parasitology* **45**, 141–151.

162. Tram, par C., Guilon, J. C., and Chouroulinkov, I. (1970). Isolement de mycoplasmes de foetus de rats obtenus par césarienne aseptique. *C. R. Seances Soc. Biol. Sas Fil.* **164**, 2470–2471.

163. Tunnicliff, R. (1916). Streptothrix in bronchopneumonia of rats similar to that of rat-bite fever. *J. Infect. Dis.* **19**, 767–772.

164. Tvedten, H. W., Whitehair, C. K., and Langham, R. F. (1973). Influence of vitamins A and E on gnotobiotic and conventionally maintained rats exposed to *Mycoplasma pulmonis*. *J. Am. Vet. Med. Assoc.* **163**, 605–612.

165. Val, F., Fournier, M., and Pariente, R. (1971). Bronchial lymphoepithelial nodules in the rat. Definition and morphological characteristics in optical and electron microscopy. *Biomedicine* **26**, 130–137.

166. Vasenius, H., and Tiainen, O. A. (1966). The etiology and therapy of labyrinthitis in laboratory animals. *Z. Versuchtierkd.* **8**, 351–356.

167. Vogelzany, A. A. (1975). The survival of *Mycoplasma pulmonis* in drinking water and in other materials. *Z. Versuchstierkd.* **17**, 240–246.

168. Vrolijk, H., Verlinde, J. D., and Braams, W. G. (1957). Virus pneumonia (snuffling disease) in laboratory rats and mice. *Antonie van Leeuwenhoek* **23**, 173–183.

169. Ward, J. R., and Cole, B. C. (1970). Mycoplasmal infections of laboratory animals. *In* "The Role of Mycoplasmas and L-Forms of Bacteria in Disease" (J. T. Sharp, ed.), pp. 212–239. Thomas, Springfield, Illinois.

170. Ward, J. R., and Jones, R. S. (1962). The pathogenesis of mycoplasmal arthritis in rats. *Arthritis Rheum.* **5**, 163–175.

171. Weisbroth, S. H. (1972). Pathogen-free substrates for gerontologic research: Review, sources, and comparison of barrier-sustained versus conventional laboratory rats. *Exp. Gerontol.* **7**, 417–426.

171a. Westerberg, S. C., Smith, C. B., Wiley, B. B., and Jensen, C. (1972). Mycoplasma-virus interrelationships in mouse tracheal organ cultures. *Infect. Immun.* **5**, 840–846.

172. Whittleston, P. (1967). Mycoplasma in enzootic pneumonia of pigs. *Ann. N.Y. Acad. Sci.* **143**, 271–276.

173. Whittlestone, P. (1974). Isolation techniques for mycoplasmas from animal diseases. *In* "Mycoplasmas of Man, Animals, Plants and Insects" (J. B. Bové and J. F. Duplan, eds.), pp. 143–151. INSERM, Paris.

174. Whittlestone, P., Lemcke, R. M., and Olds, R. J. (1972). Respiratory disease in a colony of rats. II. Isolation of *Mycoplasma pulmonis* from the natural disease, and the experimental disease induced with a cloned culture of this organism. *J. Hyg.* **70**, 387–409.

175. Wills, L., and Mehta, M. M. (1930). Studies in pernicious anemia of pregnancy. IV. The production of pernicious anemia (*Bartonella* anemia) in intact albino rats by deficient feeding. *Indian J. Med. Res.* **18**, 663–681.

176. Wolgom, W. H., and Warren, J. (1938). A pyogenic filterable agent in the albino rat. *J. Exp. Med.* **68**, 513–520.

177. Wolgom, W. H., and Warren, J. (1939). The nature of a pyogenic filterable agent in the white rat. *J. Hyg.* **39**, 266–272.

178. Woolcock, P. R., Czekalowski, J. W., and Hall, D. A. (1973). Studies on proteolytic activity of mycoplasmas: The preparation and properties of gelatinolytic enzymes from strains of *Mycoplasma arthritidis*. *J. Gen. Microbiol.* **78**, 23–32.

179. Young, C., and Hill, A. (1974). Conjunctivitis in a colony of rats. *Lab. Anim.* **8**, 301–304.

Chapter 11

Viral Diseases

Robert O. Jacoby, Pravin N. Bhatt, and Albert M. Jonas

I. INTRODUCTION

The number of viruses known to be naturally infectious for laboratory rats is small, and most cause inapparent infections which usually are detected by serological monitoring (Table I).

Table I

Viruses of Laboratory Rats

I. DNA viruses
  A. Viruses which cause naturally occurring infections in rats
    1. Rat parvoviruses (RV, H-1 virus, and related viruses)
      Natural host: Rat
      Signs: Usually asymptomatic; may have decreased litter size or decreased numbers of litters; neonatal deaths; occasional runting, jaundice or ataxis in sucklings or weanlings; neurological disease or sudden death in adults
      Lesions: Usually none; may be hemorrhages and necrosis in central nervous system, testes or elsewhere; cerebellar hypoplasia and/or hepatic necrosis and fibrosis in sucklings or weanlings; resorption sites in uterus; intranuclear inclusion bodies in brain, liver, endothelium, or elsewhere
      Epizootiology: Persistent and latent; highly infectious, spreads rapidly; transmission usually horizontal via oral or respiratory route but some strains can be transmitted vertically; virus may be excreted in feces, urine, and milk
      Diagnosis: Clinical signs and/or lesions, if present; detect antibody by HAI, CF, NT or FA tests;[a] isolate virus in primary rat embryo cultures; inoculate parvovirus-free sucklings with suspect tissues and test serum for antibody 2 to 4 weeks later
      Differentiate from: Sendai virus infection (production losses); chemical intoxication, tumors, trauma, genetic abnormalities
      Treatment: None
      Control: Destroy colony and repopulate from virus-free stock, disinfect facilities and equipment; institute routine serological surveillance to detect infection
    2. Cytomegalovirus
      Natural host: Rat (primarily wild rats)
      Signs: None
      Lesions: Usually none; intranuclear inclusion bodies in enlarged cells of salivary or lacrimal glands; mild nonsuppurative adenitis
      Epizootiology: Not well studied
      Diagnosis: Lesions with inclusion bodies; test serum for NT antibody have been detected in rats; virus has not been isolated from rats
      Treatment: None
      Control: Not studied
  B. Viruses which may infect rats[b]
    1. Minute virus of mice
      Natural host: Mouse
      Significance: Latent parvovirus of mice, but low titers of HAI antibody have been detected in rats; virus has not been isolated from rats
    2. Mouse adenovirus
      Natural host: Mouse
      Significance: CF antibodies have been detected in rats, but virus has not been isolated from rats
II. RNA Viruses
  A. Viruses which cause naturally occurring infections in rats
    1. Sialodacryoadenitis virus (SDAV)
      Natural host: Rat

Table I (*Continued*)

      Signs: Photophobia; keratoconjunctivitis; red-brown (porphyrin-containing) tears staining skin around eyes and nares; sneezing; enlarged salivary glands; cervical edema
      Lesions: Acute rhinitis; necrosis and inflammation of submaxillary and parotid salivary glands and lacrimal glands; squamous metaplasia of glands during repair; cervical edema; cervical lymph node hyperplasia; transient thymic atrophy; keratoconjunctivitis with occasional megaloglobus; interstitial pneumonia not reported, but cannot be ruled out
      Epizootiology: Acute, self-limiting infection; nonimmune rats of all ages susceptible; high morbidity; low mortality; transmitted by aerosol; virus excreted for about 1 week then CF and NT antibody detectable in serum; virus antigenically related to rat coronavirus and mouse hepatitis virus; reinfection not reported, but cannot be ruled out
      Diagnosis: Clinical signs and/or lesions; test for serum antibody by CF or NT tests; isolate virus in primary rat kidney cell cultures (must differentiate from RCV)
      Differentiate from: Rat coronavirus infection; murine respiratory mycoplasmosis; Sendai virus infection; cytomegalovirus infection; eye irritation from chemical fumes (e.g. ammonia); bacterial infections
      Treatment: Usually none; eye lesions may be treated symptomatically
      Control: Quarantine rats for 6 to 8 weeks during outbreak, stop breeding, quarantine rats from infected colonies for 30 days before placing them among susceptible rats; maintain rats in barrier facility; institute routine serological surveillance to detect infection
    2. Rat coronavirus (RCV)
      Natural host: Rat
      Signs: Usually asymptomatic; may be occasional salivary gland enlargement and cervical edema
      Lesions: Acute rhinitis; mild focal interstitial pneumonia; occasional necrotizing sialoadenitis
      Epizootiology: Same as SDAV
      Diagnosis: Clinical signs if present; lesions; detect serum antibody by CF and NT tests; isolate virus in primary rat kidney cell cultures (must differentiate from SDAV)
      Differentiate from: SDAV infection; murine respiratory mycoplasmosis; Sendai virus infection
      Treatment: None
      Control: As described for SDAV
    3. Sendai virus (parainfluenza 1)
      Natural host: Rat, mouse, hamster, guinea pig
      Signs: Usually asymptomatic; may be rough haircoat, dyspnea, production losses in breeding colonies
      Lesions: Bronchopneumonia and interstitial pneumonia with necrosis of bronchial and/or bronchiolar epithelium; upper respiratory tract usually not involved; peribronchial lymphocytic infiltrates may persist for months
      Epizootiology: Acute, self-limiting; nonimmune rats of all ages susceptible; high morbidity, low mortality; transmitted by aerosol; antibody detectable in serum by 7 to 14 days postinfection; chronology of virus excretion and stability of immunity not established for rats
      Diagnosis: Clinical signs, if present; lesions; detect serum antibody by CF test; HAI and NT tests also available; isolate virus in BHK-21 cells or embryonated eggs
      Differentiate from: Murine respiratory mycoplasmosis; coronavirus infection; RV infection (production losses; intrauterine resorption sites)

(*continued*)

Table I (Continued)

    Treatment: None
    Control: Quarantine rats for 4 to 8 weeks during outbreak including cessation of breeding; quarantine rats from infected colonies for 30 days before placing them among susceptible rats; institute routine surveillance of vendor rats for infection

  B. Viruses which may infect rats
    1. Reovirus 3
    Natural host: Many mammals
    Significance: HAI antibodies have been detected in rats, but virus has not been isolated
    2. Pneumonia virus of mice (PVM)
    Natural host: Mouse
    Significance: HAI and NT antibodies have been detected in rats, but virus has not been isolated
    3. Mouse encephalitis virus
    Natural host: Mouse
    Significance: HAI and NT antibodies have been detected in rats but virus has not been isolated

III. Unclassified viruses or viruslike agents
  A. MHG Virus
    Natural host: Rat; other species?
    Significance: Enterovirus-like neurotropic agent; can cause neurological signs and necrotizing inflammation of brain in experimentally infected rats; may be antigenically related to mouse encephalomyelitis virus; epizootiology unknown
  B. Rat submaxillary gland (RSMG) virus
    Natural host: Rat.
    Significance: Latent, vertically-transmissible agent isolated from submaxillary gland; no signs or lesions; induces HAI antibody; unrelated antigenically to rat coronaviruses or cytomegalovirus; must differentiate isolates from coronaviruses, cytomegaloviruses
  C. Novy virus
    Natural host: Rat
    Significance: Extremely stable filterable agent recovered from rat blood. Agent and its pathogenicity not well studied
  D. Viruslike pneumotropic agents (enzootic bronchiectasis agent; gray lung virus; wild rat pneumonia agent)
    Natural host: Rat; mouse?
    Significance: Incompletely characterized agents associated with pneumonias of laboratory and/or wild rats.

^aHAI, hemagglutination inhibition; CF, complement fixation; NT, neutralization; FA, fluorescent antibody.

^bMousepox virus may be mildly infectious for experimentally inoculated rats, but natural infections have not been reported and evidence of natural immunoreactivity to the virus has not been detected.

Agents from three families of viruses are especially well disseminated among rat colonies: rat virus (RV) and related strains (Parvoviridae), sialodacryoadenitis virus (SDAV) and rat coronavirus (RCV) (Coronaviridae), and Sendai virus (parainfluenza 1) (Paramyxoviridae). Rat parvoviruses normally induce latent infections, but, under conditions to be discussed, they may be lethal or cause vascular, neurological, and hepatic lesions. Sialodacryoadenitis virus causes nonlethal, self-limiting disease characterized clinically by oculonasal accumulations of lacrimal porphyrin pigment, photophobia, and cervical swelling from enlarged salivary glands and

morphologically by necrotizing inflammation in the upper respiratory tract, salivary glands, and lacrimal glands. There is recent evidence that keratoconjunctivitis, sometimes resulting in permanent eye damage, can also accompany SDAV infection. Rat coronavirus may cause sialoadenitis, but appears to be more pneumotropic than SDAV. It can induce mild interstitial pneumonia in adult rats and can cause fatal pneumonias in sucklings. Sendai virus has received increased recognition as an important virus of rats, and current evidence indicates that natural infection causes pneumonia.

The impact of other viruses naturally infectious for rats has not been thoroughly studied. Humoral antibodies to minute virus of mice (MVM), pneumonia virus of mice (PVM), reovirus 3, mouse encephalomyelitis virus, and murine adenovirus have been detected in rats. The corresponding viruses, however, are currently considered nonpathogenic for rats. Similarly, cytomegalovirus inclusions have been found in rat tissues, but are not associated with clinical disease or significant lesions. Several other viruses [MGH virus, rat submaxillary gland virus (RSMG), and Novy virus] and viruslike agents (enzootic bronchiectasis, gray lung, wild rat pneumonia) have been reported in rats, but their significance is unclear. Therefore, this chapter concentrates primarily on rat parvoviruses, rat coronaviruses, and Sendai virus. The viruses are reviewed individually, but the reader is encouraged to peruse Section VII for broad approaches to the detection, diagnosis, and control of virus infections in rat colonies. Oncogenic viruses of rats, potential or confirmed, are not discussed.

II. DNA VIRUSES

A. Parvoviruses (Rat Virus, H-1 Virus, and Minute Virus of Mice)

1. Rat Virus (RV) and H-1 Virus

a. General. Parvoviruses are extremely small (18–26 nm) icosahedral viruses that are remarkably stable to heat, acid, lipid solvents, and prolonged exposure to room temperature. Some are defective and require adenovirus "helpers" to replicate, but rat parvoviruses are not defective. Parvoviruses have been isolated from rodents, birds, dogs, cats, pigs, cattle, nonhuman primates, and humans. They appear to have a predilection for tissues containing rapidly dividing cells and may induce latent or subclinical infections. Recent reviews offer additional background information (134,142).

Rat virus (RV) is the type species for the genus *Parvovirus* and was the first virus isolated of the family Parvoviridae. The literature on rat parvoviruses is extensive, but much of it pertains to experimentally induced infections in hamsters and several other species, as well as in rats. Since this report centers on natural parvovirus infections of rats, only experimentally induced infections which contribute to understanding of natural infections will be included.

b. History.   Rat virus was isolated from several tumor-bearing rats by Kilham and Olivier (78) during studies to determine if rats harbored a polyoma-like virus. Rat virus seemed nonpathogenic for newborn and weanling rats, suckling and weanling mice, suckling hamsters, and adult rabbits and guinea pigs. They also found natural RV neutralizing (NT) and hemagglutination inhibiting (HAI) antibodies in serum of conventional rats and germfree rats. This suggested that RV was not only naturally infectious for rats but also could be transmitted vertically. Toolan (148,149,155) isolated a second parvovirus from a human tumor cell line (HEp-1) that had been passaged in rats and named it H-1 virus. Dalldorf (37) found a third parvovirus-like agent in rat-passaged human tumor HEp-3 and named it H-3 virus (also known as OLV). Moore (110) showed that H-1 virus was antigenically distinct from RV, whereas Dalldorf's virus was antigenically related to RV, but it did not appear to share antigens with H-1. Nevertheless, RV, H-1, and H-3 were similar biochemically and physically. Furthermore, each could, under suitable conditions, induce severe developmental abnormalities, particularly in suckling hamsters, despite the fact that they usually caused latent persistent infections in weaned rats.

Additional strains of RV-related and H-1-related rat parvoviruses have been characterized since the pioneering work of Kilham, Toolan, and Dalldorf. Most are antigenically related to RV (Table II). There is some controversy about whether H-1 virus is of human or rat origin. It is clear, however, that natural infections with RV and H-1 viruses are widespread in laboratory rats and in wild rats.

c. Properties.   i. Physiochemical.   Rat virus and H-1 virus are single-stranded DNA viruses (62,102,139). They

Table II

Antigenic Type of Some Rat Parvovirus Isolates

Antigenic type	Virus	Source of original isolate	Reference
RV	RV	Rat tumor	78
	H-3	HEp-3 cells	37
	X-14	Rat tumor	124
	L-5	Rat tumor	86
	HB	Human placenta	151
	SpRV	Rat tissue	73
	HER	Rat tissue	40
	HHP	Rat tissue	94
	Kirk	Detroit-6 cells human serum	16
H-1	H-1	HEp-1 cells	155
		Rat tissue	74
	H-T	Human placenta	151

range in size from 18 to 30 nm depending on conditions of purification and measurement and appear to have 32 capsomeres (69,142,158). Rat virus has a bouyant density in CsCl of about 1.40 gm/ml, and its molecular weight is approximately 6.6×10^6 daltons (138). Rat virus and H-1 virus retain infectivity and hemagglutinating activity after exposure to ether, chloroform, or alcohol and are stable at pH's from 2 to 11 at 4°, 25°, and 37°C (47,78). They also are remarkably temperature resistant. Rat virus can remain infectious after exposure to 80°C for up to 2 h, for more than 6 months at −40°C, and for up to 60 days at 40°C (162). The stability of RV and H-1 virus to variations of temperature and pH may, however, depend on virus strain and supporting medium (18,86). These viruses are inactivated by ultraviolet light at room temperature (25,155). They are, however, resistant to ultrasonication, RNase, DNase, papain, trypsin, and chymotrypsin—properties which facilitate preparation of purified virus from infected cell cultures (20,46,97,154,158).

ii. Hemagglutination and hemadsorption.   Hemagglutination is a common property of rat parvoviruses and is the simplest method to detect them. Guinea pig erythrocytes are preferred to detect RV, H-1 virus, and H-3 virus. The test can be run at 4°, 25°, or 37°C (78,110). Rat parvoviruses agglutinate erythrocytes of other species to variable degrees, and it has been suggested that differential agglutination be used to identify virus strains (153,154). Since the hemagglutinin is cell associated, cell debris should be pretreated with deoxycholate, receptor-destroying enzyme, or, preferably, alkaline buffer (48). Portella (126) showed that RV, H-1, and H-3 caused adsorption of guinea pig erythrocytes to virus-infected cells, but this technique is not generally used as a diagnostic test.

iii. Antigens.   Neutralizing (NT), hemagglutination inhibition (HAI), and complement fixation (CF) tests have shown that rat parvoviruses consist of two major antigenic groups (36,86,110,126,151,154). One group includes RV, H-3, X-14, and HER viruses, and the other includes H-1 and HT viruses (Table II). Some cross-reactivity has been detected by immunofluorescence (FA) among RV, H-1 virus, and MVM (36,54).

d. In Vitro Cultivation.   Rat virus and H-1 replicate well in primary monolayer cultures of rat embryo cells (78). Hamster embryo cultures support viral growth less well, and cells of mouse, chicken, calf, or human origin are unsuitable for RV, but the H viruses can grow in human, monkey, rat, and hamster cells [reviewed by Siegl (142)]. Primary cultures should be prepared from rats free of RV and H-1 virus infection, and all cell lines should be checked for latent contamination with parvoviruses.

Rat parvovirus-infected cultures usually develop cytopathic

changes characterized by intranuclear inclusions and necrosis (78,110,156). The inclusions can be detected with standard histological stains. The course and severity of cytopathic effects (CPE) depend on several factors in addition to the target cells used. For example, high doses of virus may cause CPE in several days, whereas cultures for endpoint titrations should be held for up to 3 weeks (86,110,156). The sensitivity of methods used to detect CPE vary. Cole and Nathanson (29) detected antigen to the HER strain of RV using immunofluorescence by 5 to 6 h and infectious virus by 20 h after infection with a large dose (100 $TCID_{50}$/cell) of RV. Viral antigen was first detected in the cytoplasm and then in the nucleus before cell necrosis appeared. Bernhard *et al.* (8) showed that margination of chromatin occurred in some cells infected with RV by 36 to 48 h. Mayor and Ito (98) detected X-14 viral antigen with immunofluorescence by 12 h postinfection and hemagglutination (HA) activity and infectious virus by 48 h. Margolis and Kilham (89) reported that RV multiplication was enhanced in dividing cultured cells. Similarly, studies by Cole and Nathanson (29) suggested that dividing BHK-21 cells supported RV replication better than nondividing cells. Analogous results were reported for H-1 virus grown in WI-26 human diploid cells (81). The morphology of viral replication and inclusion body formation have been described in detail (8,38,99,126).

e. Host Range.   Laboratory and wild rats appear to be the only natural hosts for rat parvoviruses. There is still debate as to whether the H viruses also may be naturally infectious for humans (154). Rat parvoviruses are, however, infectious for several species by experimental inoculation. Rat virus and H-1 virus can infect neonatal, suckling, and adult rats and hamsters (70,73,78,111,112,149,155). Experimental RV infection also has been reported in neonatal mice (40,96), *Praomys* (*Mastomys*) *natalensis* (129), and newborn kittens (72). H-1 virus infection has been induced in newborn and adult rhesus monkeys (*Macaca mulatta*) and in human volunteers (152). Attempts to induce RV infection in newborn rhesus monkeys were unsuccessful (154).

f. Clinical Disease.   Kilham and Margolis (73) were first to report natural RV disease. They had purchased 13-day pregnant rats from a commercial breeder, and the embryos were harvested to prepare primary tissue cultures. Embryos cultured from one rat with three uterine resorption sites contained RV. Rat virus also was isolated from three undersized pups in a second litter of eleven, and one other pup from the litter developed ataxia from cerebellar hypoplasia and was euthanatized at 45 days. A third litter consisted of two small pups. They developed severe jaundice shortly after birth and were killed at 4 days; the dam had nine intrauterine resorption sites. A fourth pregnant rat was killed at gestation day 16. One

of 10 fetuses was abnormal with a mottled placenta, and RV was isolated from the fetus.

The strain of RV isolated from these rats was designated SpRV. It proved highly infectious for pregnant and nonpregnant rats and could be isolated from tissues, milk, and feces after experimental inoculation. Furthermore, it crossed the placenta effectively and caused fetal death or teratogenesis when inoculated early in gestation, and postnatal cerebellar destruction and hepatitis when inoculated late in gestation. Simultaneous intraperitoneal (ip) and intracerebral (ic) inoculation of newborns with SpRV also caused cerebellar and liver lesions and pups usually died by 8 days. One control suckling developed cerebellar disease from contact infection.

Nevertheless, natural RV disease among newborn and suckling rats is sporadic (73), despite the high prevalence of RV infection among commercial and institutional colonies. In contrast, reduced litter size, runted litters, or decreased production with increased numbers of resorption sites in uteri of breeding females should be considered potential signs of parvovirus infection. RV infection of nonimmune pregnant dams in a production colony can be expressed initially by decreased litters or litter size followed by a progressive return to normal production as immunity to RV builds.

Parvovirus infection in adult rats is usually asymptomatic, but latent infection can be activated by immunosuppression (40). We have, however, followed a natural enzootic of severe RV infection in a colony of Sprague-Dawley rats (66). Two groups of nonimmune male weanling rats, numbering 100 and 50, respectively, were introduced to a colony known to be latently infected with RV. Two weeks later most of the newly introduced rats developed ruffled haircoats, dehydration, and cyanotic scrotums. Twenty-five rats from the first group and seven from the second group died. Rat virus was isolated from affected rats and hemorrhagic lesions compatible with RV infection were detected (see Section II, A, 1, g).

The spectrum of clinical signs associated with RV infection likely depends on the strain of virus involved in a given outbreak. Strains capable of infecting the gravid uterus of nonimmune dams may affect production. Strains restricted to horizontal transmission probably produce largely silent infections, but may occasionally provoke acute fatal disease in neonates or adult rats. The transmission of parvoviruses is discussed in more detail in Section II, A, 1, h.

g. Pathology.   Parvoviruses can be distributed widely in tissues of infected rats, but they most commonly cause lesions in tissues containing mitotically active cells, such as the developing liver and cerebellum. Since resistance to these agents usually develops during the first week or two of life, much information about the pathogenesis of parvovirus lesions stems from study of infected pregnant dams, neonates, or young sucklings. Rat parvoviruses also have been studied extensively in heterologous hosts, particularly suckling hamsters, in which they not only cause cerebellar lesions but also induce severe mongoloid developmental abnormalities and dental deformities. The reader is referred to several reviews for details about these lesions (37,89,154).

The cerebellar lesions in pups, either naturally or experimentally infected with RV, are characterized by development of intranuclear inclusions in rapidly dividing cells of the external germinal layer of cerebellum (Fig. 1) and then by necrosis (Fig. 2) (73). Since cells of the external granular layer migrate to form the internal granular layer, their depletion aborts cerebellar development and results in granuloprival cerebellar hypoplasia with ataxia (Fig. 3).

The liver lesions also begin with intranuclear inclusions in hepatocytes (Fig. 4), and less frequently in Kupffer cells, biliary epithelium, endothelial cells (Fig. 5), and connective tissue cells (73). Ensuing cytopathic changes are characterized by nuclear pyknosis, cytoplasmic eosinophilia, ballooning degeneration, dissociation of affected cells from adjacent cells and finally cell lysis. Other changes may include hepatocytic bile retention and pigmented casts (ostensibly bile) in renal tubules.

Margolis *et al.* (94) reported that fresh isolates of RV and H-1 virus were more virulent for experimentally infected newborn rats than tissue culture passaged strains. Furthermore, rats inoculated as sucklings (3 to 8 days) rather than as neonates

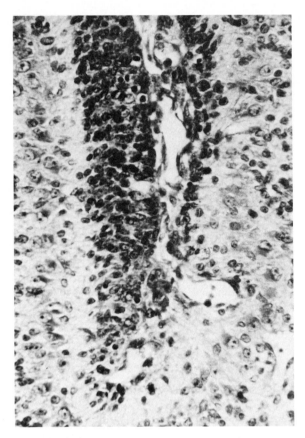

Fig. 2. Cerebellum from a 4-day-old rat naturally infected with RV. The external germinal layer has been partially destroyed. The leptomeninges are moderately infiltrated with mononuclear cells. [Courtesy of Dr. Kilham and Dr. Margolis (73); and the *American Journal of Pathology.*]

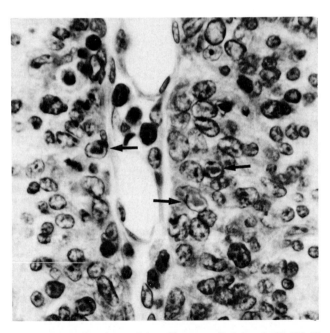

Fig. 1. Cerebellum from a 4-day-old rat naturally infected with RV. The inclusion body phase is shown. The central fissure is bordered by the external germinal layer, several cells of which contain intranuclear inclusion bodies (arrows). [Courtesy of Dr. Kilham and Dr. Margolis (73); and the *American Journal of Pathology.*]

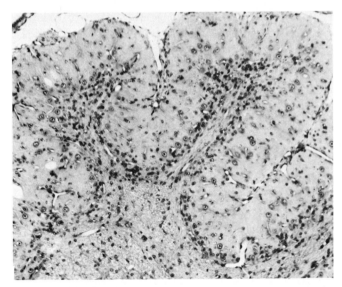

Fig. 3. Cerebellum from a 37-day-old rat inoculated intracerebrally at birth with SpRV. There is severe granuloprival hypoplasia and haphazard distribution of Purkinje cells. [Courtesy of Dr. Kilham and Dr. Margolis (73); and the *American Journal of Pathology.*]

had less severe infections, even with virulent strains. Other workers also have found that acute lethal disease, often accompanied by hemorrhage, was most easily induced in neonatally inoculated rats (111,116). Kilham and Margolis (74) found that typical cerebellar lesions developed only in rats inoculated before day 4. Hepatitis with jaundice but without fulminating generalized disease occurred in sucklings inoculated between days 6 and 12. Sucklings inoculated after day 12 developed only mild self-limiting hepatitis. The only sign of infection in adults was seroconversion.

Inclusions appeared in liver by 24 h after infection and persisted in some rats for up to 3 weeks. Persistence of virus in liver was attributed to increased mitotic activity postpartum which lasts up to 6 weeks. Ruffolo *et al.* (137) supported this view by showing that the susceptibility of adult rat liver to H-1 virus could be reestablished by partial hepatectomy or by carbon tetrachloride hepatotoxicity. Postnecrotic progression and repair of liver lesions occurred in several stages (94): (a) giant cell formation and polyploidy (Fig. 6); (b) biliary hyperplasia

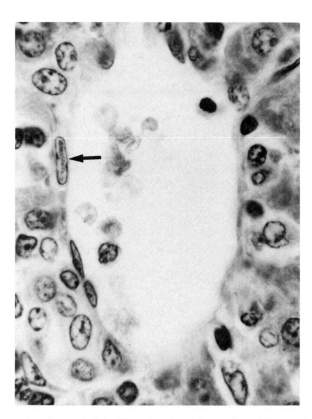

Fig. 5. An endothelial cell in an hepatic vein has an elongated intranuclear RV inclusion body (arrow). [Courtesy of Dr. Kilham and Dr. Margolis (73); and the *American Journal of Pathology.*]

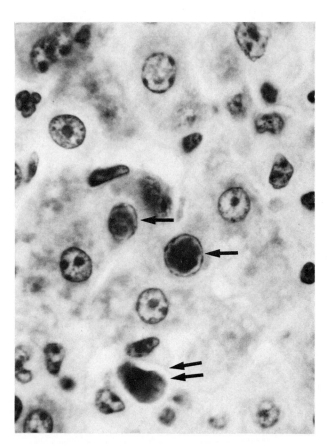

Fig. 4. Liver of a 4-day-old rat with natural RV infection. Several phases of inclusion body formation (arrows) are demonstrated in hepatocytes. A rounded eosinophilic necrotic hepatic cell (double arrows) is also shown. [Courtesy of Dr. Kilham and Dr. Margolis (73); and the *American Journal of Pathology.*]

(Fig. 7), adenomatoid distortion of lobular architecture, and formation of blood-filled spaces (peliosis hepatitis) (Fig. 8); and (c) postnecrotic stromal collapse, fibrosis, and nodular hyperplasia (Fig. 9). Peliosis hepatitis has been produced in several strains of suckling rats and is presumably a sequel of RV-induced hepatic necrosis (7).

Hemorrhagic encephalomyelopathy was first detected in about 2% of more than 1500 adult Lewis rats given a single dose of 100–225 mg/kg of cyclophosphamide (40,112). Lesions developed 10 to 25 days after treatment and consisted of multiple hemorrhages in medulla and spinal cord (Fig. 10). They were manifested clinically by paralysis. An agent, designated HER virus, was isolated from affected rats, and subsequently was shown to be a strain of RV. Hemorrhagic encephalomyelopathy was produced experimentally by intracerebral inoculation of suckling rats. It also occurred after intraperitoneal inoculation of virus into young adult rats given cyclophosphamide. Typical neurological disease and hemorrhagic lesions appeared in 1 to 4 weeks, but in only 20% of inoculated rats. Margolis and Kilham (91) showed that similar lesions could be induced by parenteral inoculation with several strains of RV and with H-1 virus (Figs. 11 and 12). Cole *et al.*

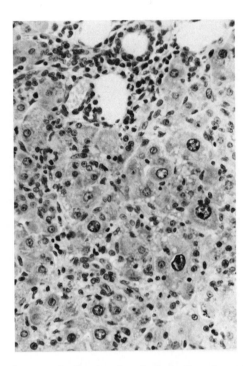

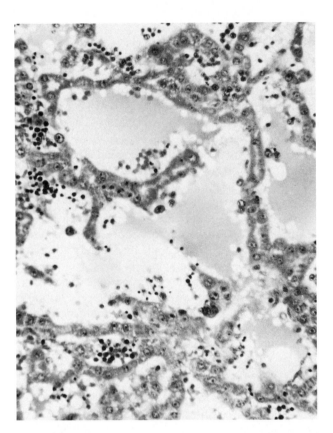

Fig. 6.   Spectrum of cell and nuclear size in the liver of a rat inoculated with RV. There is also biliary hyperplasia and nonsuppurative portal hepatitis. Rats from this study were inoculated as neonates or sucklings. Lesions were detected in rats 16 to 43 days old. [Courtesy of Dr. Margolis, Dr. Kilham, and Dr. Ruffolo (94); and *Experimental and Molecular Pathology*.]

Fig. 8.   Large vascular lakes (peliosis hepatitis) in the liver of a rat surviving RV hepatitis. Except for a rare endothelial cell or Kupffer cell only plates of hepatic cells line the cysts. [Courtesy of Dr. Margolis, Dr. Kilham, and Dr. Ruffolo (94); and *Experimental and Molecular Pathology*.]

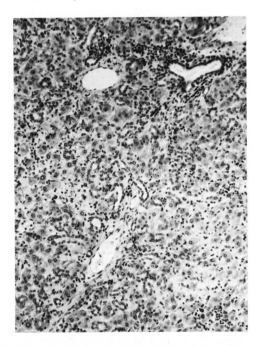

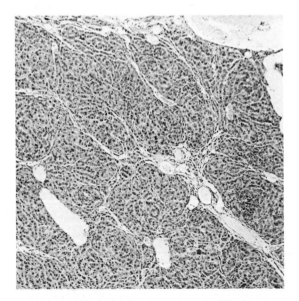

Fig. 7.   Liver of a 19-day-old rat inoculated at 8 days of age with H-1 virus. Numerous, small, dilated, hyperplastic biliary ducts extend from portal areas deep into hepatic lobules. Hypocellular zones are remnants of necrotic foci. [Courtesy of Dr. Margolis, Dr. Kilham, and Dr. Ruffolo (94); and *Experimental and Molecular Pathology*.]

Fig. 9.   Nodular hyperplasia and stromal collapse in the liver of a rat surviving RV hepatitis. Rat was inoculated at 8 days with HHP strain of RV and necropsied at 61 days. [Courtesy of Dr. Margolis, Dr. Kilham, and Dr. Ruffolo (94); and *Experimental and Molecular Pathology*.]

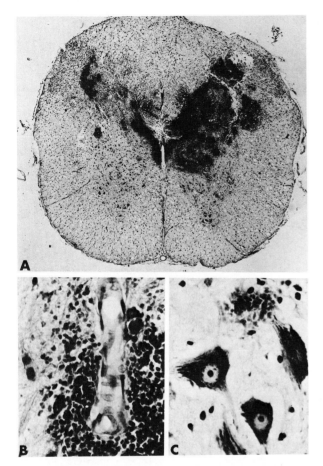

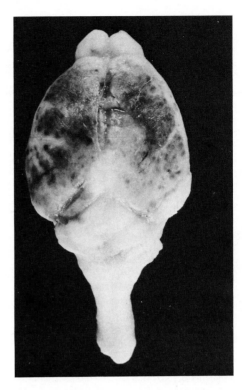

Fig. 11. Hemorrhagic encephalopathy in a suckling rat inoculated with RV. [Courtesy of Dr. Margolis and Dr. Kilham (91); and *Laboratory Investigation.*]

Fig. 10. Hemorrhagic myelopathy. Adult Lewis rat perfused at onset of paralysis 13 days after a single intraperitoneal dose of cyclophosphamide 100 mg/kg. Lumbar spinal cord. (A) Massive hemorrhage in both gray and white matter. (B) Area of typical bland hemorrhage in white matter, surrounding perfused vessel. (C). Small hemorrhage (top of frame) in anterior gray horn. [Courtesy Drs. N. Nathanson, G. A. Cole, G. W. Santos, R. A. Squire, and K. O. Smith (112); and the *American Journal of Epidemiology.*]

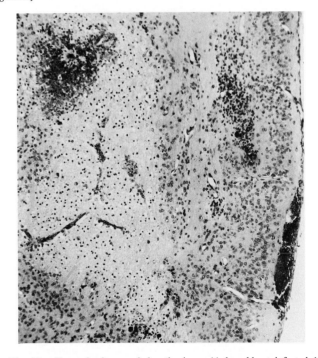

Fig. 12. Hemorrhagic encephalopathy in an 11-day-old rat infected 19 days previously with H-1 virus via intracerebral inoculation of the mother with H-1 virus. Hemorrhage, thrombosis, and malacia are evident. [Courtesy of Dr. Margolis and Dr. Kilham (91); and *Laboratory Investigation.*]

(30) reported that large doses of HER inoculated intracerebrally into newborn rats produced cerebellar destruction with hemorrhages throughout brain and spinal cord, whereas small doses produced jaundice and runting. Sucklings infected after day 10 remained well.

We have recently observed hemorrhagic lesions with thrombosis, and necrosis in the testicles and epididymis of weanling rats with naturally occurring RV infection (Figs. 13 and 14) (see Section II, A, 1, f.).

The hemorrhagic lesions appear to result from viral-induced injury to vessel walls. Inclusion bodies have been demonstrated in the vascular endothelium of small blood vessels and in arteriolar smooth muscle before extravasation of blood begins (Fig. 15) and hemorrhage is accompanied by vascular thrombosis and liquifaction necrosis (infarction) in the central

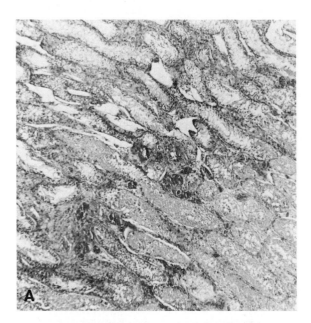 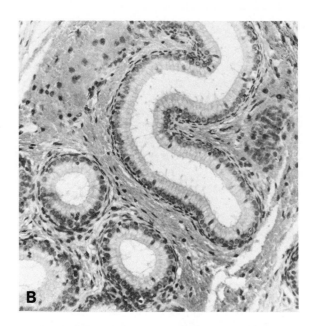

Fig. 13. Typical lesions in the testicle and epididymis from a natural outbreak of RV infection in young adult rats. (A) Testicle with coagulation necrosis of seminiferous tubules and interstitial hemorrhage. (B) Epididymis with interstitial hemorrhage.

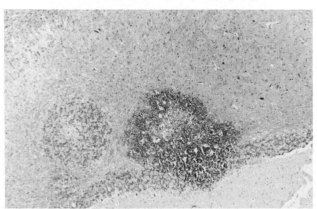

Fig. 14. Focal hemorrhage and malacia in the cerebellum of a young adult rat with naturally occurring RV infection.

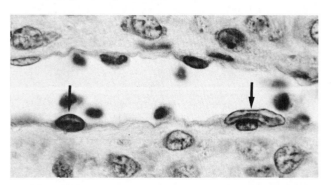

Fig. 15. Intranuclear inclusions in three endothelial cells (arrows). The rodlike form with surrounding halo is especially well shown in one cell cut in its long axis. [Courtesy of Dr. Margolis and Dr. Kilham (91); and *Laboratory Investigation.*]

nervous system (Fig. 16) (30,91). Baringer and Nathanson (5) found that platelet–fibrin aggregates attached perferentially to RV-infected cells and suggested that endothelial injury and hemorrhage followed activation of clotting near infected cells rather than from a direct cytolytic effect of RV. There is indirect evidence indicating RV may also interfere with blood coagulation. Margolis and Kilham (91) hypothesized that RV could cause coagulative disorders and promote hemorrhagic lesions through infection of hematopoietic tissues. They demonstrated, in a retrospective histological study (92), that RV inclusions can be found in megakaryocytes, and earlier reports by Portella (126) showed that RV had an affinity for erythrocytes. Corroborative coagulation studies on RV-infected rats have not been reported.

Since rat parvoviruses have a broad tissue tropism, clinical signs and lesions other than those described may occur. It seems likely that lesions of natural parvovirus infection will depend, as do their experimental counterparts, on the age and strain of the host and on the virulence and tropism of the strain of virus.

h. Epizootiology. Serological surveys show clearly that natural infections with RV and H-1 virus are common among laboratory and wild rats. Robey *et al.* (130) found that about 85% of more than 350 sera collected from Sprague-Dawley rats from three commercial breeders and from their own colony had HAI antibody to RV. Kilham (71) surveyed several populations of wild rats near Hanover, New Hampshire, and detected HAI antibody to both RV and H-1 virus. The proportion

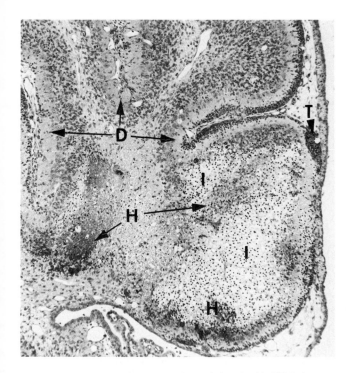

Fig. 16. Cerebellum of a 10-day-old rat infected with RV 7 days previously. There is incomplete destruction of the external germinal layer (D) a thrombosed superficial vessel (T), zones of hemorrhage (H), and infarction (I) of a folium. [Courtesy of Dr. Kilham and Dr. Margolis (91); and *Laboratory Investigation.*]

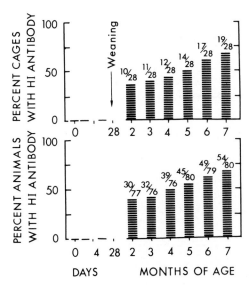

Fig. 17. Cohort study of RV infection in a closed colony. Frequency of hemagglutination inhibiting (HAI) antibody by age. After weaning, males and females were separated; cages refers to the littermates of one sex. Passively acquired maternal antibody is excluded. [Courtesy of Dr. N. Nathanson (131); and the *American Journal of Epidemiology.*]

of rats with antibody to each virus varied independently for each population. Robinson and co-workers (131) made an extensive seroepizootiological study of RV in a closed colony of 2000 McCollum rats with an annual production of about 3000 rats. Forty percent of rats without maternal antibody seroconverted to RV by 7 weeks of age, and 67% had anti-RV antibody by 7 months (Fig. 17). If several pups in a litter sustained infection, most cagemates developed antibody (Fig. 18). Passive antibody titers disappeared by 2 months. About 70% of 1000 juvenile rats were susceptible to infection at any given time, whereas 30% were actively immune (Fig. 19). Several attempts to isolate RV from immune rats were unsuccessful. Seroconversion to H-1 virus was not detected, but about half of the litters had transient maternal antibody to H-1 virus. These studies demonstrated that RV is readily transmitted among rats held in confined quarters and that infections are perpetuated by sustained introduction of susceptible hosts.

The epizootiological studies suggest that oral and perhaps respiratory infection are primarily responsible for natural horizontal transmission of RV. The studies of Kilham and Olivier (78) and Kilham and Margolis (73) indicated that natural vertical transmission can occur. Experimental studies from several laboratories support these impressions. Kilham and Margolis (74) inoculated pregnant rats orally with RV on the eleventh

day of gestation and showed that transplacental and fetal infection with fetal death occurred although dams remained asymptomatic. They also found that RV replication was widespread in tissues of pregnant dams and that dams developed viremia which contributed to placental infection. These observations parallel those of natural RV infection where signs may be limited to fetal resorption and reduced litter size (see Section II, A, 1, f). In contrast, H-1 virus infected rat fetuses only after intraperitoneal inoculation but not after oral inoculation of pregnant mothers. These workers suggested that in rats H-1 virus is more likely restricted to horizontal transmission primarily among sucklings and that adults are susceptible to clinical disease only if they are debilitated by intercurrent infections or leukemia. Lipton *et al.* (84) reported that viremia and virus infection of intestine, lung, liver, spleen, brain, and kidney followed intragastric inoculation of adult rats with the HER strain of RV. Furthermore, virus was excreted in feces but not in urine for up to 12 days, and inoculated rats were infectious for contact cohorts for 15 to 20 days. They were also able to infect rats by intranasal inoculation. Novotny and Hetrick (116) did find RV in urine of suckling rats inoculated parenterally as newborns with high doses (10^6 LD_{50}) of RV. Dams and siblings of infected sucklings developed HAI antibodies to RV which indicated that horizontal infection had ensued. Vertical transmission of RV was also confirmed, since litters born 2 to 5 days after intraperitoneal or intravenous (iv) inoculations of mothers had 100% mortality, whereas rats inoculated at least 2 weeks before conception delivered normal litters. Furthermore, litters of infected dams developed disease

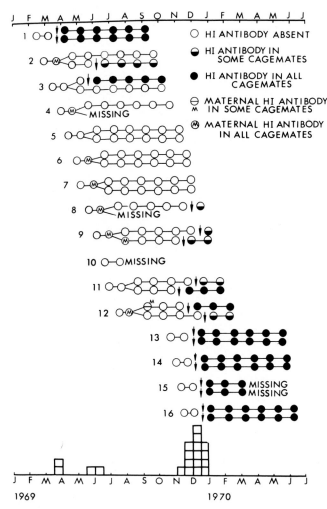

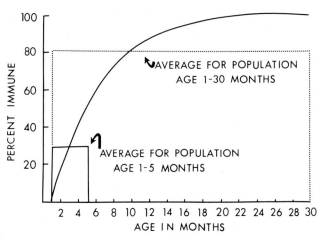

Fig. 19. Dynamics of RV in a population of juvenile rats (1 to 5 months old). The total population at any given time was about 1000 animals. Every month 250 rats entered the group, and an equal number were removed. Results are compared with those obtained from testing rats 1 to 30 months old. [Courtesy of Dr. N. Nathanson (131); and the American Journal of Epidemiology.]

Fig. 18. Cohort study of RV infection in a closed colony. Litters of rats bled periodically from 1 day to 27 weeks of age. Each litter (identified by number) is represented by a single line to age 3 weeks (when sexes were segregated) and by a double line (males above) thereafter. Date of infection of each group of cagemates is represented by an arrow and by a square on the histogram. [Courtesy of Dr. N. Nathanson (131); and the American Journal of Epidemiology.]

a single nonlethal dose (150 mg/kg) of cyclophosphamide (40). The repository of latent RV is unknown, but Margolis and Kilham (91) suggested that, since RV favors rapidly dividing cells, the gastrointestinal tract or hematopoietic system are likely sites. In addition, factors that contribute to RV latency and the role of latency in H-1 virus infection are not understood. Nevertheless, it must be assumed that seropositive rats are latently infected and can serve as a potential source of infection to all susceptible rats in a colony. Furthermore, since some strains of rat parvoviruses can be transmitted vertically, even caesarian-derived or germfree rats can not be assumed free of infection until they have been tested serologically.

i. Diagnosis. The diagnosis of parvovirus infection should be approached on four levels: (a) clinical, (b) morphological, (c) serological, and (d) virological.

Clinical signs in adult rats include neurological disturbances, sudden death, and hemorrhagic diatheses (e.g., scrotal hemorrhage and cyanosis). The incidence of clinical signs among infected adults is likely to be low, but may be exacerbated by immunological crippling. In breeding colonies, production may decrease, or neurological deficits and jaundice may occur among sucklings.

Morphological evidence of parvovirus infection is more difficult to detect since viruses are usually nonpathogenic for adults. An increase in fetal resorption sites in gravid uteri may occur. Spontaneous lesions rarely develop in young animals, but histological examination for persisting inclusion bodies occasionally may be rewarding. Other lesions in suckling and adult rats have been described in Section II, A, 1, g.

Inapparent infection is detected most easily by serological testing. Four serological tests are available (HAI, CF, NT, and

when nursed by normal mothers, whereas normal litters nursed on infected mothers remained well. Kilham and Margolis (73) showed that RV may be excreted briefly in milk by rats inoculated with RV in late pregnancy. Dams infected on the first postpartum day had RV in milk by 24 h and could excrete virus in milk up to 12 days and up to 5 days after HAI antibody appeared in their serum.

There is additional evidence that RV can persist as a latent infection in the presence of high titers of HAI antibody. Robey *et al.* (130) isolated RV from five clinically normal laboratory rats, all of which had HAI antibody to RV, and the rat with the highest antibody titer had the most widespread infection. Latent RV infection also was activated in clinically normal rats by

FA), but only the first three distinguish reliably between RV and H-1 infection (36,110,151,154). Hemagglutination inhibition tests are the simplest and least expensive to perform (131). The use of kaolin to remove nonspecific inhibitors from test serum has been recommended (49,110,131), but we and others have not found this necessary (31).

Serological evidence of infection should be confirmed, whenever possible, by viral isolation (see Section II, A, 1, d). Virus may be found in blood early in disease (viremia can persist for 2 to 10 days after experimental infection) (74). Fetuses, placenta, hematopoietic tissue, or gastrointestinal tract should, however, yield virus more consistently. Confirmation of RV infection in our laboratory is made by inoculation of test material into primary rat embryo cultures and into RV-free suckling rats. Cultures are checked for hemagglutinin production and CPE. The identity of isolates is confirmed by HAI assay with anti-RV and anti-H-1 virus immune serum. Sera are collected from inoculated rats after 2 to 4 weeks and are tested for anti-RV HAI antibody.

j. Differential Diagnosis. Rat virus infection may be subclinical or may cause lesions in the central nervous system, liver, vascular system, testes, and fetal death or resorption. Sporadic illness affecting these organs must be carefully evaluated for infection by other viruses, chemical toxicity (e.g. dicoumarin toxicity), neoplastic diseases including leukemia or pituitary adenomas, trauma, and genetic abnormalities such as hydrocephalus. Reduction in litters or litter size may be associated with Sendai virus and possibly with environmental factors. Since RV-infected colonies are common, specific antibodies per se only indicate subclinical infection has occurred. Associations between infection and clinical signs or lesions must be carefully made to confirm a cause and effect relationship. For example, dual viral infections may occur. Further, environmental or experimental factors may activate latent RV, while simultaneously causing lesions unrelated to RV infection.

k. Control. Since rat parvoviruses cause latent infections and may be transmitted vertically and/or horizontally, procedures to control infection must be planned carefully and adhered to stringently. Once a colony has been infected, infection can be eliminated effectively only by destroying the colony and repopulating from parvovirus-free stock. For experimental colonies, this usually requires purchase of rats from vendors who have adequate serological monitoring programs. For production colonies, it means rederiving breeding stock by caesarian section, nursing sucklings on parvovirus-free foster dams, and testing weanlings for antibody before they are used for breeding. Physical facilities for rederived stock should include pathogen-free barriers with personnel locks and proper air balancing to prevent reinfection. Periodic serological surveillance also is essential to detect possible reinfection as early as possible.

l. Interference with Research. The effects of rat parvoviruses on biological research are only partially understood and are potentially serious. For example, they can contaminate transplantable neoplasms (78,150) and established cell lines (49,163). Therefore, it is reasonable to assume, because of their predilection for rapidly dividing cells, that they could alter the growth characteristics of affected cells. Furthermore, parvoviruses are pathogenic for fetal and suckling rats and hamsters, so inadvertent inoculation, especially of young animals, with virus-contaminated cells could cause fatal infection. Immunosuppression also can activate latent RV infection in older rats (40), so rats subjected to immunosuppression or severe stresses, such as intercurrent infections or surgical trauma, may be at risk. Conversely, there is recent evidence that RV can suppress proliferative responses of infected lymphocytes to concanavalin A, phytohemagglutinin, or allogeneic lymphoid cells (25,26). If infected adult rats are used for experiments unaffected by latent viral infections, parvovirus infection can be tolerated, because it is unlikely that rats will develop clinical signs or lesions. Since all factors relevant to latent infection have not been identified, a decision to work with infected animals should be made cautiously. Finally, since these viruses are so hardy, instruments should be thoroughly disinfected after contacting infected animals to reduce the spread of infection.

2. Minute Virus of Mice (MVM)

This is a widely disseminated latent virus of mice. Experimentally inoculated neonatal rats developed viremia and mild necrosis of ependyma and choroid plexus, and infected cells contained intranuclear inclusion bodies. Virus also was detected on postinoculation day 6 in brain, liver, intestine, and urine (76). Low titers ($\leq$ 1 : 40) of HAI antibody to MVM have been found in wild rats (76) and in laboratory rats (119). Titers were reduced, however, by pretreating sera with receptor-destroying enzyme and attempts, to recover MVM from more than 100 seropositive rats proved unsuccessful. Therefore, the significance of anti-MVM antibody in rat serum is unresolved. Clinical signs or lesions due to natural MVM infection of rats have not been reported, and MVM has not been isolated from rats.

B. Herpesvirus (Cytomegalovirus)

Cytomegalovirus is the only known herpesvirus of rats. Similar viruses have been described for many species including humans, nonhuman primates, mice, guinea pigs, and rats

(133). They have physical and chemical characteristics typical of the Herpetoviridae, including a penchant for latency. As with other cytomegaloviruses, rat cytomegalovirus appears to have a predilection for salivary glands and has also been found in lacrimal glands (87). Typical lesions include cytomegaly, intranuclear inclusion body formation in acinar or ductal epithelium, and mild nonsuppurative interstitial inflammation (79,87,128). Kuttner and Wang (79) also showed that salivary glands of affected wild rats contained infectious virus by transmitting disease by intraglandular and ic inoculation of several "young" rats with emulsions of submaxillary gland.

The epizootiology of rat cytomegalovirus has not been well studied, but virus and/or antibody have been detected in wild rats in several widely separated areas of the world (79,128). The potential the virus has for interfering with research has not been evaluated. Infection is usually diagnosed by histological examination. Anti-viral antibody can be detected by NT test (128), but rapid serological tests are not available. Rat cytomegalovirus can be grown in primary rat fibroblasts, rat kidney cells, and hamster kidney cells (6,128).

C. DNA Virus Which May Infect Rats

Mouse Adenovirus

We have found that some rat sera contain CF antibodies to mouse adenovirus, but clinical disease or lesions attributable to infection of rats with this virus have not been detected, and virus has not been recovered from rats.

III.   RNA VIRUSES

A.   Coronaviruses (Sialodacryoadenitis Virus and Rat Coronavirus)

1.   General

Coronaviruses are lipid solvent-labile, pleomorphic, 60 to 220 nm particles with characteristic clublike projections (corona) uniformly arranged on their surfaces. They multiply in the cytoplasm and mature by budding through endoplasmic reticulum. Coronaviridae are fairly species specific and have been identified as etiologic agents in diseases of humans, pigs, bovines, rats, mice, dogs, chickens, and turkeys. They generally infect the gastrointestinal tract and its associated glandular organs and/or the respiratory tract. Reviews of coronavirus biology are available (17,103). Two strains of coronavirus have been identified as important pathogens of laboratory rats; sialodacryoadenitis virus (SDAV) (15) and rat coronavirus (RCV) (120).

2.   History

In 1961, Innes and Stanton reported two outbreaks of clinical disease in weanling rats characterized by cervical edema and "red tears" (59). They described the morphology of the disease in considerable detail and named it sialodacryoadenitis from the characteristic lesions: inflammation and edema of salivary and lacrimal glands. Hunt (58) described a similar disease of young rats, but inflammation was restricted to the intraorbital lacrimal glands and was accompanied by keratoconjunctivitis. Innes and Stanton suggested an infectious agent caused the disease, and Hunt detected acidophilic intranuclear inclusion bodies in affected Harderian glands and in conjunctival mucosa, but viral isolations were not attempted in either study. Ashe and co-workers (3,4) isolated a transmissible cytopathic viral agent from the submaxillary glands of gnotobiotic rats that hemagglutinated rabbit erythrocytes. Ashe's virus apparently was not associated with clinical signs or lesions in infected rats (see Section IV,B). Jonas et al. (67) induced sialodacryoadenitis in germfree rats by intranasal inoculation of an ultrafiltrate of diseased submaxillary salivary gland. Virus particles were detected in ducts of submaxillary glands from experimentally infected rats by electron microscopy, but attempts to isolate an agent in vitro were initially unsuccessful. However, when neonatal mice were inoculated ic with submaxillary gland homogenate they developed severe neurological deficits and died in 3 to 6 days. Brain homogenates from affected mice caused sialodacryoadenitis in intranasally inoculated rats. Bhatt and co-workers (15) subsequently isolated a virus from salivary glands of affected rats by inoculation of neonatal mice and primary rat kidney (PRK) monolayer cultures. The isolate had serological and physicochemical characteristics of a coronavirus. It was lethal for infant mice after ic inoculation but not for weanling mice. Mouse brain-passaged virus induced sialodacryoadenitis in susceptible rats.

In 1964 Hartley and associates found that some rat sera contained antibody to mouse hepatitis virus (MHV) (51). They suggested that an agent antigenically related to MHV could elicit anti-MHV antibody in rats. Parker et al. (120) offered support for Hartley's theory by isolating a coronavirus antigenically related to MHV from lungs of infected but clinically normal rats. Parker's virus (RCV) was subsequently shown to be antigenically related to both SDAV and MHV (15), and the current view is that SDAV and RCV may be different strains of one rat coronavirus. The biology of each virus is discussed separately in each of the following sections.

3.   Viral Characteristics

a. SDAV. The diameter of SDAV particles, determined by ultrafiltration, is 100 to 220 nm. Jonas et al. (67) described a 60- to 70-nm particle in ductal epithelium of experimentally

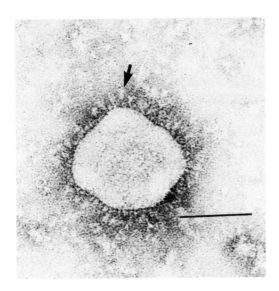

Fig. 20. Negative stained preparation of SDAV. Note projections typical of coronaviruses (arrow). Bar represents 100 nm. ×200,000. (Courtesy of Dr. M. Lipman.)

infected submaxillary glands by electron microscopy. Preliminary ultrastructural studies of SDAV propagated *in vitro* indicate it has the typical morphology of a coronavirus (Fig. 20) and is about 114 nm in diameter (83). The differences in size reported probably reflect differences in the methods of measurement and the source of virus as well as the pleomorphism characteristic of coronaviruses. Sialodacryoadenitis virus is sensitive to lipid solvents, but it is relatively stable at acid pH (3.0). It also is stable in 3% fetal bovine serum (in phosphate-buffered saline) at 4°C for up to 7 days, at 37°C for up to 3 h, and at 56°C for less than 5 min. It can be stored at −60°C for more than 7 years but loses infectivity rapidly if stored at −20°C. The virus replicates intracytoplasmically and replication is not affected by 5-bromodeoxyuridine (BUdR). Detailed morphological studies of viral maturation and release have not been reported. The virus does not hemagglutinate rabbit, guinea pig, or goose erythrocytes at 4°, 25° or 37°C (15).

Sialodacryoadenitis virus is closely related antigenically to

Table III

Results of Cross-Complement-Fixation Tests with SDAV and MHV and Their Respective Immune Sera[a]

Antigen	Antisera		
	SDAV-A	SDAV-B	MHV
SDAV	128/≥256[b]	64/128	160/128
MHV	32/32	16/16	80/≥64

[a] After Bhatt *et al.* (15) with modification.
[b] Highest dilution of serum reacting with the lowest dilution of antigen divided by the highest dilution of antigen reacting with the lowest dilution of serum.

Table IV

Results of Cross-Neutralization Tests with SDAV and RCV and Their Respective Immune Sera[a]

Antigen	Species immunized	Antibody titer versus			
		SDA		RCV	
		1	2	1	2
SDA	Mice	1:253	1:452	1:67	1:100
RCV	Rats	1:67	1:100	1:284	1:272

[a] After Bhatt *et al.* (15) with modification.

RCV, to MHV (Tables III and IV) and a human coronavirus (OC38), but its antigenic relationship to other coronaviruses has not been tested. The virus is not antigenically related to the pox-, herpes-, areno-, paramyxo-, rhabdo-, reo- or togavirus groups (15).

b. RCV. Rat coronavirus also is a typical coronavirus morphologically and, by negative staining, measures 50–118 nm in diameter including the corona. Like SDAV, it is destroyed by ether, chloroform, or heating to 56°C for 30 min. It does not hemagglutinate mouse, chicken, guinea pig, sheep, or human erythrocytes at 4°, 22°, or 37°C (120).

4. *In Vitro* Cultivation

a. SDAV. Sialodacryoadenitis virus causes a characteristic CPE in PRK monolayer cultures with formation of multinucleated giant cells (15). Cultures can be prepared from SDAV-free or infected rats, since SDAV does not infect kidney. Cultures are most sensitive if inoculated by 1 week after seeding. Older cultures appear to be less sensitive to viral replication, and CPE may not develop. Virus can be detected in the cytoplasm by immunofluorescence within 12 h after inoculation. Cytopathic effects appear by 24 h and extensive lysis occurs shortly after. Sialodacryoadenitis virus does not replicate in BHK-21, VERO, or HEp-2 cell lines; a line of polyoma-transformed mouse cells (Py-A1/N); or mouse cell line NCTC 1469 which supports growth of MHV (15). Sialodacryoadenitis virus can be plaqued in PRK monolayers only after serial *in vitro* passage.

b. RCV. Methods for and results of cultivation of RCV resemble those described for SDAV (120).

5. Host Range

a. SDAV. Early work showed that SDAV is pathogenic for in inoculated adult rats and for ic inoculated infant mice (15,61,67). Recent evidence indicates that SDAV also is infectious and pathogenic for weanling mice (12). Contact-exposed mice developed NT antibody to SDAV, whereas in inoculated mice developed NT and CF antibody. The virus was recovered from the respiratory tract for up to 7 days postinfection, and mice developed interstitial pneumonia. Anti-SDAV and anti-MHV antibody also has been found among retired breeder mice from colonies thought to be free of MHV. Since SDAV and MHV are antigenically related, SDAV infection should be considered if unexpected or unexplained seroconversions to MHV occur in mouse colonies. Seroconversions to MHV from infection of mice or rats exposed to human coronaviruses (e.g., carried by animal technicians) also should be considered (Hartley *et al.*, 1964) but has not been studied.

Extensive host range studies of SDAV have not been done, but preliminary trials with several strains of rats and mice suggest that various strains of SDAV may vary in infectivity and antigenicity (14). For example, during spontaneous outbreaks, WAG/Rij rats developed severe clinical disease, whereas DA rats developed primarily subclinical disease. Furthermore, some strains of mice developed both CF and NT antibody following experimental SDAV infection, whereas others produced only NT antibody. Conversely, one strain of SDAV induced only NT antibody in a given mouse strain whereas a second strain of SDAV induced both CF and SN antibody. These variations are important for interpretation of diagnostic and epizootiological data.

b. RCV. Host range studies with RCV also have been limited. Rat coronavirus is infectious for rats and induces seroconversion to RCV, MHV, and SDAV (11,15,120). Its pathogenicity varies with strain and age but is greatest for suckling rats. For example, mortality among intranasally inoculated Fischer 344 rats less than 48 h old approached 100%, whereas comparable Wistar rats had only 10 to 25% mortality. Furthermore, deaths among Fischer 344 sucklings occurred 6 to 12 days after infection, whereas Wistar sucklings usually died after 12 days. Resistance to mortality, however, among even highly susceptible sucklings, increased rapidly so that rats inoculated after 7 days of age had nonfatal respiratory disease and weanlings were asymptomatic (120). The pathogenicity and infectivity of RCV for other species have not been reported.

6. Clinical Disease

a. SDAV. Susceptible rats can be infected at any age, but clinical disease usually occurs in one of two patterns: endemic infection of breeding colonies or explosive outbreaks among nonimmune rats exposed to virus as weanlings or adults. In the former setting, adults may have clinical signs, but more commonly they are immune. Therefore, clinical disease develops among susceptible sucklings and is characterized by so-called "winking and blinking" associated with acute inflammation of

the eye and adnexae. Signs are transient (1 week or less) among individual sucklings, but affected animals will be prevalent among the suckling population as long as newly susceptible litters are available to become infected. In the latter situation, either new, SDAV-susceptible rats are placed in a room with infected rats or an infected rat(s) is placed in a room housing nonimmune weanlings or adults. Generally, within 1 week, the susceptible population will develop typical signs of SDAV infection. For individual rats they include cervical swelling (edema) with palpable enlargement of submaxillary salivary glands, sneezing or repeated wiping of the external nares with the forepaws, photophobia and nasal, and ocular discharges which are often red-tinged due to a high content of porphyrin pigment. Clinical signs last about 1 week. They may be mild or severe, and all signs do not occur in every infected rat. This last point is especially significant, since a single subclinically infected rat placed in a susceptible colony can initiate a full enzootic episode.

Keratoconjunctivitis has been associated with several natural outbreaks of SDAV (80,161) and may be the only clinically detectable evidence of SDAV infection. Signs and lesions commonly begin by the time of weaning, but also can occur in adults. They include photophobia, lacrimation, circumcorneal flush, diffuse corneal opacities, corneal ulcers, pannus, hypopyon, and hyphema. Lesions usually resolve completely in 1 to 2 weeks, but chronic active keratitis and megaloglobus may develop in some rats. The morbidity of eye lesions during an acute outbreak of SDAV infection varies from 0 to 100%, but is usually 10 to 30%. The prevalence of eye lesions seems greater among breeding colonies chronically infected with SDAV (65). The severity of lesions also may vary among strains of rats. In our experience, inbred Lewis and WAG/Rij rats are more susceptible to SDAV-associated eye disease than DA rats or outbred CD rats (65). Weisbroth and Peress (161) found that the spontaneously hypertensive strain TAC:SHR/N also was highly susceptible. The pathology of the eye lesions is discussed in greater detail in Section III, A, 7.

b. RCV. Rat coronavirus infection is subclinical in postweaned rats. Nonfatal respiratory disease can occur in intranasally inoculated sucklings, and intranasally infected susceptible neonates may die (11,120).

7. Pathology

a. SDAV. The lesions of SDAV infection have been described in detail by several groups of workers (59,61,67,80,161).

Gross Lesions of SDAV infection usually are restricted to mixed or serous salivary glands, lacrimal glands, cervical lymph nodes, thymus, and occasionally lung. Submaxillary and parotid salivary glands and cervical lymph nodes are un-

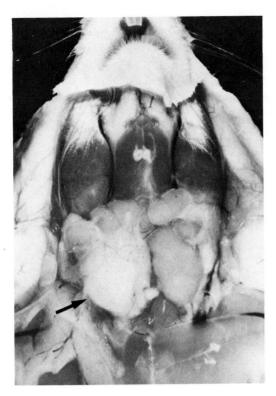

Fig. 21. Swollen pale submaxillary gland (arrow) in a rat inoculated intranasally with SDAV 5 days previously. The cervical lymph nodes are also moderately enlarged.

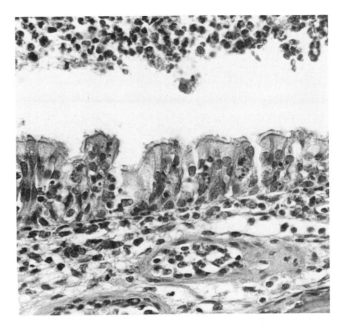

Fig. 22. Nasal mucosa from a rat inoculated intranasally with SDAV 2 days previously. There is focal necrosis of epithelium and the lamina propria is infiltrated by lymphoid cells and neutrophils. Inflammatory exudate is present in the meatus. [Courtesy of Dr. R. O. Jacoby, Dr. P. N. Bhatt, and Dr. A. M. Jonas; and *Veterinary Pathology*.]

ilaterally or bilaterally enlarged, pale yellow to white, and edematous (Fig. 21), although they may have red spots or a red tinge if they are congested. Periglandular connective tissue usually is severely edematous. The exorbital and intraorbital lacrimal glands also may be swollen, and the Harderian gland may be flecked with yellow-gray foci. Brown-red mottling of the Harderian gland is due to its normal content of porphyrin pigment and should not be interpreted as a lesion. The thymus may be small and the lungs may be spotted with small grey foci.

Histological changes of SDAV infection are found in the respiratory tract, salivary glands, thymus, cervical lymph nodes, eye, exorbital and intraorbital lacrimal glands, Harderian gland and other ocular adnexae.

Nasopharyngeal lesions include multifocal necrosis of respiratory epithelium with inflammatory edema of the lamina propria (Figs. 22 and 23). Nasal meatuses may contain neutrophils, mucus, and necrotic cell debris. Focal necrosis of glands in the lamina propria also occurs. Mild, nonsuppurative tracheitis with focal necrosis of mucosal epithelium also may occur, and hyperplastic peribronchial lymphoid nodules develop. Pneumonia has not been reported in either naturally or experimentally infected adult rats, but interstitial pneumonia may occur in naturally infected sucklings (65).

Salivary gland lesions (submaxillary and parotid) begin as necrosis of ductular epithelium which rapidly progresses to diffuse acinar necrosis (Fig. 24). Ultrastructural and immunofluorescence studies of experimentally infected rats have confirmed that SDAV has a predilection for salivary ductal epithelium (Fig. 25) (61,67). Moderate to severe interstitial inflammatory edema develops and glandular architecture is rapidly effaced (Fig. 26). Inflammatory edema and occasional focal hemorrhage are prominent in periglandular connective

Fig. 23. Nasal mucosa with fluorescing SDAV antigen in the cytoplasm of epithelial cells. [Courtesy of Dr. R. O. Jacoby, Dr. P. N. Bhatt, and Dr. A. M. Jonas; and *Veterinary Pathology*.]

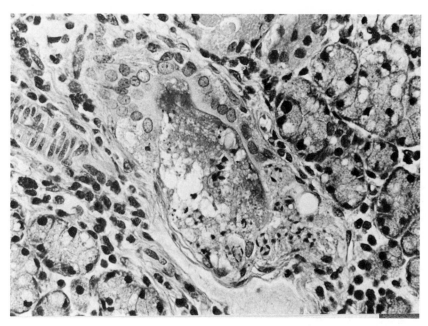

Fig. 24. Submaxillary gland 4 days after intranasal inoculation of SDAV. There is necrosis of a large salivary duct and periductular inflammation. Adjacent acini have early signs of degeneration characterized by formation of intracytoplasmic vacuoles. [Courtesy of Dr. R. O. Jacoby, Dr. P. N. Bhatt, and Dr. A. M. Jonas; and *Veterinary Pathology*.]

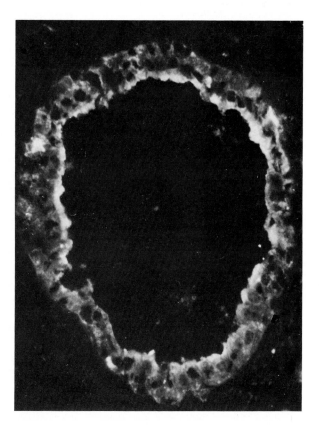

Fig. 25. A large salivary duct from a rat infected with SDAV. SDAV antigen in the cytoplasm of ductal epithelial cells demonstrated by immunofluorescence.

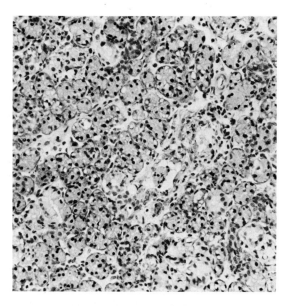

Fig. 26. Submaxillary gland from a rat infected intranasally with SDAV 5 days previously. There are advanced degenerative changes including acinar necrosis, interstitial edema, and inflammation. [Courtesy of Dr. R. O. Jacoby, Dr. P. N. Bhatt, and Dr. A. M. Jonas; and *Veterinary Pathology*.]

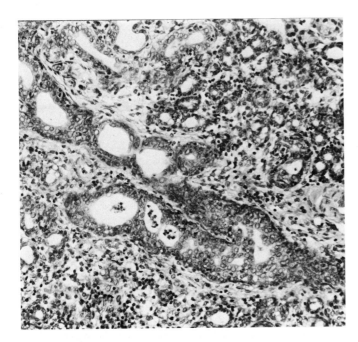

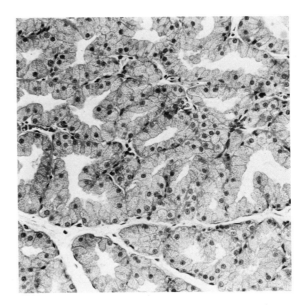

Fig. 27. Submaxillary gland 9 days after intranasal inoculation with SDAV. There is prominent squamous metaplasia of ducts. Note regenerating acini at upper right. [Courtesy of Dr. R. O. Jacoby, Dr. P. N. Bhatt, and Dr. A M. Jonas; and *Veterinary Pathology*.]

Fig. 28. Normal rat Harderian gland. [Courtesy of Dr. R. O. Jacoby, Dr. P. N. Bhatt, and Dr. A. M. Jonas; and *Veterinary Pathology*.]

tissue. Repair begins 5 to 7 days after infection and is characterized by prominent squamous metaplasia of ductular epithelium and by proliferation of hyperchromatic regenerating acinar cells (Fig. 27). The sublingual salivary gland which is exclusively of the mucus-secreting type, is not affected.

Lacrimal gland lesions develop in essentially the same pattern as salivary gland lesions, i.e., necrosis and inflammation is followed by regeneration with squamous metaplasia of ducts or tuboalveolar units (harderian gland) and acinar regeneration (Figs. 28 and 29).

Squamous metaplasia in all glands subsides by 30 days postinfection, and the cytoarchitecture is restored to normal because basement membrane is not destroyed during necrotic and inflammatory phases of the disease.

Cervical lymph nodes may be hyperplastic with focal necrosis and perinodal edema, and focal necrosis of thymus with widening of interlobular septums is common.

After experimental intranasal inoculation of adult rats, viral replication and lesions occur first in the respiratory tract, then in the salivary glands and exorbital glands, and finally in the intraorbital lacrimal glands (Fig. 30) (61). It is common, however, in either spontaneous or experimental infection, to encounter a varied distribution of lesions. For example, a rat may have marked submaxillary sialoadenitis, whereas the parotid gland may remain normal or dacryoadenitis may occur without sialoadenitis. Furthermore, lesions may be unilateral or bilateral. The factors responsible for these variations have not been

identified. Also it is not clear how virus spreads from the respiratory tract to the salivary glands, since attempts to detect viremia or retrograde infection of salivary excretory ducts were not successful (61).

Several groups have reported keratoconjunctivitis in rats with naturally occurring SDAV infection, but eye lesions have

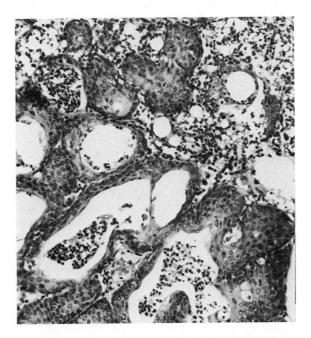

Fig. 29. Harderian gland from a rat inoculated with SDAV 10 days previously. There is prominent squamous metaplasia indicating repair, but some necrosis and inflammation remains. [Courtesy of Dr. R. O. Jacoby, Dr. P. N. Bhatt, and Dr. A. M. Jonas; and *Veterinary Pathology*.]

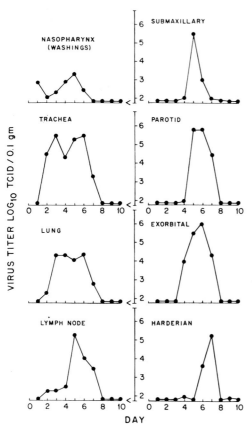

Fig. 30.   Titers of virus in tissues of germfree rats 10 days after intranasal inoculation of $10^{4.0}$ TCID$_{50}$ of SDAV. [Courtesy of Dr. R. O. Jacoby, Dr. P. N. Bhatt, and Dr. A. M. Jonas; and *Veterinary Pathology*.]

Fig. 31.   Weanling rat with SDAV-associated keratitis. The cornea is opaque and ulcerated. There is keratinaceous debris on the surface of the cornea and hypopyon.

b. RCV.   Parker and colleagues (120) showed that RCV induced lethal interstitial pneumonia in intranasally inoculated newborn rats. Lesions developed by 4 days postinoculation and

produced experimentally (52,80,161). Lai *et al.* (80) recently described an outbreak of keratoconjunctivitis in a closed colony of inbred Lewis rats. Lesions were most prominent in weanlings and included focal or diffuse interstitial keratitis, corneal ulceration, synechia, hypopyon, hyphema, and conjunctivitis (Figs. 31–34). Lesions resolved in most rats by 6 weeks of age, but about 6% of affected rats developed chronic keratitis with megaloglobus and lenticular and retinal degeneration. Similar lesions in adult rats, excluding chronic sequellae, were reported by Weisbroth and Peress (161). Affected rats from both outbreaks had a high incidence of dacryoadenitis and serum antibody to SDAV. In the Yale study (80) SDAV was recovered from the respiratory tract of some rats, but virus was not detected in the eyes either by tissue culture isolation techniques or by immunofluorescence. It has been suggested that inflammation and necrosis of lacrimal glands during SDAV infection results in impedence to flow of lacrimal fluids, proptosis, and keratitis sicca. Ostensibly, normal conjunctival bacteria could proliferate under these conditions and increase the severity of lesions (80,161).

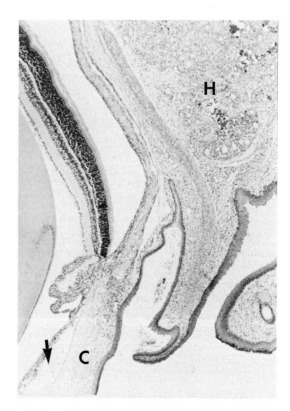

Fig. 32.   SDAV-associated keratoconjunctivitis. The Harderian gland (H) is effaced. The cornea (C) and conjunctival mucosa are diffusely infiltrated with inflammatory cells. There are also inflammatory cells in the anterior chamber (arrow). The retina is artifactually detached.

included hyperemia, mononuclear cell infiltrates, focal atelectasis, and emphysema. Sialodacryoadenitis was not observed.

Bhatt and Jacoby (11) have described rhinotracheitis and focal interstitial pneumonia in adult germfree rats inoculated intranasally with RCV. Virus was recovered from the respiratory tract for 7 days (Fig. 35), and viral antigen was detected in mucosal epithelium of the nasopharynx and in alveolar septae of some rats. Upper respiratory lesions were nearly identical to those described for experimental SDAV infection. Gross pulmonary lesions were limited to scattered red-brown to gray foci. Histologically, peribronchial lymphoid cell hyperplasia was detected by day 5. Alveolar septa contained mononuclear cells and neutrophils, and inflammatory cells occupied some adjacent alveolar spaces. Pneumocytes, macrophages, and edema fluid also filled alveolar spaces in some lungs. Lesions were mild and focal and subsided by day 7 (Fig. 36). Salivary gland lesions were rare, but, when present, were identical to those caused by SDAV. Dacryoadenitis was not detected.

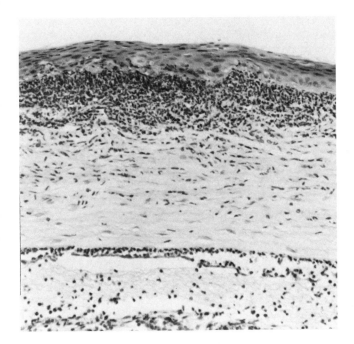

Fig. 33. Cornea from a weanling rat with SDAV-associated keratoconjunctivitis. The stroma is mildly edematous and diffusely infiltrated with neutrophils, but the heaviest accumulations are in the superficial layers. The anterior chamber contains inflammatory exudate.

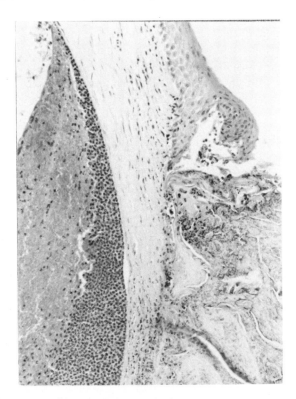

Fig. 34. Corneal ulceration in SDAV-associated keratoconjunctivitis. Masses of keratinized, necrotic cell debris have accumulated over the ulcer. There is also severe hyphema.

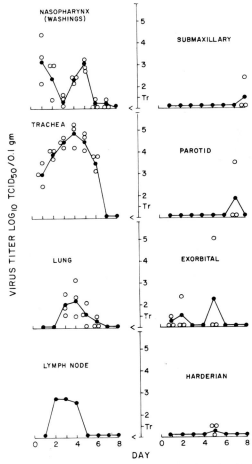

Fig. 35. Titers of virus in tissues of germfree rats inoculated intranasally with $10^{4.6}$ TCID$_{50}$ of RCV. [Courtesy of Dr. P. N. Bhatt and Dr. R. O. Jacoby; and *Archives of Virology.*]

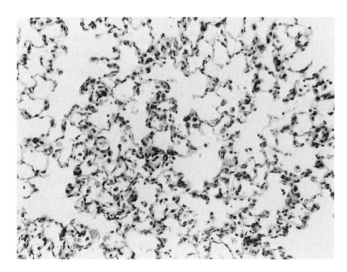

Fig. 36. Lung from a rat 5 days after intranasal inoculation with RCV. There is interstitial pneumonia characterized by infiltration of alveolar septae with mononuclear cells. [Courtesy of Dr. P. N. Bhatt and Dr. R. O. Jacoby; and *Archives of Virology.*]

Table V

Serum Antibody Titers of Germfree Rats Inoculated Intranasally with SDAV[a]

Test	Titer	No. of rats with titer days after inoculation									
		1	2	3	4	5	6	7	8	9	10
Complement fixation	<1:10	3	3	3	3	3	3	2	2		
	1:10							1			
	1:20								1	2	1
	1:40									1	1
	>1:40										1
Neutralization	<1:10	3	3	3	3	3	3				
	1:10							3	3	1	1
	1:20									2	
	1:40										2
	>1:40										

[a] After Jacoby *et al.* (61) with modification. Three rats sampled each day.

8. Epizootiology

a. SDAV. Sialodacryoadenitis virus infection is nonfatal, but is highly contagious and spreads easily and rapidly among susceptible rats in a colony room by contact, aerosol, or fomite. Morbidity is normally highest among late suckling or weanling rats, but susceptible rats of any age can be infected. Infected rats excrete virus from the respiratory tract for about 7 days, at which time anti-SDAV antibody is first detectable in serum by either NT or CF tests (Fig. 30 and Table V) (61). There is no evidence for a carrier state, so recovered rats should be considered free of virus and immune. Therefore, propagation of virus in a colony depends on continuous introduction of susceptible rats. The possibility remains, however, that as with some human coronavirus infections (109) rats immune to a strain of SDAV could be reinfected with the same strain or with an antigenically different strain. There also is no evidence that SDAV can be vertically transmitted, but this point has not been adequately tested.

The incidence of SDAV infection cannot be determined reliably by clinical signs, since infection may remain subclinical. Therefore, serological methods are preferred. Serum titers of CF and NT antibody should be determined simultaneously, since recent studies of natural and experimental SDA indicate that CF antibody titers rise during outbreaks then decline relatively rapidly, whereas NT titers increase during infection, but decline more slowly (Fig. 37) (14). Therefore, CF antibody titers appear be a better marker for current or recent infection, whereas NT antibody titers are a more reliable indicator of previous infections.

Sialodacryoadenitis virus infection is probably widespread among rat colonies. We recently tested retired breeders from six vendors, and all had anti-SDAV antibody (10). Nevertheless, confirmation that antibody detected during a given outbreak is due exclusively to SDAV infection may be difficult to obtain, since RCV and SDAV are closely related antigenically, and seroconversions for both viruses and to MHV follow infection by either RCV or SDAV. Bhatt *et al.* (15) have shown, however, that antibody titers to the homologous virus are usually higher than those to the heterologous, antigenically related viruses (Tables III and IV). This finding, coupled with documentation of clinical signs and lesions can improve the accuracy of epizootiological studies.

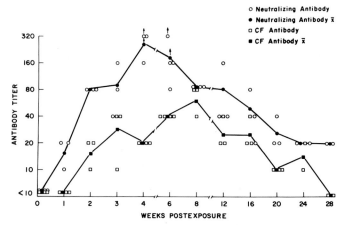

Fig. 37. Serum antibody titers to SDAV in specific pathogen-free rats after a single intranasal inoculation of virus. Rats were kept in germfree isolators for this study. Arrows = titer equal to or greater than indicated value; X̄ = average titer.

b. RCV. The epizootiology of RCV has not been reported in detail aside from Parker *et al.* (120) who demonstrated that RCV-infected rats developed anti-MHV antibody. Rat coronavirus is cleared from tissues of experimentally inoculated rats by postinoculation day 7, and there is no evidence for a carrier state. It may be assumed, for the present, that RCV is highly infectious and that it follows epizootiological patterns similar to those of SDAV.

9. Diagnosis of SDAV and RCV Infection

Infections of SDAV or RCV can be diagnosed on the basis of clinical signs, lesions, and serological profiles of NT and CF antibody and confirmed by isolation of the causative virus. Active SDAV infection should not be diagnosed solely on the basis of serological data, since anti-SDAV antibody can persist in previously infected rats for many weeks (see Section III, A, 8). The duration of anti-RCV antibody in rat serum after infection has not been reported. The viruses can be demonstrated in tissues by immunofluorescence for about 7 days after exposure and can be isolated in PRK cultures or by ic inoculation of neonatal mice (15).

It is difficult to differentiate RCV from SDAV infection virologically or serologically since the viruses are similar antigenically, physicochemically, and in their *in vitro* growth characteristics. As noted previously, however, if CF and NT antibody titers to SDAV and RCV are compared, titers to the homologous virus will likely be slightly higher than those to the heterologous virus (Table III) (15). Comparison of experimental infections with SDAV and RCV have revealed additional differences (Table VI). First, clinical signs of rhinitis and sialodacryoadenitis are common during SDAV infection and keratitis may occur, whereas RCV infection is asymptomatic. Second, RCV causes mild interstitial pneumonia in adults, whereas SDAV does not. Third, RCV replicates poorly in salivary and lacrimal glands and only rarely produces lesions, whereas SDAV is highly pathogenic for these tissues. Nevertheless, suitable caution should be maintained in differentiating these infections since the experimental data described were obtained solely from CD rats. Mitigating factors such as strain of virus, strain of rat, and age remain to be examined.

10. Differential Diagnosis

Clinically, cervical swelling from inflammatory edema and salivary gland enlargement is virtually pathogonomic for rat coronavirus infection and is more characteristic of SDAV infection than of RCV infection. Porphyrin-tinged nasoocular exudates can accumulate during SDAV infection, but they also occur in murine respiratory mycoplasmosis. Ammonia fumes,

Table VI

Comparison of the Major Features of Experimental Infection with SDAV and RCV in Adult Germfree CD Rats[a,b]

Feature	SDAV	RCV
Clinical signs		
Photophobia	Yes	No
Sneezing	Yes	No
Cervical swelling	Yes	No
Viral replication		
Respiratory system	Yes	Yes
Salivary glands	Yes	Trace
Lacrimal glands	Yes	Trace
Lesions		
Acute rhinotracheitis	Yes	Yes
Focal interstitial pneumonia	No	Yes
Sialoadenitis	Yes	Trace
Dacryoadenitis	Yes	No
Antibody response		
Complement fixing	Yes	No[c]
Neutralizing	Yes	Yes

[a] After Bhatt and Jacoby (11) with modification.
[b] See text for additional discussion.
[c] Up to 8 days postinoculation. Complement fixing antibody can occur in rats tested at later times.

especially from urine-soaked bedding, can cause acute inflammation of the eye and nose resembling SDAV infection.

Morphologically, sialoadenitis and dacryoadenitis are characteristic of rat coronavirus infections. Salivary and lacrimal gland lesions may, however, be mild and transient or may not develop, especially during RCV infection, and detectable lesions may involve only the respiratory tract. They must then be differentiated from infections caused by *Mycoplasma pulmonis,* Sendai virus, or pathogenic bacteria. Lesions associated with these agents are generally more severe than those caused by coronaviruses, and have been described elsewhere* (21,82). Dual infections, for example, with SDAV and Sendai virus, may also complicate interpretation of lesions unless adequate serological and virological tests are performed.

Infection with SDAV also must be differentiated from rat cytomegalovirus infection. The latter is clinically silent, lesions are usually mild and are characterized by enlarged salivary ductal epithelial cells with intranuclear inclusions (79). Hunt (58) detected intranuclear inclusions in the Harderian glands of rats with dacryoadenitis which may have been virus related, but inclusions have not been found in confirmed natural or experimental SDAV infection. Bacterial keratocon-

*"A Guide to Infectious Diseases of Mice and Rats," A Report of the Committee on Laboratory and Animal Diseases, Institute of Laboratory Animal Resources. Natl. Research Council, National Academy of Sciences, Washington, D.C., 1971.

junctivitis can also occur in rats, but SDAV must be eliminated as an underlying cause (53,164).

11.   Control of SDAV and RCV Infection

Rat colonies can be maintained free of coronaviruses if they are kept under rigid barrier conditions and are handled by personnel familiar with proper disinfection procedures for specific pathogen-free facilities. The key to effective control in infected colonies stems from recognition that infected rats shed virus for about 7 days and then are immune and that latent infections do not occur. Since SDAV spreads rapidly through a susceptible colony, all rats in a room should be infected and immune in 3 to 5 weeks. Thus, infected holding or experimental colonies should be quarantined for at least 4 weeks and preferably 6 to 8 weeks after infection is detected. In production colonies, breeding should cease for 6 weeks and weanlings should be removed from the room, since sucklings and weanlings of infected dams are particularly susceptible to infection. The quarantine period may be reduced by increasing contact exposure among susceptible rats. If susceptible rats are eliminated (e.g., weanlings and rats introduced from other colonies) the infection will disappear (14). However, once immune rats are replaced by susceptible rats, the opportunity for infection again increases. In addition, the possibility of reinfection with the same strain or with a different strain of coronavirus has not been ruled out. The immune status of the colony and of any rats to be introduced can be established by serological monitoring. There are no published reports of effective vaccination protocols for SDAV or RCV infection.

Rat coronaviruses are quite labile, so routine disinfection of facilities and equipment will destroy environmental sources of infection. There is no evidence that RCV or SDAV are communicable to humans. It is not known whether human coronaviruses can infect rats.

12.   Interference with Research

Rat coronavirus infections may hamper, if only transiently, studies on the respiratory system or salivary glands of rats. Furthermore, infected rats may be a greater risk for inhalation anesthesia, since excess mucus produced during acute coronavirus rhinitis can obstruct major airways. Eye research may be hampered by SDAV-associated keratoconjunctivitis, and chronically affected rats are obviously unsuitable for study. The eye lesion may be particularly troublesome for long-term studies such as those required for toxicological programs (80). Finally, since both SDAV and RCV are primary pathogens for the respiratory tract, it has been suggested they may act as initiators or as copathogens in murine respiratory mycoplasmosis (61,120).

B.   Paramyxoviruses [Sendai Virus (Parainfluenza 1)]

1.   General

Sendai virus was isolated originally from mice in Japan (44), but it can infect hamsters and guinea pigs (127). In mice, it causes silent respiratory infections, particularly in adults, but severe or even fatal bronchopneumonia and interstitial pneumonia can occur in sucklings and in genetically susceptible adults (13,118,121,122). The infectivity and pathogenicity of Sendai virus for rats is not well characterized, and there are few reports of natural outbreaks. Serological studies from our laboratory indicate, however, that Sendai virus is widely disseminated in rats from commercial breeders. (Rats from six of nine colonies had anti-Sendai virus antibody.) Therefore, we feel that Sendai virus infection is common among rats.

2.   Properties

Sendai virus is a pleomorphic, filamentous virus with a lipid coat derived from host cell plasma membrane. Therefore, it is readily inactivated by organic solvents such as ether. It can also be inactivated by ultraviolet light and is unstable at temperatures above 37°C.

Sendai virus has hemolytic activity and two major surface antigens: a hemagglutinin and a neuramidase. It also has a nonhemagglutinating internal CF antigen. Sendai virus is antigenically related to parainfluenza types 2 and 3, but can be differentiated from them serologically by CF, HAI, or NT tests. The virus grows well in the amnion and allantois of embryonated eggs and in a variety of cell cultures, including human and monkey kidney and in perfused rat lung organ cultures (19). We prefer to propogate Sendai virus in BHK-21 cells to minimize the danger of contamination with latent passenger viruses which are particularly troublesome in primary monkey kidney cultures (13). Sendai virus induces syncytial giant cells in culture as is typical for other parainfluenza viruses. The biochemical, physical, antigenic, and cultural characteristics of Sendai virus has been described at length by others (27,33,44,60).

3.   Clinical Disease

Spontaneous Sendai virus infection of rats appears to follow patterns described for mice (121). In our experience and in the experience of others, infection is usually asymptomatic (21), but it may be associated with signs of pneumonia (88). Makino and colleagues (88) also observed a decrease in average litter size and retarded growth of young rats during a cyclic epizootic of Sendai virus infection. Jonas (65) has also noticed transient decreases in production during active Sendai infection in a large breeding colony. Intranasally inoculated rats developed extensive pulmonary lesions with respiratory difficulty, anorexia, and starry haircoats within 1 week after exposure. Tyrrell and Coid (157) found that clinical disease in experimentally infected weanlings was self-limiting and that mortality was negligible. Contact exposed rats seroconverted, but remained asymptomatic.

Coid and Wardman (28) examined the effects of Sendai virus on pregnant rats. Nine of 12 rats exposed to aerosols of virus at 4 to 5 days of pregnancy resorbed all embryos, but virus was recovered from the conceptus of only 1 of 14 additional rats. Each dam developed respiratory distress, inappetence, and a rough haircoat within 1 week postinfection, and virus was recovered from lungs of all rats. The authors concluded that resorption was probably related to systemic distress from respiratory disease in the dams, rather than to direct virus infection of embryos.

4. Pathology

The pathogenesis and lesions of natural and experimental Sendai virus infection have been well described for mice (2,132,160), but only limited information on rats is available. Burek and co-workers (21) found that naturally infected rats seroconverted to Sendai virus and developed multifocal interstitial pneumonia. Unfortunately, limited attempts to isolate Sendai virus from affected rats were unsuccessful. Lesions included perivascular and peribronchial lymphocytic and plasmacytic infiltrates; focal necrosis and hyperplasia of bronchiolar epithelium with infiltration by lymphocytes and neutrophils; and accumulation of mucus, inflammatory cells, and necrotic cell debris in some airways. In some rats, interstitial inflammation was accompanied by syncytial giant cell formation. Rats younger than 8 months seemed predisposed to severe interstitial lesions compared to older rats in which peribronchial and perivascular inflammation were prominent and interstitial lesions were less severe. The frequency and severity of interstitial and perivascular lesions subsided over a 7-month period, and antibody titers to Sendai virus also decreased. Peribronchial lesions persisted in many rats, however, for at least 7 months postinfection.

They also reported that rats seroconverted to pneumonia virus of mice (PVM) as well as to Sendai virus. Since PVM can cause interstitial pneumonia in mice, they could not eliminate the possibility that PVM contributed to the Sendai-associated lesions they described in rats.

Sendai virus lesions have been produced in rats experimentally, but macroscopic descriptions were sketchy and histological findings were not reported (28,157). An example of an early lesion in a germfree rat inoculated intranasally with Sendai virus is shown in Fig. 38.

5. Epizootiology

Epizootiological studies of Sendai virus infection must be

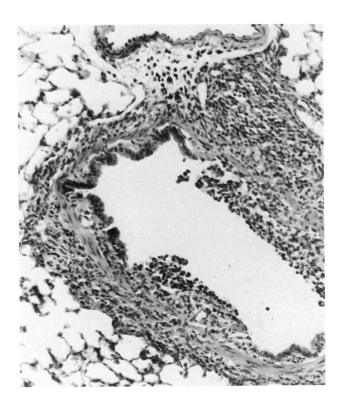

Fig. 38. Necrotizing bronchitis in a germfree rat inoculated intranasally with Sendai virus.

expanded before its natural history in rats is well defined, but based on the few reports available and our own experience, it seems likely that infection of rats will follow patterns reported previously for mice (13,121).

Sendai virus is highly infectious and can be expected to disseminate widely and rapidly through a colony. A typical outbreak was reported by Makino and co-workers (88) in a breeding colony of 500 rats. Infection spread rapidly as determined by development of HAI antibody in exposed rats, but clinical signs of respiratory disease and retarded growth lasted for several weeks during each episode. Furthermore, Sendai virus was isolated from lungs of affected rats. Outbreaks recurred at 8- to 10-month intervals among rats born after each preceding outbreak had subsided. Susceptible rats remained seronegative until the colony was reinfected. Hemagglutination inhibition antibody titers to Sendai virus increased during each outbreak and were detectable for at least 1 year postinfection. Burek *et al.* (21) reported an epizootic of Sendai virus infection in breeding and aging colonies of WAG/Rij rats. Rats seroconverted and had interstitial pneumonia, but they remained asymptomatic. Attempts to isolate virus from six seropositive rats were unsuccessful.

6. Diagnosis and Differential Diagnosis

Sendai virus infection may provoke clinical signs, such as

ruffled haircoats, inappetence, respiratory distress, growth retardation, or decreased litter size. Sendai virus infection should also be considered if deaths occur during routine anesthesia or if unexplained shifts in immunological responses occur (45). More commonly, however, infection remains subclinical, so laboratory diagnostic tests are required for diagnosis. Detection of anti-Sendai antibody in serum provides strong evidence for infection. There are HAI and NT tests available, but the CF test appears to be most sensitive (118,141). Since a single antibody survey may indicate past or current infection, diagnostic procedures should include histopathological examination for pneumonia characterized by epithelial necrosis in bronchi and bronchioles and, when possible, virus isolation from lung during acute stages of infection. Virus can be isolated in embryonated eggs, primary monkey kidney cells (121), or BHK-21 cells, but in our laboratory the last is preferred. If CPE does not appear by 1 week, it has been recommended that cultures be checked for Sendai antigen by hemadsorption with guinea pig erythrocytes (27). In our experience, however, BHK-21 cultures that are CPE negative are never hemadsorption positive.

Sendai virus infection of rats must be differentiated from respiratory infection of rats due to coronaviruses, *Mycoplasma,* and bacteria. It is helpful to remember that uncomplicated Sendai virus infection involves the lungs, but not the upper respiratory tract, whereas coronavirus and *Mycoplasma* infections regularly induce rhinitis. Furthermore, in contrast to Sendai virus infection, coronavirus infection is not associated with necrotizing bronchitis and bronchiolitis, and *Mycoplasma* infection produces chronic inflammation of the lung culminating in bronchiectasis and bronchiolectasis (11,61,82). It has been suggested that PVM may cause pneumonia in rats, but this has not been confirmed (21). Since Sendai virus infection also has been associated with embryonic resorption and retarded growth, RV infection must also be ruled out.

7. Control

Based on studies in mice (13,121), it is likely that Sendai virus infection in rats is self-limiting. Persistence of infection in a colony probably depends on continued introduction of susceptible animals, especially sucklings and weanlings. Thus, infection should be controlled by halting introduction (including breeding) of susceptible rats into a colony for 4 to 8 weeks. This will allow infection to run its course in exposed rats.

Rats can be kept Sendai virus-free in suitable barrier facilities serviced by technical personnel schooled in the control of infectious diseases. Filter lids for animal boxes have, in our experience, neither prevented nor contained Sendai virus outbreaks. Furthermore, since Sendai virus infection is prevalent among mice and hamsters, susceptible rats should not be housed with infected mice and hamsters if they must remain

Sendai virus-negative. If valuable animals from an infected colony must be placed in a Sendai virus-free colony, seropositive animals should be selected and quarantined for 30 days in a laminar air flow unit or other appropriate isolation unit. Production colonies, experimental colonies, and, if possible, animal vendors should be monitored serologically for Sendai virus infection at regular intervals (e.g., semiannually). Additional discussion on control of virus infections in rat colonies is found in Section VII.

8. Interference with Research

Sendai virus can cause pneumonia in rats, but infection is frequently subclinical. Therefore, any experimental procedure involving the respiratory system, from anesthesia to inhalation toxicology, could be at risk during active infection. In addition, histological interpretation of experimentally induced pulmonary lesions may be complicated by residual lesions of Sendai viral pneumonia. Tissue harvests from inapparently infected rats also present a danger of tissue culture contamination or inadvertent infection of recipient rats subsequently inoculated with virus-infected material. There is recent evidence that selected immunological responses of rats may be suppressed during Sendai virus infection (45) and that the virus can induce numerous long-lasting immunological changes in mice (69a). Sendai virus also should be considered a potential copathogen or aggrevating factor in other respiratory diseases of rats. Finally, infected rats may disseminate infection to other susceptible rodents.

C. RNA Viruses Which May Infect Rats [Reovirus 3, Pneumonia Virus of Mice (PVM), and Mouse Encephalomyelitis Virus]

1. General

These viruses are not known to be naturally pathogenic for rats. Subclinical infections may occur, however, since antibodies to them have been detected in rat sera.

2. Reovirus 3

Mammalian reoviruses consist of three serotypes: 1, 2, and 3. They cross-react by CF and immunofluorescence assays, but can be separated antigenically by NT and HAI tests (135). Reovirus 3 is a natural pathogen of mice and produces a syndrome in sucklings characterized clinically by runting, oily skin, jaundice, conjunctivitis, hair loss, neurological disturbances, and sporadic deaths and pathologically by focal necrosis of liver, pancreas, heart, and brain with nonsuppurative encephalitis (32,143–145). Spontaneous reovirus infection has not been described for rats, but we and others (10,31) have detected HAI antibody to reovirus 3 in rat serum. Experimental infections of rats with reoviruses 1 and 3 have been reported as described in Section VI.

3. Pneumonia Virus of Mice (PVM)

This virus is closely related to respiratory syncytial virus of humans. It produces silent infections in mice. However, virulent, mouse lung-passaged strains can cause severe interstitial pneumonia after intranasal inoculation in mice (56,57). It can be cultivated in vitro in BHK-21 cells and produces CPE in about 1 week (50). Neutralizing and HAI antibodies have been detected in rats (55), but clinical signs or lesions have not been reported. Burek et al. (21) detected seroconversions to PVM among rats sustaining Sendai virus-associated interstitial pneumonia and suggested that PVM may have contributed to the development of pneumonia.

4. Mouse Encephalomyelitis Virus

Mouse encephalomyelitis virus is a picornavirus that causes persistent latent infection of mice, but which can occasionally cause a paralytic syndrome similar clinically and morphologically to human poliomyelitis (24,147). There are no published reports documenting infection of rats either by isolation of virus or by detection of serum antibody. We and others (10; J. C. Parker, personal communication) have, however, detected occasional low titers of HAI or NT antibody to the GD VII strain of mouse encephalomyelitis virus in rats. The biological significance of these antibodies is unknown.

IV. UNCLASSIFIED VIRUSES OR VIRUSLIKE AGENTS

A. MHG Virus

McConnell and co-workers (101) isolated a neurotropic agent from adult Sprague-Dawley rats that was biologically and physically similar to an enterovirus. A clinical disease characterized by circling, incoordination, tremors, torticollis, and high mortality was produced in suckling rats and mice by ic inoculation. Virus was isolated from brain, lung, and intestine. Lesions included hydrocephalus, brain edema, neuronal destruction, and gliosis in the gray matter of spinal cord and brainstem and mild encephalitis. The agent was nonhemag-

glutinating with erythrocytes from a variety of species, but it reacted at low titer (up to 1 : 16) with antibody to mouse encephalomyelitis virus (GD VII strain) by CF assay.

Interestingly, these workers also found CF antibody to their agent in human sera. Additional characterization of the agent has, to our knowledge, not been reported.

B. Rat Submaxillary Gland (RSMG) Virus

Ashe and co-workers (4) isolated a cytopathic agent from submaxillary glands of 74 of 97 clinically normal conventional, monoinfected, or germfree rats and designated it RSMG virus. It measured approximately 50 to 300 nm by filtration techniques and was not sensitive to lipid solvents (in contrast to cytomegalovirus). It was, inactivated, however, at pHs below 5, after exposure to bactericidal light or by heating to 60°C for 30 min. The RSMG virus was originally isolated by inoculating salivary gland extracts into primary monolayer cultures of rabbit kidney and was passed repeatedly in similar cultures. Cytopathic effect developed in 2 to 5 days, and cultures were rapidly destroyed. Efforts to grow RSMG virus in cultured cells from monkeys, rats, hamsters, mice, and chickens were unsuccessful. Virus could be adapted to HeLa cells and to human skin cells. Clinical signs did not develop in experimentally inoculated suckling, weanling, or adult rats, mice, or hamsters.

Infected salivary glands were histologically normal, but contained a hemagglutinin which was first detected at 8 weeks, was apparently specific for rabbit erythrocytes, and was ostensibly virus associated. Neutralizing antibody to RSMG virus and HAI antibody to the hemagglutinin were detected in sera of affected rats, the latter antibody being detected in rats as young as 3 weeks. The hemagglutinin was not inhibited by antisera to RV, H-1 virus, MVM, or reovirus 3. Additional studies of RSMG virus have not been reported, but Ashe's work indicates that it is a persistent latent virus that can probably be transmitted vertically and that is not related to other sialotropic rat viruses such as RCV, SDAV, or rat cytomegalovirus (3).

C. Novy Virus

Jordan et al. (68) recovered a filterable agent from virus-infected rat blood stored in glycerin at about 9°C. Remarkably, the blood had been collected by Novy 35 years prior to Jordan's reisolation (117). It was subsequently shown to cause fatal infection of ic- or ip-inoculated weanling rats and mice, but lesions were not described. Further work with this agent has not been reported.

D. Viruslike Pneumotropic Agents (Enzootic Bronchiectasis Agent, Gray Lung Virus, and Wild Rat Pneumonia Agent)

1. Enzootic Bronchiectasis

Nelson *et al.* [reviewed by Nelson (114)] found that suspensions of pneumonic lung from *Mycoplasma*-free rats caused chronic pneumonia in intranasally inoculated mice and rats. The putative etiological agent was filterable, but was not isolated. Nelson indicated that it may be a copathogen with *M. pulmonis* in chronic respiratory disease of rats, but this view has not been confirmed.

2. Gray Lung Virus

Andrews and Glover (1) found a pneumotropic agent in mice inoculated with bovine and human material. It caused red-gray consolidation of lungs due to interstitial pneumonia and pulmonary edema. Vrolijk *et al.* (159) described similar naturally occurring lesions in laboratory and wild rats. A viruslike agent infectious for mice was isolated from affected animals, but it has not been characterized in laboratory rats.

3. Wild Rat Pneumonia Agent

Nelson (113) also recovered an agent from lungs of several wild rats which produced interstitial pneumonia in mice. He was not able to isolate the etiological agent, but suggested it was related to the agent of gray lung pneumonia (114).

The association of these agents with rat pneumonias needs further definition. None has been adequately characterized *in vitro* or *in vivo,* and serological tests to detect them have not been developed.

V. EXPERIMENTAL VIRAL INFECTIONS

A. General

Excluding studies of natural infections, rats have been used sparingly as experimental animals for *in vivo* virological investigations compared to mice and hamsters. Examples of experimental infections of rats with heterologous viruses are listed in Table VII. Three viruses have been selected for additional comment either because they induce interesting lesions in inoculated rats or because they are natural pathogens for other rodents.

Table VII

Examples of Experimental Infections of Rats with Heterologous Viruses

Virus	Experimental disease	Reference
Borna	Slow virus infection	100
		115
Herpes simplex	Encephalitis, hepatitis, skin lesions	95
		125
		146
Lymphocytic choriomeningitis	Cerebellar hypoplasia, retinopathy	104–108
		85
Measles	Encephalitis	23
		140
Reovirus 1	Hydrocephalus	75
		90
Rubella	Congenital defects of heart, eye, other organs	34
		35
SV40	Retinopathy	42
Toga	Encephalitis	39

B. Lymphocytic Choriomeningitis (LCM) Virus

Wild mice are the natural hosts for LCM virus, an arenavirus which may occur as a latent infection of mice and hamsters and as an acute infection of guinea pigs. The pathobiology of LCM infection has been studied extensively, since in mice it is a prototype for cell-mediated immunological injury to the nervous system of animals infected as adults and a prototype for immune complex-mediated glomerular disease in adult mice infected as fetuses or neonates [reviewed by Lehmann-Grube (81a)]. LCM infection has not been detected in laboratory rats, but infection can be induced experimentally. Monjan and colleagues (104,105,107,108) have induced cerebellar "hypoplasia" and retinopathy in ic inoculated suckling rats. Cerebellar hypoplasia was secondary to necrosis of the cortex which was most severe in rats inoculated 4 days postpartum. Interestingly, cerebellar lesions were prevented by immunosuppressing rats with anti-lymphocytic serum (106) and were elicited by adoptive immunization of infected rats with LCM virus-immune splenic cells (105). Therefore, the pathogenesis is similar to that observed for LCM disease in mice in that cell-mediated immunity to LCM virus is required to cause acute lesions. The retinal lesions were characterized by progressive destruction of all retinal layers and modest inflammation, and the presence of LCM virus in all layers of retina. The retinal lesions were also inhibited by immunosuppression (108).

Löhler and associates (85) showed that the WE strain of LCM virus caused widespread infection of the brain after ic inoculation of adult Sprague-Dawley rats. Lesions were characterized by lymphocytic infiltration of meninges, choroid plexus, and paraventricular areas. They closely resembled changes seen in brains of inoculated adult mice. Recently, Zinkernagel *et al.* (165) have shown that rats sensitized with live LCM virus develop potent cell-mediated immune responses to the virus.

C. Reoviruses 1 and 3

The infectivity and pathogenicity of reoviruses for naturally infected rats are believed to be low; however, suckling rats inoculated ic and ip with reovirus 1 at 3 days of age develop encephalitis. Some rats also develop hydrocephalus secondary to viral infection of ependymal cells inoculated with reovirus 3. Pregnant rats (75,90) developed viremia in the presence of maternal serum antibodies, and transient nonfatal infection of fetuses occurred. When virus was inoculated directly into fetuses, however, there was a high rate of death and resorption (77,93).

D. Mousepox Virus (Ectromelia)

We are not aware of adequately documented studies showing that mousepox virus produces natural infections of rats. Burnet and Lush (22) found that rats inoculated intranasally with large doses of virus developed inapparent infection of olfactory mucosa. Neutralizing antibody was detected in serum of infected rats. Rats inoculated intradermally developed either a minute papule at the injection site or no lesions, but HAI antibodies appeared in serum (41). Intraperitoneally inoculated rats also developed antiviral antibody.

VI. COLLECTION OF SAMPLES FOR SEROLOGICAL TESTS AND VIRUS ISOLATION

A. Serum for Antibody Titrations

The demonstration of anti-viral antibody in serum is good evidence for infection with the homologous virus or with an antigenically related virus. Serological testing, in fact, often provides the first evidence of asymptomatic or latent viral infections and, in laboratories with limited capabilities in tissue culture or pathology, may be the only method to detect infection. Sampling protocols must reflect the population under study and the goals of the surveillance program. Consideration must be given to the number, age, and sex of rats to be tested for each sampling period, to the geographical distribution of rats selected from a colony or a vendor shipment, and to the frequency of sampling. These variables are discussed further in Section VII.

Blood can be collected at necropsy by cardiac, intraaortic, or intravenacaval puncture and should be allowed to clot at room temperature. If an animal must be saved, smaller samples can be obtained under ether anesthesia by aseptic cardiac puncture or by retroorbital bleeding with a heparinized capillary tube. Use of nonheparinized tubes frequently results in premature clotting. Heparinized samples (plasma) should be held in ice, separated shortly after collection, and assayed immediately or frozen as described below. Clotted blood samples can be held overnight at 4°C and centrifuged and 0.5- to 1.0-ml aliquots of serum should be stored at −20°C or lower until tested. We recommend that samples not be pooled. Frozen or refrigerated samples may be packed for shipment as described at the end of Section VI, B. If serum samples are collected with reasonable care (aseptically) and rapid shipment is assured, they can be shipped unrefrigerated. It is best, however, to discuss shipping details with the respective testing laboratory.

B. Tissues for Virus Isolation

Clinical signs, serological profiles, lesions, and immunofluorescence staining can incriminate a particular virus and thus narrow the selection of tissues for virus isolation. Conversely, there is considerable overlap in the spectrum of tissues susceptible to the common rat viruses, so it is better to "overcollect" than to "undercollect." Timing is also critical for successful isolations during acute, self-limiting infection such as those caused by SDAV or Sendai virus, since rats may harbor virus before lesions develop or only during early or florid stages of disease. Attempts to isolate infectious virus once seroconversion has occurred may therefore be unsuccessful. On the other hand, isolation of latent viruses such as RV may be difficult without proper culturing techniques, regardless of when tissues are harvested. As a rule fetal, suckling, or weanling rats with low levels of antibody should be selected (Table VIII).

Regardless of the tissues to be harvested, aseptic techniques should be followed. The pelt should be wet with an antiseptic solution (e.g., 70% ethyl alcohol), but care must be taken not to contaminate the tissues so that virus is not inactivated. Instruments should be autoclaved or flame-sterilized (after immersion in 95% or absolute alcohol) and allowed to cool at room temperature. Small pieces of tissue 0.5–1.0 cm² should be placed in labeled sterile vials (2 dram screwcap glass vials are adequate) and held on ice until they are frozen. Clotted or anticoagulant-treated blood also can be stored in vials.

Tissues should be stored at −60°C or below. If deep cold storage must be delayed, tissues may be held overnight in ice. Storage of tissues at −20°C should be avoided, since virus titers may drop quickly. If tissues are triturated before freezing, antibody-free protein (e.g., fetal bovine serum) should be added to the suspension to a final concentration of 50% to

Table VIII

Tissues Recommended for Isolation of Viruses from Naturally Infected Rats[a]

Virus	Tissues
I. Common viruses	
Parvoviruses[b]	Lesions, spleen, liver, intestine
Rat coronavirus[c]	Nasal wash, lung
Sendai virus	Nasal wash, lung
Sialodacryoadenitis virus[d]	Nasal wash, submaxillary salivary gland, Harderian gland
II. Other viruses	
Adenovirus[e]	Intestines, mesenteric lymph nodes
Cytomegalovirus	Submaxillary salivary gland, saliva
Mouse encephalomyelitis virus[e]	Intestines, central nervous system
Pneumonia virus of mice[e]	Nasal wash, lung
Rat submaxillary gland virus	Submaxillary gland
Reovirus 3[e]	Intestines

[a] See footnote, p. 301.
[b] Select young rats with low titers of HAI antibody.
[c] See Fig. 35.
[d] See Fig. 30.
[e] Has not been isolated from rats, but antibody has been detected in rat serum. See text for details.

prevent inactivation of virus. Tissues stored in dry ice should be placed in heat-sealed glass ampules, since CO_2 can infiltrate tightened screwcap vials and may inactivate virus.

If facilities for serological testing or virus isolation are not available locally, samples can be shipped in sealed ampules encased in rigid cardboard carriers placed in a styrofoam or other suitable insulated shipping carton filled with dry ice. Alternatively, serum can be shipped in containers containing ice cubes or crushed ice. Containers should be clearly marked according to federal regulations,* and the receiving laboratory should be notified of the time and mode of shipping in advance.

VII. GENERAL COMMENTS ON THE DETECTION, DIAGNOSIS, AND CONTROL OF VIRAL INFECTIONS

A. Detection

A healthy laboratory rat has been defined as one free of all currently detectable known viruses, with a defined microbial flora and maintained in a protective barrier.† In practice, how-

* "Transportation of Hazardous Materials," Vol. 4, No. 1. National Institutes of Health Guide for Grants and Contracts, United States Department of Health, Education and Welfare, Washington, D.C., 1975.
† "Long-Term Holding of Laboratory Rodents," A Report of the Committee on Long-Term Holding of Laboratory Rodents, Institute of Laboratory Animal Resources. *ILAR News* **19**, No. 4 (1976).

ever, investigators usually use rats with a varied history of exposure to viruses, so documentation of exposure of infection should be integrated into quality assurance programs for breeding and experimental colonies.

Recognition of viral infection may be relatively simple if mortality or characteristic clinical signs occur, but detection is more difficult if infection is asymptomatic or latent. Detection procedures must, therefore, be designed to encompass all possibilities so that preventive measures can be instituted to reduce the negative impact of infection on animal-related research.

Infectious disease can be detected by intramural routine surveillance programs. "Routine surveillance" is, however, a general term which, depending on the needs and judgment of the professional staff, may imply minimal effort or a thorough multidimensional program. Minimal surveillance procedures may be limited to clinical spot checks of production or research colonies or newly received vendor rats with follow-up clinical checks. This may detect active infections such as SDAV, but will usually miss latent infections such as those caused by rat parvoviruses. Effective surveillance must, therefore, include adequate diagnostic laboratory support in virology, serology, and pathology as well as thorough clinical evaluations (63,64).

Signs of disease may be reported to the veterinary staff by veterinary assistants or animal health technicians, animal care technicians, research technicians, or the investigator. The role of investigators in the early recognition of disease can be extremely helpful, since they often have unique opportunities to observe effects of potential viral infection through unexplained changes of responses in experimental procedures. Increased mortality, decreased survival times, increased or unexplained anesthesia-associated deaths, altered metabolic or immunological responses, or variations in tumor growth *in vivo* or *in vitro* all may be potentially associated with viral infection.

The detection of subclinical or latent infections depends on well conceived, statistically valid sampling procedures, since it is impossible to test every animal. Sampling protocols must consider the age, sex, and number of rats to be tested for each sampling period, the distribution of rats selected from a colony room or vendor, and the frequency of sampling. As a rule, we recommend testing equal numbers of males and females at 90 days of age and also retired breeders at 9 to 12 months of age. The number of rats tested may vary according to the goals of the program, but we suggest that a minimum of 10 rats per age group from a given room be screened twice yearly for a total of 40 rats annually. For intramural production colonies or long-term holding colonies, no more than 1 rat per cage should be included in a single sample population, but blood or tissues from all rats selected should be collected on the same day. In addition, rats should be chosen from as many different racks or shelves in a room as possible. A useful technique, especially for intramural sampling, is to place sentinel weanling rats in separate cages among other rats in a room and test them when

Table IX

Sample Size Required to Detect at Least One Infected Animal in a Population for a Given Expected Incidence of Infection[a-c]

No. of animals in sample	Incidence of infection in population (%)
29	10
14	20
9	30
6	40
5	50
4	60
3	70
2	80
2	90

[a] Confidence limit = 95%.

[b] See text for additional explanation.

[c] From formulas contributed by Dr. C. White, Yale University School of Medicine and described in "Long-term Holding of Laboratory Rodents." A report of the Committee on Long-Term Holding of Laboratory Rodents, Institute of Laboratory Animal Resources. *ILAR News*, **19**, No. 4 (1976).

they reach the appropriate age.

Vendor sampling is more difficult to control, but if the source of the rats (specific room or area) is requested, reliable data can be obtained. We recommend testing only 90-day-old vendor-derived rats according to the protocol outlined above, unless screening of retired breeders can be justified.

The recommendations for sample size were derived from a statistical formula provided by White and extrapolated to Table IX. For example, from the table one can infer that if a population of rats with a 30% incidence of viral infection is properly sampled, there is a 95% probability that at least one infected rat will be detected in a sample of nine rats from that population. The sampling protocol must include a thorough serological and pathological evaluation. The reader is referred to Table IX for additional information.

B. Diagnosis

Accurate diagnosis is critical for control of infection and for evaluating the impact of infection on research, not only with respect to requirements of individual investigators but also with respect to hazards infection may present to other animals. Seroconversion or increasing titers of antibody to a particular viral antigen over several weeks is commonly accepted as good presumptive evidence of active or recent viral disease. Antibody alone, however, indicates only that exposure to a viral agent or to an antigenically related agent may have occurred. For example, it was known that rats developed humoral antibody to MHV (51); however, signs or lesions of MHV infection were never seen in rats. Parker *et al.* (120) and Bhatt *et al.*

(15) showed subsequently that rat antibodies to MHV were elicited by rat coronaviruses (SDA and RCV), viruses which are antigenically related to MHV. Thus, antibody induced by one virus reacted with all three viruses. Similarly, isolation of a virus per se during a suspected outbreak may be misleading, since passenger viruses, multiple viral infections, or latent viruses with no influence on the problem under investigation may occur. Therefore the significance of serological and virological findings must be assessed after evaluating clinical signs, lesions, and epizootiological data. Furthermore, before a final diagnosis is confirmed, the veterinarian must critically evaluate information about an outbreak and compare it to his or her knowledge about the suspected infection derived from personal experience or from the scientific literature. Diagnosis of viral infection must be combined with a search for its source. For example, rats lose colostrum-derived antibody progressively and, around the time of weaning, can become susceptible to viruses to which they were passively immune. Therefore, although infection may be present in a vendor colony, it is possible that weanling rats can be infected by an "in-house" source after arrival when passive immunity has decayed. It may be helpful in such situations to test whether rats received from a vendor become seronegative and remain free of infection when held in isolation for several weeks.*

C. Control

Effective control of viral infection requires that the veterinary clinician have thorough knowledge of the epizootiology of the etiological agent. For example, the clinician must be aware of the influence viruses may have on research so that control procedures are neither overzealous nor inadequate. The clinician must also consider the implications of spread from a colony where the impact of the virus may be negligible to a colony where its impact may be major and engender significant losses in time and money.

Generally, infections such as those caused by SDAV or Sendai virus can be eliminated, as discussed in previous sections, by appropriate quarantining procedures. On the other hand, persistent latent infections, such as those due to RV and which can occasionally be transmitted vertically, may require that all stock be killed, that rooms and equipment be thoroughly decontaminated, and that colonies be restocked with virus-free rats.

*The mouse antibody production (MAP) test has been used to detect viruses in mouse tissues and has been useful in detecting viral contamination of tumors (136). The test is based on the principle that inoculation of test tissues into a nonimmune recipient will elicit antibody to viruses in the inoculum. This test can be adapted for rats and is particularly valuable when dealing with latent infections such as those caused by RV.

Spread of air-borne virus from infected to susceptible animals can be reduced by methods ranging from (a) improved air handling procedures, such as maintaining infected rooms under constant negative pressure, to (b) reducing animal populations, separating cages, and covering cage tops with filter lids, to (c) use of more sophisticated air moving equipment, such as mass air flow rooms, laminar air flow racks, or Horsfall-type cabinets. Technicians should be instructed about proper animal handling procedures and more frequent cage changes to reduce ammonia levels.

Adequate surveillance programs to prevent entry of infected animals is a major priority. Rats received from other institutions should be quarantined for 30 days and tested, at least serologically, before introducing them to holding rooms. Portable isolators, like those used for gnotobiotic work are satisfactory for this purpose. Tissues for animal inoculation, such as tumors, should also be tested for passenger viruses prior to their use in animal rooms.

In summary, evaluation of a viral disease and recommendations for its control depend on many factors, including (a) detection of a problem, (b) accurate diagnosis of the disease, (c) complete knowledge of its epizootiology, (d) knowledge of the research using the rats being evaluated, (e) thorough assessment of the risk infection implies for other colonies or research projects, and (e) the economic and logistical feasibility of implementing adequate control measures successfully.

REFERENCES

1. Andrewes, C. H., and Glover, R. E. (1945). Grey lung virus: An agent pathogenic for mice and other rodents. *Br. J. Exp. Pathol.* **26**, 375–387.
2. Appell, L. H., Kovatch, R. M., Reddecliff, J. M., and Gerone, P. J. (1971). Pathogenesis of Sendai virus infection in mice. *Am. J. Vet. Res.* **32**, 1835–1841.
3. Ashe, W. K. (1969). Properties of the rat submaxillary gland virus hemagglutinin and antihemagglutinin and their incidence in apparently healthy gnotobiotic and conventional rats. *J. Gen. Virol.* **4**, 1–7.
4. Ashe, W. K., Scherp, H. W., and Fitzgerald, R. J. (1965). Previously unrecognized virus from submaxillary glands of gnotobiotic and conventional rats. *J. Bacteriol.* **90**, 1719–1729.
5. Baringer, J. R., and Nathanson, N. (1972). Parvovirus hemorrhagic encephalopathy of rats. Electron microscopic observations of the vascular lesions. *Lab. Invest.* **27**, 514–522.
6. Benyesh-Melnick, M. (1969). Cytomegaloviruses. *In* "Diagnostic Procedures for Viral and Rickettsial Infections" (E. H. Lenette and N. J. Schmidt, eds.), 4th ed., pp. 701–732. Am. Public Health Assoc., New York.
7. Berg, V. V., and Scotti, R. M. (1967). Virus-induced peliosis hepatitis in rats. *Science* **158**, 377–378.
8. Bernhard, W., Kasten, F. H., and Chany, C. H. (1963). Etude cytochimique et ultrastructurale de cellules infectées par le virus K des rat et le virus H-1. *C.R. Hebd. Seances Acad. Sci.* **257**, 1566–1569.
9. Berthiaume, L., Joncas, J., and Pavilanis, V. (1974). Comparative structure, morphogenesis and biological characteristics of the respiratory syncytial (RS) virus and the pneumonia virus of mice (PVM). *Arch. Gesamte Virusforsch.* **45**, 39–51.
10. Bhatt, P. N. (1977). Unpublished data.
11. Bhatt, P. N., and Jacoby, R. O. (1977). Experimental infection of adult axenic rats with Parker's rat coronavirus. *Arch. Virol.* **54**, 345–352.
12. Bhatt, P. N., Jacoby, R. O., and Jonas, A. M. (1977). Respiratory infection in mice with sialodacryoadenitis virus, a coronavirus of rats. *Infect. Immun.* **18**, 823–827.
13. Bhatt, P. N., and Jonas, A. M. (1974). An epizootic of Sendai infection with mortality in a barrier-maintained mouse colony. *Am. J. Epidemiol.* **100**, 222–229.
14. Bhatt, P. N., and Jonas, A. M. (1977). Epizootiology of experimental and natural infection with sialodacryoadenitis virus—a coronavirus of rats. In preparation.
15. Bhatt, P. N., Percy, D. H., and Jonas, A. M. (1972). Characterization of the virus of sialodacryoadenitis of rats: A member of the coronavirus group. *J. Infect. Dis.* **126**, 123–130.
16. Boggs, J. D., Melnick, J. L., Conrad, M. E., and Felsher, B. F. (1970). Viral hepatitis: Clinical and tissue culture studies. *J. Am. Med. Assoc.* **214**, 1041–1046.
17. Bradburne, A. F., and Tyrrell, D. A. (1971). Coronaviruses of man. *Prog. Med. Virol.* **131**, 373–403.
18. Brailovsky, C. (1966). Recherches sur le virus K du rat (Parvovirus Ratti). 1. Une méthode de titrage par plaques et son application à l'étude du cycle de multiplication du virus. *Ann. Inst. Pasteur, Paris* **110**, 49–59.
19. Braun, P., and Henley, J. O. (1971). Growth of parinfluenza 1 in isolated perfused rat lungs. *Proc. Soc. Exp. Biol. Med.* **136**, 374–376.
20. Breese, S. S., Howatson, A. F., and Chang, C. (1964). Isolation of virus-like particles associated with Kilham rat-virus infection of tissue cultures. *Virology* **24**, 598–603.
21. Burek, J. D., Zurcher, C., Van Nunen, M. C. J., and Hollander, C. F. (1977). A naturally occurring epizootic caused by Sendai virus in breeding and aging rodent colonies. II. Sendai virus infection in rats. *Lab. Anim. Sci.* **27**, 963–971.
22. Burnet, F. M., and Lush, D. (1936). Inapparent (subclinical) infection of the rat with the virus of infectious ectromelia of mice. *J. Pathol. Bacteriol.* **42**, 469–476.
23. Byington, D. P., and Burnstein, T. (1973). Measles encephalitis produced in suckling rats. *Exp. Mol. Pathol.* **19**, 36–43.
24. Callisher, C. H., and Rowe, W. P. (1966). Mouse hepatitis, Reo-3 and the Theiler viruses. *Natl. Cancer Inst., Monogr.* **20**, 67–75.
25. Campbell, D. A., Manders, E. K., Oehle, J. R., Bonnard, G. D., Oldham, R. K., and Herberman, R. B. (1977). Inhibition of *in vitro* lymphoproliferative responses by *in vivo* passaged rat 13762 mammary adenocarcinoma cells. I. Characteristics of inhibition and evidence for an infectious agent. *Cell. Immunol.* **33**, 364–377.
26. Campbell, D. A., Staal, S. P., Manders, E. K., Bonnard, G. D., Oldham, R. K., Salzman, L. A., and Herberman, R. B. (1977). Inhibition of *in vitro* lymphoproliferative responses by *in vivo* passaged rat 13762 mammary adenocarcinoma cells. II. Evidence that Kilham rat virus is responsible for the inhibitory effect. *Cell. Immunol.* **33**, 378–391.
27. Chanock, R. M. (1969). Parainfluenza viruses. *In* "Diagnostic Procedures for Viral and Rickettsial Infections" (E. H. Lenette and N. J. Schmidt, eds.), 4th ed., Chapter 12, pp. 434–456. Am. Public Health Assoc., New York.
28. Coid, C. R., and Wardman, G. (1971). The effect of parainfluenza type 1 (Sendai) virus infection on early pregnancy in the rat. *J. Reprod. Fertil.* **24**, 39–43.

29. Cole, G. A., and Nathanson, N. (1969). Immunofluorescent studies of the replication of rat virus (HER strain) in tissue culture. *Acta Virol. (Engl. Ed.)* **13**, 515–520.

30. Cole, G. A., Nathanson, N., and Rivet, H. (1970). Viral hemorrhagic encephalopathy of rats. II. Pathogenesis of central nervous system lesions. *Am. J. Epidemiol.* **91**, 339–350.

31. Collins, M., and Parker, J. C. (1977). Personal communication.

32. Cook, I. (1963). Reovirus type 3 infection in laboratory mice. *Aust. J. Exp. Biol. Med. Sci.* **41**, 651–660.

33. Cook, M. K., Andrews, B. E., Fox, H. H., Turner, H. C., James, W. D., and Chanock, R. M. (1959). Antigenic relationships among the "newer" myxoviruses (parainfluenza). *Am. J. Hyg.* **69**, 250–254.

34. Cotlier, E. (1972). Rubella in animals and experimental ocular aspects of congenital rubella. *Int. Ophthalmol. Clin.* **12**, 137–146.

35. Cotlier, E., Fox, J., Bohigian, G., Beaty, C., and Dupré, A. (1968). Pathogenic effects of rubella virus on embryos and newborn rats. *Nature (London)* **217**, 38–40.

36. Cross, S. S., and Parker, J. C. (1972). Some antigenic relationships of the murine parvoviruses: Minute virus of mice, rat and H-1 virus. *Proc. Soc. Exp. Biol. Med.* **139**, 105–108.

37. Dalldorf, G. (1960). Viruses and human cancer. *Bull. N.Y. Acad. Med.* [2] **36**, 795–803.

38. Dawe, C. J., Kilham, L., and Morgan, W. D. (1961). Intranuclear inclusions in tissue cultures infected with rat viruses. *J. Natl. Cancer Inst.* **27**, 221–235.

39. El Dadah, A. N., Nathanson, N., and Sarsitis, R. (1967). Pathogenesis of West Nile encephalitis in mice and rats. I. Influence of age and species on mortality and infection. *Am. J. Epidemiol.* **86**, 765–775.

40. El Dadah, A. N., Nathanson, N., Smith, K. O., Squire, R. A., Santos, G. W., and Melby, E. C. (1967). Viral hemorrhagic encephalopathy of rats. *Science* **156**, 392–394.

41. Fenner, F. (1949). Mouse-pox (infectious ectromelia of mice): A review. *J. Immunol.* **63**, 341–370.

42. Friedman, A. H., Bellhorn, R. W., and Henkind, P. (1973). Simian virus 40-induced retinopathy in the rat. *Invest. Ophthalmol.* **12**, 591–595.

43. Fukumi, H., and Nishikawa, F. (1961). Comparative studies of Sendai and HA2 viruses. *Jpn. J. Med. Sci. Biol.* **14**, 109–120.

44. Fukumi, H., Nishikawa, F., and Kitoyama, T. (1954). A pneumotropic virus from mice causing hemagglutination. *Jpn. J. Med. Sci. Biol.* **7**, 345–363.

45. Garlinghouse, L. E., and Van Hoosier, G. L. (1978). Studies on adjuvant-induced arthritis, tumor transplantability and serologic response to bovine serum albumin in Sendai virus-infected rats. *Am. J. Vet. Res.* **39**, 297–300.

46. Greene, E. L. (1964). Ph.D. Thesis, Cornell University, Ithaca, New York.

47. Greene, E. L. (1965). Physical and chemical properties of H-virus. I. pH and heat stability of the hemagglutinating property. *Proc. Soc. Exp. Biol. Med.* **118**, 973–975.

48. Hallauer, C., and Kronauer, G. (1962). Nachweis eines nicht identifizierten Hämagglutinins in manschlichen Tumorzellstämmen. *Arch. Gesamte Virusforsch.* **11**, 754–756.

49. Hallauer, C., Kornauer, G., and Siegl, G. (1971). Parvoviruses as contaminants of permanent human cell lines. *Arch. Gesamte Virusforsch.* **35**, 80–90.

50. Harter, D. H., and Choppin, P. W. (1967). Studies on pneumonia virus of mice (PVM). I. Replication in baby hamster kidney cells and properties of the virus. *J. Exp. Med.* **126**, 251–266.

51. Hartley, J. W., Rowe, W. P., Bloom, J. J., and Turner, H. C. (1964). Antibodies to mouse hepatitis virus in human sera. *Proc. Soc. Exp. Biol. Med.* **115**, 414–418.

52. Heywood, R. (1973). Some clinical observations on the eyes of Sprague-Dawley rats. *Lab Anim.* **7**, 19–27.

53. Hill, A. (1974). Experimental and natural infection of the conjunctiva of rats. *Lab. Anim.* **8**, 305–310.

54. Hoggan, D. M. (1971). Small DNA viruses. *In* "Comparative Virology" (K. Maramorosch and E. Kurstak, eds.), Chapter 2, pp. 43–79. Academic Press, New York.

55. Horsfall, F. L., and Curnen, E. C. (1946). Studies on pneumonia virus of mice (PVM). II. Immunological evidence of latent infection with the virus in numerous mammalian species. *J. Exp. Med.* **83**, 43–64.

56. Horsfall, F. L., and Ginsberg, H. S. (1951). The dependence of the pathological lesion upon the multiplication of pneumonia virus of mice (PVM). Kinetic relation between the degree of viral multiplication and the extent of pneumonia. *J. Exp. Med.* **93**, 139–150.

57. Horsfall, F. L., and Hahn, R. G. (1940). A latent virus in normal mice capable of producing pneumonia in its natural host. *J. Exp. Med.* **71**, 391–408.

58. Hunt, R. D. (1963). Dacryoadenitis in the Sprague-Dawley rat. *Am. J. Vet. Res.* **24**, 638–641.

59. Innes, J. R. M., and Stanton, M. (1961). Acute disease of the submaxillary and Harderian glands (sialodacryoadenitis) of rats with cytomegaly and no inclusion bodies. *Am. J. Pathol.* **38**, 455–468.

60. Jackson, G. G., and Muldoon, R. L. (1976). Animal paramyxoviruses—parainfluenza 1. *In* "Viruses Causing Common Respiratory Infections in Man" (E. H. Kass, ed.), pp. 94–99. Univ. of Chicago Press, Chicago, Illinois.

61. Jacoby, R. O., Bhatt, P. N., and Jonas, A. M. (1975). The pathogenesis of sialodacryoadenitis in gnotobiotic rats. *Vet. Pathol.* **12**, 196–209.

62. Jamison, R. M., and Mayor, H. D. (1965). Acridine orange staining of purified rat virus strain X-14. *J. Bacteriol.* **90**, 1486–1488.

63. Jonas, A. M. (1973). The role of veterinary diagnostic support laboratories in a research animal colony. *In* "Research Animals in Medicine" (L. T. Harmison, ed.), DHEW Publ. No. (NIH)72-333, pp. 1055–1060. USDHEW, Washington, D.C.

64. Jonas, A. M. (1976). The research animal and the significance of a health monitoring program. *Lab. Anim. Sci.* **26**, 339–344.

65. Jonas, A. M. (1977). Unpublished observations.

66. Jonas, A. M., Coleman, G. L., Jacoby, R. O., and Bhatt, P. N. (1977). Unpublished observations.

67. Jonas, A. M., Craft, J., Black, D. L., Bhatt, P. N., and Hilding, D. (1969). Sialodacryoadenitis in the rat (a light and electron microscopic study). *Arch. Pathol.* **88**, 613–622.

68. Jordan, R. T., Nungester, W. J., and Preston, W. S. (1953). Recovery of the Novy rat virus. *J. Infect. Dis.* **93**, 124–129.

69. Karasaki, S. (1966). Size and ultrastructure of the H-viruses as determined with the use of specific antibodies. *J. Ultrastruct. Res.* **16**, 109–122.

69a. Kay, M. M. B. (1978). Long term subclinical effects & parainfluenza (SENDAI) infection on immune cells of aging mice. *Proc. Soc. Exp. Biol. Med.* **158**, 326–331.

70. Kilham, L. (1961). Rat virus (RV) infections in the hamster. *Proc. Soc. Exp. Biol. Med.* **106**, 825–829.

71. Kilham, L. (1966). Viruses of laboratory and wild rats. *Natl. Cancer Inst., Monogr.* **20**, 117–135.

72. Kilham, L., and Margolis, G. (1965). Cerebellar disease in cats induced by inoculation of rat virus. *Science* **148**, 244–245.

73. Kilham, L., and Margolis, G. (1966). Spontaneous hepatitis and cerebellar hypoplasia in suckling rats due to congenital infection with rat virus. *Am. J. Pathol.* **49**, 457–475.

74. Kilham, L., and Margolis, G. (1969). Transplacental infection of rats and hamsters induced by oral and parenteral inoculations of H-1 and rat

viruses (RV). *Teratology* **2,** 111–123.

75. Kilham, L., and Margolis, G. (1969). Hydrocephalus in hamsters, ferrets, rats, and mice following inoculations with reovirus type 1. I. Virologic studies. *Lab. Invest.* **21,** 183–188.

76. Kilham, L., and Margolis, G. (1970). Pathogenicity of a minute virus of mice (MVM) for rats, mice and hamsters. *Proc. Soc. Exp. Biol. Med.* **133,** 1447–1452.

77. Kilham, L., and Margolis, G. (1973). Pathogenesis of intrauterine infections in rats due to reovirus type 3. I. Virologic studies. *Lab. Invest.* **28,** 597–604.

78. Kilham, L., and Olivier, L. (1959). A latent virus of rats isolated in tissue culture. *Virology* **7,** 428–437.

79. Kuttner, A. G., and Wang, S. H. (1934). The problem of the significance of the inclusion bodies found in the salivary glands of infants and the occurrence of inclusion bodies in the submaxillary glands of hamsters, white mice, and wild rats (Peiping). *J. Exp. Med.* **60,** 773–791.

80. Lai, Y. L., Jacoby, R. O., Bhatt, P. N., and Jonas, A. M. (1976). Keratoconjunctivitis associated with sialodacryoadenitis in rats. *Invest. Ophthalmol.* **15,** 538–541.

81. Ledinko, N., Hopkins, S., and Toolan, H. W. (1969). Relationship between potentiation of H-1 growth of human adenovirus 12 and inhibition of the "helper" adenovirus by H-1. *J. Gen. Virol.* **5,** 19–31.

81a. Lehmann-Grube, ed. (1973). Lymphocytic choriomeningitis virus and other arenaviruses. Springer-Verlag, New York.

82. Lindsey, J. R., Baker, H. J., Overcash, R. G., Cassell, G. H., and Hunt, C. E. (1971). Murine chronic respiratory disease. Significance as a research complication and experimental production with *Mycoplasma pulmonis. Am. J. Pathol.* **64,** 675–708.

83. Lipman, M., Bhatt, P. N. (1977). Unpublished observations.

84. Lipton, H. G., Nathanson, N., and Hodous, J. (1973). Enteric transmission of parvoviruses: Pathogenesis of rat virus infection in adult rats. *Am. J. Epidemiol.* **6,** 443–446.

85. Löhler, J., Schwerdemann, G., and Lehmann-Grube, F. (1973). LCM disease of the adult rat: Morphological alterations of the brain. *In* "Lymphocytic Choriomeningitis Virus and Other Arenaviruses" (F. Lehmann-Grube, ed.), pp. 217–231. Springer-Verlag, Berlin and New York.

86. Lum, G. S., and Schreiner, A. W. (1963). Study of virus isolated from a chloroleukemic Wistar rat. *Cancer Res.* **23,** 1742–1747.

87. Lyon, H. W., Christian, J. J., and Mitler, C. W. (1959). Cytomegalic inclusion disease of lacrimal glands in male laboratory rats. *Proc. Soc. Exp. Biol. Med.* **101,** 164–166.

88. Makino, S., Seko, S., Nakao, H., and Midazuki, K. (1972). An epizootic of Sendai virus infection in a rat colony. *Exp. Anim.* **22,** 275–280.

89. Margolis, G., and Kilham, L. (1965). Rat virus, an agent with an affinity for the dividing cell. *In* "Slow, Latent and Temperate Virus Infections," (D. C. Gadjusek, C. J. Gibbs, and M. Alpers, eds.), NINDB Monogr. 2, pp. 361–367. U. S. Department of Health, Education and Welfare.

90. Margolis, G., and Kilham, L. (1969). Hydroencephalus in hamsters, ferrets, rats and mice following incoulations with reovirus type 1. II. Pathologic studies. *Lab. Invest.* **21,** 189–198.

91. Margolis, G., and Kilham, L. (1970). Parvovirus infections, vascular endothelium and hemorrhagic encephalopathy. *Lab. Invest.* **22,** 478–488.

92. Margolis, G., and Kilham, L. (1972). Rat virus infection of megakaryocytes: A factor in hemorrhagic encephalopathy. *Exp. Mol. Pathol.* **16,** 326–340.

93. Margolis, G., and Kilham, L. (1973). Pathogenesis of intrauterine infections in rats due to reovirus type 3. II. Pathologic and fluorescent antibody studies. *Lab. Invest.* **28,** 605–613.

94. Margolis, G., Kilham, L., and Ruffolo, P. R. (1968). Rat virus disease as an experimental model of neonatal hepatitis. *Exp. Mol. Pathol.* **8,** 1–20.

95. Marks, M. I., and Carpenter, S. (1973). Experimental animal model for encephalitis due to herpes simplex virus. *J. Infect. Dis.* **128,** 331–334.

96. Matsuo, V., and Spencer, H. J. (1969). Studies on the infectivity of rat virus (RV) in BALB/C mice. *Proc. Soc. Exp. Biol. Med.* **130,** 294–299.

97. May, P., and May, E. (1970). The DNA of Kilham rat virus. *J. Gen. Virol.* **6,** 437–439.

98. Mayor, H. D., and Ito, M. (1968). The early detection of picodnavirus X-14 by immunofluorescence. *Proc. Soc. Exp. Biol. Med.* **129,** 684–686.

99. Mayor, H. D., and Jordan, E. L. (1966). Electron microscopic study of the rodent "picodnavirus" X-14. *Exp. Mol. Pathol.* **5,** 580–589.

100. Mayr, A., and Danner, K. (1974). Persistent infections caused by Borna virus. *Infection* **2,** 64–69.

101. McConnell, S. J., Juxsoll, D. L., Garner, F. M., Spertzel, R. O., Warner, A. R., and Yager, R. H. (1964). Isolation and characterization of a neurotropic agent (MHG Virus) from adult rats. *Proc. Soc. Exp. Biol. Med.* **115,** 362–367.

102. McGeoch, D. J., Crawford, L. V., and Follett, E. A. C. (1970). The DNA's of three parvoviruses. *J. Gen. Virol.* **6,** 33–40.

103. McIntosh, K. (1974). Coronaviruses: A comparative review. *Curr. Top. Microbiol. Immunol.* **63,** 85–129.

104. Monjan, A. A., Cole, G. A., Gilden, D. H., and Nathanson, N. (1973). Pathogenesis of cerebellar hypoplasia produced by lymphocytic choriomeningitis virus infection of neonatal rats. I. Evolution of disease following infection at 4 days of age. *J. Neuropathol. Exp. Neurol.* **32,** 110–124.

105. Monjan, A. A., Cole, G. A., and Nathanson, N. (1973). Pathogenesis of LCM disease in the rat. *In* "Lymphocytic Choriomeningitis Virus and Other Arenaviruses" (F. Lehman-Grube, ed.), pp. 195–206. Springer-Verlag, Berlin and New York.

106. Monjan, A. A., Cole, G. A., and Nathanson, N. (1974). Pathogenesis of cerebellar hypoplasia produced by lymphocytic choriomeningitis virus infection of neonatal rats: Protective effect of immunosuppression with anti-lymphoid serum. *Infect. Immun.* **10,** 499–501.

107. Monjan, A. A., Gilden, D. H., Cole, G. A., and Nathanson, N. (1970). Cerebellar hypoplasia in neonatal rats caused by lymphocytic choriomeningitis virus. *Science* **171,** 194–196.

108. Monjan, A. A., Silverstein, A. M., and Cole, G. A. (1972). Lymphocytic choriomeningitis virus-induced retinopathy in newborn rats. *Invest. Ophthalmol.* **11,** 850–856.

109. Monto, A. S., and Lim, S. K. (1974). The Tecumseh study of respiratory illness. VI. Frequency of relationship between outbreaks of coronavirus infection. *J. Infect. Dis.* **129,** 271–276.

110. Moore, A. E. (1962). Characteristics of certain viruses isolated from transplantable tumors. *Virology* **18,** 182–191.

111. Moore, A. E., and Nicastri, A. D. (1965). Lethal infection and pathological findings in AXC rats inoculated with H-virus and RV. *J. Natl. Cancer Inst.* **35,** 937–947.

112. Nathanson, N., Cole, G. A., Santos, G. W., Squire, R. A., and Smith, K. O. (1970). Viral hemorrhagic encephalopathy of rats. I. Isolation, identification and properties of the HER strain of rat virus. *Am. J. Epidemiol.* **91,** 328–338.

113. Nelson, J. B. (1949). Observations on a pneumotropic virus obtained from wild rats. *J. Infect. Dis.* **84,** 21–31.

114. Nelson, J. B. (1967). Respiratory infections of rats and mice with emphasis on indigenous Mycoplasms. *In* "Pathology of Laboratory Rats and Mice" (E. Cotchin and F. J. C. Roe, eds.), pp. 251–294.

Blackwell, Oxford.

115. Nitzschke, E. (1963). Untersuchungen über die experimentella Bornavirus-Infektion bei der Ratte. *Zentralbl. Veterinaermed.* **10**, 470–527.

116. Novotny, J. F., and Hetrick, F. M. (1970). Pathogenesis and transmission of Kilham rat virus infection in rats. *Infect. Immun.* **2**, 298–303.

117. Novy, F. G., Perkins, W. A., Chambers, R., and DeKruif, P. H. (1953). The rat virus. *J. Infect. Dis.* **93**, 111–123.

118. Parker, J. C., Whiteman, M. D., and Richter, C. B. (1978). Susceptibility of inbred and outbred mouse strains to Sendai virus and prevalence of infection in laboratory rodents. *Infect. Immun.* **19**, 123–130.

119. Parker, J. C., Collins, M. J., Cross, S. S., and Rowe, W. P. (1970). Minute virus of mice: Prevalence, epidemiology and occurrence as a contaminant of transplanted tumors. *J. Natl. Cancer Inst.* **45**, 305–310.

120. Parker, J. C., Cross, S. S., and Rowe, W. P. (1970). Rat coronavirus (RCV): A prevalent, naturally occurring pneumotropic virus of rats. *Arch. Gesamte Virusforsch.* **31**, 293–302.

121. Parker, J. C., and Reynolds, R. K. (1968). Natural history of sendai virus infection in mice. *Am. J. Epidemiol.* **89**, 112–125.

122. Parker, J. C., Tennant, R. W., Ward, T. G., and Rowe, W. P. (1964). Enzootic Sendai virus infections in mouse breeder colonies within the United States. *Science* **146**, 936–938.

123. Parker, J. C., Tennant, R. W., Ward, T. G., and Rowe, W. P. (1965). Virus studies with germfree mice. I. Preparation of serologic diagnostic reagents and survey of germfree and mono contaminated mice for indigenous murine viruses. *J. Natl. Cancer Inst.* **34**, 371–380.

124. Payne, F. E., Shellabarger, C. J., and Schmidt, R. W. (1963). A virus from mammary tissue of rats treated with X-ray or methylcholanthrene (MC). *Proc. Am. Assoc. Cancer Res.* **4**, 51.

125. Percy, D. H., and Hatch, L. A. (1975). Experimental infection with herpes simplex virus type 2 in newborn rats: Effects of treatment with iododeoxyuridine and cytosine arabinoside. *J. Infect. Dis.* **132**, 256–261.

126. Portella, O. B. (1963). Hemadsorption and related studies on the hamster-osteolytic viruses. *Arch. Gesamte Virusforsch.* **14**, 277–305.

127. Profeta, M. L., Lief, F. S., and Plotkin, S. A. (1969). Enzootic Sendai infection in laboratory hamsters. *Am. J. Epidemiol.* **89**, 216–324.

128. Rabson, A. S., Edgecomb, J. H., Legallais, F. Y., and Tyrell, S. A. (1969). Isolation and growth of rat cytomegalovirus *in vitro. Proc. Soc. Exp. Biol. Med.* **131**, 923–927.

129. Rabson, A. S., Kilham, L., and Kirschstein, R. L. (1961). Intranuclear inclusions in *rattus* (Mastomys) *natalensis* infected with rat virus. *J. Natl. Cancer Inst.* **27**, 1217–1223.

130. Robey, R. E., Woodman, D. R., and Hetrick, F. M. (1968). Studies on the natural infection of rats with the Kilham rat virus. *Am. J. Epidemiol.* **88**, 139–143.

131. Robinson, G. W., Nathanson, N., and Hodous, J (1971). Sero-epidemiological study of rat virus infection in a closed laboratory colony. *Am. J. Epidemiol.* **94**, 91–100.

132. Robinson, T. W. E., Cureton, R. J. R., and Heath, R. B. (1968). The pathogenesis of Sendai virus infection in the mouse lung. *J. Med. Microbiol.* **1**, 89–95.

133. Roizman, B., and Furlong, D. (1974). The replication of herpesviruses. *Compr. Virol.* **3**, 229–403.

134. Rose, J. A. (1974). Parvovirus reproduction. *Compr. Virol.* **3**, 1–62.

135. Rosen, L. (1969). Reoviruses. *In* "Diagnostic Procedures for Viral and Rickettsial Infections" (H. E. Lenette and N. J. Schmidt, eds.), 4th ed., pp. 354–363. Am. Public Health Assoc., New York.

136. Rowe, W. P., Hartley, J. W., and Huebner, R. J. (1962). Polyoma and other indigenous mouse viruses. *In* "The Problems of Laboratory Animal Disease" (R. J. C. Harris, ed.), pp. 131–142. Academic Press, New York.

137. Ruffolo, P. R., Margolis, G., and Kilham, L. (1966). The induction of hepatitis by prior partial hepatectomy in resistant adult rats injected with H-1 virus. *Am. J. Pathol.* **49**, 795–824.

138. Salzman, L. A., and Jori, L. A. (1970). Characterization of the Kilham rat virus. *J. Virol.* **5**, 114–122.

139. Salzman, L. A., White, W. L., and Kakefuda, T. (1971). Linear, single-stranded DNA isolated from Kilham rat virus. *J. Virol.* **7**, 830–835.

140. Schumacher, H. P., Albrecht, P., and Tauraso, N. M. (1972). The effect of altered immune reactivity on experimental measles encephalitis in rats. *Arch. Gesamte Virusforsch.* **37**, 218–229.

141. Sever, J. L. (1962). Application of a microtechnique to viral serological investigations. *J. Immunol.* **88**, 320–329.

142. Siegl, G. (1976). The parvoviruses. *Virol. Monog.* **15**, 000–000.

143. Stanley, N. F., Dorman, D. C., and Ponsford, J. (1953). Studies on the pathogenesis of a hitherto undescribed virus (hepato-encephalomyelitis) producing unusual symptoms in suckling mice. *Aust. J. Exp. Biol. Med. Sci.* **31**, 147–180.

144. Stanley, N. F., Dorman, D. C., and Ponsford, J. (1954). Studies on the hepatoencephalomyelitis virus (HEV). *Aust. J. Exp. Biol. Med. Sci.* **32**, 543–562.

145. Stanley, N. F., Leak, P. J., Walter, M. N., and Jaske, R. A. (1964). Murine infection with reovirus. II. The chronic disease following reovirus type 3 infection. *Br. J. Exp. Pathol.* **45**, 142–149.

146. Tanaka, S., and Southam, C. M. (1965). Zoster-like lesions from herpes simplex virus in newborn rats. *Proc. Soc. Exp. Biol. Med.* **120**, 56–59.

147. Theiler, M. (1937). Spontaneous encephalomyelitis of mice, a new virus disease. *J. Exp. Med.* **65**, 705–719.

148. Toolan, H. W. (1960). Production of a "mongolian-idiot" like abnormality in hamsters. *Fed. Proc. Fed. Am. Soc. Exp. Biol.* **19**, 208.

149. Toolan, H. W. (1960). Experimental production of mongoloid hamsters. *Science* **131**, 1446–1448.

150. Toolan, H. W. (1961). A virus associated with transplantable human tumors. *Bull. N. Y. Acad. Med.* [2] **37**, 305–310.

151. Toolan, H. W. (1964). Studies on H-viruses. *Proc. Am. Assoc. Cancer Res.* **5**, 64.

152. Toolan, H. W. (1966). Susceptibility of the rhesus monkey (*Macaca mulatta*) to H-1 virus. *Nature (London)* **209**, 833–834.

153. Toolan, H. W. (1967). Agglutination of the H-viruses with various types of red blood cells. *Proc. Soc. Exp. Biol. Med.* **124**, 144–146.

154. Toolan, H. W. (1968). The picodnaviruses: H, RV and AAV. *Int. Rev. Exp. Pathol.* **6**, 135–180.

155. Toolan, H. W., Dalldorf, G., Barclay, M., Chandra, S., and Moore, A. E. (1960). An unidentified, filtrable agent isolated from transplanted human tumors. *Proc. Natl. Acad. Sci. U.S.A.* **46**, 1256–1259.

156. Toolan, H. W., and Ledinko, N. (1965). Growth and cytopathogenicity of H-viruses in human and simian cell cultures. *Nature (London)* **208**, 812–183.

157. Tyrrell, D. A. J., and Coid, C. R. (1970). Sendai virus infection of rats as a convenient model of acute respiratory infection. *Vet. Rec.* **86**, 164–165.

158. Vasquez, C., and Brailovsky, C. (1965). Purification and fine structure of Kilham's rat virus. *Exp. Mol. Pathol.* **4**, 130–140.

159. Vrolijk, H., Verlinde, J. D., and Braoams, W. G. (1957). Virus pneumonia (snuffling disease) in laboratory rats and wild rats due to an agent probably related to grey lung virus of mice. *Antonie van Leeuwenhoek* **23**, 173–183.

160. Ward, J. M., Houchens, D. P., and Collins, M. J. (1976). Naturally occurring Sendai virus infection of athymic nude mice. *Vet. Pathol.* **13**, 36–46.

161. Weisbroth, S. H., and Peress, N. (1977). Ophthalmic lesions and dac-

ryoadenitis: A naturally occurring aspect of sialodacryoadenitis of the laboratory rat. *Lab. Anim. Sci.* **27,** 466–473.

162. Whitman, J. E., and Hetrick, F. M. (1967). Purification of the Kilham rat virus. *Appl. Microbiol.* **15,** 62–66.

163. Wozniak, J., and Hetrick, F. (1969). Persistent infection of a rat nephroma cell line with Kilham rat virus. *J. Virol.* **4,** 313–314.

164. Young, C., and Hill, A. (1974). Conjunctivitis in a colony of rats. *Lab. Anim.* **8,** 301–304.

165. Zinkernagel, R. M., Althage, A., and Jensen, F. C. (1977). Cell-mediated immune response to lymphocytic choriomeningitis and vaccinia virus in rats. *J. Immunol.* **119,** 1242–1247.

Chapter 12

Parasitic Diseases

Chao-Kuang Hsu

I. INTRODUCTION

The laboratory rat has been found to serve as host for a wide variety of protozoan and metazoan parasites. Although infections due to these organisms rarely result in death, there is no doubt that some of them can cause serious clinical disease. Perhaps even more important, however, is the fact that the majority can occur as clinically silent infections which insidiously alter the host. These subtle effects may influence the validity and reproducibility of experimental results in fields such as immunology, hematology, radiobiology, toxicology, and oncology. Consequently, only rats free of parasitic infections are entirely suitable as research subjects.

Due to differences in their ecological niches, wild rats (*Rattus norwegicus*) and laboratory rats have parasitic loads which greatly differ. Because of advances attributable to gnotobiology and improved colony management practices in recent years, it is uncommon to find many different parasites in a colony of laboratory rats. This chapter will review those parasites most likely to be encountered in contemporary rat stocks. The information has been arranged according to organ system(s) involved as well as parasite taxonomy. Each parasite is presented with a brief morphologic description, its life cycle, pathobiology, diagnosis, treatment, prevention, and control. More detailed information and illustrations on some species may be found in a number of review articles and books (25,56,68,78,84,106).

Treatment of parasite infections deserves a special word of caution. Although various treatment regimens recommended in the past have been incorporated in subsequent pages of this text, it is now well established that insecticides (20,76) and various anthelminthics (13,93,97) have profound effects on host physiological changes and immune responses. Many additional drugs and environmental chemicals (67,99) have been shown to influence body chemistry in yet other ways. It has been shown that niridazole suppresses cellular immunity and delayed hypersensitivity and enhances tumor metastasis (21,61,71). Some anthelminthics, such as hycanthone, metronidazole, and niridazole, are carcinogenic in rodents (21,39). Dichlorovos, an organophosphate compound, is well known as a cholinesterase inhibitor. Thiabendazole has a nonimmunosuppressive antiinflammatory property in rodents (97). Levamisole has been extensively used as a broad-spectrum anthelminthic in both human and veterinary medicine. It is an immuno-modulating agent and affects host defense mechanisms in a number of ways (13,93). Levamisole restores phagocytosis by macrophages and T lymphocyte functions when they are defective. It has little or no effect on a primary invasion by viruses, bacteria, or tumor cells. Levamisole was shown to be a potent inhibitor of mammalian alkaline phosphatases at low concentrations (12). In the rat, aldrin and dieldrin decrease the incidence of mammary and lymphatic tumors, increase the metabolism and excretion of steroid hormones, and induce microsomal enzyme activity (20). Aramite, an ascaricide, was shown to be carcinogenic in the rat (76). Thus, it follows that rats (including breeding colonies) to be used in serious research should not be treated chemically, but by methods to exclude the parasite entirely. Usually, this means cesarean derivation and barrier maintenance as the only satisfactory alternative.

II. PARASITES OF PARENTERAL SYSTEMS

A. Flagellates

Both *Trypanosoma lewisi* and *Trypanosoma cruzi* are hemoflagellates commonly found in wild rats. Natural infection of the trypanosomes in animal hosts requires the presence of suitable arthropod vector(s). A well-managed rat colony should be free of trypanosome infections if it is a wild rodent- and vermine-proof facility. However, experimental infection of rats with *T. lewisi* has been widely used as an animal model for studying important human and ruminant trypanosomiasis (see this volume, Chapter 11).

Trypanosoma lewisi

a. Description and Life Cycle. In host blood the trypanosome has a long, slender body tapering to a point at the posterior end, measuring from 25 to 36 μm in total length. A simple flagellum arising from the kinetoplast runs along the body and is free at the anterior end.

The rat flea *Nosopsyllus fusciatus* serves as the usual arthropod host. *Trypanosoma lewisi* multiplies rapidly in the alimentary tract of the flea. Rats become infected by ingestion of fleas or their feces.

b. Host. *Trypanosoma lewisi* is host specific, occurring only in *Rattus*. It cannot be transmitted to man, mouse, guinea pig, or other laboratory animals.

c. Pathobiology. *Trypanosoma lewisi* is a relatively rare parasite of the laboratory rat. It is essentially nonpathogenic and occurs as a short-term infection in the blood. Clinical disease may develop in rats under experimental conditions such as irradiation, splenectomy, immunosuppression, and adrenalectomy (27).

d. Clinical Symptoms. Even in heavily infected rats, there are no apparent symptoms. However, a significantly higher body weight gain of between 10 and 30% has been reported (59).

e. Diagnosis. Diagnosis is established by the demonstration of trypanosomes in blood smears. Subinoculation of suspected blood into uninfected young animals is a useful confirmatory step.

f. Treatment. Rats can be cured by giving spirotrypan, 200 mg/kg, intravenously (40).

g. Prevention and Control. Prevention of the infection in the animal colony should consist of elimination of fleas, other ectoparasites, and wild rodents.

B. Sporozoa

For the convenience of discussion, *Hepatozoon, Toxoplasma, Sarcocystis, Frenkelia,* and *Encephalitozoon* will be included in the blood protozoan group, even though their taxonomic position is still uncertain in some cases.

1. *Hepatozoon muris*

a. Description and Life Cycle. Miller (1908) first described this parasite from a sick laboratory rat colony in Washington, D.C. (63). Schizogony takes place in the parenchymal cells of the liver. The schizonts measure 30–35 by 25–28 μm, and give rise to 20 to 30 merozoites. Gametogony occurs in lymphocytes, monocytes or, occasionally, in granulocytes. Further development requires the spiny rat mite, *Laelaps echidninus,* which sucks the blood of the rat. Fertilization and sporogony then take place in the gut and hemocoele of the mite. Rats become infected by eating the mite, not by its bite.

b. Host. The parasite is apparently common and nonpathogenic in wild Norway and Black rats throughout the world. *Hepatozoon muris* occurs in *Rattus* only. Wild rats in Washington, D.C. have been reported to have natural infections (77).

c. Pathobiology. *Hepatozoon muris* may be pathogenic, although slight infections are asymptomatic in laboratory rats. A mortality of 50% has been reported in rats heavily infected by the parasite (27). The infection induces remarkable leukocytosis and monocytosis and may also cause splenomegaly, hepatic degeneration, anemia, and emaciation in rats.

d. Diagnosis. Detection of the infection can be made by the demonstration of gametocytes from leukocyte preparations or schizogonic stages in tissue sections of liver.

e. Treatment. No effective treatment is known.

f. Prevention and Control. The infection is transmitted by mites and wild rats which should be prevented from entering or coming into contact with the laboratory animal colony. Foster reported that *H. muris* is absent from cesarean-derived, barrier-sustained animal colonies (26).

2. *Toxoplasma gondii*

a. Description and Life Cycle. *Toxoplasma gondii* is a very ubiquitous coccidian parasite. It occurs in parenteral tissues of many mammals, reptiles, and avian species. Infection of this parasite is common in man and animals although they do not, in most cases, show clinical systems. Perrin described an episode of spontaneous toxoplasmosis in the laboratory rat (73). This is an important zoonotic parasite believed to be transmissible between man and animal.

The typical coccidian life cycle of *T. gondii* only recently has been established (28,30,48). Its definitive host is the cat and other Felidae. Asexual developmental stages (Fig. 1) (i.e.,

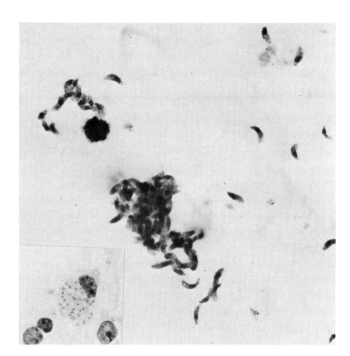

Fig. 1. *Toxoplasma gondii* tachyzoites (trophozoites) from peritoneal fluid. Giemsa. ×800. Lower left insert, bradyzoites (cystic form).

proliferative and cystic forms) occur in the tissues of rats and other intermediate animal hosts. Cats become infected by ingestion of the infected intermediate host tissues, by carnivorism, or by ingesting sporulated oocysts in feces of other Felidae. The prepatent period is 20 to 24 days after ingestion of oocysts; 3 to 5 days after ingestion of bradyzoites, the tissue cyst stage; and 5 to 10 days after ingestion of tachyzoites, the tissue proliferative stage (52). The enteroepithelial multiplicative stage (schizogony), gametocyte stage (gametogony), and oocyst stage may all occur in the intestine of Felidae. The patent period is about 14 days during which millions of oocysts are shed. Oocysts can survive for long periods of time under humid conditions and are resistant to acids, alkalai, and most disinfectants.

Transplacental infection of *T. gondii* has been reported in either naturally occurring or experimentally induced infections in man, rats, mice, hamsters, guinea pigs, mink, dogs, pigs, sheep, and, rarely, cats (52).

b. Host. The definitive host of *T. gondii* is the cat or other felids. Intermediate animal hosts include man, mice, rats. guinea pigs, hamsters, other rodents, dogs, other domestic animals, nonhuman primates, and zoo animals.

c. Pathobiology. *Toxoplasma gondii* is an intracellular parasite affecting all organs, but primarily the reticuloendothelial system. The organisms can be found in brain, lung, heart, muscle, placenta, liver, spleen, and lymph nodes. Tissue cysts are formed inside the cell as the host develops both cellular and humoral immunity. Tissue cysts may persist for many months, years, or for the life of the host. Ordinarily, they do not provoke inflammatory host responses. When a cyst does rupture, this will elicit an inflammatory response characterized by lymphocyte infiltration. Tissue cysts are most numerous in the central nervous system and are found more often in gray than white matter.

Infection of *T. gondii* in newborn or young rats usually results in acute clinical manifestation with fatal pneumonia (58). However, adult rats usually are refractory and develop chronic encephalitis. Therefore, rats are commonly used to maintain *T. gondii* in the laboratory. Histologic changes are relatively infrequent and minor in rats and calves, while they are more frequent and distinct in cats, dogs, rabbits, guinea pigs, ferrets, pigs, lambs, and chimpanzees (8).

The chronic infection of animals with *T. gondii* has been shown to confer resistance against unrelated mammary gland tumors and leukemia (43), virus (79), bacteria (82), protozoa (83), and fungi (35). It delays the development of liver tumors induced by a chemical carcinogen (32).

d. Diagnosis. Spontaneous toxoplasmosis in laboratory rats has not been reported frequently. Rats acquire this infection usually through tissue or cell inoculations or ingestion of oocysts from felid feces. The diagnosis of *T. gondii* infection is based on the demonstration of parasites in tissue section or in stained cell preparations; or based on immunological tests (complement fixation, direct or indirect hemagglutination, Sabin-Feldman dye test, immunofluorescent antibody, toxoplasmin skin test, lymphocyte transformation, Elisa; etc.); or by animal inoculation. Suspected brain or other tissue homogenates are inoculated into *Toxoplasma*-free mice or hamsters. The peritoneal exudates or tissues are then examined for the presence of *Toxoplasma*. Differentiation of *Toxoplasma* from other tissue protozoa has been reviewed by Frenkel (27) and Pakes (70).

e. Treatment. Frenkel (27) recommends pyrimethamine (30 mg/100 gm in food) and/or sulfonamides (i.e., sulfadiazine, 60 mg/100 ml of drinking water or 100 mg in food). The combined use of these two drugs results in a synergism (23) that will inhibit the multiplication of *T. gondii* and give the most effective chemotherapy. However, this treatment regimen cannot erradicate the established infection in a given colony. More satisfactory treatments for this infection need to be developed. For prolonged treatment of rats with both pyrimethamine and sulfadiazine, one has to use folinic acid to prevent toxicity from these drugs.

f. Prevention and Control. Laboratory rats acquire the infection either by the ingestion of *Toxoplasma* oocysts in feline feces, by cannibalism, by experimental inoculation of infected cells or tissues, or through transplacental transmission. Oocysts are resistant to most disinfectants and environments. Obviously, cats should never be allowed to enter or be housed in rat rooms. If representative animals are proved to be infected with *Toxoplasma,* the colony should be destroyed because the infection can persist through transplacental infection. An additional consideration is the zoonotic potential of rat toxoplasmosis (see Chapter 15).

Oocysts are resistant to environmental temperatures. They are susceptible to drying, however, and are killed by heating to 55°C for 30 min. Iodine, 10% ammonia water, ammonium sulfide, and 5% dichlorobenzol are effective oocystocidal agents (51).

3. *Sarcocystis muris*

a. Description and Life Cycle. *Sarcocystis muris* occurs in the muscle of the wild Norway rat, black rat and mice (25). Tuffery and Innes (95) mentioned that in the 1920s–1930s, *Sarcocystis* infection was prevalent in laboratory rats and mice, but it is now rare to find any infected laboratory rodents (33). Twort and Twort reported *Sarcocystis* infection in 14% of laboratory mice used for cancer research (96). Various species of *Sarcocystis* occurs in different animal species. In rats, the elongated cysts (or Miescher's tubes) of *S. muris* have a thin smooth wall, are not compartmented and are several millimeters long. Trophozoites are slightly curved and measure 9 to 15 by 2.5 to 3 μm.

The life cycle of *Sarcocystis* is believed to be similar to that of *Toxoplasma* (56). Transmission is presumably by ingestion of sarcocystis-infected meat or fecal oocysts. *Sarcocystis* has not been found in Cesarean-derived, barrier-maintained colonies (26).

b. Host. *Sarcocystis muris* occurs in the Norway rat, black rat, and the mouse.

c. Pathobiology. *Sarcocystis muris* is nonpathogenic.

d. Diagnosis. Diagnosis is based on the demonstration of the cysts in tissue.

e. Treatment, Prevention, and Control. Treatment, prevention and control are not known.

4. *Frenkelia*

a. Description and Life Cycle. *Frenkelia* sp. was first (and only very recently) reported in the brain of laboratory-reared Fischer 344 rats by Hayden *et al.* (41). The cysts are large, thin-walled, and multilobulated with many compartments outlined by fine interlacing septae. Each compartment contains numerous crescent-shaped, periodic acid-Shiff (PAS)-positive bradyzoites. The cysts are indistinguishable from those of *Sarcocystis*.

The life cycle of *Frenkelia* is not yet well known. Frenkel (28) and Rommel and Krampitz (81) reviewed its life cycle in the bank vole, *Clethrionomys glareolus*.

b. Hosts. *Frenkelia* occurs in rats, voles (*Microtus agrestris*), meadow mice (*Microtus modestus*), muskrats (*Ondatra zibethica*), lemmings (*Lemmus lemmus*), red-backed mice (*Clethrionomys glareolus*), and buzzards (*Buteo buteo*).

c. Pathobiology. Rats show no clinical syndromes even when heavily infected. Large cysts are found in the cerebrum, cerebellum, brainstem, and cervical spinal cord (41). The cysts induce inflammatory changes characterized by granulomas, giant cell formation, perivascular and meningeal infiltrates, and gliosis. Oral infections of mice, hamsters, guinea pigs, and rabbits with cysts were unsuccessful.

d. Diagnosis. Diagnosis is based on the histological demonstration of cysts in the brain.

e. Treatment, Prevention, and Control. Nothing is known about transmission, treatment, or prevention.

5. *Encephalitozoon (Nosema) cuniculi*

a. Description and Life Cycle. *Encephalitozoon cuniculi* is frequently found in the brain and kidney of apparently normal rabbits, rats, and mice. It usually causes no clinical signs but induces a focal granulomatous encephalitis (Fig. 2) and nephritis which may complicate experimental results. The significance of the widespread infection in laboratory animals should be recognized by investigators because undesirable effects on cancer research and drug assay studies are well documented (3,50,66,74,75,96).

Encephalitozoon cuniculi is a microsporidian parasite. The spores are oval, about 1.5×2.5 μm in size, and usually form clusters in the brain, macrophages, peritoneal exudates, kidney, heart muscle, pancreas, liver, spleen, and other organs. The spores have a capsule which is poorly stained with hematoxylin, but well stained with Gram, Giemsa, or Goodpasture's carbol fuchsin stains. The spore contains a nucleus near one end, and a polar filament forms 5 to 6 coils at the other end (70). *Encephalitozoon cuniculi* is disseminated via the bloodstream, either free or within macrophages. Infections last for more than 1 year in the brain of mice, rats, rabbits, and hamsters (28). The life cycle is direct. The mode of transmission is through the ingestion of spores or perhaps via urine or carnivorism. Intrauterine transmission has been suggested in gnotobiotic mice, rabbits, and dogs.

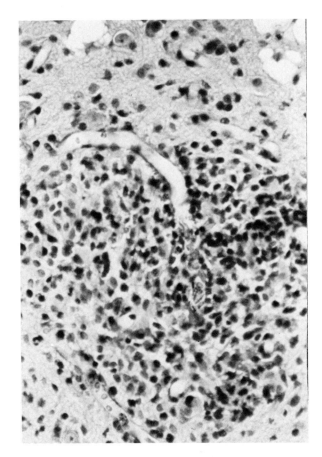

Fig. 2. *Encephalitozoon cuniculi.* Focal granulomatous encephalitis induced by the rupture of a spore in the brain. Hematoxylin and eosin. ×600.

b. Host. *Encephalitozoon cuniculi* infections have been reported in rats, mice, hamsters, guinea pigs, multimammate mice (*Mastomys*), rabbits, cottontails, dogs, and man.

c. Pathobiology. The infections in rats are inapparent. Attwood and Sutton found granulomatous encephalitis throughout the brain in 21% of 365 adult laboratory rats (3). The granulomas ("glial nodules") consisted of collections of activated glial cells surrounded by lymphocytes, without giant cells and necrosis. They demonstrated perivascular spread of encephalitozoa in the brain. The intact cysts did not induce an inflammatory response. They demonstrated encephalitozoa in interstitial infiltrates in rat kidneys.

Encephalitozoon cuniculi can grow in some tumor cell lines. Petri reported the infection in 25% of tumor cells of a Yoshida transplantable ascites sarcoma (74). The infected tumor cells became less pathogenic than usual and did not give rise to solid tumors after a subcutaneous inoculation in rats.

d. Diagnosis. Thin smears prepared from ascitic fluid of suspected animals should be air-dried, fixed with methanol, stained with Giemsa, and examined for spores. Normally, only a small portion of peritoneal mononuclear cells are infected, although the infection rate can be increased by immunosuppressive treatments. In tissue sections, one can demonstrate spores or granulomatous inflamation using an appropriate special stain. The lesions should be distinguished from toxoplasma (70).

e. Treatment, Prevention, and Control. No effective treatment is known. Sporadic recognition of the infection is experienced in apparently well-managed colonies. Tumor cell lines should be examined regularly for *E. cuniculi.*

C. Nematodes

1. *Trichinella spiralis*

a. Description and Life Cycle. The natural hosts of *Trichinella spiralis* are carnivores, although it occurs in a wide host spectrum of mammalian species. The pig, rats, and man are commonly infected. The adult male and female worms (Fig. 3) occur in the small intestine where they live for only

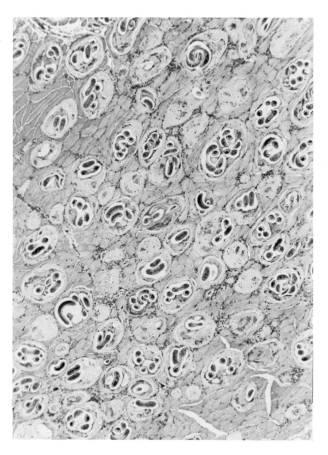

Fig. 4. *Trichinella spiralis* larvae encysting in the muscle section. ×300.

approximately 6 weeks. The male worm (1.4–1.6 mm long and 40 μm wide) is more slender than the female (3–4 mm by 60 μm). The infective larvae encyst in skeletal muscle (Fig. 4).

Infection occurs by eating raw or poorly cooked meat. Adult worms appear in the anterior small intestine of the rat within 24–36 h after infection. After copulation the female worm deposits larvae in the submucosa, and they pass via the lymphatics to the bloodstream and thence are carried all over the body. The larvae penetrate into skeletal muscle where they encyst and may remain alive for years.

b. Pathobiology. Rats usually show no clinical signs except in very heavily infected cases. During the adult stages in the intestine, there may be diarrhea, enteritis, abdominal pain, and weakness. During the larval migration and invasive stages, there may be muscular pain, weakness, pneumonia, and difficulty in breathing and eating. Eosinophilia and high levels of immunoglobulin production (including IgE) are characteristic of this infection.

c. Diagnosis. Definitive diagnosis is made by finding the larvae by microscopic examination of diaphragm or other

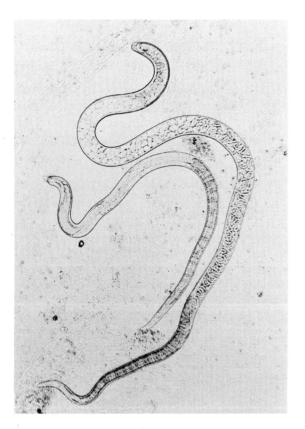

Fig. 3. Female (right) and male (left) adult *Trichinella spiralis.* The gravid female worm contains numerous larvae. ×100.

skeletal muscle. Squash preparations of unfixed muscle are satisfactory for this purpose.

d. Treatment, Prevention, and Control. The laboratory rat is unlikely to be infected by the parasite except for experimental purposes. Rats have been commonly used for experimental trichinosis studies. No satisfactory treatment is available for elimination of encysted larvae. Prevention and control depend on not eating raw or poorly cooked meat.

2. *Capillaria hepatica*

a. Description and Life Cycle. *Capillaria hepatica* is found commonly in the liver of wild rats and, less frequently, in mice, squirrels, muskrats, hares, beavers, dogs, pigs, some species of monkey, and man. The infection rate is usually very high in rats. Luttermoser (60) reported an incidence of 86% in 2500 wild adult rats. Adult worms are slender. The eggs are brownish, barrel-shaped, and possess a thick double wall, of which the outer one is distinctly pitted. At each end of the egg there is a plug (operculum) which does not bulge beyond the outline of the outer wall.

Capillaria hepatica has a direct life cycle but involves a transport host. The natural host acquires the infection by ingesting infective embryonated eggs. These hatch in the cecum, and the larvae penetrate the intestinal mucosa and enter the portal vein reaching the liver where they mature within 3 weeks. The adult worms live for short periods of time but deposit large numbers of unembryonated eggs in the liver. These eggs do not develop but remain in the liver for a long time. When the infected mammal is eaten by a carnivorous animal, eggs are then discharged and passed out in the feces. Eggs reach the soil where they develop to embryonated eggs and become infective in 2 to 6 weeks under suitable conditions.

b. Pathobiology. Although the parasite has low pathogenicity in rats, it can cause significant damage to the liver which becomes enlarged. Accumulations of eggs in liver are usually visible as irregular yellow-gray spots or streaks on the surface. Histologically, the liver architecture is distorted by granulomatous foci and abscesslike lesions. One may see parasites or eggs in the central portion of the lesions (Fig. 5). The granulomas consist of an amorphous center with parasites surrounded by eosinophils, palsma cells, macrophages, epithelioid cells, and multinucleated giant cells. In more advanced stages there are extensive areas of scarring containing large numbers of eggs. Mice are usually more seriously affected by this nematode than rats.

c. Diagnosis. Final diagnosis is based only on the demonstration of parasites or eggs in the liver. Parasites and eggs are not found in feces.

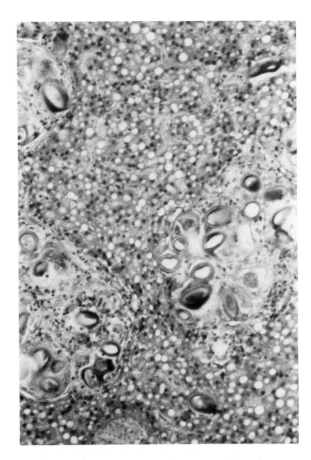

Fig. 5. Capillaria hepatica. Abscess formation and granulomatous reaction to the eggs of *C. hepatica* in the portal areas. Eosinophils, macrophages, and giant cells may be observed in the lesion. ×240.

d. Treatment, Prevention, and Control. No satisfactory treatment is known. *Capillaria hepatica* is rare in well-managed laboratory rat colonies. Wild rodents should be isolated from stable colonies.

3. *Trichosomoides crassicauda*

a. Description and Life Cycle. *Trichosomoides crassicauda* occurs in the urinary bladder, ureter, and renal pelvis of wild and laboratory rats. This parasite is frequently found in conventional laboratory rat colonies, but is absent from cesarian-derived, barrier-sustained colonies (7,10,17,25,104, 105). The small male worms, 1.3 to 3.5 mm long, live and are parasitic in the uterus or vagina of the adult female. The latter is 10 to 9 mm long. The male worm has no spicule, bursa, or trace of copulatory organ. The eggs are oval, brown, and thick-shelled with an operculum at each end, measuring 55 to 70 by 30 to 35 μm (Fig. 6). They are embryonated when laid.

The life cycle of *T. crassicauda* is direct. Infection is by ingestion of embryonated eggs which are passed in the urine.

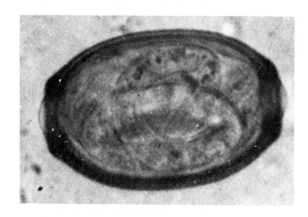

Fig. 6. Embryonated *Trichosomoides crassicauda* egg. ×450. (Courtesy of Dr. S. H. Weisbroth.

The primary means of transmission in the laboratory rat is from adult females to offspring prior to weaning (104). The ingested eggs hatch in the stomach, and the larvae penetrate the stomach wall to migrate via the bloodstream to the lungs, kidney, ureter, and thence to the urinary bladder. The complete life cycle takes 8 to 9 weeks after ingestion.

b. Pathobiology. The anterior end of the adult female embeds within the transitional epithelium of the bladder. Microscopically there is no inflammatory response to the parasite. However, dead parasites may act as a nidus for calculus formation (34). This parasite has been suggested as associated with bladder tumors and urinary calculi (17,90). Migrating larvae in the lung may produce multifocal granulomas (49) and eosinophilia (1). The presence of *Trichosomoides* in certain colonies apparently still constitutes a serious problem in the study of the relationship between bladder cancer and saccharin or cyclamates using laboratory rats as a model (46).

c. Diagnosis. Diagnosis depends on the demonstration of characteristic eggs in the urine, or female worms in the bladder (by dissecting microscope or histological section).

d. Treatment, Prevention, and Control. Treatments with 0.2% nitrofurantoin in food for 6 weeks (17) or single subcutaneous injections of 100 mg/gm methyridine (105) are effective. The infection can be eliminated from a rat colony by cesarean-derivation and good colony management.

4. *Gongylonema neoplasticum*

a. Description and Life Cycle. *Gongylonema neoplasticum* is apparently common in wild rodents but rarely found in laboratory rats. It occurs in the tongue and anterior (squamous) portion of the stomach of the rat and mouse. This spirurid worm is characterized by having the cephalic and esophageal regions ornamented with numerous cuticular plaques irregularly arranged in longitudinal rows on the dorsal and ventral parts of the body (68). The life cycle involves intermediate hosts which include cockroaches, mealworms, and fleas. In these hosts, eggs hatch and the larvae develop and encyst in muscle. Infection occurs by ingestion of the infected intermediate host.

b. Pathobiology. The nematode parasitizes the wall of the stomach and can cause gastric ulcers. The parasitic lesions resemble a carcinomalike proliferation of the gastric mucosa. Although *G. neoplasticum* was initially thought to contribute to the induction of stomach tumors, later studies failed to reproduce neoplastic lesions by the parasites. It was found that the earlier reported neoplastic lesions were due to the original rats having unwittingly been fed a vitamin A-deficient diet (44).

D. Cestodes

1. *Cysticercus fasciolaris* (Syn. *Strobilocercus fasciolaris*)

a. Description and Life Cycle. Cysticercus fasciolaris is the larval form of adult tapeworm, *Taenia taeniaeformis*, which is frequently found in the small intestine of the cat and other carnivores. This parasite occurs commonly in the liver of rats, mice, and other rodents (Fig. 7). The strobilocercus can reach a considerable size, 6–12 cm. It lies loosely coiled and encapsulated in a thick-wall cyst. The cysticercus comprises an extruded scolex followed by a segmented strobila and terminates in a relatively small bladder, so that the whole larva looks like a small adult tapeworm.

The infection of rats occurs by ingestion of the embryonated eggs which are shed in cat feces. The oncospheres emerge in the small intestine and make their way to the liver where they grow rapidly and develop into the infective cysticercus in 30 days. Transmission to the definitive host is by ingestion of infected liver. Following ingestion the strobila and bladder of the larva are digested away and the scolex, which is all that remains, attaches itself to the wall of the small intestine, and strobilation proceeds. The worms mature in a few weeks.

b. Pathobiology. The cysticercus, even in large numbers, appears to be relatively nonpathogenic in the rat. The lesion is characterized by the white to clear cyst measuring up to several centimeters. Usually one or two cysts occur per animal. Microscopically, the parasite may induce the proliferation of connective tissue forming a thin circular fibrous capsule. The strobilocercus has incomplete segmentations with little internal structures except for smooth muscle and calcareous bodies (34). Occasionally the cysticercus may be associated with the

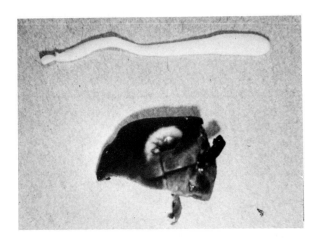

Fig. 7. Relaxed *Cysticercus fasciolaris* obtained from a cyst (white area) of the liver.

development of hepatic sarcoma, especially in older animals about 12 to 15 months after infection (53,85). It appears that certain strains of rats are more susceptible to the induction of the tumors by the parasite than other strains. One cyst usually produces sufficient immunity to prevent reinfection.

c. Diagnosis. Definitive diagnosis of infection in rats is based on the demonstration of the characteristic cyst in the liver.

d. Treatment, Prevention, and Control. No satisfactory treatment is known. Effective prevention and control include proper management of the animal facility to eliminate cross-contamination between rodents and cats, proper handling and sanitization of food and bedding, and control of wild rodents.

E. Unclassified Protozoa

1. *Pneumocystis carinii*

a. Description and Life Cycle. *Pneumocystis carinii* is an ubiguitous microorganism and has world-wide distribution. It is a major etiologic agent of pneumonitis and causes significant morbidity and mortality in human patients with conditions such as immunosuppression, malnutrition, debilitation, immunodeficiency, malignancy, or other generalized disorders (2,80). The prevalence of *P. carinii* is presumably high as a latent, subclinical infection. The taxonomical status of *P. carinii,* either belonging to a fungus or a sporozoa (protozoan), has not yet been established. It lacks lysosomes, Golgi apparatus, and the conoid apparatus. It does not feed by phagocytosis.

Recently, Frenkel (29) proposed that pneumocystis from human origin be regarded as *Pneumocystis jiroveci* n.sp. and

that from the rat be designated *P. carinii* on the basis of serological differences and species specificity, although they do not have significant morphological differences.

Pneumocystis organisms occur in the lung, rarely in other organs, of a variety of mammal hosts including man, chimpanzee, marmosets, dogs, cats, horses, sheep, goats, pigs, rabbits, hares, guinea pigs, rats, and mice. They are found in the foamy, honeycomb materials filling the alveolar space of the infected lungs resulting in pulmonary insufficiency.

The entire life cycle of *P. carinii* is simple and occurs within alveoli of the lung. Four morphologically distinguished entities can be demonstrated in the infected lung by light and electron microscopy. They are trophozoites, precysts, cysts, and intracystic bodies. Based on the studies by Barton and Campbell (6), Vávra and Kucera (98), Campbell (14), Frenkel (29), and Seed and Aikawa (86), the morphological features of these parasitic forms are described as the following.

i. Trophozoites. The growth forms are small (1–5 μm), uninucleated, polymorphic (ovoid to ameboid or small and large forms), and are limited by a double-layer membrane or pellicle. They may possess tubular extensions; filopodia or pili extended from the surface. They contain mitochondria, free ribosomes, endoplasmic reticulum, vacuoles, granules, and globules. Trophozoites probably feed on metabolites in the alveolar cavity. Trophozoites may undergo several reproductions through a cell division resembling binary fission before cyst forms develop. By light microscopy, usually only the stained nucleus is seen. Trophozoites may be present in large and small forms. The tubular surface filopodial extensions of trophozoites often intermesh with similar structures from other adjacent organisms, with filopodia of pneumocytes (host epithelial cells) and with pseudopodial extensions of lung macrophages, consequently forming an intraaveolar clustering aggregate composed of various forms. Trophozoites play a primary contributing role in the pathogenicity of the infection by adhering to the aveolar wall, which causes the blockage of aveolar–capillary gas exchange.

ii. Precysts. Precyst form (5 μm) is the intermediate stage between the trophozoite and cyst forms. Precysts are oval, smooth surfaced, and without pseudopodial tubular extension. Several nuclear chromatin masses bounded by plasmalemma are embedded in the cytoplasm. Precysts have a thick cell wall.

iii. Cysts. The cyst stage is characterized by a thick, three-layered cyst wall which apparently lacks transport enzymes. The filopodial tubular extensions may be present but they do not communicate with the cytoplasma and are not as prominent as those in trophozoites. The cyst cytoplasma contains several, up to eight, nuclear masses or intracystic bodies. The cyst wall stains deeply with methenamine silver, PAS, and

Gram techniques. The cyst stage is most likely the source of infection. The cyst forms appear to be highly antigenic.

iv. Intracystic bodies. In the cyst, the developing intracystic organisms undergo development and transformation into large oval extracystic trophozoites that escape from the cyst and multiply, probably asexually, or develop into precysts or cysts. The intracystic bodies possess nucleus, mitochondria, endoplasmic reticulum, vacuoles, and glycogen particles. Usually eight intracystic bodies are formed. They are best seen when stained with Giemsa or methenamine silver.

b. Pathobiology. *Pneumocystis carinii* naturally is widely distributed in many different species. Under normal conditions, the organism is either not pathogenic or has low pathogenicity for the host. *Pneumocystis* pneumonia usually occurs in patients with impairment of host resistance or in premature and debilitated infants during the early months of life (2 to 6 months). It appears that many humans have been subclinically infected during their lifetime.

Certain colonies of rats are commonly infected with *P. carinii* (31). Animals from the infected colony develop clinical manifestations when they are immunosuppressed with corticosteroids or other immunosuppressive agents. This system has been used as an animal model for studies of human *Pneumocystis* infection. Rats developed a low-grade infection early in life (10–16 weeks old). The transmission is probably via the inhalation of infective cysts that are expelled during exhalation or coughing.

Pneumocystis organisms grow extracellularly and slowly in animals and man. In rats, the organisms will fill the aveolar space in approximately 6–8 weeks under a proper immunosuppressed condition. In the infected lung, the filopodial surface structures of trophozoites facilitate their attachment to the epithelial cells of the alveolar wall and inhibit their phagocytosis by macrophages. The presence of filopodial extensions determine the pathogencity of the organism. Consequently, the adhering clusters of organisms on the alveolar wall causes the alveolar capillary blockage which results in the blood gas exchanges of the alveoli as well as pulmonary insufficiency. Phagocytosis by macrophages prevents the accumulation of *Pneumocystis* in alveoli, since the clusters of organisms are cleared in the hosts after the withdrawal of immunosuppressive agents or after chemotherapy (with either pentamidine or sulfadiazine-pyrimethamine treatment), even in a hypercosticoid animal.

Apparently many immunological factors of the host control the clinical manifestations of *Pneumocystis* infection. Antibodies of IgG, IgM, complement, T lymphocytes, and macrophages are involved. A deficiency in any one of the above factors may result in clinical pneumocystosis. Successful experimental infection and environmental contamination of nude mice with *P. carinii* have been recently demonstrated (102).

The gross pathological changes of infected lungs are similar regardless of species. The lungs are usually enlarged and solid with a firm rubber consistency. The cut surface is homogeneous and shows multiple irregular gray, brownish, or pink areas of consolidation. Histologically, the pulmonary alveoli may be extended and filled with a homogeneous homeycombed foamy material (Fig. 8) containing organisms which are best seen with the periodic acid–Schiff (PAS) or methenamine silver staining techniques (Fig. 9). The foamy exudates are thought to represent a mixture of disintegrating parasites and antigen-antibody complexes. There are no or minimal cell response in the alveoli. The alveolar septa are usually thickened and infiltrated by inflammatory cells that predominantly are plasma cells, lymphocytes, and histocytes with fewer numbers of neutrophils or eosinophils involved.

c. Clinical Symptoms. Subclinical latent infections of *Pneumocystis* are common in many colonies (31). Normal rats show no signs of clinical symptoms unless they are immunosuppressed with corticosteroids (35–40 mg/kg/day) or cyclophosphamide (Cytoxan). These immunosuppresive regimens also favor complicating bacterial and fungal infections. The *Pneumocystis* pneumonia is characterized by the pulmonary alveolar filling with numerous cysts and trophozoites that are embedded in a foamy matrix. Clinical signs include loss of weight, cyanotic, rough hair coat, and rapid labored breathing.

d. Diagnosis. The diagnosis of pneumocystosis is primarily determined by the demonstration of eight nuclei cyst forms or trophozoites in the lung tissues or lung impression smears stained with methanamine silver or Giemsa techniques. One needs to differentiate other fungal infections. Serological tests such as complement fixation (CF) and fluorescent antibody methods are also very useful.

e. Treatment. The combined use of sulfadiazine (100 to 200 mg per 100 gm food) and pyrimethamine (30 mg per 100 gm food) gives an effective treatment. However, this regimen results in the arrested maturation of granulocytes in the bone marrow of the rats. Sulfadiazine inhibits the biosynthesis of folic acid, and pyrimethamine interferes with the conversion of folic to folinic acid. This toxicity could be antagonized with folinic acid supplement. Either drug alone is less effective than the synergistic treatment. The stilbamidine compounds such as hydroxystilbamidine and pentamidine are also effective. Hydroxystilbamidine (3 to 5 mg per kg. iv, for 7 to 10 days) is recommended for rats because it has no side effects, whereas pantamidine is much more toxic in rats than in man.

f. Control. Cesarean derivation, good colony management, and selection of *Pneumocystis*-free animals as breeders

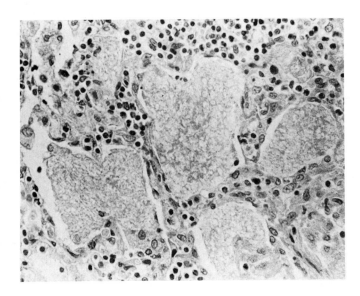

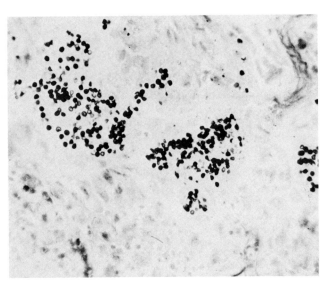

Fig. 8. *Pneumocystis carinii* in lung of a rat. Distention of alveoli by foamy, mucinous material, characteristic of *P. carinii* disease in man and animals. Many plasmalike cells are present, and alveolar septae are widened. Hematoxylin and eosin. ×250. (Courtesy of Dr. J. R. Lindsey.)

Fig. 9. Methenamine silver stained lung as in Fig. 8 demonstrating *P. carinii* organisms (black) in the alveolar space. ×250. (Courtesy of Dr. J. R. Lindsey.)

may be the proper control measures. However, many rat colonies of cesarean derivation origins were reported to be subclinically infected. Periodically treating the animals may minimize the infection in the colony.

III. PARASITES OF THE ALIMENTARY SYSTEM

A. Flagellates

1. *Giardia muris*

Giardia muris occurs in the anterior small intestine of mice, rats, hamsters, and various wild rodents. It is common in laboratory rats and mice. The trophozoites are 7–13 by 5–10 μm, pear-shaped, and bilaterally symmetrical (Fig. 10). They attach themselves to intestinal epithelial cells by means of concave sucking discs. The anterior end is broadly rounded with a large sucking disc, and the posterior end is drawn out. There are two anterior nuclei, two blepharoplasts, and four pairs of flagellae. Trophozoites can be easily recognized in a wet preparation of intestinal contents by their characteristic rolling and tumbling movements. Cysts are ellipsoidal (15 × 7 μm), have thick walls, have four nuclei each, and are found in the large intestine and feces.

Giardia muris proliferates by binary and multiple fission. It is not pathogenic or has low pathogenicity, although it can cause digestive disturbance, malnutrition, or even ulcers when

present in large numbers. Transmission is usually by ingestion of cysts.

Treatment with chloroquine, quinacrine, or amodiaquin is effective. Control depends on proper sanitization. Cysts can be killed by 2.5% phenol or Lysol or by temperatures above 50°C.

2. *Spironucleus (Hexamita) muris*

Spironucleus muris (54a) commonly occurs in the small intestine of rats, mice, hamsters, and other wild rodents. It occurs especially in the crypts of Lieberkuhen. It is an elongated, pear-shaped, and bilaterally symmetrical flagellate, measuring 7–9 by 2–3 μm. It has two anterior nuclei, two longitudinal axostyles, and four pairs of flagellae arising from a pair of blepharoplasts situated between the paired nuclei. It multiplies by longitudinal fission. There are no cysts, although one may find pseudocysts in the crypts. It is transmitted by oral ingestion of trophozoites. These organisms are more pathogenic to stressed or young than to old animals, and the former may develop a chronic infection. Infected animals show rough hair coat, depression, weight loss, listlessness, distended abdomen, diarrhea, and sometimes death. The parasite causes duodenitis accompanied by catarrhal fluid accumulation and cellular infiltration in the small intestine (62,101). Histologically, the crypts of Lieberkühn and lumina between villi are distended with accumulation of *Spironucleus* organisms (Fig. 11). Hexamitiasis (Spironucleosis) causes a number of research variables including changes of macrophage metabolism (54), in-

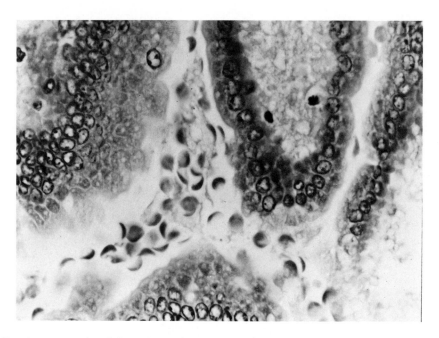

Fig. 10. *Giardia muris.* Round or crescent-shaped *G. muris* organisms are present between intestinal villi of a portion of small intestine. Hematoxylin-phloxin–saffron (HPS). ×550. (Courtesy of Dr. M. J. van Zwieten and Dr. C. Zurcher.)

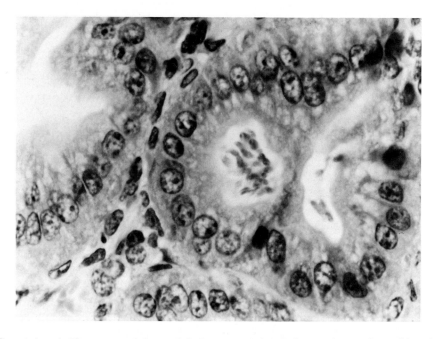

Fig. 11. *Spironucleus (Hexamita) muris.* The presence of characteristic *S. muris* organisms in the crypt lumen of a small intestine. Hematoxylin-phloxin–saffron (HPS). ×880. (Courtesy of Dr. M. J. van Zwieten and Dr. C. Zurcher.)

creased sensitivity to irradiation (65), increased mortality in cadmium-exposed mice (22), and shortened life span of athymic mice (11). It can be diagnosed by microscopic examination of intestinal smears or scrapings from anterior small intestinal mucosa or Peyer's patches in saline. The parasite can be differentiated from *Giardia* spp. and trichomonads by its small size and rapid horizontal movement. Treatment with dimetridazole, nitrofurantoin, or oxytetracycline proved ineffective (62). Control of the infection depends on proper animal husbandry and sanitation.

3. *Hexamastix muris*

Hexamastix muris occurs as a nonpathogenic protozoan of the cecum in the rat, hamster, and other rodents. The incidence rate of this infection in laboratory rats in not known. The parasite has a piriform body (9 × 7 μm) with an anterior nucleus and cytostome, a pelta, a conspicuous axostyle, a prominent parabasal body, five anterior flagella, and a trailing flagellum (55). Cysts are not formed.

4. *Chilomastix bettencourti*

Chilomastix bettencourti occurs in the cecum of the rat, mice, hamster, and wild rodents and is common in some colonies (25). The trophozoites of this nonpathogenic flagellate are piriform, with an anterior nucleus, a large cystostomal groove, three anterior flagella, a short fourth flagellum which undulates within the cytostomal cleft, and a cytoplasmic fibril along the anterior end and sides of the cytostomal groove (55). The cysts are usually lemon shaped and contain a nucleus and all organelles of the trophozoites.

5. *Tritrichomonas muris*

Tritrichomonas muris (Fig. 12) is nonpathogenic and occurs in the cecum, colon, and small intestine of the rat, mouse, hamster, and wild rodents (55). It is pear shaped and measures 16–26 by 10–14 μm. It has an anterior vesicular nucleus. Anterior to the nucleus is the blepharoplast from which arise three anterior flagella and a posterior flagellum which is attached to the body by means of an undulating membrane which continues posteriorly as a free flagellum. The trichomonal is supported by a stiff rodlike axostyle which protrudes terminally like a tail. It has large oval-shaped nucleus, slitlike cytostome, and a sausage-shaped parabasal body in the anterior region. Reproduction is asexual by binary fission. The organism swims with a characteristic wobbly movement. Cysts are not formed. Transmission is by ingestion of trophozoites passed in feces.

6. *Tritrichomonas minuta*

This nonpathogenic trichomonad occurs in the cecum and colon of the rat, mouse, hamster, and marmot. It measures 4–9 by 2–5 μm and is similar to *T. muris*.

7. *Tetratrichomonas microti*

This nonpathogenic flagellate is common in the cecum of the mouse, rat, hamster, vole, and other wild rodents in North America. It is 4–9 μm long. The organism has four anterior flagellae, a posterior free-trailing flagellum, and a pelta. It has been experimentally transmitted from the mouse to the labora-

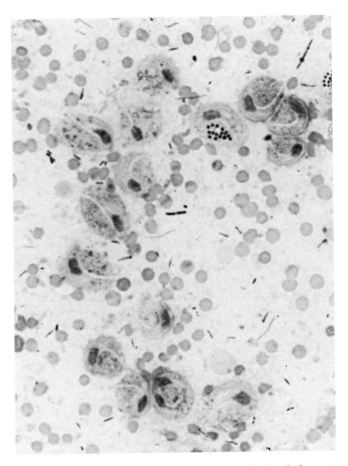

Fig. 12. *Tritrichomonas muris* trophozoites from the intestinal content smear. Giemsa. ×800.

tory rat, guinea pig, ground squirrel (*Citellus citellus*), dog, and cat (55).

8. *Pentatrichomonas homonis*

Pentatrichomonas homonis occurs commonly in the cecum and colon of the mouse, rat, hamster, dog, cat, macaques, apes, and man (55). This nonpathogenic organism can be cross-transmitted among these species. Its body is piriform, measuring 8–20 by 3–14 μm, and ordinarily has five anterior flagellae plus one posterior flagellum. Levine (55) described its detailed morphology.

B. Sporozoa

1. *Eimeria nieschulzi*

Eimeria nieschulzi is common in the small intestine of the wild rat but uncommon in the laboratory rat. *Eimeria* spp. do not occur in cesarean-derived, barrier-sustained colonies. Oocysts of *E. nieschulzi* are ellipsoidal to ovoid, measuring 16–26 by 13–21 μm. A detailed description of this coccidium is given by Levine and Ivens (57).

Oocysts sporulate in 3 days with the formation of four sporocysts, each with two sporozoites. Rats become infected after ingestion of sporulated oocysts. Sporozoites escape from the oocysts and enter the intestinal epithelium in which they undergo four generations of asexual schizogony producing thousands of merozoites. These merozoites enter epithelial cells and form microgametocytes and macrogametes. After fertilization, macrogametes become oocysts. Oocysts are then passed out in the feces. The prepatent and patent periods of *E. nieschulzi* are 7 and 5 days, respectively. It is a self-limiting infection without reinfection. Mice are not susceptible to infection.

Eimeria nieschulzi primarily affects young rats and causes diarrhea, weakness, emanciation, and death in young animals less than 6 months old. Perard (72) reported that this coccidium caused considerable loss among rats in the Pasteur Institute.

Diagnosis is based on the histological demonstration of the organism in the intestine or the identification of oocysts in the feces. Since coccidiosis is self-limiting, control primarily depends on good sanitization.

2. *Eimeria miyairii*

This coccidium occurs in the small intestine of wild rats, but its incidence in laboratory rats is not known. For further information the reader is referred to the monograph by Levine and Ivens (57).

3. *Eimeria separata*

This slightly pathogenic coccidium is apparently common in the cecum and colon of wild rats. The oocyst is ellipsoidal and measures 10–19 by 10–17 μm Levine and Ivens (57) gave a full description of this species. Its life cycle is similar to that of *E. nieschulzi*. Diagnosis and control are similar to that described for *E. nieschulzi*.

C. Amoebae

Entamoeba muris

Entamoeba muris is common in the cecum and colon of rats, mice, hamsters, and many wild rodents (55) throughout the world. It occurs in high incidence in conventional laboratory rat, mouse, and hamster colonies (64). *Entamoeba muris* resembles *Entamoeba coli*. Its trophozoites are large, 8–30 μm long. Its oocysts are 9–20 μ mand contain up to eight nuclei (Fig. 13). A food vacuole often is present in the cyst. The cyst

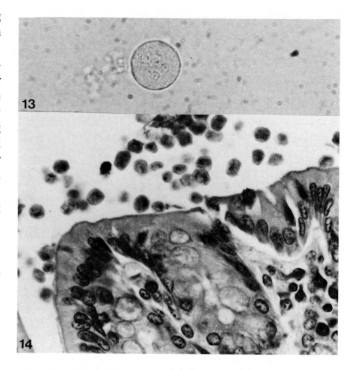

Fig. 13. *Entamoeba muris* cyst. ×800.

Fig. 14. *Entamoeba muris* trophozoites in the lumen of a large intestine. Hematoxylin–phloxin–saffron (HPS). ×550. (Courtesy of Dr. M. J. van Zwieten and Dr. C. Zurcher.)

produces eight amoebae when ingested by a new host. Trophozoites multiply by binary fission. Normally these amoeba live in the lumen of the gut (Fig. 14) where they feed on the particles of food and bacteria. *Entamoeba muris* can be cross-transmitted among rats, mice, and hamsters. *Entamoeba muris* is nonpathogenic, but is undesirable in research animals.

D. Ciliates

Balantidium coli

Balantidium coli is a large ciliate that commonly occurs in the cecum and colon of pigs, nonhuman primates, man, and occasionally the rat, hamster, and dog (55). It trophozoites are ovoid, 30–150 × 150 × 25 μm, with subterminal tubular mouth and a body covered with cilia. It moves rapidly. Its cysts are sperical to ovoid and 40 to 50 μm in diameter. There are many food vacuoles containing starch. Multiplication is by transverse binary fission and conjugation.

Balantidium coli is usually nonpathogenic. It may sometimes be a secondary invader of the lesions initiated by pathogenic bacteria or viruses. It produces hyaluronidase which might enlarge the lesions by attacking the ground substance between cells (55). *Balantidium coli* is pathogenic in man and nonhuman primates and may be a primary pathogen. It causes diarrhea or dysentery and ulcerative enteritis similar to the lesions caused by *Entamoeba histolytica.* Transmission is by ingestion of cysts. Balantidiosis can be recognized easily by microscopic examination of intestinal contents for the organism (Fig. 15) or by histological examination of intestine. Good sanitary measures and proper colony management should prevent balantidial infections.

E. Nematodes

1. *Syphacia muris* (Rat Pinworm)

a. Description and Life Cycle. *Syphacia muris* is an oxyurid nematode that commonly occurs in the cecum and colon of the laboratory rat, wild rat, and occasionally laboratory mouse. It closely resembles *Syphacia obvelata,* and it has been confused with this species for many years (25,47). This parasite can be transmitted from rats to mice if they are housed in the same room (47).

The adult worms are cylindrical with rounded anterior region and posterior tapering to a long, slender, and sharply pointed tail. The mouth is surrounded by three distinct lips. The esophagus is a typical oxyurid three-club shaped with a prebulbar swelling and a posterior globular valvulated bulb. The male is about 1.2 to 1.3 mm long and 100 μm wide. The anterior mamelon is near the middle of the body. The tail of the male is bent ventrally. The male worm has a single, long, prominent spicule and a gubernaculum. The female is 2.8 to 4.0 mm long, and its vulva is in the anterior quarter of the body. The egg (Fig. 16) is vermiform, slightly flattened on one side and measures 72–82 by 25–36 μm averaging 75 $\times$ 29 μm.

The life cycle of *S. muris* is direct (47,91). The eggs are deposited by the female on the perianal area of the host or in the colon (Fig. 17). The eggs embryonate and become infective in a few hours. Infection of the rat occurs in three ways: (1) direct ingestion of eggs by licking the perianal region of an infected animal; (2) ingestion of food, water, or fomites contaminated with infective eggs (the prepatent period being about 15 days following ingestion of embryonated eggs); and (3) retroinfection, when embryonated eggs hatch in the perianal region and larvae migrate by way of anus back into the colon

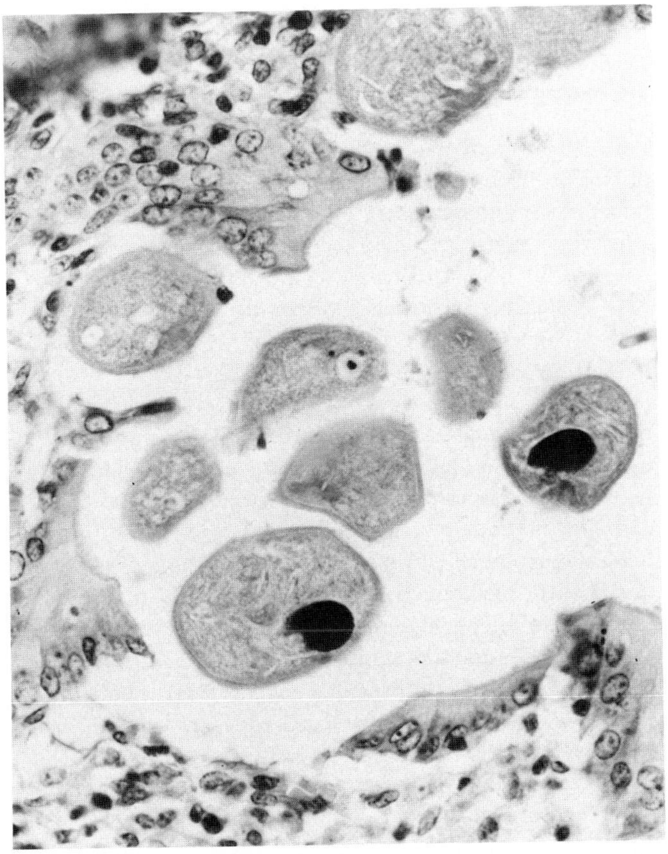

Fig. 15. *Balantidium coli* organisms in the lumen of the large intestine. Hematoxylin and eosin. $\times$400.

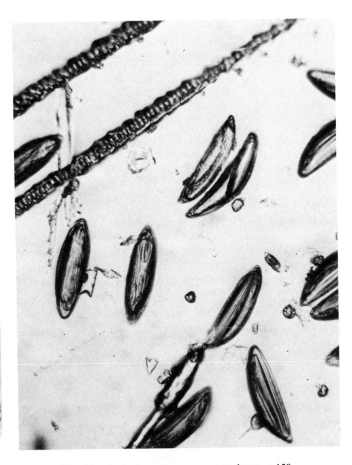

Fig. 16. *Syphacia muris* eggs on an anal tape. $\times$150.

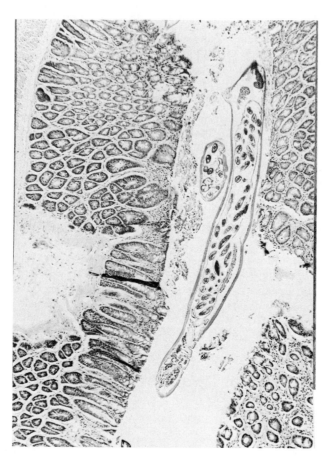

Fig. 17. Adult *Syphacia muris* section from the intestine. The worm contains many eggs. Hematoxylin and eosin. ×20.

and cecum where they grow and moult becoming sexually mature in 6 to 7 days. After fertilization, the male worm dies and is passed with the feces. The gravid female worms migrate to the rectum on the twelfth day and deposit eggs.

b. Pathobiology. This nematode usually does not induce clinical symptoms or microscopic lesions, although its pathogenic effects are probably similar to those of *S. obvelata.* It may be associated with poor general health, slow weight gain and growth rate, diarrhea, intestinal impaction, intussusception, and rectal prolapse, especially when animals are under stresses of shipping and drastic changes in diet.

Rats are the usual host of *S. muris.* Mice also can be infected if they are housed in the same room as infected rats. The infection of this nematode is difficult to eradicate from the colony. The parasite may not cause any serious ill effects on its host but can certainly affect research and may influence experimental results.

c. Diagnosis. Since the nematode deposits its characteristic eggs in the perianal region, the diagnosis of infection is

based on the demonstration of eggs on perianal tapes, anal swabs, in feces, or the demonstration of adults in the large intestine.

d. Treatment, Prevention, and Control. Treatment and control of *S. muris* are the same as for *S. obvelata* in mice. Some drugs are effective in eliminating adult worms, but are not efficient in clearing immature worms (19). Thus repeated treatment is necessary. Effective treatments include piperazine adipate at a rate of 50 mg/ml in drinking water for 3 days (37); piperazine citrate at a rate of 3 gm/liter in drinking water in a three-week regime of treatment (treat animals for 7 days, followed by 7 days without treatment, and then 7 days of treatment again (45); pyrvinium pamoate given orally at a concentration of 3 mg/liter of drinking water or 1.2 mg per kg of food over a period of 30 days (9); trichlorfon at a rate of 1.75 mg/ml of water for 14 days (88); and dichlorovos at a rate of 0.5 mg/gm of food for 1 day (100). All inhabitants of a single room should be treated at one time, and strict hygienic measures should be instituted to remove and disinfect contaminated fomites.

Control of *S. muris* infection is accomplished by cesarean derivation of animals and barrier housing methods.

2. *Syphacia obvelata*

a. Description and Life Cycle. *Syphacia obvelata* is primarily the mouse pinworm, although it can occasionally infect rats in the laboratory (47). It is found commonly in the laboratory mouse and hamster and occurs in the cecum and colon of the host. It is believed that many reports in the past have mistakenly identified *S. muris* as *S. obvelata* of the rat (25).

The males are 1.1–1.5 mm long and 120–140 μm wide (122 μm average), with a long tail with a distinct spicule and gubernaculum. The mouth is surrounded by three simple lips. The esophagus has a club-shaped corpus, very short isthmus, and bulb. The anterior end has small cervical alae. The posterior end is bent ventrally. The females are 3.4–5.8 mm long and 0.24–0.4 mm wide with a tail 530–675 μm long. The eggs are thin-shelled, crescent shaped, flattened on one side, and contain an undifferentiated embryo *in utero.* Eggs measure 111–153 by 33–55 μm with a mean of 134 × 36 μm (47).

The life cycle of *S. obvelata* is direct and very similar to that of *S. muris* (15a). Unembryonated eggs are deposited by the females on the perianal region. Infection is either by oral ingestion of infective eggs or by retroinfection (the direct migration of larvae from the anal area forward to the cecum and colon).

b. Pathobiology, Diagnosis, Treatment, and Control. These subjects are similar to those described for *S. muris.*

3. *Aspicularis tetraptera*

a. Description and Life Cycle. *Aspicularis tetraptera* commonly occurs in the cecum and colon of mouse and rat. This oxyurid nematode is morphologically similar to, but readily distinguished from *S. obvelata*. The males are 2–4 mm long and 120–190 μm wide with a short conical tail measuring 117–169 μm long. Both spicule and gubernaculum are absent. The females are 3–4 mm long and 215–275 μm wide with a conical tail of 445–605 μm long. The eggs are symmetrically ellipsoidal, 70–98 by 29–50 μm with a mean of 86 × 37 μm, (Fig. 18).

The life cycle of *A. tetraptera* is direct but distinct from that of *Syphacia* species in that the unembryonated eggs are passed entirely in the feces and are not found in the perianal region of the host. Chan (16) studied its life cycle in mice and found that *A. tetraptera* eggs develop and become infective in 5–8 days at 27°C. The eggs are highly resistant to adverse environmental factors (such as disinfectants, desiccation, and cold) but are relatively susceptible to heat. Infection is by ingestion of infective eggs. Larvae hatch and develop in the posterior colon and then migrate anteriorly and develop to maturity in the proximal colon. They remain in the lumen of the intestine and do not invade the mucosa. The prepatent period is 23 days.

b. Pathobiology. *Aspicularis tetraptera* does not appear to be pathogenic. Its pathobiologic effects may be the same as *Syphacia* spp. in rodents.

c. Diagnosis. Diagnosis is by finding the characteristic eggs in the feces or adult worms in the intestines. Sampling by the anal tape technique is of no value.

d. Treatment, Prevention, and Control. The chemotherapy of *A. tetraptera* is the same as that for *S. muris* and *S. obvelata*.

The use of thiabendazole at a level of 0.3% in food is effective in removing rodent pinworms (94).

The prevention and control of the oxyurid infection are based on cesarean derivation, effective treatment, and strict sanitation and husbandry.

4. *Heterakis spumosa*

a. Description and Life Cycle. *Heterakis spumosa* is common in the cecum and colon of wild rats but rare in the laboratory rat (68). The male worms are 6–10 mm long and 200–260 μm wide with a distinct spicule. The males have a preanal sucker with a chitinous rim. The females are 7–13 mm long by 680–740 μm wide. The eggs have a thick-walled, mammillated shell and measure 55–60 by 40–55 μm (Fig. 19).

The life cycle is direct. The eggs are passed in the feces and embryonate to become infectious in 14 days under optimal conditions. Infection is by ingestion of infective eggs which hatch in the stomach. The larvae then migrate to the cecum and colon where they develop and mature. Eggs appear in the feces 26 to 47 days after infection (89).

b. Pathobiology. *Heterakis spumosa* is nonpathogenic for rats or mice.

c. Diagnosis. Diagnosis is based on the demonstration of eggs in the feces or adults in the cecum.

d. Treatment, Prevention, and Control. For the treatment of *H. spumosa* in the rat, Habermann and Williams (38) recommended 1 gm of phenothiazine with 10 ml of molasses

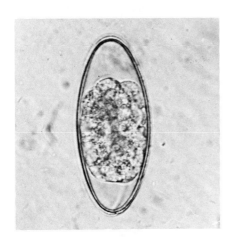

Fig. 18. Aspicularis tetraptera eggs. ×400.

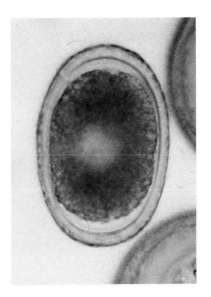

Fig. 19. Heterakis spumosa eggs. (Courtesy of Dr. A. M. Allen.)

added to 20 gm of feed (38). Prevention and control are by appropriate sanitation and good facility management.

F. Cestodes

1. *Hymenolepis nana*

a. Description and Life Cycle. *Hymenolepis nana* occurs in the small intestine of mice, rats, hamsters, nonhuman primates, and humans. This parasite is commonly known as the dwarf tapeworm, although the adults vary greatly in size ranging from 7 to 100 mm in length, it averages 20–40 mm long by 1 mm wide. The scolex is globular and bears four suckers and a retractible muscular rostellum encircled at the anterior end by a ring of 20–27 small hooklets. The strobila consists of a chain of segments.

Mature proglottids (measure 0.8 mm in width and 0.2 mm in length) contain three globular testes, a single globular ovary and a compound vitelline gland. Gravid segments contain 100–200 eggs which are liberated by disintegration of the terminal segments and are passed out in the feces. The egg (Fig. 20) is oval and measures 40–62 μm by 30–55 μm. The egg shell is thin and colorless. The embryo is spherical and thin walled with a knob at each pole, from which six fine filaments emerge. The onchosphere (hexacanth embyro) possesses three pairs of small hooks.

The life cycle of *H. nana* may be direct or indirect (without or with an intermediate host) as well as by autoinfection. In the direct life cycle, embryonated eggs are ingested by the definitive host and hatch in the small intestine, where the onchosphere emerges and penetrates into villi to develop into a cysticercoid larva in 4 to 5 days. The larva then reenters the lumen of the gut, and the scolex evaginates and attached to the mucosa (Figs. 21–23). Mature worms then develop in 10 to 11 days. Therefore the direct life cycle requires 14 to 16 days. The adults live for only a few weeks in the host. The direct cycle induces a certain degree of immunity which can prevent autoinfection.

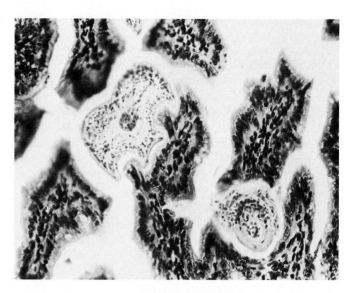

Fig. 21. *Hymenolepis nana*. Scolices are visible between intestinal villi. A scolex (center) attaches to a villus by two of its suckers. Hematoxylin and eosin. ×125. (Courtesy of Dr. J. R. Lindsey.)

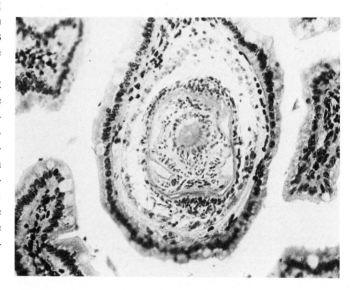

Fig. 22. *Hymenolepis nana*. An intestinal villus with a developing intermediate stage (cysticercoid). Parts of the scolex, including the hooks and suckers, are visible. Hematoxylin and eosin. ×125. (Courtesy of Dr. J. R. Lindsey.)

In the indirect life cycle, proper intermediate hosts such as grain beetles (*Tenebrio molitor* and *T. obscurus*) and fleas (*Pulex irritans, Ctenocephalus canis,* and *Xenopsylla cheopis*) are required. The embryonated eggs are ingested by the arthropod intermediate hosts. In the gut, the egg hatches, and the onchosphere penetrates the hemocoele and develops into a cysticercoid larva; this process is completed at metamorphosis. The definitive host becomes infected by ingestion of the in-

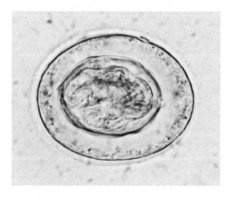

Fig. 20. *Hymenolepis nana* egg. ×500.

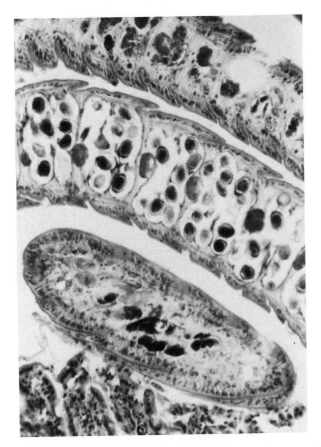

Fig. 23. Portions of strobila of adult *Hymenolepis nana* tapeworms in intestinal lumen Hematoxylin and eosin. ×200.

fected arthropod. The indirect route of transmission confers little host immunity and permits autoinfection to occur.

Autoinfection by *H. nana* does occur in animals that have no or little immunity to the tapeworm. The eggs hatch within the small intestine of the same host in which they are produced (42). The embryos develop into mature worms without leaving the original host.

b. Pathobiology. *Hymenolepis nana* is a common parasite of the laboratory rodents, pathogenic only when large numbers of worms are present. Heavy infection in young animals causes retarded growth, weight loss, intestinal occlusion and impaction, and death (25). In less severe cases, it causes catarrhal enteritis with little inflammatory response. Chronic inflammation, abscesses, and focal granulomatous lymphadenitis of mesenteric lymph nodes have been reported (34,87). The eggs are relatively susceptible to the environment and cannot survive long outside the host under ordinary conditions.

Hymenolepis nana is a public health hazard and can infect laboratory personnel who work around infected rodents; therefore, proper precautions should be taken.

c. Diagnosis. Diagnosis is based on the demonstration of the eggs in the feces or the adults in the intestine at necropsy.

d. Treatment, Prevention, and Control. Effective treatment with a single dose of niclosamide at 100 mg/kg orally (36) or quinacrine hydrochloride orally at 75 mg/kg have been reported (5).

Hymenolepis nana can be transmitted among animals in many ways. An effective prevention and control program should include cesarean derivation, barrier husbandry techniques, insect and rodent control, quarantine of acquired animals, and high standards of management.

2. *Hymenolepis diminuta*

a. Description and Life Cycle. *Hymenolepis diminuta* occurs in the small intestine of rats, mice, hamsters, monkey and man. The parasite is commonly known as the rat tapeworm. *Hymenolepis diminuta* may be distinguished from *H. nana* by (a) its greater size, being 20–60 mm in length and 3–4 mm in width; and (b) its having small pear-shaped scolex bearing four deep suckers and having an unarmed rostellum. The proglottids are similar to those of *H. nana* being broader than they are long and containing three oval testes. The gravid segments break away from the strobila and are passed in the feces. The eggs (Fig. 24) are spherical, measure 60–88 μm by 52–81 μm and contain a hexacanth embryo which also possess three pairs of small hooks. However, the embryo does not have

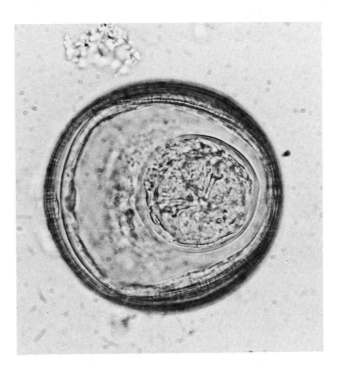

Fig. 24. *Hymenolepis diminuta* egg. ×600.

polar filaments. The eggs are more resistant to the environment than those of *H. nana*.

The life cycle of *H. diminuta* is always indirect and requires arthropods such as the four beetle (*Tenebrio molitor* or *Tribolium confusum*), fleas (*Nosopsyllus fasciatus*), or moths as an intermediate host. The eggs are ingested by intermediate hosts and hatch, and the onchosphere migrates to the hemocoele, where it develops into a cysticercoid. Infection of the definitive host is by ingestion of the infected arthropod. The larva is liberated from the cysticercoid, and the scolex evaginates and attaches itself to the intestinal mucosa where they develop into adult worms in 19 to 21 days.

b. Pathobiology. Hymenolepis diminuta is pathogenic only when large numbers are present, and this seldom occurs. When this does occur, it causes an acute catarrhal enteritis or a chronic enterocolitis with lymphoid hyperplasia. Infected animals develop immunity. The eggs are more resistant to disinfectants or environment than those of *H. nana* and can survive in the feces for up to 6 months.

c. Diagnosis. Diagnosis is based on the finding of eggs or proglottids in the feces or the adult worms in the intestine.

d. Treatment, Prevention, and Control. Treatment, prevention, and control of *H. diminuta* are the same as recommended for *H. nana*.

Fig. 25. Polyplax spinulosa adult male louse. ×40.

IV. PARASITES OF THE INTEGUMENT

A. Lice

1. *Polyplax spinulosa*

a. Description and Life Cycle. Polyplax spinulosa (the spined rat louse) is the common louse of both laboratory and wild rats. It occurs in the body fur especially on the midbody, shoulder, and neck. *Polyplax spinulosa* is a blood sucking louse. It is slender, yellow-brown, and 0.6–1.5 mm long with a brown tinge (Figs. 25 and 26). The head is rounded with two five-segmented antennae. Eyes are absent. The body is thickset. The ventral thoracis plate is pentagonal. The abdomen has about seven lateral plates on each side and seven to thirteen dorsal plates. The third segment of the male antenna is provided with a pointed apophysis. The eggs are elongated and are laid and fastened to the hair near the skin (Fig. 27). They have a conelike operculum with a row of pores along the cone. Nymphs resemble adults but are smaller and paler.

Eggs hatch in 5 to 6 days, and the nymphs emerge from the operculum and develop to first-stage nymphs. They then undergo three moltings and become adults. Each nymph resembles adults, but is paler in color and is lacking the reproductive organs. The whole life cycle takes place on the body fur and is completed in 26 days. Adults survive 35 to 28 days. Transmission between animals is by direct contact.

b. Pathobiology. Polyplax spinulosa sucks the blood of the host. Infected animals usually show unthrifty appearance, constant scratching and irritation, restlessness, anemia, and debilitation. It is the vector of *Hemobartonella muris*, *Rickettsia typi*, *Trypanosoma lewisi*, *Borrellia duttoni*, and *Brucella brucei*. The lice move slowly and usually do not leave their host.

c. Diagnosis. Diagnosis depends on finding and identifying the adult lice, nymphs or eggs on the fur.

d. Treatment and Control. Infested rats can be effectively treated for lice with several insecticides as sprays, dusts or dips. They should be applied both to the animals and bedding. Dusts containing 0.25 to 0.5% lindane, 3 to 5% malathion, 10% methoxychlor, 0.5 to 1% rotenone, or a silica dust (Dri-Die 67), with or without pyrethrins, are effective. Sprays or

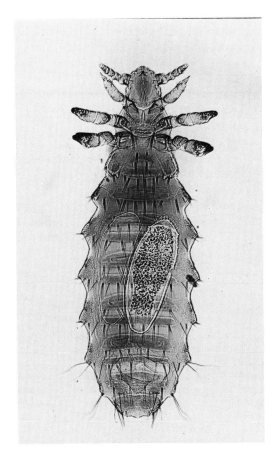

Fig. 26. *Polyplax spinulosa* adult female louse. ×60.

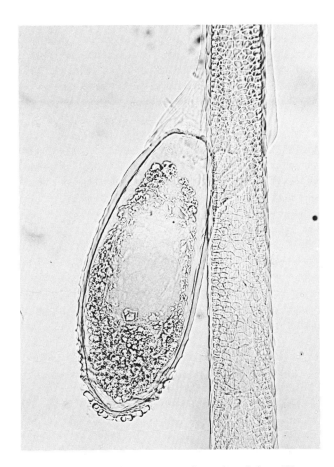

Fig. 27. *Polyplax spinulosa* egg fastened to a hair. ×150.

dips containing 0.03 to 0.06% Diazinon, 0.1 to 0.25% malathion, 0.5% methoxychlor, or a 2% aqueous suspension of a wettable powder containing 15% butylphenoxyisopropyl chloroethyl sulfite (Aramite) are effective treatment (56).

Louse infestation can be prevented and controlled by proper quarantine procedures, effective treatment, and by cesarean derivation and barrier techniques.

B. Mites

1. *Radfordia ensifera (Myobia ratti)*

a. Description and Life Cycle. *Radfordia ensifera,* the fur mite of rat, is a pelage-inhabiting mite that commonly occurs on wild and laboratory rats. It is closely related to *Myobia musculi* and *Radfordia affinis* of mice. Morphologically, it can be distinguished readily by the claws on the tarsi of the second pair of legs of *R. ensifera.* They are paired and equal, whereas those of *R. affinis* are paired and subequal; *M. musculi* has only a single empodial claw on these segments.

The life cycle, transmission, pathological effects, treatment

and control are unknown but are probably similar to those of *Myobia musculi* of the mice as outlined by Flynn (25).

2. *Notoedres muris*

a. Description and Life Cycle. *Notoedres muris,* the ear mange mite of the rat, occurs in the skin and ears of wild and laboratory rats (24,25). It is a short-legged, roundish white mite. The females measure 400 × 350 μm and resemble those of *Sarcoptes scabiei* but differ by the dorsal position of the anal opening and by the absence of heavy dorsal spines, cones, and triangular scales (4). The males and immature forms resemble those of *S. scabiei.* Both males and females bear suckers on unointed pedicels on the first two pairs of legs. The third and fourth pairs of legs are short and do not protrude beyond the edge of the body. The male also bears suckers on the fourth pair of legs.

The female mites burrow into the deeper layers of the epidermis laying eggs at regular intervals in the stratum corneum. The larvae hatch in 4 or 5 days and either remain in the maternal burrow or emerge to excavate a pocket on the surface in which to molt; two nymphal stages follow, each of which may form its own molting pocket. The second-stage nymphs molt and develop into the adult male and female approximately 16 to 18 days after oviposition. The males emerge from their pockets to search for the females; copulation takes place in the burrow. The whole life cycle from egg to adult takes 19 to 21 days. Transmission is by direct contact.

b. Pathobiology. *Notoedres muris* causes a scabies-like mange in the rat. The lesions are characterized by red papules, reddened and thickened skin, thick horny crusts, warty excrescences, and erythematous vesicles on the hairless or sparsely haired regions of the body such as ear pinnae, nose, face, head, tail, and, sometimes, external genitalia and legs. In severe cases, the outer ear may be completely eroded.

The mites are usually restricted to the stratum corneum but occasionally they penetrate this layer, giving rise to a more severe skin reaction and serum exudation. Epidermal cells proliferate and gradually cornify (103). The inflammation is localized and characterized by an influx of neutrophils and lymphocytes.

c. Diagnosis.   Diagnosis is by finding and identifying the mites in deep skin scrapings from the infected lesions.

d. Treatment, Prevention, and Control.   Effective treatments include topical application of 0.1 to 0.25% lindane solution once or twice, or dipping in a 2% aqueous suspension of a wettable powder containing 15% butylphenoxyisopropyl chlorethyl sulfite (Aramite-15W). Where *Notoedres muris* infections are severe, crusts and papules should be softened and removed before treatment. Cook found dibutyl phthalate painted directly on the ears effective in eliminating ear mange (18).

The prevention and control of *N. muris* infection includes proper quarantine of all newly acquired rats or the establishment of a rat colony by means of gnotobiotic technique with barrier maintenance.

3. *Laelaps echidninus*

a. Description and Life Cycle.   *Laelaps echidninus* is the "spiny rate mite." It is common on wild rats and occasionally found on laboratory rats and mice. The mites live in the bedding and come out at night to feed on abraded skin and body fluids; they sometimes suck blood. The are long-legged and globular in shape. The females are about 1 mm long with sclerotized, reddish brown shields. Morphology of the mite is described in detail elsewhere (92).

Owen studied the life cycle of *L. echidninus* (69). The larvae are born alive, do not feed, and molt into protonymphs in 10 to 13 h. The protonymphs feed and molt in 3 to 11 days to deutonymphs, which also feed and molt in 3 to 9 days. The

females feed shortly after molting and produce larvae within 5 days, sometimes by parthenogenesis. The whole life cycle takes at least 16 days, and females can live 2–3 months if they feed.

b. Pathobiology.   *Laelaps echidninus* is nonpathogenic. However, it sucks host blood or feeds on body fluid. It is the natural vector of *Hepatozoon muris*.

c. Diagnosis.   Diagnosis is by finding and identifying the mites on the rat or in the bedding or cage crevices.

d. Treatment, Prevention, and Control.   The treatments described for other rodent mites may be applied to this mite. Prevention and control involve effective wild rodent control and high standards in hygienic practices.

REFERENCES

1. Ahlquist, J., Rytömaa, T., and Borgmästar, H. (1962). Blood eosinophilia caused by a common parasite in laboratory rats. *Acta Haematol.* **28,** 306–312.
2. Arean, V. M. (1971). Pulmonary pneumocystitis. *In* "Pathology of Protozoal and Helminthic Diseases with Clinical Correlation" (R. A. Marcial-Rojas, ed.), pp. 291–317. Williams & Wilkins, Baltimore, Maryland.
3. Attwood, H. D., and Sutton, R. D. (1965). *Encephalitozoon* granuloma in rats. *J. Pathol. Bacteriol.* **89,** 735–738.
4. Baker, E. W., Evans, T. M., Gould, D. J., Hull, W. B., and Keegan, J. L. (1956). "A Manual of Parasitic Mites of Medical or Economic Importance," Tech. Publ. Natl. Pest Control Assoc., Inc., New York.
5. Balazs, T., Hatch, A. M., Gregory, E. R. W., and Grice, H. C. (1962). A comparative study of hymenolepicides in *Hymenolepis nana* infestation of rats. *Can. J. Comp. Med. Vet. Sci.* **26,** 160–162.
6. Barton, E. G., and Campbell, W. G. (1969). *Pneumocystis carinii* in lungs of rats treated with cortisone acetate. Ultrastructural observations relating to the life cycle. *Am. J. Pathol.* **54,** 209–217.
7. Bell, E. P. (1968). Disease in a caesarian-derived albino rat colony in a conventional colony. *Lab. Anim.* **2,** 1–17.
8. Beverley, J. K. A. (1976). Toxoplasmosis in animals. *Vet. Rec.* **99,** 123–127.
9. Blair, L. S., and Thompson, P. E. (1969). Effects of pyrvinium pamoate in the ration or drinking water of rats against the pinworm *Syphacia muris*. *Lab. Anim. Care* **19,** 639–643.
10. Bone, J. F., and Harr, J. R. (1967). *Trichosomoides crassicauda* infection in laboratory rats. *Lab. Anim. Care* **17,** 321–326.
11. Boorman, G. A., Lina, P. H. C., Zurcher, C., and Nieuwerkerk, H. T. M. (1973). *Hexamita* and *Giardia* as a cause of mortality in congenitally thymus-less (nude) mice. *Clin. Exp. Immunol.* **15,** 623–627.
12. Borgers, M. (1973). The cytochemical application of new potent inhibitors of alkaline phosphatases. *J. Histochem. Cytochem.* **21,** 812–824.
13. Brugmans, J., and Symoens, J. (1977). The effects of levamisole on host defense mechanisms: A review. *Modulation Host Immune Resistance Prev. Treat. Induced Neoplasia, 1974.* Fogarty Int. Cent. Proc. No. 28, DHEW Publ. No. (NIH)77-893, pp. 3–15.
14. Campbell, W. G. (1972). Ultrastructure of *Pneumocystis* in human lung. *Arch. Pathol.* **93,** 312–324.

15. Chan, K.-F. (1952). Life cycle studies on the nematode *Syphacia obvelata. Am. J. Hyg.* **56,** 14–21.

16. Chan, K.-F. (1955). The distribution of larval stages of *Aspicularis tetraptera* in the intestine of mice. *J. Parasitol.* **41,** 529–532.

17. Chapman, W. H. (1964). The incidence of a nematode, *Trichosomoides crassicauda,* in the bladder of laboratory rats. Treatment with nitrofurantoin and preliminary report of their influence on urinary calculi and experimental bladder tumors. *Invest. Urol.* **2,** 52–57.

18. Cook, R. (1956). Murine ear mange: The control of *Psorergates simplex* infestation. *Br. Vet. J.* **112,** 22–25.

19. Cook, R. (1969). Common diseases of laboratory animals. *In* "The I.A.T. Manual of Laboratory Animal Practice and Techniques" (D. J. Short and D. P. Woodnott, eds.), 2nd ed., pp. 160–215. Thomas, Springfield, Illinois.

20. Deichmann, W. B. (1970). Tumorigenicity of aldrin, dieldrin and endrin in the albino rat. *Ind. Med.* **39,** 37–45.

21. Doedhar, S. D., Lee, V. W., Chiang, T., Mahmoud, A. F., and Warren, K. S. (1976). Effects of the immunosuppressive drug niridazole in isogeneic and allogenic mouse tumor systems in vivo. *Cancer Res.* **36,** 3147–3150.

22. Exon, J. H., Patton, N. M., and Koller, L. D. (1975). Hexamitiasis in cadmium-exposed mice. *Arch. Environ. Health* **30,** 463–464.

23. Eyles, E. E. (1963). Chemotherapy of toxoplasmosis. *Exp. Chemother.* **1,** 641–655.

24. Flynn, R. J. (1960). *Notoedres muris* infestation of rats. *Proc. Anim. Care Panel* **10,** 69–70.

25. Flynn, R. J. (1973). "Parasites of Laboratory Animals." Iowa State Univ. Press, Ames.

26. Foster, H. L. (1963). Specific pathogen-free animals. *In* "Animals for Research: Principles of Breeding and Management" (W. Lane-Petter, ed.), pp. 110–137. Academic Press, New York.

27. Frenkel, J. K. (1971). Protozoal diseases of laboratory animals. *In* "Pathology of Protozoal and Helminthic Diseases" (R. A. Marcial-Rojas, ed.), pp. 319–369. Williams & Wilkins, Baltimore, Maryland.

28. Frenkel, J. K. (1973). Toxoplasmosis: Parasite, life cycle, pathology, and immunology. *In* "The Coccidia" (D. M. Hammond and P. L. Cong, eds.), pp. 343–410. Univ. Park Press, Baltimore, Maryland.

29. Frenkel, J. K. (1976). *Pneumocystis jiroveci* n.sp. from Man: Morphology, Physiology and Immunology in relation to pathology. *Natl. Cancer Inst., Monogr.* **43,** 13–27.

30. Frenkel, J. K., Dubey, J. P., and Miller, N. C. (1970). *Toxoplasma gondii* in cats: Fecal stages identified as coccidian oocysts. *Science* **167,** 893–896.

31. Frenkel, J. K., Good, J. T., and Shults, J. A. (1966). Latent pneumocystis infection of rats, relapse, and chemotherapy. *Lab. Invest.* **15,** 1559–1577.

32. Frenkel, J. K., and Reddy, J. K. (1977). Induction of liver tumor by 3'-methyl-dimethylaminoazobenzene (3'-Me-DAB) in rats chronically infected with *Toxoplasma* or *Besnoitia. RES, J. Reticuloendothel. Soc.* **21,** 61–68.

33. Garner, F. M., Innes, J. R. M., and Nelson, D. H. (1967). Murine neuropathology. *In* "Pathology of Laboratory Rats and Mice" (E. Cotchin and F. J. C. Roe, eds.), pp. 295–348. Blackwell, Oxford.

34. Garner, F. M., and Patton, C. S. (1971). Helminthic diseases of laboratory animals (rats and mice). *In* "Pathology of Protozoal and Helminthic Diseases" (R. A. Marcial-Rojas, ed.), pp. 960–969. Williams & Wilkins, Baltimore, Maryland.

35. Gentry, L. O., and Remington, G. S. (1971). Resistance against *Crytococcus* conferred by intracellular bacteria and protozoa. *J. Infect. Dis.* **123,** 22–31.

36. Gonnert, R., and Schraufstatter, E. (1960). Experimentelle Untersuchungen mit *N*-(2'-chlor-4'-nitrophenyl)-5-chlorsalicylamidcinem neuen Bandwurmmittel. I. Mitteilung: Chemotherapeutische versuche. *Arzneim.-Forsch.* **10,** 881–884.

37. Habermann, R. T., and Williams, F. P. (1957). The efficacy of some piperazine compounds and slytomycin in drinking water for the removal of oxyurids from mice and rats and a method of critical testing of anthelminthics. *Proc. Anim. Care Panel* **7,** 89–97.

38. Habermann, R. T., and Williams, F. P. (1958). The identification and control of helminths in laboratory animals. *J. Natl. Cancer Inst.* **20,** 979–1010.

39. Haese, W. H., Smith, D. L., and Bueding, E. (1973). Hycanthone-induced hepatic changes in mice infected with *Schistosoma mansoni. J. Pharmacol. Exp. Ther.* **186,** 430–440.

40. Hawking, F. (1963). Chemotherapy of Trypanosomiasis. *Exp. Chemother.* **1,** 129.

41. Hayden, D. W., King, N. W., and Murthy, A. S. K. (1976). Spontaneous *Frenkelia* infection in a laboratory-reared rat. *Vet. Pathol.* **13,** 337–342.

42. Heyneman, D. (1961). Studies on helminth immunity. III. Experimental verification of autoinfection from cystercercoids of *Hymenolepis nana* in the white mouse. *J. Infect. Dis.* **109,** 10–18.

43. Hibbs, J. B., Jr., Lewis, H. L., Jr., and Remington, J. S. (1971). Resistance to murine tumors conferred by chronic infection with intracellular protozoa, *Toxoplasma gondii* and *Besnoitia jellisoni. J. Infect. Dis.* **124,** 587–592.

44. Hitchcock, C. R., and Bell, E. T. (1952). Studies on the nematode parasite, *Gonglylonema neoplasticum,* and avitaminosis A in the forestomach of rats: Comparison with Fibiger's results. *J. Natl. Cancer. Inst.* **12,** 1345–1387.

45. Hoag, W. C. (1961). Oxyuriasis in laboratory mouse colonies. *Am. J. Vet. Res.* **2,** 150–153.

46. Homburger, F. (1977). Saccharin and cancer. *N. Engl. J. Med.* **297,** 560–561.

47. Hussey, K. L. (1957). *Syphacia muris* vs. *S. obvelata* in laboratory rats and mice. *J. Parasitol.* **43,** 555–559.

48. Hutchison, W. M., Durachie, J. F., Siim, J. C., and Work, K. (1970). Coccidian-like nature of *Toxoplasma gondii. Br. Med. J.* **1,** 142–144.

49. Innes, J. R. M., Garner, F. M., and Stookey, J. L. (1967). Respiratory disease in rats. *In* "Pathology of Laboratory Rats and Mice" (E. Cotchin and F. J. C. Roe, eds.), pp. 229–257. Blackwell, Oxford.

50. Innes, J. R. M., Zeman, W., Frenkel, J. K., and Borner, G. T. (1962). Occult endemic encephalitozoonosis of the central nervous system of mice. *J. Neuropathol.* **21,** 519.

51. Ito, S., Tsunoda, K., Shimada, K., Taki, T., and Matsui, T. (1975). Disinfectant effects of several chemicals against *Toxoplasma* oocysts. *Jpn. J. Vet. Sci.* **37,** 229–234.

52. Jones, S. R. (1973). Toxoplasmosis: A review. *J. Am. Vet. Med. Assoc.* **163,** 1038–1042.

53. Jones, T. C. (1967). Pathology of the liver of rat and mice. *In* "Pathology of Laboratory Rats and Mice" (E. Cotchin and F. J. C. Roe, eds.), pp. 1–23. Blackwell, Oxford.

54. Keast, D., and Chesterman, F. C. (1972). Changes in macrophage metabolism in mice heavily infected with *Hexamita muris. Lab. Anim.* **6,** 33–39.

54a. Kunstyr, I. (1977). Infectious form of *Spironucleus (Hexamita) muris:* Banded cysts. *Lab. Anim.* **11,** 185–188.

55. Levine, N. D. (1973). "Protozoan Parasites of Domestic Animals and of Man," 2nd ed. Burgess, Minneapolis, Minnesota.

56. Levine, N. D. (1974). Diseases of laboratory animals-parasitic. *In* "CRC Handbook of Laboratory Animal Science" (E. C. Melby and N. H. Altman, eds.), Vol. II, pp. 289–327. CRC Press, Cleveland, Ohio.

57. Levine, N. D., and Ivens, V. (1965). "The Coccidian Parasites (Protozoa, Sporozoa) of Rodents." Univ. of Illinois Press, Urbana.

58. Lewis, W. P., and Markell, E. K. (1958). Acquisition of immunity to toxoplasmosis by the newborn rat. *Exp. Parasitol.* **7**, 463.

59. Lincicome, D. R., Rossman, R. M., and Jones, W. C. (1963). Growth of rats infected with *Trypanosoma lewisi. Exp. Parasitol.* **14**, 54.

60. Luttermoser, G. W. (1936). A helminthological survey of Baltimore house rats (*Rattus norvegicus*). *Am. J. Hyg.* **24**, 350.

61. Mahmoud, A. A. F., and Warren, K. S. (1974). Anti-inflammatory effects of tartar emetic and niridazole: Suppression of schistosome egg granuloma. *J. Immunol.* **112**, 222–228.

62. Meshorer, A. (1969). Hexamitiasis in laboratory mice. *Lab. Anim. Care* **19**, 33–37.

63. Miller, W. W. (1908). *Hepatazoon perniciosum* (n.g., n.sp.); a hemogregarine pathogenic for white rats; with a description of the sexual cycle in the intermediate host, a mite (*Laelaps echidninus*). *Bull. U.S. Hyg. Lab.* **46**, 1.

64. Mudrow-Reichenow, L. (1956). Spontanes vorkommen von Amöben und Ciliaten bei Laboratoriumstieren. *Z. Tropenmed. Parasitol.* **7**, 198–211.

65. Myers, D. D. (1973). Sensitivity to X-irradiation of mice infected with *Hexamita muris. 24th Annu. Meet. Am. Assoc. Lab. Anim. Sci.* Abstract No. 22.

66. Nelson, J. B. (1962). An intracellular parasite resembling a microsporidian associated with ascites in Swiss mice. *Proc. Soc. Exp. Biol. Med.* **109**, 714–717.

67. Newberne, P. M. (1975). Influence on pharmacological experiments of chemicals and other factors in diets of laboratory animals. *Fed. Proc., Fed. Am. Soc. Exp. Biol.* **34**, 209–218.

68. Oldham, J. N. (1967). Helminths, ectoparasites and protozoa in rat and mice. *In* "Pathology of Laboratory Rats and Mice" (E. Cotchin and F. J. C. Roe, eds.), pp. 641–679. Blackwell, Oxford.

69. Owen, B. L. (1956). Life history of the spiny rat mite under artificial conditions. *J. Econ. Entomol.* **49**, 702–703.

70. Pakes, S. P. (1974). Protozoal diseases. *In* "Biology of the Laboratory Rabbit" (S. H. Weisbroth, R. E. Flatt, and A. L. Kraus, eds.), pp. 263–286. Academic Press, New York.

71. Pelley, R. P., Pelley, R. J., Stavitsky, A. B., Mahmoud, A. A. F., and Warren, K. S. (1975). Niridazole, a potent long-acting suppressant of cellular hypersensitivity. IV. Minimal suppression of antibody responses. *J. Immunol.* **115**, 1477–1482.

72. Perard, C. (1926). Sur la coccidiose du sat. *Bull. Acad. Vet. Fr.* **102**, 120–124.

73. Perrin, T. L. (1943). *Toxoplasma* and *Encephalitozoon* in spontaneous and in experimental infections of animals: A comparative study. *Arch. Pathol.* **36**, 568–578.

74. Petri, M. (1965). A cytolytic parasite in the cells of transplantable, malignant tumors. *Nature (London)* **205**, 302–303.

75. Petri, M. (1966). The occurrence of *Nosema cuniculi* (*Encephalitozoon cuniculi*) in the cells of transplantable, malignant ascites tumors and its effect upon tumor and host. *Acta Pathol. Microbiol. Scand.* **66**, 13.

76. Popper, H., Sternberg, S. S., Oser, B. L., and Oser, M. (1960). The carcinogenic effect of aramite in rats—a study of hepatic nodules. *Cancer* **13**, 1035–1046.

77. Price, E. W., and Chitwood, B. G. (1931). Incidence of internal parasites in wild rats in Washington, D.C. *J. Parasitol.* **18**, 55.

78. Ratcliffe, H. L. (1967). Metazoan parasites of the rat. *In* "The Rat in Laboratory Investigation" (E. J. Farris and J. Q. Griffith, eds.), pp. 502–514. Hafner, New York.

79. Remington, J. S., and Merigan, T. C. (1969). Resistance to virus challenge in mice infected with protozoa or bacteria. *Proc. Soc. Exp. Biol. Med.* **131**, 1184–1188.

80. Robbins, J. B., Devita, V. T., Jr., and Dutz, W., eds. (1976). "Symposium on *Pneumocystis carinii* Infection," Natl. Cancer Inst. Monogr. No. 43, DHEW Pub. No. (NIH) 76-930. Natl. Inst. Health, Bethesda, Maryland.

81. Rommel, M., and Krampitz, H. E. (1975). Contributions to the life cycle of *Frenkelia*. I. The identity of *Isospora buteonis* of *Buteo buteo* and *Frenkelia* sp. (*F. clethrionomyobuteonis* spec.n.) of *Clethrionomys glareolus. Berl. Muench. Tieraerztl. Wochenschr.* **88**, 338–340.

82. Ruskin, J., and Remington, J. S. (1968). Immunity and intracellular infection: Resistance to bacteria in mice infected with a protozoan. *Science* **160**, 72–74.

83. Ruskin, J., and Remington, J. S. (1969). A role for the macrophage in acquired immunity to phylogenetically unrelated intracellular organisms. *Antimicrob. Agents Chemother.* pp. 474–477.

84. Sasa, M., Tanaka, H., Fukin, M., and Takata, A. (1962). Internal parasites of laboratory animals. *In* "The Problems of Laboratory Animal Diseases" (R. J. C. Harris, ed.), pp. 195–214. Academic Press, New York.

85. Schwabe, C. W. (1955). Helminth parasites and neoplasia. *Am. J. Vet. Res.* **16**, 485.

86. Seed, T. M., and Aikawa, M. (1977). Pneumocystis. *In* "Parasitic Protozoa" (J. P. Kreier, ed.), Vol. 4, pp. 329–357. Academic Press, New York.

87. Simmons, M. L., Richter, C. B., Franklin, J. A., and Tennant, R. W. (1967). Prevention of infectious diseases in experimental mice. *Proc. Soc. Exp. Biol. Med.* **126**, 830–837.

88. Simmons, M. L., Williams, H. E., and Wright, E. B. (1965). Therapeutic value of the organic phosphate Trichlorfon against *Syphacia obvelata* in inbred mice. *Lab. Anim. Care* **15**, 382–385.

89. Smith, P. E. (1953). Life history and host parasite relations of *Heterakis spumosa*, a nematode parasite in the colon of the rat. *Am. J. Hyg.* **57**, 194–221.

90. Smith, V. S. (1946). Are vesical calculi associated with *Trichosomoides crassicauda*, the common bladder nematode of rats. *J. Parasitol.* **32**, 142–149.

91. Stahl, W. B. (1963). Studies on the life cycle of *Syphacia muris*, the rat pinworm. *Keio J. Med.* **12**, 55–60.

92. Strandtmann, R. W., and Mitchell, C. J. (1963). The laelaptine mites of the *Echinolaelaps* comples from the Southwest Pacific area (Acarina: Mesostigmata). *Pac. Insects* **5**, 541–576.

93. Symoens, J. (1977). Levamisole, an antianergic chemotherapeutic agent: An overview. *In* "Control of Neoplasia by Modulation of the Immune System" (M. A. Chirigos, ed.), pp. 1–24. Raven, New York.

94. Taffs, L. F. (1976). Further studies on the efficacy of thiabendazole given in the diet of mice infected with *H. nana, S. obvelata* and *A. tetraptera. Vet. Rec.* **99**, 143–144.

95. Tuffery, A. A., and Innes, J. R. M. (1963). Diseases of laboratory mice and rats. *In* "Animals for Research" (W. Lane-Petter, ed.), pp. 47–108. Academic Press, New York.

96. Twort, J. M., and Twort, C. C. (1932). Disease in relaation to carcinogenic agents among 60,000 experimental mice. *J. Pathol. Bacteriol.* **35**, 219.

97. Van Arman, G. G., and Campbell, W. C. (1975). Anti-inflammatory activity of thiabendazole and its relation to parasitic disease. *Tex. Rep. Biol. Med.* **33**, 303–311.

98. Vavra, J., and Kucera, K. (1970). *Pneumocystis carinii* Delanoe, its ultrastructure and ultrastructural affinities. *J. Protozool.* **17**, 463–483.

99. Vesell, E. S., Lang, C. M., White, W. J., Passananti, G. T., Hill, R. N., Clemens, T. L., Lia, D. L., and Johnson, W. D. (1976). Environmental and genetic factors affecting response of laboratory animals to drugs. *Fed. Proc., Fed. Am. Soc. Exp. Biol.* **35**, 1125–1132.

100. Wagner, J. E. (1970). Control of mouse pinworms, *Syphacia obvelata,* utilizing Dichlorvos. *Lab. Anim. Care* **20,** 39–44.

101. Wagner, J. E., Doyle, R. E., Ronald, N. C., Garrison, R. G., and Schmitz, J. A. (1974). Hexamitiasis in laboratory mice, hamsters and rats. *Lab. Anim. Sci.* **24,** 349–354.

102. Walzer, P. D., Schnelle, V., Armstrong, D., and Rosen, P. P. (1977). Nude mouse: A new experimental model for *Pneumocystis carinii* infection. *Science* **197,** 177–179.

103. Watson, D. P. (1962). On the immature and adult stages of *Notoedres alepis* (Railliet and Lucet, 1893) and its effect on the skin of the rat. *Acarologia* **4,** 64–77.

104. Weisbroth, S. H., and Scher, S. (1971). *Trichosomoides crassicauda* infections of a commercial rat breeding colony. I. Observations on the life cycle and propagation. *Lab. Anim. Care* **21,** 54–61.

105. Weisbroth, S. H., and Scher, S. (1971). *Trichosomoides crassicauda* infections of a commercial rat breeding colony. I. Drug screening for anthelmintic activity and field trials with methyridine. *Lab. Anim. Sci.* **21,** 213–219.

106. Wenrich, D. H. (1967). Protozoan parasites of the rat. *In* ''The Rat in Laboratory Investigation'' (E. J. Farris and J. Q. Griffith, eds.), pp. 486–501. Hafner, New York.

Chapter 13

Neoplastic Diseases

Norman H. Altman and Dawn G. Goodman

I. INTRODUCTION

A. General Considerations

Despite the fact that millions of rats are used every year in biomedical research, the incidence of spontaneous tumors in most strains is poorly documented. There are many reasons for this void in our existing knowledge, and some of these will be addressed briefly. Bullock and Rohdenberg (25) reviewed the early literature and they state the following in their introduction.

> Although spontaneous tumors of the rat have ceased to be a pathological novelty, the number of cases recurring in this species is, nevertheless, small as compared with that of neoplasms in the mouse. In the experience of McCoy [123] ... the rate of incidence, considering benign and malignant growth collectively, was only about one in one thousand, and Woolley and Wherry ... have reported a rate not much higher. Yet it must not be forgotten that the tumors recorded by these authors occurred in wild rats, of which probably fewer survive to reach the cancer age than is the case with white mice.... Malignant tumors of connective tissue origin, comprising 86 percent of all malignant growths, arose in the liver in 5 percent of the cases, and in 90 percent of these *Cysticercus fasciolaris* was demonstrable in the tumor.

These observations dramatically point out some of the problems in interpreting the literature on spontaneous neoplasms in rats. This confusion extends well after 1916; significant improvement was not noted until the 1960s. Inbred strains of rats were not available, and the genetic background of most animals was unknown. Many animals died of infectious or parasitic disease at a relatively young age and, therefore, did not reach the "cancer age." The association of the parasite *Cys-*

ticercus fasciolaris with hepatic sarcomas is worthy of special note because many investigators have considered these as "spontaneous tumors." We prefer the philosophy of Bullock and Curtis (24) who state the following in their paper.

> The object of the present communication is to consider the tumors which have arisen in the rats of this colony without any direct etiological relationship to *Cysticercus fasciolaris*. For these neoplasms the term "Spontaneous Tumors" will be employed to distinguish them from the experimentally induced *Cysticercus* tumors, though we realize that a spontaneous tumor is only one of which we do not at present know the causative agent.

While this optimistic quotation assumes that ultimately, the etiology of all tumors will be discovered, it vividly illustrates that the definition of "spontaneous" constantly changes. As investigators recognize the effects of hormones, diet, environment, etc., on biological systems, more tumors will be removed from the "spontaneous" category as defined by Bullock and Curtis.

B. Dietary Factors

Diet has a profound effect on the incidence of spontaneous tumors. It is beyond the scope of this chapter to provide an in depth discussion of the relationship between dietary constituents and tumor induction. However, there are some factors which must be mentioned since they are relevant to the interpretation of published data.

The number of calories and the source of these calories in rodent diets is extremely important. When rats were continuously fed a low caloric diet which retarded their growth, the number of spontaneous tumors was decreased and the age of tumor onset was delayed (165,175). These investigators also placed growth-retarded rats on control diets at varying ages and found that some types of tumors increased in incidence, but others did not. However, none of the rats on the early caloric restriction diets attained the tumor incidence of the controls. Caloric restrictions also can inhibit the growth of some transplanted tumors (209).

The protein–calorie ratio is an important factor in tumor induction under *ad libitum* as well as isocaloric feeding conditions. Chronic marginal protein undernutrition has been reported to predispose rats to early tumor occurrence and high morbidity (164,165,167), whereas protein overnutrition increased the susceptibility to urinary bladder papillomas (167). Other studies also have shown a variety of dietary effects on tumor incidence (58,75).

There are few published reports on the effects of vitamins and minerals on spontaneous tumor incidence in rats (99). However, experimental studies have shown that both vitamins and minerals may enhance or inhibit the induction of experimental tumors (19,34,90,108,155,162,182,187). The report by Jacobs and Huseby (94) states that the rats were fed a standard commercial laboratory diet and yet 23% of the animals had enlarged pituitary glands (2–3×) and many had gross enlargements of the thyroid glands. Were these changes related to the diet? Axelrad and Leblond (3) demonstrated that rats fed an iodine-deficient diet had an increased incidence of thyroid and pituitary gland tumors. These reports only emphasize the necessity for basic research on dietary requirements in order to standardize experimental diets.

C. Hormones

It is well known that a delicate hormonal balance exists in the body, and disturbances of this balance can result in neoplasia. Experimentally, the introduction of exogenous hormones can induce a variety of tumors in rats, particularly endocrine and mammary gland tumors (101). Tumors of nonendocrine organs such as the liver also can be significantly influenced by hormones. Rodent diets may also be inadvertently contaminated with hormones or contain feed ingredients which have hormonal effects and, therefore, influence the ''spontaneous'' tumor incidence (180). Another example of hormonal interaction is the increased incidence of adrenal tumors after gonadectomy (76,84).

D. Environmental Factors

We are only beginning to appreciate the effects of many environmental factors on tumor incidence. Temperature, light quality and quantity, noise, gaseous contaminants (e.g., ammonia) can all have profound effects on metabolism and tumor induction (59). This group is probably one of the most significant experimental variables, and yet one of the least understood or appreciated.

E. Genetic Factors

Chapter 1 of this volume (also see this volume, Chapter 3) describes the contributions of early investigators toward developing inbred strains of rats. While our knowledge of inbred rat strains is not as broad as with mouse strains, many of the same concepts are valid. The incidence of spontaneous tumors varies between strains and their F_1 hybrids (42). It is just as important to recognize that there are differences in tumor incidence between strains from the same source raised under different conditions. These differences may be greater than those between strains (115,180). We have attempted to distinguish the origin of strains wherever possible; however, a great deal of confusion still exists in the literature. An example of this confusion is the Sprague-Dawley rat which is a random-bred rat but is commonly mistaken for an inbred strain. Several references in addition to this text are available (48,89a). For example, Jacobs and Huseby (94) report neoplasms from inbred Fischer rats obtained from Dr. W. F. Dunning. However, Dr. Dunning maintains four sublines of Fischer rats, each with different characteristics, and the subline provided is not mentioned in the Jacobs and Huseby paper.

F. Microbiological Status

We are moving from an era where experimental animals had a variety of intercurrent diseases and unknown flora and fauna into a period of using ''defined'' animals. It is generally agreed that healthy animals are vital to the success of biomedical research programs. However, there is little consistency in the quality of animals currently being utilized, since scientists are just beginning to appreciate the importance of this area. A classification of animals based on their microbiological status recently has been published (98).

The few publications on spontaneous tumors in germfree rats suggest these animals may have fewer tumors than their conventional counterparts (147,149,172). The majority of the tumors observed were of mammary or hematopoietic origin. Since the effect of many chemical carcinogens can be altered by manipulating the intestinal flora (20,130), standardization of this complex flora may be vitally important in experimental studies.

We have attempted to review as many original papers as possible in order to exclude questionable ''spontaneous'' tumors. For example, some investigators have used animals in studies with X radiation and various chemical treatments and then concluded that the tumors observed were not experimentally induced. We have excluded many of these tumors from our tables since we prefer to follow the philosophy of Bullock and Curtis as previously discussed.

There are two different types of tables in this chapter. One type lists tumors by organ and includes a reference for each report. We have excluded other information about the tumor-bearing rats, e.g., strain, sex, and age, since it would have made the tables unwieldy. These data can readily be obtained from the references. Where there was a significant strain incidence, the tumors were excluded from the tables on organ incidence and listed in Table XVI on strain incidence. We have attempted to be as inclusive as possible using the sources available to us. We have also interpreted the tumor designations in the foreign literature in light of currently acceptable nomenclature. For example, "epithelioma" is commonly used to designate a carcinoma, and "malpighian epithelioma" denotes a squamous cell carcinoma. The modern translated terms have been used in the tables with appropriate explanations provided in the text.

II. INTEGUMENTARY SYSTEM

Spontaneous tumors of the skin, adnexae (excluding the mammary gland), and subcutis are uncommon in rats.

A. Epithelial Tumors

Epithelial tumors have been reported sporadically in most strains of rats (Table I) and usually are found around the face, tail, or paws. The most common types include papillomas, squamous cell carcinomas, and basal cell carcinomas. Other than tumors of the specialized adnexal glands described below, adnexal tumors are rare, although a few trichoepitheliomas have been reported. Eight malignant melanomas have been reported in BN/Bi rats of both sexes in animals greater than 26 months of age (26). Other than this, melanomas appear to be extremely rare in laboratory rats.

Papillomas are small raised papilliform proliferations composed of connective tissue fronds covered by squamous epithelium which is usually thickened by varying degrees of hyperkeratosis. Squamous cell carcinomas are characterized by a proliferation of squamous epithelium extending downward into the dermis resulting in solid nests of epithelial cells of varying degrees of differentiation. In more differentiated tumors, there is a relatively normal maturation sequence with small basal cells at the periphery adjacent to the stroma and maturation to squamous epithelium occurring toward the center of the epithelial nests. When central keratinization of these

islands is prominent, these nests are known as "keratin pearls." Mitotic figures are usually common, especially in the more anaplastic forms. These tumors are locally invasive and may metastasize to regional lymph nodes and elsewhere.

Basal cell carcinomas also arise from squamous epithelium and are thought to be derived from the basal or germinal layer (*stratum germinativum*). The tumors are usually composed of small closely packed cells with scant cytoplasm and uniform oval nuclei. The cells may be arranged in ribbons, cords, or small or large nests. Occasionally, there may be some degree of maturation toward sebaceous glands or hair follicles. Those tumors, composed primarily of hair folliclelike structures, are referred to as trichoepitheliomas and are thought to be derived from the basal layer of the hair follicle. Basal cell tumors and, to a lesser degree, trichoepitheliomas may be locally invasive but do not metastasize.

Table I

Spontaneous Tumors of the Integument in Rats

Tumor	Number of animals	Reference
Epithelial tumors		
Papilloma	2	24
		37
	1	153
	3	42
	4	69
	2	164
	1	102
	4	115
	3	166
	1	116
Squamous papilloma	1	151
	4	26
Squamous cell carcinoma	24	24
		37
	1	1
	1	153
	7	42
	5	164
	1	189
	1	15
	5	115
	13	166
	10	173
	1	26
	1	203
Adenoma	2	102
	1	166
Sebaceous gland adenoma	7	177
	1	151
	1	116
Sebaceous gland carcinoma	1	42
	1	58
		59
	1	194

(continued)

Table I (*Continued*)

Tumor	Number of animals	Reference
Trichoepithelioma	1	189
Adenoacanthoma	1	24
		37
	1	42
Basal cell tumor	1	15
	1	116
	2	24
		37
	1	175
Carcinoma	5	69
	4	177
Carcinosarcoma	7	24
		37
	1	42
Melanoma	1	42
	8	79
Zymbal's gland		
Squamous papilloma	1	189
Adenoma	5	177
Squamous cell carcinoma	3	189
Zymbal's gland carcinoma	3	200
Preputial/clitoral gland		
Myoepithelioma	1	94
Carcinoma	2	24
		37
Mesenchymal tumors		
Myoepithelioma	1	116
Mesodermal tumor	2	36
Leiomyoma	2	166
Rhabdomyosarcoma	1	116
Lipoma	1	25
	1	153
	3	42
	1	175
	1	177
	4	115
	3	132
	1	116
Lipoma or fibrolipoma	3	24
Fibrolipoma	1	102
Lipoblastoma	1	199
Liposarcoma	1	115
Myxolipoma	1	151
Myxoma	4	142
Myxosarcoma	2	153
	1	132
Fibromyxoma	1	206
Dermatofibrosarcoma	1	36
Fibroma	20	25
	22	24
	7	153
	10	42
	4	175
	1	36
	2	189
	7	94

Table I (*Continued*)

Tumor	Number of animals	Reference
	4	177
	1	102
	12	115
	10	132
	1	151
	1	116
	1	26
Fibrosarcoma	10	153
	74	42
	2	175
	2	142
	1	59
		58
	1	189
	2	177
	12	115
	3	132
	1	151
	7	166
	14	173
Sarcoma	11	25
	67	24
	15	42
	1	39
	1	59
		58
	4	94
	3	177
	5	115
	1	199

B. Specialized Adnexal Gland Tumors

1. Zymbal's Glands

These are large modified sebaceous glands which surround the external ear canal (Fig. 1). Spontaneous tumors of these glands are uncommon but are readily induced by a variety of systemically administered carcinogens (145,189,191). Grossly, Zymbal's gland tumors usually present as an ulcerated mass within or just below the external ear canal. They may become so large as to involve the entire side of the face and neck. Histologically, these tumors may be either sebaceous or squamous or admixtures thereof (Figs. 2 and 3). Frequently, squamous cell carcinomas arise from the squamous epithelium of the ducts or of the external ear canal. These tumors are highly aggressive locally but rarely metastasize.

2. Preputial (Clitoral) Glands

These are modified sebaceous glands located subcutaneously in the ventral inguinal region with stratified squamous epithelial lined ducts emptying into the preputial (clitoral) sac (Fig.

338   NORMAN H. ALTMAN AND DAWN G. GOODMAN

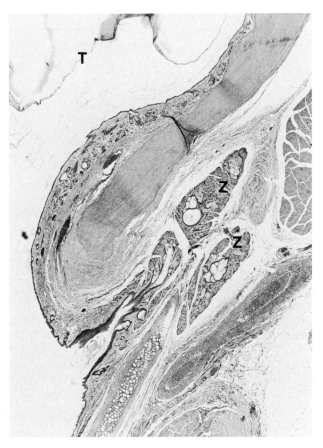

Fig. 1. Normal Zymbal's gland. Lobes of the gland located at the base of the ear (Z). Tympanic membrane (T). ×38.

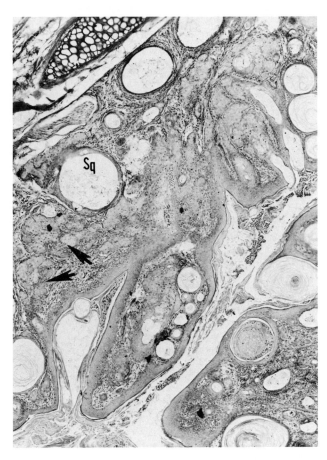

Fig. 2. Zymbal's gland tumor. Note both the sebaceous (arrows) and squamous elements (Sq). ×54.

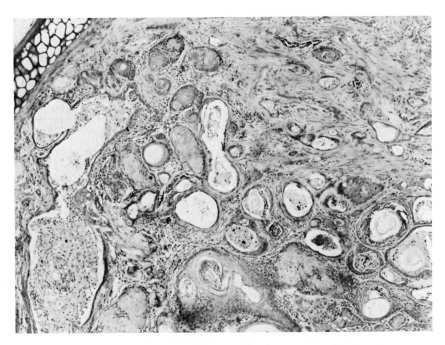

Fig. 3. Zymbal's gland tumor. Higher magnification of tumor in Fig. 2. ×80.

Table II

Spontaneous Tumors of the Musculoskeletal System in Rats

Tumor	Number of animals	Reference
Muscle		
Mixed	1	23
Giant cell sarcoma	1	94
Myoblastic sarcoma	1	69
Myosarcoma	1	203
Rhabdomyosarcoma	1	33
Skeletal		
Osteoma	1	24
	1	181
	1	115
	1	203
Osteosarcoma	3	25
	2	24
	6	153
	1	1
	1	69
	1	115
	1	173
	1	114
	3	203
Chrondrosarcoma	1	172

Guérin (69) reported a myoblastic sarcoma, but no further details were available. Coleman *et al.* (33) reported a subscapular rhabdomyosarcoma in an aged Fischer male rat.

Bullock and Rohdenburg (25) reviewed the sparse literature on spontaneous bone tumors in rats and they found three osteosarcomas. Bullock and Curtis (24) reported one osteoma and two osteosarcomas in a series of 489 rats. These rats were infected with *Cysticercus fasciolaris,* but the bone tumors were thought to be spontaneous and not parasite induced. Ratcliffe (153) described six osteosarcomas from a colony of 468 Wistar-derived rats. Three of these were located in the femur. Single cases of osteosarcoma were reported by Arai (1) and Guérin (69). Schulze (181) reported an osteoma of the cranial cavity in a female Sprague-Dawley rat from necropsies of 1040 animals. An osteoma and an osteosarcoma were reported by Mackenzie and Garner (115) in a study of 535 Charles River, Sprague-Dawley-derived rats. An osteosarcoma of the vertebral column was reported (173) in one Fischer female out of a group of 192 females. The age of the animal was not given, but the tumor was described as having a bizarre cellular pattern and irregular foci of calcification. An osteosarcoma in a WI/TEN rat has recently been reported (114).

IV. RESPIRATORY SYSTEM

Spontaneous pulmonary tumors are rare in rats (Table III).

Bullock and Curtis (24) made the observation that the lung seemed to be relatively insusceptible to neoplasia even in older rats. They observed four spontaneous pulmonary tumors out of 521 primary neoplasms. Dunning and Curtis (42) reported pulmonary sarcomas in seventeen Copenhagen rats and one Fischer rat. Saxton *et al.* (175) described two pulmonary adenomas in 498 rats. Olcott (142) reported a sarcoma of the lung in a 20-month-old Sherman-Mendel rat. Horn and Stewart (82) published a case report of a papillary pulmonary adenoma in a 12-month-old Marshall 520 rat. Guérin (69) reported eight carcinomas and two connective tissue tumors of the "pleuropulmonary system." Crain (36) reported 18 primary pulmonary tumors, mostly malignant lymphomas, in 786 Wistar-derived rats. Schardein *et al.* (177) reported a bronchogenic carcinoma in an 11-month-old Sprague-Dawley-derived rat. Snell (189) reported two bronchogenic polyps and three carcinomas in 488 rats of various inbred strains. Mackenzie and Garner (115) published a comparison of neoplasms in rats of six sources raised in seven laboratories. Out of 2082 rats, they found twelve primary pulmonary tumors. Préjean *et al.* (151) reported four tumors in 360 Sprague-Dawley rats. Morii and Fujii (132) reported a squamous cell carcinoma in one Sprague-Dawley male rat out of 108 rats. Maekawa and Odashima (116) reported a single papilloma in the trachea and a lymphoma in the lungs out of 264 ACI/N rats. Sacksteder (172) described four pulmonary carcinomas in germfree Fischer rats.

Table III

Spontaneous Tumors of the Respiratory System in Rats

Tumor	Number of animals	Reference
Papilloma (trachea)	1	116
Bronchogenic polyps	2	189
Papillary adenoma	1	82
	2	175
Adenoma	5	194
Adenocarcinoma	2	194
Squamous cell carcinoma	1	24
	1	132
Bronchogenic carcinoma	1	177
	12	115
	4	151
	1	215
Carcinoma	8	69
	1	39
	4	172
	3	189
Fibroma	1	69
Fibrosarcoma	2	36
Fibrocystic sarcoma	1	69
Osteochondrosarcoma	1	24
Sarcoma	18	42
	1	142

V. CARDIOVASCULAR SYSTEM

In general, spontaneous tumors of the cardiovascular system are uncommon in rats.

A. Vascular Tumors

Vascular tumors, including hemangiomas, hemangiosarcomas (endotheliomas, hemangioendotheliomas, malignant endotheliomas), lymphangiomas, and lymphangiosarcomas, have been reported (Table IV). These tumors are seen most frequently in the subcutaneous tissues, liver, and spleen, although they have been reported as occurring in a wide variety of sites.

Hemangiomas are composed of vascular spaces of variable size usually filled with blood and lined by a single layer of endothelial cells. The vascular spaces may be capillary in size or large and irregularly shaped (cavernous). These lesions are considered by some to be vascular malformations and not true neoplasms. Hemangiosarcomas are malignant tumors of vascular endothelial cells. Vascular spaces also are present in these tumors, although they are very irregular in size and shape, sometimes only being slit or cleftlike spaces in otherwise solid sheets of endothelial cells. The cells are very pleomorphic, oval to spindle-shaped, with hyperchromatic and often bizzare-shaped nuclei. Mitotic figures are frequent. The tumors are locally invasive and may metastasize widely. While rare, hepatic hemangiosarcomas are thought to arise from the sinusoidal lining cells or Kupffer cells and so may be referred to as Kupffer cell sarcomas. A peculiar feature of these tumors is the entrapment of one or more small nests of hepatocytes by the proliferating endothelial cells so that a cluster of hepatocytes is surrounded by a single or multiple layer of neoplastic endothelium (Fig. 8). These frequently project into blood-filled spaces as papillary projections.

B. Heart Tumors

Primary tumors of the heart are exceedingly rare in most strains of rats. However, in NZR/Gd albino rats, a strain originating in New Zealand, there is approximately 20% incidence of atriocaval mesotheliomas in animals of both sexes over 1 year of age (63). Except for pituitary tumors, atriocaval tumors are the most common type of neoplasm seen in this strain. These tumors arise in the wall of the right atrium or occasionally in the wall of the inferior vena cava at the level of the diaphragm.

The atrial tumors are located within the pericardial sac and are lobulated, firm, yellow-white masses. They infiltrate the atrial wall and extend into adjacent structures. Metastases to other thoracic viscera have been seen in some cases. Micros-

Table IV

Spontaneous Tumors of the Cardiovascular System in Rats

Tumor	Number of animals	Reference
Heart		
Endocardial disease	40	17
Endocardial fibromatosis	1	151
Endocardial proliferation	4	116
	1	55
Atrial mesothelioma	1	185
Atriocaval epithelial mesothelioma	18	63
Fibroma	4	116
Fibrosarcoma	2	115
	1	26
	1	203
	1	24
Anitschow cell sarcoma (treated animal)	1	133
Sarcoma	1	115
Rhabdomyoma	1	166
Blood vessels		
Hemangioma	4	42
	2	69
	1	101
	1	164
	2	189
	1	13
	7	115
	1	132
	2	151
	2	203
	4	199
	2	26
Angioma	7	25
Hemangiosarcoma	1	25
	1	42
	1	142
	2	189
	1	177
	3	13
	2	115
	2	116
	1	203
Histioendothelial sarcoma	1	69
Lymphatics		
Lymphangioma	1	24
	1	142
	3	164
	1	189
	5	115
	4	203
Lymphangiosarcoma	1	115

copically, these tumors consist of alveolar structures lined by flattened to low cuboidal epithelium as well as solid nests of tumor cells with variable amounts of fibrous stroma. Mucin is present within some tumor cells as well as within the alveolar spaces. It is postulated that these tumors may arise from mesothelial nests entrapped during embryological develop-

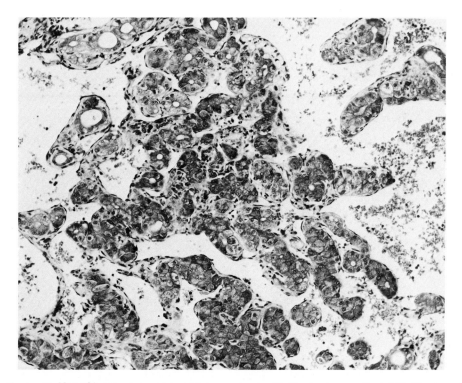

Fig. 8. Kupffer cell sarcoma. Nest of hepatocytes are trapped between strands of infiltrating tumor cells. ×140. (Courtesy of J. Strandberg.)

ment. Sharma (185) reported a morphologically similar tumor.

Other primary tumors of the heart have been reported sporadically (Table IV). Boorman *et al.* (17) described 40 cases of an unusual endocardial disease occurring in three strains (WAG/Rij, CIVO and BN/Rij) strains of rats. These lesions, usually in the left ventricle, are seen in rats of both sexes greater than 21 months of age. Microscopically there is marked thickening of the endocardium due to proliferation of spindle cells beneath the endocardium. The lesion may be sharply demarcated from the underlying myocardium or extend into it. Large polypoid tumorlike masses filling the left ventricle have been described in six cases. Extracardiac lesions have not been reported. The nature of these lesions is controversial, and some feel they represent a spectrum from hyperplasia to neoplasia. Similar lesions have been seen by Goodman ranging from subendocardial spindle cell proliferation (Figs. 9 and 10) to apparent locally invasive sarcomas. Proliferative endocardial lesions have been reported by others (Table IV). The fibromas described by Maekawa and Odashima (116) in ACI/N rats were all subendocardial. These may represent the same disease process described by Boorman *et al.* (17).

VI. DIGESTIVE SYSTEM

Primary tumors of the digestive tract are uncommon except in the WF strain which has a high incidence of colonic adenocarcinomas.

A. Oral Cavity

A review of the older literature uncovered only one primary tumor of the oral cavity, an epithelioma of the tongue (25) (Table V). Isolated cases of dental origin tumors have been reported: two odontomas (24), one adamantinoma (153), one adamantinoma (42), and one ameloblastic fibroodontoma (15). Guérin (69) reported twelve squamous cell carcinomas and 1 unspecified carcinoma in the oropharynx.

B. Salivary Glands

One submaxillary salivary gland tumor was found in the older literature; however, the histological type was not stated (25). Fifteen additional primary salivary gland tumors subsequently have been reported (Table V). It is of interest that all but one of these tumors was malignant, and there were no reports of mixed tumors. A recent review described a number of salivary gland tumors; however, it is not possible to separate spontaneous from induced tumors (61).

C. Esophagus

Tumors of the esophagus are exceedingly rare with only two cases of a squamous cell carcinoma reported to date (36,168a).

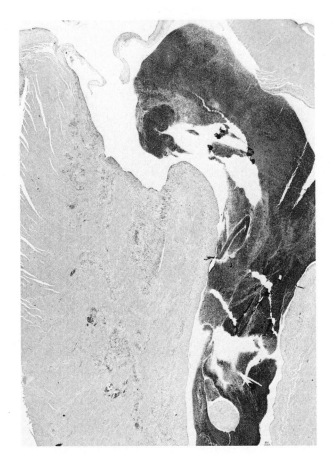

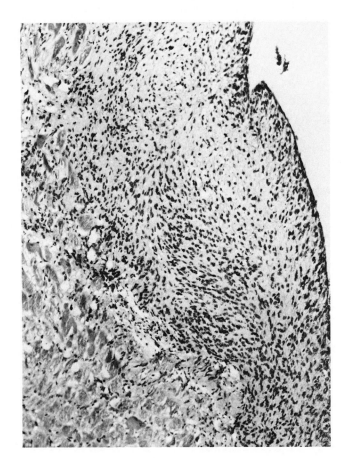

Fig. 9. Endocardial proliferation. Note the thickened endocardium and nests of proliferating spindle cells extending into the myocardium. ×130.

Fig. 10. Endocardial proliferation. Large sheets of proliferating spindle cells are present on the endocardial surface of the heart and extend into the myocardium. Note trapped myocardial fibers. ×140.

Table V

Spontaneous Tumors of the Digestive System in Rats

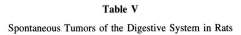

Tumor	Number of animals	Reference
Oral cavity		
Odontoma	2	24
Adamantinoma	1	153
	1	42
Ameloblastic fibroodontoma	1	15
Squamous cell carcinoma	12	69
	1	25
Carcinoma	1	69
Salivary glands		
Sarcoma (submaxillary gland)	1	25
Carcinoma	1	153
	5	69
	1	58
	2	202
	1	177
	1	116
Fibroma	1	116
Fibrosarcoma	1	116
	1	173
Myxoma	1	142

Table V (*Continued*)

Tumor	Number of animals	Reference
Esophagus		
Squamous cell carcinoma	1	36
	1	169
Stomach		
Papilloma	1	42
	1	175
	2	69
	1	107
	2	169
	1	151
	1	173
Adenomatous cystic tumor	2	69
Adenocarcinoma	1	42
	1	58
Mucoid carcinoma	1	36
Carcinoma	1	69
	1	58
Squamous cell carcinoma	1	140
	2	42

(*continued*)

Table V (*Continued*)

Tumor	Number of animals	Reference
	1	175
	1	202
	3	115
	1	151
Fibrosarcoma	1	35
	1	115
Sarcoma	6	42
	1	69
	1	189
Small Intestine		
Adenoma	1	115
	1	116
Adenocarcinoma	2	42
	1	202
	1	177
	1	208
	6	115
	1	116
Undifferentiated carcinoma	1	13
Fibroma	1	153
	2	42
	1	69
	1	115
Fibrosarcoma	1	115
Undifferentiated sarcoma	1	115
Carcinosarcoma	1	189
Cecum		
Adenoma	2	159
	1	189
Carcinoma	1	58
	2	28
Fibroma	1	189
Colon		
Adenocarinoma	2	211
	2	36
	1	58
	1	28
	1	124
	2	35
	1	65
	(See text)	77
	(See text)	126
	1	173
	28	125
Fibroma	1	42
Sarcoma	1	42
Liver		
Hepatocellular tumors[a]	1	41a
	1	136
	1	69
	2	35
	1	94
	14	72
	1	177
	12	115
	2	172
	1	33

Table V (*Continued*)

Tumor	Number of animals	Reference
Cholangioma	1	35
	1	69
Cholangiocarcinoma	1	35
Cystadenoma	5	41a
Adenocarcinoma	1	41a
Fibroma	2	25
Fibrosarcoma	1	115
Hepatic sarcomas	34	25
	9	1
	28	58
	1	35
Undifferentiated spindle cell tumor	1	36

[a] Based on the description in the literature it was not usually possible to separate the benign and malignant lesions.

D. Stomach

Spontaneous tumors of either the squamous or glandular portions of the stomach are rare. An early investigator observed papillary growths in the forestomach of rats infested with *Gongylonema neoplasticum* and concluded that the parasites caused tumors (49). Bullock and Rohdenburg (25) were able to reproduce the gastric lesions with nonspecific mechanical and chemical irritation. They also concluded that the pulmonary lesions which Fibiger observed were patches of squamous metaplasia and not metastatic tumor. Both lesions were reproduced by feeding a vitamin A-deficient diet (143). Many additional studies have reinforced these observations (103). A later study demonstrated that infection with *Gongylonema neoplasticum* coupled with a vitamin A deficiency produced the gastric lesions observed by Fibiger (78). The few reports of spontaneous stomach tumors, listed in Table V, represent a wide spectrum of histological types with the majority being malignant.

E. Small Intestine

Primary tumors of the small intestine also are rare. Twenty-two cases in the literature are tabulated in Table V. Many small tumors probably are missed because the intestine is not routinely opened and closely examined during necropsy.

F. Cecum and Colon

Primary tumors of the cecum and colon consist of scattered case reports (Table V) except for the two instances listed below.

There appears to be a high incidence of colon carcinoma in the WF strain of rat (126). The tumors initially developed as focal thickenings of the colonic wall and continued to progress to well-differentiated tubular adenocarcinomas. Occasionally, metastases to regional lymph nodes were found. Tumors were noted as early as 3 months of age. An interesting outbreak of spontaneous colonic adenocarcinomas was reported by Heslop (77). In one year 44 of 314 AS rats, predominantly young males, were noted to have tumors. However, during the next 4 years, not a single case was found in 2524 rats of the same strain.

G. Liver

The nomenclature of liver tumors is confusing since they have variable histological patterns and many diagnoses have been made only by gross examination. The term "hepatoma" does not indicate whether the tumor is benign or malignant and its use is to be discouraged. Terminology has been further confused by the use of the term "hyperplastic nodule" in the literature on experimental hepatocarcinogenesis for a discrete circumscribed nodule of proliferating hepatocytes which compress the surrounding parenchyma (Fig. 11). There is a large body of evidence to suggest these lesions are neoplasms which are part of the carcinogenic process. A conference on hepatic neoplasms in rats (192) recommended the term "neoplastic nodule" for these lesions since it is difficult to predict, on a morphologic basis, which of these will progress to cancer. The term hepatocellular adenoma may be preferable with the understanding that a portion of such lesions may progress to cancer. The term hepatocellular carcinoma should always be used for malignant tumors of hepatocyte origin.

The preponderance of reported hepatic sarcomas deserves clarification. It is well known that *Cysticercus fasciolaris,* the larval stage of the cat tapeworm (*Taenia taeniformis*), can induce hepatic tumors in rats. These generally are sarcomas and, unfortunately, many have been classified as spontaneous neoplasms. To our knowledge, hepatic sarcomas have only been reported from conventional rat colonies which have harbored this parasite. Several other authors agree with our interpretation (24,35). Many interesting studies have been conducted on the transmission of the tapeworm egg and the pathogenesis of the induced tumors (40). This is probably a nonspecific foreign body effect and not an effect specifically induced by the parasitic cyst. There are two lines of evidence in support of this idea. One is that spontaneous, nonparasitic sarcomatous tumors (other than vascular tumors) have never been reported in the rat liver. Additionally, the implantation of many foreign bodies in different sites in rats leads to the development of mesenchymal tumors (21). The hepatic tumors have been tabulated in Table V. Vascular tumors of the liver are reported in the cardiovascular section.

H. Pancreas

Tumors of the exocrine pancreas appear to be rare in rats based on published surveys (161,171). The literature review by Rowlatt and Roe (171) describes three reports of exocrine adenomas and eleven exocrine adenocarcinomas. These au-

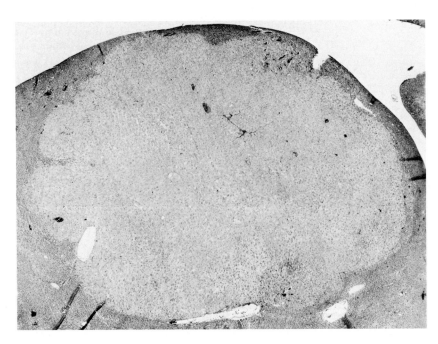

Fig. 11. Hepatic neoplastic nodule. The proliferating hepatocytes are compressing and invading the adjacent parenchyma. ×135.

Table VI

Spontaneous Tumors of the Pancreas in Rats

Tumor type	Number of animals	Reference
Endocrine		
Islet cell adenoma	1	5
	1	69
	1	207
	3	93
	24	58
	4	163
	1	202
	2	8
	2	189
	6	171
	42	115
	7	13
	6	173
Islet cell carcinoma	1	58
	5	171
	1	173
Exocrine		
Adenoma	1	105
	2	171
	7	115
	1	151
Adenocarcinoma	1	113
	1	212
	1	183
	1	207
	1	66
	1	58
	1	189
	2	115

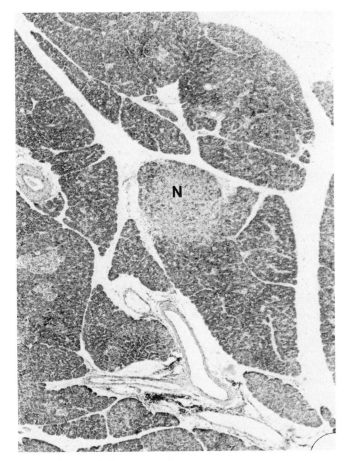

Fig. 12. Pancreatic nodular hyperplasia. Proliferating nodule (N) is lighter than surrounding parenchyma. ×138.

thors concluded that the tumors were essentially spontaneous in origin even though most of the rats had been treated with various chemicals or irradiated. We disagree with this interpretation, and, therefore, Table VI contains only the tumors from control animals. These include one exocrine adenoma and seven exocrine adenocarcinomas. Two additional cases of multiple exocrine adenomas in Chester Beatty rats were also reported (169,170,171).

Nodular hyperplasia of exocrine cells is a common finding in older rats, and in many cases it is difficult to distinguish from benign adenomas (Fig. 12).

Grossly, pancreatic exocrine adenomas tend to be discrete, nonencapsulated white masses which may be single or multiple. Microscopically, adenomas are composed of well-differentiated zymogen-granule-containing cells arranged similarly to the normal pancreas. Ducts and islets are absent. Pancreatic exocrine adenocarcinomas are less differentiated than adenomas, and they usually spread by extension to the liver, spleen, and omentum. Abdominal seeding can also occur, but thoracic metastases are rare.

VII. ENDOCRINE SYSTEM

Spontaneous tumors of the endocrine glands occur with a moderate to high frequency in most rat strains. We must regard the reported incidence of tumors as conservative since many of the endocrine tumors are small and easily missed at necropsy. Thompson and Hunt (201) compared the incidence rates of some tumors as determined by serial sections versus single random section. They demonstrated a significant increase in tumor detection in many endocrine glands by multiple sectioning. For example, the number of C cell tumors of the thyroid gland increased from 9 to 55. This study dramatically demonstrates the necessity for thorough histopathological examinations when attempting to define tumor incidence in a population.

There is a delicate physiological balance within the glands in the endocrine system and between endocrine glands and target organs. Any upset in this balance usually manifests itself in several organs. If a tumor occurs in one endocrine organ, it is highly probable that other endocrine organs will have hyperplastic or neoplastic changes. It is not always possible to distinguish between multiple primary effects (e.g., dietary mineral levels) from secondary hormonal interactions.

We have avoided lengthy discussions on the separation of hyperplasia from neoplasia. This can be particularly difficult with endocrine organs; however, it is beyond the scope of this chapter to present detailed morphological criteria.

A. Pituitary Gland

Tumors of the pituitary gland are extremely common in rats. The tumor incidence increases proportionately with the age of the animal, and in some colonies it is the main factor limiting their life span. Females have a higher incidence than males. Diet can have a significant effect on pituitary tumors. Rats given a high caloric diet *ad libitum* had an extremely high incidence of spontaneous tumors (167,175). Under restricted dietary regimens, tumor prevalence was directly related to the protein level in the diet. The lowest tumor incidence was obtained when both caloric and protein intake were depressed. A large volume of literature has accumulated concerning the relationship of endogenous and exogenous hormones and the induction of pituitary tumors (56). In addition, the relationship between the pituitary, other endocrine tumors, and organ tumors (e.g., mammary) is coming into clearer focus (56,92,101,67).

A major problem arises in interpreting the "classic" pathological descriptions of pituitary tumors which have been used in the past in light of current knowledge of this organ. The pituitary is composed of six cell types secreting seven or more hormones (56). The current classification has been based on a combination of light microscopy, immunohistochemical stains, and electron microscopy. The older description of acidophils, basophils, or chromophobes may be useful if one does not ascribe a functional significance to them based on morphological criteria. As Furth *et al.* (56) have stated, "Cells after depletion of their hormones appear chromophobic, and many functional tumors that do not produce enough hormone appear to be "chromophobic" in hematoxylin–eosin (H–E) preparations and may be detectable only by electron microscopy." We have developed the following discussion using classic descriptions as morphological indicators to provide some comparison with published strain data.

Chromophobe Adenoma

Tumors of this type constitute the vast majority of pituitary tumors in rats. They vary from microscopic foci to nodules, 0.5 cm or larger. Grossly, the tumors are soft with an irregular surface, and many have prominent hemorrhagic areas (Fig.

13). They generally are well circumscribed and compress adjacent brain tissue, occasionally causing hydrocephalus. Frank invasion of brain tissue is uncommon. Microscopically, the tumors consist of large polygonal cells with prominent vesicular nuclei and abundant eosinophilic cytoplasm. The cells are arranged in nests, cords, or sheets separated by vascular sinusoids (Fig. 14). In some tumors these vascular sinusoids may be a major component. Compression of adjacent pituitary tissue and a uniform or single cell population help delineate these tumors. Other less common morphological patterns have been described, ranging to frank anaplastic carcinomas. Metastases are rare (91).

Strains with a high incidence are included in Table XVI (See later). Less common pituitary tumors are listed in Table VII. We have not listed all of the reported pituitary tumors which do not fit into the above categories since many reflect bias, e.g., young animals, incomplete necropsies, or noninbred strains.

B. Thyroid Gland

It is difficult to assess accurately the incidence of thyroid tumors in rats since this gland usually is not carefully examined at necropsy and the reported data are contradictory. As with many endocrine organs, small tumors may be overlooked. This

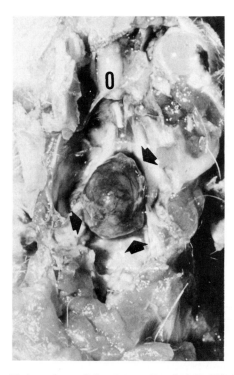

Fig. 13. Pituitary chromophobe adenoma (dorsal view). This large tumor (arrows) compressed the brain producing hydrocephalus. (O) olfactory bulb. (Courtesy of A. M. Allen.)

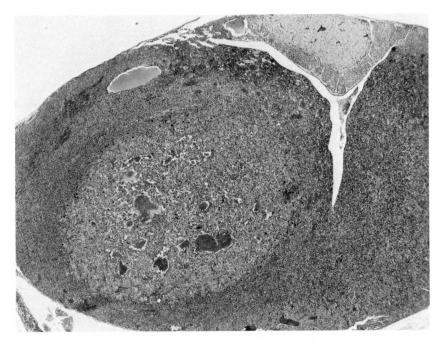

Fig. 14. Pituitary chromophobe adenoma. This tumor contains many dilated vascular sinuses. ×38.

is particularly important when one considers that the large tumors, clearly evident on gross examination, are usually found in older animals.

The most prevalent thyroid neoplasm found in rats has been described as a "clear cell" or "light cell" tumor (3,16,111,135,178,179,196). They originate from parafollicular or C cells in the thyroid gland (210). Parafollicular cells secrete the hormone calcitonin, and this hormone has been identified in variable amounts in C cell tumors (15). The tumors begin as gradual increases in C cells until small groups are recognized as circumscribed nests or nodules. As the tumors enlarge, they compress and entrap thyroid follicles (Fig. 15). These can be focal or multicentric. Tumor cells are

round to oval with moderate amounts of pale granular cytoplasm. Nuclei are pale and round with delicate chromatin (Fig. 16). Mitoses are rare. Eventually some C cell tumors invade the capsule, and a small percentage will metastasize. Sass *et al.* (173) reported metastatic parafollicular cell tumors in the lymph node and lung of Fischer rats.

Tumors of follicular origin have been less commonly reported than parafollicular cell tumors. The follicular cells of the rat thyroid are extremely sensitive to dietary iodine levels. Thyroid tumors can be induced by feeding low iodine diets (3,90), and one report noted significant enlargement of the thyroid glands in rats fed a commercial diet (94). Both follicular adenoma and carcinomas have been reported (Table VIII).

C. Parathyroid Gland

Parathyroid hyperplasia secondary to chronic renal disease is commonly observed in older rats. However, tumors of the parathyroid gland are extremely rare. The few reported cases are listed in Table VIII.

D. Adrenal Gland

Spontaneous tumors of the adrenal gland occur frequently in rats. However, there is a considerable variation in incidence between strains (Table IX).

Table VII

Spontaneous Tumors of the Endocrine System in the Rat (Pituitary)[a]

Tumor type	Number of animals	Reference
Basophilic adenoma	3	69
	26	68
	3	50
Acidophilic adenoma	2	1
	2	50
	3	9
	10	116
Chromophobe carcinoma	2	116
Craniopharyngioma	1	50
Granular cell myoblastoma	1	197
Fibrosarcoma	1	151

[a]Excluding chromophobe adenomas.

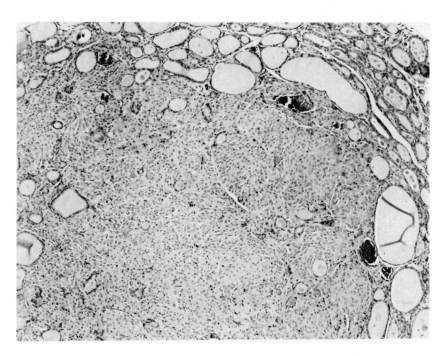

Fig. 15.   Thyroid C cell tumor. Note the proliferating nests of parafollicular cells and trapped thyroid follicles. ×80.

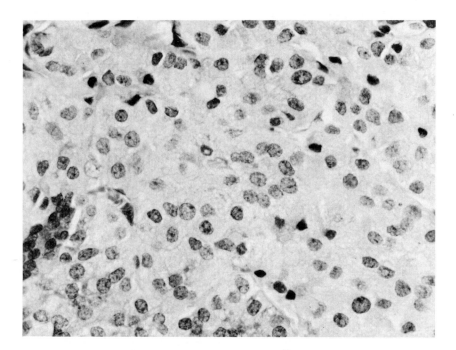

Fig. 16.   Thyroid C cell tumor. Tumor nests are composed of large round to oval cells with abundant granular cytoplasm. Nests are separated by delicate strands of collagen. ×540.

1. Adrenal Cortex

Most of the cortical tumors in rats are adenomas (Table IX). They begin as single or multiple circumscribed nodules composed of hypertrophied and hyperplastic cortical cells which compress adjacent parenchyma. As the tumors expand, they gradually distort and less commonly, invade the capsule. The tumors consist of large polygonal cells with eosinophilic cytoplasm which is occasionally vacuolated. Nuclei are hyperchromatic and variable in size (Fig. 17). Hemorrhage, nec-

Table VIII

Spontaneous Tumors of the Thyroid and Parathyroid Gland in Rats

Tumor type	Number of animals	Reference
Thyroid		
Parafollicular adenomas (C cell tumors)	2	202
	49	115
	11	11
	3	33
Adenoma (unspecified type)	2	25
	6	1
	87	69
	2	36
	1	102
	5	151
	1	116
Papillary cystadenoma	8	25
Papillary adenocarcinoma	1	94
Follicular adenoma	20	177
	1	33
Follicular adenocarcinoma	1	38
	2	115
	5	151
Small cell carcinoma	1	100
Sarcoma	2	25
Parathyroid		
Adenoma	1	1
	2	69
	1	177
	2	13
	5	115
	2	151

Table IX

Spontaneous Tumors of the Adrenal Gland in Rats

Tumor type	Number of animals	Reference
Cortex		
Adenoma	1	1
	1	153
	2	42
	1	69
	2	101
	91	177
	38	115
	2	116
	25	173
	2	33
Adenocarcinoma	1	87
	2	116
	3	173
Myelolipoma	1	177
Sarcoma	1	25
Medulla		
Pheochromocytoma	8	69
	3	202
	7	177
	7	13
	8	173
Medullary adenoma	36	115

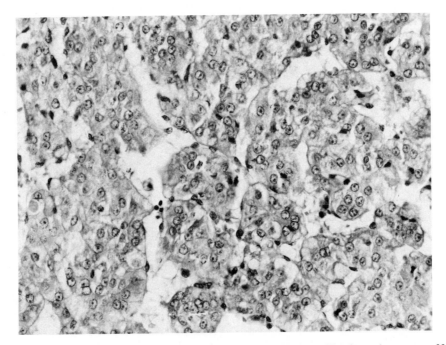

Fig. 17. Adrenal cortical adenoma. Nests of tumor cells are separated by dilated vascular spaces. ×330.

rosis, and vascular dilation are common. Cortical adenocarcinomas are distinguished by their cellular atypia, increased mitotic activity, local invasion, and metastases. Many of these tumors are transplantable (30).

Hyperplastic nodules in the cortex, capsule, and pericapsular area are common in older rats, and they should be carefully examined before diagnosing neoplasia.

2. Adrenal Medulla

The medullary tumors reported in rats are mainly pheochromocytomas (60). Grossly, when the adrenal is bisected, they appear as circumscribed light nodules. Microscopically, the cells resemble normal medullary cells which are arranged in nests or cords (Fig. 18). Cytoplasm usually is scanty, and it gives a positive chromaffin reaction. Occasionally such tumors will invade adjacent structures, particularly veins, and will metastasize. Medullary hyperplasia has also been reported (216).

E. Pancreatic Islets

Tumors of the exocrine pancreas are rare and have been discussed under the digestive system. Tumors of the islets have been reported more frequently, but they still must be considered uncommon (Table IV). Grossly, islet cell tumors are single or multiple, circumscribed, and reddish-brown on section. Microscopically, they are composed of polygonal cells arranged in cords along sinusoidal spaces. The cells have a scanty amount of lightly eosinophilic cytoplasm and small hyperchromatic nuclei. Islet cell carcinomas are distinguished from adenomas by capsular invasion and metastases

VIII. HEMATOPOIETIC SYSTEM

Tumors of the hematopoietic system have been reported in inbred and random-bred rats. Several strains have a relatively high incidence of tumors, although none approaches the incidences seen in certain mouse strains. There are no known rat leukemia viruses, although lymphocytic neoplasms have been induced experimentally by various mouse leukemia viruses.

Hematopoietic tumors in rats are not well classified, and in the older literature there is some confusion due to the diversity of terms used, often for the same entity. Swaen and Van Heerde (198) classified these tumors according to the predominant cell type and gave detailed descriptions. Myelofibrosis and mast cell leukemia have been seen only as induced lesions in rats. Reticulum cell sarcoma, stem cell leukemia, monocytic leukemia, lymphosarcoma, lymphoid leukemia, plas-

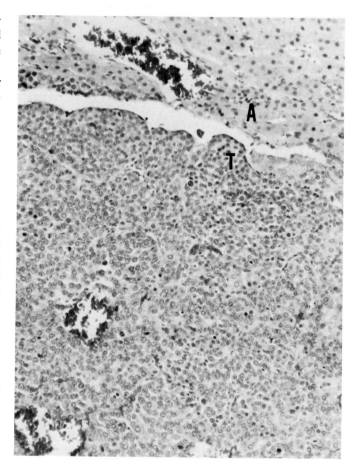

Fig. 18. Pheochromocytoma. Nests of tumor cells (T) are compressing adjacent adrenal cortex (A). ×138.

macytomas (''immunocytomas''), granulocytic sarcoma and leukemia, erythroleukemia, and unclassified leukemia all have been reported as spontaneous tumors in the rat.

Lymphomas, lymphocytic leukemias, and reticulum cell sarcomas (histiocytic lymphomas) occur sporadically (Table X). Many of these apparently arise in mesenteric or ileocecal lymph nodes, in the cecal wall, or in peribronchial lymphoid tissue of the lung. Typhlitis and peribronchiolitis are common in older rats, particularly those housed in conventional colonies, and these must be differentiated from lymphomas. Myelocytic tumors and erythroleukemia have been reported in the rat (Table X) but are not well described. Other than these, a few strains of rats have a relatively high incidence of specific types of spontaneous hematopoietic tumors as described below.

A. Leukemia of Fischer Rats

The mononuclear cell leukemia seen in aged F344 and W/Fu rats, although not precisely characterized, has been referred to

Table X

Spontaneous Tumors of the Lymphoreticular System in Rats

Tumor	Number of animals	Reference
Thymus		
Thymoma	90	24
		37
	3	153
	527	42
	10	193
	2	175
	20	69
	2	164
	1	189
	11	147
	19	166
	7	139
	30	214
	6	116
	1	199
	5	203
Lymphosarcoma	1	149
	1	115
	25	203
Fibrosarcoma	1	175
Lymphoreticular tumors		
Leukemia	6	101
	14	166
	12	139
	21	199
Lymphatic leukemia	3	116
Lymphatic leukosis	6	203
Lymphoid leukemia	8	69
Lymphocytic leukemia	2	85
Lymphoma	4	160
Malignant lymphoma	1	142
	2	101
	1	202
	34	164
Lymphosarcoma	32	69
	4	189
	44	12a
		79
		26
	9	115
	1	151
Lymphoreticular tumors	4	102
	91	173
Reticulum cell sarcoma	1	97
	1	189
	7	115
	14	166

Table X (*Continued*)

Tumor	Number of animals	Reference
	6	203
Reticulum cell sarcoma (not lung)	10	139
Reticuloendothelial tumors	7	69
Reticuloses (liver)	14	70
Unclassified	5	203
Lung		
Lymphoma	18	36
Lymphosarcoma	91	175
	10	115
	1	116
Reticulum cell sarcoma	12	24
		37
	35	136
	1	115
	45	139
Histiocytic sarcoma	13	37
		69
Spleen		
Fibroma	1	166
Fibrosarcoma	2	166
Sarcoma	1	42
Mesenteric lymph nodes or mesentery		
Lymphoma	1	153
	23	36
Lymphosarcoma	215[a]	118
	1	175
	4	115
Reticulum cell sarcoma	78	24
		37
	14	47
	1	115
Reticulosarcoma	11	59
Myeloid tumors		
Myeloid leukemia	6	175
	6	69
	1	132
	1	116
	2	203
Myelogenous leukemia	1	142
Chronic myeloid leukemia	1	1
Granulocytic leukemia	11	209a
Myeloma	1	152
Myelocytoma	1	177
Myelocytic sarcoma	2	151
Erythroblastic leukemia	1	69

[a] Majority in mesenteric lymph nodes.

as monocytic, myelomonocytic, unclassified, or Fischer rat leukemia. It occurs in older animals, with an incidence of 17% in W/Fu rats and 25% in F344 rats (127–129). Grossly, splenomegaly is consistent. Hepatomegaly, lymphadenopathy, and involvement of other organs is variable. Microscopically the disease appears to begin in the marginal zone of the spleen (Fig. 19), eventually involving the entire spleen. There is usually infiltration of liver sinusoids (Fig. 20), lymph nodes, and,

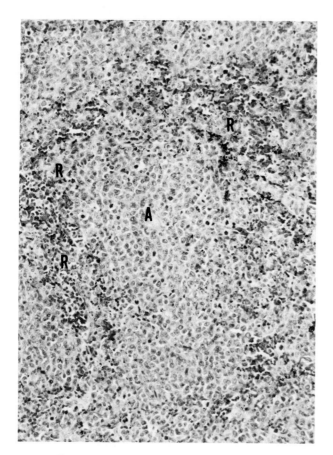

Fig. 19. Fischer rat leukemia (spleen). Large sheets of proliferating tumor cells can be seen around central artery (A). The red pulp (R) is diminished and compressed. ×220.

to a lesser extent, other organs. Bone marrow involvement occurs in most animals.

Clinically, detection of atypical mononuclear cells in the peripheral blood is usually the first evidence of the disease. Leukocytosis, with counts up to 180,000/mm³, and up to 90% leukemic cells occurs. Hemolytic anemia is often present. The leukemic cells are large mononuclear cells 15 to 20 nm in diameter with oval or slightly lobulated nuclei, one to two nucleoli, and abundant blue cytoplasm often with azurophilic granules demonstrable by Wright–Giemsa stains. Histochemical studies for peroxidase, esterase, alkaline and acid phosphatase, and periodic acid-Schiff and oil red O stains are negative. Thus, a specific classification of these cells has not been made. In tissue sections the cells resemble undifferentiated or stem cells with moderate amounts of cytoplasm and large oval or slightly indented nuclei. Cytoplasmic granules are difficult to identify.

B. Stem Cell Leukemia

Spontaneous leukemia in young rats is unusual. However, Richter *et al.* (158) reported a stem cell leukemia occurring as early as 81 days of age and most under 1 year of age in approximately 2% of their CR:CD(SD) colony. Grossly, hepatosplenomegaly and anemia were present. Microscopically, infiltration of bone marrow, spleen, and liver was consistently present, with the bone marrow being the most extensively involved. Lymph node and other organ involvement was inconsistent. In several cases, there was CNS involvement.

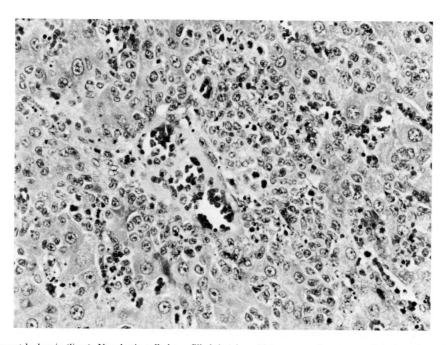

Fig. 20. Fischer rat leukemia (liver). Neoplastic cells have filled the sinusoidal spaces and compressed the hepatic cords. ×350.

The cells were large (20 to 40 μm) in diameter with large oval or occasionally indented vesicular nuclei with a small rim of cytoplasm without granules when stained with Giemsa. The tumor is readily transplantable, but the etiology is unknown.

C. Granulocytic Leukemia

WN rats over 18 months of age have been reported to have a 14% incidence of granulocytic leukemia (189). Detailed descriptions of these tumors were not available.

D. Plasmacytomas ("Immunocytomas")

Plasma cell tumors are generally rare and can occur in a variety of sites. However, in the LOU/C/Wsl strain of rats, the incidence of these tumors is high; 32% in males and 16% in females (133). These tumors originate in the ileocecal lymph nodes and are confined to the ileocecal region. Seventy-five percent of the tumors secrete monoclonal immunoglobulin, and a variety of immunoglobulin types have been identified. These tumors are transplantable (for further information, see Volume II, Chapter 9.)

E. Thymomas

Thymomas also have been reported in various rat strains, usually in a low incidence (Table X). However, in reviewing the literature it is not always clear whether these represent true thymomas or lymphomas originating in the thymus. Pollard and Kajima (147) reported 11 of 49 aged germfree Wistar rats with large thymic tumors. These tumors were restricted to the thymus and were composed of small lymphocytes. Yamada *et al.* (214) reported thymomas in 54% (14/26) male and 39% (16/41) of female Buffalo rats. Grossly, the thymus was markedly enlarged, weighing more than 2.0 gm. Microscopically, the tumors were composed of both lymphocytes and reticular epithelial cells in various proportions. Other studies have reported both lymphosarcomas and mixed epithelial and lymphomatous tumors of the thymus.

IX. URINARY SYSTEM

Primary tumors of the urinary tract are, in general, uncommon in the rat (Table XI). There is one report of familial renal adenomas in Wistar rats (44), and the BN/BiRij strain of rats is reported to have a relatively high incidence of ureter and urinary bladder tumors (14). Other than these, most urinary tract tumors are reported as sporadic cases in various strains.

Table XI

Spontaneous Tumors of the Urinary System in Rats

Tumor	Number of animals	Reference
Kidney		
Lipomatous hamartoma	1	36
Hamartoma	1	190
	2	115
Papilloma	1	25
Papilloma, pelvis	2	160
Transitional cell papilloma	2	203
Hypernephroma	1	1
Lipoma	4	190
Liposarcoma	1	58
		59
	1	203
Nephroblastoma	6	24
		37
	1	110
	4	153
	4	42
	1	142
	1	69
	1	36
	1	4
	1	202
	1	190
	1	83
	1	177
	1	117
	1	62
	1	132
	4	144
	1	203
Adenoma	10	25
	1	175
	1	69
	1	36
	1	164
		190
	1	164
	2	164
Adenoma (familial)	25	44
Adenoma (multiple)	1	175
Adenocarcinoma	1	138
		190
	1	116
Transitional cell carcinoma	7[a]	203
Squamous cell carcinoma	1	42
	1	203
Carcinoma	8	25
	1	24
	1	69
	1	190
	3	203
Fibroadenoma	7	25
Mixed tumor	1	190
Fibrosarcoma	1	153
	2	58
		59

(continued)

Table XI (*Continued*)

Tumor	Number of animals	Reference
Sarcoma	1	24
		37
	3	69
	1	189, 190
	1	203
Urinary bladder		
Papilloma	1	25
	5	42
	2	115
	58	166
	10	116
Papillomatosis	1	69
	1	139a
Transitional cell papilloma	2	115
	10	116
	19	203
Transitional cell carcinoma	2	116
	2	173
	17[b]	203
Carcinoma	1	25
Urethra		
Carcinoma	2	173

[a] Four also in urinary bladder.
[b] Four also in pelvis of kidney.

A. Kidney

Nephroblastomas, renal adenomas and adenocarcinomas, mixed tumors, lipomas, sarcomas, and hamartomas all have been described in the rat. Nephroblastomas (embryonal nephroma, Wilm's tumors, adenosarcoma) have previously been reported to be the most frequent type of renal tumor (83). However, renal tubular neoplasms appear to be more common (Table XI) (74).

Nephroblastomas tend to occur in younger animals. The mean age is less than 1 year, but tumors have been found in animals up to 2 years of age. There was no strain or sex predilection. Grossly, these tumors are firm and yellow to gray. Microscopically, they consist of primitive tubular and glomerular structures and islands of epithelial cells scattered through a connective tissue stroma (Fig. 21). The epithelial structures range from poorly defined nests of cells with indistinct cytoplasm and oval nuclei to well-defined tubules with distinct lumens lined with cuboidal to columnar epithelium which occasionally have brush borders. Structures resembling rudimentary glomeruli are also seen. Occasionally, nests of transitional cell or squamous epithelium are present. There usually is a loose fibrovascular stroma, the cells of which may vary from immature fibroblasts to well-differentiated fibrous tissue. In some areas, the stromal cells blend indistinctly into the nests of epithelium. The stroma may occasionally contain skeletal muscle or cells resembling smooth muscle (24). These tumors frequently resemble fetal kidney. Metastases to the lungs have been reported in one case (117). Nephroblastomas have been transplanted successfully (190).

Renal tubular adenomas and adenocarcinomas have been sporadically reported in the literature in small numbers (Table XI). Familial renal adenomas have been reported in Wistar rats (44). The tumors occurred in both males and females which ranged in age from 6 to 16 months. They were frequently bilateral and multiple. Grossly, the tumors were located in the cortex and varied in size up to $8 \times 6 \times 4$ mm, were yellow to gray or white and either cystic or solid. Microscopically, the tumors were composed of variable sized cuboidal to polygonal cells with either eosinophilic or basophilic cytoplasm and prominent oval nuclei. Mitoses are occasionally found. The cells resemble tubular epithelium and are arranged in solid nests, tubular, or acinar patterns or formed papillary projections into cystic spaces. Stroma was scanty. Adenomas generally are well circumscribed. Renal adenomas reported in other strains are similar in appearance.

Renal carcinomas generally are larger and more anaplastic, with numerous mitotic figures and invasion of adjacent renal parenchyma. Metastases have occasionally been reported. Renal carcinomas have been transplanted successfully (138).

Transitional cell papillomas and carcinomas arising from the renal pelvis have been reported (Table XI) and are similar in appearance to those arising elsewhere in the urinary tract. Papillary proliferations of the urinary epithelium have been reported in association with calculi in the pelvis.

Mesenchymal tumors, including fibrosarcomas, sarcomas, and lipomatous tumors have been reported sporadically as arising in the kidney. The lipomatous tumors include those diagnosed as lipoma, lipomatous hamartoma, hamartoma, mixed tumor (both benign and malignant), and liposarcoma. These tumors consist of adipose, stromal, and epithelial components in variable amounts with the adipose cells generally predominating (Fig. 22). These cells may be mature fat cells or relatively undifferentiated lipoblasts depending on the nature of the tumor. Occasionally, groups of tumor cells may be seen which appear to be forming tubularlike structures, leading some to believe that these tumors are true mixed tumors with both an epithelial and a mesenchymal component. Entrapped glomeruli and tubules, sometimes hyalinized, may be scattered throughout the tumor. Large cystic spaces lined with flattened epithelium may also be present. Large, thin-walled vessels are common in the larger tumors, as are areas of thrombosis and hemorrhage. Smaller tumors usually involve the medulla or corticomedullary region; larger ones usually involve the cortex as well and often extend beyond the renal capsule. Metastases are rare. Snell (190) has described the occurrence of a mixed tumor which had a lipomatous component as well as transitional type epithelium.

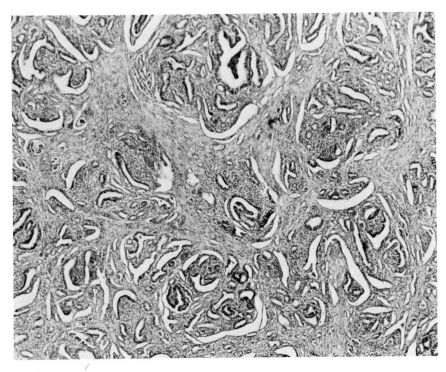

Fig. 21. Nephroblastoma. Note the loose connective tissue stroma and the primitive glomeruli formed by the epithelial elements. ×138.

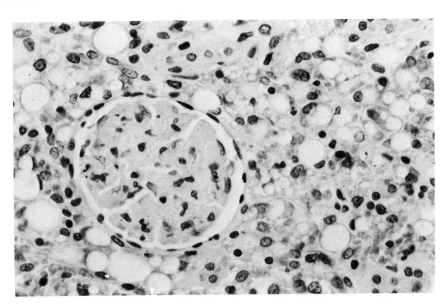

Fig. 22. Renal mixed tumor. Neoplastic cells, many with lipid globules, have surrounded the glomerulus. ×330.

B. Urinary Bladder, Ureter, and Urethra

In most strains of rats, spontaneous tumors of the ureter, urinary bladder, and urethra are uncommon, although transitional cell papillomas and carcinomas have occasionally been reported (Table XI and Fig. 23). In the strain Chbb:THOM(SPF) rat from Germany, there is an 8% inci-dence of epithelial urinary bladder tumors, about one-half of which are malignant (203). However, in the Brown Norway (BN/BiRij) rat, there is a high incidence of epithelial tumors of the ureter and urinary bladder (14). The incidence of ureter tumors is 20% in females and 6% in males, whereas in the bladder, 28% of the males and 2% of the females have tumors. The bladder tumors were generally papillary with a narrow

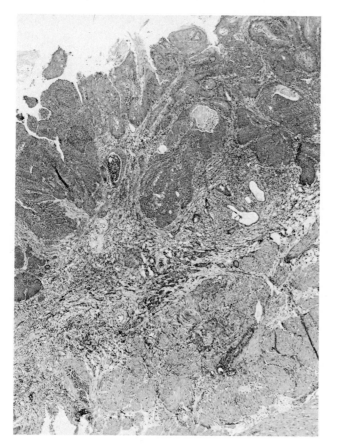

Fig. 23. Transitional cell carcinoma. The tumor cells have invaded the mucosa, submucosa, and muscularis. ×34.

connective tissue stalk covered by markedly thickened transitional epithelium with some degree of pleomorphism (Fig. 24). In some tumors, there were areas of squamous cell differentiation. Invasion of the submucosa and muscularis occurred in a few cases, and metastases in one. Tumors of the ureters were similar in appearance but had a much greater tendency toward squamous differentiation with keratinization. A few appeared to be well differentiated squamous cell carcinomas. Several of these tumors were invasive and metastatic. Many of these tumors were associated with urolithiasis, the significance of which is unknown.

Two carcinomas of the urethra have been reported, both in F344 rats (173). One, a carcinosarcoma, was very anaplastic with marked pleomorphism and both epithelial and mesenchymal components.

X. FEMALE REPRODUCTIVE SYSTEM

A. Ovary

Spontaneous tumors of the ovary are uncommon in most strains of rats. The most common type is the granulosa or granulosa-theca cell tumor. Snell (189) reports that approximately 33% of OM rats over 18 months old develop granulosa cell tumors. These tumors have been reported sporadically in a variety of other strains. Granulosa-theca cell tumors are similar to those described in other species. The tumors are usually uni-

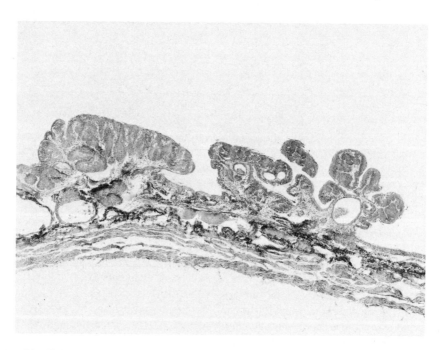

Fig. 24. Urinary bladder papillomas. Note the multiple papillomas extending from the mucosa. × 38.

lateral. Occasionally, they are associated with endometrial hyperplasia, suggesting a functional tumor. Microscopically, the tumors vary greatly. Usually, the granulosa cells are arranged in solid nests which may have cystic centers, thus resembling normal Graffian follicles. Stroma is scanty, forming a delicate network around the solid nests of cells. Adenomatous and trabecular patterns can also be present. In some tumors, thecal cells forming whorled and nodular patterns predominate. Luteinization is a variable feature. Occasionally, these tumors will metastasize (189), and some have been transplanted (88).

Adenomas, cystadenomas, adenocarcinomas, cystadenocarcinomas, fibromas, lipomas, hemangiomas, sarcomas, and other miscellaneous tumors of the ovary also have been reported (Table XII). Engle (46) described tubular adenomas in the ovaries of aged Wistar rats. Tubules lined by cells resembling Sertoli cells of the testis were also found in ovaries of old Wistar rats, and the adenomas were thought to arise from these structures.

B. Uterus

Uterine tumors are more common than ovarian tumors, although still relatively infrequent. Most of these are endometrial stromal tumors which have variously been termed adenomatous polyps, angiomatous polyps, fibrovascular polyps, stromal polyps, or endometrial stromal polyps. In approximately 66% of female F344 rats and 33% of M520 female rats over 21 months of age, polypoid tumors of endometrial origin have been reported (189). In these strains, the tumors reportedly are more frequent in virgin females than in breeders. There are reports of such tumors occurring sporadically in other strains. These tumors are polypoid masses protruding into the uterine lumen, usually with a slender stalk. They may involve one or both uterine horns. Microscopically, they are composed of a variably cellular stroma covered by glandular epithelium which is continuous with the endometrial lining epithelium (Figs. 25 and 26). There may be small epithelial nests and acini within the stroma. In addition, these tumors are frequently highly vascular with numerous variable-sized endothelial-lined spaces. Jacobs and Huseby (94) have successfully transplanted nine of these tumors; four grew as sarcomas, and five retained their original appearance. Endometrial stromal sarcomas, composed of poorly differentiated spindle cells, are occasionally seen.

A variety of other uterine tumors also have been reported (Table XII). These include benign and malignant neoplasms of both epithelial and mesenchymal origin, the most common being adenocarcinomas, leiomyomas, and leiomyosarcomas. The adenocarcinomas frequently are anaplastic, with invasion of the uterine wall. Squamous metaplasia may be seen in these

Table XII

Spontaneous Tumors of the Female Reproductive Tract in Rats

Tumor	Number of animals	References
Ovary		
Adenoma	5	175
	3	46
	31	69
	2	8
	1	29
	2	115
	5	203
Adenocarcinoma	1	42
	1	36
	2	115
	1	132
Carcinoma	3	153
	1	69
	1	181
	4	203
Cystadenocarcinoma	3	24
		37
	Sporadic	189
Granulosa-theca cell tumor	1	86
	3	58
		59
	1	181
	1	202
	(See text)	189
	1	29
	1	13
	5	115
	1	116
	4	203
	3	79
		26
Fibroma	1	24
		37
	2	175
	1	69
	3	36
Sarcoma	2	24
		37
	2	69
	1	29
	1	115
Osteosarcoma	1	69
Fibromyoma	1	69
Malignant fibroadenoma	1	181
Lipoma	1	177
Mesonephroma	1	58
		59
Seminoma	1	69
Uterus		
Endometrial stroma polyp	15	69
	(See text)	189
	17	94
	4	177
	2	151
	5	116

(continued)

Table XII (*Continued*)

Tumor	Number of animals	Reference
	9	26
Papillomatosis	3	69
Papilloma	1	24
		37
Adenomatosis	1	69
Adenoma	1	102
Fibroadenoma	2	25
Adenocarcinoma	4	24
		37
	14	153
	3	36
	11	58
		59
	4	189
	19	116
	2	173
	6	203
Adenoacanthoma	1	24
		37
Carcinoma *in situ*	1	58
		59
Carcinoma	1	24
		37
	1	71
	2	42
	13	69
	23	181
Squamous cell carcinoma	10	24
		37
	6	203
Carcinosarcoma	3	69
	4	24
		37
Fibroma	2	42
	2	181
	4	115
Fibrosarcoma	1	153
	1	115
	1	203
Fibromyoma	8	69
	5	58
		59
Leiomyoma	4	24
		37
	2	36
	1	101
	6	12a
	2	115
	1	132
	1	116
	6	203
	1	26
Leiomyosarcoma	1	202

Table XII (*Continued*)

Tumor	Number of animals	Reference
	2	115
	1	151
	2	116
	1	173
	2	203
Myoma	1	181
	1	189
Myosarcoma	2	24
		37
	1	142
	2	58
		59
Uterine (endometrial stromal) sarcomas	4	94
Myxosarcoma	1	153
Sarcoma	1	188
	21	24
		37
	16	42
	2	69
	5	181
	1	115
	1	203
	1	27
Mesodermal tumor	7	36
Benign mesenchymal tumors	3	153
Uterine tumor	2	199
Cervix and Vagina		
Squamous cell carcinoma	1	173
	7	27
		26
Leiomyoma	4	27
		26
Sarcoma	43	27
		26
Polyp, cervix	1	189
Fibrovascular polyp, cervix	1	202
Fibroendothelioma, cervix	1	58
		59
Fibrosarcoma, cervix	1	115
Squamous cell carcinoma, vagina	1	153
Carcinoma, vagina	2	58
Papilloma, vagina	2	116
Stromal polyp, vagina	1	151
Fibroma, vagina	2	153
	1	115
	1	203
Myxoma, vagina	1	104
Myxosarcoma, vagina	1	153
Sarcoma, vagina	1	24
	1	69
Vulva		
Squamous cell carinoma	2	25

tumors. Occasionally, squamous cell carcinomas occur in the uterus. Leiomyomas and leiomyosarcomas are similar to those found elsewhere.

C. Cervix and Vagina

Cervical and vaginal tumors are generally rare in the rat. In a

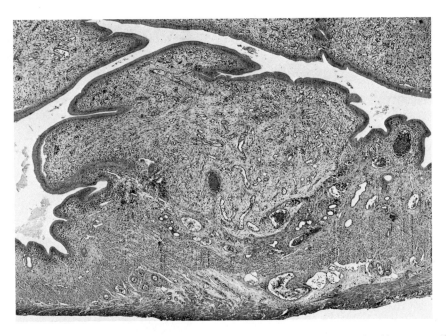

Fig. 25. Endometrial stromal tumor. Most of these tumors extend into the uterus as polypoid masses. ×40.

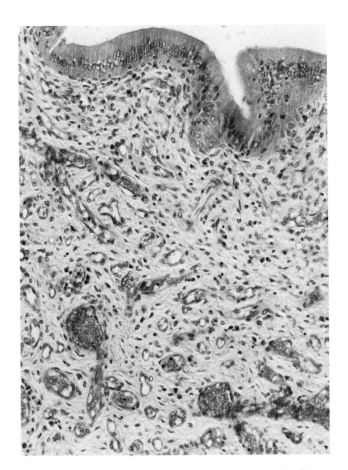

Fig. 26. Endometrial stromal tumor. Note the loose connective tissue matrix and the extensive network of dilated blood vessels. × 220.

literature review by Billups (10) squamous cell papillomas and carcinomas, polyps, fibromas, fibrosarcomas, and myxosarcomas were reported (Table XII). Burek *et al.* (27) reported a 19% incidence of cervical and vaginal tumors in 252 BN/Bi female rats greater than 26 months of age. Of the 252 tumors, 43 were sarcomas of various types (primarily leiomyosarcomas or round cell sarcomas). Many of the sarcomas had variable histologic patterns. Four leiomyomas and seven squamous cell carcinomas were seen also. Five rats with squamous cell carcinomas also had sarcomas. Most of the tumors were highly invasive and several metastasized.

Clitoral gland tumors are described in Section II.

XI. MAMMARY GLAND

Mammary tumors are probably the most common spontaneous tumors found in the rat. They occur with a frequency up to 50% in many strains, and there are scattered reports of up to 90% incidence (See Table XVI). Sher (186) has a comprehensive literature tabulation of the incidence of mammary tumors in a variety of strains. Mammary tumors occur primarily in females, only 16% occurring in males (141). Age also affects the incidence of these tumors. They are rare before 1 year of age and increase markedly in incidence after about 18 months of age.

Mammary tumors are sensitive to hormonal stimuli. As mentioned, they occur primarily in females. Administration of estrogen will increase the incidence of carcinomas and cause them to appear at an earlier age. Prolonged growth hormone administration will increase the incidence of fibroadenomas. There also is an association between the occurrence of induced anterior pituitary tumors which produce prolactin and the occurrence of fibroadenomas and cystic ducts of the mammary gland. Spontaneous mammary and pituitary tumors often are found in the same animal as well, which further suggests that these tumors may be related. Many studies have demonstrated the effect of hypophysectomy, ovariectomy, or the administration of excess hormones and the induction of mammary tumors by carcinogens (217).

In addition to hormones, a variety of chemical carcinogens, as well as ionizing radiation, can induce both benign and malignant mammary tumors of various types. Unlike the mouse, there does not appear to be a virus associated with the induction of mammary tumors in the rat.

The vast majority of mammary tumors (217) are fibroadenomas. Pure adenomas are uncommon. Adenocarcinomas comprise less than 10% of all spontaneous mammary tumors. Fibromas and other mesenchymal tumors, both benign and malignant, are occasionally seen in the mammary gland and are probably coincidental subcutaneous tumors. It is possible some fibromas arise as fibroadenomas in which there is such an abundance of connective tissue produced that the epithelial component is not detected.

Mammary tissue in the rat extends from the axillary to the inguinal region on either side of the ventral midline so that mammary tumors may be found anywhere in this area. Grossly, mammary tumors are firm, smooth lobulated masses usually freely moveable in the subcutaneous tissues. They are usually slow growing and may attain a very large size (Fig. 27), occasionally becoming as large as the rat. The larger tumors may interfere with the ability of the rat to move around or even to eat. In addition, they often become ulcerated and infected, thus further contributing to the general debility and eventual demise of the animal.

As mentioned previously, fibroadenomas are the most common mammary tumor in the rat. Microscopically, these lesions are proliferations of both duct epithelium forming small tubules, clefts, and acini and of the periductular connective tissue (Fig. 28). Either component may predominate. The epithelium is usually one to two cell layers thick and may exhibit secretory activity. Although these tumors may become quite large, they do not metastasize. However, a certain percentage are transplantable. When transplanted, they may maintain their original morphology or there may be overgrowth of one element or the other (43). On occasion, there may be malignant transformation, to either a sarcoma or carcinoma in a fibroadenoma (45).

Adenomas and cystadenomas occasionally have been reported but are uncommon (217). These tumors are well circumscribed masses composed of well-differentiated epithelial

Fig. 27. Mammary tumor. This gross photograph illustrates a large mammary tumor in the left axilla and a smaller tumor in the right flank. (Courtesy of A. M. Allen.)

cells proliferating to form papillary, acinar, or tubular structures with a delicate connective tissue stroma.

Carcinomas of various types have also been reported, although in a much lower incidence than fibroadenomas. These tumors include adenocarcinomas, papillary carcinomas, comedocarcinomas, and squamous cell carcinomas. Although these tumors are histologically malignant with piling up, pleomorphism and anaplasia of the epithelium, and a high mitotic index, they remain fairly well circumscribed and are only moderately invasive (Fig. 29). They metastasize infrequently. Metastasis, when it occurs, is usually late in the course of the disease, and many animals die or are killed before this occurs.

XII. MALE REPRODUCTIVE SYSTEM

A. Testis

The incidence of tumors of the testis varies greatly between

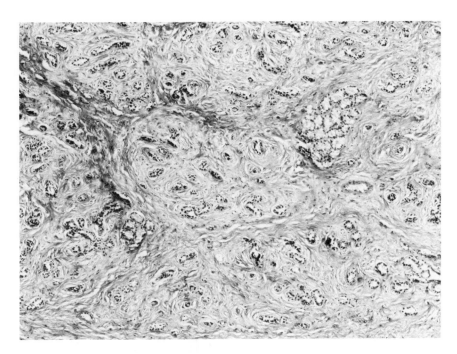

Fig. 28. Mammary fibroadenoma. Note the dense collagenous stroma separating nests of secretory and nonsecretory acini. ×90.

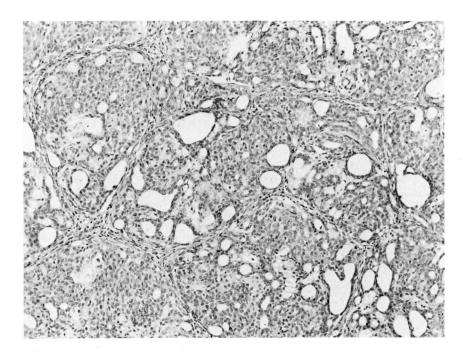

Fig. 29. Mammary adenocarcinoma. Tumor cells are arranged in large nests with differentiation into acini. ×130.

strains of rats and between colonies of the same strain. The vast majority of testicular tumors are interstitial cell tumors. Tumors arising from seminiferous tubules are exceedingly rare (Table XIII).

Interstitial cell (Leydig cell) tumors are the most common type of testicular tumor in rats. They occur with a high inci-

dence, increasing with age in certain strains (Table XIII). Grossly, these tumors are discrete, soft, and yellow to brown with areas of hemorrhage. They may be multiple within a testis and frequently are bilateral. Microscopically, these tumors are composed of large eosinophilic polyhedral cells with central vesicular nuclei and foamy or vacuolated cytoplasm (Fig. 30),

Table XIII

Spontaneous Tumors of the Male Reproductive Tract in Rats

Tumor	Number of animals	Reference
Testis		
Seminoma	1	24
	1	37
	1	173
	3	134
	1	199
Embryonal carcinoma	1	175
	1	134
Carcinoma	1	134
Teratoma	2	134
Sertoli cell tumor	1	134
Scrotal sac		
Lipoma	1	26
Lipofibroma	1	37
Fibrosarcoma	1	24
	2	153
Sarcoma	1	24
	3	42
Mesothelioma	Sporadic	189
	1	121
Seminal vesicles		
Carcinoma	1	25
Adenocarcinoma	1	132
	1	203
Carcinosarcoma	1	53
Epididymus		
Scirrhous carcinoma	1	203
Penis		
Sarcoma	1	37
	2	42
Accessory sex ducts		
Papillary adenocarcinoma	1	94
Prostatic tumors		
Adenoma	1	166
	1	195
Adenocarcinoma	1	41
	1	166
	8	148
	7	184
Carcinoma	2	69
Fibroma	1	203
Fibrosarcoma	1	175
	1	166
Sarcoma	3	42

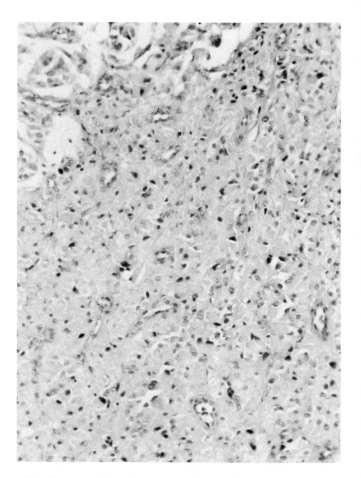

Fig. 30. Interstitial cell tumor. Atrophic seminiferous tubules can be seen in the upper left. ×138.

tumors have been transplanted, and some appear to produce androgen and/or estrogen (95,109,168). In addition, a transplantable Leydig cell tumor of the Fischer rat causes hypercalcemia (157).

Mesenchymal tumors, primarily fibromas, and fibrosarcomas of the testis and/or scrotal sac have been reported occasionally (Table XIII). An interesting lesion seen sporadically in several strains is a papillary mesothelioma involving the tunica vaginalis and genital omentum. These may be microscopic in size or may involve the entire tunica, appearing as variable-sized, pearly white nodules. Occasionally there is involvement of the peritoneal cavity. Microscopically, these tumors are composed of papillary fronds of hyaline connective tissue covered with single to several cell layers of plump, cuboidal, basophilic mesothelial cells (Fig. 31) (64).

B. Accessory Sex Glands

A single scirrhous carcinoma of the epididymus has been reported by Tilov *et al.* (203) in an Chbb:THOM(SPF) rat.

the vacuoles containing lipids. The cells are arranged in solid sheets or may form an organoid pattern. Variable-sized cysts filled with pink homogenous material or blood are frequently present. Often, there is compression of adjacent seminiferous tubules which are usually atrophic. Although these tumors may become extremely large, most appear benign. Interstitial cell

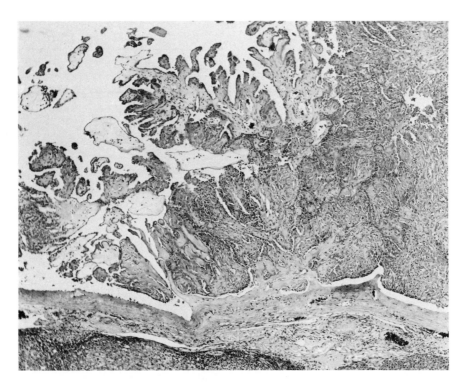

Fig. 31. Mesothelioma. This polypoid tumor originated in the tunica vaginalis. ×50. (Courtesy of J. Strandberg.)

Tumors of the accessory sex glands are relatively rare (Table XIII). Carcinomas of the seminal vesicles have been reported by Bullock and Rohdenburg (25), Morii and Fujii (132) in a Sprague-Dawley rat, and by Tilov *et al.* (203) in a Chbb: THOM(SPF) rat. Flexner and Jobling (53) described a transplantable tumor of uncertain origin attached to the left seminal vesicle of a male albino rat. The original tumor apparently had both carcinomatous and sarcomatous elements (52).

Prostatic tumors have been reported sporadically (Table XIII). In more recent studies where animals survive past 2 years of age, prostatic carcinomas have been seen more frequently. Eight adenocarcinomas in a group of 52 aged germfree Wistar rats have been reported (146,148). Seven of the eight tumor-bearing animals were over 30 months of age. Most of the tumors metastasized widely, and three have been transplanted successfully.

Shain *et al.* (184) reported adenocarcinomas of the ventral prostate in 7 of 41 (16%) AxC (ACI) rats 34 to 37 months old. Focal hemorrhage was the only grossly visible lesion. Microscopically, the tumors were intraluminal except for one with stromal invasion and were composed of anaplastic epithelial cells filling the lumens and forming cribiform and glandular patterns within these masses.

Tumors of bulbourethral (Cowper's) glands, urethral glands, coagulating glands, and ampullary glands have not been reported. The preputial glands are similar to the clitoral glands. (Tumors of these glands have been described in Section II.)

XIII. NERVOUS SYSTEM

Few primary tumors of the central or peripheral nervous system have been reported. This is probably a reflection of the fact that the nervous system is not fully examined during most necropsy procedures. Many early investigators failed to open the calvarium, and, even today, full brain and spinal column examinations are not performed routinely in rats. The tumors listed in Table XIV reflect the type and incidence of neurogenic tumors reported in the literature, and the most frequent types will be described briefly. Metastases from all neurogenic tumors are rare (Table XIV).

A. Astrocytomas

These are the most commonly reported brain tumors in rats. As with most neurogenic tumors in rats, there may be few if any clinical signs observed. Grossly, the larger tumors are evident when the brain is sectioned. They are soft and blend into the adjacent parenchyma. Histopathologically, the tumors vary from well-differentiated to extremely anaplastic (glioblastoma multiforme). There is usually no clear distinction between the tumor and the surrounding brain tissue. The vast majority occur in the cerebrum of young male rats. Fourteen astrocytic tumors out of a total of 20 brain tumors from the

Table XIV

Spontaneous Tumors of the Nervous System in Rats

Tumor	Number of animals	Reference
Brain		
Astrocytoma	6	57
	1	18
	2	177
	10	137
	2	151
	22	51
	2	173
	1	196a
Oligodendroglioma	2	69
	3	57
	1	18
	3	137
	3	51
	1	116
	1	196a
Glioma	4	69
	2	89
	14	57
	15	115
	5	137
	2	196a
Glioblastoma multiforme	1	175
	4	122
Ganglioneuroma	2	51
Meningioma	1	69
	1	58
	1	89
	13	137
	3	51
	1	173
Complex glio-ependymal	1	69
Ependymoma	1	96

Table XIV (*Continued*)

Tumor	Number of animals	Reference
	6	57
	1	177
	2	115
	4	51
Pinealoma	1	202
	1	54
Medulloblastoma	6	57
Granular cell tumors	1	173
	12	80
Spinal Cord		
Ependymoma	1	89
Polymorphic cell sarcoma	1	89
	1	119
Nerve		
Neurilemmoma-auditory	1	94
Neurilemmoma-trigeminal	2	89
Ganglioneuroma-optic	1	24
Glioma-optic	2	51
Neuroma-optic	1	42
Neurofibroma-skin	1	177
Others		
Ganglioneuroma-adrenal	1	7
	3	204
	2	115
	2	51
Neurofibroma-adrenal	1	177
Neurofibrosarcoma-retroperitoneal	1	177
Neuroblastoma-olfactory	1	58
Paranglioma-aorta	2	58
Paranganglioma-heart	1	58
Aortic body tumor	1	205

examination of 41,000 Sprague-Dawley rats were reported (120,122). There was no analysis of sex or age incidences. Fitzgerald *et al.* (51) reported 22 astrocytomas out of 31 brain tumors from 7803 outbred rats. A few mixed tumors containing astrocytes and oligodendroglia were noted in the series; however, they were classified as astrocytomas. Mackenzie and Garner (115) reported 15 gliomas, but they did not classify them further.

B. Oligodendrogliomas

These tumors are less common than astrocytomas and generally occur in older rats. Histologically, they are characterized by monotonous sheets of uniform cells with small dark nuclei and abundant cytoplasm imparting a honeycomb appearance to the tumor.

C. Ganglioneuroma

Ganglioneuromas have been reported in the central nervous system and from peripheral ganglia associated with the adrenal gland. Grossly, these tumors appear as soft, solitary, circumscribed masses. Microscopically, they contain clusters of well-differentiated ganglion cells in an abundant stroma of nerve fibers and Schwann cells (Fig. 32).

D. Meningioma

These are the most frequently reported mesenchymal tumors in the nervous system, but their incidence is still exceedingly low. Grossly, they can be nodular or plaquelike, with local infiltration of the brain substance. Histologically, meningiomas in rats have several distinctive patterns. The syncytial

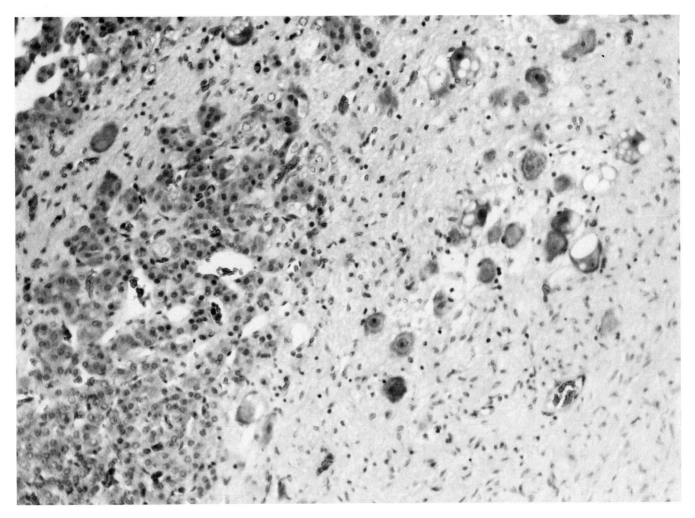

Fig. 32. Ganglioneuroma—adrenal gland. Tumor containing large ganglion cells and nerve tissue is invading the adjacent adrenal gland (arrows). ×225.

pattern is composed of two cell types. One is a large cell with abundant eosinophilic cytoplasm and a large prominent, vesicular nucleus. The second cell type is smaller with a tendency to be elongated, and it contains a hyperchromatic nucleus and sparse cytoplasm. The fibroblastic pattern has typical elongated spindle cells arranged in whorls, often around blood vessels. The nuclei in these cells are oval, hyperchromatic, and contain scanty cytoplasm.

E. Granular Cell Tumors

Other than the paper by Hollander and Snell (80), only one other report of a granular cell tumor in the nervous system of rats has been published (173). Hollander's group described twelve granular cell tumors in several strains, primarily aged BN/Bi rats. They arose in the leptomeninges, cerebrum, and cerebellum. All of the tumors were solitary, well-circumscribed, and light pink or yellowish.

Histologically, the tumors were composed of sheets or nests of large, closely packed round to oval cells with abundant cytoplasm (Fig. 33). The cytoplasm contained lightly eosinophilic granules which were PAS-positive and diastase resistant. These tumors are similar morphologically to granular cell tumors seen elsewhere in the body in a variety of species. They speculated that the tumors may have all developed from the leptomeninges. These tumors have been mistaken for gemnistocytic astrocytomas.

Granular cell tumors have been reported in a variety of other species from many organs including the lung, tongue and muscle. The most recent evidence suggests that granular cell tumors at all sites are neurogenic in origin from Schwann cells (2,6,22).

Spontaneous miscellaneous tumors for the rat are presented in Table XV. Table XVI lists reported tumor incidences by rat strain.

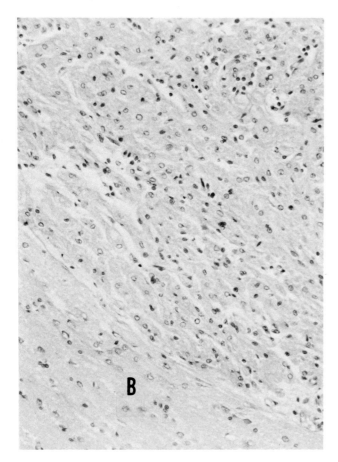

Fig. 33. Granular cell tumor. Note compression of adjacent brain tissue (B). ×220.

Table XV

Spontaneous Miscellaneous Tumors in Rats

Tumor	Number of animals	Reference
Pleura		
Mesothelioma	1	15
	1	194
Mediastinum		
Fibrosarcoma	1	153
Lymphoblastoma	51	58
Teratoma	2	166
Abdomen Cavity		
Mesothelioma	1	115
	5	173
	1	116
	2	26
Lipoma	1	151
	3	33
Liposarcoma	1	94
Teratoma	1	176
Sarcoma	7	25
	4	42
	6	69
Ovary		
Mesothelioma	1	58
		59
Site unknown		
Fibroma	22	164
	10	203
Fibroblastic tumors	9	69
Fibrosarcoma	4	69
	28	164
	7	203
Lipoma	3	69
	1	101
	2	164
	6	203
Liposarcoma	3	69
Sarcoma	12	203

Table XVI

A Listing of Reported Tumor Incidences by Rat Strain

Strain	Organ	Tumor type[a]	Incidence (%)	Age (months)	Sex	Comments	References
ACI	Anterior pituitary	—	15				48
	Prostate	Adenocarcinoma	17	34	M		184
	Testis	Interstitial cell tumor	25	Older	M		48
ACI/N	Adrenal medulla	—	5–10	>18		Breeders	73
							116
			6	>24	F		116
			16	>30	M		116
	Anterior pituitary	—	15–40	>18		Primarily chromophobe adenomas and primarily found in females	73
		—	6	>22	M		116
		—	21	>19	F		116

(continued)

Table XVI (*Continued*)

Strain	Organ	Tumor type[a]	Incidence (%)	Age (months)	Sex	Comments	References
	Mammary gland	Fibroadenomas, adenomas, carcinomas	11	>24	F	Two-thirds are benign	116
	Testis	Interstitial cell tumors	20	12–18	M		73
			46	>22	M		116
			85	>18	M		73
	Uterus	—	8–12	>18	F	Primarily adenocarcinomas but also endometrial polyps and smooth muscle tumors	73 116
BN/Bi	Adrenal cortex	Adenoma	12	33	M		79
			22	28	F		79
		Carcinoma	5	32	F		79
	Anterior pituitary	Adenoma	23	28	F		79
			8	21	M		79
	Cervix and vagina	—	20	>24	F	Primarily sarcomas, some squamous cell carcinomas	27
	Hematopoietic system	Lymphosarcoma	13	25	M		79
			3	28	F		79
	Mammary gland	Fibroadenoma	11	31	F		79
		Carcinoma	3	30	F		79
	Pancreatic islet	Adenoma	10	35	M		79
	Urinary bladder	Transitional cell carcinoma	23	30	M		13
			5	27	F		13 79
	Ureter	Transitional cell papillomas, carcinomas and squamous cell carcinomas	20		F		12
			6		M		12
BUF	Adrenal cortex	—	25	Older			48
	Anterior pituitary	—	30	Older			48
	Thymus	Thymoma	54		M		214
			39		F		214
	Thyroid	Parafollicular cell (C cell) tumors	25	>24		Increase with age	112
BUF/N	Adrenal cortex	—	30–70	>18			73
	Adrenal medulla	—	5–40	>18			73
	Anterior pituitary	—	5–20	<18			73
	Anterior pituitary	—	55–75	>18			73
	Uterus	—	2	>18	F	Virgin	73
COP	Thymus	Thymoma	High				48
F344	Adrenal cortex	—	11	>16	M		173
		—	6	>24	F		173
	Adrenal medulla	Pheochromocytoma	37	>16	M	27% Bilateral	94
			12	>14	F		94
			4	>18	M		33
	Anterior pituitary	Adenoma	24		M	Increase with age	173
			36		F	Increase with age	173
			11	>16	M		94
			3	>14	F		94
	Anterior pituitary	Chromophobe adenoma	15	>18	M		33
	Hematopoietic system	Leukemia	31		M	Increase with age	129 173
			21		F	Increase with age	129 173
	Mammary gland		41		F		173

(*continued*)

Table XVI (*Continued*)

Strain	Organ	Tumor type[a]	Incidence (%)	Age (months)	Sex	Comments	References
			23		M		173
	Pancreas	Islet cell adenoma	6	>18	M		33
	Testis	Interstitial cell tumors	60–90	>18	M	Frequently bilateral	32
	Thyroid	Parafollicular cell (C cell)	22	>24		Increase with age	112
		Parafollicular	9	>16	M		94
			2	>14	F		94
			12	>24	M		173
			10	>16	F		173
			3	>25	—		31
	Uterus	Polypoid tumors	21	22	F	Primarily benign	94
F344/N	Adrenal cortex	—	Up to 10	>18			73
	Adrenal medulla	—	10–45	>18			73
	Anterior pituitary	—	25	>18			73
	Testis	Interstitial cell tumors	70	<18	M	Nonbreeders	73
			30	<18	M	Breeders	73
			100	>18	M		73
	Uterus	—	75	>18	F	Virgin	73
			33	>18	F	Breeders	73
LOU/C/Wsl	Ileocecal lymph node	Plasmacytoma	32		M	May secrete immunoglobulins	131
			16		F		
M520	Adrenal medulla	—	21–25				48
	Uterus	Endometrial tumors	30	>21	F		189
M520/N	Adrenal cortex	—	20–45	>18			73
	Adrenal medulla	—	60–85	>18			73
	Anterior pituitary	—	20–40	>18			73
	Testis	Interstitial cell tumors	35	>18	M	Nonbreeders	73
	Uterus	—	12–50	>18	F		73
OM	Adrenal cortex	—	70				48
	Anterior pituitary	Chromophobe adenoma	60		M	Increases with age	174
			30		F	Increases with age	174
	Hematopoetic system	Reticulum cell sarcoma	15	>12			175
	Mammary gland	—	26–52		F		186
	Ovary	—	21–25		F		48
	Thyroid	Parafollicular cell (C cell) tumors	33	>24		Increases with age	112
OM[b]	Anterior pituitary	Chromophobe adenoma	29	17–24	F		213
			12	18–30	M		213
OM[c]		Chromophobe adenoma	5	7–13	M	None observed under 7 months of age	174
			52	14–20	M	—	174
			61	>20	M	—	174
			13	14–20	F	None observed under 13 months of age	174
			30	>20	F	—	174
OM/N	Adrenal cortex	—	50–70	<18			73
			75–95	>18			73
	Anterior pituitary	—	15–20	>18			73
	Mammary gland	—	25–30		F		73
	Ovary	—	20–25		F		73
	Uterus	—	10–15	>18	F		73
R (Amsterdam)	Anterior pituitary	Adenoma	28	25–29	F		106
WAG/Rij	Adrenal cortex	Adenoma	31	32	F		79
			7	21	M		79

(*continued*)

Table XVI (*Continued*)

Strain	Organ	Tumor type[a]	Incidence (%)	Age (months)	Sex	Comments	References
	Anterior pituitary	Adenoma	68	31	F		79
			96	21	M		79
	Mammary gland	Fibroadenomas	14	30	F		79
	Mammary gland	Adenocarcinomas	8	29	F		79
	Testis	Mesothelioma	5	22	M		79
	Thyroid	Parafollicular cell (C cell) tumors	40	31	F		79
			25	22	M		79
W	Adrenal medulla	Pheochromocytoma	86	>24	M		60
			76	>24	F		60
	Thyroid	Parafollicular cell	19	>24			112
W	Prostate	Adenocarcinoma	13	>30	M		146
W[d]	Anterior pituitary	Chromophobe adenoma	68	17–28	F		213
WF	Anterior pituitary	—	High			Many are mammotropic	48
	Hematopoietic system	Leukemia	High				48
	Mammary gland	—	21	20	F		101
W/FU	Anterior pituitary	Adenoma	27	>17	F		101
WN	Anterior pituitary	—	21–25	Older			48
	Mammary gland	—	30–50				48
WN/N	Adrenal cortex	—	8–12	>18			73
	Adrenal medulla	—	25–50	>18			73
	Anterior pituitary	—	40–93	>18		Increases with age	73
	Mammary gland	—	30–50		F	Benign and malignant tumors	73

[a]Many references do not list specific tumor types occurring at a given site. In such cases, this column is left blank.

[b]Obtained from Drs. Osborne and Mendel but published as the Vanderbilt strain.

[c]Yale strain.

[d]Originated from Wistar colony.

ACKNOWLEDGMENTS

The authors would like to thank the following for their assistance: Dr. Sannie Y. Demaray (editorial); Mrs. Joan E. O'Brien and Mrs. Virginia Salisbury (secretarial); and Mr. Jack Romaine and Mr. Robert Nye (photography).

REFERENCES

1. Arai, M. (1940). Uberdie spontanen Geschwulste bei weissen Ratten (Spontaneous tumors in white rats). *Gann* **34**, 137–145.
2. Ashburn, L. L., and Rogers, R. C. (1952). Myoblastoma, neural origin. *Am. J. Clin. Pathol.* **22**, 440–448.
3. Axelrad, A. A., and Leblond, C. P. (1955). Induction of thyroid tumors in rats by a low iodine diet. *Cancer* **8**, 339–367.
4. Babcock, V. I., and Southham, C. M. (1961). Transplantable renal tumor of the rat. *Cancer Res.* **21**, 120–131.
5. Bagg, H. J., and Hagopian, F. (1939). The functional activity of the mammary gland of the rat in relation to mammary cancer. *Am. J. Cancer* **35**, 175–187.
6. Bangle, R. J. (1953). An early granular cell myoblastoma confined within a small peripheral myelinated nerve. *Cancer* **6**, 790–793.
7. Barofsky, I., Matalka, E., and Russfield, A. B. (1970). A ganglio-neuroma in the adrenal medulla of a rat bearing a preoptic anterior hypothalmic lesion. *Cancer Res.* **30**, 2913–2916.
8. Berdjis, C. C. (1963). Protracted effects of repeated doses of x-ray irradiation in rats. *Exp. Mol. Pathol.* **2**, 157–172.
9. Bielschowsky, F. (1954). Functional acidophilic tumors of the pituitary of the rat. *Br. J. Cancer* **8**, 154–160.
10. Billups, L. H. (1976). Spontaneous tumors in laboratory animals—Rat. *In* "Handbook of Laboratory Animal Science" (E. C. Melby and N. H. Altman, eds.), Vol. 3, pp. 227–252. CRC Press, Cleveland, Ohio.
11. Bollmann, R., and Pearse, A. G. (1974). Calcitonin secretion and APUD characteristics of naturally occurring medullary thyroid carcinomas in rats. *Virchows Arch. B* **15**, 95–105.
12. Boorman, G. A., Burek, J. D., and Hollander, C. F. (1977). Spontaneous urothelial tumors in BN/Bi Ri rats. *Am. J. Pathol.* **88**, 251–254.
12a. Boorman, G. A., and Hollander, C. F. (1972). Occurrence of spontaneous cancer with aging in an inbred strain of rats. *TNO Nieuws* **27**, 692–695.
13. Boorman, G. A., and Hollander, C. F. (1973). Spontaneous lesions in the female WAG/Rij (Wistar) rat. *J. Gerontol.* **28**, 152–159.
14. Boorman, G. A., and Hollander, C. F. (1974). High incidence of spontaneous urinary bladder and ureter tumors in the Brown Norway rat. *J. Natl. Cancer Inst.* **52**, 1005–1008.
15. Boorman, G. A., and Hollander, C. F. (1976). Animal model: Medullary carcinoma of the thyroid in the rat. Human disease: Medullary carcinoma of the thyroid. *Am. J. Pathol.* **83**, 237–240.
16. Boorman, G. A., van Noord, M. J., and Hollander, C. F. (1972).

Naturally occurring medullary thyroid carcinoma in the rat. *Arch. Pathol.* **94,** 35–41.

17. Boorman, G. A., Zurcher, C., Hollander, C. F., and Feron, V. J. (1973). Naturally occurring endocardial disease in the rat. *Arch. Pathol.* **96,** 39–45.

18. Bots, G. T. A. M., Kroes, R., and Feron, V. J. (1968). Spontaneous tumors of the brain in rats. A report of 3 cases. *Pathol. Vet.* **5,** 290–296.

19. Brada, Z., Altman, N. H., and Bulba, S. (1975). Effect of cupric acetate on ethionine metabolism. *Cancer Res.* **35,** 3172–3180.

20. Brada, Z., Bulba, S., and Cohen, J. (1975). The metabolism of ethionine in rats. *Cancer Res.* **35,** 2674–2683.

21. Brand, K. G., Buoen, L. C., Johnson, K. H., and Brand, I. (1975). Etiological factors, stages, and the role of the foreign body in foreign body tumorigenesis: A review. *Cancer Res.* **35,** 279–286.

22. Budzilovich, G. N. (1968). Granular cell "myoblastoma" of vagus nerve. *Acta Neuropathol.* **10,** 162–165.

23. Bullock, F. D., and Curtis, M. R. (1922). A metastasizing condro-rhabdo-myosarcoma of the rat. *J. Cancer Res.* **7,** 195–207.

24. Bullock, F. D., and Curtis, M. R. (1930). Spontaneous tumors of the rat. *J. Cancer Res.* **14,** 1–115.

25. Bullock, F. D., and Rohdenburg, G. L. (1917). Spontaneous tumors of the rat. *J. Cancer Res.* **2,** 39–60.

26. Burek, J. D., and Hollander, C. F. (1977). Incidence patterns of spontaneous tumors in BN/Bi rats. *J. Natl. Cancer Inst.* **58,** 99–105.

27. Burek, J. D., Zurcher, C., and Hollander, C. F. (1976). High incidence of spontaneous cervical and vaginal tumors in an inbred strain of rats. *J. Natl. Cancer Inst.* **57,** 549–551.

28. Burn, J. I., Sellwood, R. A., and Bishop, M. (1966). Spontaneous carcinoma of the colon of the rat. *J. Pathol. Bacteriol.* **91,** 253–254.

29. Carter, R. L. (1968). Pathology of ovarian neoplasms in rats and mice. *Eur. J. Cancer* **3,** 537–543.

30. Cohen, A. J., Furthand, J., and Buffett, R. F. (1957). Histologic and physiologic characteristics of hormone-secreting transplantable adrenal tumors in mice and rats. *Am. J. Pathol.* **33,** 631–651.

31. Cockrell, B. Y., and Garner, F. M. (1975). Thyroid tumors in the Fischer rat. *Lab. Invest.* **32,** 420.

32. Cockrell, B. Y., and Garner, F. M. (1976). Interstitial cell tumor of the testis in rats. *Comp. Pathol. Bull.* **8,** 2–4.

33. Coleman, G. L., Barthold, S. W., Osbaldiston, G. W., Foster, S. J., and Jonas, A. M. (1977). Pathological changes during aging in barrier-reared Fischer 344 male rats. *J. Gerontol.* **32,** 258–278.

34. Copeland, D. H., and Salman, W. D. (1946). The occurrence of neoplasms in the liver, lungs, and other tissues of rats as a result of prolonged choline deficiency. *Am. J. Pathol.* **22,** 1059–1079.

35. Cotchin, E., and Roe, F. J. C., eds. (1967). "Pathology of Laboratory Rats and Mice." Blackwell, Oxford.

36. Crain, R. C. (1958). Spontaneous tumors in the Rochester strain of the Wistar rat. *Am. J. Pathol.* **34,** 311–335.

37. Curtis, M. R., Bullock, F. D., and Dunning, W. F. (1931). A statistical study of the occurrence of spontaneous tumors in a large colony of rats. *Am. J. Cancer* **15,** 67–121.

38. Davey, F. R., and Moloney, W. C. (1970). Postmortem observations on Fischer rats with leukemia and other disorders. *Lab. Invest.* **23,** 327–334.

39. Davis, R. K., Stevenson, G. T., and Busch, K. A. (1956). Tumor incidence in normal Sprague-Dawley female rats. *Cancer Res.* **16,** 194–197.

40. Dunning, W. F. (1958). Evidence from experimentally induced neoplasms relative to the mutation hypothesis of the origin of malignant disease. *Ann. N.Y. Acad. Sci.* **71,** 1114–1123.

41. Dunning, W. F. (1963). Prostate cancer in the rat. *Natl. Cancer Inst., Monogr.* **12,** 351–369.

41a. Dunning, W. F., and Curtis, M. R. (1930). Spontaneous tumours of the rat. *J. Cancer Res.* **14,** 1–115.

42. Dunning, W. F., and Curtis, M. R. (1946). The respective roles of longevity and genetic specificity in the occurrence of spontaneous tumors in the hybrids between two inbred lines of rats. *Cancer Res.* **6,** 61–81.

43. Dunning, W. F., Curtis, M. R., and Maun, M. E. (1945). Spontaneous malignant mixed tumors of the rat, and the successful transplantation and separation of both components from a mammary tumor. *Cancer Res.* **5,** 644–651.

44. Eker, R. (1954). Familial renal adenomas in Wistar rats. A preliminary report. *Acta Pathol. Microbiol. Scand.* **34,** 554–562.

45. Emge, L. A. (1938). Sarcomatous degeneration of transplantable mammary adenofibroma of the white rat. *Arch. Pathol.* **26,** 429–440.

46. Engle, E. T. (1946). Tubular adenomas and testis-like tubulis of the ovaries of aged rats. *Cancer Res.* **6,** 578–582.

47. Farris, E. J., and Yeakel, E. H. (1944). Spontaneous and transplanted reticulum cell sarcomas in Wistar rats. *Am. J. Pathol.* **20,** 773–781.

48. Festing, M., and Staats, J. (1973). Standardized nomenclature for inbred strains of rats. Fourth listing. *Transplantation* **16,** 221–245.

49. Fibiger, J. (1913). Untersuchungen über eine nematode (*Spiroptera* sp. n.) und deren fähigkeit, papillomatöse und carcinomatöse geschwulstbidungen im magen der ratte hervorzurufen. *Z. Krebsforsch.* **13,** 217–280.

50. Fitzgerald, J. E., Schardein, J. L., and Kaump, D. H. (1971). Several uncommon pituitary tumors in the rat. *Lab. Anim. Sci.* **21,** 581–584.

51. Fitzgerald, J. E., Schardein, J. L., and Kurtz, S. M. (1974). Spontaneous tumors of the central nervous system in albino rats. *J. Natl. Cancer Inst.* **52,** 265–273.

52. Flexner, S., and Jobling, J. W. (1907). Infiltrating and metastasising sarcoma of the rat. *J. Am. Med. Assoc.* **48,** 420.

53. Flexner, S., and Jobling, J. W. (1910). Studies upon a transplantable rat tumor. Originally regarded as sarcoma; probably a teratoma from which an adenocarcinoma developed. *Monogr. Rockefeller Inst. Med. Res.* **1**(pt. 1).

54. Frauchiger, E., O'Hara, P. J., and Shortridge, E. H. (1966). Pinealome bei Tieren. *Schweiz. Arch. Tierheilkd.* **108,** 368–372.

55. Frith, C. H., Farris, H. E., and Highman, B. (1977). Endocardial fibromatous proliferation in a rat. *Lab. Anim. Sci.* **27,** 114–115.

56. Furth, S., Nakane, P., and Pasteels, J. C. (1976). Tumours of the pituitary gland. *In* "Pathology of Tumours in Laboratory Animals" (V. S. Turusov, ed.), IARC Sci. Publ. No. 5, Vol. 1, Part 2. IARC, Lyon.

57. Garner, J. M., Innes, J. R. M., and Nelson, D. H. (1967). Murine neuropathology. *In* "Pathology of Laboratory Rats and Mice" (E. Cotchin and F. J. C. Roe, eds.), pp. 000–000. Blackwell, Oxford.

58. Gilbert, C., and Gillman, J. (1958). Spontaneous neoplasms in the albino rat. *S. Afr. J. Med Sci.* **23,** 257–272.

59. Gilbert, C., Gillman, J., Loustalot, P., and Lutz, W. (1958). The modifying influence of diet and the physical environment on spontaneous tumor frequency in rats. *Br. J. Cancer* **7,** 565–593.

60. Gillman, J., Gilbert, C., and Spence, I. (1953). Phaeochromocytoma in the rat. Pathogenesis and collateral reactions and its relation to comparable tumors in man. *Cancer* **6,** 494–511.

61. Glucksmann, A., and Cherry, C. P. (1973). Tumours of the salivary glands. *In* "Pathology of Tumours in Laboratory Animals" (V. S. Turusov, ed.), IARC Sci. Publ. No. 5, Vol. 1, Part 1, pp. 75–86. IARC, Lyon.

62. Gohlke, R., and Schmidt, P. (1971). Spontaneous nephroblastomas in laboratory rats. *Z. Versuchstierkd.* **13,** 326–337.

63. Goodall, C. M., Christie, G. S., and Hurley, J. V. (1975). Primary epithelial tumour in the right atrium of the heart and inferior vena cava in NZR/Gd inbred rats: Pathology of 18 cases. *J. Pathol.* **116,** 239–251.

64. Gould, D. H. (1977). Mesotheliomas of the tunica vaginalis propria and peritoneum in Fischer rats. *Vet. Pathol.* **14**, 372-379.

65. Grasso, P., and Creasy, M. (1969). Carcinoma of the colon in a rat. *Eur. J. Cancer* **5**, 415-419.

66. Griem, W. (1957). Uber einen fall von pankreaskarzinom bei liner ratte. *Berl. Muench. Tieraerztl. Wochenschr.* **21**, 451-452.

67. Griepentrog, F. (1964). Spontaneous pituitary tumors as a frequent phenomenon in white rats used in the laboratory. *Beitr. Pathol. Anat. Allg. Pathol.* **130**, 40-50.

68. Griesbach, W. E. (1967). Basophil adenomata in the pituitary glands of 2-year old male Long-Evans rats. *Cancer Res.* **27**, 1813-1818.

69. Guérin, M. (1954). "Tumerus spontanées des animaux de laboratorie (Spuris-Rat-Poule)." Book, Legrand, Paris.

70. Guérin, M., Chouroulinkov, I., and Guillon, J. C. (1969). Rat lymphomas (reticulosis). *Bull. Cancer (Paris)* **56**, 309-320.

71. Guérin, M., and Guérin, P. (1934). Epithelioma of rat uterus report on transplantation studies. *Bull. Assoc. Fr. Cancer* **23**, 623-646.

72. Guimaraes, J. P., and Ballini-Kerr, I. (1968). Atypical cellular areas and the origin of the spontaneous hepatoma in the rat. *Rev. Bras. Pesquis. Med. Biol.* **1**, 143-146.

73. Hansen, C. T., Judge, F. J., and Whitney, R. A. (1974). "Catalogue of NIH Rodents," DHEW Publ. No. (NIH) 74-606. Natl. Inst. Health, Washington, D.C.

74. Hard, G. C., and Grasso, P. (1976). Nephroblastoma in the rat: Histology of a spontaneous tumor, identify with respect to renal mesenchyma neoplasms, and a review of previously recorded cases. *J. Natl. Cancer Inst.* **57**, 323-329.

75. Hegsted, D. M. (1975). Relevance of animal studies to human disease. *Cancer Res.* **35**, 3537-3539.

76. Heiman, J. (1944). Spontaneous tumors of the adrenal cortex in castrated male rat. *Cancer Res.* **4**, 430-432.

77. Heslop, B. F. (1969). Cystic adenocarcinoma of the ascending colon in rats occurring as a self-limiting outbreak. *Lab. Anim.* **3**, 185-195.

78. Hitchcock, C. R., and Bell, E. T. (1952). Studies on the nematode parasite, Gongylonema neoplasticum (spiroptera neoplasticum), and avitaminosis A in the forestomach of rats: Comparison with Fibiger's results. *J. Natl. Cancer Inst.* **12**, 1345-1387.

79. Hollander, C. F. (1976). Current experience using the laboratory rat in aging studies. *Lab. Anim. Sci.* **26**, 320.

80. Hollander, C. F., Burek, J. D., Boorman, G. A., Snell, K. C., and Laqueur, G. L. (1976). Granular cell tumors of the central nervous system of the rat. *Arch. Pathol. Lab. Med.* **100**, 445-447.

81. Hollander, C. F., and Snell, K. C. (1976). Tumours of the adrenal gland of the rat. *In* "Pathology of Tumours in Laboratory Animals" (V. S. Turusov, ed.), IARC Sci. Publ. No. 5, Vol. 1, Part 2, pp. 273-294. IARC, Lyon.

82. Horn, H. A., and Stewart, H. L. (1952). Spontaneous pulmonary tumor in the rat: Report of a lesion. *J. Natl. Cancer Inst.* **12**, 743-749.

83. Hottendorf, G. H., and Ingraham, K. J. (1968). Spontaneous nephroblastomas in laboratory rats. *J. Am. Vet. Med. Assoc.* **153**, 826-829.

84. Houssay, B. A., Houssay, A. B., Cardeza, A. F., Fogua, V. G., and Pinto, R. M. (1953). Adrenal tumors in gonadectomized rats. *Acta Physiol. Lat. Am.* **3**, 125-130.

85. Iglesias, R., and Mardones, E. (1956). Twenty generations of a spontaneous transplantable, functional granulosa-cell tumor in the rat, 1949 to 1955. *Cancer* **9**, 740-748.

86. Iglesias, R., and Mardones, E. (1956). Two instances of spontaneous transplantable leukemias in AXC rats. *Proc. Am. Assoc. Cancer Res.* **2**, 121.

87. Iglesias, R., and Mardones, E. (1958). Spontaneous and transplantable functional tumour of the adrenal cortex in the AXC rat. *Br. J. Cancer* **7**, 20.

88. Iglesias, R., Sternberg, W. H., and Segaloff, A. (1950). A transplanta-ble functional ovarian tumor occurring spontaneously in a rat. *Cancer Res.* **10**, 668-673.

89. Innes, J. R. M., and Borner, G. (1961). Tumors of the central nervous system of rats: With reports of two tumors of the spinal cord and comments on posterior paralysis. *J. Natl. Cancer Inst.* **26**, 719-735.

89a. Institute of Laboratory Animal Resources (1975). "Animals for Research," 9th ed. Natl. Acad. Sci., Washington, D.C.

90. Isler, H., Leblond, C. P., and Axelrad, A. A. (1958). Influence of age and of iodine intake on the production of thyroid tumors in the rat. *J. Natl. Cancer Inst.* **21**, 1065-1081.

91. Ito, A. (1976). Animal model: Pituitary tumors in rats—human dis: Pituitary tumors. *Am. J. Pathol.* **83**, 423-426.

92. Ito, A., Moy, P., Kaunitz, H., Kortwright, K., Clarke, S., Furth, J., and Meites, J. (1972). Incidence and character of the spontaneous pituitary tumors in strain CR and W/Fu male rats. *J. Natl. Cancer Inst.* **49**, 701-711.

93. Ives, M., and Dack, G. M. (1957). Safety of inside enamel coatings used in food cans. *Food Res.* **22**, 102-109.

94. Jacobs, B. B., and Huseby, R. A. (1967). Neoplasms occurring in aged Fischer rats with special reference to testicular, uterine and thyroid tumors. *J. Natl. Cancer Inst.* **39**, 303-309.

95. Jacobs, B. B., and Huseby, R. A. (1968). Transplantable Leydig cell tumors in Fischer rats: Hormone responsivity and hormone production. *J. Natl. Cancer Inst.* **41**, 1141-1153.

96. Janisch, W., and Osske, G. (1966). Spontaneous ependymoma of the brain in a rat. *Zentralbl. Allg. Pathol. Pathol. Anat.* **109**, 520-523.

97. Jenne, F. S. (1941). Reticulum cell sarcoma of the rat transferred through twelve successive passages in animals of related stock. *Cancer Res.* **1**, 407-409.

98. Jonas, A. M., chairman (1976). Long term holding of laboratory rodents. *ILAR News* **19**, 1-25.

99. Kanesawa, M., and Schroeder, H. A. (1969). Lifetime studies on the effect of trace elements on spontaneous tumors in mice and rats. *Cancer Res.* **29**, 892-895.

100. Kihlstrum, I. M., and Clements, G. R. (1969). Spontaneous pathologic findings on Long-Evans rats. *Lab. Anim. Sci.* **19**, 710-715.

101. Kim, U., Clifton, K. H., and Furth, J. (1960). A highly inbred line of Wistar rats yielding spontaneous MA/MMO somatotropic pituitary and other tumors. *J. Natl. Cancer Inst.* **24**, 1031-1055.

102. Kinkel, H. J. (1971). Spontaneous tumors in Sprague-Dawley rats. *Z. Versuchstierkd.* **13**, 97-100.

103. Klein, A. J., and Palmer, W. L. (1941). Experimental gastric carcinoma: A critical review with comments on the criteria of induced malignancy. *J. Natl. Cancer Inst.* **1**, 559-584.

104. Klein-Szanto, A. J. P., Conti, C. J., and Cartagenova, R. E. (1974). Ultrastructure of a vaginal myxoma of a rat. *Vet. Pathol.* **11**, 289-296.

105. Koletsky, S., and Gustafson, G. E. (1955). Whole-body radiation as a carcinogenic agent. *Cancer Res.* **15**, 100-104.

106. Kwa, H. G., van der Gugten, A. A., and Verhofstad, F. (1969). Radio immunoassay of rat prolactin. Prolactin levels in plasma of rats with spontaneous pituitary tumours, primary. *Eur. J. Cancer* **5**, 571-579.

107. Laqueur, G. L. (1965). The induction of intestinal neoplasms in rats with the glycoside cycasin and its aglycone. *Virchows Arch. Pathol. Anat. Physiol.* **340**, 151-163.

108. Leblond, C. P., Isler, H., and Axelrad, A. (1956). Induction of thyroid tumors by a low-iodine diet. *Proc. Can. Cancer Res. Conf.* **2**, 248-266.

109. Lehmann, H., Lutzen, L., and Deutsch, H. (1975). Hormonal activity in testes in aged rats with Leydig cell hyperplasia. *Beitr. Pathol.* **156**, 109-116.

110. Lillie, R. D., and Engle, J. L. (1935). Renal adenosarcoma in a white rat. *Arch. Pathol.* **19**, 687-689.

111. Lindsay, S., and Nichols, C. W., Jr. (1969). Medullary thyroid car-

cinoma and parathyroid hyperplasia in rats. *Arch. Pathol.* **88**, 402–406.

112. Lindsay, S., Nichols, C. W., Jr., and Chaikoff, I. L. (1968). Naturally occurring thyroid carcinoma in the rat. Similarities to human medullary carcinoma. *Arch. Pathol.* **86**, 353–364.

113. Loeb, L. (1902). Further investigations in transplantation of tumors. *J. Med. Res.* **8**, 44–73.

114. Machado, E. A., and Beauchene, R. E. (1976). Spontaneous osteogenic sarcoma in the WI/Ten rat: A case report. *Lab. Anim. Sci.* **26**, 98–100.

115. Mackenzie, W. F., and Garner, F. M. (1973). Comparison of neoplasms in six sources of rats. *J. Natl. Cancer Inst.* **50**, 1243–1257.

116. Maekawa, A., and Odashima, S. (1975). Spontaneous tumors in ACI/N rats. *J. Natl. Cancer Inst.* **55**, 1437–1445.

117. Magnusson, G. (1969). Spontaneous nephroblastomas in the rat. *Z. Versuchstierkd.* **11**, 293–297.

118. Maisin, J., Maldague, P., Deckers, C., and Hong-Que, P. (1963). Le leukosarcome du rat. *Symp. Lymph Tumours Afr.*, pp. 341–354.

119. Mawdesley-Thomas, L. E. (1967). A polymorphic-cell sarcoma of the rat spinal cord. *Lab. Anim.* **1**, 31–34.

120. Mawdesley-Thomas, L. E. (1974). Spontaneous tumors of the central nervous system in the rat. *Vet. Pathol.* **11**, 454.

121. Mawdesley-Thomas, L. E., and Hague, P. H. (1970). Mesothelioma of the tunica vaginalis in a rat. *Lab. Anim.* **4**, 29–36.

122. Mawdesley-Thomas, L. E., and Newman, A. J. (1974). Some observations on spontaneously occurring tumors of the central nervous system of Sprague-Dawley delivered rats. *J. Pathol.* **112**, 107–116.

123. McCoy, G. W. (1909). A preliminary report on tumors found in wild rats. *J. Med. Res.* **21**, 282–296.

124. Mitra, S. K. (1966). Carcinoma of the colon in a rat. *Br. J. Cancer* **20**, 399–401.

125. Miwa, M. (1976). Spontaneous colon tumors in rats. *J. Natl. Cancer Inst.* **56**, 615–621.

126. Miyamoto, M., and Takizawa, S. (1975). Brief communication: Colon carcinoma of highly inbred rats. *J. Natl. Cancer Inst.* **55**, 1471–1472.

127. Moloney, W. C. (1974). Primary granulocytic leukemia in the rat. *Cancer Res.* **34**, 3049–3057.

128. Moloney, W. C., Boschetti, A. E., and King, V. (1969). Observations on leukemia in Wistar and Wistar Furth rats. *Cancer Res.* **29**, 938–946.

129. Moloney, W. C., Boschetti, A. E., and King, V. P. (1970). Spontaneous leukemia in Fischer rats. *Cancer Res.* **30**, 40–43.

130. Moore, W. E. C., and Holdeman, C. V. (1975). Discussion of current bacteriological investigations of the relationships between intestinal flora, diet and colon cancer. *Cancer Res.* **35**, 3418–3420.

131. Moriame, M., Beckers, A., and Bazin, H. (1977). Decrease in the incidence of malignant ileo-caecal immunocytoma in LOU/C rats after surgical removal of ileocaecal lymph nodes. *Cancer Lett.* **3**, 139–143.

132. Morii, S., and Fujii, T. (1973). Spontaneous tumors in Sprague-Dawley JCL rats. *Exp. Anim.* **22**, 127–138.

133. Morris, H. P., Wagner, B. P., Ray, F. E., Snell, K. C., and Stewart, H. L. (1961). Comparative study of cancer and other lesions of rats fed *N, N'*2,7-fluoroenylene bisacetamide or *N*-2-fluorenyl acetamide. *Natl. Cancer Inst., Monogr.* **5**, 1–53.

134. Mostofi, F. K., and Bresler, V. M. (1976). Tumours of the testis. *In* "Pathology of Tumours in Laboratory Animals" (V. S. Turusov, ed.), IARC Sci. Publ. No. 6, Vol. 1, Part 2, pp. 135–151. IARC, Lyon.

135. Napalkov, N. P. (1976). Tumours of the thyroid gland. *In* "Pathology of Tumours in Laboratory Animals" (V. S. Turusov, ed.), IARC Sci. Publ. No. 5, Vol. 1, Part 2, pp. 239–272. IARC, Lyon.

136. Nelson, A. A., and Morris, J. M. (1941). Reticulum cell lymphosarcoma in rats. *Arch. Pathol.* **31**, 578–584.

137. Newman, A. J., and Mawdesley-Thomas, L. E. (1974). Spontaneous tumours of the central nervous system of laboratory rats. *J. Comp. Pathol.* **84**, 39–50.

138. Nicholson, G. W. (1913). A transplantable carcinoma of the kidney of a white rat. *J. Pathol. Bacteriol.* **17**, 329–346.

139. Nifatov, A. P., and Koshurnikova, N. A. (1973). Spontaneous tumors in Wistar rats. *Vopr. Onkol.* **19**, 83–86.

139a. Niggeschulze, A., and Tilov, T. (1975). Kasuistischer beitrag zur urolithiasis bei der ratte. *Z. Versuchstierkd.* **17**, 315–319.

140. Nino, F. L., and Jorg, M. E. (1933). Carcinosis de Fiebiger a localizaciones multiples en una rara cruzado. *Prensa Med. Argent.* **20**, 883–893.

141. Noble, R. L., and Cutts, J. H. (1959). Mammary tumors of the rat: A review. *Cancer Res.* **19**, 1125–1139.

142. Olcott, C. T. (1950). A transplantable nephroblastoma (Wilms' tumor) and other spontaneous tumors in a colony of rats. *Cancer Res.* **10**, 625–628.

143. Passey, R. D., Leese, A., and Knox, J. C. (1936). Bronchiectasis and metaplasia in the lung of the laboratory rat. *J. Pathol. Bacteriol.* **42**, 425–34.

144. Pittermann, W., and Deerberg, F. (1974). Spontaneous nephroblastomas in SPF Han rats. *Berl. Muench. Tieraerztl. Wochenschr.* **87**, 178–180.

145. Pliss, G. B. (1973). Tumours of the auditory sebaceous glands. *In* "Pathology of Tumours in Laboratory Animals" (V. S. Turusov, ed.). IARC Sci. Publ. No. 5, Vol. 1, Part 1, pp. 23–30. IARC, Lyon.

146. Pollard, M. (1973). Spontaneous prostate adenocarcinomas in aged germfree Wistar rats. *J. Natl. Cancer Inst.* **51**, 1235–1241.

147. Pollard, M., and Kajima, M. (1970). Lesions in aged germfree Wistar rats. *Am. J. Pathol.* **61**, 25–36.

148. Pollard, M., and Luckert, P. H. (1975). Transplantable metastasizing prostate adenocarcinomas in rats. *J. Natl. Cancer Inst.* **54**, 643–647.

149. Pollard, M., and Teah, B. A. (1963). Spontaneous tumors in germfree rats. *J. Natl. Cancer Inst.* **31**, 457–465.

150. Pradhan, S. N., Chung, E. B., Ghosh, B., Paul, B. D., and Kapadia, G. J. (1974). Potential carcinogens. I. Carcinogenicity of some plant extracts and their tanin-containing factions in rats. *J. Natl. Cancer Inst.* **52**, 1579–1582.

151. Prejean, J. D., Peckham, J. C., Casey, A. E., Griswold, D. P., Weisburger, E. K., and Weisburger, J. H. (1973). Spontaneous tumors in Sprague-Dawley rats and Swiss mice. *Cancer Res.* **33**, 2768–2773.

152. Rask-Nielsen, R. (1938). Experimental studies on a transplantable aleukemic myelomatosis in white rats. *Acta Pathol. Microbiol. Scand.* **15**, 285–300.

153. Ratcliffe, H. L. (1940). Spontaneous tumors in two colonies of rats of the Wistar Institute of Anatomy and Biology. *Am. J. Pathol.* **16**, 237–254.

155. Reddy, J., Svoloda, D., Azarnoff, D., and Dawar, R. (1973). Cadmium-induced Leydig cell tumors of rat testis: Morphologic and cytochemical study. *J. Natl. Cancer Inst.* **51**, 891–903.

157. Rice, B. F., Roth, L. M., Cole, F. E., MacPhee, A. A., Davis, K., Ponthier, R. L., and Sternberg, W. H. (1975). Hyperplasia and neoplasia. *Lab. Invest.* **33**, 428–439.

158. Richter, C. B., Estes, P. C., and Tennant, R. W. (1972). Spontaneous stem cell leukemia in young Sprague-Dawley rats. *Lab Invest.* **26**, 419–428.

159. Roberts, D. C. (1960). Tumours of the large intestine of mammals other than man. *In* "Cancer of the Rectum" (C. E. Dukes, ed.), pp. 43–58. Livingstone, Edinburgh.

160. Roe, F. J. C. (1941). Spontaneous tumours in rats and mice. *Food Cosmet. Toxicol.* **3**, 707–720.

161. Roe, F. J. C., and Roberts, J. D. B. (1973). Tumours of the pancreas. *In* "Pathology of Tumours in Laboratory Animals" (V. S. Turusov, ed.), IARC Sci. Publ. No. 5, Vol. 1, Part 1, pp. 141–150. IARC, Lyon.

162. Roger, A. E., and Newberne, P. M. (1975). Dietary effects on chemi-

cal carcinogenesis in animal models for colon and liver tumors. *Cancer Res.* **35,** 3427-3431.

163. Rosen, V. J., Castanera, T. J., Jones, D. C., and Kimeldorf, D. J. (1961). Islet-cell tumors of the pancreas in irradiated and nonirradiated rat. *Lab. Invest.* **10,** 608-616.

164. Ross, M. H., and Bras, G. (1965). Tumor incidence patterns and nutrition in the rat. *J. Nutr.* **87,** 245-260.

165. Ross, M. H., and Bras, G. (1971). Lasting influence of early caloric restriction on prevalence of neoplasms in the rat. *J. Natl. Cancer Inst.* **47,** 1095-1113.

166. Ross, M. H., and Bras, G. (1973). Influence of protein under- and overnutrition on spontaneous tumor prevalence in the rat. *J. Nutr.* **103,** 944-963.

167. Ross, M. H., Bras, G., and Ragbeer, M. S. (1970). Influence of protein and caloric intake upon spontaneous tumor incidence of the anterior pituitary gland of the rat. *J. Nutr.* **100,** 177-189.

168. Roth, L. M., Spurlock, B. O., Sternberg, W. H., and Rice, B. F. (1970). Fine structure of transplantable Leydig cell tumors of the rat. *Am. J. Pathol.* **60,** 137-152.

168a. Rowlatt, U. F. (1967). Neoplasms of the alimentary canal of rats and mice. *In* "Pathology of Laboratory Rats and Mice" (E. Cotchin and F. J. C. Roe, eds.), pp. 57-84. Blackwell, Oxford.

169. Rowlatt, U. F. (1967). Pancreatic neoplasms of rats and mice. *In* "Pathology of Laboratory Rats and Mice" (E. Cotchin and F. J. C. Roe, eds.), pp. 85-103. Blackwell, Oxford.

170. Rowlatt, U. F. (1967). Spontaneous epithelial tumors of the pancreas of mammals. *Br. J. Cancer* **21,** 82-107.

171. Rowlatt, U. F., and Roe, F. J. C. (1967). Epithelial tumors of the rat pancreas. *J. Natl. Cancer Inst.* **39,** 18-32.

172. Sacksteder, M. R. (1976). Occurrence of spontaneous tumors in the germfree F344 rat. *J. Natl. Cancer Inst.* **57,** 1371-1373.

173. Sass, B., Rabstein, L. S., Madison, R., Nims, R. M., Peters, R. L., and Kelloff, G. J. (1975). Incidence of spontaneous neoplasms in F344 rats throughout the natural life-span. *J. Natl. Cancer Inst.* **54,** 1449-1456.

174. Saxton, J. A., and Graham, J. B. (1944). Chromophobe adenoma-like lesions of rat hypophysis. *Cancer Res.* **4,** 168.

175. Saxton, J. A., Sperling, G. A., Barnes, L. L., and MacCay, C. M. (1948). The influence of nutrition upon the incidence of spontaneous tumors of the albino rat. *Acta Unio. Int. Cancrum.* **6,** 423-431.

176. Schardein, J. L., and Fitzgerald, J. E. (1977). Teratoma in a Wistar rat. *Lab. Anim. Sci.* **27,** 114.

177. Schardein, J. L., Fitzgerald, J. E., and Kaump, D. H. (1968). Spontaneous tumors in Holtzman-source rats of various ages. *Pathol. Vet.* **5,** 238-252.

178. Schilling, B. F., Frohberg, H., and Oettel, H. (1967). On the incidence of naturally occurring thyroid carcinoma in the Sprague-Dawley rat. *Ind. Med. Surg.* **36,** 678-684.

179. Schilling, B. F., Frohberg, H., and Oettel, H. (1967). Spontaneous tumors of the thyroid gland of the laboratory rat. *Krebsforsch.* **70,** 161-164.

180. Schoental, R. (1974). Role of podophyllotoxin in the bedding and dietary zearalenone on incidence of spontaneous tumors in laboratory animals. *Cancer Res.* **34,** 2419-2420.

181. Schulze, E. (1960). Spontantumoren der schadelhohle und genitalorgane bei Sprague-Dawley und Bethesda-Black-Ratten. *Z. Krebsforsch.* **64,** 78-82.

182. Schwartz, M. K. (1975). Role of trace elements in cancer. *Cancer Res.* **35,** 3481-3487.

183. Scott, K. G., Bostich, W. C., Shimkin, M. B., and Hamilton, J. G. (1949). Distribution of globulin-bound thyroid fractions of iodine in normal and tumorous animals after intravenous administration using radioiodine as a tracer. *Cancer* **22,** 692-696.

184. Shain, S. A., McCullough, B., and Segaloff, A. (1975). Spontaneous adenocarcinoma of the ventral prostate of aged AXC rats. *J. Natl. Cancer Inst.* **55,** 177-180.

185. Sharma, R. N. (1977). Naturally occurring mesothelioma in the right atrium of the heart of an albino rat. *Bull. Soc. Pharm. Environ. Pathol.* **5,** 2-4.

186. Sher, S. P. (1972). Mammary tumors in control rats: Literature tabulation. *Toxicol. Appl. Pharmacol.* **22,** 562-588.

187. Shubik, P. (1975). Potential carcinogenicity of food additives and contaminants. *Cancer Res.* **35,** 3475-3480.

188. Singer, C. (1913-1914). A transplantable sarcoma arising in the uterus of a rat. *J. Pathol. Bacteriol.* **18,** 495-497.

189. Snell, K. C. (1965). Spontaneous lesions of the rat. *In* "The Pathology of Laboratory Animals" (W. E. Ribelin and J. R. McCoy, eds.), pp. 241-302. Thomas, Springfield, Illinois.

190. Snell, K. C. (1967). Renal disease of the rat. *In* "Pathology of Laboratory Rats and Mice" (E. Cotchin and F. J. C. Roe, eds.), pp. 105-147. Blackwell, Oxford.

191. Squire, R. A., Goodman, D. G., Valerio, M. G., Frederickson, T., Levitt, M. H., Strandberg, J. D., Lingeman, C. H., Dawe, C. J., and Harshbarger, J. (1978). "Tumors in Pathology of Laboratory Animals" (K. Bernewski, F. M. Garner, and T. C. Jones, eds.), pp. 1051-1262. Springer-Verlag, Berlin and New York.

192. Squire, R. A., and Levitt, M. H. (1975). Report of a workshop on classification of specific hepatocellular lesions in rats. *Cancer Res.* **35,** 3214-3223.

193. Stoerk, H., Guérin, M., and Guérin, P. (1946). Tumeurs lymphoepitheliales du thymus chez le rat. *Bull. Assoc. Fr. Cancer* **33,** 141-151.

194. Stula, E. F. (1975). Naturally occurring pulmonary tumors of epithelial origin in Charles River—CD rats. *Bull. Soc. Pharm. Environ. Pathol.* **3,** 3-11.

195. Stupfel, M., Romary, F., Magnier, M., Tran, M. H., and Moutet, J. P. (1973). Life long exposure of SPF rats to automotive exhaust gas. *Arch. Environ. Health* **26,** 264-269.

196. Stux, M., Thompson, B., Isler, H., and Leblond, C. P. (1961). The "light cells" of the thyroid gland in the rat. *Endocrinology* **68,** 292-308.

196a. Sumi, N., Stavrou, D., Frohberg, H., and Jochmann, G. (1976). The incidence of spontaneous tumors of the CNS of Wistar rats. *Arch. Toxicol.* **35,** 1-13.

197. Swaen, G. J. V., and Becker, A. E. (1968). Granular-cell myoblastoma of the pituitary gland in an old rat. A case report. *Lab. Anim.* **2,** 41-43.

198. Swaen, G. J. V., and Van Heerde, P. (1973). Tumours of the haematopoietic system. *In* "Pathology of Tumours in Laboratory Animals" (V. S. Turusov, ed.), IARC Sci. Publ. No. 5, Vol. 1, Part 1, pp. 185-214. IARC, Lyon.

199. Takizawa, S., and Miyamoto, M. (1976). Observations on spontaneous tumors in Wistar Furth strain rats. *Hiroshima J. Med. Sci.* **25,** 89-98.

200. Tannenbaum, A., and Maltoni, C. (1962). Neoplastic response of various tissues to the administration of urethane. *Cancer Res.* **22,** 1105-1112.

201. Thompson, S. W., and Hunt, R. D. (1963). Spontaneous tumors in the Sprague-Dawley rat and incidence rates of some types of neoplasms as determined by serial section versus single section techniques. *Ann. N.Y. Acad. Sci.* **108,** 832-845.

202. Thompson, S. W., Huseby, R. A., Fox, M. A., Davis, C. L., and Hunt, R. D. (1961). Spontaneous tumors in the Sprague-Dawley rat. *J. Natl. Cancer Inst.* **27,** 1037-1057.

203. Tilov, T., Kollmer, H., Weisse, I., and Stotzer, H. (1976). Spontan auftretende Tumoren des Rattenstammes Chbb:Thom (SPF) (Occurrence of spontaneous tumours in the rat strain Chbb:THOM(SPF)). *Arzneim.-Forsch.* **21,** 45-50.

204. Todd, G. C., Pierce, E. C., and Clevinger, W. G. (1970). Ganglioneuroma of the adrenal medulla in rats: A report of 3 cases. *Pathol. Vet.* **7,** 139-144.

205. Trevino, G. S., and Nessmith, W. B. (1972). Aortic body tumor in a white rat. *Vet. Pathol.* **9,** 243.

206. Ueberberg, H. (1968). Uber ein spontanes fibromyxom der haut bei des ratte (spontaneous fibromyxma of the skin in a rat). *Z. Versuchstierkd.* **10,** 222-225.

207. Walpole, A. L., Roberts, D. C., Rose, F. C., Hendry, J. A., and Homer, R. F. (1954). Cytotoxic agents. IV. The carcinogenic actions of some monofunctional ethyleneimine derivatives. *Br. J. Pharmacol. Chemother.* **9,** 306-323.

208. Wells, G. A. H. (1971). Mucinous carcinoma of the ileum in the rat. *J. Pathol.* **103,** 271-275.

209. White, F. R. (1961). The relationship between underfeeding and tumor formation, transplantation, and growth in rats and mice. *Cancer Res.* **21,** 281-290.

209a. Wilens, S. L., and Sproul, E. E. (1936). Spontaneous leukemia in the rat. *Am. J. Pathol.* **12,** 249-258.

210. Williams, E. D. (1966). Histogenesis of medullary carcinoma of the thyroid. *J. Clin. Pathol.* **19,** 114-118.

211. Willis, R. A. (1935). Carcinoma of the intestine in rats. *J. Pathol. Bacteriol.* **40,** 187-188.

212. Wilson, R. H., De Eds, F., and Cox, A. J., Jr. (1941). The toxicity and carcinogenicity activity of 2-acetaminofluorene. *Cancer Res.* **1,** 595-608.

213. Wolfe, J. M., Bryan, W. R., and Wright, A. W. (1938). Histologic observations on the anterior pituitaries of old rats with particular reference to the spontaneous appearance of pituitary adenomata. *Cancer* **34,** 352-372.

214. Yamada, S., Masuko, K., Ito, M., and Nagayo, T. (1973). Spontaneous thymoma in Buffalo rats. *Gann* **64,** 287-291.

215. Yang, Y. H., and Grice, H. C. (1965). Mucinous bronchiolar carcinoma of the rat lung: A case report. *Can. J. Comp. Med. Vet. Sci.* **29,** 15-17.

216. Yeakel, E. H. (1947). Medullary hyperplasia of the adrenal gland in aged Wistar albino and gray Norway rats. *Arch. Pathol.* **44,** 71-77.

217. Young, S., and Hallowes, R. C. (1973). Tumours of the mammary gland. *In* "Pathology of Tumours in Laboratory Animals" (V. S. Turusov, ed.), IARC Sci. Publ. No. 5, Vol. 1, Part 1, pp. 31-74. IARC, Lyon.

Chapter 14

Lesions Associated with Aging

Miriam R. Anver and Bennett J. Cohen

Old age is a hospital that takes in all diseases.

German proverb

THE LABORATORY RAT, VOL. I

I. INTRODUCTION

Whether or not there is a unitary concept underlying the aging process, the cause of death of an aged animal rarely is attributed to an entity identifiable as old age per se. Rather, as in the proverb above, old animals—man, rats, and others—develop a multiplicity of nonneoplastic and neoplastic lesions. Such lesions, singly or in combination, often are severe enough to produce mortality. Differentiation between spontaneous age-associated lesions and true lesions of aging is a challenge to researchers in experimental gerontology. The purpose of this chapter is not to determine whether such differentiation is indeed possible but instead to describe major age-associated lesions in laboratory *Rattus norvegicus*. Many of these lesions are common to all rat strains and stocks although their incidence and time of onset vary.

In theory, interpretation of age-associated lesions in rats should be easier than in human or pet animal populations because of tighter control of environmental influences. However, upon closer examination of the laboratory environment, it is possible to identify a large number of variables, many of which have effects on rat lesion development. Such items as husbandry, intercurrent infectious disease, nutrition, breeding status, and genetic background of rats should always be considered in interpretation of one's own data or when trying to make correlations between studies from different laboratories. Some of these variables are further discussed in Volume II, Chapter 7, and in Section III of this chapter.

II. MAJOR AGE-ASSOCIATED LESIONS

By the time an aged conventional or specific pathogen-free (SPF) rat is submitted for necropsy examination, it usually has major lesions in several organ systems. Prominent among these are neoplasms which are often multiple (see Chapter 13). The most common nonneoplastic lesions associated with aging are chronic nephritis (nephrosis), polyarteritis (periarteritis) nodosa, myocardial degeneration, skeletal muscle degeneration and radiculoneuropathy (8,10,12,16,17,19,25,43,43a,103,110). The possible relationship of these lesions to nutrition will be discussed in Section III.

A. Chronic Renal Disease

Chronic renal disease is a significant cause of morbidity and mortality in many rat strains and stocks. The condition has many synonyms: chronic nephritis, chronic nephrosis, chronic progressive nephrosis, glomerulosclerosis, glomerulonephritis, and nephropathy (4,8,26,32,36,43a,44,61,103). The syn-

drome is most common in rats fed a commercial laboratory pelleted diet *ad libitum*. In many rat stocks, there is a significant correlation between chronic renal disease and nutrition (see Section III). Renal disease in rats has been comprehensively reviewed by Gray (43a).

Kidney lesions have been described in rats of Sprague-Dawley (SD) (8,16,36,85), Fischer (F344) (25), Wistar (WI) (4,32,52,83,103), BD 1 (WI derived) (61), Marshall 520 (103), and other lineages. It is of interest that WAG/Rij (WI derived) and BN/Bi/Rij inbred rats fed commercial rodent diets (protein concentration 22–25%) *ad libitum* in The Netherlands do not have significant renal disease (12,18). Similar absences of the disease were reported by Snell (103) for Osborne-Mendel (OM) and Buffalo (BUF) rats at the U.S. National Cancer Institute. Conventional, SPF and germfree (GF) rats have developed chronic renal disease.

Gross and light microscopic renal lesions have been well characterized by several workers (4,8,36,43a,103). End-stage kidneys, as seen in rats greater than 1 year of age, are bilaterally enlarged, nodular or granular, yellow, pitted, and contain multiple cortical cysts which may be as large as 3 mm diameter (Fig. 1). Histologically, the glomeruli have thickened basement membranes in the capillary tufts and Bowman's capsules, eosinophilic deposits, adhesions between the parietal and visceral layers of Bowman's capsule, and partial to complete hyalinization or sclerosis. Tubules in the cortex and medulla are dilated to cystic with flattened epithelium and proteinaceous casts in the lumens. Other tubules may be collapsed and atrophic. The cortical interstitium is fibrotic and infiltrated by lymphocytes and other mononuclear cells (Fig. 2).

Although chronic renal disease can cause clinical illness and death in older rats, its pathogenesis starts at an early age with the initial lesions in glomeruli. In rats 3–6 months old, these lesions consist of an eosinophilic PAS-positive thickening of the glomerular mesangial matrix (8,16,26,52,61). Only scattered glomeruli are affected initially.

By the time rats reach 12–14 months, a certain proportion of them (about 50%) have proteinuria > 20 mg/dl (8,26,44). On electrophoresis of urinary protein, the principal early proteins in Berg's (8) SD-derived rats were α_2- and β-globulins. Older rats excreted principally albumin. Couser and Stilmant (26) found albumin excreted primarily by 3- and 6-month-old Crl:CD(SD)BR rats followed by a "progressive decrease in selectivity," until by age 24 months all classes of serum protein were present in the urine. In any case, protein excretion in the urine has been reported to rise from < 5 mg/dl in rats of weaning to 6 month of age to > 20 mg/dl in a majority of rats 24 months and older. Following the onset of proteinuria, glomerulosclerosis develops (8,26,27,44). Initially, glomerulosclerosis is focal and segmental within glomeruli, later becoming more diffuse with partial to total obliteration of glomerular tufts (7,26,44,52,61).

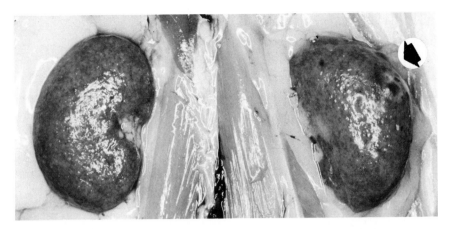

Fig. 1. Kidney with diffuse chronic renal disease. Roughened, nodular capsular surface and multiple cortical cysts (arrow). Crl:CD(SD)BR, 30 months.

On electron microscopy, the primary lesion is thickening of the glomerular basement membrane (26,44,52,61), ranging from two to three times normal at 18 months (44) to as much as five to fifteen times normal at 24 months (26) (Fig. 3). There also is focal membrane wrinkling and collapse (26,61). Some segments of the glomerular basement membrane contain linear areas of electron density within the membrane but not on either side (26). Epithelial podocytes have fusion of the foot processes and may, in more advanced disease, contain dense granular material, lipids, vacuoles, and protein droplets (26,44,61).

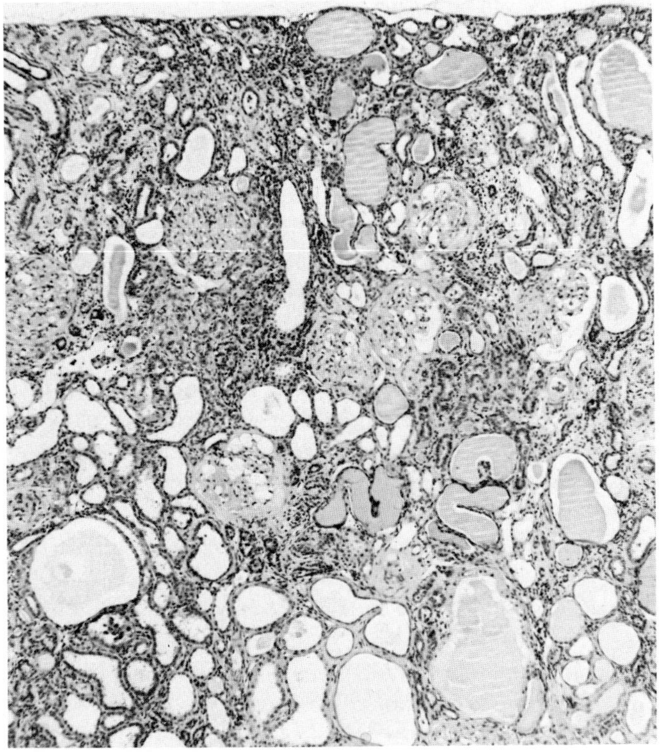

Lesions progress in glomerular tufts until there is narrowing or total obliteration of capillary lumens with the visceral epithelium adhering sclerotic capillary loops to the parietal layer of Bowman's capsular epithelium. Mesangial matrix increases, and electron-dense, finely granular material containing lipid is present on the subendothelial side of the basement membrane (26).

In the proximal tubular epithelium of 10-month-old SD rats with glomerulosclerosis, Gray *et al.* (44) described ultrastructural lesions of atrophy, increase in electron-dense cytoplasmic bodies, granular cytoplasmic material, and cytoplasmic degradation. Hirokawa (52) theorized that renal tubular and interstitial lesions in his WI rats were secondary to and caused by the excessive leakage of protein through diseased glomeruli.

Immunofluorescence studies on aging rat kidneys detected the presence of immunoglobulins in glomeruli (26,27,52). In their study of Crl:CD(SD)BR rats, Couser and Stilmant (26,27) found IgM in a coarse granular or diffuse pattern. Small amounts of IgM were present in the mesangium of 50% of glomeruli of rats 3 and 6 months old. Nonproteinuric 12-month-old rats resembled the younger animals. Proteinuric 12- and 24-month-old rats had a marked increase in mesangial IgM deposits involving the entire stalk of most glomeruli (Fig. 4). The IgM deposits did not fix complement. There were minimal to no IgG deposits and nonspecific generalized staining for other immunoglobulins. Complement (C3) was deposited along tubular basement membranes but not in glomeruli. Bolton *et al.* (11a) had similar findings in SD rats but could not correlate the quantity of deposited IgM with proteinuria.

WI rats studied by Hirokawa (52) showed with aging a progressive increase of IgG deposition in the glomerular basement membrane and mesangium. Some of the glomeruli also had

Fig. 2. Kidney with diffuse chronic renal disease. Numerous sclerotic glomeruli, dilated tubules with proteinaceous casts and interstitial fibrosis Hematoxylin and eosin. ×40. Crl:CD(SD)BR, 30 months.

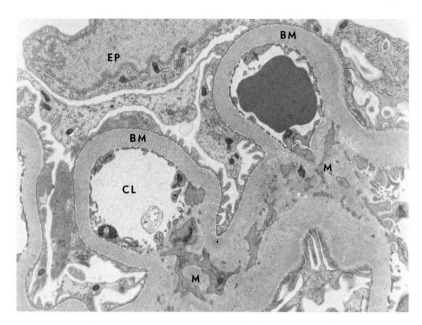

Fig. 3. Electron micrograph of glomerular lesions from rat with proteinuria. Marked diffuse increase in thickness of basement membrane (BM) and focal epithelial cell (EP) foot process fusion. No electron dense deposits within the basement membrane. CL, capillary lumen; M, mesangium. ×9700. [From Couser and Stilmant (26).] Crl:CD(SD)BR, 24 months.

Moloney leukemia virus antigen, but it was associated only with a portion of the immune complex deposits.

Couser and Stilmant (26,27) and Hirokawa (52) agreed that glomerulosclerosis of aging rats was not an age-related autoimmune phenomenon. Couser and Stilmant (26) did not detect circulating IgM or IgG antibodies to renal or other tissue antigens either in serum or eluates of kidneys from aged rats with prominent glomerular lesions. Antigens within the immune complex were thought to be of exogenous origin, vari-

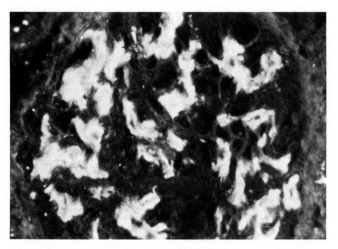

Fig. 4. Kidney of proteinuric rat. Immunofluorescent micrograph of glomerulus with diffuse mesangial IgM deposition. No basement membrane staining is present. ×475. [From Couser and Stilmant (26).] Crl:CD(SD)BR, 12 months.

able in type, and present in small amounts (26,27,52). Complex deposition in the mesangium was theorized to represent increased mesangial phagocytosis rather than a primary insult, with possible secondary contribution to glomerulosclerosis either by sheer volume or by reducing the mesangial phagocytic capability. Bolton *et al.* (11a) suggest that there is indirect evidence for participation of the cellular immune system in the pathogenesis of glomerulosclerosis.

Prolactin also has been implicated as an important factor in the etiology of chronic renal disease. Male OFA (Sandoz,Ltd) SPF rats given exogenous prolactin had more severe renal disease than controls. Another experimental group given bromocriptin, a compound which inhibits endogenous prolactin, had a significantly lower incidence of renal disease, decrease in relative and absolute kidney weight, and less severe lesions than untreated controls. Prolactin is theorized to lower urinary pH and decrease urine volume, thereby decreasing the solubility of urinary protein. In proteinuric rats, formation of protein casts in the ducts is thus facilitated (88).

Sequelae of chronic renal disease in rats are variable, depending on age and strain. In affected rats older than 12 months, any or all of the following can occur: elevated blood urea nitrogen (BUN) and creatinine, polydypsia, lowered serum albumin/globulin ratio, hypercholesterolemia, hypertension, fibrous osteodystrophy, hydrothorax, and ascites. In female WI rats 30 months old, glomerular filtration rates were depressed by 25% compared with those of 12- and 24-month-old animals (45). Most severe sequelae of chronic renal disease are in rats older than 24 months (8,26,36,44,52,61,103).

B. Myocardial Degeneration and Fibrosis

Aged conventional and SPF rat strains and stocks, but not germfree rats (83), develop myocardial degeneration and fibrosis (8,12,25,33,108,120). The incidence is generally higher in males than females. Onset is usually after 1 year of age with females developing lesions later in life than males (8). Maximum incidences in both sexes can be as high as 60–80% (8,12) with the most striking increase occurring after 18 months (33).

Microscopic lesions consist of myocardial atrophy, degeneration, and necrosis; condensation fibrosis of the stroma; and a variable inflammatory infiltrate of lymphocytes, Anitschkow cells, and macrophages (Fig. 5) (8,22,25,33,120). Most frequently involved are the papillary muscles and their attachment sites in the wall of the left ventricle (8,12). Other portions of myocardium which may be affected are the heart base, interventricular septum, and papillary muscle attachments in the right ventricle, apex of the heart, and adjacent to the coronary arteries (8,12). Grossly, these lesions are not obvious except for very advanced cases in which the myocardium appears white (8).

In two studies involving F344 rats or an unspecified rat strain or stock, myocardial lesions in aged animals seemed secondary to chronic interstitial myocarditis of young rats, an inflammatory lesion with an incidence greatest in rats younger than 4–6 months of age (25,33). Other workers have not implicated myocarditis as a precursor to myocardial degeneration and fibrosis but instead suggest the lesions result from chronic ischemia caused by coronary arteriosclerosis (33).

Ultrastructurally, marked changes occur in aged rat myocardium with the most severe lesions after 18 months of age. There are accumulations of residual bodies in mitochondria at myocardial nuclear poles (visible on light microscopy as lipofuscin pigment), increased numbers of lysosomes, lysosomal degradation of mitochondria, increased numbers of glycogen granules and membrane-bound lipid droplets in the sarcoplasm, and the presence of collagen fibers. Fibrosis correlates well with histochemical findings of increased hydroxyproline content in the left ventricle of aged rats (107,108).

Clinically, aged rats do not have signs of cardiac insufficiency. Heart size relative to body weight increases with age (9), principally due to left ventricular hypertrophy (8,64). WI rats at 24 months of age had little difference in cardiac performance compared to 6- and 12-month-old rats (64). Aged SD-derived rats of Berg (6) had minor electrocardiogram alterations.

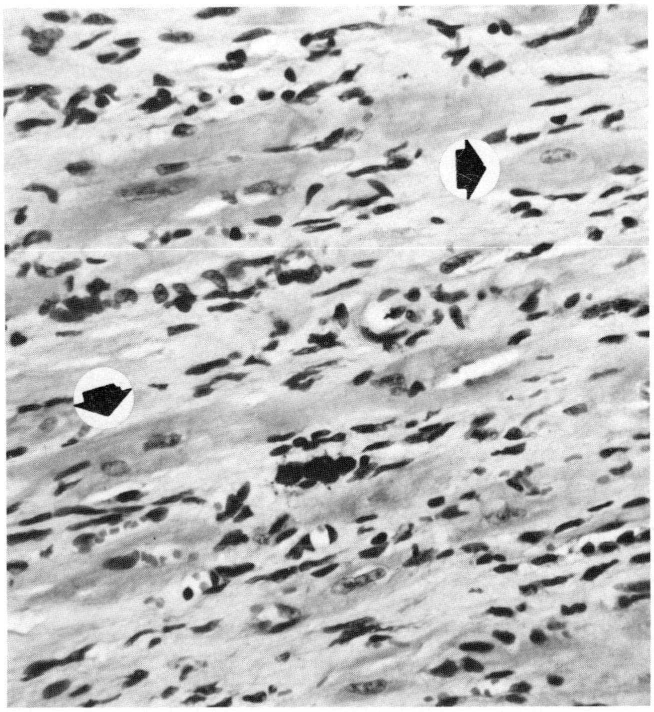

C. Radiculoneuropathy

Spinal nerve root degeneration (radiculoneuropathy, degenerative myelopathy) has a late onset, with lesions most extensive in rats greater than 24 months of age (10,43). The incidence in rats older than 24 months is generally high, ranging from 75 to 90% (10,43). Berg *et al.* (10) reported that SD-derived male rats developed lesions earlier than females (500 days as compared to 700 days), whereas Gilmore's (43) study on Crl:CD(SD)BR rats did not reveal sex differences in time of onset.

Lesions have been reported in the following strains or stocks: Crl:CD(SD)BR, SD derived, WI SPF, BN/Bi/Rij, WAG/Rij and F_1 (BN/Bi/Rij × WAG/Rij) (10,19,43,110). Differences in incidence of paralysis have occurred in two inbred strains (BN/Bi/Rij and WAG/Rij—2%) and their F_1 hybrid (21%) (19).

Clinical disease, such as disturbances in motor function, posterior paresis, and paralysis, have been correlated by investigators (19,110) with skeletal muscle atrophy and degeneration. Others state that the two conditions are unrelated and that the nervous lesions are clinically silent (10,43,46) (also see Section II, D).

Histologic lesions are present in the cauda equina, ventral spinal nerve roots, white matter of the spinal cord (especially the ventral and lateral funiculi and sciatic and brachial plexuses), and the lower brainstem (10,19,43,110). Less often

Fig. 5. Heart with myocardial degeneration, atrophy, fibrosis, and mononuclear infiltrate. Few remaining myocardial fibers with cross striations (such as at arrows). Hematoxylin and eosin. ×200. Crl:CD(SD)BR, 35 months.

and/or less severely affected are dorsal spinal nerve roots and dorsal funiculi of the spinal cord (19). In BN/Bi/Rij and WAG/Rij rats, spinal cord lesions were most severe in the midcervical to midthoracic cord and less severe caudal to the fifth thoracic vertebra to the lumbosacral cord. Pathologic changes in ventral nerve roots, however, became more extensive caudal to T3 (19). SD-derived rats did not have cord lesions localized to specific areas (10).

In ventral spinal nerve roots, lesions consist of myelin sheath swelling, segmental demyelination, infiltration of macrophages into damaged fibers, and focal Schwann cell proliferation. Advanced lesions are characterized by axonal degeneration and loss with formation of cystic spaces containing cholesterol crystals, gitter cells, and hemosiderin (Fig. 6) (10,19,43,110). Spinal cord lesions are essentially similar; they also are usually bilaterally symmetrical (19).

The pathogenesis of radiculoneuropathy is still in question. Berg (8) states that there is no obvious link with nutrition since caloric restriction had no effect on the incidence of myelin degeneration. Burek *et al.* (19) suggested a relationship between the nerve root lesions and multiple vertebral bone and disc lesions also found in their aged rats (also see Section IV, J). Berg *et al.* (10) and Gilmore (43) did not describe vertebral bone or disc changes.

D. Skeletal Muscle Degeneration and Atrophy

Skeletal muscle degeneration of the hindlimbs of rats has an onset time similar to that of radiculoneuropathy, i.e., after 24 months of age. Clinical signs can range from difficulty in use of the hindlegs to posterior paresis, paralysis of the hindlimbs, loss of tail control, and urinary incontinence or failure to empty the urinary bladder. Animals with severe signs often have weight loss. Grossly, muscles of the hind quarters, especially the gastrocnemius and adductor, are atrophic, brownish, and flabby (8,19,110).

Microscopically, there is a decrease in the diameter of individual muscle fibers and increased prominence of sarcolemmal nuclei due to hypertrophy and/or hyperplasia (8,110). Berg (8) describes an early loss of cross-striations. Van Steenis and Kroes (110) note that striations are preserved until lesions become advanced. In the later stages, the sarcoplasm undergoes hyaline or granular degeneration and fragmentation (Fig. 7). In the final stages, only collapsed sarcolemmal sheaths remain, and there is adipose tissue replacement of damaged muscle fibers. Inflammatory cells are not present at any stage of the process, and muscle fibers are not uniformly affected (8,110).

Concerning the relationship between skeletal muscle lesions and radiculonueropathy, several workers (19,110) consider

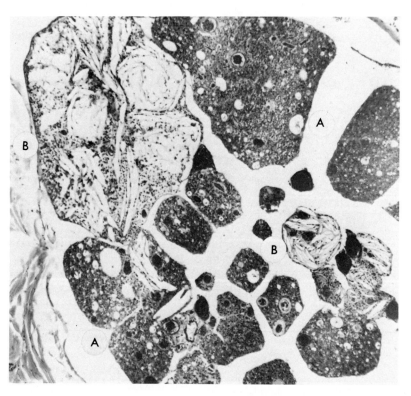

Fig. 6.   Radiculoneuropathy of the cauda equina. Early lesions (A) of myelin sheath swelling and axon degeneration. Advanced lesions (B) consist of demyelination, cholesterol clefts, axon loss, and gitter cells. Hematoxylin and eosin. ×80. Crl:CD(SD)BR, 36 months.

skeletal muscle lesions to be neurogenic atrophy secondary to nerve root and spinal cord lesions. Berg (8), on the other hand, could delay the onset of skeletal muscle lesions by caloric restriction, whereas radiculoneuropathy incidence remained unchanged. Pollard and Kajima (83) did not see any skeletal muscle lesions (examination of nerve roots and spinal cord were not reported) in WI germfree rats fed a specially compounded diet.

Ultrastructural lesions were present in motor end plates of the levator ani muscle in 30-month-old as compared to 2-month-old rats (Czechoslovakia, of an unspecified strain or stock). Affected muscle had sarcolemmal thickening, proliferation of subsarcolemmal tubules and smooth endoplasmic reticulum, increased ribosomal activity, and enlargement of the smooth endoplasmic reticulum cisternae. Disorganization and disintegration of the myofilaments with autophagic vacuoles occurred initially at the periphery with sparing of the center of muscle fibers. Lesions such as axonal degeneration and synaptic Schwann cell changes were not present. Thus, the authors felt the muscle changes were more characteristic of senile rather than neurogenic muscular atrophy (46).

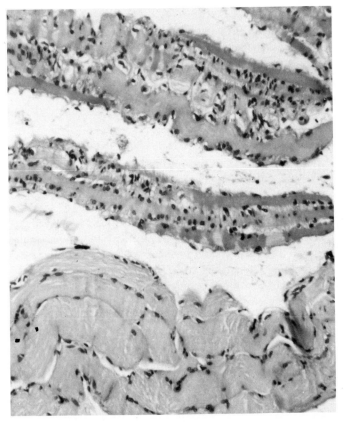

Fig. 7. Skeletal muscle atrophy and degeneration with proliferation of sarcolemmal nuclei. Less affected muscle with normal-sized fibers at bottom. Hematoxylin and eosin. ×200. Crl:CD(SD)BR, 36 months.

In 24-month-old WI and Donryu rats, the decrease in skeletal muscle mass of the hind quarters was due to a decrease in volume of white muscle fibers and a decrease in number of red muscle fibers (106).

E. Polyarteritis Nodosa

Polyarteritis (periarteritis) nodosa (PAN) is an inflammatory lesion of unknown etiology which involves muscular arteries, especially the pancreatic, mesenteric, and spermatic vessels. Hepatic, coronary, ovarian, uterine, cerebral, adrenal, and other arteries can also be affected (8,75,102). SD rats and stocks derived from them have a low incidence of renal arterial disease (8,124), while such lesions are seen more frequently in Mendel-Sherman (121), August rats (76), and other strains (102). Incidence of PAN is higher in males than females (8,124), but the condition, unlike chronic renal disease, does not reach a 100% incidence even in males.

A single artery may have lesions ranging from acute to chronic. In the acute stage, the arterial tunica media and adventitia undergo degeneration, fibrinoid necrosis, and disruption of the internal elastic lamellae accompanied by a marked infiltration of polymorphonuclear leukocytes (PMNs) and mononuclear cells (Fig. 8). Mural thrombosis occurs on the damaged intima. In subacute lesions, the inflammatory cells are PMNs, eosinophils, lymphocytes, and plasma cells. Fibrosis occurs in the adventitia and media, and there is intimal proliferation. Vessels with chronic PAN have marked fibrosis of the wall (Fig. 9) and occluding or recanalized thrombi. Some vessels have stenotic lumens while others have aneurysms filled with blood or thrombus.

Gross lesions of chronic PAN are especially obvious in the pancreatic and mesenteric vessels. Affected arteries are enlarged, thick-walled, gray to red, hard, nodular and tortuous and have aneurysmal outpouching (Fig. 10) (8,76,102,121,124).

Berg's (8) clean conventional male SD-derived rats had a 60% incidence of PAN at 900 days. Other investigators have reported a 14–17% incidence in both sexes of SD rats at average age 699 days (124) and 43% in an unspecified strain of stock of rats at average age 498 days (102). Lesions have not been reported in all strains or stocks of aged rats. Crl:CD(SD)BR rats have developed PAN (24), but the condition has not been demonstrated in germfree WI rats (83). Opie et al. (76) found an extremely low incidence of mesenteric PAN (about 7%) in conventional aged WI rats.

According to Wexler (116), the frequency of breeding during the first 12–14 months of life influences the age of onset of PAN. SD-derived males siring as many as nine consecutive litters died of other causes (myocardial infarcts, diabetes, hypertension) with only a 5% incidence of PAN. Females outlived the males and at 12–14 months of age had the following

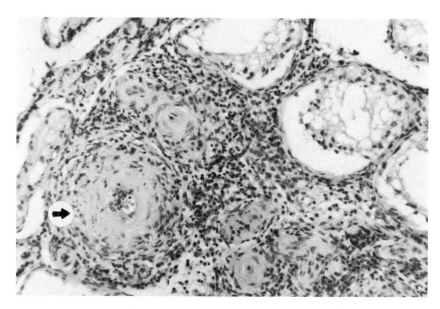

Fig. 8.   Acute polyarteritis nodosa of testicular arteries. Fibrinoid medial necrosis (arrows); medial and adventitial inflammatory cells, predominantly PMNs. Degeneration of adjacent testicular tubules. Hematoxylin and eosin. ×100. Crl:CD(SD)BR, 34 months.

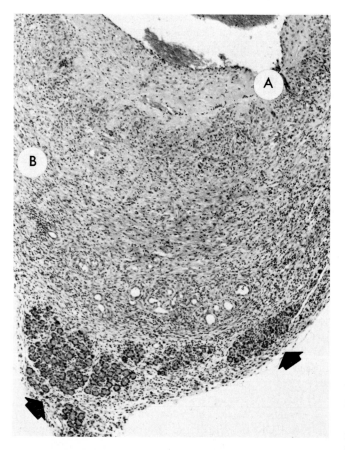

Fig. 9.   Chronic polyarteritis nodosa of pancreatic artery. Subintimal thickening and fibrinoid necrosis (A); marked medial thickening by fibrosis, PMNs, and mononuclear cells (B). Atrophic pancreatic acini in tunica adventitia (arrow). Hematoxylin and eosin. ×40. Crl:CD(SD)BR, 25 months.

incidences of PAN: 3–8% after the fourth and fifth consecutive pregnancy, 12% after the sixth, 21% after the seventh, 40% after the eighth, and 86% following the ninth. Virgin controls did not develop PAN until 30–36 months of age. Such clearcut differences may not exist in all strains of rats (see Volume II, Chapter 7).

III.   INFLUENCE OF DIET ON LONGEVITY AND AGE-ASSOCIATED LESIONS

The interrelationship of nutrition and aging is complex. There are no simple answers to questions such as what and how much old rats should be fed. It is difficult to compare nutritional studies even when "identical" diets are used because of the complex interactions among dietary components and because ingredients of the commercial and semisynthetic diets commonly used in nutritional research can vary from batch to batch. These considerations indicate that the results of feeding studies should be interpreted cautiously and conservatively particularly in relation to nutritional effects on longevity and the incidence of lesions. Despite this cautionary note, nutrition is an essential element in the life history of rats intended for use in gerontologic research and must be controlled adequately. In this section, the effects of selected nutritional factors on age-related lesions and longevity are considered. The role of nutrition in experimental murine pathology is discussed more fully in this volume, Chapter 6.

Fig. 10. Chronic polyarteritis nodosa of mesenteric arteries. Segmental arterial enlargement, thickening, tortuosity, and aneurysmal dilatation. Multiple discrete gray nodules (arrows) are implanted renal carcinoma. Small intestine at A. Crl:CD(SD)BR, 29 months.

A. Nutrition and Obesity

Most laboratory rats are fed commercially formulated pelleted rations *ad libitum* throughout life. The nutrient composition of the rations is based largely on National Research Council specifications of nutrient requirements for the rat (73). These specifications promote rapid early growth, vigor, and reproduction. However, a diet that emphasizes growth criteria in its composition may be unsatisfactory in respect to the late development of diseases and to longevity. For example, it may promote obesity and reduce longevity in some rat stocks (74). The Crl:CD(SD)BR outbred male rat can weigh as much as 900–1000 gm and exhibit heavy concentrations of abdominal fat by 20–24 months of age or older when fed a standard commercial pasteurized diet *ad libitum* (22). On the other hand, the F344 inbred rat fed the same diet tends to plateau in weight after about 12 months of age, remains relatively stable thereafter, and does not become obese (25). Nevertheless, obesity and, as shown in Section III, B and C, reduced longevity are possible consequences of feeding standard commercial diets *ad libitum* to some stocks or strains of aging rats. Dietary modification, caloric restriction, or both may be required if obesity is a problem in a specific aging research project.

B. Caloric Intake

The relationship of caloric intake to longevity has been the subject of numerous investigations since McKay *et al.* (68,69) first demonstrated increased life expectancy in rats restricted in growth throughout postweaning life by having been fed diets complete except for calories. More recent studies on the increase or decrease in life expectancy associated with dietary alterations are summarized in Table I. They confirm the earlier studies showing a direct relationship between caloric intake and life expectancy.

Caloric restriction also has been shown to reduce the incidence of certain lesions in aging rats, or to delay their onset. For example, onset of myocardial degeneration in SD rats was delayed significantly, and the incidence at 800 days was significantly lower when food intake was restricted throughout life by 33% or 46% compared with the *ad libitum* intake of age matched controls (101). Caloric restriction between 21 and 70 days of age in Crl:CD(SD) rats resulted in a reduction in tumor incidence of 14–83% depending on the tumor type (91). This reduction occurred even though followed by *ad libitum* feeding which normally is conducive to a relatively high incidence of spontaneous neoplasms (92). Life-long restriction of the intake of isocaloric semisynthetic diets that varied in the percentage composition of protein, carbohydrate, and fat resulted in a significant reduction in chronic nephritis (15). Nolen (74) did not find great differences in the overall incidence of age-related diseases between restricted and *ad libitum*-fed rats, but lesions seemed to appear later in life in the restricted rats.

C. Dietary Composition

There is strong evidence that the composition of a diet, as well as caloric intake, can influence the incidence and severity

Table I

Effects of Dietary Alterations on Longevity

Rat stock	Type of diet	Feeding regimen	Longevity compared with controls	Reference	Remarks
Random-bred Norway	Natural ingredients	Access to food limited to 6 hr/day	158 days longer	30	Increased longevity attributed to greater physical activity of underfed rats
Simonson (SD)	Semisynthetic	60–80% of *ad libitum* intake of controls	170 days longer	74	
Crl:CD(SD)	Semisynthetic	Caloric restriction 21–70 days of age	Restricted rats lived longer	90	44% of the restricted rats were alive after the last *ad libitum*-fed rat died at 924 days
Crl:CD(SD)	Commercial laboratory chow	60–70% of *ad libitum* intake of controls	Restricted rats lived longer	89	Mortality risk was greater in *ad libitum*-fed rats at 300 days and older than in restricted rats
Crl:CD(SD)	Semisynthetic	Caloric restriction 21–70 days	Up to 500 days longer	91	
Crl:CD(SD)	Semisynthetic	Self-selected intake	Longevity greater in rats with low intake	93,94	
Crl:CD(SD)	High caloric	20% fat versus 5% or 10% fat	20% fat diet was life shortening	50	Decreased life expectancy was related to increased body weight. Nature of dietary fat was not a factor

of lesions and longevity. Berg (8) and Hirokawa (52) have shown that rats fed commercial laboratory diets *ad libitum* can develop light or electron microscopic evidence of chronic nephritis as early as 60 days of age, although there are no gross lesions at this time. Male Crl:CD(SD) rats maintained throughout postweaning life on a commercial rat diet exhibited severe chronic nephritis at 795 days of age; rats fed semisynthetic diets had a lower incidence and a delayed appearance of the disease (16). This led Bras and Ross (16) to the concept that the dietary regimen early in life has an important bearing on chronic nephritis later in life, and that the initiating mechanisms can be altered by dietary means. It is also interesting that in this study, none of the rats fed the commercial diet survived longer than 1100 days, while substantial numbers of rats fed semisynthetic diets lived 1500 days or more. Pollard (81) retrospectively compared the incidence of chronic nephritis in two groups of germfree WI rats. Six of 16 rats older than 24 months were found to have had nephritis in 1961, while none of another group of 41 rats were affected in 1969. The only difference between the groups was the diet, which in 1961 was prepared from semirefined components while in 1969 it contained natural unprocessed grains (58).

The protein content or intake of the diet appears to be an important factor in the incidence of chronic nephritis (15,62,63,109). In general, it appears that a high level of dietary protein is associated with a high incidence of chronic nephritis. For example, the index of chronic nephritis was 36, 81 and 100%, respectively, in groups of 250 Crl:CD(SD) male

rats fed isocaloric semisynthetic diets containing 10, 22, or 51% vitamin-free casein as the protein source. Furthermore, the disease was more severe in rats fed the highest level of protein (15). The glomerular immunoglobulins discussed in Section II,A do not, however, appear to be related to diet, since Hirokawa (52) was unable to demonstrate serum antibodies against diet pellets in old WI rats with age-associated kidney disease.

Both protein under- or overnutrition has been shown to modify spontaneous neoplasm incidence in Crl:CD(SD) rats (92,95). However, the interactions among dietary intake, protein–calorie ratio, tumor incidence, and tumor type are complex. No single diet, when fed throughout postweaning life, uniformly reduces or increases the prevalence of every type of neoplasm.

Finally, longevity is influenced by the level of dietary protein and the age at which a given level is ingested. "Hooded" female rats fed a low (4%) protein diet *ad libitum* starting at 4 months of age, after having been fed a diet containing 12% biologically useful protein up to that time, lived an average of 980 days compared with 763 days for a control group that was fed the 12% protein diet throughout life (71). Whether caloric intake was identical for the two groups was not specified, but the results imply that life expectancy can be extended by reducing protein intake during adult life after early full feeding. This implication is borne out in recent experiments by Ross and Bras (93,94) involving the effects of dietary self-selection in Crl:CD(SD) male rats. Three complete isocaloric diets were provided, differing only in their content of casein (10%, 22%, 51%) and sucrose (30%, 59%, 70%). Other rats were fed one of the diets *ad libitum* as a fixed regimen. Complex interactions were demonstrated among the levels of dietary protein, carbohydrate, total food intake, and the age of the rats during which significant correlations occurred. With respect to dietary protein, self-selection of a low protein diet early in life resulted in a shorter life span than did self-selection of a high protein diet. The pattern was reversed in older rats, where self-selection of a low protein diet resulted in greater longevity.

D. Interpretive Comments

It is evident that the nutritional maintenance of aging rats to promote longevity and minimize the incidence of obesity, chronic nephritis, neoplasms, and other age-associated lesions is a complex problem. The standard practice of feeding commercially available pelleted rations *ad libitum* is convenient and may be adequate for some aging studies. However, this feeding method is not conducive to the greatest longevity or freedom from chronic diseases. Currently available information suggests that life-long control of nutritional intake and dietary composition is desirable for rats to be used in gerontologic studies. No single "aging diet" is likely to meet the needs of all investigators, but various combinations of protein and caloric restriction, starting after rats have peaked in growth, may provide the most satisfactory approach to long-term nutritional maintenance. Additional research is needed to establish how nutritional factors affect the potential for longevity in the rat and its susceptibility to age-associated lesions.

IV. OTHER AGE-ASSOCIATED LESIONS

A. Cardiovascular System

1. Heart

Most cardiac lesions of rats seem to be age-associated. They generally appear in rats greater than 12 months old and their incidence increases strikingly after 18 months of age (33). The most prominent morphologic change, myocardial degeneration and fibrosis, is discussed in Section II, B.

Intracardiac thrombi are found occasionally. They are most common in the left atrium but also have been reported in the right atrium and left ventricle. The thrombi are firm, gray-red, and laminated. Older thrombi are white and smooth, organized by fibrous connective tissue which extends through the endocardium and underlying myocardium (8,33). There is no evidence of concomitant bacterial infection.

Valvular endocardiosis (thickening of the heart valves by myxomatous connective tissue) is present in aged rats (33). It apparently causes no clinical signs and is essentially similar to the lesion commonly seen in aged dogs.

Endocardial and subendocardial proliferative lesions have been reported by several investigators. The most comprehensive study was by Boorman *et al.* (14). They described 40 cases in two rat strains (WAG/Rij and BN/Bi/Rij) and one outbred WI derived stock (CIVO). The condition usually appeared between 25 and 30 months of age. Early lesions consisted of subendocardial undifferentiated mesenchymal cells (Fig. 11) and scattered lymphocytes. In more advanced stages, there were deep subendocardial accumulations of fibroblasts and collagen. Mononuclear cells and nuclear debris were present on the endocardial surface. Spindle cells extended into the myocardium, which had focal degenerative changes. Mitotic figures also were present within the reaction which in six cases, became so extensive that large tumorlike masses filled some of the heart chambers. The left ventricle was most often affected. Similar histologic lesions diagnosed as cardiac fibromas were found in four ACI/N rats (67). Chronic auriculitis described by Fairweather (33) also may be related as is en-

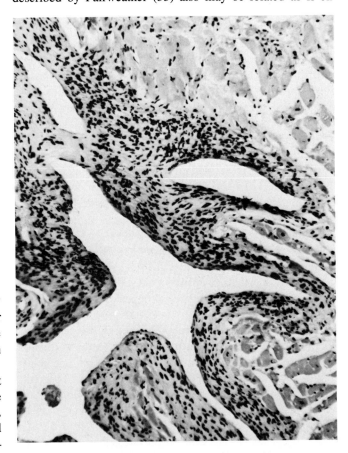

Fig. 11. Heart (ventricle) with subendocardial proliferation of spindle-shaped cells. Hematoxylin and eosin. ×100. Crl:CD(SD)BR, 34 months.

docardial fibrosis seen by Wexler (115).

Heart weight relative to body weight of the rat increases with advancing age. The ratios are highest in females who also show an increase in heart size in association with hypertension (9).

2. Vessels

Polyarteritis nodosa is discussed in Section III,E. Arteriosclerosis occurs in the aorta, carotid, and cerebral arteries of retired breeders. Whether this is an age-associated lesion is questionable. There appears to be a stronger correlation with breeding frequency since intensively bred rats (siring or whelping 5–6 litters in a 9–12 month period) have a high incidence of arterial lesions when examined at the end of their reproductive life. Such lesions are rare in virgin rats or occur much later in life, e.g., 36 months as compared to 12. This association is similar to the incidence and onset time of polyarteritis nodosa reported by Wexler (116). Investigations on retired breeder rats have been reviewed by Fairweather (33).

Aortic and carotid lesions most resemble Mönkeberg's medial sclerosis. Aortic tunica media may have any combination of the following: basophilic degeneration, accumulations of mucopolysaccharides, necrosis, marked calcification, and cartilagenous and osseous metaplasia. Grossly, severely calcified aortas are white, rigid, and have a "bamboo stick" appearance with irregularities in shape. Carotid lesions are not obvious grossly and are greatly attenuated in the cranial portions of the arteries. Microscopically, they resemble aortic changes. Cerebral arterial lesions are even less severe, less prevalent, and do not lead to any central nervous system clinical disease. The majority of studies of arteriosclerosis of the large elastic arteries were on SD or SD-derived rats fed commercial diets (33,42,54,87,119,122). Electron micrographs of the aortic tunica interna of breeder rats (albinos from Australia, strain or stock unspecified) showed intimal changes compatible with early atherosclerosis (40).

Arteriosclerosis, including atherosclerosis, also occurs in coronary arteries of many strains and stocks of aged rats. Similar to vascular pathology already reviewed, lesions are exacerbated and time of onset reduced in repeatedly bred rats. The distribution of coronary arterial lesions plus strain and sex differences are comprehensively discussed by Wexler (115).

Coronary arteries have both medial and intimal changes. Medial lesions consist of smooth muscle edema and hypertrophy sometimes causing stenosis of the lumen, basophilic degeneration, and mucopolysaccharide vacuolization. Intimal lesions are internal elastic membrane disruption, endothelial hyperplasia, and subintimal mucopolysaccharide accumulations. These areas can form plaques which extend into the lumen. Atheromas with lipids and cholesterol clefts, as seen in humans, do not form. As well as myocardial degeneration and

fibrosis seen, however, by numerous workers in the absence of significant coronary arterial disease, acute subendocardial infarctlike lesions have been linked with coronary arteriosclerosis (115).

Spontaneous hypertension occurs not only in rats specifically bred to serve as a model of this human disease (55) but also is a spontaneous disease of aged rats. Berg (8) found hypertension in 40% of SD-derived rats older than 800 days. In another study, Berg and Harmison (9) found that the incidence of hypertension was greater in males than females and increased with advancing age. Rapp (85) states that there is a relationship between chronic renal disease and systolic arterial blood pressure and also between polyarteritis nodosa and blood pressure. However, renal arterial lesions seldom occur in rats despite numerous parenchymal lesions in the kidney. Willgram and Ingle (122) provide one of the few descriptions of atherosclerosis in renal arteries of aging female retired breeders similar to that seen in the coronaries.

B. Genital System

1. Male

a. Testes. The principal lesions in the testes of old rats are seminiferous tubular degeneration (Fig. 4), interstitial edema, and neoplasms, especially interstitial (Leydig) cell tumors (22,25,57). Polyarteritis nodosa of testicular arteries (Fig. 8) also is seen commonly in some rat stocks (8,22) (see Section II,E). Morphological changes in testes were studied in 24 to 26-month-old FW 49 Biberach albino rats that had been reared under SPF conditions and fed a fortified steam-sterilized pelleted ration *ad libitum* (66). Testes weights ranged from 1.5 to 13.5 gm. The testes were soft and light yellow, and many contained amber-colored fluid. Microscopically, 24 of 90 rats exhibited degenerative or atrophic seminiferous tubular changes, while 70 of 90 had interstitial cell tumors. In some instances, degeneration of tubular epithelium appeared secondary to interstitial cell tumors. Giant cells or bizarre-shaped cells also were present in the tubules. Giant cells also can be seen in testes of some rat stocks unassociated with neoplasia (24). Dystrophic calcification may occur in degenerating tubules.

b. Prostate. Small eosinophilic to basophilic concretions are a common microscopic finding in acinar lumens of prostate from male rats over 12 months old (Fig. 12) (22). Their significance is not clear. Acute or chronic prostatitis occurs commonly, but usually is not extensive. Occasionally, however, abscess formation ensues (22,25). Ross and Bras (94) reported an increase in prostatitis in aging Crl:CD(SD) rats whose dietary intake was self-selected compared with rats on fixed dietary regimens. Prostatic hyperplasia, consisting largely of

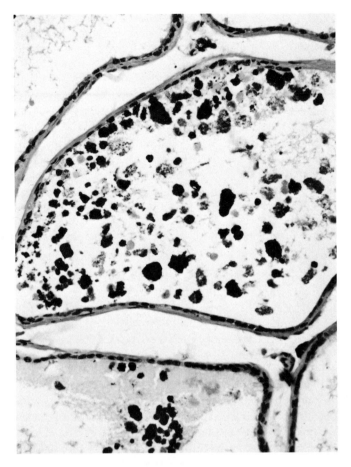

Fig. 12. Basophilic concretions in acinar lumens of prostate. Hematoxylin and eosin. ×250. Crl:CD(SD)BR, 34 months.

stratification of acinar epithelium, also has been observed in Crl:CD(SD)BR rats 24 months old and older (22). There was no indication that the hyperplasis was precancerous, although Pollard (82) and Shain *et al.* (99) have reported spontaneous prostatic adenocarcinomas of aged rats in which areas of glandular hypertrophy and hyperplasia were present.

c. Preputial Glands. The functional significance of the preputial glands is not clear. It has been suggested that they are a source of pheromones which influence sexual behavior of female rats (31). In mice, preputial secretion has been related to the synchronization of estrus (20).

Unilateral or bilateral acute or chronic preputial adenitis was common in rats over 12 months of age (22,31). Ducts of the gland were sometimes markedly distended with inspissated secretion and necrotic debris. Abscesses also were found. According to Franks (37), preputial gland abscesses often are subacute or chronic and show a granulomatous reaction. Giant cells can be present. Similar inflammatory lesions have not been reported in aging rats in the female counterpart of the preputial glands, the clitoral glands.

2. Female

Existing information on morphological and functional changes in the ovary of aging rats has been well summarized by Russfield (96). Meites *et al.* (70) have described a relationship among abnormal estrous cyclicity; morphologic changes in uterus, ovaries, and mammary glands; and pituitary lesions. Earlier (53), they showed that in aging female Long-Evans (LE) rats, there was a gradual reduction in the total number of ova and ovulations, and estrous cycles became irregular and finally ceased. Three major patterns were demonstrated after cessation of estrous cycles in 20-month-old or older rats: a progression from constant estrus with constant vaginal cornification to irregular pseudopregnancies to anestrus. The reason why some old rats developed pseudopregnancy after constant estrus was not clear.

In SD or LE rats which were 20 months old or older and in constant estrus, ovarian follicles were well developed or cystic but no corpora lutea were present. Uterine glands were abundant and lined by columnar epithelium, a morphologic change which Meites *et al.* (70) thought was suggestive of estrogen stimulation. Mammary glands were well developed. The pituitary gland was normal appearing but weighed up to 17 mg compared with 10 mg in 3- to 4-month-old mature females in estrus. Abundant acidophils were present. In pseudopregnant rats, ovaries contained numerous corpora lutea; uterine glands had an appearance suggestive of progesterone secretion; mammary glands were poorly developed; and pituitaries contained small tumors and hemorrhages. The anestrous state occurred most often in the oldest rats. They had atrophic ovaries, a small aglandular uterus, highly stimulated mammary glands, and, almost always, a large pituitary tumor. Meites *et al.* (70) believe that a gradual failure occurs in the hypothalamo-pituitary mechanisms controlling LH release and ovulation, and that the ovaries themselves are not the reason for cessation of estrous cycles in old female rats. However, no hypothalamic lesions have as yet been described.

As far as other changes in female reproductive organs are concerned, Wolfe *et al.* (123) reported a gradual transformation of reticulum into collagen in the stroma of the uterus, cervix, and vagina of aging rats. Hydrometra, pyometra, and cystic endometrial hyperplasia also have been described (37,123).

C. Urinary System

The most intensively studied urinary tract lesions are those involving the kidney. Other than chronic renal disease, urolithiasis has been described in the rat involving both the renal pelvis and the urinary bladder. Clinical signs may be absent. However, if obstruction occurs within the pelvis, hydronephrosis may develop. Occlusion of the neck of the bladder by uroliths can cause acute dilatation, uremia, and death.

Renal calculi are reported to be common in SD-derived and Osborne-Mendel (OM) rats (8). Calculi of the urinary bladder occur in SD-derived (8), BN/Bi/Rij (13), and ACI/N rats (67). In the last two strains, bladder cancers also were present: transitional cell papillomas, transitional cell carcinomas, and squamous cell carcinomas. There seemed to be a definite association between the presence of calculi and neoplasia. BN/Bi/Rij rats also had neoplastic lesions in the ureter. Females had more ureteral tumors, while males had a high neoplasia incidence in the bladder. Another strain (WAG/Rij) kept under identical husbandry conditions as the BN/Bi/Rij rats did not develop calculi or neoplasia. The BN/Bi/Rij rat also has a 30% incidence of congenital hydronephrosis (13,23).

In various other strains and stocks—WI (86), F344 (25,28), Crl:CD(SD)BR (22)—any type of urolithiasis is rare. Slonaker-Addis rats have unilateral (right) hydronephrosis in the absence of calculi (98).

Under conventional conditions, ascending urinary tract infections can be found more frequently in older rats. These probably reflect the duration of exposure to pathogens and probably are not age-related lesions.

D. Digestive System

1. Mouth

Old rats are subject to overgrowth of incisor teeth just as young rats are. This type of malocclusion, as well as broken or missing teeth, has been noted with moderate frequency in Crl:CD(SD)BR male rats over 24 months of age (22). Undoubtedly it occurs in other stocks and strains also. Occasionally, the long incisors penetrate the soft tissues of the upper or lower jaw (Fig. 13) or the soft palate, and acute or chronic inflammation and abscess formation ensues. Such lesions can lead to inanition through mechanical interference with food intake.

2. Stomach

Crl:CD(SD)BR male rats exhibit gastric glandular atrophy, crypt dilatation and submucosal fibrosis with increasing frequency as they age (Fig. 14) (22). Burek (18) has noted similar changes in the WAG/Rij, BN/Bi/Rij, and F₁ hybrid of these two strains. The significance of these changes in gastric morphology is not clear, but they are not often seen in young rats. It is possible that these lesions may correlate with gastric function since the volume of gastric secretion and hydrochloric acid output are decreased in Wof:(WI) rats older than 12 months (60). Apparently the lesions are not present in all stocks, since Berg (8) reported that the stomach of SD rats was

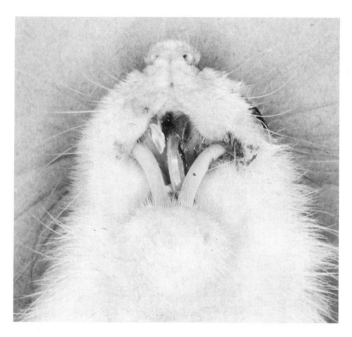

Fig. 13. Malocclusion with broken upper incisor and overgrown lower incisors. Crl:CD(SD)BR, 34 months.

unaltered with advancing age. Fundic mucosal hyperplasia has been demonstrated in aging Mendel-Sherman rats to a variable extent (8). Gland dilatation and submucosal fibrosis in the hyperplastic areas are similar to those described above.

Burek (18) occasionally has observed small punctate erosions and ulcers of the glandular stomach and more frequently has seen large, sometimes perforating ulcers of the squamous portion of the stomach in WAG/Rij, BN/Bi/Rij, and F₁ hybrid rats from these two strains. Frequency clearly increased with age.

3. Intestines

Spontaneous age-associated, nonneoplastic lesions of the small intestine, cecum, and colon are rare. Duodenal ulcers accompanied by acute diffuse mucosal inflammation and/or atrophy have been produced in aging rats subjected to pantothenic acid deficiency (8).

4. Pancreas—Exocrine

Acinar atrophy and fibrosis and, sometimes, cavitations occur with increasing frequency in various rat stocks and strains as aging progresses, and the severity of the lesion also appears to be age-related (1,8,12,22,25,59). Microscopically the acini are small, and zymogen granules are absent. Vacuolization of acinar cells may accompany atrophic changes. The acini are transformed into ductlike structures lined by cuboidal or flattened epithelial cells (Fig. 15) and there is sometimes an

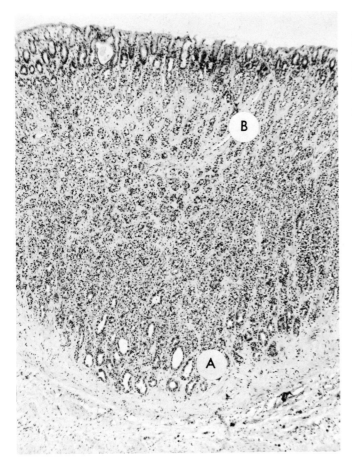

Fig. 14. Stomach with mucosal and submucosal fibrosis. Glands in the fibrotic areas are dilated (A) or atrophic (B). Mast cells and eosinophils in fibrotic areas adjacent to muscularis mucosae. Hematoxylin and eosin. ×40. Crl:CD(SD)BR, 34 months.

associated lymphocytic infiltrate (12). Fat cells often are prominently interspersed within areas of pancreatic lobular atrophy (22).

5. Liver

Cholangiofibrosis (105) is an age-associated lesion with differing incidences in various rat stocks and strains. It occurs with increasing frequency and severity in F344 inbred rats after 18 months of age. The incidence in 24-month-old F344 rats has been reported as 98% (25). It occurs to a lesser extent in the outbred Crl:CD(SD)BR male rat (22), and it is uncommon in the WAG/Rij and BN/Bi/Rij inbred strains, although cystic bile ductules are seen (18). Microscopically, the lesion consists of an increase in portal bile ductules with associated mild fibrosis (Fig. 16). A few mononuclear cells often are seen, but large infiltrates of inflammatory cells usually are not present. The cause and functional significance of the lesion are not clear.

Cystic and telangiectatic changes in liver are common age-associated lesions in rats. The pathogenesis of many of these lesions is unclear, and opinions differ as to their origin and progression. Hepatic cysts in aging rats often are multiple and vary from small to very large. Microscopically, they may be lined by flat to cuboidal cells and can compress adjacent parenchyma. Many appear to contain mucopolysaccharide material (22) (Fig. 17). Burek (18) also has noted mucopolysaccharide-containing hepatic cysts in his colonies of aging BN/Bi/Rij and WAG/Rij inbred rats. Boorman and Hollander (12) rarely encountered hepatic cysts in young female WAG/Rij rats, but such cysts occurred in 30% of rats over 36 months of age. These investigators speculated that the cystic areas were an end stage of massive hepatic necrosis; however, no clear-cut progression of lesions was demonstrated.

A progressive increase with age in hepatic telangiectasis was reported in WI female rats (39). Almost no telangiectasis was observed in 9-month-old rats, but the incidence was 16 and 50% in 13- and 18-month-old rats, respectively. In males, the incidence was 10–15% higher than in females. Cohen *et al.* (22) also have noted telangiectatic foci in livers from

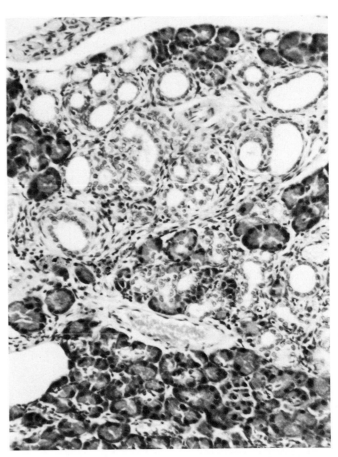

Fig. 15. Pancreatic acinar atrophy and fibrosis. Normal pancreatic lobule at bottom. Hematoxylin and eosin. ×100. Crl:CD(SD)BR, 18 months.

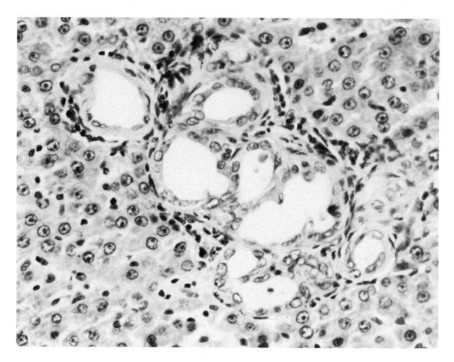

Fig. 16. Liver with cholangiofibrosis (bile ductule proliferation, periductal mononuclear cells, and fibrosis). Hematoxylin and eosin. ×250. Crl:CD(SD)BR, 36 months.

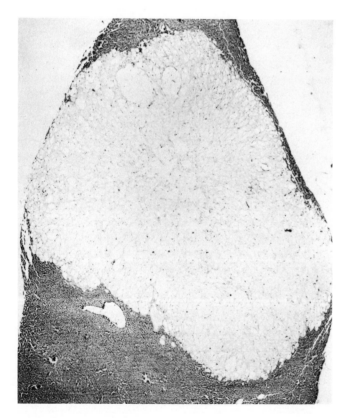

Fig. 17. Liver with cystic area containing mucopolysaccharide material. Edges of the area are well circumscribed from surrounding parenchyma, but there is no lining or cyst wall. Hematoxylin and eosin. ×15.8. Crl:CD(SD)BR, 30 months.

Crl:CD(SD)BR male rats more than 25 months old (Fig. 18). The lesion has not been observed in young rats.

Foci or areas of cellular alteration or proliferation are seen with moderate frequency in aging rat livers. Squire and Levitt (105) have pointed to the difficulties in classifying such lesions and to their significance with respect to ultimate progression to cancer. Detailed discussion of this matter is beyond the scope of this section, but in view of the widespread use of rats in long-term drug carcinogenicity testing, it clearly is essential to characterize and classify spontaneously arising hepatic lesions accurately.

E. Endocrine System

The most dramatic lesion of the endocrine system of aged rats is adenomatosis involving cells of the so-called APUD (amine precursor uptake and decarboxylation) series (80): adrenal medulla, parafollicular (C) cells, anterior pituitary, pancreatic islet cells. These neoplasms are discussed in detail in this volume, Chapter 13.

Other endocrine lesions have less effect on the general health of old rats and usually are incidental findings.

1. Adrenal

Aged rat adrenals occasionally have dilatation of cortical sinusoids, sometimes accompanied by thrombosis (8,22,25).

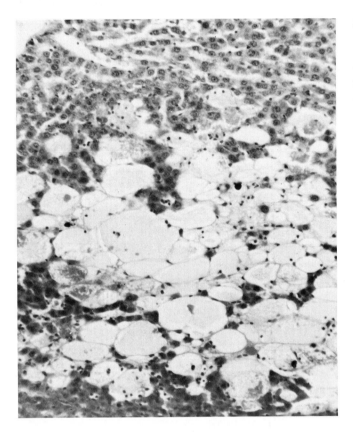

Fig. 18. Liver with focal telangiectasis. Spaces are filled with homogenous or fibrillar material and/or erythrocytes. Hepatocytes between the spaces are compressed and atrophic. Hematoxylin and eosin. ×100. Crl:CD(SD)BR, 25 months.

Focal degeneration and necrosis of cortical cells also have been reported (8,96). The adrenal capsule becomes thickened by hyalinized connective tissue (96).

A common finding in the adrenal zona fasciculata is circumscribed nodules of plump lipoid-containing cells. It is theorized that these are derived from spindle-shaped subcapsular "A" cells (96). Such nodules have been termed adenomas by Boorman and Hollander (12) and included in their tumor series. These investigators also report vascular dilatation, hemorrhage, and necrosis occurring within the "adenomas." Coleman *et al.* (25) found that aged rats with pituitary tumors had more focal and diffuse adrenal cortical vacuolization than did those with normal pituitaries. For further discussion of adrenal neoplasms, see this volume, Chapter 13.

2. Thyroid

Cysts lined by stratified squamous epithelium are reported by one investigator to be common in aged rats (96). Older animals may have marked variation in follicle size and colloid content. Some distended follicles also may contain basophilic debris and calcific concretions (22).

3. Parathyroid

Mitotic activity occurs within the rat parathyroid glands to an advanced age. Parathyroid hyperplasia in aged rats generally is regarded as secondary renal hyperparathyroidism (8,96). Some strains of aged rats, however, have minimal parathyroid lesions while others, such as LE, have incidences of up to 40% (96).

4. Pituitary

Colloid cysts are not uncommon in the anterior pituitary of rats 20 months and older (8,22). A marked increase in reticular connective tissue also occurs in aged rats (96). In aged female breeder rats, there is degranulation of acidophils and proliferation of chromophobe-type cells. On histochemistry, the latter cells are often found to produce prolactin. Chromophobe cells have a spectrum of proliferative lesions ranging from hyperplasia to neoplasia (adenomas) (96). (Also, see this volume, Chapter 13).

5. Pancreas (Islets)

Lesions in the exocrine pancreas are described in Section IV,D. While secondary atrophy of islets may occur in pancreatic tissue with extensive exocrine damage, a more common finding is islet cell hyperplasia (47,48,117,118). This lesion has been reported both in virgin (47,48) and breeder (117,118) SD rats. Nonbreeder males 40 weeks and older had a higher incidence of islet hyperplasia (33–60%) than females (5% maximum) (48).

Beta cells of the islets become both hypertrophic and hyperplastic; granularity is normal or increased. Alpha cells are present in the islet margins. Intra- and periinsular fibrosis occurs, and islets coalesce to form giant conglomerates and, eventually, tumors.

Glucose tolerance tests of SD nonbreeder rats 10–12 months old are abnormal even when islet histology is not (47), with all values elevated except the fasting and 30-min blood glucose levels. The abnormal pattern is further exaggerated in rats with islet lesions. This functional change in islets preceding microscopic abnormalities is theorized to be due to a discrepancy between insulin demand and availability to tissues.

Wexler and Fischer's (118) breeder rats had hyperglycemia 1 hr after glucose injection as compared to virgin controls of comparable age. Fasting blood glucose levels were similar in both groups. Wexler and Fischer (117) also found lesions increased with breeding frequency especially in females. They state that islet cell changes in breeders can be correlated with the intensity of atherosclerosis.

Clinical identification of rats with abnormal blood glucose values could be extremely important in studies on glucose metabolism in a system unrelated to the pancreas, where investigators may not immediately realize the possible effects of islet lesions on their research (21).

F. Respiratory System

Murine respiratory mycoplasmosis is the most common disease problem in conventionally reared rats (see this volume, Chapter 10). Lindsey *et al.* (65) and Cohen and Anver (21) have cited examples of the devastating effects of this widespread disease on aging research. With present day technology, it is possible to rear and maintain aging rats free of chronic upper respiratory infection, pneumonia, and most other infectious diseases (35,56). For example, Crl:CD(SD)BR male rats and barrier-reared F344 rats of both sexes were free of respiratory tract inflammation at 24 months of age and older (24,25). Thus, research in aging need not be complicated by using rats infected with *Mycoplasma pulmonis*.

Pulmonary foam cells (foamy macrophages, alveolar histiocytes) are seen often in the lungs of rats, including rats that are free of mycoplasmosis (Fig. 19). Alveolar histiocytosis is characterized by subpleural accumulations of cells containing free fatty acids, cholesterol, and phospholipids (34). The lesion has been observed to occur in rats of various ages (17,41,125). Flodh *et al.* (34) demonstrated that accumulations of foam cells were more frequent and pronounced in 22-month-old conventional outbred SD rats than in 6- or 9-month-old animals. They speculated that the cells may aid in maintaining a steady state of lipid metabolism in the body, while Yang *et al.* (125) suggested that they reflect a degenerative change that may also be influenced by aging. However, the significance and role of the cells is not well understood.

G. Hemic and Lymphatic Systems

Alteration of the immune system is considered to be an important factor in the pathogenesis of aging in mammals. With aging, normal immune function decreases, and there is an increase in autoimmune phenomena. Such immunologic changes are best investigated by functional studies on lymphocytes (114).

Morphologic changes do, however, occur in spleen and lymph nodes. In the spleen of aged rats, the T and B cell zones in the Malpighian corpuscles become less discrete, and germinal centers are rare to absent. The red pulp, which is compact in younger rats, develops a more sinusoidal arrangement with an increase in hemosiderin-filled macrophages (2,7). Megakaryocytes may be present in the spleen of both adult and aged rats (49). Splenic extramedullary hematopoiesis also occurs and can be quite extensive, especially in tumor-bearing animals (8).

Despite a decrease in immune competence in aged rats (11), lymph node changes are not marked. Deep cervical lymph nodes in WI and gray Norway rats greater than 700 days (3) had several different appearances: hyperplastic with germinal

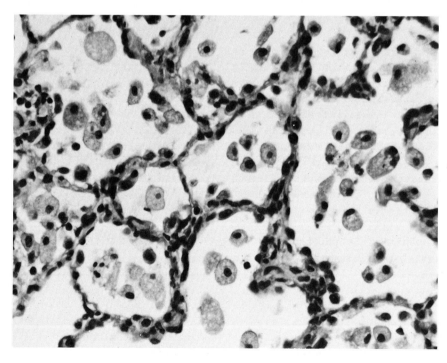

Fig. 19. Lung with many foamy histiocytes in alveolar lumens. Hematoxylin and eosin. ×250. Crl:CD(SD)BR, 26 months.

centers and many plasma cells or cystic with medullary cavitation and cortical compression. Macrophages containing pigment or phagocytized debris were present in the medulla of the cystic nodes. Cervical nodes from aged rats also had capsular and trabecular fibrosis with focal mast cell infiltration.

Leukemoid reactions have been reported in rats with prostatic carcinoma (82).

H. Central Nervous System and Special Senses

1. Central Nervous System

Radiculoneuropathy and associated degenerative myelopathy of the spinal cord are the most clinically significant lesions of the central nervous system (see Section II,C). Other than this syndrome, central nervous system pathology is absent or minimal in aged rats. Lesions which are present apparently cause no functional impairment.

Membranous bodies with a median size of 6 μm have been described by Vaughn (111). These occur in dendrites of cerebral cortical neurons of both SD and SD-derived stocks. They are found most frequently in rats 27–28 months old (the oldest group studied) but also are present sporadically in rats 12 months but not 3 months old. Even in the 28-month-old animals, they are not numerous. On electron microscopy, the bodies are composed principally of layered sheets of membranes. They also contain cytoplasm and empty vacuoles. The source of the membranes has not been ascertained.

Although sometimes difficult to differentiate from tissue processing artifacts, focal or generalized vacuolization of the neuropil and adjacent to neurons occurs in aged rats (22,25,38). The vacuoles do not stain with hematoxylin and eosin, and there is no accompanying cellular reaction (22). The incidence of vacuolization can be quite high, e.g., up to 80% in 30- to 33-month-old F344 rats (25).

2. Eyes

Retinal lesions of photoreceptor degeneration can occur in albino rats of any age exposed to fluorescent or incandescent light levels comparable to those which may be found in laboratories or animal rooms (e.g., 70 fc). The degree of damage is dependent on light intensity, environmental and body temperature, and exposure time (77,78). In albino rats (stock or strain unspecified) experimentally exposed to high intensity light levels (150–350 fc) at 39°C environmental temperature, animals 11–14 weeks, 16 weeks, and 24 weeks old had more severe thinning and disruption of photoreceptors than did those between 3 and 10 weeks old (79).

Even at standard animal room light levels, the incidence of spontaneous retinal degeneration in SD-derived rats at 2 years

of age was 11.6% compared to 0.3% in rats less than 6 months old. The lesion also occurred in rats as young as 3 months. This is somewhat similar to findings of von Sallmann and Grimes (112,113) in inbred and outbred OM and SHR(WI) rats. These animals, housed in rooms with cyclic 40–50 fc room illumination and 5–12 fc cage illumination, developed retinal degeneration after 9 months of age. The onset time and the severity of lesions were extremely variable. In general, the OM animals had more advanced disease than did the SHR rats.

In all rat stocks and strains studied, lesions generally consisted of fragmentation of the outer segments of the photoreceptors, disappearance of the inner segments, and a reduction in width of the outer nuclear layer. Karyorrhexis or other frank damage to the photoreceptor nuclei was not detected nor did debris from the fragmented outer layer accumulate. Involvement was focal or diffuse with the central retina more often affected than the peripheral portions (97,112).

It is theorized that gradual photoreceptor attrition with age in rats may be in fact light-induced although there may also be a genetic component determining susceptibility (113).

Nuclear and posterior capsule cataracts are reported in rats greater than 1 year old (51). Cataract incidence is usually less than 10% (25,51). Heywood's (51) SD rats were from a conventional colony which had enzootic sialodacryoadenitis, but the cataracts in F344 SFP rats did not seem secondary to viral infection (25).

Sherman rats (Lederle colony) also developed progressive bilateral lens changes so that by 22–24 months of age, one or more lens abnormalities were detectable with a biomicroscope. Histologic lesions occurred in the anterior and posterior cortices, consisting of fiber swelling or destruction, accumulation of fine granular proteinaceous material, fiber fragmentation, focal lysis, and aggregates of balloon cells. Associated lenticular epithelium had nuclear degeneration. Lens nuclei developed focal sclerosis (5).

I. Musculoskeletal System

Osteoarthritis occurs sporadically in the tibiotarsal joints and medial femoral condyles of aged rats (8,104). Other joints also are affected but less frequently. In Sokoloff's (104) study, lesions were noninflammatory, consisting of chondromucoid degeneration of articular cartilage, erosion of cartilage, and eburnation of underlying bone. Berg's (8) rats had a clinical syndrome more suggestive of an infectious process with acute swelling progressing to chronic periarticular and articular inflammation.

Chronic spondylitis also affects old rats. Spinal column lesions are proliferative exostoses with bridging and osteophytes on the dorsal and/or ventral surface of the vertebral canal (19,104). Vertebral disc lesions described in BN/Bi/Rij,

WAG/Rij, and an F_1 hybrid consist of degeneration (most severe in F_1) and occasional disc prolapse or rupture with cord compression (19).

Aseptic necrosis of bone of old rats involves epiphyses of both limbs and trunk. Lesions are seen after 1 year of age; bones most frequently affected are those of the knee joint and the vertebrae. In the latter instance, there may be associated thoracolumbar kyphosis (19,104).

In aged rats with advanced chronic renal disease, secondary renal hyperparathyroidism may develop (103). Bone lesions consist of resorption of bone, osteoclasis, and replacement with fibrous connective tissue (fibrous osteodystrophy). Bone lesions secondary to hyperparathyroidism can be reproduced by partial (5/6) nephrectomy. Animals thus nephrectomized at weaning develop typical advanced chronic renal disease in the remaining one-sixth kidney, with most severe fibrous osteodystrophy occurring between weeks 27 and 45 of the experiment (72,100). Feeding a high phosphorus diet also increases bone resorption (29).

Decubital ulcers on the plantar surface of the hindfeet—similar to "hock sores" of rabbits—have been seen in aged obese male Crl:CD(SD)BR rats housed on wire. Granulation tissue and phlegmon can extend to the tarsal and metatarsal bones with development of chronic periostitis and osteitis (22).

J. Other

1. Integument

Cutaneous neoplasms are common in aged rats (see Chapter 13). The fur of old albino rats, especially males, tends to become yellow and coarse (8).

2. Lipofuscins

These golden-brown granular pigments are primarily associated with the aging process of cells uncomplicated by local or environmental pathologic factors. This is in contrast to ceroid, a pigment closely related chemically but associated principally with nutritional deficiencies especially vitamin E. Both pigments are found intracytoplasmically. Synonyms for lipofuscin are age pigment, wear and tear pigment, and pigment of brown atrophy. Rats, as well as other mammals including humans, accumulate lipofuscin in various cells as they age. Most prominently affected are the nervous system, cardiac and skeletal muscle, and liver. Other organs and tissues may also contain pigment. Ultrastructurally, lipofuscin is a heterogenous material composed of vacuoles, dense granules, lipid droplets, and laminated bodies; the entire pigment granule is surrounded by a simple membrane indicative of its lysosomal location. A comprehensive discussion of lipofuscin and ceroid in rats and other species is provided by Porta and Hartroft (84).

REFERENCES

1. Andrew, W. (1944). Senile changes in the pancreas of Wistar Institute rats and of man with special regard to the similarity of locule and cavity formation. *Am. J. Anat.* **74,** 97–125.
2. Andrew, W. (1946). Age changes in the vascular architecture and cell content in the spleens of 100 Wistar Institute rats including comparisons with human material. *Am. J. Anat.* **79,** 1–74.
3. Andrew, W., and Andrew, N. V. (1948). Age changes in the deep cervical lymph nodes of 100 Wistar Institute rats. *Am. J. Anat.* **82,** 105–165.
4. Andrew, W., and Pruett, D. (1957). Senile changes in the kidneys of Wistar Institute rats. *Am. J. Anat.* **100,** 51–79.
5. Balazs, T., Ohtake, S., and Noble, J. F. (1970). Spontaneous lenticular changes in the rat. *Lab. Anim. Care* **20,** 215–219.
6. Berg, B. N. (1955). The electrocardiogram in aging rats. *J. Gerontol.* **10,** 420–423.
7. Berg, B. N. (1965). Spontaneous nephrosis with proteinuria, hyperglobulinaemia, and hypercholesterolemia in the rat. *Proc. Soc. Exp. Biol. Med.* **119,** 417–420.
8. Berg, B. N. (1967). Longevity studies in rats. II. Pathology of aging. In "Pathology of Laboratory Rats and Mice" (E. Cotchin and F. J. C. Roe, eds.), pp. 749–786. Davis, Philadelphia, Pennsylvania.
9. Berg, B. N., and Harmison, C. R. (1955). Blood pressure and heart size in aging rats. *J. Gerontol.* **10,** 416–419.
10. Berg, B. N., Wolf, A., and Simms, H. S. (1962). Degenerative lesions of spinal roots and peripheral nerves in aging rats. *Gerontologia* **6,** 72–80.
11. Bilder, G. E. (1975). Studies on immune competence in the rat: Changes with age, sex and strains. *J. Gerontol.* **30,** 641–646.
11a. Bolton, W. K., Benton, F. R., Maclay, J. G., and Sturgill, B. C. (1976). Spontaneous glomerular sclerosis in aging Sprague-Dawley rats. *Am. J. Pathol.* **85,** 277–300.
12. Boorman, G. A., and Hollander, C. F. (1973). Spontaneous lesions in the female WAG/Rij (Wistar) rat. *J. Gerontol.* **28,** 152–159.
13. Boorman, G. A., and Hollander, C. F. (1974). High incidence of spontaneous urinary bladder and ureter tumors in the Brown Norway rat. *J. Natl. Cancer Inst.* **52,** 1005–1008.
14. Boorman, G. A., Zurcher, C., and Hollander, C. F. (1973). Naturally occurring endocardial disease in the rat. *Arch. Pathol.* **96,** 39–45.
15. Bras, G. (1969). Age associated kidney lesions in the rat. *J. Infect. Dis.* **120,** 131–135.
16. Bras, G., and Ross, M. H. (1964). Kidney disease and nutrition in the rat. *Toxicol. Appl. Pharmacol.* **6,** 247–262.
17. Bullock, B. C., Banks, K. L., and Manning, P. J. (1968). Common lesions in the aged rat. In "The Laboratory Animal in Gerontological Research," pp. 62–82. Natl. Acad. Sci. Publ. 1591, Washington, D.C.
18. Burek, J. D. (1978). "Pathology of Aging Rats." CRC Press, Inc., West Palm Beach, Florida.
19. Burek, J. D., van der Kogel, A. J., and Hollander, C. F. (1976). Degenerative myelopathy in three strains of aging rats. *Vet. Pathol.* **13,** 321–331.
20. Chipman, R. K., and Albrecht, E. D. (1974). The relationship of the male preputial gland to the acceleration of oestrus in the laboratory mouse. *J. Reprod. Fertil.* **38,** 91–96.
21. Cohen, B. J., and Anver, M. R. (1976). Pathological changes during aging in the rat. "Spec. Rev. Exp. Aging Res. Prog. in Biol." (M. F. Elias, B. E. Eleftheriou, and P. K. Elias, eds.), pp. 379–403. Ear Inc., Bar Harbor, Maine.
22. Cohen, B. J., Anver, M. R., and Ringler, D. H. (1979). Age-associated lesions in barrier reared outbred albino male rats. (In preparation.)
23. Cohen, B. J., deBruin, R. W., and Kort, W. J. (1970). Heritable

hydronephrosis in a mutant strain of Brown Norway rats. *Lab. Anim. Care* **20**, 489–493.

24. Cohen, B. J., and deVries, M. J. (1969). Giant cells in rat testes after accidental and experimental thermal injury. *Pathol. Eur.* **4**, 336–344.

25. Coleman, G. L., Barthold, S., Osbaldiston, G., Foster, S., and Jonas, A. M. (1977). Pathological changes during aging in barrier reared Fischer 344 male rats. *J. Gerontol.* **32**, 258–278.

26. Couser, W. G., and Stilmant, M. M. (1975). Mesangial lesions and focal glomerular sclerosis in the aging rat. *Lab. Invest.* **33**, 491–501.

27. Couser, W. G., and Stilmant, M. M. (1976). The immunopathology of the aging rat kidney. *J. Gerontol.* **31**, 13–22.

28. Davey, F. R., and Moloney, W. C. (1970). Postmortem observations on Fischer rats with leukemia and other disorders. *Lab. Invest.* **23**, 327–334.

29. Draper, H. H., Ten-Lin, S., and Bergan, J. G. (1972). Osteoporosis in aging rats induced by high phosphorus diets. *J. Nutr.* **102**, 1133–1142.

30. Drori, D., and Folman, Y. (1976). Environmental effects on longevity in the male rat: Exercise, mating, castration and restricted feeding. *Exp. Gerontol.* **11**, 25–32.

31. Ekstrom, M. E., and Ewald, P. E. (1975). "Chronic Purulent Preputial Gland Adenitis in the Male Laboratory Rat," Abstr. No. 10, Publ. 75-2. Am. Assoc. Lab. Anim. Sci., Joliet, Illinois.

32. Elema, J., Koudstall, J. D., Lamberts, H. B., and Arends, A. (1971). Spontaneous glomerulosclerosis in the rat. *Arch. Pathol.* **91**, 418–425.

33. Fairweather, F. A. Cardiovascular disease in rats. *In* "Pathology of Laboratory Rats and Mice" (E. Cotchin and F. J. C. Roe, eds). pp. 213–227. Davis, Philadelphia, Pennsylvania.

34. Flodh, H., Magnusson, G., and Magnusson, O. (1974). Pulmonary foam cells in rats of different ages. *Z. Viersuchstierkd.* **16**, 299–312.

35. Flynn, R. J. (1972). Development and maintenance of laboratory rodents to meet the needs of aging research. *In* "Development of the Rodent as a Model System of Aging" (D. Gibson, ed.), DHEW Publ. No. (NIH) 72-121, pp. 13–17. DHEW, Bethesda, Maryland.

36. Foley, W. A., Jones, D. C. L., Osborn, G. K., and Kimeldorf, D. J. (1964). A renal lesion associated with diuresis in the aging Sprague-Dawley rat. *Lab. Invest.* **13**, 439–450.

37. Franks, L. M. (1967). Normal and pathological anatomy and histology of the genital tract of rats and mice. *In* "Pathology of Laboratory Rats and Mice" (E. Cotchin and F. J. C. Roe, eds.). pp. 469–499. Davis, Philadelphia, Pennsylvania.

38. Garner, F. M., Innes, J. R. M., and Nelson, D. H. (1967). Murine neuropathology. *In* "Pathology of Laboratory Rats and Mice" (E. Cotchin and F. J. C. Roe, eds.), pp. 295–348. Davis, Philadelphia, Pennsylvania.

39. Gellatly, J. B. M. (1967). Discussion in Pathology of the liver of rats and mice. *In* "Pathology of Laboratory Rats and Mice" (E. Cotchin and F. J. C. Roe, eds.), p. 22. Davis, Philadelphia, Pennsylvania.

40. Gerrity, R. G., and Cliff, W. J. (1972). The aortic tunica intima in young and aging rats. *Exp. Mol. Pathol.* **16**, 382–402.

41. Giddens, W. E., and Whitehair, C. K. (1969). The peribronchial lymphocytic tissue in germfree, defined-flora, conventional and chronic murine pneumonia-affected rats. *In* "Germfree Biology" (E. A. Mirand and N. Back, eds.), pp. 75–84. Plenum, New York.

42. Gilman, T., and Hathorn, M. (1959). Sex incidence of vascular lesions in ageing rats in relation to previous pregnancies. *Nature (London)* **183**, 1139–1140.

43. Gilmore, S. A. (1972). Spinal nerve root degeneration in aging laboratory rats: A light microscopic study. *Anat. Rec.* **174**, 251–257.

43a. Gray, J. E. (1977). "Chronic Progressive Nephrosis in the Albino Rat," pp. 115–144. CRC Critical Reviews in Toxicology, CRC Press, Inc., West Palm Beach, Florida.

44. Gray, J. E., Weaver, R. N., and Purmalis, A. (1974). Ultrastructural observations of chronic progressive nephrosis in the Sprague Dawley

rat. *Vet. Pathol.* **11**, 153–164.

45. Gregory, J. G., and Barrows, C. H. (1969). The effect of age on renal functions of female rats. *J. Gerontol.* **24**, 321–323.

46. Gutmann, E., Hanzlikova, V., and Vyskocil, F. (1971). Age changes in cross striated muscle of the rat. *J. Physiol. (London)* **219**, 331–343.

47. Hajdu, A., Herr, F., and Rona, G. (1968). The functional significance of a spontaneous pancreatic islet change in aged rats. *Diabetologia* **4**, 44–47.

48. Hajdu, A., and Rona, G. (1967). Morphological observations on spontaneous pancreatic islet changes in rats. *Diabetes* **16**, 108–110.

49. Hardy, J. (1967). Haematology of rats and mice. *In* "Pathology of Laboratory Rats and Mice" (E. Cotchin and F. J. C. Roe, eds.), pp. 501–536. Davis, Philadelphia, Pennsylvania.

50. Harman, D. (1971). Free radical theory of aging: Effect of the amount and degree of unsaturation of dietary fat on mortality rate. *J. Gerontol.* **26**, 451–457.

51. Heywood, R. (1973). Some clinical observations on the eyes of Sprague-Dawley rats. *Lab. Anim.* **7**, 19–27.

52. Hirokawa, H. (1975). Characterization of age associated kidney disease in Wistar rats. *Mech. Ageing Dev.* **4**, 301–316.

53. Huang, H. H., and Meites, J. (1975). Reproductive capacity of aging female rats. *Neuroendocrinology* **17**, 289–295.

54. Ingle, D. J., and Baker, B. L. (1953). A consideration of the relationship of experimentally produced and naturally occurring pathologic changes in the rat to the adaptation diseases. *Recent Prog. Horm. Res.* **8**, 143–169.

55. Institute of Laboratory Animal Resources. Committee on Care and Use of Spontaneously Hypertensive (SHR) Rats. (1976). Spontaneous hypertensive (SHR) rats: Guidelines for breeding, care, and use. *ILAR News* **19** (3), 1–20.

56. Institute of Laboratory Animal Resources. Committee on Long Term Holding of Laboratory Rodents. (1976). Long term holding of laboratory rodents. *ILAR News* **19**, (4), 1–25.

57. Jacobs, B. B., and Huseby, R. A. (1967). Neoplasms occurring in aged Fischer rats with special reference to testicular, uterine and thyroid tumors. *J. Natl. Cancer Inst.* **39**, 303–309.

58. Kellogg, T. F., and Wostmann, B. S. (1969). Stock diet for colony production of germfree rats and mice. *Lab. Anim. Care* **19**, 812–814.

59. Kendrey, G., and Roe, F. J. C. (1969). Histopathological changes in the pancreas of laboratory rats. *Lab. Anim.* **3**, 207–220.

60. Kowalewski, K. (1976). Effect of age and sex on histamine-induced gastric secretion in laboratory rats. *Lab. Anim. Sci.* **26**, 562–565.

61. Kraus, B., and Cain, H. (1974). Über eine spontane nephropathie bei Wistarratten. Die licht- und elektronenmikroskopischen Glomerulumveranderungen. *Virchows Arch. A* **363**, 343–358.

62. Lalich, J. J., and Allen, J. R. (1971). Protein overload nephropathy. II. Ultrastructural study. *Arch. Pathol.* **91**, 372–382.

63. Lalich, J. J., Faith, G. C., and Harding, G. E. (1970). Protein overload nephropathy in rats subjected to unilateral nephrectomy. *Arch. Pathol.* **89**, 548–559.

64. Lee, J. C., Karpelles, L. M., and Downing, S. E. (1972). Age related changes of cardiac performance in male rats. *Am. J. Physiol.* **222**, 432–438.

65. Lindsey, J. R., Baker, H. J., Overcash, R. G., Kassell, G. H., and Hart, E. D. (1971). Murine chronic respiratory disease. *Am. J. Pathol.* **64**, 675–716.

66. Lutzen, L., and Ueberberg, H. (1973). A study on morphological changes in the testes of old albino rats. *Beitr. Pathol.* **149**, 377–385.

67. Maekawa, A., and Odashima, S. (1975). Spontaneous tumors in ACI/N rats. *J. Natl. Cancer Inst.* **55**, 1437–1445.

68. McKay, C. M., Crowell, M. E., and Maynard, L. A. (1935). The effect of retarded growth upon the length of life span and upon the ultimate body size. *J. Nutr.* **10**, 63–79.

69. McKay, C. M., Maynard, L. A., Sperling, G., and Barnes, L. L. (1939). Retarded growth, life span, ultimate body size and age changes in the albino rat after feeding diets restricted in calories. *J. Nutr.* **18,** 1–13.

70. Meites, J., Huang, H. H., and Simpkins, J. W. (1978). Recent studies on neuroendocrine control of reproductive senescence in rats. *In* "The Aging Reproductive System" (E. L. Schneider, ed.), pp. 213–235. Raven, New York.

71. Miller, D. S., and Payne, P. R. (1968). Longevity and protein intake. *Exp. Gerontol.* **3,** 231–234.

72. Morrison, A. B. (1962). Experimentally induced chronic renal insufficiency in the rat. *Lab. Invest.* **11,** 321–332.

73. National Research Council Subcommittee on Laboratory Animal Nutrition (1978). Nutrient requirements of the laboratory rat. *In* "Nutrient Requirements of Laboratory Animals," 3rd ed., pp. 7–37, NAS-NRC, Washington, D.C.

74. Nolen, G. A. (1972). Effect of various restricted dietary regimes on the growth, health, and longevity of albino rats. *J. Nutr.* **102,** 1477–1493.

75. Okamoto, K., Aoki, K., Nosaka, S., and Fukushima, M. (1964). Cardiovascular disease in the spontaneously hypertensive rat. *Jpn. Circ. J.* **28,** 943–952.

76. Opie, E. L., Lynch, C. J., and Tershakovec, M. (1970). Sclerosis of the mesenteric arteries of rats. Its relation to longevity and inheritance. *Arch. Pathol.* **89,** 306–313.

77. O'Steen, W. K., and Anderson, K. V. (1971). Photically evoked responses in the visual system of rats exposed to continuous light. *Exp. Neurol.* **30,** 525–533.

78. O'Steen, W. K., and Anderson, K. V. (1972). Photoreceptor degeneration after exposure of rats to incandescent illumination. *Z. Zellforsch. Mikrosk. Anat.* **127,** 306–313.

79. O'Steen, W. K., Anderson, K. V., and Shear, C. R. (1974). Photoreceptor degeneration in albino rats: Dependency on age. *Invest. Ophthalmol.* **13,** 334–339.

80. Pearse, A. G. E. (1969). The cytochemistry and ultrastructure of polypeptide hormone-producing cells of the APUD series and the embryologic, physiologic and pathologic implications of the concept. *J. Histochem. Cytochem.* **17,** 303–313.

81. Pollard, M. (1971). The germfree rat. *In* "Pathobiology Annual" (H. L. Ioachim, ed.), pp. 83–94. Appleton, New York.

82. Pollard, M. (1973). Spontaneous prostate adenocarcinomas in aged germfree Wistar rats. *J. Natl. Cancer Inst.* **51,** 1235–1241.

83. Pollard, M., and Kajima, J. (1970). Lesions in aged germfree Wistar rats. *Am. J. Pathol.* **61,** 25–32.

84. Porta, E. A., and Hartroft, W. S. (1969). Lipid pigments in relation to aging and dietary factors (lipofuscins). *In* "Pigments in Pathology" (M. Wolman, ed.), pp. 191–235. Academic Press, New York.

85. Rapp, J. P. (1973). Age-related pathologic changes, hypertension, and 18-hydroxydeoxy-corticosterone in rats selectively bred for high or low juxtaglomerular granularity. *Lab. Invest.* **28,** 343–351.

86. Ratcliffe, H. L. (1949). Spontaneous diseases of laboratory rats. *In* "The Rat in Laboratory Investigation" (E. J. Farris and J. Q. Griffiths, eds.), 2nd ed., pp. 528–529. Lippincott, Philadelphia, Pennsylvania.

87. Renaud, S. (1962). Calcification in the renal-cardiovascular system of female breeder rats. *Br. J. Exp. Pathol.* **43,** 387–391.

88. Richardson, B., and Luginbuhl, H. (1976). The role of prolactin in the development of chronic progressive nephropathy in the rat. *Virchows Arch. A* **370,** 13–19.

89. Ross, M. H. (1967). Life expectancy modification by change in dietary regimen of the mature rat. *Proc. Int. Cong. Nutr., 7th, 1966.* Vol. 5, pp. 35–38.

90. Ross, M. H. (1972). Length of life and caloric intake. *Am. J. Clin. Nutr.* **25,** 834–838.

91. Ross, M. H., and Bras, G. (1971). Lasting influence of early caloric restriction on prevalence of neoplasms in the rat. *J. Natl. Cancer Inst.* **47,** 1095–1113.

92. Ross, M. H., and Bras, G. (1973). Influence of protein under- and over-nutrition on spontaneous tumor prevalence in the rat. *J. Nutr.* **103,** 944–963.

93. Ross, M. H., and Bras, G. (1974). Dietary preference and diseases of age. *Nature (London)* **250,** 263–265.

94. Ross, M. H., and Bras, G. (1975). Food preference and length of life. *Science* **190,** 165–167.

95. Ross, M. H., Bras, G., and Ragbeer, M. S. (1970). The influences of protein and caloric intake upon spontaneous tumor incidence of the anterior pituitary gland of the rat. *J. Nutr.* **100,** 177–189.

96. Russfield, A. B. (1967). Pathology of the endocrine glands, ovary and testes of rats and mice. *In* "Pathology of Laboratory Rats and Mice" (E. Cotchin and F. J. C. Roe, eds.), pp. 391–467. Davis, Philadelphia, Pennsylvania.

97. Schardien, J. L., Lucas, J. A., and Fitzgerald, J. E. (1975). Retinal dystrophy in Sprague Dawley rats. *Lab. Anim. Sci.* **25,** 323–326.

98. Sellers, A. L., Rosenfeld, S., and Friedman, N. B. (1960). Spontaneous hydronephrosis in the rat. *Proc. Soc. Exp. Biol. Med.* **104,** 512–515.

99. Shain, S., McCullough, B., and Segaloff, A. (1975). Spontaneous adenocarcinomas of the ventral prostate of aged A × C rats. *J. Natl. Cancer Inst.* **55,** 177–180.

100. Shimamura, T., and Morrison, A. (1971). Secondary hyperparathyroidism in rats with an experimental chronic renal disease. *Exp. Mol. Pathol.* **15,** 345–353.

101. Simms, H. S. (1967). Longevity studies in rats. I. Relation between life span and age of onset of specific lesions. *In* "Pathology of Laboratory Rats and Mice" (E. Cotchin and F. J. C. Roe, eds.), pp. 733–747. Davis, Philadelphia, Pennsylvania.

102. Skold, B. H. (1961). Chronic arteritis in the laboratory rat. *J. Am. Vet. Med. Assoc.* **138,** 204–207.

103. Snell, K. C. (1967). Renal disease of the rat. *In* "Pathology of Laboratory Rats and Mice" (E. Cotchin and F. J. C. Roe, eds.), pp. 105–147. Davis, Philadelphia, Pennsylvania.

104. Sokoloff, L. (1967). Articular and musculoskeletal lesions of rats and mice. *In* "Pathology of Laboratory Rats and Mice" (E. Cotchin and F. J. C. Roe, eds.), pp. 373–390. Davis, Philadelphia, Pennsylvania.

105. Squire, R. A., and Levitt, M. H. (1975). Report of a workshop on classification of specific hepatocellular lesions in rats. *Cancer Res.* **35,** 3214–3223.

106. Tauchi, H., Yoshioka, T., and Kobayashi, H. (1971). Age change of skeletal muscles of rats. *Gerontologia* **17,** 219–227.

107. Tomanek, R. J., and Karlsson, O. L. (1973). Myocardial ultrastructure of young and senescent rats. *J. Ultrastruct. Res.* **42,** 201–220.

108. Travis, D. F., and Travis, A. (1972). Ultrastructural changes in the left ventricular rat myocardial cells with age. *J. Ultrastruct. Res.* **39,** 124–148.

109. Tucker, S. M., Mason, R. L., and Beauchene, R. E. (1976). Influence of diet and feed restriction on kidney function of aging male rats. *J. Gerontol.* **31,** 264–270.

110. van Steenis, G., and Kroes, R. (1971). Changes in the nervous system and musculature of old rats. *Vet. Pathol.* **8,** 320–332.

111. Vaughn, D. W. (1976). Membranous bodies in cerebral cortex of aging rats: Electron microscope study. *J. Neuropathol. Exp. Neurol.* **35,** 152–166.

112. von Sallmann, L., and Grimes, P. (1972). Spontaneous retinal degeneration in mature Osborne-Mendel rats. *Arch. Ophthalmol.* **88,** 404–411.

113. von Sallmann, L., and Grimes, P. (1974). Retinal degeneration in

mature rats. Comparison of the disease in an Osborne-Mendel and a spontaneously hypertensive Wistar strain. *Invest. Ophthalmol.* **13,** 1010–1015.

114. Walford, R. L. (1974). Immunologic theory of aging: Current status. *Fed. Proc., Fed. Am. Soc. Exp. Biol.* **33,** 2020–2027.

115. Wexler, B. C. (1964). Spontaneous coronary arteriosclerosis in repeatedly bred male and female rats. *Circ. Res.* **14,** 32–43.

116. Wexler, B. C. (1970). Co-existent arteriosclerosis, PAN, and premature aging. *J. Gerontol.* **25,** 373–380.

117. Wexler, B. C., and Fischer, C. T. (1963). Hyperplasia of the islets of Langerhans in breeder rats with arteriosclerosis. *Nature (London)* **200,** 33–37.

118. Wexler, B. C., and Fischer, C. T. (1963). Abnormal glucose tolerance in repeatedly bred rats with arteriosclerosis. *Nature (London)* **200,** 133–136.

119. Wexler, B. C., and True, C. W. (1963). Carotid and cerebral arteriosclerosis in the rat. *Circ. Res.* **12,** 659–665.

120. Wilens, S. L., and Sproul, E. E. (1938). Spontaneous cardiovascular disease in the rat. I. Lesions of the heart. *Am. J. Pathol.* **14,** 177–200.

121. Wilens, S. L., and Sproul, E. E. (1938). Spontaneous cardiovascular disease in the rat. II. Lesions of the vascular system. *Am. J. Pathol.* **14,** 201–216.

122. Wilgram, G. F., and Ingle, D. J. (1959). Renal-cardiovascular pathologic changes in aging female breeder rats. *Arch. Pathol.* **68,** 690–703.

123. Wolfe, J. M., Burack, E., Lensing, W., and Wright, A. W. (1942). The effects of advancing age on the connective tissue of the uterus, cervix and vagina of the rat. *Am. J. Anat.* **70,** 135–165.

124. Yang, Y. H. (1965). Polyarteritis nodosa in laboratory rats. *Lab. Invest.* **14,** 81–88.

125. Yang, Y. H., Yang, C. Y., and Grice, H. C. (1966). Multifocal histiocytosis in the lungs of rats. *J. Pathol. Bacteriol.* **92,** 559–561.

Chapter 15

Health Hazards for Man

Estelle Hecht Geller

I. INTRODUCTION

"Man and the rat will always be pitted against each other as implacable enemies." This quote from Hans Zinsser's classic "Rats, Lice and History" (122) may well apply to the wild rat but hardly to its laboratory counterpart. However, there is always the danger that diseases will be introduced from wild rats that may be present in the laboratory as experimental animals or as invaders in inadequately rodent-proofed facilities. One also must be aware of the possible susceptibility of immunodeficient persons to agents not usually pathogenic for man, which may be carried by laboratory animals.

In the preparation of this chapter, letters were sent to American and Canadian institutions using research animals, requesting anecdotal information on disease problems in personnel resulting from association with laboratory rats (42).* Of 51 responding, nine were not using rats. Fifteen of the remaining 42 reported no problems of which the laboratory directors were aware. No laboratory reported any infection transmitted from a rat to man. Paradoxically perhaps, most of this chapter is concerned with infectious diseases which may be contracted from laboratory and/or wild rats.

II. ALLERGY

Twenty-three of the responding laboratories (42) reported problems with allergic reactions to rats, including minor urticarial response to contact with rats' claws, dermal erythema wherever a rat tail touched, extensive dermatitis of the hands, rhinitis, conjunctivitis, and bronchial asthma. One technician developed a severe conjunctivitis after her eye had been exposed to rat urine. When she was skin tested with rat dander, she developed acute anaphylactic shock. Many allergic individuals had to change jobs, and a few scientists were forced to change careers. One researcher became an Information Scientist. Another allergic investigator left the university where she had been working because of the apparent presence of rat dander throughout the building, even beyond the animal quarters. A pharmaceutical company reported that 11 of 24 employees directly involved with laboratory animals were allergic, predominantly to rats. Two institutions which noted no problems

*I acknowledge with thanks the responses of L. Arrington, L. Belbeck, E. Bernstein, D. Brooks, D. Clifford, P. Conran, K. Davis, P. Day, S. DeSalva, W. Dieterich, D. D'Innocenzo, E. Dolensek, W. Dorward, F. Enzie, E. Feenstra, C. Frith, T. Grafton, M. Hedge, J. Knill, K. Kohlstadt, A. Leash, L. Lewis, T. Liem, L. Lord, A. Macklin, D. Martin, D. McKay, R. Miller, R. Minner, G. Moore, T. Murchison, J. Pick, H. Rowsell, J. Russell, A. Sanders, P. Saunders, L. Serrano, C. Shannon, L. Slaughter, O. Soave, R. Steelman, A. Stilson, J. Sullivan, M. Sutton, C. Thayer, E. Truitt, B. Trum, J. Tufts, E. Watkins, F. Wazeter, and B. Zook.

routinely screen prospective employees for allergy to laboratory animals and were not aware of any conversions during employment.

In a recent study of laboratory animal dander allergy as an occupational disease (69), 15% of 1300 workers having direct daily contact with laboratory animals were found to be affected. Of these, 56.5% had become sensitized to rats. Most had a personal or family history of atopy and developed symptoms within 3 years, generally much sooner that the relatively few without such history. Another survey for allergic reactions in 474 laboratory workers who handle animals found that 23% developed one or more symptoms within 12 hr of contact (107). In this study, a family history of allergy was present in only 22% of those with symptoms and in 19% of those without symptoms.

Face masks; antihistamine, decongestant, and other drugs; and hyposensitization are useful in dealing with the allergy problem, but change of occupation is often necessary (81,84). A unique complete "costume" which enabled one highly allergic investigator to work with rodents has been described (45). A ventilated cage/rack system was developed recently which confines and eliminates the animal dander directly from the cages and rack into the room's exhaust system (58a). This is available commercially.

III. BITES

Four of the 41 responding laboratories reported rat bites resulting in localized infections. One of these bites damaged sensory nerves in a surgeon's finger, necessitating reparative microneurosurgery (42).

Throughout the world, bites of wild rats can be serious problems. It is estimated that over 14,000 people in urban areas of the United States are bitten by rats annually, babies and other helpless persons maimed, and at least 6000 cases of disease, transmitted directly by their bites or indirectly by wild rats, occur each year (24). Bite wounds should be cleaned thoroughly, and tetanus booster injections given if immunization is not current.

IV. BACTERIAL DISEASES

A. Rat-Bite Fever

This is the most common zoonosis of laboratory rats in the United States, with cases continuing to be reported in spite of the widespread (but not total) use of hygienic methods of experimental animal production and management in recent years

(19,25,55,91). Rat-bite fever is the designation for two similar but distinct diseases usually following a bite or other contact with rats or rat-infested areas, caused by either of two gram-negative bacteria: *Spirillum minus (Spirillum morsus muris, Spirochaeta morsus muris)* and *Streptobacillus moniliformis (Streptothrix muris ratti, Actinomyces muris ratti, Haverhillia multiformis, Actinobacillus muris)*.

Almost all of the infections of laboratory personnel have been caused by the latter. This disease, Haverhill or streptobacillary rat-bite fever, is characterized by a short incubation period of 2–10 days, usually prompt healing of the bite wound, regional lymphadenitis in about 25% of the cases, fever, chills, myalgia, sore throat, arthritis, and a morbilliform, pink, petechial rash. Complications such as abscesses, anemia, mastoiditis, endocarditis, and pericarditis may occur. Penicillin and other antibiotic therapy usually is effective (10,68,91), but a few deaths due to endocarditis and pericarditis have been reported in treated patients, perhaps because of inadequate dosage (18,74,87,88,91,116). The organism may be cultured from patient's blood or local lesions when grown on enriched media, or identified by animal inoculation (10,91). (See this volume, Chapter 9.)

Streptobacillus moniliformis is found in the nasopharynx of apparently healthy rats from where it is injected into the bite wound. The disease also may be acquired via contact with a rat-infested environment, possibly due to airborne spread (10). Surveys of small numbers of rats have demonstrated the bacteria in 0/15, 7/10, 2/20, and 7/14 of laboratory rats (43,66,68,103) and 4/6 of wild rats (103).

Since 1942, at least 21 cases have been reported due to bites or other association with wild rats (15,18,33,63,74,75,76, 91,98,102,115), and fifteen due to bites of laboratory rats (13,15,19,25,44,46,48,49,55,66,68,88,91). Immediate disinfection of the wound is, of course, recommended, although in at least one case it did not prevent development of the disease (68).

Sodoku, rat-bite fever caused by *Spirillum minus,* is characterized by a longer incubation period (2–3 weeks or more), an indurated lesion at the site of the initially healed bite wound, regional lymphadenitis in 50% of the cases, generalized symptoms such as a relapsing fever, chills, myalgia, and a purple to red-brown macular rash (10,91). Complications such as endocarditis can occur (54). Penicillin and other antibiotics are effective treatments (10,91). Diagnosis is confirmed by demonstration of the spirilla via dark-field microscopic examination of blood or peritoneal fluid of a mouse or guinea pig inoculated 2 weeks earlier with blood or wound fluid from the patient or by (impractical) serologic means.

Spirillum minus may be found in the blood of apparently healthy rats, and it may be associated with a conjunctivitis. Objective evidence is lacking as to the mode of transfer; however, it is believed to enter bite wounds by way of the nasolac-

rimal duct or from an injury to the rat's buccal mucosa sustained during biting (93). The organism may also be found in rat urine (57). Various surveys of wild rats report an infection rate of 0–25% (10,51,52,57,113). At least four cases in man (25,35,51,54) have been reported since 1944, only one of which (25) was caused by the bite of a laboratory rat.

Considering the large number of bites inflicted on people by wild (24) and laboratory rats, the incidence of rat-bite fever in man is very low.

B. Leptospirosis

Leptospirosis is a widespread, often inapparent infection of wild rats and other mammals. In some colonies of laboratory rats, it has been the source of a number of cases of disease, including fatalities, in laboratory personnel.

Thirteen species of the genus *Rattus* are listed as harboring 23 serotypes of *Leptospira* (110). Human infections occur world-wide as a result of rodent-associated leptospirosis (28,53,65). In several surveys, it appeared that a higher percentage, up to 51%, of trapped *R. norvegicus* than of other species are infected, usually with serotype *L. icterohemorrhagiae* (26,28,53,77,78). In the United States, of 119 reported cases of human leptospirosis with onset in 1975, rodents were the probable source of infection in 11% (20). Of 146 cases occurring between 1955 and 1960 in which a probable infecting cause could be ascertained, 13% had been exposed to rats (97). In another report (50), of 191 such traceable cases between 1949 and 1961, 31% were related to contact with rats. The majority of patients infected with *Leptospira icterohemorrhagiae* had been exposed to rats.

Leptospirosis in colonies of experimental rats, with transmission to humans, has been reported from several European laboratories including at least ten cases of disease in man in about 20 years of observation (67); three cases, one fatal, in a 4-year period; seventeen cases, two fatal, in a 2-year period; five deaths out of fourteen cases in yet another report (82); and one case in a 14-year period (85).

In one survey, 68% of 215 laboratory rats were positive for leptospirosis in one or more tests: 88.8% of these were serotype *L. icterohemorrhagiae*; the others were *L. ballum, L. australis, L. hebdomanis,* and *L. tarassovi* (82). *Leptospira pomona* was found in another laboratory (60). Concurrent with testing the laboratory rats, 534 people working with them were checked serologically, and 54 of these had antibodies (82). None of fifteen laboratory rats from three commercial sources in the United States were found to be serologically positive to twelve serotypes of *Leptospira* (43).

Laboratory rat colonies may be initially contaminated with leptospirosis from wildlife reservoirs, from colonies of laboratory mice which may harbor *Leptospira ballum,* and from other laboratory rat colonies (3,29,39,67,82,101). After an early bacteremia, the organisms become localized in the kidneys and are shed in the urine of carrier animals. Infection through contact during mating or via lacerated skin seems to be the means of spread within a rat colony (67). The disease may be eliminated by treatment with chlortetracycline (100), dihydrostreptomycin (96), serologic testing followed by destruction of reactors, or sacrifice of the entire colony and replacement with caesarian-derived stock (67). Suckling young of infected dams are protected by maternal antibodies and can remain free of infection if reared in a Leptospira-free environment after weaning (60).

Diagnosis of leptospirosis in the rat may be made by culture of kidney and urine in special media (EMJH, Fletcher's, Noguchi's, etc.) or by animal inoculation and subsequent culture. Serologic techniques used include macroscopic slide agglutination tests, microscopic slide agglutination tests using killed or live antigens, complement fixation tests, indirect fluorescent antibody (IFA) test, and indirect hemagglutination test (28,41,53,99,104,105). Findings in surveys of rat populations may vary according to the test. Two surveys in Canada found 31 and 9% of trapped R. norvegicus positive for Leptospira icterohemorrhagiae by culture methods, and 60 and 28% of the same groups positive by the agglutination-lysis test (77). In a study of wild rats in Hawaii, 1.2 times as many positives were found by kidney culture as by the macroscopic slide agglutination test (53). A survey of trapped wild urban rats in Georgia found that culture of urine or kidney was superior to either the macroscopic slide agglutination or IFA tests (104). The agglutination test correlated better with culture, but the IFA demonstrated the greatest number of those that were positive by only one method. Such differences may be due to several factors, such as recent infections in which organisms are shed before detectable antibodies appear, recovered animals in which organisms are no longer shed but antibodies remain, carrier animals which may continue to shed leptospires after detectable antibodies decline, and possible measurement of different immune reactions by the slide agglutination and IFA tests (104). In a Roumanian survey of laboratory rats, in which 146 or 68% were found to be positive in one or more of four tests, 44 were examined by all methods. The greatest number of positives was obtained with the IFA test, followed by the microscopic agglutination test using live antigen, and next by the dark-field examination of fresh kidney suspensions. Kidney and liver cultures were all negative (82).

C. Salmonellosis

Infection with hundreds of types of Salmonella occurs world-wide in many species of vertebrates, including the rat

which most frequently harbors S. typhimurium and S. enteritidis (21,23,26,34,77,118). These two are also the first and third most common serotypes reported in man in the United States. Salmonella typhimurium accounted for 28% of common salmonellae reported from human sources in 1975 and 18% of those from nonhuman sources (21). Salmonellosis can be a serious disease of laboratory rats. Rats so infected may become chronic carriers (47,67). It is not prevalent in most colonies, although a recent report from the Soviet Union indicated that 38% of a group of laboratory rats were infected with Salmonella, and stressed it as a possible hazard for man (61). It is also possible that laboratory personnel can infect animals, which may reinfect man (67). However, no cases of humans contracting salmonellosis from laboratory rats have been reported.

D. *Bordetella bronchiseptica* Infection

Bordetella bronchiseptica is found in the respiratory tracts of wild and laboratory rats where it may be associated with pneumonia (58,106). It has also been associated with a whooping coughlike disease in humans (16), but there is no evidence of transmission from rat to man or vice versa.

E. Clostridial Infections

Most herbivores and man are believed to carry *Clostridium tetani,* and virtually all mammals carry *Clostridium perfringens* in their intestinal tracts (90). Therefore, a risk, albeit a small one, of contracting tetanus exists when working with rats through contamination of a puncture wound. Animal handlers should keep their immunizations against tetanus current. Culture of feces and colon of fifteen laboratory rats from three sources revealed no *C. tetani* (43), but apparently others have reported finding it in rat feces (16).

F. Tularemia

Rattus spp. are not significant sources of human infection with tularemia, caused by *Francisella tularensis*. They are susceptible to infection but have a low sensitivity, with probably insufficient bacteremia to infect the blood-sucking arthropod vectors, such as ixodid ticks (species of *Amblyomma, Dermacentor, Haemaphysalis,* and *Ixodes*), some tabanid flies, and *Aedes* mosquitoes. However, since the disease is also transmitted by direct contact, there is a remote danger from the rat. Laboratory rats have not been involved in reported human tularemia (56,83).

G. Pasteurelloses

Rats are believed to be resistant to disease caused by *Pasteurella multocida,* although they may carry the organism (16,58,111). In one case of human infection following the bite of a laboratory rat, the organism was not isolated from the suspect animal when an attempt was made some time after the bite (7).

Pasteurella pneumotropica can cause widespread, usually inapparent infections in laboratory rats, where it may be harbored in the upper respiratory and alimentary tracts. No human infections related to contact with rats have been reported (9,58).

H. Staphylococcal Infection

Although rats are believed to be relatively resistant to *Staphylococcus* organisms of human phage type, they have been found to infect specific pathogen-free (SPF) laboratory rats. These organisms were presumably introduced by laboratory personnel. There were no silent carriers among these rats as there were among some of the other laboratory species studied (11,67).

I. Streptococcal Infection

Streptococcus (Pneumococcus, Diplococcus) pneumoniae of three serotypes (II, III, XIX) has been found to be widely distributed in conventional laboratory rats in which it may cause inapparent or overt infection. The organism is carried in the nasoturbinates of asymptomatic animals (117). There is no evidence that it is acquired from or transmitted to humans.

J. Pseudotuberculosis

Yersinia pseudotuberculosis is a questionable pathogen of laboratory rats (58,112), and, although rare in man, it can cause serious and fatal infections (16,112). Lesions due to this organism have been found in wild rats (78), and it has been isolated from rats with no apparent lesions. Experimentally inoculated rats with no visible lesions or disease may excrete the bacteria in feces and urine over a period of several weeks (70).

K. Plague

The introductory quote from ''Rats, Lice and History'' continues: ''and the rat's most potent weapons against mankind

have been its perpetual maintenance of the infectious agents of plague and typhus fever.'' Plague, the Black Death of the fourteenth century, caused by *Yersinia (Pasteurella) pestis,* killed many millions of people in three pandemics. It is primarily a septicemic, usually fatal, disease of rodents. It may be spread to man by direct contact, but usually is transmitted by rat fleas (*Xenopsylla cheopis, Nosopsyllus fasciatus,* and others), which regurgitate the bacteria when they bite (27,30,36,122). Laboratory rats have not been involved except experimentally.

V. RICKETTSIAL DISEASES

A. Murine Typhus

Murine or endemic typhus, due to *Rickettsia typhi (mooseri),* is an inapparent infection in the rat which is transmitted to man via feces of infected rat fleas (*Xenopsylla cheopis* and *Nosopsyllus fasciatus*) and probably the louse (*Polyplax spinulosa*) which contaminate the insect's bite wounds. It has been postulated that the disease can also be contracted by ingestion of food contaminated with the urine of infected rats. *Rickettsia typhi* is believed by some to be the parent organism of *R. prowazeki,* the causative agent of the more serious epidemic louse-borne typhus which is transmitted directly from man to man (6,16,17,119,122).

B. Other Rickettsial Diseases

Other rickettsial diseases with which the rat is associated are tick-transmitted boutonneuse fever, North Asian tick typhus, possibly Rocky Mountain spotted fever, mite-transmitted scrub typhus (tsutsugamushi), and rickettsial pox. Although the known host of the last is the house mouse, the vector mite, *Liponyssoides (Allodermanyssus) sanguineus,* is also found on the rat, suggesting that it too might be a source of infection (17,36).

VI. VIRAL DISEASES

A. Exotic Viral Diseases

The rat does not seem to be a significant host for many viral zoonoses. It probably does not carry Lassa fever, since 0/50

Rattus rattus trapped in an epidemic area were negative for the virus. However, another sometime laboratory rodent, *Mastomys natalensis,* the multimammate mouse or rat, was found to be the natural reservoir of this pathogen (79,120). *Rattus rattus* may be a minor carrier of the New Jersey serotype of vesicular stromatitis virus (109).

B. Lymphocytic Choriomeningitis

The rat has not been reported to be a source of human infection with lymphocytic choriomeningitis (LCM) as have other wild and laboratory rodents (1,5,14). Immature rats are susceptible to experimental disease, but older animals appear to be resistant (64). One of 28 laboratory rats had a positive complement fixation test for LCM, in a vivarium in which LCM was present in hamsters and had been transmitted to humans (14). In a group of rodents trapped in an Argentinian city, two *Rattus rattus* were found to have antibodies to LCM (71).

C. Rabies

Rabies is not endemic in rodents. Only four or five of 25,000 rodents examined annually in the United States have the disease. No case of human rabies in the United States has been traced to a rodent. Therefore, it is not recommended that people bitten by rodents receive antirabies treatment as a matter of course, unless the animal has laboratory-confirmed rabies (22). However, a case in which a rabid rat bit the face of a sleeping boy has been noted (89). In the German Federal Republic during 1961–1967, eighteen rats and three mice were reported as infected with rabies. People were bitten twenty times by the animals, and these authors believe that they are a dangerous source of infection to man (94). Laboratory rats are not a likely source of human exposure.

D. Sendai

Sendai virus, found in some laboratory rats (58), has been reported (as influenza D or parainfluenza 1 virus) to cause pneumonitis in newborn children and an influenzalike disease in adults (8). In a survey done to determine the potential of infections with murine viruses, antibodies to Sendai virus were found in persons associated with laboratory animals, but also in those who had no contact (108), probably due to cross-reaction with parainfluenza 2 (107a).

VII. PROTOZOAN DISEASES

A. Amebiasis

Entamoeba histolytica has been found in wild Norway rats throughout the world. It has not been reported to occur specifically in laboratory rodents (36,40,80). Rats, which are believed to be only incidental hosts of this parasite, are easily infected with human strains and can be carriers for long periods of time. Thus, it is conceivable that they may transmit amebiasis to man (80).

B. Encephalitozoonosis

Encephalitozoon (Nosema) cuniculi is found world-wide in brains, kidneys, and other tissues of wild and domesticated lagomorphs and rodents, including some laboratory rats (36,95). There is some belief that it can infect man, and at least two cases, one in an immunologically deficient infant, have been attributed to it or to similar organisms (62,72,73). The mode of natural horizontal transmission has not been definitely determined, but since it has been found in the kidneys of infected animals, it may be spread by way of the urine.

C. *Pneumocystis carinii* Infection

Pneumocystis carinii exists in many laboratory rat colonies, apparently as a saprophyte in the animals' lungs. It can become clinically evident, producing the typical pneumonia, when the rat is immunosuppressed with corticosteroids (4,37,38,114). The organism is believed, although without proof, to be species specific (37,114), and attempts to transmit it from heavily infected rat lungs to cortisone-treated mice and hamsters have failed (37). It has been transmitted from cortisone-treated rats (and from infected human lung tissue) to the genetically athymic "nude" mouse by inoculation, close contact, and probably the airborne route (114). Immunodeficient humans should be considered at risk. A suggestion of evidence for its possible transmission from rat to man is the observation that abundant *Pneumocystis* organisms were found in rodents from homes of many index patients in certain hospital ward epidemics (2).

VIII. HELMINTH INFECTIONS

Hymenolepsis nana is a common parasite of wild rats, may be found in laboratory rats, and is pathogenic for man. Since it is one cestode that does not require an intermediate invertebrate host, it can be transmitted to laboratory personnel via the animal's feces (36).

IX. ARTHROPOD INFESTATIONS

A. Flea Infestations

Fleas are rather uncommon on laboratory rats, but not on wild rats. *Xenopsylla cheopis,* the oriental rat flea; *Leptopsylla segnis,* the mouse flea; and *Nosopsyllus fasciatus,* the northern rat flea, do bite man. In addition to their being intermediate hosts for *Hymenolepis nana* and *H. diminuta* and vectors of plague and typhus, their bites can be irritating and allergenic (36,121).

B. Mite Infestations

Ornithonyssus (Liponyssus) bacoti, the tropical rat mite, is found in many rat colonies. It feeds on the host, lives in close proximity, and will attack man, causing local irritation and sometimes an allergic dermatitis. It can transmit murine typhus and plague (36,121).

Liponyssoides (Allodermanyssus) sanguineus, the house mouse mite, is also found on the wild Norway rat but has not been reported in laboratory rat colonies. It is the vector of *Rickettsia akari,* the cause of rickettsial pox, and its bite can cause a rash in man (36,121).

X. MYCOSES

Trichophyton mentagrophytes is a dermatophyte which frequently infects wild and laboratory rodents and other mammals and is capable of causing ringworm in man (59,92). Infected laboratory rats may have irregularly defined areas of alopecia, with or without scaling, crusts, or erythema, or carry the spores with no visible lesions (31,86). Clinically affected conventional rats, which later apparently recovered spontaneously, have been reported (86). Other workers have noted scaly patches of alopecia due to *T. mentagrophytes* in 20% of their rat colony (32). These investigators cultured the fungus from sixteen of 271 apparently normal rats and again from 237 of 900 asymptomatic conventional rats, but none from 900 "pathogen-free" rats (31). Transmission to laboratory workers has been reported (12,32,86).

XI. CONCLUSION

In a well-managed animal facility, and if handled with reasonable care and proper personal hygiene, the normal laboratory rat is a safe animal with which to work. It harbors relatively few organisms pathogenic to man which can be transmitted directly. With an appropriate barrier about the animal colony, preventing entry of vermin and wild rodents and contamination from other laboratory animals and man, there should be little risk.

The greatest problem seems to be sensitization of personnel to rat allergens. With an appropriate barrier about the person, e.g., mask and gloves (and perhaps a nonallergic predisposition), this problem can be diminished.

REFERENCES

1. Ackermann, R. (1973). Epidemiologic aspects of lymphocytic choriomeningitis in man. *In* "Lymphocytic Choriomeningitis Virus and Other Arenaviruses" (F. Lehmann-Grube, ed.), pp. 233–237. Springer-Verlag, Berlin and New York.
2. Araujo, F. G., and Remington, J. S. (1975). *Pneumocystis carinii* infection. *In* "Diseases Transmitted from Animals to Man" (W. T. Hubbert, W. F. McCulloch, and P. R. Schnurrenberger, eds.), pp. 801–803. Charles C Thomas, Springfield, Illinois.
3. Barkin, R. M., Guckian, J. C., and Glosser, J. W. (1974). Infection by *Leptospira ballum:* A laboratory-associated case. *South. Med. J.* **67,** 155.
4. Barton, E. G., and Campbell, W. G. (1969). *Pneumocystis carinii* in lungs of rats treated with cortisone acetate. *Am. J. Pathol.* **54,** 209–236.
5. Baum, S. G., Lewis, A. M., Rowe, W. P., and Huebner, R. J. (1966). Epidemic nonmeningitic lymphocytic-choriomeningitis-virus infection. An outbreak in a population of laboratory personnel. *N. Engl. J. Med.* **274,** 934–936.
6. Benenson, A. S. (1970). "Control of Communicable Diseases in Man." Am. Public Health Assoc., New York.
7. Bergogne-Berezin, E., Christol, D., Zechorsky, N., and Bonfils, S. (1972). Human *Pasteurella* infections from bites: Epidemiologic survey in a laboratory environment. *Nouv. Presse Med.* **44,** 2953–2957.
8. Bernkopf, H. (1964). Influenza. *In* "Zoonoses" (J. van der Hoeden, ed.), pp. 394–395. Elsevier, Amsterdam.
9. Biberstein, E. L. (1975). Miscellaneous pasteurelloses. *In* "Diseases Transmitted to Man" (W. T. Hubbert, W. F. McCulloch, and P. R. Schnurrenberger, eds.), pp. 139–146. Thomas, Springfield, Illinois.
10. Biberstein, E. L. (1975). Rat bite fever. *In* "Diseases Transmitted from Animals to Man" (W. T. Hubbert, W. F. McCulloch, and P. R. Schnurrenberger, eds.), pp. 186–190. Thomas, Springfield, Illinois.
11. Blackmore, D. K., and Francis, R. A. (1970). The apparent transmission of staphylococci of human origin to laboratory animals. *J. Comp. Pathol.* **80,** 645–651.
12. Blank, F. (1955). Dermatophytes of animals transmissible to man. *Am. J. Med. Sci.* **229,** 302–316.
13. Borgen, L. O. (1948). Infection with *Actinomyces muris ratti* after a rat bite. *Acta Pathol. Microbiol. Scand.* **25,** 161–166.

14. Bowen, G. S., Calisher, C. H., Winkler, W. G., Kraus, A. L., Fowler, E. H., Garman, R. H., Fraser, D. W., and Hinman, A. R. (1975). Laboratory studies of a lymphocytic choriomeningitis virus outbreak in man and laboratory animals. *Am. J. Epidemiol.* **102**, 233–240.

15. Brown, T. M., and Nunemaker, J. C. (1942). Rat-bite fever. A review of American cases with reevaluation of its etiology: Report of cases. *Bull. Johns Hopkins Hosp.* **70**, 201–328.

16. Bruner, W. D., and Gilespie, J. H. (1973). "Hagan's Infectious Diseases of Domestic Animals." Cornell Univ. Press, Ithaca, New York.

17. Burgdorfer, W. (1975). Rickettsialpox. Rocky mountain spotted fever. Murine (flea-borne) typhus fever. Boutonneuse fever. North Asian tick typhus. Scrub typhus. *In* "Diseases Transmitted from Animals to Man" (W. T. Hubbert, W. F. McCulloch, and P. R. Schnurrenberger, eds.), pp. 393–414 and 418–429. Thomas, Springfield, Illinois.

18. Carbeck, R. B., Murphy, J. F., and Pratt, E. M. (1967). Streptobacillary rat-bite fever with massive pericardial effusion. *J. Am. Med. Assoc.* **201**, 703–704.

19. CDC Morbidity and Mortality Weekly Report (1974). Rat-bite fever. Vol. 23, pp. 357–358. Center for Disease Control, U.S. Department of Health, Education, and Welfare, Atlanta, Georgia.

20. CDC Morbidity and Mortality Weekly Report. (1977). Leptospirosis—United States. Vol. 26, pp. 21–22. Center for Disease Control, U.S. Department of Health, Education, and Welfare, Atlanta, Georgia.

21. CDC Salmonella Surveillance (1976). "Annual Summary 1975," Rep. No. 126. Center for Disease Control, U.S. Department of Health, Education, and Welfare, Atlanta, Georgia.

22. CDC Veterinary Public Health Notes (1976). Rodent rabies. July, p. 2. Center for Disease Control, U.S. Department of Health, Education, and Welfare, Atlanta, Georgia.

23. Clarenburg, A. (1964). Salmonellosis. *In* "Zoonoses" (J. van der Hoeden, ed.), pp. 133–161. Elsevier, Amsterdam.

24. Clinton, J. M. (1969). Rats in urban America. *Public Health Rep.* **84**, 1–7.

25. Cole, J. S., Stoll, R. W., and Bulger, R. J. (1969). Rat-bite fever. Report of three cases. *Ann. Intern. Med.* **71**, 979–981.

26. Davis, D. E. (1951). The relation between the level of population and the prevalence of *Leptospira, Salmonella,* and *Capillaria* in Norway rats. *Ecology* **32**, 465–468.

27. Davis, D. H. S., Hallett, A. F., and Isaacson, M. (1975). Plague. *In* "Diseases Transmitted from Animals to Man" (W. T. Hubbert, W. F. McCulloch, and P. R. Schnurrenberger, eds.), pp. 147–173. Thomas, Springfield, Ill.

28. Diesch, S. L., and Ellinghausen, H. C. (1975). Leptospiroses. *In* "Diseases Transmitted from Animals to Man" (W. T. Hubbert, W. F. McCulloch, and P. R. Schnurrenberger, eds.), pp. 436–462. Thomas, Springfield, Illinois.

29. Diesch, S. L., Glosser, J. W., Hanson, L. E., Morter, R. L., Smith, R. E., and Stoenner, H. G. (1976). Leptospirosis of domestic animals. *U.S. Dep. Agric., Agric. Inf. Bull.* **394.**

30. Dinger, J. E. (1964). Plague. *In* "Zoonoses" (J. van der Hoeden, ed.), pp. 58–73. Elsevier, Amsterdam.

31. Dolan, M. M., and Fendrick, A. J. (1959). Incidence of *Trichophyton mentagrophytes* infections in laboratory rats. *Proc. Anim. Care Panel* **9**, 161–164.

32. Dolan, M. M., Kligman, A. M., Kobylinski, P. G., and Motsavage, M. A. (1958). Ringworm epizootics in laboratory mice and rats: Experimental and accidental transmission of infection. *J. Invest. Dermatol.* **30**, 23–25.

33. Dolman, C. E., Kerr, D. E., Chang, H., and Shearer, A. R. (1951). Two cases of rat-bite fever due to *Streptobacillus moniliformis. Can. J. Public Health* **42**, 228–241.

34. Edwards, P. R., and Galton, M. M. (1967). Salmonellosis. *Adv. Vet. Sci.* **11**, 2–63.

35. Floch, H., Constant, Y., and Destombes, P. (1950). Sur le sodoku en Guyane française et son traitement par la streptomycine. *Bull. Soc. Pathol. Exot.* **43**, 406–410.

36. Flynn, R. J. (1973). "Parasites of Laboratory Animals." Iowa State Univ. Press, Ames.

37. Frenkel, J. K., Good, J. T., and Schultz, J. A. (1966). Latent *Pneumocystis* infection of rats, relapse, and chemotherapy. *Lab. Invest.* **15**, 1559–1577.

38. Frenkel, J. K., and Havenhill, M. A. (1963). The corticoid sensitivity of golden hamsters, rats, and mice. Effects of dose, time, and route of administration. *Lab. Invest.* **12**, 1204–1220.

39. Friedmann, C. T., Spiegel, E. L., Aaron, E., and McIntyre, R. (1973). Leptospirosis *ballum* contracted from pet mice. *Calif. Med.* **118**, 51–52.

40. Fulton, J. D., and Joiner, L. P. (1948). Natural amoebic infections in laboratory rodents. *Nature (London)* **161**, 66–68.

41. Galton, M. M., Menges, R. W., Shotts, E. B., Nahmias, A. J., and Heath, C. W. (1962). Leptospirosis. Epidemiology, clinical manifestations in man and animals, and methods in laboratory diagnosis. *U.S., Public Health Serv. Publ.* **951.**

42. Geller, E. H. (1976). Personal communications.

43. Geller, E. H., Nusbaum, K. E., and Szilagyi, G. (1977). Unpublished observations.

44. Gilbert, G. L., Cassidy, J. F., and Bennett, N. McK. (1971). Rat-bite fever. *Med. J. Aust.* **2**, 1131–1134.

45. Goldberg, M. (1972). Adequate protection. *Science* **178**, 1151.

46. Griffith, R. L., and McNaughton, D. W. (1953). Report of rat-bite fever due to *Streptobacillus moniliformis. Public Health Rep.* **68**, 947–948.

47. Habermann, R. T., and Williams, F. P. (1958). Salmonellosis in laboratory animals. *J. Natl. Cancer Inst.* **20**, 933–941.

48. Hamburger, M., and Knowles, H. C. (1953). *Streptobacillus moniliformis* infection complicated by acute bacterial endocarditis. *Arch. Intern. Med.* **92**, 216–220.

49. Hayes, E. R., Kidd, E. G., and Cowan, D. W. (1950). Rat bite fever due to *Streptobacillus moniliformis. J.-Lancet* **70**, 394–395.

50. Heath, C. W., Alexander, A. D., and Galton, M. M. (1965). Leptospirosis in the United States. *N. Engl. J. Med.* **273**, 857–864 and 915–922.

51. Heisch, R. B. (1950). A case of rat-bite fever in Kenya Colony and the discovery of rodents naturally infected with *Spirillum minus. J. Trop. Med. Hyg.* **53**, 33–39.

52. Herrenberger, R. (1954). A contribution to the study of "sodoku" in North Africa. *Trop. Dis. Bull.* **51**, 270.

53. Higa, H. H., and Fujinaka, I. T. (1976). Prevalence of rodent and mongoose leptospirosis on the island of Oahu. *Public Health Rep.* **91**, 171–177.

54. Hitzig, W. M., and Liebesman, A. (1944). Subacute endocarditis associated with infection with a *Spirillum. Arch. Intern. Med.* **73**, 415–424.

55. Holden, F. A., and MacKay, J. C. (1964). Rat-bite fever—An occupational hazard. *Can. Med. Assoc. J.* **91**, 78–81.

56. Hopla, C. E. (1974). The ecology of tularemia. *Adv. Vet. Sci. Comp. Med.* **18**, 25–53.

57. Humphreys, F. A., Campbell, A. G., Driver, M. W., and Hatton, G. N. (1950). Rat-bite fever. *Can. J. Public Health* **41**, 66–71.

58. Institute of Laboratory Animal Resources (1971). "A Guide to Infectious Diseases of Mice and Rats." Natl. Acad. Sci., Washington, D.C.

58a. James, P. D. (1977). Ventilated animal racks: A practical system.

AALAS Publ. **77-3,** 104. American Association for Laboratory Animal Science.

59. Kaplan, W. (1967). Epidemiology and public health significance of ringworm in animals. *Arch. Dermatol.* **96,** 404–408.

60. Kemenes, F., and Temesvari, E. (1967). A new leptospiral serotype in albino rat breeds. *Acta Vet. Acad. Sci. Hung.* **17,** 257–261.

61. Kuzina, R. F., and Lifshitz, Y. I. (1974). Laboratory animals as a source of spread of salmonellae. *Zh. Mikrobiol., Epidemiol. Immunobiol.* **51,** 126–128.

62. Lainson, R., Garnham, P. C., Killick-Kendrick, R., and Bird, R. G. (1964). Nosematosis, a microsporidial infection of rodents and other animals, including man. *Br. Med. J.* **2,** 470–472.

63. Lambe, D. W., Jr., McPhedran, A. M., Mertz, J. A., and Stewart, P. (1973). *Streptobacillus moniliformis* isolated from a case of Haverhill fever. *Am. J. Clin. Pathol.* **60,** 854–860.

64. Lehmann-Grube, F. (1971). Lymphocytic choriomeningitis virus. *Virol. Monogr.* **10,** 88, 118.

65. Lindenbaum, I., and Eylan, E. (1970). Human infections with *Leptospira ballum* in Israel. *Isr. J. Med. Sci.* **6,** 403–407.

66. Lominski, I. R., Henderson, A. S., and McNee, J. W. (1948). Rat-bite fever due to *Streptobacillus moniliformis*. *Br. Med. J.* **2,** 510–511.

67. Loosli, R. (1967). Zoonoses in common laboratory animals. *In* "Husbandry of Laboratory Animals" (M. L. Conalty, ed.), pp. 307–325. Academic Press, New York.

68. Lubsen, N., van der Plaats, A., and Wolthuis, F. (1950). A case of rat bite fever caused by *Streptobacillus moniliformis*. *Nederland. Tijdschr. Geneesk* **94,** 102–106.

69. Lutsky, I., and Neumann, I. (1975). Laboratory animal dander allergy. I. An occupational disease. *Ann. Allergy* **35,** 201–205.

70. Mair, N. S. (1973). Yersiniosis in wildlife and its public health implications. *J. Wildl. Dis.* **9,** 64–71.

71. Maiztegu, J. (1972). Activity of lymphocytic choriomeningitis virus (LCM) in the endemic area of Argentine hemorrhagic fever (AHF). I. Serologic studies in rodents captured in the city of Pergamino. *Medicina (Buenos Aires)* **32,** 131–137.

72. Margileth, A. M., Strano, A. J., Chandra, R., Neafie, R., Blum, M., and McCulty, R. M. (1973). Disseminated nosematosis in an immunologically comprised infant. *Arch. Pathol.* **95,** 145–150.

73. Matsubayashi, H., Koike, T., Mikata, I., Takei, H., and Hagiwara, S. (1959). A case of *Encephalitozoon*-like body infection in man. *Arch. Pathol.* **67,** 181–187.

74. McCormack, R. C., Kaye, D., and Hook, E. W. (1967). Endocarditis due to *Streptobacillus moniliformis*. *J. Am. Med. Assoc.* **200,** 77–79.

75. McDermott, W., Leask, M., and Benoit, M. (1945). *Streptobacillus moniliformis* as a cause of subacute bacterial endocarditis: Report of a case treated with penicillin. *Ann. Intern. Med.* **23,** 414–423.

76. McGill, R. C., Martin, A. M., and Edmunds, P. N. (1966). Ratbite fever due to *Streptobacillus moniliformis*. *Br. Med. J. 1,* 1213–1214.

77. McKiel, J. A., Rappay, D. E., Cousineau, J. G., Hall, R. R., and McKenna, H. E. (1970). Domestic rats as carriers of leptospires and salmonellae in eastern Canada. *Can. J. Public Health* **61,** 336–340.

78. Mesina, J. E., and Campbell, R. S. (1975). Wild rodents in the transmission of disease to animals and man. *Vet. Bull.* **45,** 87–96.

79. Monath, T. P., Newhouse, V. F., Kemp, G. E., Setzer, H. W., and Cacciapuoti, A. (1974). Lassa virus isolation from *Mastomys natalensis* rodents during an epidemic in Sierra Leone. *Science* **185,** 263–265.

80. Neal, R. A. (1951). The duration and epidemiological significance of *Entamoeba histolytica* infections in rats. *Trans. R. Soc. Trop. Med. Hyg.* **45,** 363–370.

81. Neumann, I., and Lutsky, I. (1976). Laboratory animal dander allergy. II. Clinical studies and the potential protective effect of disodium cromoglycate. *Ann. Allergy* **36,** 23–29.

82. Nicolescu, M., Borsai, L., and Alamita, I. (1973). Leptospirosis in albino rats. *Arch. Roum. Pathol. Exp. Microbiol.* **32,** 171–177.

83. Olsen, P. F. (1975). Tularemia. *In* "Diseases Transmitted from Animals to Man" (W. T. Hubbert, W. F. McCulloch, and P. R. Schnurrenberger, eds.), pp. 191–223. Thomas, Springfield, Illinois.

84. Patterson, R. (1964). The problem of allergy to laboratory animals. *Lab. Anim. Care* **14,** 466–469.

85. Plesko, I., Hruzik, J., and Hlavata, Z. (1975). Occupational leptospiroses in laboratory workers. *Cesk. Epidemiol., Mikrobiol., Immunol.* **24,** 77–83.

86. Povar, M. L. (1965). Ringworm (*Trichophyton mentagrophytes*) infection in a colony of albino Norway rats. *Lab. Anim. Care* **15,** 264–265.

87. Priest, W. S., Smith, J. M., and McGee, C. J. (1947). Penicillin therapy of subacute bacterial endocarditis. *Arch. Intern. Med.* **79,** 333–359.

88. Prouty, M., and Schaefer, E. L. (1950). Periarteritis nodosa associated with ratbite fever due to *Streptobacillus moniliformis* (erythema arthriticum epidermicum). *J. Pediatr.* **36,** 605–613.

89. Public Health Service (1956). Rabid rat attacks boy. *J. Am. Vet. Med. Assoc.* **128,** 148.

90. Rosen, M. N. (1975). Clostridial infections and intoxications. *In* "Diseases Transmitted from Animals to Man" (W. T. Hubbert, W. F. McCulloch, and P. R. Schnurrenberger, eds.), pp. 251–262. Thomas, Springfield, Illinois.

91. Roughgarden, J. W. (1965). Antimicrobial therapy of rat-bite fever. *Arch. Intern. Med.* **116,** 39–54.

92. Rowsell, H. C. (1963). Mycotic infections of animals transmissible to man. *Am. J. Med. Sci.* **245,** 333–344.

93. Ruys, A. C. (1964). Rat bite fevers. *In* "Zoonoses" (J. van der Hoeden, ed.), pp. 233–239. Elsevier, Amsterdam.

94. Scholz, V., and Weinhold, E. (1969). Epidemiology of rabies in rats and mice. *Berl. Muench. Tieraerztl. Wochenschr.* **82,** 255–257.

95. Shadduck, J. A., and Pakes, S. P. (1971). Encephalitozoonosis (nosematosis) and toxoplasmosis. *Am. J. Pathol.* **64,** 657–671.

96. Stalheim, O. H. (1973). Chemical aspects of leptospiroses. *Crit. Rev. Microbiol.* **2,** 423–456.

97. Steele, J. H. (1960). Epidemiology of leptospirosis. *J. Am. Vet. Med. Assoc.* **136,** 247–252.

98. Steen, E. (1951). Rat-bite fever. Report of a case with examination of *Haverhillia moniliformis*. *Acta Pathol. Microbiol. Scand.* **28,** 17–26.

99. Stoenner, H. G. (1957). The laboratory diagnosis of leptospirosis. *Vet. Med.* **52,** 540–542.

100. Stoenner, H. G., Grimes, E. F., Thrailkill, F. B., and Davis, E. (1958). Elimination of *Leptospira ballum* from a colony of Swiss albino mice by use of chlortetracycline hydrochloride. *Am. J. Trop. Med. Hyg.* **7,** 423–426.

101. Stoenner, H. G., and MacLean, D. (1958). Leptospirosis (*ballum*) contracted from Swiss albino mice. *Arch. Intern. Med.* **101,** 606–610.

102. Stokes, J. F., Gray, I. R., and Stokes, E. J. (1951). *Actinomyces* endocarditis treated with chloramphenicol. *Br. Heart J.* **13,** 247–251.

103. Strangeways, W. I. (1933). Rats as carriers of *Streptobacillus moniliformis*. *J. Pathol. Bact. ol.* **37,** 45–51.

104. Sulzer, C. R., Harvey, T. W., and Galton, M. M. (1968). Comparison of diagnostic technics for the detection of leptospirosis in rats. *Health Lab. Sci.* **5,** 171–173.

105. Sulzer, C. R., and Jones, W. L. (1973). Evaluation of a hemagglutination test for human leptospirosis. *Appl. Microbiol.* **26,** 655–657.

106. Switzer, W. P., Mare, C. J., and Hubbard, E. D. (1966). Incidence of *Bordetella bronchiseptica* in wildlife and man in Iowa. *Am. J. Vet. Res.* **27,** 1134–1136.

107. Taylor, G., Davies, G. E., Altounyan, R. E., Brown, H. M., Frankland, A. W., Smith, J. M., and Winch, R. (1976). Allergic reactions to

laboratory animals. *Nature (London)* **260,** 280.

107a. Tennant, R. W. (1977). Personal communication.

108. Tennant, R. W., Reynolds, R. K., and Layman, K. R. (1967). Incidence of murine virus antibody in humans in contact with experimental animals. *Exp. Hematol.* **14,** 76 (Oak Ridge National Laboratory).

109. Tesh, R. B., Peralta, P. H., and Johnson, K. M. (1969). Ecologic studies of vesicular stromatitis virus. *Am. J. Epidemiol.* **90,** 255-261.

110. van der Hoeden, J. (1964). Leptospirosis. *In* "Zoonoses" (J. van der Hoeden, ed.), pp. 240-274. Elsevier, Amsterdam.

111. van der Hoeden, J. (1964). Pasteurellosis. *In* "Zoonoses" (J. van der Hoeden, ed.), pp. 50-53. Elsevier, Amsterdam.

112. van der Hoeden, J. (1964). Pseudotuberculosis. *In* "Zoonoses" (J. van der Hoeden, ed.), pp. 54-57. Elsevier, Amsterdam.

113. Varela, G., and Roche, E. (1954). Natural infections of rats with rat-bite fever. *Trop. Dis. Bull.* **51,** 269.

114. Walzer, P. D., Schelle, V., Armstrong, D., and Rosen, P. P. (1977). Nude mouse: A new experimental model for *Pneumocystis carinii* infection. *Science* **197,** 177-179.

115. Waterson, A. P., and Wedgwood, J. (1953). Rat-bite fever. Report of a case due to *Actinomyces muris. Lancet* **1,** 472-473.

116. Wedding, F. S. (1947). Actinomycotic endocarditis. *Arch. Intern. Med.* **79,** 203-227.

117. Weisbroth, S. H., and Freimer, E. H. (1969). Laboratory rats from commercial breeders as carriers of pathogenic pneumococci. *Lab. Anim. Care* **19,** 473-478.

118. Williams, L. P., and Hobbs, B. C. (1975). Enterobacteriaceae infections. *In* "Diseases Transmitted from Animals to Man" (W. T. Hubbert, W. F. McCulloch, and P. R. Schnurrenberger, eds.), pp. 33-109. Thomas, Springfield, Illinois.

119. Wolff, J. W. (1964). Flea-borne rickettsioses. Endemic typhus. *In* "Zoonoses" (J. van der Hoeden, ed.), pp. 288-290. Elsevier, Amsterdam.

120. World Health Organization (1975). Biological hazards associated with Mastomys. *WHO Chron.* **29,** 241.

121. Yunker, C. E. (1964). Infections of laboratory animals potentially dangerous to man: Ectoparasites and other arthropods, with emphasis on mites. *Lab. Anim. Care* **14,** 455-465.

122. Zinsser, H. (1935). "Rats, Lice and History." Little, Brown, Boston, Massachusetts.

Appendix 1

Selected Normative Data

Henry J. Baker, J. Russell Lindsey, and Steven H. Weisbroth

Adult weight	
Male	300–400 gm
Female	250–300 gm
Life Span	
Usual	2.5–3 years
Maximum reported	4 yrs. 8 mo.
Surface area	0.03–0.06 cm²
Chromosome number (diploid)	42
Water consumption	80–110 ml/kg/day
Food consumption	100 gm/kg/day
Body temperature	99.5°F, 37.5°C
Puberty	
Male	50 ± 10 days
Female	50 ± 10 days
Breeding season	None
Gestation	21–23 days
Litter size	8–14 pups
Birth weight	5–6 gm
Eyes open	10–12 days
Weaning	21 days
Heart rate	330–480 beats/min
Blood pressure	
Systolic	88–184 mm Hg
Diastolic	58–145 mm Hg
Cardiac output	50 (10–80) ml/min
Blood volume	
Plasma	40.4 (36.3–45.3) ml/kg
Whole blood	64.1 (57.5–69.9) ml/kg
Respiration frequency	85.5 (66–114)/min
Tidal volume	0.86 (0.60–1.25) ml

(continued)

(Continued)

Minute volume	0.073 (0.05–0.101) ml
Stroke volume	1.3–2.0 ml/beat
Plasma	
pH	7.4 ± 0.06
CO_2	22.5 ± 4.5 mM/liter
CO_2 pressure	40 ± 5.4 mm Hg
Leukocyte counts	
Total	$14 (5.0–25.0) \times 10^3/\mu$l
Neutrophils	22 (9–34)%
Lymphocytes	73 (65–84)%
Monocytes	2.3 (0–5)%
Eosinophils	2.2 (0–6)%
Basophils	0.5 (0–1.5)%
Platlets	$1240 (1100–1380) \times 10^3/\mu$l
Packed cell volume	46%
Red blood cells	$7.2–9.6 \times 10^6$/mm^3
Hemoglobin	15.6 gm/dl
Maximum volume of single bleeding	5 ml/kg
Urine	
pH	7.3–8.5
Specific gravity	1.04–1.07

Appendix 2

Drugs and Dosages

Sam M. Kruckenberg

Table I

Drug	Classification	Dosage	Comment	Reference[a]
Acetylsalicylic acid	Analgesic	450 mg/kg po	Analgesic dose	22
α-Prodine	Analgesic	1.5–6 mg/kg sc	Analgesic activity dosages	33
Aminopyrineone	Analgesic	25–100 mg/kg sc	Analgesic activity dosages	33
Codeine	Analgesic	6.25–25 mg/kg sc	Analgesic activity dosages	33
Codeine	Analgesic	20 mg/kg sc	Hypnotic, analgesic, and antitussive drug	34
d-Propoxyphene	Analgesic	25 mg/kg ip	Analgesic dosage	34
Levorphan	Analgesic	0.125–0.5 mg/kg sc	Analgesic activity dosages	33
Meperidine	Analgesic	6.25–25 mg/kg sc	Analgesic activity dosages	33
Morphine	Analgesic	1.5–6 mg/kg sc	Analgesic activity dosages	33
Nalorphine	Analgesic	1.0–2.1 mg/kg sc	Analgesic—in writing tests	34
Nalorphine	Analgesic	5 mg/kg im, sc	Analgesic ED_{50}	34
Phenazocine	Analgesic	0.25–0.5 mg/kg sc	Analgesic dosage	34
Phenylbutazone	Analgesic	25–100 mg/kg sc	Analgesic activity dosages	33
Sodium salicylate	Analgesic	25–100 mg/kg sc	Analgesic activity dosages	33
α-Chloralose	Anesthetic	55 mg/kg ip	Anesthetic dose	22
Anesthetic drugs	Anesthetic	Various drugs and dosages	14 Drugs, dosages, routes and references, p. 19	1
Chloral hydrate[b]	Anesthetic	300 mg/kg ip	Anesthetic dose	22
Hexobarbital	Anesthetic	100 mg/kg ip	Anesthetic dose	15
Inactin	Anesthetic	160 mg/kg ip	Anesthetic dose for 150–400 gm rat	8
Innovar-Vet	Anesthetic	0.2–0.3 ml/kg im	Anesthetic dose	17
Innovar-Vet	Anesthetic	0.13 ml/kg im	Analgesia, sedation and tranquilization	23
Ketamine + atropine	Anesthetic	44 mg/kg ketamine im	0.04 mg/kg atropine given with ketamine	44
Ketamine hydrochloride	Anesthetic	44 mg/kg im	Surgical anesthesia	46
Methitural	Anesthetic	120 mg/kg ip	Anesthetic dosage	34
Pentobarbital-chlorpromazine	Anesthetic	25 mg/kg Chl. 20 mg/kg Pen.	Chloropromazine 30 min before pento-barbital	5

(continued)

Table I (*Continued*)

Drug	Classification	Dosage	Comment	Reference[a]
Pentobarbital	Anesthetic	30–40 mg/kg ip	Adult rat dosage	15
Pentobarbital	Anesthetic	10–30 mg/kg ip	Dosage for rats under 50 gm	5
Carbon dioxide	Anesthetic gaseous	Gaseous anesthetic	Quick knockdown for short-term procedures	15
Halothane	Anesthetic gaseous	1% with oxygen	No rats died when oxygen was given with gas	18
Methoxyflurane (Metofane)	Anesthetic gaseous	Inhalation	Multiple rat anesthetic apparatus	28
Atropine	Anesthetic (pre-)	0.04 mg/kg im	Used with ketamine to prevent salivation	46
Atropine	Anesthetic (pre-)	None given	Rat has a high atropinase level	25
Atropine	Anesthetic (pre-)	1–5 mg/kg sc	Preanesthetic	34
Atropine sulfate	Anesthetic (pre-)	2.5 mg/rat sc	Give 30 min before gas anesthesia	15
Chlordiazepoxide	Anesthetic (pre-)	10 mg/kg po	Preanesthetic dosage	10
Mikedimide	Anesthetic–antagonist	1 mg/mg of pentobarbital	Barbiturate antagonist–respiratory stimulant	15
Meperidine	Anesthetic-narcotic	2–16 mg/kg ip	Analgesic dosage	34
Morphine	Anesthetic-narcotic	1.25–20 mg/kg/hr iv	Addictive dose	19
Morphine sulfate	Anesthetic-narcotic	20 mg/kg	Narcotic	14
Opium	Anesthetic-narcotic	0.0025 mg/kg	Narcotic analgesic expectorant dosage	34
Levallorphan	Anesthetic-narcotic antagonist	6 mg/kg sc	Narcotic antagonist	34
Levorphanol	Anesthetic-narcotic antagonist	3–5 mg/kg sc	Narcotic antagonist	34
Methadone	Anesthetic-narcotic antagonist	1 mg/kg ip	Dosage to stop shakes in morphine addiction	34
Phenobarbital	Anesthetic–sedative	15 mg/kg ip	Sedative dosage	34
Aminopyrine	Antiinflammatory	20 mg/kg ip	Antipyretic dosage	39
Aminopyrine	Antiinflammatory	20–40 mg/kg ip	Antiinflammatory	39
Benzydamine	Antiinflammatory	15–45 mg/kg sc	Analgesic and antiinflammatory	39
Chlorpromazine	Antiinflammatory	0.5–4 mg/kg iv	Antiedema dosages	31
Cortisone	Antiinflammatory	25 mg/kg sc	Prevent inflammation caused by brewers yeast	36
Indomethacin	Antiinflammatory	0.1–0.5 mg/rat	Inhibition of cotton pellet granuloma	39
Indomethacin	Antiinflammatory	4 mg/kg po	Analgesic and antiinflammatory drug	34
Methotrimeprazine	Antiinflammatory	0.1–1 mg/kg iv	Antiedema dosages	31
Methylpromazine	Antiinflammatory	0.5–4 mg/kg iv	Antiedema dosages	31
Oxyphenbutazone	Antiinflammatory	100 mg/kg po	Protective action in nicotinic acid deficiency	10
Phenylbutazone	Antiinflammatory	7.5–15 mg/kg sc	Antiinflammatory dosage	39
Phenylbutazone	Antiinflammatory	30 mg/kg po	Antiinflammatory dosage	39
Phenylbutazone	Antiinflammatory	100 mg/kg po daily	Inhibited primary arthritic lesions	29
Promethazine	Antiinflammatory	2–8 mg/kg iv	Antiedema dosages	31
Trimeprazine	Antiinflammatory	0.25–1 mg/kg iv	Antiedema dosages	31
6-Mercaptopurine	Antiinflammatory	2 mg/kg	Antiinflammatory in experimental arthritis	34
Chloramphenicol palmitate	Antimicrobial-antibiotic	20–50 mg/kg po	Treat once or twice a day	35
Chloramphenicol succinate	Antimicrobial-antibiotic	6.6 mg/kg im	Treat once or twice a day	35
Chlortetracycline	Antimicrobial-antibiotic	0.1% of water	Effective against *Corynebacterium kutscheri*	12
Chlortetracycline	Antimicrobial-antibiotic	6–10 mg/kg im	Treat once or twice a day	35
Chlortetracycline	Antimicrobial-antibiotic	0.025% of diet	Did not eliminate rhinitis, ear infection, etc.	13
Gentamicin	Antimicrobial-antibiotic	4.4 mg/kg im	Treat one or two times a day	35
Oxytetracycline	Antimicrobial-antibiotic	6–10 mg/kg IM	Treat one or two times a day	35
Penicillin g-procaine	Antimicrobial-antibiotic	20,000 units/kg po	Treat once daily	35
Tetracycline	Antimicrobial-antibiotic	0.1–.5% of diet	14-Day treatment for labrynthitis	34
Tetracycline	Antimicrobial-antibiotic	15–20 mg/kg po	Treat two or three times a day	35
Tylosin	Antimicrobial-antibiotic	2–4 mg/kg im	Treat once or twice a day	35
Amphotericin B	Antimicrobial-fungal	6.25 mg/kg sc daily	Fungal prevention in cortisoned rats	12

(*continued*)

Table I (*Continued*)

Drug	Classification	Dosage	Comment	Reference[a]
Griseofulvin	Antimicrobial–fungal	0.03% of feed	Fungal prevention in cortisoned rats	12
Pimaricin	Antimicrobial–fungal	100–200 ppm of feed	Eliminated yeast flora	26
Nitrofurantoin	Antimicrobial–furacin	12.5 mg/kg B.I.D. for 3–7 day	Treatment of *Proteus* or *E. coli* bladder infection	34
Sulfadiazine	Antimicrobial–sulfa	220 mg/kg then 110 mg/kg	Initial dose twice maintainance dose	34
Sulfadimethoxine	Antimicrobial–sulfa	20–50 mg/kg po	Treat once daily	35
Sulfamerazine	Antimicrobial–sulfa	0.025% of diet	Eliminated rhinitis, middle ear infection, etc.	13
Sulfamerazine	Antimicrobial–sulfa	50–80 mg/kg po	May be used in drinking water also	35
Sulfaquinoxaline	Antimicrobial–sulfa	0.025–0.1% of water	Therapeutic dose	35
Sulfaquinoxaline	Antimicrobial–sulfa	0.05% of diet	Therapeutic dose	35
Triple sulfas	Antimicrobial–sulfa	0.2% of water	Treatment for pneumonia	6
Bithional	Antiparasitic	35 mg/kg	Anthelmintic dosage for *Fasciola*	34
Carbarsone	Antiparasitic	0.05% of feed	Slight suppression of *Pneumocystis*	12
Carbon tetrachloride	Antiparasitic	800–1600 mg/kg po ip	Anthelmintic	34
Chloroquine	Antiparasitic	0.012% of feed	Ineffective against *Pneumocystis*	12
Dichlorvos (DDVP)	Antiparasitic	0.05–0.1% of diet	Used for 3–5 days not effective *T. crassicauda*	47
Emetine	Antiparasitic	5 mg/kg	Treatment for amoebiasis	34
Hexachloroethane	Antiparasitic	0.6–1.2 mg/kg po	Anthelmintic dosage	34
Hexachlorophene	Antiparasitic	10 mg/kg	Anthelmintic dose	34
Hydroxystilbamidine	Antiparasitic	0.17–1 mg/rat daily	Kills pneumocystis	12
Lead arsenate	Antiparasitic	0.5% of diet	Treatment of *Hymenolepis nana*	34
Malathion	Antiparasitic	0.125% dip	For control of external parasites	34
Methoxychlor	Antiparasitic	Topical application	Up to 2% dust for external parasites	34
Methyridine	Antiparasitic	100 mg/kg ip	Anthelmintic dose	34
Methyridine	Antiparasitic	200 mg/kg	Single dose effective against *Trichosomides*	32
Methyridine	Antiparasitic	65–125 mg/kg/day/4 days po	0.15% of drinking water for 4 days	34
Metronidazole	Antiparasitic	10–40 mg/day/rat	Slightly effective in killing *Pneumocystis*	12
Metronidazole	Antiparasitic	125 mg/kg ip	Trichomonacide	34
Niclosamide	Antiparasitic	100 mg/kg/day for 21 days	Effective against *Hymenolepis nana* and *H. diminuta*	16
Niclosamide	Antiparasitic	50–100 mg/kg po	Treatment for *Hymenolepis diminuta* infection	34
Niclosamide	Antiparasitic	30–90 mg/kg po in feed	Treat 3 days/off 3 days/treat 3 days in feed	35
Niridazole	Antiparasitic	100 mg/kg sc	Ineffective against *T. crassicauda*	47
Niridazole	Antiparasitic	0.05% of diet	Ineffective against *T. crassicauda*	47
Nitrofurantoin	Antiparasitic	0.2% of feed	Eradication of *Trichosomoides crassicauda*	4
Pentamidine isethionate	Antiparasitic	20 mg/kg sc daily	Causes tissue necrosis at injection site	12
Phenothiazine	Antiparasitic	1000 mg/kg po	Treatment for *Heterakis spumosa* infection	41
Piperazine	Antiparasitic	0.2% of water	Ineffective against *T. crassicauda*	47
Piperazine	Antiparasitic	200 mg/100 ml water	Anthelmintic dose for pinworm infections	34
Piperazine citrate	Antiparasitic	500–1000 mg/kg po-water	Treat week-off week-treat week	35
Pyrantel	Antiparasitic	50 mg/kg po	Ineffective against *T. crassicauda*	47
Pyrantel	Antiparasitic	6 mg/kg	Anthelmintic	34
Resorantel	Antiparasitic	20 mg/kg po	Treatment for *Hymenolepis diminuta* infection	34
SKF 29044	Antiparasitic	0.5% of diet-14 days	Partially effective against *T. crassicauda*	47
Tetramisole	Antiparasitic	40 mg/kg po	Partially effective against *T. crassicauda*	47
Tetramisole	Antiparasitic	12–25 mg/kg in water	Partially effective against *T. crassicauda*	47
Thiabendazole	Antiparasitic	100 mg/kg po	Ineffective against *T. crassicauda*	47
Thiabendazole	Antiparasitic	200 mg/kg for 5 days po	Anthelmintic	34
Thiabendazole	Antiparasitic	50 mg/kg ip	Partially effective against *T. crassicauda*	47

(*continued*)

Table I (*Continued*)

Drug	Classification	Dosage	Comment	Reference[a]
Pyrimethazine plus sulfadiazine	Antiparasitic–protozoal	0.03% of feed	Pneumocystosis treatment with sulfadiazine	12
Sulfadiazine plus pyrimethamine	Antiparasitic–protozoal	0.1% of feed	Pneumocystosis treatment with pyrimethazine	12
Metyrapone	Body function test	66 mg/kg ip	Pituitary and hypothalmic function test	34
Inulin	Body function test—kidney	90 mg/kg iv	Urinary excretion test for kidney function	8
Amaranth (azorubin s)	Body function test—liver	200 mg/kg iv	Plasma disappearance test for liver function	20
Bilirubin	Body function test—liver	25 mg/kg iv	Plasma disappearance test for liver function	20
Chlorothiazide	Body function test—liver	40 mg/kg iv	Plasma disappearance test for liver function	20
Dibromophthalein (DBSP)	Body function test—liver	96 mg/kg iv	Plasma disappearance test for liver function	20
Indocyanine green (ICG)	Body function test—liver	1–50 mg/kg iv	Plasma disappearance test for liver function	20
Ouabain octohydrate	Body function test—liver	5 mg/kg iv	Plasma disappearance test for liver function	20
PAEB	Body function test—liver	10 mg/kg iv	Procaine amide ethyl bromide liver function	20
Phenol red	Body function test—liver	30 mg/kg iv	Plasma disappearance test for liver function	20
Probenecid	Body function test—liver	75 mg/kg iv	Plasma disappearance test for liver function	20
Succinylsulfathiazole	Body function test—liver	40 mg/kg iv	Plasma disappearance test for liver function	20
Sulfobromophthalein BSP	Body function test—liver	120 mg/kg iv	Plasma disappearance test for liver function	20
Taurocholic acid-sodium	Body function test—liver	100 mg/kg iv	Plasma disappearance test for liver function	20
Iodine	Body function—thyroid	Iodine deficient diet	Triiodthyronine (T_3) and protein-bound iodine	45
Bretylium	CNS	5 mg/kg iv	Norepinephrine release inhibitor	1
Nervous system drugs	CNS and *peripheral* drugs	Various drugs and dosages	Review off peripheral nervous system drugs	3
Imipramine	CNS antidepressant	40 mg/kg po	Protective action in nicotinic acid deficiency	10
Imipramine	CNS antidepressant	20–40 mg/kg po	Depressant dosage	34
Imipramine	CNS antidepressant	Up to 10 mg/kg ip	Antidepressant	34
Dichloroisoproterenol	CNS beta blocking agent	0.1 mg/kg iv	Beta blocking agent	1
Nervous system drugs	CNS drugs	Various drugs and dosages	Review of CNS drugs	2
Mephenesin	CNS muscle relaxant	100 mg/kg po	Muscle relaxant	10
d-Tubocurarine tubarine	CNS nerve transmission	0.04–0.06 mg/kg iv	Neuromuscular junction affector	40
Decamethonium	CNS nerve transmission	1.25 mg/kg iv	Neuromuscular junction affector	40
Dimethine or Metubine	CNS nerve transmission	0.009 mg/kg iv	Neuromuscular junction affector	40
Laudolissin	CNS nerve transmission	1.05 mg/kg iv	Neuromuscular junction affector	40
Atropine	CNS parasympatholytic	10 mg/kg po 3 mg/kg sc	Anticholinergic parasympathathetic blockade	1
Acetylcholine	CNS parasympathomimetic	0.01 mg/kg iv	Direction acetylcholine sensitive sites	1
Dextroamphetamine	CNS stimulant	5 mg/kg po	CNS stimulant	10
Dextroamphetamine	CNS stimulant	1–2 mg/kg sc	Sympathomimetic and CNS stimulant	34
Pentylenetetrazol	CNS stimulant	50 mg/kg ip	Convulsive dose	34
Dihydrotestosterone	Hormone	125 μg/rat/day	Replacement therapy following castration	9
Epinephrine	Hormone		Sympathomimetic or adrenergic drug	34
Estradiol	Hormone	1–5 μg sc im	Estrogenic hormone	34
Estradiol benzoate	Hormone	.5–167 μg/rat/day	Replacement therapy following castration	9
Estradiol benzoate	Hormone	0.1–1 μg sc	Ovulation dosage for immature rats	48

(*continued*)

Table I (*Continued*)

Drug	Classification	Dosage	Comment	Reference[a]
Estrone	Hormone	20 μg im 200 units	Estrogenic hormone	34
Insulin	Hormone	Not over 3 units sc	Antidiabetogenic	34
Norethindrone	Hormone	15 μg oral	Progesteroid oral contraceptive	34
Oxytocin	Hormone	1 unit im or sc	Therapeutic dose	34
Parathyroid hormone	Hormone	30 units/kg ip		34
Pregnant mare's serum	Hormone	20 iu sc	Causes ovulation in immature rats	27
Progesterone	Hormone	3 mg sc im + estradiol	Causes eclampsia in pregnant rat	34
Prostaglandins	Hormone	0.5–1 μg iv	Three types of prostaglandins used	14
Testosterone	Hormone	125 μg/rat/day	Replacement therapy following castration	9
Testosterone	Hormone	0.5–1 mg/kg im sc	Androgenic hormone	34
Thyroxine	Hormone	30 μg/kg im sc		34
Adrenalectomized rat	Hormone replacement Rx	Various drugs and dosages	Hormone maintenance and replacement dosage, p. 278	1
Castrated rat	Hormone replacement Rx	Various drugs and dosages	Hormone maintenance and replacement dosage, p. 279	1
Hypophysectomized rat	Hormone replacement Rx	Various drugs and dosages	Hormone maintenance and replacement dosage, p. 277	1
Ovariectomized rat	Hormone replacement Rx	Various drugs and dosages	Hormone maintenance and replacement dosage, p. 279	1
Pancreatectomized rat	Hormone replacement Rx	Various drugs and dosages	Hormone maintenance and replacement dosage, p. 277	1
Parathyroidectomized rat	Hormone replacement Rx	Various drugs and dosages	Hormone maintenance and replacement dosage, p. 280	1
Thyroidectomized rat	Hormone replacement Rx	Various drugs and dosages	Hormone maintenance and replacement dosage, p. 280	1
Thyroparathyroidectomized	Hormone replacement Rx	Various drugs and dosages	Hormone maintenance and replacement dosage, p. 280	1
Dihydrotachysterol	Hormone synthetic	0.25–2 mg/rat sc po	Parathyroid-like hormone	34
Medroxyprogesterone	Hormone synthetic	300 μg sc	ED_{50} in delaying parturition	34
Acetylsalicylic acid	Miscellaneous	300 mg/kg	Analgesic and antipyretic	34
α-methyl-p-tyrosine	Miscellaneous	200 mg/kg po, 50 mg/kg ip	Catecholamine synthesis inhibitor	34
Apomorphine	Miscellaneous	1–1.5 mg/kg iv	Use to cause experimental agressiveness	34
Benadryl	Miscellaneous	100 mg/kg po	Toxic to rats at this dosage	10
Betaine hydrochloride	Miscellaneous	180–400 mg/kg po	Lipotropic agent and hydrochloric acid releaser	34
Carbon tetrachloride	Miscellaneous	800–1600/week po ip	Give weekly for liver necrosis	34
Chlorcyclizine HCl	Miscellaneous	25 mg/kg ip	Microsomal enzyme inducer	21
Cyclophosphamide	Miscellaneous	50–100 mg/kg ip	Antineoplastic drug	34
Diphenylhydantoin	Miscellaneous	20 mg/kg ip	Anticonvulsant	34
Evan's blue dye	Miscellaneous	2 mg/rat iv	Dosage to study vascular changes	31
Folinic acid	Miscellaneous	0.38–0.54 mg/kg daily	To treat pyrimethazine-sulfadiazine toxicity	12
γ-Chlordan	Miscellaneous	50 mg/kg ip	Microsomal enzyme inducer	21
Glycidol	Miscellaneous	15 mg/kg	Antifertility agent	34
Hydralazine	Miscellaneous	2 mg/kg ip	Hypotensive agent	34
Hydroxyurea	Miscellaneous	250 mg/kg ip	Antineoplastic drug	34
Iodoacetamide	Miscellaneous	50–100 mg/100 ml water	Ulcerogenic agent	34
Iproniazid	Miscellaneous	25–100 mg/kg ip	Monoamine oxidase inhibitor	34
Isoproterenol	Miscellaneous	Various dosages and routes	Adrenergic or sympathomimetic drug	34
Isosorbide	Miscellaneous	1–2 gm/kg	Osmotic diuretic	34
Lysergide-LSD	Miscellaneous	5–300 μg po ip	Psychopharmaceutical	34
Methotrexate	Miscellaneous	0.65–1 mg/kg po or ip	Antineoplastic drug	34
Methoxamine	Miscellaneous	2–3 mg/rat	Sympathomimetic or adrenergic drug	34
Metopirone (SU-4885)	Miscellaneous	10 mg 2×/day for 42 days	Causes hypertension	7
Minerals	Miscellaneous	Various dosages and routes	Chapter on the nutrient requirements of rats	30

(*continued*)

Table I (*Continued*)

Drug	Classification	Dosage	Comment	Reference[a]
Nicotinamide	Miscellaneous	10 mg/kg po	Protective action in nicotinic acid deficiency	10
Norethandrolone	Miscellaneous	0.0002–0.004% of feed	Anabolic agent	34
Nutrients	Miscellaneous	Various dosages and routes	Chapter on the nutrient requirements of rats	30
Oubain	Miscellaneous	0.05–0.09 mg/kg iv	Dosage to increase cardiac output	34
Pentazocine	Miscellaneous	2 mg/kg sc	Analgesic	34
Phenacetin	Miscellaneous	200 mg/kg	Antipyretic action lasts at least 4 hrs	34
Phenobarbital	Miscellaneous	75 mg/kg ip for 4 days	Microsomal enzyme inducer	20
Phentolamine	Miscellaneous	25 mg/kg ip	Alpha adrenergic blocking agent	34
Phenylbutazone sodium	Miscellaneous	125 mg/kg ip	Microsomal enzyme inducer	21
Physostigmine	Miscellaneous	0.02 mg/kg iv	Parasympathomimetic and anti-cholinesterase drug	34
Physostigmine	Miscellaneous	0.1–1 mg/kg im	Parasympathomimetic and anti-cholinesterase drug	34
Pilocarpine	Miscellaneous	0.5 mg/kg iv	Parasympathomimetic and cholinergic drug	34
Proadifen	Miscellaneous	25–50 mg/kg ip	Drug potentiator—liver microsome inhibitor	34
Procarbazine	Miscellaneous	30–250 mg/kg ip	Antineoplastic drug	34
Procyclidine	Miscellaneous	15 mg/kg	Anticholinergic	34
Promethazine	Miscellaneous	12.5 mg/kg ip, 1–5 mg/kg sc	Antihistamine	34
Prostaglandins	Miscellaneous	5–10 μg/kg	Experimental parenteral dosage	34
Ribaminol	Miscellaneous	25–50 mg/kg po	Learning and memory enhancer—experimental	34
Spironolactone	Miscellaneous	5–10 mg/rat B.I.D. sc or po	Diuretic—Aldosterone antagonist	34
Streptokinase	Miscellaneous	90,000 units po	Antiinflammatory	34
Teratogenic agents	Miscellaneous	Varous dosages and routes	649 drugs and agents tested—teratology	37
Thio-TEPA	Miscellaneous	1.8–8.4 mg/kg iv, ip	Antineoplastic drug	34
Tolbutamide	Miscellaneous	250–2000 mg/day	Hypoglycemic agent	34
Tranylcypromine	Miscellaneous	2.5–10 mg/kg ip po	Monoamine oxidase inhibitor	34
Tremorine	Miscellaneous	5–20 mg/kg po iv ip im sc	Parasympathomimetic or cholinergic	34
Triamterene	Miscellaneous	30 mg/kg po	Diuretic—naturetic	34
Tripelennamine	Miscellaneous	6 mg/kg iv	Antihistamine	34
Tyramine	Miscellaneous	0.5 mg/kg iv	Vasopressor drug	34
Uridine	Miscellaneous	10 mg/kg iv	Nucleoside	34
Vinblastine	Miscellaneous	0.4–0.65 mg/kg iv	Antineoplastic	34
Vitamins	Miscellaneous	Various dosages and routes	Chapter on the nutrient requirements of rats	30
2-Thiouracil	Miscellaneous	0.01% of ration	Thyroid inhibitor	34
6-Mercaptopurine	Miscellaneous	20–100/rat/day po or ip	Antineoplastic drug	34
Methylcholanthrene	Miscellaneous carcinogen	100–600 mg/kg	Treat weekly for 50–60 days for carcinogen	34
Zedalin-streptozotocin	Miscellaneous carcinogen	15:85 mixture of 50 mg/kg iv	To produce experimental renal tumors	34
Carrageenin	Miscellaneous irritant	0.05 ml of a 1% solution id	Edema and erythema following hindpaw injection	39
Compound 48/80	Miscellaneous irritant	0.1 ml id of 10 μg/ml	Hindpaw edema producing dosage	31
Croton oil and air	Miscellaneous irritant	25 ml air + oil sc	Granuloma pouch method for inflammation	24
Dextran	Miscellaneous irritant	0.1 ml id of 60 μg/ml	Hindpaw edema producing dosage	31
Egg white	Miscellaneous irritant	0.1 ml id of 0.5% solution	Hindpaw edema producing dosage	31
Formalin	Miscellaneous irritant	1 mg/kg of a 1% solution ip	Peritonitis and pleuritis following ip injection	39
Formalin	Miscellaneous irritant	0.1 ml of a 3% solution id	Edema and erythema following hindpaw injection	39
Histamine	Miscellaneous irritant	0.1 ml id of 1 mg/ml solution	Hindpaw edema producing dosage	31

(*continued*)

Table I (*Continued*)

Drug	Classification	Dosage	Comment	Reference[a]
Mustard	Miscellaneous irritant	0.1 ml of 2.5% solution	Edema and erythema following hindpaw injection	43
Phenylquinone	Miscellaneous irritant	0.25 ml of 0.02% solution	Peritonitis and pleuritis following ip injection	38
Silver nitrate	Miscellaneous irritant	0.2 ml of a 1% solution	Silver nitrate injected under Achilles tendon	39
5-Hydroxytryptamine	Miscellaneous irritant	0.1 ml id of 0.5 μg/ml	Hindpaw edema producing dosage	31
Atropine	Miscellaneous poison Rx	17 mg/kg + oximes 30 mg/kg	Treatment for organophosphate poisoning	34
Pralidoximes	Miscellaneous poison Rx	30 mg/kg atropine 17 mg/kg	Cholinesterase reactivator in organophosphate	34
Aminopterin	Miscellaneous—cytostatic	0.4–0.8 mg/kg po 3×/week	Dose prevented body weight gain	12
Azathioprine	Miscellaneous—cytostatic	0.02% of feed	Dose prevented body weight gain	12
Chlorambucil	Miscellaneous—cytostatic	8 mg/kg po 3×/week	Dose prevented body weight gain	12
Cyclophosphamide	Miscellaneous—cytostatic	20–40 mg/kg po 3×/week	Dose prevented body weight gain	12
Nitrogen mustard	Miscellaneous—cytostatic	0.2–0.4 mg/kg ip 3×/week	Dose prevented body weight gain	12
Vinblastine	Miscellaneous—cytostatic	0.25–0.50 mg/kg ip 3×/week	Dose prevented body weight gain	12
6-Mercaptopurine	Miscellaneous—cytostatic	75 mg/kg po 3×/week	Dose prevented body weight gain	12
Aldosterone	Steroid	20–40 μg/rat sc	Given B.I.D. to produce hypertension	11
Corticosterone	Steroid	40 mg/kg po 10 mg/kg im	Protective action in nicotinic acid deficiency	10
Cortisone	Steroid	0.25–1.25 mg/day sc, im, po	Antiinflammatory and glucocorticoid	34
Cortisone acetate	Steroid	135 mg/kg sc 2–3×/week	Dosage to "cortsone-condition" rats	12
Deoxycorticosterone	Steroid	40 mg/kg sc	Protective action in nicotinic acid deficiency	10
Dexamethazone	Steroid	0.05 mg/kg sc	Protective action in nicotinic acid deficiency	10
Hydrocortisone	Steroid	5–20 mg/kg po, im, or sc	Antiinflammatory and glucocorticoid	34
Prednisolone	Steroid	2–10 mg/kg im, sc, po	Immunosuppressant dosage	34
Triamcinolone	Steroid	0.01 mg/kg sc	Protective action in nicotinic acid deficiency	10
Chlordiazepoxide	Tranquilizer	3–10 mg/kg ip	CNS depressant and tranquilizer	34
Chlorpromazine	Tranquilizer	20 mg/kg po	Protective action in nicotinic acid deficiency	10
Guanethidine	Tranquilizer	5 mg/kg iv	Norepinephrine depleting agent	1
Haloperidol	Tranquilizer	0.35 mg/kg sc	Tranquilizer	34
Meprobamate	Tranquilizer	100 mg/kg po	Tranquilizer	10
Meprobamate	Tranquilizer	15–150 mg/kg ip im	Tranquilizer	34
Methotrimeprazine	Tranquilizer	10 mg/kg im	Tranquilizer	34
Oxazepam	Tranquilizer	20 mg/kg	Tranquilizer	34
Perphenazine	Tranquilizer	1 mg/kg po	ED_{50} for tranquilization	34
Phenylcyclidine HCl	Tranquilizer	10 mg/kg ip	Excited with ataxia	42
Phenylcyclidine HCl	Tranquilizer	50 mg/kg ip	Cataleptoid with tremors	42
Phenylcyclidine HCl	Tranquilizer	2 mg/kg ip	Slightly ataxic	42
Promazine	Tranquilizer	0.5–1 mg/kg im	Tranquilizer	34
Reserpine	Tranquilizer	5 mg/kg sc	Norepinephrine depleting agent	1
Reserpine	Tranquilizer	1 mg/kg po	Stimulates corticosteroid output	10
Reserpine	Tranquilizer	10 mg/kg po	Toxic dose to rat	10
Reserpine	Tranquilizer	0.4 mg/kg iv	Norepinephrine depleting agent	1
Reserpine	Tranquilizer	5 mg/kg po	Stimulates corticosteroid output	10
Thalidomide	Tranquilizer	150 mg/kg po	Given during pregnancy but few malformations	10

[a] Key to references:

1. Barnes, C. D., and Eltherington, L. G. (1973). "Drug Dosage in Laboratory Animals," Rev. Ed. Univ. of California Press, Berkeley, California.
2. Bowen, J. M. (1976). Drugs acting on the central nervous system. In "Handbook of Laboratory Animal Science" (E. C. Melby, Jr. and N. H. Altman, eds.), Vol. 3, pp. 65–95. CRC Press, Cleveland, Ohio.
3. Bowen, J. M., and Butrram, J. M. (1976). Drugs affecting the peripheral nervous system. In "Handbook of Laboratory Animal Science" (E. C. Melby, Jr. and N. H. Altman, eds.), Vol. 3, pp. 96–118. CRC Press, Cleveland, Ohio.

Table I (*Continued*)

4. Chapman, W. H. (1964). The incidence of a nematode, *Trichosomoides crassicauda* in the bladder of laboratory rats: Treatment with nitrofurantoin and preliminary report of the influence on the urinary calculi and experimental bladder tumor. *Invest. Urol.* **2**, 52–57.

5. Charles River Digest (1966). ''Anesthesia in Small Laboratory Animals,'' Vol. 5, No. 3. Charles River Breed. Lab., Inc., Wilmington, Massachusetts.

6. Charles, R. T., and Rees, O. (1958). Use of sulphonamides in the treatment of pleuro-pneumonia-like organisms in rats. *Nature (London)* **181**, 1213.

7. Colby, D., Skelton, F. R., and Brownie, A. C. (1970). Metopirone-induced hypertension in the rat. *Endocrinology* **86**, 620–628.

8. Danielson, R. A., and Schmidt-Nielsen, B. (1972). Recirculation of urea analogs from renal collecting ducts of high- and low-protein-fed rats. *Am. J. Physiol.* **223**, 130–139.

9. Feder, H. H., Naftolin, F., and Ryan, K. J. (1974). Male and female sexual responses in male rats given estradiol benzoate and 5α-ancrostan-17β-1,3-one propionate. *Endocrinology* **94**, 136–141.

10. Fratta, I. D. (1969). Nicotinamide deficiency and thalidomide: Potential teratogenic disturbances in Long-Evans rats. *Lab. Anim. Care* **19**, 727–732.

11. Fregly, M. J., Kim, K. J. and Hood, C. I. (1969). Development of hypertension in rats treated with aldosterone acetate. *Toxicol. Appl. Pharmacol.* **15**, 229–243.

12. Frenkel, J. K., Good, J. T., and Schultz, J. A. (1966). Latent *Pneumocystis* infection of rats, relapse, and chemotherapy. *Lab Invest.* **15**, 1559–1576.

13. Habermann, R. T., Williams, F. P., McPherson, C. W., and Every, R. R. (1963). The effect of orally administered sulfamerazine and chlortetracycline on chronic respiratory disease in rats. *Lab. Anim. Care* **13**, 28–40.

14. Hedge, G. A., and Hanson, S. D. (1972). The effects of prostaglandins on ACTH secretion. *Endocrinology* **91**, 925–933.

15. Hoar, R. M. (1965). Anesthetic technics of the rat and guinea pig. *In* ''Experimental Animal Anesthesiology'' (D. C. Sawyer, ed.), pp. 325–344. USAF School of Aerospace Medicine, Brooks Air Force Base, Texas.

16. Hughes, H. C., Jr., Barthel, C. H., and Lang, C. M. (1973). Niclosamide as a treatment for *Hymenolepis nana* and *Hymenolepsis diminuta* in rats. *Lab. Anim. Care* **23**, 72–73.

17. Jones, J. B., and Simmons, M. L. (1968). Innovar-Vet as an intramuscular anesthetic for rats. *Lab. Anim. Care* **18**, 642–643.

18. Kaczmarczyk, G., and Reinhardt, H. W. (1975). Arterial blood gas tensions and acid–base status of Wistar rats during thiopental and halothane anesthesia. *Lab. Anim. Sci.* **25**, 184–190.

19. Khazan, N., and Colasanti, B. (1972). Protracted rebound in rapid eye movement sleep time and electroencephalogram voltage output in morphine dependent rats upon withdrawal. *J. Pharmacol. Exp. Ther.* **183**, 23–30.

20. Klassen, C. D. (1970). (A). Effects of phenobarbital on the plasma disappearance and biliary excretion of drugs in rats. *J. Pharmacol. Exp. Ther.* **175**, 289–300.

21. Klaassen, C. D. (1970). (B). Plasma disappearance and biliary excretion of sulfobromophthalein and phenol-3,6-dibromphthalein disulfonate after microsomal enzyme induction. *Biochem. Pharmacol.* **19**, 1241–1249.

22. Latt, R. H. (1976). Drug dosages for laboratory animals. *In* ''Handbook of Laboratory Animal Science'' (E. C. Melby, Jr. and N. H. Altman, eds.), Vol. 3, pp. 561–568. CRC Press, Cleveland, Ohio.

23. Lewis, G. E., and Jennings, P. B. (1972). Effective sedation of laboratory animals using Innovar-Vet. *Lab. Anim. Sci.* **22**, 430–432.

24. Llaurado, J. G. (1961). The effects of some 21-methyl-substituted corticoids on inflammation, liver glycogen and electrolyte-regulating activity in the rat. *Acta Endocrinol. (Copenhagen)* **38**, 137–150.

25. Lumb, W. V. (1965). Pre- and postanesthetic agents. *In* ''Experimental Animal Anesthesiology'' (D. C. Sawyer, ed.), pp. 48–66. USAF School of Aerospace Medicine, Brooks Air Force Base, Texas.

26. Manten, A., and Hoogerheide, J. C. (1958). The influence of a new antifungal antibiotic, pimaricin, on the yeast flora of the gastrointestinal tract of rats and mice during tetracycline administration. *Antibiot. Chemother. (Washington, D.C.)* **8**, 381–386.

27. McCormack, C. E., and Meyer, R. K. (1968). Evidence for the release of ovulating hormone in PMS-treated immature rats. *Proc. Soc. Exp. Biol. Med.* **128**, 18–23.

28. Molello, J. A., and Hawkins, K. (1968). Methoxyflurane anesthesia of laboratory rats. *Lab. Anim. Care* **18**, 581–583.

29. Newbould, B. B. (1965). Suppression of adjuvant-induced arthritis in rats with 2-butoxycarbonylmethylene-4-oxothiazolidine. *Br. J. Pharmacol. Chemother.* **24**, 632–640.

30. ''Nutrient Requirements of Laboratory Animals'' (1972). 2nd rev. ed., No. 10. N.A.S., Washington, D.C.

31. Parratt, J. R., and West, G. B. (1958). Inhibition by various substances on oedema formation in the hind-paw of the rat induced by 5-hydroxytryptamine, histamine, dextran, eggwhite and compound 48/80. *Br. J. Pharmacol. Chemother.* **13**, 65–70.

32. Peardon, D. L., Tufts, J. M., and Eschroeder, H. C. (1966). Experimental treatment of laboratory rats naturally infected with *Trichosomoides crassicauda*. *Invest. Urol.* **4**, 215–219.

33. Randall, L. O., and Selitto, J. J. (1957). A method for measurement of analgesic activity of inflamed tissue. *Arch. Int. Pharmacodyn. Ther.* **111**, 409–419.

34. Rossoff, I. F. (1974). ''Handbook of Veterinary Drugs.'' Springer-Verlag, Berlin and New York.

35. Schuchman, S. M. (1974). Individual care and treatment of mice, rats, guinea pigs, hamsters and gerbils. *Curr. Vet. Ther.* **5**, 610–614.

36. Selitto, J. J., and Lowell, O. R. (1954). Screening of antiinflammatory agents in rats. *Fed. Proc., Fed. Am. Soc. Exp. Biol.* **13**, 403–404.

37. Shepard, T. H. (1973). ''Catalog of Teratogenic Agents.'' Johns-Hopkins Univ. Press, Baltimore, Maryland.

38. Siegmund, E., Cadmus, R., and Go, L. (1957). A method for evaluating both non-narcotic and narcotic analgesics. *Proc. Soc. Exp. Biol. Med.* **95**, 729–731.

39. Silvestrini, B., Garan, A., Pozzatti, C., and Cioli, V. (1965). Pharmacological research on benzydamine—A new analgesic-anti-inflammatory drug. *Arzneim.-Forsch.* **16**, 59–63.

40. Spector, S. (1956). ''Handbook of Biological Data.'' Saunders, Philadelphia, Pennsylvania.

41. Steward, J. S. (1955). Anthelmentic studies. *Parasitology* **45**, 231–241.

42. Stoliker, H. E. (1965). The physiologic and pharmacologic effects of sernylan: A review. *In* ''Experimental Animal Anesthesiology'' (D. C. Sawyer, ed.), pp. 148–184. USAF School of Aerospace Medicine, Brooks Air Force Base, Texas.

Table I (*Continued*)

43. Stucki, J. C., and Thompson, C. R. (1962). Anti-inflammatory activity of amidines of substituted triphenylethylenes. *Toxicol. Appl. Pharmacol.* **4,** 362–384.

44. Stunkard, J. A., and Miller, J. C. (1974). An outline guide to general anesthesia in exotic species. *Vet. Med. Small Anim. Clin.* **69,** 1181–1186.

45. Volpert, E. M., and Werner, S. C. (1972). Serum triiodothyronine concentration in the iodine-deficient rat. *Am. J. Anat.* **135,** 187–190.

46. Weisbroth, S. H., and Fudens, J. H. (1972). Use of ketamine hydrochloride as an anesthetic in laboratory rabbits, rats, mice, and guinea pigs. *Lab. Anim. Sci.* **22,** 904–906.

47. Weisbroth, S. H., and Scher, S. (1971). *Trichosomoides crassicauda* infection of a commercial rat breeding colony. 2. Drug screening for anthelmentic activity and field trials with methyridine. *Lab. Anim. Sci.* **21,** 213–219.

48. Ying, S., and Greep, R. O. (1971). Effect of a single low dose of estrogen on ovulation, pregnancy, and lactation in immature rats. *Fertil. Steril.* **22,** 165–169.

[b]Important complications have been reported from use of this drug in rats (Fleischman, R. W., McCracken, D., and Forbes, W. (1977). *Lab. An. Sci.* **27,** 238–243).

Subject Index